Advanced Technologies for Position and Navigation under GNSS Signal Challenging or Denied Environments

Advanced Technologies for Position and Navigation under GNSS Signal Challenging or Denied Environments

Editors

Changhui Jiang
Yuwei Chen
Qian Meng
Panlong Wu
Bing Xu
Lianwu Guan
Wang Gao
Zeyu Li

MDPI • Basel • Beijing • Wuhan • Barcelona • Belgrade • Manchester • Tokyo • Cluj • Tianjin

Editors

Changhui Jiang
Nanjing University of Aeronautics and Astronautics
Nanjing
China

Yuwei Chen
Finnish Geospatial Research Institute
Masala
Finland

Qian Meng
Southeast University
Nanjing
China

Panlong Wu
Nanjing University of Science and Technology
Nanjing
China

Bing Xu
The Hong Kong Polytechnic University
Hong Kong
China

Lianwu Guan
Harbin Engineering University
Halbin
China

Wang Gao
Southesast University
Nanjing
China

Zeyu Li
Shandong University of Science and Technology
Qingdao
China

Editorial Office
MDPI
St. Alban-Anlage 66
4052 Basel, Switzerland

This is a reprint of articles from the Special Issue published online in the open access journal *Remote Sensing* (ISSN 2072-4292) (available at: https://www.mdpi.com/journal/remotesensing/special_issues/GNSS_signal).

For citation purposes, cite each article independently as indicated on the article page online and as indicated below:

LastName, A.A.; LastName, B.B.; LastName, C.C. Article Title. *Journal Name* **Year**, *Volume Number*, Page Range.

ISBN 978-3-0365-7630-5 (Hbk)
ISBN 978-3-0365-7631-2 (PDF)

Contents

remote sensing

MDPI

Article

Low-Cost Single-Frequency DGNSS/DBA Combined Positioning Research and Performance Evaluation

Shengliang Wang [1,2], Xianshu Dong [1,*], Genyou Liu [2], Ming Gao [2,3], Wenhao Zhao [2,3], Dong Lv [2,3] and Shilong Cao [2,3]

1 College of Mining Engineering, Taiyuan University of Technology, Taiyuan 030024, China; wangshengliang@tyut.edu.cn
2 State Key Laboratory of Geodesy and Earth's Dynamics, Innovation Academy for Precision Measurement Science and Technology, Chinese Academy of Sciences, Wuhan 430077, China; liugy@whigg.ac.cn (G.L.); gm2015@apm.ac.cn (M.G.); zhaowenhao@apm.ac.cn (W.Z.); lvdong@apm.ac.cn (D.L.); caoshilong@apm.ac.cn (S.C.)
3 College of Earth and Planetary Sciences, University of Chinese Academy of Sciences, Beijing 100049, China
* Correspondence: dxshu520@163.com

Abstract: In recent years, low-cost single-frequency GNSS receivers have been widely used in many fields such as mass navigation and deformation monitoring; however, due to the poor signal quality of low-cost patch antennae, it is difficult for carrier phase real-time kinematic (RTK) technology to fix the integer ambiguity. Differential GNSS (DGNSS) positioning with pseudorange can effectively meet the high robustness and reliability requirements for the submeter to the meter level positioning accuracy of UVA/vehicle/aerospace users. To improve the DGNSS positioning accuracy and reliability of low-cost single-frequency GNSS receivers in complex environments, we propose a differential barometric altimetry (DBA)-assisted DGNSS positioning algorithm, which solves the DGNSS observation equations jointly and rigorously with the Earth ellipsoidal constraint equations constructed by the DBA altitude. The DBA altitude accuracy at different baseline lengths was evaluated in detail, and the DGNSS positioning performance of the single-frequency low-cost u-blox receiver NEO-M8T with a patch antenna and DGNSS/DBA combined positioning performance with the BMP280 barometer was analyzed by several sets of static and dynamic experiments under different environments. The results show that the single-frequency NEO-M8T receiver with patch antenna DGNSS positioning accuracy is submeter level in the static environment and drops to meter level in the dynamic environment. GPS+BDS dual system has higher positioning accuracy than single GPS or single BDS. DGNSS/DBA combination has higher positioning accuracy than DGNSS, especially the root mean square error (RMSE) can be improved by 30% to 80% in the U direction and slightly improved in the N and E directions. This study can provide an effective solution reference for various applications of low-cost sensor fusion positioning in the mass consumer market.

Keywords: differential GNSS; DBA; low-cost; combined positioning

Citation: Wang, S.; Dong, X.; Liu, G.; Gao, M.; Zhao, W.; Lv, D.; Cao, S. Low-Cost Single-Frequency DGNSS/DBA Combined Positioning Research and Performance Evaluation. *Remote Sens.* **2022**, *14*, 586. https://doi.org/10.3390/rs14030586

Academic Editors: Changhui Jiang, Yuwei Chen, Qian Meng, Panlong Wu, Bing Xu, Lianwu Guan, Wang Gao and Zeyu Li

Received: 18 December 2021
Accepted: 24 January 2022
Published: 26 January 2022

1. Introduction

With the continuous development and improvement of the Global Navigation Satellite System (GNSS), the number of visible satellites has been greatly increased, which effectively improves the positioning accuracy, reliability, and availability. However, in complex environments such as under trees, urban canyons, tunnels, deep mine pits, and indoors, etc., GNSS signal attenuation is severe, observation quality is poor, or the visible satellites are insufficient and there is the serious multipath effect. These drawbacks will lead to decreased positioning accuracy or even unable to locate, which greatly limit the availability and application of GNSS [1,2]. Combining GNSS with other multi-source sensor technologies to realize the complementary advantages of each system and improve the location-based services (LBS) accuracy of user terminals in harsh scenarios has become a major research hotspot in the field of navigation.

Single-frequency low-cost GNSS receivers, such as u-blox series, SkyTraq S2525F8, etc., are widely used in various industries, providing solutions for surveying and mapping applications [3], landslide deformation monitoring [4,5], pedestrian navigation, vehicle tracking [6], and small unmanned aerial vehicles (UAV) navigation [7]. Low-cost GNSS receivers are smaller in size and mass, but their hardware performance is inferior to that of geodetic GNSS receivers. For example, lower signal-to-noise ratio (SNR), poor observed values, and more frequent satellite out-of-lock and observation data are missing. The main reason is that the observation quality of the low-cost patch antenna is poor. Low-cost GNSS receivers for standard point positioning (SPP) can only obtain meter-level positioning accuracy, which means positioning accuracy and reliability will significantly reduce with positioning errors up to tens of meters in complex urban environments [8]. The single-frequency low-cost u-blox receiver for GNSS RTK positioning has a low fixed rate of carrier phase ambiguity in practical applications due to the poor observed data quality [9].

Compared with GNSS RTK positioning to achieve centimeter-level positioning accuracy, the DGNSS positioning with code pseudorange can only achieve 1–2 m positioning accuracy [10]. However, DGNSS is simpler to implement and can avoid positioning failure caused by RTK ambiguity fixed incorrectly. DGNSS can be widely used in many fields, for example, in the location of mobile devices [11,12], marine navigation, and in coastal navigation and in dynamic vessel positioning [10,13], in hydrography for positioning of acoustic systems [14], in autonomous vehicle positioning [15,16], and civil aviation during precision approach procedures [17]. DGNSS is currently the most widely used augmentation system around the world.

The idea behind DGNSS operation lies in determination of the error related to pseudorange observations and calculated comparing the actual value received by the GNSS receiver and the true value calculated using the satellite and the reference station antenna coordinates. This difference, referred to as a pseudorange correction, is transmitted to users who use a GNSS receiver and take it into account in the positioning process [18]. They can be divided into so-called local-area DGNSS (LADGNSS) services for small areas, such as a relatively small area of several dozen to several hundred square kilometers, and wide-area DGNSS (WADGNSS) services for larger areas such as an entire continent or even worldwide. The positioning accuracy achieved by LADGNSS method is 1–3 m and it decreases with increasing distance between a user and the single reference station [19]. WADGNSS can extend the service area using a few geosynchronous equatorial orbit (GEO) satellites and overcome the error due to the spatial decorrelation, as in, for example, the Wide-Area Augmentation System (WAAS, USA), European Geostationary Navigation Overlay Service (EGNOS, Europe), and MTSAT Satellite Augmentation System (MSAS, Japan). This system is used to obtain a meter-level accuracy over a large region while using a fraction of the number of reference stations [10].

Most studies and analyses of DGNSS positioning with low-cost GNSS receivers have used geodetic antennas. For example, an Flächen Korrektur parameter (FKP)-DGPS algorithm [19] is studied as a new augmentation method for the low-cost GPS receivers by integrating the conventional DGPS correction with the modified FKP correction to mitigate the positioning error due to the spatial decorrelation. Single-frequency DGPS aided low-cost inertial navigation system (INS) positioning [20] was studied to achieve the real-time high-frequency state output with decimeter position accuracy and centimeter velocity accuracy. However, there is still little research on DGNSS for low-cost GNSS receivers with low-cost patch antenna and fusion positioning with other low-cost sensors.

Height constraint is an effective method to improve the GNSS positioning accuracy, for example, due to the insufficient number of available satellites, the BDS-1 has used electronic maps as height constraints to improve the users' positioning accuracy [21]. This method is more complicated to implement and it is difficult to promote its application. The barometric altimetry is low-cost, independent of environmental restrictions, and can be used both indoors and outdoors. Low-cost barometer altimetry-assisted GNSS positioning navigation is also widely used in the field of aviation flights and smartphone navigation, etc. [22–25]. The basic principle of barometric altimetry is to use the physical phenomenon that the atmospheric pressure on the Earth's surface gradually decreases with increasing height, but, due to the irregular changes of atmospheric pressure, the altitude error directly calculated by a single barometer is as high as tens or even hundreds of meters, which cannot be used as a constraint to improve GNSS positioning accuracy. Similar to DGNSS, the accurate height can be obtained by barometric correction compensation or differential barometric altitude (DBA). That is, using the property that local atmospheric pressure changes are similar, a barometer is placed at the reference station and another barometer is used as a mobile station to determine its high-precision relative altitude by the differential equation. The altitude accuracy and reliability obtained by DBA mode are high, which can effectively constrain other technology to improve positioning and navigation accuracy.

The user altitude obtained by the DBA system was applied as a virtual satellite in the China Area Positioning System (CAPS), and the construction of independent earth ellipsoidal constraint equations by users' altitude can effectively solve the insufficient number of CAPS satellites and improve its 3D positioning accuracy and availability [26–28]. The DBA system also be applied to mobile cellular network positioning for accurately determining the user's height, and reduced 3D positioning to planar positioning which can obtain more desirable positioning accuracy [29]. Mobile cellular base stations can be used as DBA reference stations and transmitted the relevant atmosphere pressure data to the user side through mobile communication networks, achieving GNSS SPP/DBA combined 3D positioning with the altitude accuracy better than 1 m [30]. Inertial/barometric altitude can be fused to measure vertical velocity and height with velocity root mean squared error (RMSE) between 0.04 to 0.24 m/s and RMSE in height between 5 to 68 cm [31]. In addition, a barometer installed on a wearable device can measure vital signs such as blood pressure by detecting the position and orientation of the human body, thus providing a better telemedicine solution for precision medicine [32,33].

At present, to our knowledge, little research has been reported on the DGNSS/DBA combined positioning with low-cost GNSS receivers and a patch antenna, and there is also a lack of research and analysis on the theoretical methods and application effects of DBA, which is worth further study. In this study, we firstly propose a DGNSS/DBA combined positioning algorithm. Second, the DBA altitude accuracy at different baseline lengths is evaluated in detail. Then, the DGNSS performance of single-frequency low-cost NEO-M8T receiver and the accuracy and reliability of DGNSS/DBA combined positioning with low-cost BMP280 barometer are fully evaluated through actual measurement data.

This manuscript is organized as follows: Section 1 is the introduction. Section 2 is the mathematical model of DGNSS/DBA combined positioning, which contains Section 2.1 about DGNSS positioning observation equations; Section 2.2 about the principle of the DBA system; and Section 2.3 on the DGNSS/DBA combined positioning algorithm. Section 3 reports the experiment results, containing Section 3.1 that introduces experimental data; Section 3.2 about DBA altitude accuracy evaluation at different baseline lengths; Sections 3.3 and 3.4 on the DGNSS/DBA combined static and kinematic vehicle positioning performance evaluation for the single-frequency low-cost NEO-M8T receiver and BMP280 barometer, respectively. Section 4 presents the discussions. Section 5 is the conclusion.

2. Mathematical Model of DGNSS/DBA Combined Positioning

2.1. DGNSS Positioning Observation Equation

GNSS receivers can simultaneously receive observation data such as code pseudorange, carrier phase, Doppler shift, and SNR. The raw observation values contain the receiver geometric position parameters, clock error as well as hardware delays parameters, and various error corrections, such as tropospheric errors and ionospheric errors. Therefore, the raw code pseudorange observation equation between satellite s and receiver i can be expressed as:

$$P_i^s = \rho_i^s + c(dt_i - dt^s) + I_i^s + T_i^s + M_i^s + \varepsilon_i^s \tag{1}$$

where P_i^s denotes the code pseudorange observation between satellite s and receiver i (Unit: m); $\rho_i^s = \sqrt{(X^s - X)^2 + (Y^s - Y)^2 + (Z^s - Z)^2}$ denotes the geometric distance between satellite s and receiver i at the moment of signal transmission, (X^s, Y^s, Z^s) and (X, Y, Z) are the satellite s and unknown receiver i antenna center position. c is the speed of light in vacuum; dt_i denotes the receiver clock error; dt^s denotes the satellite clock error; I_i^s is the ionospheric error; T_i^s is the tropospheric error; M_i^s is the multipath delay error; ε_i^s contains the code pseudorange measurement noise and other uncorrected errors.

For the short and medium baselines, the receiver clock error and satellite clock error are eliminated in the double-difference observation equation, the ionospheric and tropospheric errors can be neglected, and the DGNSS observation equation can be simplified and expressed as [34]:

$$\nabla\Delta P_{i,j}^{s_1 s_k} = \nabla\Delta\rho_{i,j}^{s_1 s_k} + \nabla\Delta\varepsilon_{i,j}^{s_1 s_k} \tag{2}$$

where $\nabla\Delta$ denotes the double-difference operator; i and j denotes the reference station receiver and mobile receiver; s_1 and s_k denote the reference and nonreference satellite, respectively.

2.2. Principle of the DBA System

Single-barometer altimetry affected by atmospheric temperature, seasonal changes, and other factors drifts up to several tens of meters within a day, with poor stability and reliability [35]. Due to the Earth's gravitational field, the space atmosphere pressure and height show a certain regular distribution. Except for the local strong convection zone, the trend of atmospheric pressure variation in the local range of several tens of kilometers shows the same physical characteristics, and the atmosphere is basically in hydrostatic equilibrium in the vertical direction. Usually, the atmospheric pressure is distributed more evenly in the horizontal direction, and the pressure difference is about 1 hPa at a distance of 100 km. Thus, the concept of "difference" can be extended to the field of barometric altimetry with the help of differential GNSS positioning, that is, by setting one (or several) barometric reference points, the barometric altimetry value of the mobile station can be corrected by the precise altitude of the reference station to compensate the influence of atmospheric physical environment changes on the altitude measurement results of the mobile station, thus improving the accuracy of the user altitude of the mobile station.

When the barometric reference station and the mobile station are within a few tens of kilometers, their latitude, gravitational acceleration, and water vapor factors have the same effect on the atmospheric pressure, so the above three errors can be neglected to obtain the simplified DBA formula [36]:

$$h = h_0 + 18{,}410\left(1 + \frac{t_m}{273.15}\right)\lg\frac{P_0}{P} \tag{3}$$

where h is the altitude of the mobile station to be found, h_0 is the known altitude of the reference station, P_0 is the pressure of the reference station, P is the pressure of the mobile station, and t_m is the average Celsius temperature between the reference station and the mobile station.

2.3. DGNSS/DBA Combined Positioning Algorithm

In this study, we propose a DGNSS/DBA combined positioning algorithm, in which the user's altitude obtained by the DBA system in Section 2.2 is used to construct the Earth ellipsoid constraint equation and solved rigorously as an independent observation jointly with the DGNSS observation equation, which is equivalent to adding a virtual satellite located at the center of the Earth [26]. Since the geodetic height is an independent variable in GNSS coordinates, an approximate ellipsoid with the altitude h from the reference ellipsoid (WGS-84) can be constructed using the user geodetic height, as shown in Figure 1.

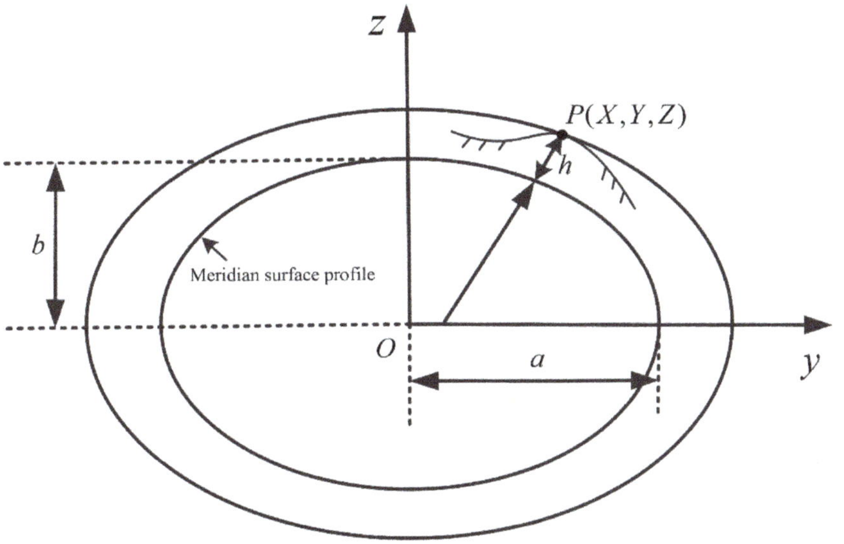

Figure 1. Approximate reference ellipsoidal meridian profile where the ground user's geodetic height is located.

At this time, when the ground user's geodetic height is not very large, the observation equation after DGNSS/DBA combination can be expressed as:

$$\begin{cases} \nabla\Delta P_{i,j}^{s_1s_k} = \nabla\Delta\rho_{i,j}^{s_1s_k} + \nabla\Delta\varepsilon_{i,j}^{s_1s_k} \\ \frac{X^2+Y^2}{(a+h)^2} + \frac{Z^2}{(b+h)^2} = 1 \end{cases} \tag{4}$$

the symbols in the DGNSS observation equation in the first line of Equation (4) are the same as Equation (2). $P(X,Y,Z)$ is the 3D coordinate of the ground user; a and b are the long and short semiaxes of the WGS-84 Earth reference ellipsoid, respectively. Since h is much smaller than the long and short semiaxes of the Earth reference ellipsoid, an approximate reference ellipsoid with a long semiaxis $a+h$ and short semiaxis $b+h$ is used instead without causing much bias [37]. To solve the Earth ellipsoid constraint equation in the second expression of Equation (4) by differential processing, the ellipsoid constraint equation is expanded in the user's approximate position (X_0, Y_0, Z_0) according to the Taylor series, and only the first-order term is retained, where the partial derivative of X is obtained as:

$$\frac{2X_0}{(a+h)^2}dX - \frac{2X_0^2}{(a+h)^3}dh - \frac{2Y_0^2}{(a+h)^3}dh - \frac{2Z_0^2}{(b+h)^3}dh = 0 \tag{5}$$

after simplification, we get:

$$\frac{\partial h}{\partial X_0} = \frac{X_0(a+h)(b+h)^3}{(X_0^2+Y_0^2)(b+h)^3 + Z_0^2(a+h)^3} \tag{6}$$

similarly, taking partial derivatives of Y and Z yields:

$$\frac{\partial h}{\partial Y_0} = \frac{Y_0(a+h)(b+h)^3}{(X_0^2+Y_0^2)(b+h)^3 + Z_0^2(a+h)^3} \tag{7}$$

$$\frac{\partial h}{\partial Z_0} = \frac{Z_0(a+h)^3(b+h)}{(X_0^2+Y_0^2)(b+h)^3+Z_0^2(a+h)^3} \quad (8)$$

let $\alpha = \partial h/\partial X_0$, $\beta = \partial h/\partial Y_0$, $\gamma = \partial h/\partial Z_0$, that is, the Earth ellipsoidal constraint equation in Equation (4) is linearized at the approximate coordinates (X_0, Y_0, Z_0) to give:

$$V_{DBA} = \alpha dX + \beta dY + \gamma dZ - dh \quad (9)$$

where $dh = h - \hat{h}$ is the altitude residual, h is the altitude obtained by the DBA system, and $\hat{h}$ is the geodetic height obtained by the users' at the approximate position. dX, dY, dZ are positional corrections of the receiver antenna center. The detailed conversion process can be found in the original literature [38].

Similarly, the DGNSS observation equation in the first expression of Equation (4) is expanded by the Taylor series at the approximate position (X_0, Y_0, Z_0), omitting higher-order terms above the first order, and combined with Equation (9) to obtain the DGNSS/DBA combined positioning error equation:

$$V = H\hat{x} - l, P \quad (10)$$

in Equation (10), the parameter estimated as $\hat{x} = \begin{bmatrix} dX & dY & dZ \end{bmatrix}^T$ contains three approximate position correction values; $V = \begin{bmatrix} V^1 & \cdots & V^i & V_{DBA} \end{bmatrix}^T$ is the residual vector, $H = \begin{bmatrix} l^1 & m^1 & n^1 \\ \cdots & \cdots & \cdots \\ l^i & m^i & n^i \\ \alpha & \beta & \gamma \end{bmatrix}$ is the coefficient matrix, $l^i = \frac{(X^s - X_0)}{\rho_j^{(0)s}} - \frac{(X^k - X_0)}{\rho_j^{(0)k}}$, $m^i = \frac{(Y^s - Y_0)}{\rho_j^{(0)s}} - \frac{(Y^k - Y_0)}{\rho_j^{(0)k}}$, $n^i = \frac{(Z^s - Z_0)}{\rho_j^{(0)s}} - \frac{(Z^k - Z_0)}{\rho_j^{(0)k}}$ are the pseudorange double-difference directional cosine, respectively. $l = \begin{bmatrix} L^1 & \cdots & L^i & dh \end{bmatrix}^T$ are the observation value vectors. $P = \begin{bmatrix} P_{DGNSS} & 0 \\ 0 & P_{DBA} \end{bmatrix}$ is the DGNSS/DBA combined positioning weight matrix. $P_{DGNSS} = Q_{DD}^{-1}$ is the a priori weight matrix of GNSS pseudorange double-difference observation equation. Q_{DD} is the GNSS pseudorange double-difference observation values covariance matrix, the stochastic model of GNSS nondifferential observations adopts the sine trigonometric function elevation angle fixed-weight model [39]. According to the error propagation law, the GNSS relative positioning variance-covariance matrix can be expressed as Q_{DD} [34]. The a priori weight P_{DBA} of the DBA can be determined based on the results of the empirical evaluation in Section 3.2. We can solve Equation (10) using the single-epoch weighted least squares method, as Equation (11):

$$\begin{cases} \hat{x} = (H^T P H)^{-1} H^T P l \\ Q_{\hat{x}\hat{x}} = (H^T P H)^{-1} \end{cases} \quad (11)$$

where $Q_{\hat{x}\hat{x}}$ is the posterior covariance matrix of the parameter $\hat{x}$. It can be found that the DBA altitude constraint is equivalent to adding a code pseudorange observation, and the condition number of the error equation coefficient matrix H becomes significantly better, the position dilution of precision (PDOP) value can be effectively reduced. However, the accuracy of the positioning solution is influenced by the accuracy of DBA altitude, i.e., if the accuracy of DBA altitude is better than DGNSS altitude, the improvement effect is obvious, otherwise, the positioning accuracy cannot be improved.

3. Experiment Results

3.1. The Introduction of Experiment Data

In the experiment, the reference station included a high-precision geodetic GNSS receiver Trimble NET R9, a Trimble Choking 59,800 antenna, and a low-cost BMP280 barometer. The mobile station consisted of a single-frequency low-cost u-blox receiver NEO-M8T, a patch antenna, and two BMP280 barometers. The mobile station was also equipped with a geodetic receiver Trimble NET R9 for data acquisition, and the postprocessed kinematic (PPK) mode of commercial software Inertial Explorer 8.70 was used to process data to obtain high-precision 3D coordinate sequences as reference values. The nominal resolution of the low-cost BMP280 barometer was 0.01 mbar (0.1 m) and the data sampling rate was set at 1 Hz. As shown in Figure 2, to prevent the effect of crosswind on the barometric pressure, the barometer was placed inside a transparent plastic box with several small holes at the top of the box. All experimental data were recorded and postprocessed by a laptop computer. The height deviation of the BMP280 barometer from the GNSS antenna phase center was compensated by data preprocessing, and the DGNSS/DBA combined positioning analysis was performed by the self-written programs.

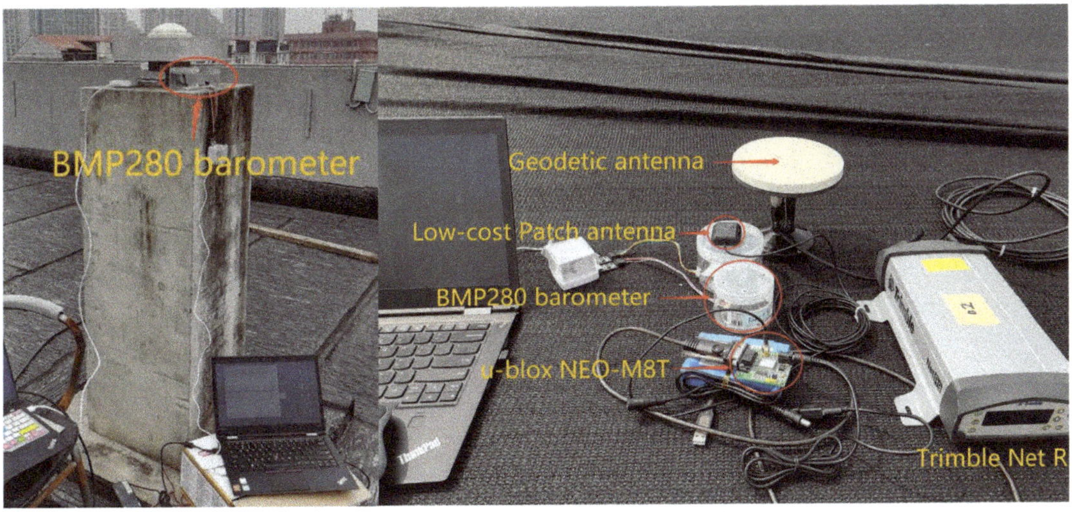

(a) (b)

Figure 2. The hardware equipment in the DBA altitude accuracy evaluation and static positioning experiment: (**a**) reference station; (**b**) mobile station.

The whole experiment was divided into three parts. The first experiment was used to evaluate the DBA altitude accuracy at different baseline lengths and to provide a priori information for subsequently combined positioning with DGNSS. The second experiment evaluated the positioning performance of low-cost single-frequency DGNSS and DGNSS/DBA combined positioning through static experiments with 65 m and 6.0 km baseline lengths. The third experiment evaluated the kinematic vehicle positioning performance of low-cost single-frequency DGNSS and DGNSS/DBA combined positioning in open and complex urban environments, respectively.

3.2. DBA Altitude Accuracy Evaluation at Different Baseline Lengths

The altitude accuracy and practical range obtained by the BMP280 barometer DBA model were evaluated through different baseline lengths outdoors. The experimental data were collected on 18 September 2020, and Figure 2 shows the hardware equipment of the reference station and mobile stations. The reference station was arranged on the observation pier on the roof of the Innovation Academy for Precision Measurement Science and Technology, Chinese Academy of Sciences (APM, CAS). Figure 3 shows five different mobile station locations at baseline lengths of 0 m, 65 m, 2.6 km (shopping mall plaza), 6.0 km (Wufu Plaza on the Yangtze River), and 10.0 km (the roof of the PET center of Tongji

Medical College), with a data acquisition time of about 30 min for each static point. Figure 4 and Table 1 show the time series and RMSE accuracy statistics of the altitude results of the DBA system under five groups of different baseline lengths.

Figure 3. Five mobile station locations at different baseline lengths for altitude accuracy evaluation of the outdoor DBA systems.

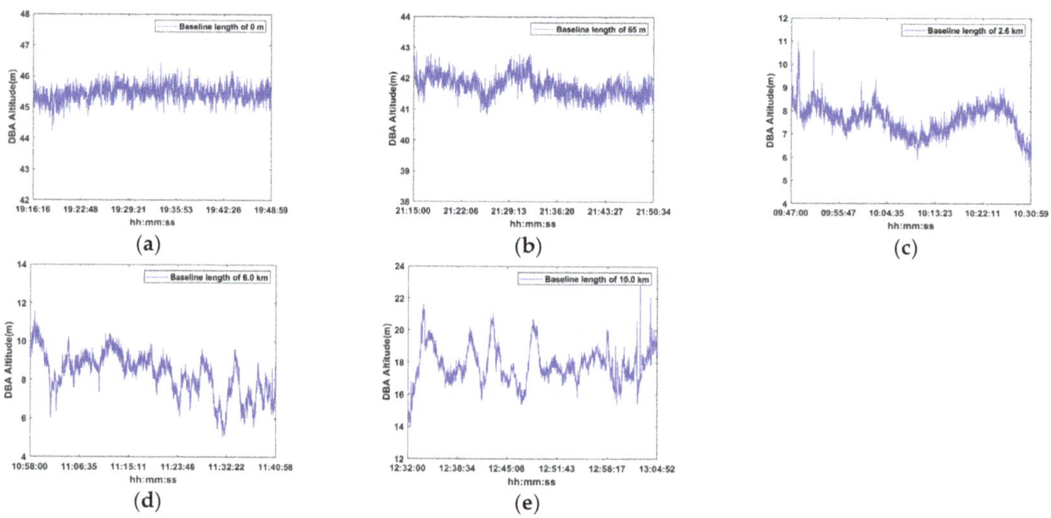

Figure 4. Time series of the DBA altitude results for five mobile station locations at different baseline lengths: **(a)** 0 m; **(b)** 65 m; **(c)** 2.6 km; **(d)** 6.0 km; **(e)** 10.0 km.

Table 1. The DBA altitude statistical results of the five different baseline lengths (Unit: m).

Baseline Length	Reference Altitude	DBA_Min	DBA_Max	Mean	Std	RMSE
0 m	45.453	44.200	46.436	45.453	0.271	0.270
65 m	41.125	40.814	42.871	41.738	0.350	0.706
2.6 km	8.645	5.626	10.994	7.656	0.649	1.182
6.0 km	8.944	5.047	11.583	8.308	1.101	1.27
10.0 km	16.542	13.91	23.67	17.86	1.16	1.76

In Table 1, DBA_min and DBA_max denote the minimum and maximum altitude results obtained by the DBA model. Mean and STD denotes the average value of DBA altitude and standard deviation. As can be seen from Figure 4 and Table 1, the outdoor DBA altitude RMSE increased gradually with an increase of baseline length, and the DBA altitude RMSE was submeter level within the 2 km baseline length and did not exceed 2 m within the 10 km baseline length. This result can provide a priori information to determine the weight matrix P_{DBA} in Equation (11) for the DGNSS/DBA combined positioning.

3.3. *DGNSS/DBA Combined Static Positioning Results*

This section mainly evaluates the static positioning performance of the single-frequency low-cost NEO-M8T receiver and BMP280 barometer DGNSS/DBA combined positioning algorithm. The experimental data were consistent with Section 3.2, and the 65 m and 6.0 km baseline length data were selected for processing and analysis. Two data processing modes, DGNSS and DGNSS/DBA, were set, and each mode was divided into single GPS, single BDS, and GPS+BDS dual systems by the satellite system.

3.3.1. Baseline Length 65 m

This experiment used 65 m short baseline static data and performed statistical analysis by setting the elevation mask angle from 10 to 40 degrees, and the prior error of the DBA system was $\sigma_{DBA} = 1.0$ m. Table 2 shows the average number of visible satellites at different elevation mask angles, and Table 3 lists the average PDOP values for DGNSS and DGNSS/DBA modes at different elevation mask angles.

Table 2. The average number of GNSS visible satellites at different elevation mask angles.

Satellite System	Elevation Mask Angles (Degree)			
	10	20	30	40
GPS	7.81	7.21	4.25	3.0
BDS	14.19	12.90	10.21	8.99
GPS+BDS	21.97	20.11	14.45	11.99

As can be seen from Table 2, the average number of visible satellites was seven to eight for GPS and 12 to 14 for BDS at a low elevation mask angle of 10 or 20 degrees. With the increase of the elevation mask angle, the number of available satellites of both GPS and BDS systems decreased significantly, the satellite space geometry distribution became worse, and the PDOP value gradually increased. The number of GPS satellites was only three at the elevation mask angles of 40 degrees, and the user receiver could not be positioned at this time, while the number of visible satellites of BDS in the China region was larger with eight to 10 visible satellites at the elevation mask angles of 30 or 40 degrees. The GPS+BDS dual system significantly increased the number of visible satellites compared to the single system, which significantly improved the satellite geometry and reduced the PDOP value. As shown in Table 3, increasing a DBA observation was equivalent to adding a virtual satellite, which improved the satellite geometry distribution and reduced the PDOP value; and the reduction of PDOP value was more significant in the environment with a higher elevation mask angle. When the elevation mask angle was 40 degrees, three GPS satellites could not complete the positioning, and adding a DBA observation ensured the availability of user receiver positioning. Figures 5–8 show the north (N)/east (E)/up (U) direction deviation sequence of the two data processing modes at the elevation mask angle 10 to 40 degrees. Each mode included single GPS, single BDS, and a GPS+BDS dual system. Table 4 shows the RMSE values in the N/E/U directions for the two data processing modes at different elevation mask angles.

Table 3. The average PDOP values for DGNSS and DGNSS/DBA modes at different elevation mask angles.

Positioning Mode	Satellite System	Elevation Mask Angles (Degree)			
		10	20	30	40
DGNSS	GPS	1.95	2.11	5.53	-
	BDS	1.49	1.81	3.02	4.66
	GPS+BDS	1.09	1.25	2.30	3.69
DGNSS/DBA	GPS	1.32	1.37	2.04	2.66
	BDS	1.16	1.27	1.53	1.79
	GPS+BDS	0.91	0.99	1.32	1.56

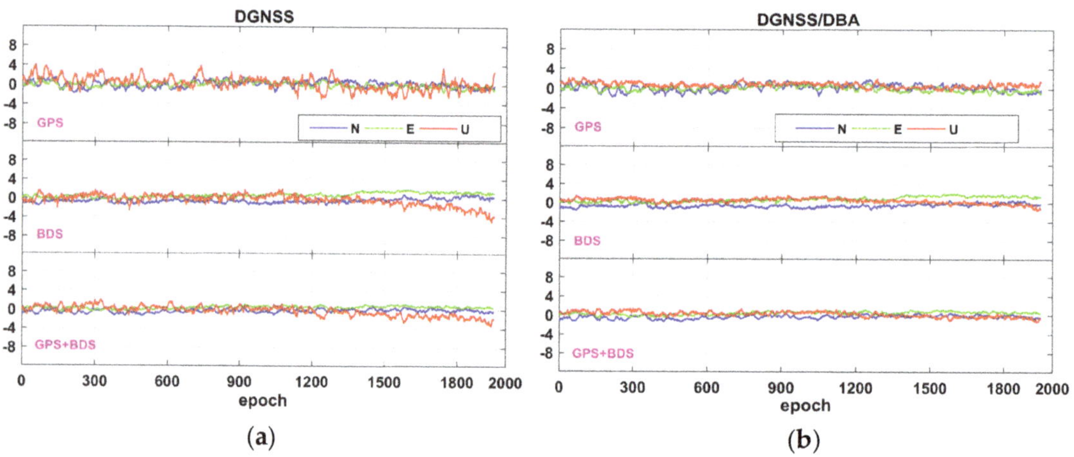

Figure 5. The deviation sequence diagram in the N/E/U directions at 10−degree elevation mask angles: (**a**) DGNSS positioning mode; (**b**) DGNSS/DBA positioning mode.

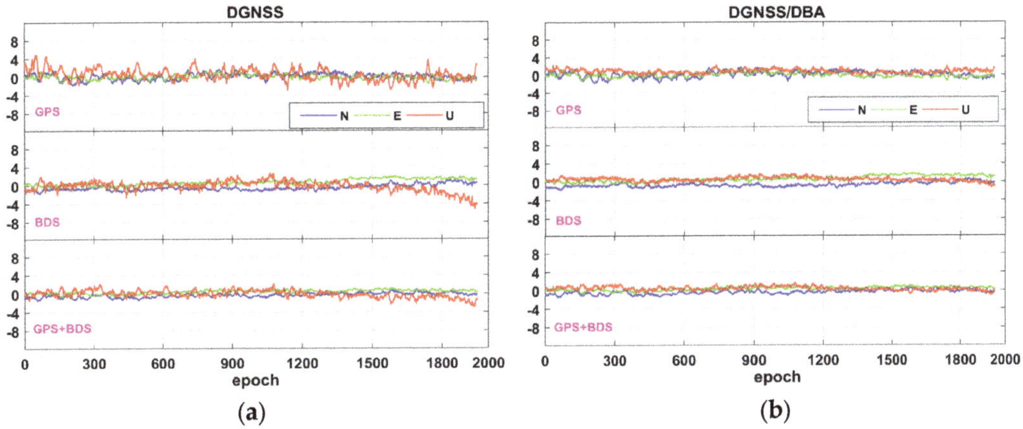

Figure 6. The deviation sequence diagram in the N/E/U directions at 20−degree elevation mask angles: (**a**) DGNSS positioning mode; (**b**) DGNSS/DBA positioning modes.

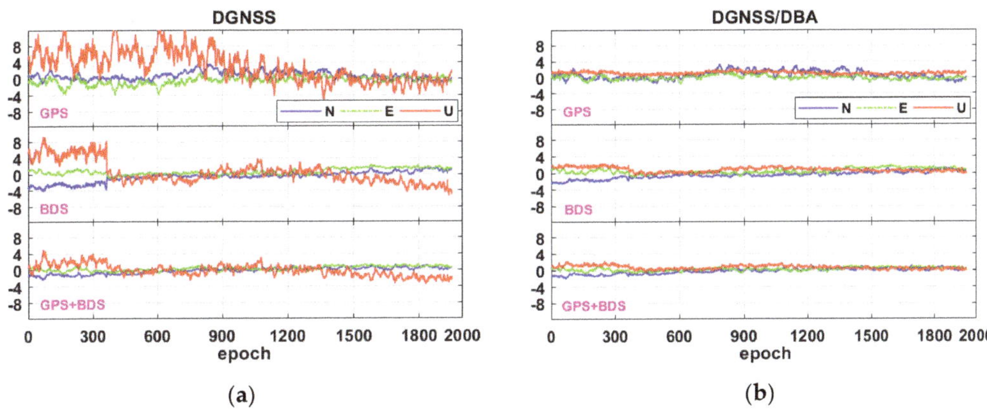

Figure 7. The deviation sequence diagram in the N/E/U directions at 30−degree elevation mask angles: (**a**) DGNSS positioning mode; (**b**) DGNSS/DBA positioning mode.

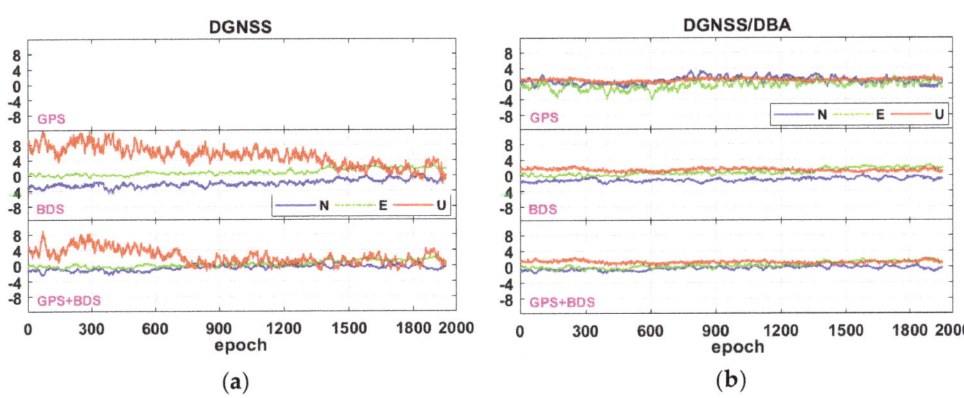

Figure 8. The deviation sequence diagram in the N/E/U directions at 40−degree elevation mask angles: (**a**) DGNSS positioning mode; (**b**) DGNSS/DBA positioning mode.

Table 4. The RMSE values of the two data processing modes in the N/E/U directions at different elevation mask angles.

Elevation Mask Angles (Degree)	RMSE Directions	DGNSS			DGNSS/DBA		
		GPS	BDS	GPS+BDS	GPS	BDS	GPS+BDS
10	N	0.82	0.72	0.58	0.76	0.69	0.57
	E	0.53	0.96	0.61	0.52	0.96	0.60
	U	1.32	1.08	0.94	0.89	0.63	0.57
20	N	0.70	0.73	0.53	0.70	0.74	0.53
	E	0.49	0.92	0.54	0.49	0.87	0.50
	U	1.38	1.14	0.85	0.94	0.72	0.62
30	N	1.24	1.33	0.71	1.16	0.94	0.65
	E	0.99	0.93	0.56	0.59	0.74	0.42
	U	4.81	2.85	1.47	1.09	0.95	0.80
40	N	-	2.21	0.89	1.23	1.20	0.59
	E	-	1.23	0.85	1.08	1.22	0.85
	U	-	5.81	3.07	1.02	1.45	1.20

Tables 2–4 and Figure 5 show that due to sufficient number of visible satellites and low PDOP values at the elevation mask angle of 10 degrees, the positioning accuracy of the DGNSS mode in the N/E/U directions could reach the decimeter to submeter level, and the RMSE of single GPS and single BDS in the N/E/U directions were 0.82/0.53/1.32 m and 0.72/0.96/1.08 m, respectively. In DGNSS/DBA mode, the RMSE of GPS/DBA and BDS/DBA in the N/E/U directions were 0.76/0.52/0.89 m and 0.69/0.96/0.63 m, respectively, which were 30% to 40% better than DGNSS in the U direction and slightly better in N and E directions. Due to the increase of available observations and better satellite geometry of the GPS+BDS dual system, the RMSE of DGNSS and DGNSS/DBA mode in the N/E/U directions were 0.58/0.61/0.94 m and 0.57/0.60/0.57 m, respectively. Both had some improvements over the single system. The results at the elevation mask angle of 20 degrees were similar to that of 10 degrees.

Tables 2–4 and Figure 7 show that the PDOP value became larger due to the fewer available observation satellites and worse satellite geometry at the elevation mask angle of 30 degrees, and the RMSE in the N/E/U directions became significantly larger than that of 10 and 20 degrees. The RMSE in the N/E/U directions were 1.24/0.99/4.81 m and 1.33/0.93/2.85 m for single GPS and single BDS, and 0.71/0.56/1.47 m for GPS+BDS dual system in DGNSS mode, respectively. Compared with the DGNSS mode, the accuracy of the N/E/U directions was significantly improved by the DGNSS/DBA combination, and the RMSE of GPS/DBA and BDS/DBA in the N/E/U directions were improved by 6.4%/40%/77.3% and 29.3%/20.4%/66.6%, and the RMSE of GPS+BDS/DBA combination in the N/E/U directions were improved by 8.5%/25%/45.6%.

When the elevation mask angle was 40 degrees, the number of available satellites for single GPS was three and they could not be located. The RMSE of single BDS was 2.21/1.23/5.81 m in the N/E/U directions, and 0.89/0.85/3.07 m for the GPS+BDS dual system (Tables 2–4 and Figure 8). In DGNSS/DBA mode, GPS/DBA met the most basic positioning requirements for four satellites and the RMSE in the N/E/U directions was 1.23/1.08/1.02 m; the RMSE of BDS/DBA in the N/E/U directions was improved by 44.3%/12.2%/82.4%, and the RMSE of the GPS+BDS/DBA combination in the N/E/U directions was improved by 33.7%/0%/60.9%, respectively.

3.3.2. Baseline Length 6.0 km

In this experiment, static data with 6.0 km baseline length were processed and analyzed, the elevation mask angle was 10 degrees, and the priori error of the DBA system was $\sigma_{DBA} = 1.5$ m. Figure 9 shows the number of GNSS visible satellites and the PDOP value sequence for both DGNSS and DGNSS/DBA modes at the baseline length of 6.0 km. The RMSE accuracy statistics and corresponding deviation sequence in the N/E/U directions for the DGNSS and DGNSS/DBA mode are shown in Table 5 and Figure 10, respectively.

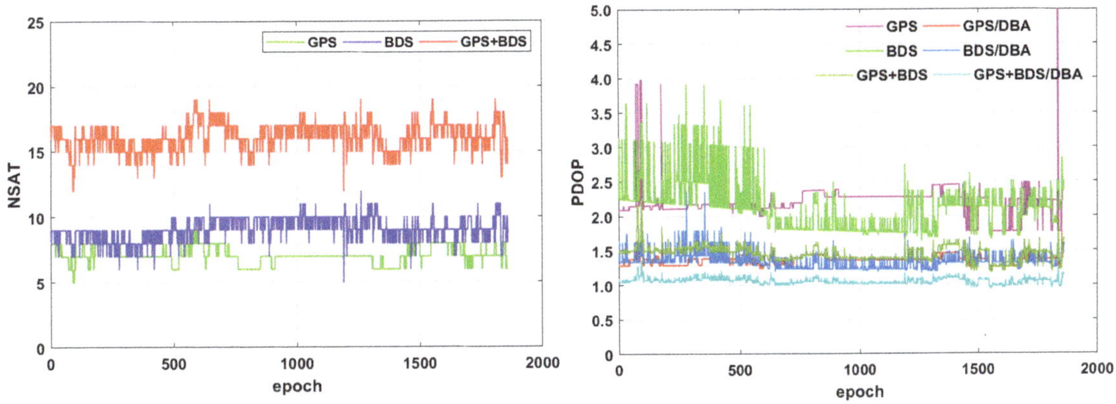

(**a**) (**b**)

Figure 9. Observation values of static experiment at the baseline length of 6.0 km: (**a**) the number of GNSS visible satellites; (**b**) the sequence of PDOP values.

Table 5. The RMSE of bias in the N/E/U directions for the DGNSS and DGNSS/DBA mode at 6.0 km baseline length.

Positioning Mode	Satellite System	N/m	E/m	U/m
DGNSS	GPS	2.23	1.12	4.41
	BDS	0.61	2.27	2.64
	GPS+BDS	0.60	1.01	2.22
DGNSS/DBA	GPS	2.13	1.11	3.91
	BDS	0.55	2.15	2.54
	GPS+BDS	0.62	0.98	2.44

Table 5 and Figures 9 and 10 show that the PDOP value of BDS was smaller than GPS in the China region due to a large number of observable satellites. The RMSE of single BDS in the N/E/U directions was 0.61/2.27/2.64 m, which was better than that of single GPS in DGNSS mode (2.23/1.12/4.41 m). The RMSE of the GPS+BDS dual system in the N/E/U directions was 0.60/1.01/2.22 m with higher positioning accuracy than single system. Compared to DGNSS mode, the RMSE in the U direction of DGNSS/DBA mode reduced by 0.5 m and 0.1 m for single GPS and single BDS, and there was also some improvement in the N and E directions. However, the GPS+BDS dual system did not improve significantly, due to the higher DGNSS accuracy and the lower DBA height accuracy at 6.0 km baseline length did not prove to be an obvious constraint.

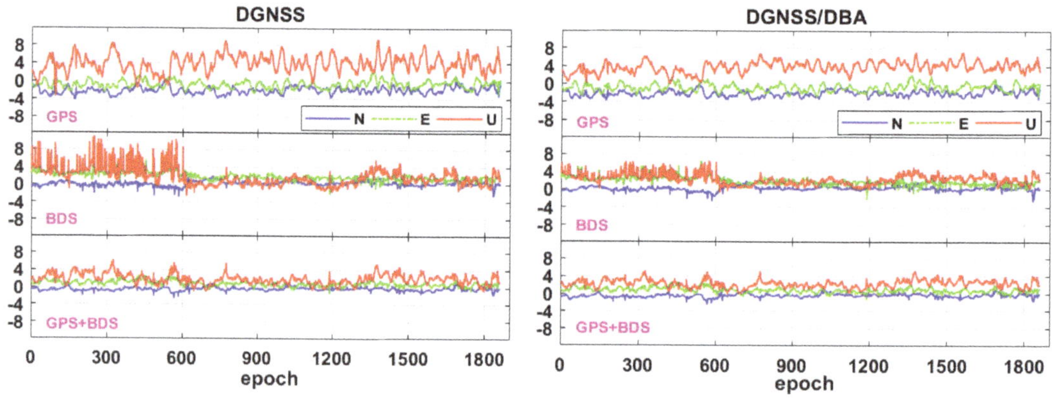

(a) (b)

Figure 10. The deviation sequence diagram in the N/E/U directions at 6.0 km baseline length: (**a**) DGNSS positioning mode; (**b**) DGNSS/DBA positioning mode.

3.4. DGNSS/DBA Combined Kinematic Vehicle Positioning Results

In this section, two kinematic vehicle experiments were designed to evaluate the dynamic positioning performance of the single-frequency low-cost NEO-M8T receiver and the BMP280 barometer DGNSS/DBA combined positioning algorithm.

The site photo of the mobile station equipment of the kinematic vehicle experiment is shown in Figure 11. The platform contained four multisystem dual-frequency geodetic GNSS antennas which were connected to Trimble Net R9 receivers to obtain the high precision reference value. The single-frequency, low-cost u-blox NEO-M8T receiver, a patch antenna and a BMP280 barometer were laid on the roof of the car. Two sets of kinematic vehicle data were collected, the first set was located in an open urban environment and the second set was in a complex urban environment.

(a) (b)

Figure 11. The mobile station hardware equipment site for kinematic vehicle experiment, including four geodetic GNSS receivers and antennas, NovAtel SPAN-FSAS GNSS/INS system, low-cost NEO-M8T receiver with a patch antenna, and a BMP280 barometer: (**a**) the experimental vehicle; (**b**) equipment setup diagram.

3.4.1. Open Urban Environment

This experiment was collected near the industrial park in Caidian District, Wuhan, China, on 20 November 2020, with a data duration of about 80 min. The area has an open urban environment with less shading and good data quality. Figure 12 shows the kinematic vehicle experiment test scene and test trajectory in the open urban environment. Figure 13 shows the number of GNSS visible satellites and the sequence of PDOP values during the kinematic vehicle experiment.

(**a**) (**b**)

Figure 12. Kinematic vehicle experiment in the open urban environment: (**a**) test scene; (**b**) test trajectory.

As can be seen in Figure 13, the number of observable satellites of GPS and BDS fluctuated greatly for the kinematic vehicle, resulting in a significant increase in PDOP values compared to a static environment. The addition of DBA observations significantly improved the satellite geometric spatial distribution and reduced the PDOP values. In the open urban environment kinematic vehicle experiment, the comparison of the low-cost BMP280 barometer DBA altitude with the Trimble NET R9 reference altitude is shown in Figure 14. It can be seen that the BMP280 barometer DBA altitude had a high consistency with the reference altitude with an RMSE of 2.10 m. This value can be used to set the a priori weight matrix in the DGNSS/DBA combined positioning process.

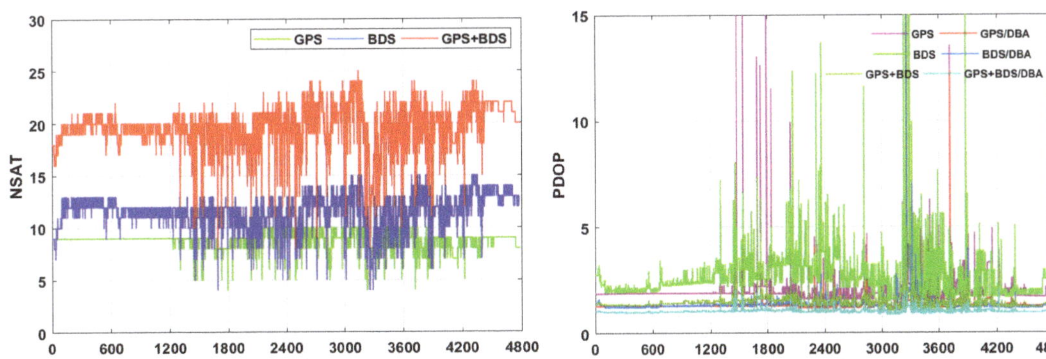

(**a**) (**b**)

Figure 13. Observation values of kinematic vehicle experiment in the open urban environment: (**a**) the number of GNSS visible satellites; (**b**) the sequence of PDOP values.

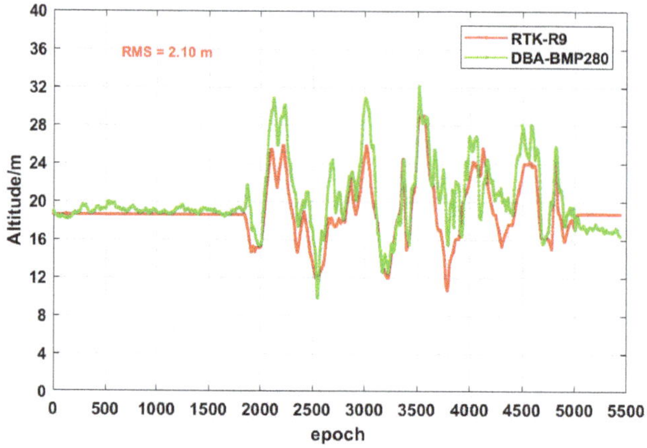

Figure 14. Comparison of the low-cost BMP280 barometer DBA altitude with the Trimble NET R9 reference altitude during the kinematic vehicle experimental in the open urban environment.

The RMSE statistical results and deviation sequence in the N/E/U directions for both DGNSS and DGNSS/DBA modes are given in Table 6 and Figure 15, respectively. The RMSE values of single GPS and single BDS in the N/E/U directions were 1.20/1.32/3.18 m and 1.58/2.11/5.02 m for DGNSS mode, respectively. The GPS/DBA, BDS/DBA, and GPS+BDS/DBA modes improved the RMSE by 40% to 60% in the U direction and increased slightly in the N and E directions, and the DBA altitude showed a good constraint effect.

Table 6. The RMSE of bias in the N/E/U directions for DGNSS and DGNSS/DBA modes during the kinematic vehicle experiment in the open urban environment.

Positioning Mode	Satellite System	N/m	E/m	U/m
DGNSS	GPS	1.20	1.32	3.18
	BDS	1.58	2.11	5.02
	GPS+BDS	1.00	1.31	2.12
DGNSS/DBA	GPS	1.11	1.33	1.34
	BDS	1.54	2.17	2.71
	GPS+BDS	1.02	1.31	1.24

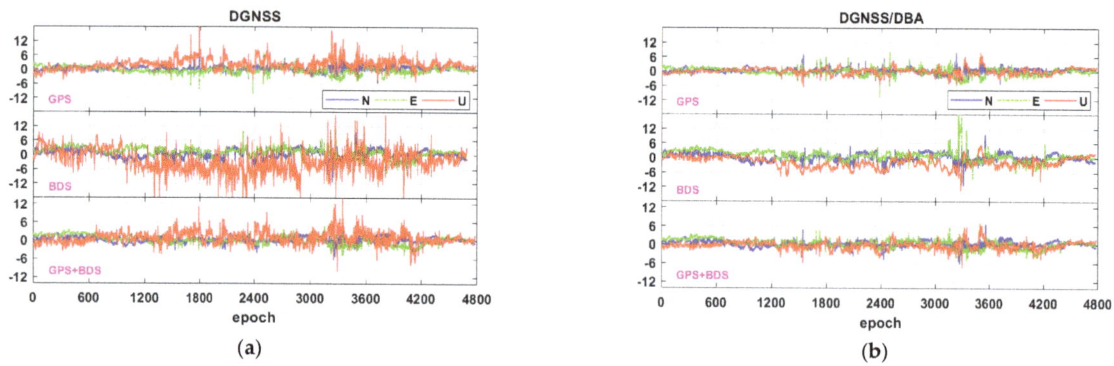

Figure 15. The deviation sequence diagram in the N/E/U directions during the kinematic vehicle experiment in the open urban environment: (**a**) DGNSS positioning mode; (**b**) DGNSS/DBA positioning mode.

3.4.2. Complex Urban Environment

This experiment was collected in Wuchang District, Wuhan, China, on November 20, 2020, with a valid data duration of about 1 h. The area is a complex urban environment, and the occlusion is relatively serious. Figure 16 shows kinematic vehicle experiment test scene and test trajectory in the complex urban environment. Figure 17 shows the number of visible GNSS satellites and the sequence of PDOP values during the kinematic vehicle experiment, and it can be seen that compared with the open urban environment, the number of GNSS visible satellites of the kinematic vehicle in the urban environment was significantly lower and the PDOP values became larger.

(**a**) (**b**)

Figure 16. Kinematic vehicle experiment in the complex urban environment: (**a**) indicative test scene; (**b**) test route.

Figure 18 shows the DBA altitude results of the low-cost BMP280 barometer during the kinematic vehicle experiment in the complex urban environment compared to the Trimble Net R9 reference altitude. It can be seen that the DBA altitude in the complex urban environment was also very consistent with the reference altitude with an RMSE of 2.19 m, which is approximately the same as the RMSE results of the DBA altitude in the open urban environment in Section 3.4.1. It indicates that the DBA altitude accuracy was reliable and stable in different environments, and could assist GNSS to improve the positioning accuracy in the complex urban environment.

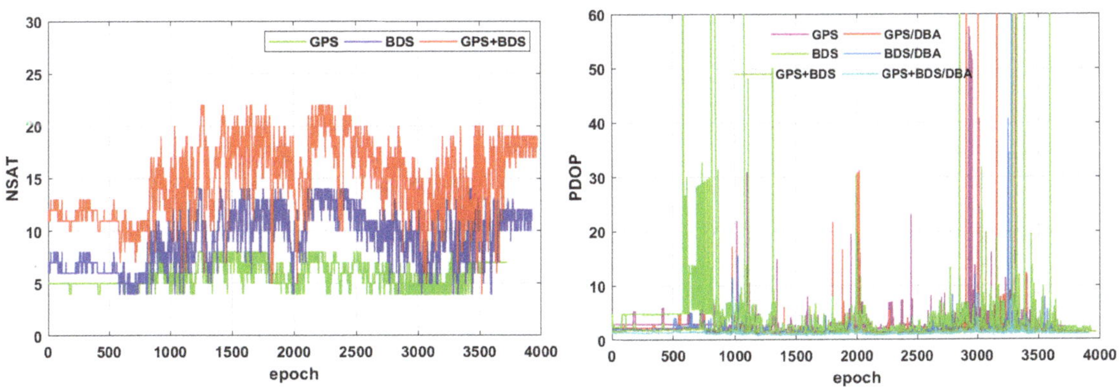

(**a**) (**b**)

Figure 17. Observation values of kinematic vehicle experiment in the complex urban environment: (**a**) number of GNSS visible satellites; (**b**) sequence of PDOP values.

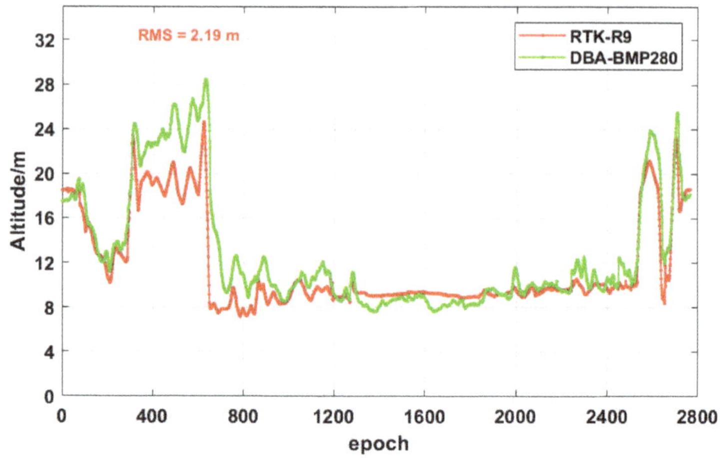

Figure 18. Comparison of the low-cost BMP280 barometer DBA altitude with the Trimble NET R9 reference altitude during the kinematic vehicle experiment in the complex urban environment.

The RMSE statistics and deviation sequence of the single-frequency, low-cost u-blox NEO-M8T and the geodetic Trimble Net R9 receiver for both DGNSS and DGNSS/DBA modes in the N/E/U directions are given in Table 7 and Figures 19 and 20. The RMSE of the low-cost u-blox NEO-M8T receiver with single GPS and single BDS in DGNSS mode were 4.33/4.69/8.35 m and 4.91/6.91/19.48 m in the N/E/U directions, respectively, and the RMSE of the GPS+BDS dual system was 3.28/5.23/8.91 m. The RMSE statistics of the geodetic Trimble Net R9 receiver in N/E direction in the complex urban environment was maintained at submeter level; the U direction was relatively poor, with RMSE not exceeding 2.7 m. The difference of RMSE between the two GNSS receivers in 3D directions was determined by their hardware performance. Compared with DGNSS mode, the RMSE of single GPS, single BDS, and GPS+BDS dual system in DGNSS/DBA mode of low-cost NEO-M8T receiver slightly worsened in the N and E directions, while the RMSE in U direction could be improved by 50% to 80%, and this improvement ratio is higher than that in the open urban environment. Similarly, for the geodetic Trimble Net R9 receiver, the RMSE of the DGNSS/DBA combination remains the same in the N/E directions compared to DGNSS mode, and the RMSE in the U direction could be improved by 30% to 60%, which verifies the advantage of DBA altitude in assisting DGNSS positioning.

Table 7. The RMSE of bias in the N/E/U directions for low-cost u-blox NEO-M8T and geodetic Trimble Net R9 receivers in the complex urban environment.

Positioning Mode	Satellite System	NEO-M8T Receiver				Trimble Net R9 Receiver			
		N/m	E/m	U/m	3D/m	N/m	E/m	U/m	3D/m
DGNSS	GPS	4.33	4.69	8.35	10.51	1.16	2.03	2.47	3.40
	BDS	4.91	6.91	19.48	21.24	1.19	0.80	2.64	3.00
	GPS+BDS	3.28	5.23	8.91	10.84	0.96	1.03	1.38	1.97
DGNSS/DBA	GPS	4.69	4.94	2.19	7.15	1.19	1.95	1.17	2.56
	BDS	5.13	7.14	4.76	9.99	1.05	0.64	1.24	1.74
	GPS+BDS	3.51	5.64	4.28	7.90	0.97	1.05	0.97	1.72

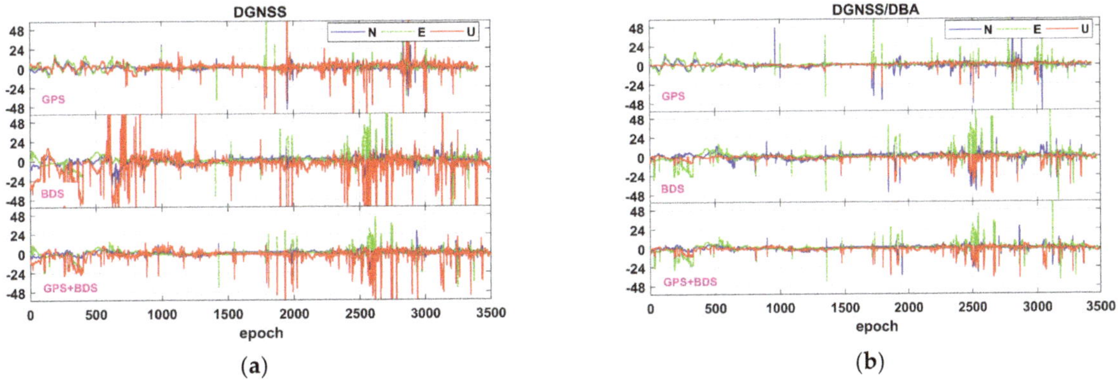

Figure 19. The deviation sequence of low−cost NEO−M8T receiver in the N/E/U directions during the kinematic vehicle experiment in the complex urban environment: (**a**) DGNSS positioning mode; (**b**) DGNSS/DBA positioning mode.

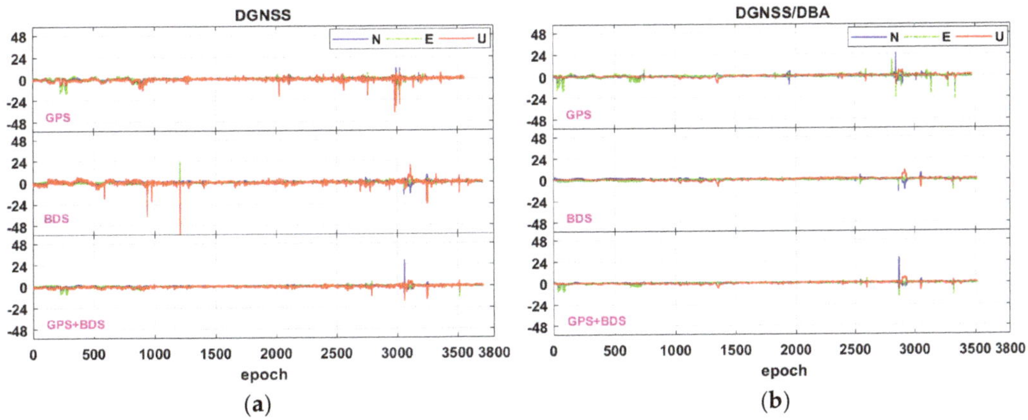

Figure 20. The deviation sequence of geodetic Trimble Net R9 receiver in the N/E/U directions during the kinematic vehicle experiment in the complex urban environment: (**a**) DGNSS positioning mode; (**b**) DGNSS/DBA positioning mode.

It can also be seen that when comparing Tables 3–5 for the static experiment with Tables 6 and 7 for the kinematic experiments, a low-cost receiver with only BDS signals provided lower RMSE values than GPS in static experiments, which was mainly due to BDS having more observation satellites and the PDOP value being lower than GPS at this time. In open and complex urban kinematic experiments, as shown in Figures 13 and 17, BDS had very large fluctuations for the number of observable satellites and a larger average PDOP value (3.02 and 4.10) than GPS (1.97 and 3.16) for both cases. This is the main reason that a low-cost receiver with only GPS signals provides lower RMSE values than with BDS in kinematic experiments.

4. Discussion

As the low-cost single-frequency GNSS receivers dominate most of the GNSS market [40], there is a strong interest in enhancing their accuracy. Low-cost DBA altitude plays a significant constraining role in improving the DGNSS positioning accuracy.

In the DBA altitude accuracy evaluation experiment, BMP280 barometers can achieve better than 2 m altitude accuracy within 10 km baseline lengths in static environments.

The DBA altitude consistent with GNSS reference altitude in Figures 14 and 18 implies that it is also reliable and stable in complex environments. Low-cost single-frequency GNSS receivers with a patch antenna have become increasingly popular due to their lower and lower price. The DGNSS positioning accuracy of single-frequency low-cost GNSS receivers can still meet the submeter positioning accuracy needed by the general public in GNSS-friendly environments. However, the RMSE in the N/E/U directions are all at the meter-level in complex urban environments since low-cost GNSS receivers have poor observation quality, and the positioning accuracy of GPS+BDS dual system is significantly improved compared to single system. There is only a single-epoch resolution algorithm rather than a filtering algorithm is used in this study. In the future, with more and more satellites available for low-cost GNSS receivers and the use of multiple filtering algorithms, DGNSS positioning accuracy is expected to be further improved.

The DGNSS/DBA combined positioning can effectively improve the DGNSS positioning accuracy and meet the demand for real-time positioning applications. The Earth ellipsoid constraint equation constructed by the DBA altitude is equivalent to adding a virtual satellite located at the center of the Earth, effectively improving the spatial geometry structure of the observation satellite. The DGNSS/DBA combined positioning improves the positioning accuracy in the U direction by 30% to 80% compared with the DGNSS positioning, while the positioning accuracy in N and E directions also has a certain improvement effect.

Nowadays, most smartphones integrate both an inexpensive GNSS chip and a barometric pressure sensor. WADGNSS services [10] and a large number of meteorological stations [26] can provide correction information to users. We can achieve higher positioning accuracy without increasing hardware costs. The applications of low-cost DGNSS/DBA, such as indoor and outdoor seamless switching positioning, car navigation, emergency mapping, LBS, and rescue, etc. are likely to increase dramatically. Subsequently, based on the combined positioning of low-cost DGNSS/DBA, the positioning performance research by integrating other sensors, such as MEMS IMU and geomagnetic, etc., will be worth further investigation.

5. Conclusions

In this study, low-cost single-frequency DGNSS/DBA combined positioning research and performance evaluation was carried out. First, a DGNSS/DBA combined positioning model is proposed. The Earth ellipsoid constraint equation act as a virtual satellite observation at the center of the Earth, effectively improving the spatial geometry structure and PDOP value of the observation satellite. The low-cost BMP280 barometer DBA altitude accuracy is evaluated by different baseline lengths, which is better than the submeter level within 2 km and better than 2 m within 10 km baseline length. In both open and complex urban environment kinematic vehicle experiments, the DBA altitude accuracy is better than 2.20 m, which indicates that the DBA system has highly reliable and stability in different environments in local area.

The low-cost single-frequency NEO-M8T receiver with a patch antenna can achieve submeter level positioning accuracy for DGNSS positioning in the N/E directions and better than 1.5 m in the U direction in a short baseline static environment; as the baseline length increases, the DGNSS positioning accuracy gradually decreases. The positioning accuracy in a kinematic vehicle environment is significantly lower than in the static environment, and the RMSE in the N/E/U directions are all at the meter level in the complex urban environment, and the positioning accuracy of both GPS+BDS dual system is significantly improved compared to single system. The DGNSS/DBA combined positioning for low-cost NEO-M8T receiver and BMP280 barometer improves the positioning accuracy in the U direction by 30% to 80% compared with the DGNSS positioning, while the positioning accuracy in the N and E directions also has a certain improvement effect.

Author Contributions: Conceptualization, S.W.; methodology, S.W.; software, S.W., M.G., W.Z., D.L. and S.C.; validation, M.G., W.Z. and D.L.; formal analysis, M.G.; investigation, S.W., M.G. and S.C.;

resources, G.L.; data curation, S.W.; writing—original draft preparation, S.W.; writing—review and editing, X.D., G.L., M.G., W.Z., D.L. and S.C.; visualization, S.W. and M.G.; supervision, X.D. and G.L.; project administration, X.D. and G.L.; funding acquisition, S.W., X.D. and G.L. All authors have read and agreed to the published version of the manuscript.

Funding: This work was jointly supported by the National Key Research Program of China "Collaborative Precision Positioning Project" (No. 2016YFB0501900) and the National Natural Science Foundation of China (Grant No. 41774017).

Acknowledgments: The authors would like to thank Gongwei Xiao and Aizhi Guo for their help in setting up the multi-sensor platform used in the kinematic vehicle experiments. Meanwhile, the authors are grateful to Chengfeng Zhang for his valuable advice in writing.

Conflicts of Interest: The authors declare no conflict of interest.

References

1. Han, H.; Wang, J.; Wang, J.; Tan, X. Performance Analysis on Carrier Phase-Based Tightly-Coupled GPS/BDS/INS Integration in GNSS Degraded and Denied Environments. *Sensors* **2015**, *15*, 8685–8711. [CrossRef] [PubMed]
2. Cristodaro, C.; Dovis, F.; Falco, G.; Pini, M. GNSS Receiver Performance in Urban Environment: Challenges and Test Approaches for Automotive Applications. In Proceedings of the International Conference of Electrical & Electronic Technologies for Automotive, Turin, Italy, 15–16 June 2017.
3. Tsakiri, M.; Sioulis, A.; Piniotis, G. The use of low-cost, single-frequency GNSS receivers in mapping surveys. *Surv. Rev.* **2016**, *50*, 46–56. [CrossRef]
4. Biagi, L.; Grec, F.C.; Negretti, M. Low-Cost GNSS Receivers for Local Monitoring: Experimental Simulation, and Analysis of Displacements. *Sensors* **2016**, *16*, 2140. [CrossRef]
5. Hamza, V.; Stopar, B.; Ambroi, T.; Turk, G.; Sterle, O. Testing Multi-Frequency Low-Cost GNSS Receivers for Geodetic Monitoring Purposes. *Sensors* **2020**, *20*, 4375. [CrossRef]
6. Long, C.; Wen, Q. Positioning performance analysis of a low-cost u-blox single-frequency multi-GNSS receiver. *GNSS World China* **2019**, *44*, 82–88.
7. Stempfhuber, W.; Buchholz, M. A precise, low-cost RTK GNSS system for UAV applications. *ISPRS—Int. Arch. Photogramm. Remote Sens. Spat. Inf. Sci.* **2012**, *XXXVIII-1/C22*, 289–293. [CrossRef]
8. Wang, Y.; Pan, S.; Guorong, Y.U.; Zhang, J. An algorithm of BDS/GPS positioning with single clock bias parameter with the consideration of inter-system bias. *Eng. Surv. Mapp.* **2019**, *28*, 7–12.
9. Lu, L.G.; Ma, L.Y.; Wu, T.T.; Chen, X.Y. Performance Analysis of Positioning Solution Using Low-Cost Single-Frequency U-Blox Receiver Based on Baseline Length Constraint. *Sensors* **2019**, *19*, 4352. [CrossRef]
10. Specht, C.; Pawelski, J.; Smolarek, L.; Specht, M.; Dabrowski, P. Assessment of the Positioning Accuracy of DGPS and EGNOS Systems in the Bay of Gdansk using Maritime Dynamic Measurements. *J. Navig.* **2019**, *72*, 575–587. [CrossRef]
11. Ji, S.; Gao, Z.; Wang, W. M-DGPS: Mobile devices supported differential global positioning system algorithm. *Arab. J. Geosci.* **2015**, *8*, 6667–6675. [CrossRef]
12. Yoon, D.; Kee, C.; Seo, J.; Park, B. Position Accuracy Improvement by Implementing the DGNSS-CP Algorithm in Smartphones. *Sensors* **2016**, *16*, 910. [CrossRef] [PubMed]
13. Chen, H.; Moan, T.; Verhoeven, H. Effect of DGPS failures on dynamic positioning of mobile drilling units in the North Sea. *Accid. Anal. Prev.* **2009**, *41*, 1164–1171. [CrossRef]
14. Lubis, M.Z.; Anggraini, K.; Kausarian, H.; Pujiyati, S. Review: Marine Seismic And Side-Scan Sonar Investigations For Seabed Identification With Sonar System. *J. Geosci. Eng. Environ. Technol.* **2017**, *2*, 166–170. [CrossRef]
15. Rathour, S.S.; Boyali, A.; Zheming, L.; Mita, S.; John, V. A map-based lateral and longitudinal DGPS/DR bias estimation method for autonomous driving. *Int. J. Mach. Learn. Comput.* **2017**, *7*, 67–71. [CrossRef]
16. Ssebazza, L.; Pan, Y.J. DGPS-based localization and path following approach for outdoor wheeled mobile robots. *Int. J. Rob. Autom.* **2015**, *30*, 13–25. [CrossRef]
17. Grunwald, G.; Bakua, M.; Cieko, A. Study of EGNOS accuracy and integrity in eastern Poland. *Aeronaut. J.* **2016**, *120*, 1275–1290. [CrossRef]
18. Specht, C.; Weintrit, A.; Specht, M. A History of Maritime Radio-Navigation Positioning Systems used in Poland. *J. Navig.* **2016**, *69*, 468–480. [CrossRef]
19. Kim, J.; Song, J.; No, H.; Han, D.; Kim, D.; Park, B.; Kee, C. Accuracy Improvement of DGPS for Low-Cost Single-Frequency Receiver Using Modified Flächen Korrektur Parameter Correction. *Int. J. Geo-Inf.* **2017**, *6*, 222. [CrossRef]
20. Yang, Y. Low-Cost Single Frequency DGPS Aided INS for Vehicle Control. In Proceedings of the 13th International Technical Meeting of the Satellite Division of The Institute of Navigation (ION GPS 2000), Salt Lake City, UT, USA, 19–22 September 2000; pp. 2299–2308.

21. Guo, R.; Su, R.R.; Liu, L.; Hu, G.M.; Chang, Z.Q. COMPASS RDSS Positioning Accuracy Analysis. In Proceedings of the China Satellite Navigation Conference (CSNC) 2014 Proceedings, Nanjing, China, 21–23 May 2014; Springer: Berlin/Heidelberg, Germany, 2014; Volume III, pp. 219–228.
22. Jan, S.S.; Gebre-Egziabher, D.; Walter, T.; Enge, P. Improving GPS-based landing system performance using an empirical barometric altimeter confidence bound. *IEEE Trans. Aerosp. Electron. Syst.* **2008**, *44*, 127–146.
23. Paces, P.; Popelka, J. Performance evaluation of two altimeters intended for Euler angles measurement. In Proceedings of the 2013 IEEE/AIAA 32nd Digital Avionics Systems Conference (DASC), New York, NY, USA, 5–10 October 2013.
24. Xia, H.; Wang, X.G.; Qiao, Y.Y.; Jian, J.; Chang, Y.F. Using Multiple Barometers to Detect the Floor Location of Smart Phones with Built-in Barometric Sensors for Indoor Positioning. *Sensors* **2015**, *15*, 7857–7877. [CrossRef]
25. Vanini, S.; Faraci, F.; Ferrari, A.; Giordano, S. Using barometric pressure data to recognize vertical displacement activities on smartphones. *Comput. Commun.* **2016**, *87*, 37–48. [CrossRef]
26. Ai, G.; Shi, H.; Wu, H.; Yan, Y.; Bian, Y.; Hu, Y.; Li, Z.; Guo, J.; Cai, X. A Positioning System based on Communication Satellites and the Chinese Area Positioning System (CAPS). *Chin. J. Astron. Astrophys.* **2008**, *8*, 611–630.
27. Shi, H.; Pei, J. The solutions of navigation observation equations for CAPS. *Sci. China Ser. G-Phys. Mech. Astron.* **2009**, *52*, 434–444. [CrossRef]
28. Ji, Y.F.; Sun, X.Y. Analysis on the positioning precision of CAPS. *Sci. China Ser. G-Phys. Mech. Astron.* **2009**, *52*, 328–332. [CrossRef]
29. Hu, Z.Q.; Zhang, L.R.; Ji, Y.F. Applications of differential barometric altimeter in ground cellular communication positioning network. *IET Sci. Meas. Technol.* **2020**, *14*, 322–331. [CrossRef]
30. Du, X.; Pei, J.; Zhang, L.; Liu, C. Differential barometric altimetry method based on mobile phone base stations. *J. Beijing Univ. Aeronaut. Astronaut.* **2013**, *39*, 83–88. [CrossRef]
31. Sabatini, A.M.; Genovese, V. A Sensor Fusion Method for Tracking Vertical Velocity and Height Based on Inertial and Barometric Altimeter Measurements. *Sensors* **2014**, *14*, 13324–13347. [CrossRef]
32. Wang, H.; Wen, Y.; Zhao, D. Differential barometric-based positioning technique for indoor elevation measurement in IoT medical applications. *Technol. Health Care* **2017**, *25*, S295–S304. [CrossRef]
33. Masse, F.; Bourke, A.K.; Chardonnens, J.; Paraschiv-Ionescu, A.; Aminian, K. Suitability of commercial barometric pressure sensors to distinguish sitting and standing activities for wearable monitoring. *Med. Eng. Phys.* **2014**, *36*, 739–744. [CrossRef]
34. Xu, G.; Xu, Y. *GPS Theory, Algorithms and Applications*, 3rd ed.; Science Press: Beijing, China, 2017.
35. Bolanakis, D.E.; Laopoulos, T.; Kotsis, K.T. A Prototype Wireless Sensor Network System for a Comparative Evaluation of Differential and Absolute Barometric Altimetry. *IEEE Aerosp. Electron. Syst. Mag.* **2015**, *30*, 20–28. [CrossRef]
36. Hu, Z. *Study on Key Technologies in Integration System of Navigation based on Multi-Sensor Information such as Differential Barometric Altimetry*; University of Chinese Academy of Sciences: Beijing, China, 2014.
37. Shi, H.-L.; Sun, X.; Li, Z. *Principle of Forwarding Satellite Navigation*; Science Press: Beijing, China, 2009.
38. Li, Z.; Hunag, J. *GPS Surveying and Data Processing*; Wuhan University Press: Wuhan, China, 2013.
39. King, B.; Bock, Y. *Documentation for the GAMIT GPS Analysis Software*; Massachusetts Institute of Technology: Cambridge, MA, USA, 1999.
40. GSA, G. *GNSS Market Report Issue 3*; European Global Navigation Satellite Systems Agency: Prague, Czech Republic, 2013; pp. 7–10.

remote sensing

MDPI

Article

LiDAR-Visual-Inertial Odometry Based on Optimized Visual Point-Line Features

Xuan He [1,2], Wang Gao [1,2,*], Chuanzhen Sheng [3,4], Ziteng Zhang [3,4], Shuguo Pan [1,2], Lijin Duan [5], Hui Zhang [1,2] and Xinyu Lu [1,2]

1 School of Instrument Science and Engineering, Southeast University, Nanjing 210096, China; hexuan@seu.edu.cn (X.H.); psg@seu.edu.cn (S.P.); amzhanghui@seu.edu.cn (H.Z.); 220213597@seu.edu.cn (X.L.)

2 Key Laboratory of Micro-Inertial Instrument and Advanced Navigation Technology, Southeast University, Nanjing 210096, China

3 State Key Laboratory of Satellite Navigation System and Equipment Technology, Shijiazhuang 050081, China; shengchuanzhen@163.com (C.S.); zzteng54@163.com (Z.Z.)

4 The 54th Research Institute of China Electronics Technology Group Corporation, Shijiazhuang 050081, China

5 Linzi District Transportation Service Center, Zibo 255400, China; lzqjtysjqdjgs@zb.shandong.cn

* Correspondence: gaow@seu.edu.cn

Abstract: This study presents a LiDAR-Visual-Inertial Odometry (LVIO) based on optimized visual point-line features, which can effectively compensate for the limitations of a single sensor in real-time localization and mapping. Firstly, an improved line feature extraction in scale space and constraint matching strategy, using the least square method, is proposed to provide a richer visual feature for the front-end of LVIO. Secondly, multi-frame LiDAR point clouds were projected into the visual frame for feature depth correlation. Thirdly, the initial estimation results of Visual-Inertial Odometry (VIO) were carried out to optimize the scanning matching accuracy of LiDAR. Finally, a factor graph based on Bayesian network is proposed to build the LVIO fusion system, in which GNSS factor and loop factor are introduced to constrain LVIO globally. The evaluations on indoor and outdoor datasets show that the proposed algorithm is superior to other state-of-the-art algorithms in real-time efficiency, positioning accuracy, and mapping effect. Specifically, the average RMSE of absolute trajectory in the indoor environment is 0.075 m and that in the outdoor environment is 3.77 m. These experimental results can prove that the proposed algorithm can effectively solve the problem of line feature mismatching and the accumulated error of local sensors in mobile carrier positioning.

Keywords: multi-sensor fusion; visual point and line feature; SLAM; LiDAR-visual-inertial odometry

Citation: He, X.; Gao, W.; Sheng, C.; Zhang, Z.; Pan, S.; Duan, L.; Zhang, H.; Lu, X. LiDAR-Visual-Inertial Odometry Based on Optimized Visual Point-Line Features. *Remote Sens.* **2022**, *14*, 622. https://doi.org/10.3390/rs14030622

Academic Editor: Giuseppe Casula

Received: 20 December 2021
Accepted: 26 January 2022
Published: 27 January 2022

1. Introduction

Multi-sensor fusion localization technology based on Simultaneous Localization and Mapping (SLAM) is a fundamental technology in the field of high-precision localization of mobile carriers [1]. The SLAM-based multi-sensor fusion system applied to mobile carriers can be divided into two core parts: the front-end, and the back-end. The function of the front-end is used to analyze the environmental fingerprint information collected by the sensors, in order to estimate the positional information of the mobile carrier in time. In addition, the change in the surrounding environment with the movement of the carrier is restored. The function of the back-end is used to obtain the final positioning results by iteratively optimizing the position estimates obtained from the front-end analysis. Depending on the sensors used in the front-end, it can be divided into methods mainly based on LiDAR and vision [2,3]. Engineers and researchers in related fields have conducted a lot of research in both directions and produced a series of research-worthy results.

The main vision-based SLAM approach, namely visual odometry (VO), has long dominated the SLAM technology field due to the lower cost of the camera compared with

LiDAR. However, pure monocular visual SLAM systems cannot recover metric scales. Thus, there is a growing trend to utilize low-cost inertial measurement units to assist monocular vision systems, which is called visual-inertial odometry (VIO). Monocular VIO provides high-quality self-motion simulation by using monocular cameras and inertial measurement unit (IMU) measurements, which has significant advantages in terms of size, cost, and power. Based on the method of feature association, visual SLAM can be classified into feature point method and direct method. The feature point-based method VIO accomplishes the inter-frame feature constraint by extracting and matching image feature points [4–6]. Therefore, rich environmental texture is required to ensure that the threshold of the number of effective feature points required for feature tracking is reached. Tracking loss of feature points is prone to occur in weak texture environments such as parking lots and tunnels, which in turn affects localization accuracy and real-time performance. The theoretical basis of the direct method-based VIO is the assumption of constant grayscale [7,8]. It only needs to capture environmental features by the changes in the grayscale image to establish constraints, which has a better real-time performance. Nevertheless, the tracking accuracy is greatly affected by environmental illumination changes. Therefore, stable and rich line feature models are required to be introduced into the front-end to provide stable and accurate feature constraints for visual back-end state estimation. In 2018, He et al. proposed PL-VIO based on point-line feature fusion, but too many optimization factors greatly limited the real-time performance in practical tests [9]. In 2020, Wen et al. proposed PLS-VIO to optimize the 6-DOF pose by minimizing the objective function and improving the line feature matching filtering strategy to reduce the probability of mismatching [10]. Although the VIO based on point-line features has a positive effect on the number of features [11,12], it still cannot solve the scale uncertainty problem of monocular cameras. The development of VIO in practical applications still has certain limitations.

As another important technical means of SLAM-based localization technology, SLAM mainly based on LiDAR is also widely used in the industry for its high resolution, high accuracy, and high utilization of spatial features. In 2016, Google proposed Cartographer, a 2D LiDAR based on particle filtering and graph optimization. In 2017, Zhang et al. proposed the LOAM for the first time, which uses the curvature of the LiDAR point cloud to register the effective point cloud features as planar points and edge points [13]. In 2018, Shan et al. proposed LeGO-LOAM based on LOAM, which uses the ground plane feature point cloud to further filter outliers from the scanned point cloud and improve the LOAM frame [2]. In 2020, Shan et al. further introduced the LIO-SAM algorithm based on the previous work, which uses IMU pre-integrated measurements to provide initial pose estimation for laser odometry [14]. In addition, a Bayesian network-based factor graph optimization framework is proposed, in which the global position is constrained by adding GPS factors, and an incremental smooth global voxel map is established. These schemes provide technical feasibility for the high-precision positioning by fusing LiDAR with other sensors.

However, due to the inherent shortcomings of the main sensing sensors, such as the limited scanning angle of LiDAR and the sensitivity of the mainly vision-based methods to light variations, these methods can hardly show excellent robustness in real-world applications. To further improve the localization performance, LiDAR-Visual-Inertial Odometry, as a multi-sensor fusion localization method, has become a research focus of SLAM with its advantages of multi-sensor heterogeneity and complementarity.

The existing LVIO multi-sensor fusion strategy can be described from the front-end and back-end perspectives. First, the front-end fusion strategy of LVIO is introduced. Generally, LiDAR acts as a feature depth provider for monocular VO as a way to improve the scale ambiguity of visual features. Meanwhile, VO performs state estimation from the extracted visual features, which is provided as the initial state for LiDAR scan matching. Therefore, the quantity and quality of visual features are closely related to the precision of state estimation of the fusion system. In existing fusion systems, the features extracted by camera are mainly point features [15,16]. Xiang et al. proposed a combination of fisheye camera and LiDAR based on a semantic segmentation model, which improved the confidence of the depth of visual features in the driving environment of unmanned vehicles [15]. Chen et al. proposed a method to construct a loopback constraint for LiDAR-visual odometry by using the Distributed bag of Words (DboWs) model in the visual subsystem, although, without introducing IMU sensors to assist in the initial positional estimation [16]. In 2021, Lin et al. proposed R2LIVE to incorporate IMU into the fused localization system, in which the LiDAR odometry is used to establish depth constraints for VIO [17]. Although the above-mentioned algorithms exhibit superior performance to the VIO based on point features, it is still difficult to extract rich and effective features in weak texture environments, which leads to the failure in LiDAR scan matching. Therefore, additional feature constraints on the LiDAR need to be added with line features that are more robust to environmental texture and luminosity variations. Visual SLAM based on point-line features has been studied but not widely applied to LVIO systems in recent years [18,19]. In 2020, Huang et al. first proposed a LVIO based on a robust point and line depth extraction method, which greatly reduces the three-dimensional ambiguity of features [18]. Zhou et al. introduced line features in the direct method-based VIO to establish data association [19]. The above-mentioned algorithms provide technical feasibility for LVIO based on point-line features.

From the perspective of the back-end fusion strategy, LVIO can be classified into two categories based on different optimization algorithms: filter-based methods and factor graph methods. Although the filtering method is a traditional technology to realize multi-sensor fusion, its principle defect of frequent reconstruction of increasing or decreasing sensors limits its application in LVIO [20]. As an emerging method in recent years, the factor graph method can effectively improve the robustness of SLAM system when a single sensor fails because of its plug-and-play characteristics. Therefore, it is widely applied to deal with such heterogeneous aperiodic data fusion problems [21]. In addition, since LVIO is in the local frame, there are inherent defects such as accumulated errors. Thus GNSS measurements need to be introduced for global correction [22–24] to realize local accuracy and global drift-free position estimation, which makes full use of their complementarity [24]. The research on adding GNSS global constraints into the local sensor fusion framework are as follows: Lin et al. modified the extended Kalman filter to realize a loose coupling between GPS measurements and LiDAR state estimation, but there is a large single linearization error to be solved [17]. In 2019, Qin et al. proposed VINS-Fusion, which uses nonlinear optimization strategies to support Camera, IMU, and GNSS [25], but it assumes that GNSS is continuous and globally convergent, which is inconsistent with reality. In any case, these strategies presented above provide numerous reliable ideas.

Generally speaking, we can conclude that the existing LVIO fusion system has two problems that deserve further exploration. First, on the premise of ensuring the real-time performance, more abundant feature constraints are needed to improve the pose estimation accuracy of LVIO. Secondly, global constraints are needed to globally optimize the LVIO local pose estimation results. To address these issues, this study presents a LiDAR-Visual-Inertial Odometry based on optimized visual point-line features. First of all, an improved line feature extraction in scale space and constraint matching strategy based on the square method are proposed, which provides richer visual feature for the front-end of LVIO. Secondly, multi-frame LiDAR point clouds were projected into the visual frame for feature depth correlation, which improves the confidence of monocular visual depth estimation. At the same time, the initial visual state estimation can be used to optimize the scan matching of LiDAR. Finally, a factor graph based on the Bayesian network was used to build the LVIO fusion system, in which the GNSS factor and loop factor are introduced to constrain LVIO globally, to achieve locally accurate and globally drift-free position estimation in the complex environment.

2. System Overview

The general framework of the LiDAR-Visual-Inertial Odometry based on optimized visual point-line features proposed in this study is shown in Figure 1. The system consists of the front-end of LiDAR-Visual-Inertial Odometry tight combination and the back-end of factor graph optimization.

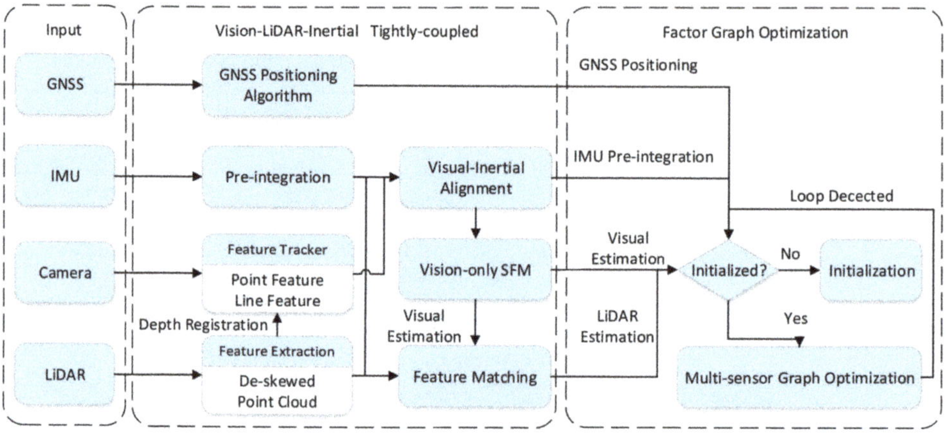

Figure 1. Overall algorithm framework, system inputs include IMU, camera, lidar and optional GNSS. IMU provides initial state correction for VIO subsystem and LiDAR-inertial odometry (LIO) subsystem, VIO and LIO systems use each other's information to improve the positioning accuracy, and GNSS signals are optionally added to the back-end to provide global constraints.

In the front-end of our algorithm, the visual odometry not only extracts point features, but also further extracts line features in the improved scale space and performs geometric constraint matching on them, which improves the number of features in the weak texture environment. Then, the feature depth provided by LiDAR point clouds performed a role in correlating the depth of monocular visual features. IMU pre-integration provides all necessary initial values, including attitude, velocity, acceleration bias, gyroscope bias, and three-dimensional feature position, for completing the initial state estimation after time alignment with a camera. If VIO initialization fails, the IMU pre-integration value is used as the initial assumption to improve the robustness of the fusion system in the texture-free environment.

After the front-end initialization is successfully realized, the back-end optimizes the factor graph by using the estimated residual of each sensor's state. IMU pre-integration, visual residual and lidar residual were added to the factor graph as local state factors for maximum a posteriori estimation. In order to further correct the cumulative error of local state estimation, the residual of GNSS single-point positioning measurements was used as the global positioning factor to add to the factor graph. Besides, when the system detects the path loop, the loop factor will be added to the factor graph to participate in the nonlinear optimization and obtain the optimal global pose estimation.

3. Front-End: Feature Extraction and Matching Tracking

3.1. Line Feature Extraction

Commonly used line feature extraction algorithms include Hough [26], LSWMS [27], EDLine [28], and LSD [29]. Weighing factors such as accuracy, real-time performance, and the need for parameter adjustment, we chose LSD to extract line features. According to the bottom parameter optimization strategy, we modified an improved LSD algorithm, and a minimum geometric constraint method to realize line feature constraint matching.

Given an N-layer Gaussian pyramid as the scale space of LSD line features, the scale ratio of images in each layer is defined to reduce or eliminate the sawtooth effect in images. After scaling the image s times, a downsampling was performed, and then the gradient was calculated for all pixels in the new image obtained after downsampling. By traversing the image and getting the gradient values of all pixels, the pixel gradient rectangle can be merged according to the density of same-sex points to obtain a rectangle-like line segment l. The density d of homogeneous points in the rectangle can be expressed as:

$$d = \frac{k}{length(l) \cdot width(l)}, d \leq D \tag{1}$$

where k is defined as the total number of pixels in the rectangle, and D is the density threshold of parity points. Different from the hypothesis in [12], a low co-location density threshold in the outdoor complex texture environment will extract a large number of invalid line features. Therefore, it is necessary to re-optimize the strategy according to the underlying parameters and select the following combinations near the original parameters ($s = 0.8$, $D = 0.7$), for real-time and accuracy experiments.

We measured the positioning accuracy by the root mean square error of absolute trajectory error (APE_RMSE). The accuracy and real-time performance of different values of s and D on the Hong Kong 0428 dataset are shown in Figure 2. The Monte Carlo method was used in this experiment. Within the parameter range that ensures the stable operation of the line feature extraction algorithm, we conducted three experiments. First of all, as shown in Figure 2a, under the premise that the original scaling times $s = 0.8$, 100 random numbers were selected in the range of $D \in (0.3, 0.9)$ to carry out the experiment of density threshold selection. Secondly, as shown in Figure 2b, we kept the original density threshold $D = 0.7$, and then selected 100 random numbers in the range of $s \in (0.4, 0.9)$, which is to select the appropriate range of scaling times s. Finally, as shown in Figure 2c, within the appropriate parameter range obtained in the previous experiments, 100 groups of parameter combinations were randomly selected for line feature extraction to obtain the optimal value.

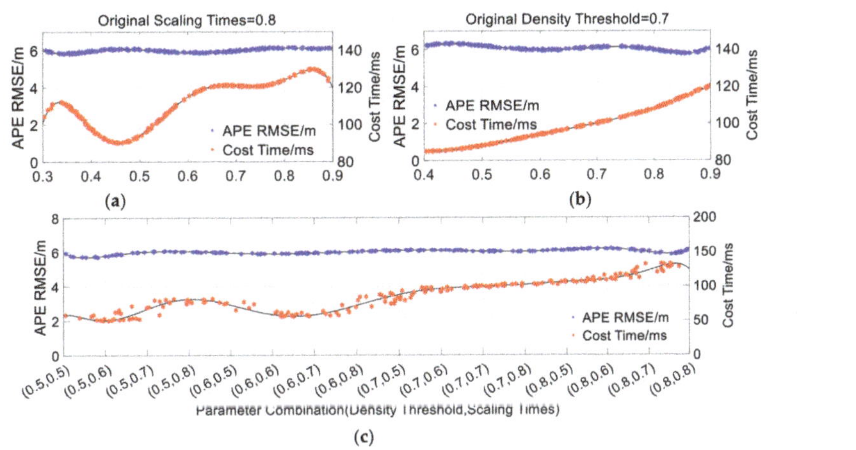

Figure 2. Underlying parameter selection. (**a**) Density threshold selection, (**b**) scaling times selection. (**c**) Experimental results by selecting the best combination of parameters. Noted that decreasing s and D will show better real-time performance with negligible loss of accuracy.

According to Figure 2c it can be seen that the operation time is shorter when the value of (s, D) is around $(0.5, 0.6)$ or around $(0.6, 0.6)$. Furthermore, we compared the accuracy of the above two groups of parameters. It can be concluded that the accuracy of line feature extraction of the former group is slightly higher than that of the latter group. Considering the accuracy and real-time, we chose $s = 0.5$, $D = 0.6$ as the parameter combination for our system.

3.2. Inter-Frame Feature Constraint Matching

Different from the neighboring line merging of different line features within the same frame in feature extraction, the least square method-based line feature constraint matching is for the same line feature pair whose angle and distance change between two consecutive frames. Considering the angle and translation changes in the same line feature pair during the carrier movement, a minimized sparse matrix model can be constructed to ensure the minimum total error in matching the line features extracted between the front and back frames.

Given a line $l^W = \left[n^{W^T}, v^{W^T}\right]^T \in R^6$ extracted from the world coordinate system, where $n^W, v^W \in R^3$ is the normal vector and direction vector, respectively, of l^W, let the transformation matrix from the world frame to camera frame be $T_C^W = \left[R_C^W, t_C^W\right]$, with R_C^W, t_C^W denoting the rotation and translation, respectively, then l^W can be expressed in Plücker coordinates within the camera frame as:

$$l^C = \begin{bmatrix} n^C \\ v^C \end{bmatrix} = T_C^W l^W = \begin{bmatrix} R_C^W & \left[t_C^W\right]_\times R_C^W \\ 0 & R_C^W \end{bmatrix} \begin{bmatrix} n^W \\ v^W \end{bmatrix} \in R^6 \tag{2}$$

It can be seen that the matching of line feature pairs in the camera frame is a 6-DOF parametric matching problem. In order to improve the accuracy and simplify the line feature matching problem, it can be simplified as a 4-DOF parameter matching optimization problem. Let all the line feature pairs obtained by matching between two consecutive frames in the camera frame be:

$$F_{ij} = \left\{ (l_i, l_j) \mid j \in [1, n] \right\} \tag{3}$$

where l_i and l_j are certain line features extracted in the previous frame and subsequent frame, respectively, n is the total number of line features in the subsequent frame.

According to the variation in the inter-frame line characteristics shown in Figure 3, the parameter matrix can be set as $e_{ij} = \left[\theta_{ij}, \mu_{ij}, \rho_{ij}, d_{ij}\right]^T$, θ_{ij} and d_{ij} are the included angle and translation distance between two consecutive frames, respectively, μ_{ij} and ρ_{ij} are the projection ratio and length ratio of the front-to-back interframe line features. Constructing the parameter matrix may establish a linear constraint matrix $A_i = \left[e_{i1}, \ldots, e_{ij}, e_{in}\right]$ of the subsequent keyframe for l_i. The target vector of the matching judgment of l_i is $m_i = \left[m_{i1}, \ldots, m_{ij}, \ldots, m_{in}\right]^T$. The value of each component is determined by the result of feature matching, where matching is 1 and non-matching is 0. If $\sum m_{in} = 1$, the linear constraint $A_i m_i = t$ will be satisfied. Therefore, the line feature matching problem can be optimized into a constrained matching equation based on least squares:

$$\min_{m_i} \lambda \|m_i\|_1 + \frac{1}{2}\|A_i m_i - t\|_2 \tag{4}$$

where λ is the weight coefficient and $t = [0, 1, 1, 0]^T$ is the constraint target vector.

Figure 3. Deviation of a line feature during the movement of the carrier. (**a**) Parallel offset (**b**) angular offset.

3.3. LiDAR-Aided Depth Correlation of Visual Features

LiDAR-aided depth correlation of visual features can effectively improve the scale ambiguity of monocular cameras. Since the LiDAR resolution is much lower than that of the camera, the use of only a single frame of sparse point cloud depth correlation will result in a large number of visual feature depth deletions [30]. Therefore, this study purposes a strategy of superimposing multi-frame sparse point cloud to obtain the depth value of the point cloud, which is used to establish the depth correlation with the visual features.

As shown in Figure 4, f_1^V is a feature point in the visual frame $\{V\}$, and $\{d_1^L, \ldots, d_m^L\}$ is a group of depth point clouds in the lidar frame $\{L\}$. Projecting d_n^L onto a unit spherical surface $\{V_g\}$ with f_1^V as the spherical center to obtain a projection point $d_n^{V_g}$:

$$d_n^{V_g} = R_L^{V_g} d_n^L + p_L^{V_g} \quad n \in [1, m] \tag{5}$$

where $R_L^{V_g}$ and $p_L^{V_g}$ are the rotation matrix and external parameter matrix of $\{L\}$ to $\{V_g\}$, respectively. Taking f_1^V as the root node to establish KD tree to search for the three closest depth points d_1, d_2, d_3 on the sphere. Then, connecting f_1^V with the camera center O and intersecting $\Delta d_1 d_2 d_3$ with O_d, we can obtain the characteristic depth of f_1^V as $f_1^V O_d$.

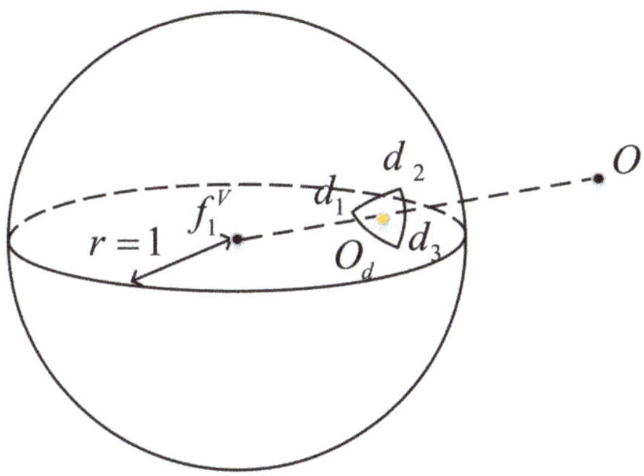

Figure 4. Association of visual feature depth.

4. Back-End: LVIO-GNSS Fusion Framework Based on Factor Graph

4.1. Construction of Factor Graph Optimization Framework

The framework of factor graph optimization based on the Bayesian network proposed in this study is shown in Figure 5. The state vector in that world frame construct according to the constraint factor shown in the figure is:

$$\mathcal{X} = \left[x_1, x_2, \ldots, x_i, \lambda_1, \lambda_2, \ldots, \lambda_p, {}_1, {}_2, \ldots, {}_l, d_1^e, d_2^e, \ldots, d_k^e, d_1^p, d_2^p, \ldots, d_k^p\right] \tag{6}$$

where $x_n = [p_n, q_n, v_n, b_a, b_g]$ represents the IMU state at the nth time, which includes the carrier position p_i, the rotation quaternion q_i and the velocity v_n obtained by IMU pre-integration in the world frame, b_a and b_g stand for the acceleration bias and the gyroscope bias in IMU body frame, respectively, λ_p represents the inverse depth of the visual point feature in the visual frame from its initial observation in the first frame, ${}_l$ represents the orthogonal frame of the visual line feature, d_k^e and d_k^p stand for the distances between the LiDAR feature points and its corresponding edge or plane feature point cloud, respectively.

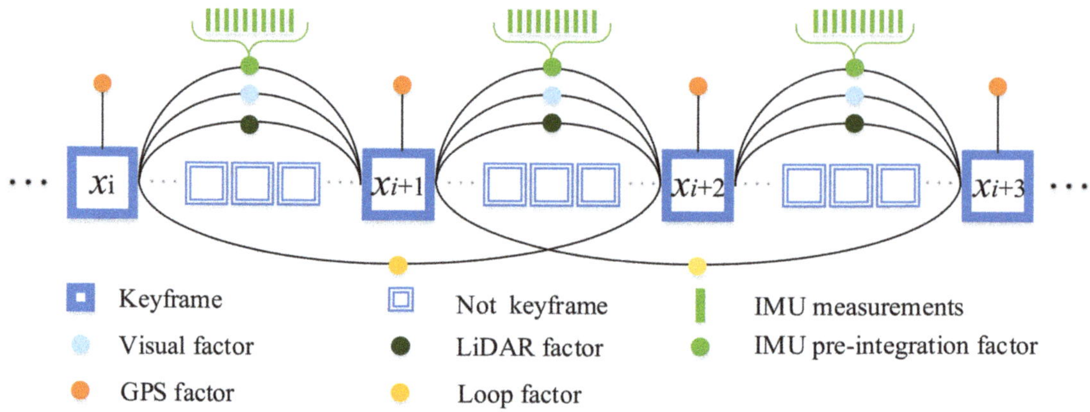

Figure 5. Factor graph optimization framework of our system. Constraints of factor graph on the keyframe maintenance include three local constraints and two global constraints.

Therefore, the Gaussian–Newton method can be used to minimize all cost functions to construct a maximum a posteriori estimation problem, to perform nonlinear optimization on the state vectors in the sliding window:

$$\begin{aligned} \min_{\mathcal{X}} \Bigg\{ & \|r_p - \mathcal{J}_p \mathcal{X}\|^2 + \sum_{k \in B} \left\| r_B\left(\hat{z}^k_{k+1}, \mathcal{X}\right) \right\|^2_{p_i} + \\ & \sum_{(i,j) \in F} \rho\left(\left\| r_f\left(\hat{z}^j_i, \mathcal{X}\right) \right\|^2_{p_c} \right) + \sum_{(i,j) \in L} \rho\left(\left\| r_l\left(\hat{z}^j_i, \mathcal{X}\right) \right\|^2_{p_c} \right) + \sum d^e_k + \sum d^p_k \Bigg\} \end{aligned} \tag{7}$$

where $\{r_p, \mathcal{J}_p\}$ contains the prior states after the marginalization in the sliding window, and $\mathcal{J}_p$ is the Jacobian matrix, $r_B\left(\hat{z}^k_{k+1}, \mathcal{X}\right)$ represents the IMU residuals, and p_i is the IMU covariance matrix; $r_f\left(\hat{z}^j_i, \mathcal{X}\right)$ and $r_l\left(\hat{z}^j_i, \mathcal{X}\right)$ represent the re-projection errors of visual point and line features, p_c is the visual covariance matrix, and ρ represents Huber norm, with specific values as follows:

$$\rho(e(s)) = \begin{cases} \frac{1}{2} e_1(s)^2 & e(s) = e_1(s), |e_1(s)| \leq \delta \\ \delta |e_2(s)| - \frac{1}{2}\delta^2 & e(s) = e_2(s), |e_2(s)| > \delta \end{cases} \tag{8}$$

The specific meaning of each sensor cost function in Formula (6) is as follows.

4.2. IMU Factor

The IMU state of the kth frame and the $k+1$th frame in the global coordinate system can be defined as:

$$\begin{aligned} x_k &= \left[p^G_{b_k}, q^G_{b_k}, v^G_{b_k} b_{ak}, b_{gk}\right] \\ x_{k+1} &= \left[p^G_{b_{k+1}}, q^G_{b_{k+1}}, v^G_{b_{k+1}} b_{ak+1}, b_{gk+1}\right] \end{aligned} \tag{9}$$

Take the IMU state of the kth frame, x_k, as an example, which includes position $p^G_{b_k}$, rotation $q^G_{b_k}$, velocity $v^G_{b_k}$, accelerometer bias b_{ak} and gyroscope bias b_{gk}.

Next, the IMU residual equation can be constructed, which is defined as:

$$r_B\left(\hat{z}_{k+1}^{k}, \mathcal{X}\right) = \begin{bmatrix} r_p \\ r_q \\ r_v \\ r_{ba} \\ r_{bg} \end{bmatrix} = \begin{bmatrix} R_G^{B_k}\left(p_{b_{k+1}}^G - p_{b_k}^G + \frac{1}{2}g\Delta t_k^2 - v_{b_k}^G \Delta t_k\right) - \hat{p}_{k+1}^k \\ 2\left[q_{b_k}^{G^{-1}} \otimes q_{b_{k+1}}^G \otimes \hat{q}_{k+1}^{k^{-1}}\right]_{xyz} \\ R_G^{B_k}\left(v_{k+1}^G + g\Delta t_k - v_k^G\right) - \hat{v}_{k+1}^k \\ b_{ak+1} - b_{ak} \\ b_{gk+1} - b_{gk} \end{bmatrix} \tag{10}$$

where $\left[r_p, r_q, r_v, r_{ba}, r_{bg}\right]^T$ represents the observation residual of IMU state between two consecutive keyframes in the sliding window, including the residual of position, rotation, velocity, accelerometer bias and gyroscope bias, $R_G^{B_k}$ represents the pose conversion matrix of the kth frame from the IMU coordinate system to GNSS global coordinate system, and $\left[\hat{p}_{k+1}^k, \hat{q}_{k+1}^k, \hat{v}_{k+1}^k\right]$ represents the IMU pre-integration value of two keyframes in the sliding window within Δt_k.

4.3. Visual Feature Factor

The visual feature factor is essentially the re-projection error of the visual feature, that is, the difference between the theoretical value projected on the image plane and the actual observation value. In order to unify the coordinate system in Section 3.3, we provide the definition of re-projection error on the unit sphere instead of the generalized image plane. Specific schematic diagrams are shown in Figures 6 and 7.

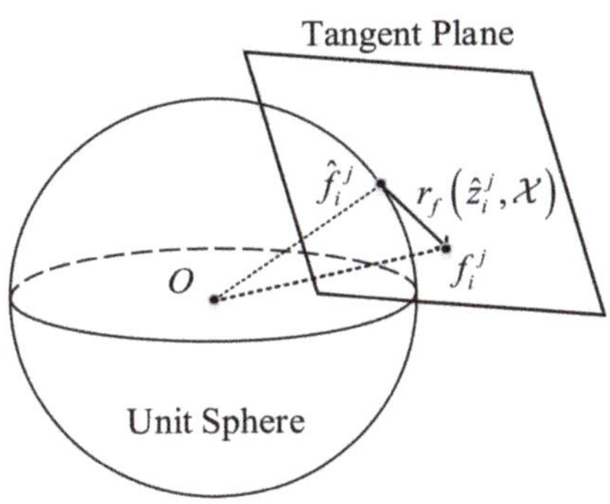

Figure 6. Re-projection error of visual point features.

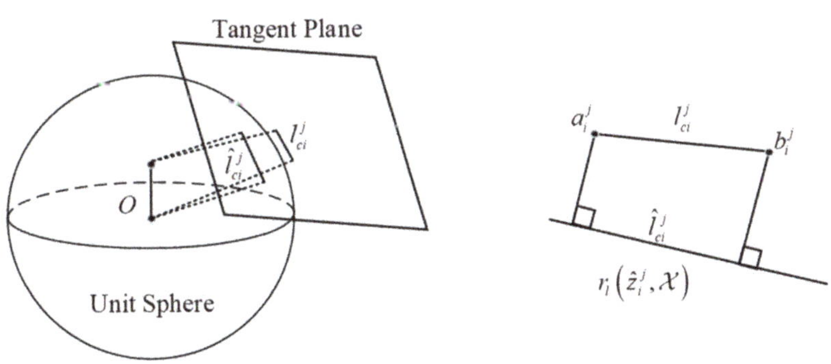

Figure 7. Re-projection error of visual line features.

4.3.1. Visual Point Feature Factor

In this study, the visual feature factors are built with reference to VINS-Mono [5]. As shown in Figure 6, the re-projection error of visual point features can be defined as the difference between the projection point on the unit spherical surface and the observation value after distortion correction. Given the ith normalized projection point $\hat{f}_i^j = \left[\hat{u}_i^j, \hat{v}_i^j, 1\right]^T$ and observation point $f_i^j = \left[u_i^j, v_i^j, 1\right]^T$ in the jth frame, we use the first observation value $f_i^j = \left[u_{i0}^j, v_{i0}^j, 1\right]^T$ in the jth frame to define the visual point feature factor as:

$$\begin{cases} r_f\left(\hat{z}_i^j, \mathcal{X}\right) = \begin{bmatrix} \hat{u}_i^j - u_i^j \\ \hat{v}_i^j - v_i^j \end{bmatrix} \\ \begin{bmatrix} u_i^j \\ v_i^j \end{bmatrix} = R_B^V\left(R_G^{B_j}\left(R_{B_0}^G\left(R_V^B \frac{1}{\kappa_i}\begin{bmatrix} u_{i0}^j \\ v_{i0}^j \end{bmatrix} + p_V^B\right) + p_{b_0}^G - p_{b_i}^G\right) - p_V^B\right) \end{cases} \tag{11}$$

where R_B^V represents the external parameter matrix between camera and IMU, which is obtained by calibration, $R_G^{B_j}$ represents the pose conversion matrix from the IMU observation in the jth frame to the global coordinate system, $R_{B_0}^G$ represents the pose conversion matrix from the global coordinate system to the initial IMU observation, κ_i stands for the inverse depth of f_j^i, p_V^B represents the displacement from the IMU coordinate system to the camera coordinate system. Finally, $p_{b_0}^G$ and $p_{b_i}^G$ represent the displacement of the first and the ith IMU observation in the global coordinate system, respectively.

4.3.2. Visual Line Feature Factor

As shown in Figure 7, similar to the visual point feature, the definition of the re-projection error of the visual line feature is as follows: Given the characteristics of a visual line in space, the end point of a line segment is the center of the sphere to construct a unit sphere. Therefore, the reprojection error is the difference between the projection line on the unit sphere and the observed value. According to Equation (2), given the observed value of the characteristic factor of the ith line in the jth frame in the camera coordinate system as $l_{c_i}^j = \left[n_{c_i}^j, v_{c_i}^j\right]^T$, the projection line is obtained by projecting it onto the unit sphere, and can be expressed as:

$$\hat{l}_{ci}^j = \begin{bmatrix} \hat{l}_1 \\ \hat{l}_2 \\ \hat{l}_3 \end{bmatrix} = \mathrm{K}n_{ci}^j \in R^6 \tag{12}$$

where K is the camera internal reference projection matrix. It can be seen from Equation (12) that the spatial coordinates of the line features projected onto the unit sphere are only related to n_c. The two end points of the observation line are a_i^j and b_i^j, then the re-projection error of the line feature can be expressed by the dotted distance from the two end points of the observation line feature to the projection line feature:

$$\begin{cases} r_l\left(\hat{z}_i^j, \mathcal{X}\right) = \left[d\left(a_i^j, \hat{l}_{ci}^j\right), d\left(b_i^j, \hat{l}_{ci}^j\right)\right]^T \\ d\left(a_i^j, \hat{l}_{ci}^j\right) = \frac{\left(a_i^j\right)^T \hat{l}_{ci}^j}{\sqrt{\hat{l}_1^2 + \hat{l}_2^2}} \\ d\left(b_i^j, \hat{l}_{ci}^j\right) = \frac{\left(b_i^j\right)^T \hat{l}_{ci}^j}{\sqrt{\hat{l}_1^2 + \hat{l}_2^2}} \end{cases} \tag{13}$$

4.4. LiDAR Factor

As mentioned in Section 3.3, after the LiDAR-assisted monocular visual depth correlation, the VIO will provide the LiDAR with visual initial positional estimates to correct the motion distortion of the LiDAR point cloud and improve the scan matching accuracy. The scanning matching error between adjacent keyframes of LiDAR involved in this study can be expressed by the distance from the feature point to the matched edge line and feature plane as:

$$\left\{\begin{array}{l} d_k^e = \frac{\left|\left(X^e_{(k+1,i)} - X^e_{(k,a)}\right) \times \left(X^e_{(k+1,i)} - X^e_{(k,b)}\right)\right|}{\left|X^e_{(k,a)} - X^e_{(k,b)}\right|} \\ d_k^p = \frac{\left|\begin{array}{c} \left(X^p_{(k+1,i)} - X^p_{(k,b)}\right) \\ \left(\left(X^p_{(k,a)} - X^p_{(k,b)}\right) \times \left(X^p_{(k,a)} - X^p_{(k,c)}\right)\right) \end{array}\right|}{\left|\left(X^p_{(k,a)} - X^p_{(k,b)}\right) \times \left(X^p_{(k,a)} - X^p_{(k,c)}\right)\right|} \end{array}\right. \tag{14}$$

where $X^e_{(k+1,i)}$ represents the edge feature point at the $k+1$th time, $X^e_{(k,a)}$ and $X^e_{(k,b)}$ are the endpoint of the edge line matched with the feature point at the kth time, $X^p_{(k+1,i)}$ represents the plane feature point at the $k+1$th time, and the feature surface matched with it at the kth time can be represented by three points $X^p_{(k,a)}$, $X^p_{(k,b)}$ and $X^p_{(k,c)}$.

4.5. GNSS Factor and Loop Factor

When the carrier moves to a GNSS signal trusted environment, GNSS factors can be added to optimize with local sensors. The time interval of two frames of GNSS observations is Δt, and given the GNSS measurements p_k^{Gg} in the global frame and p_k^{Vg} representing the observation of LVIO in the global frame, the GNSS factor can be expressed by the following observation residuals:

$$r_G\left(\hat{z}^k_{k+1}, \mathcal{X}\right) = p_k^{Vg} - p_k^{Gg} \tag{15}$$

Different from the assumption in [14] that GNSS factors are added to the system only when the GNSS measurement covariance is smaller than the LVIO measurement covariance, we noticed that the accuracy of outdoor GNSS positioning results is much higher than the LVIO local positioning results. The covariance threshold size for judging whether to add GNSS factors has little impact on the positioning accuracy. Therefore, we present that once the GNSS signal is detected by the system, the GNSS factor is added to the factor graph. In this way, even if the mobile carrier enters the GNSS rejection environment (such as the indoor parking lot or tunnel), it can also provide a more accurate initial observation value after GNSS correction. The fusion strategy of GNSS and LVIO is shown in Figure 8.

Further, considering the possible overlap of the mobile carrier travel area, i.e., the mobile carrier travels to the same position again after a period of time, we also added a loop-back detection link to establish the loopback constraint that exists between non-adjacent frames. Unlike introducing another sensor (GNSS) for global correction of the local sensor (LVIO), the loopback factor establishes the correlation between the current observed frames and the historical data by the local sensor itself to obtain a globally consistent estimate. The conditions for adding the loopback factor are similar to those of GNSS. Once the carrier motion trajectory is detected to travel to the environment passed by the history, the loop factor is added to the factor graph. By registering with the point cloud of the prior map, the historical trajectory is corrected, and the global pose estimation result with higher accuracy is obtained.

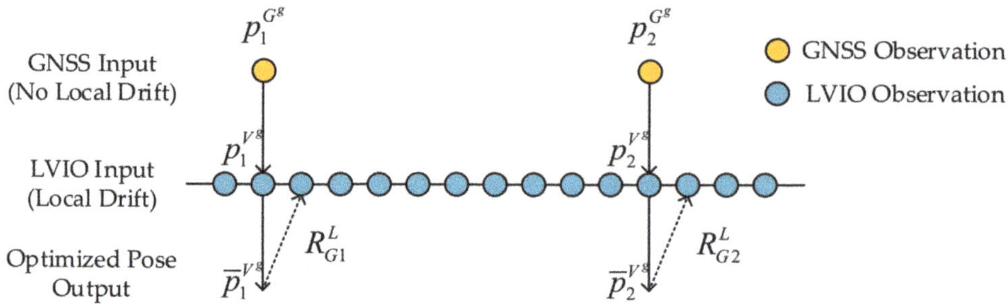

Figure 8. Fusion strategy of GNSS and LVIO. The initial rotation R_G^L of LVIO in the local frame and the global frame is set to identity matrix. GNSS provides global constraints to LVIO to correct the global position of LVIO and update R_G^L, and the new R_G^L is used for the next frame of LVIO.

5. Experimental Results

5.1. Real-Time Performance

5.1.1. Indoor Environment

For evaluating the real-time performance of our algorithm, we randomly selected the MH_01_easy dataset for indoor experiments. Since the strategy of adding line feature constraints to the VIO subsystem of our algorithm is referenced to PL-VIO, the time consumption of several threads involving line features of PL-VIO and this algorithm is compared. As shown in Figure 9, the appropriate selection of hidden parameters and the least-squares-based geometric constraint matching strategy have positive effects on real-time performance. The time cost of the line feature extraction and matching process and the line feature tracking process of the proposed algorithm is about one-third that of similar algorithms.

The time consumption of the line feature matching process is shown in Figure 9a. In the period of (110 s, 170 s), the carrier passes through the well-lit factory wall duct area. The number of line features extracted by both algorithms increases, and the corresponding time cost of line feature matching also increases with the number of line features. However, unlike PL-VIO which is significantly affected by the increase in the number of line features, the line feature matching the process time of our algorithm remains relatively stable within 1 ms. The reason is that the number of invalid line features is reduced due to the geometric constraint-based line feature matching strategy, which improves the accuracy of line feature matching between the front and current frames of the image. In the time-consuming of line feature tracking process shown in Figure 9b, it can be seen that in the initial stage (0 s, 5 s) of the visual subsystem, the line feature tracking process of the two systems takes longer. The reason is the UAV is at rest during this time and the VIO subsystem does not receive sufficient motion excitation, which leads to its incomplete initialization. After 5 seconds of initialization, the PL-VIO line feature tracking time remains stable at about 125 ms, while the time consumption of our algorithm is about 4/5 less than that of PL-VIO, about 25 ms. It has a strong positive effect on the real-time performance of the fusion system in the actual operating environment.

Although as shown in Figure 9c, the time-consuming cost of the line feature residual optimization process increases by about 10 ms, the time-consuming of the line feature tracking process is significantly reduced. Thus, the proposed method leads to a decrease in the total time cost of the three line feature-related processes in the fusion system, which still has a better real-time performance overall than before the improvement.

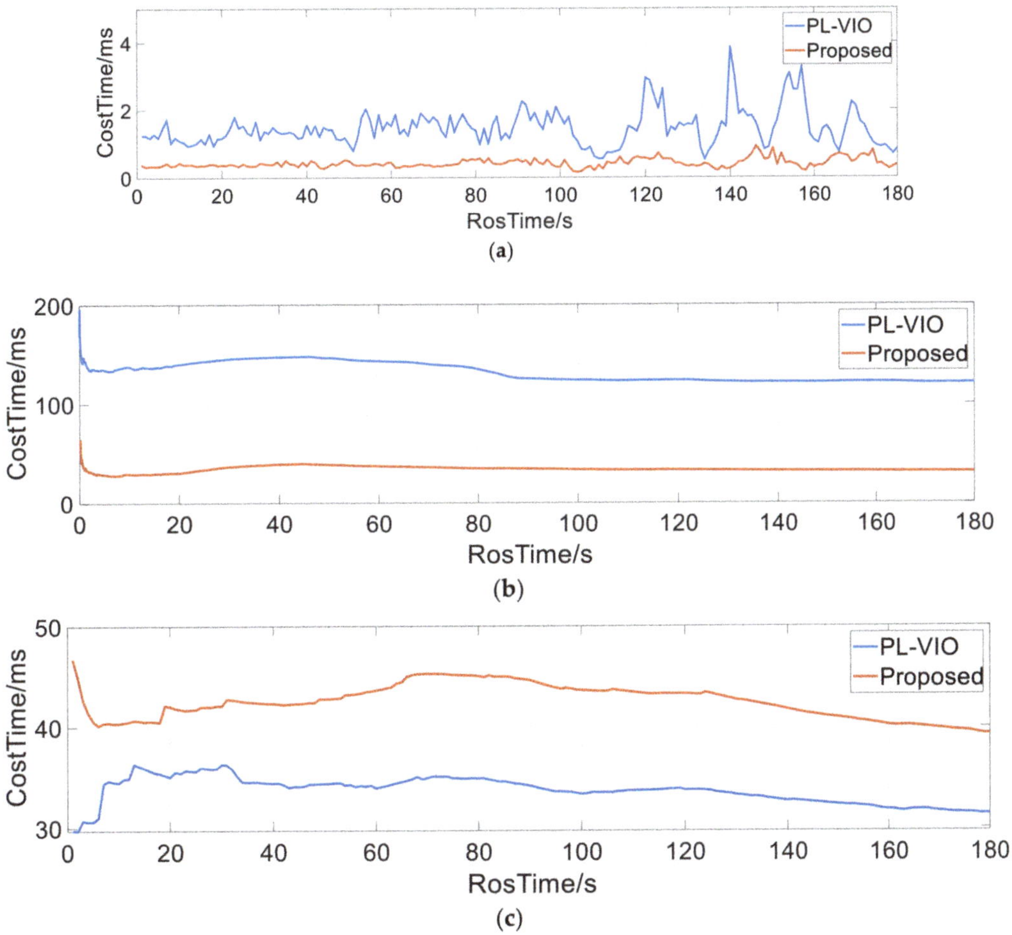

Figure 9. Real-time comparison experiment of MH_01_easy dataset. (**a**) Line feature extraction and matching process. (**b**) Line feature tracking process. (**c**) Line feature residual optimization process.

5.1.2. Outdoor Environment

Since the distribution characteristics of line features are different in indoor and outdoor environments, in order to fully evaluate the superior performance of this algorithm in terms of real-time, we selected the Hong Kong 0428 dataset for outdoor experiments. The experimental results are shown in Figure 10.

Different from the indoor environment, the outdoor environment has more complex conditions of light refraction and reflection, and the dynamic interference such as pedestrians and vehicles in the driving process of moving vehicles. The time consumption of the line feature matching process in the outdoor environment is shown in Figure 10a. It can be seen that the line feature matching time of PL-VIO in the outdoor environment is about 10 ms on average, and our algorithm still maintains the same good real-time characteristics as the indoor environment. In the line feature tracking process shown in Figure 10b, it can be seen that the line feature tracking process in the initialization phase (0 s, 5 s) of the visual subsystem is abnormally high for both systems. The same reason is that the VIO system is not provided sufficient motion excitation at the beginning of the vehicle stationary phase. It can be concluded that it is more difficult to match and track visual line features in the outdoor environment, and the time consumed for line feature tracking rises about 3–4 times compared with the indoor environment. However, the time consumed by our algorithm is still greatly shortened compared with similar algorithms, leaving more time for the optimization of a multi-sensor fusion at the back-end.

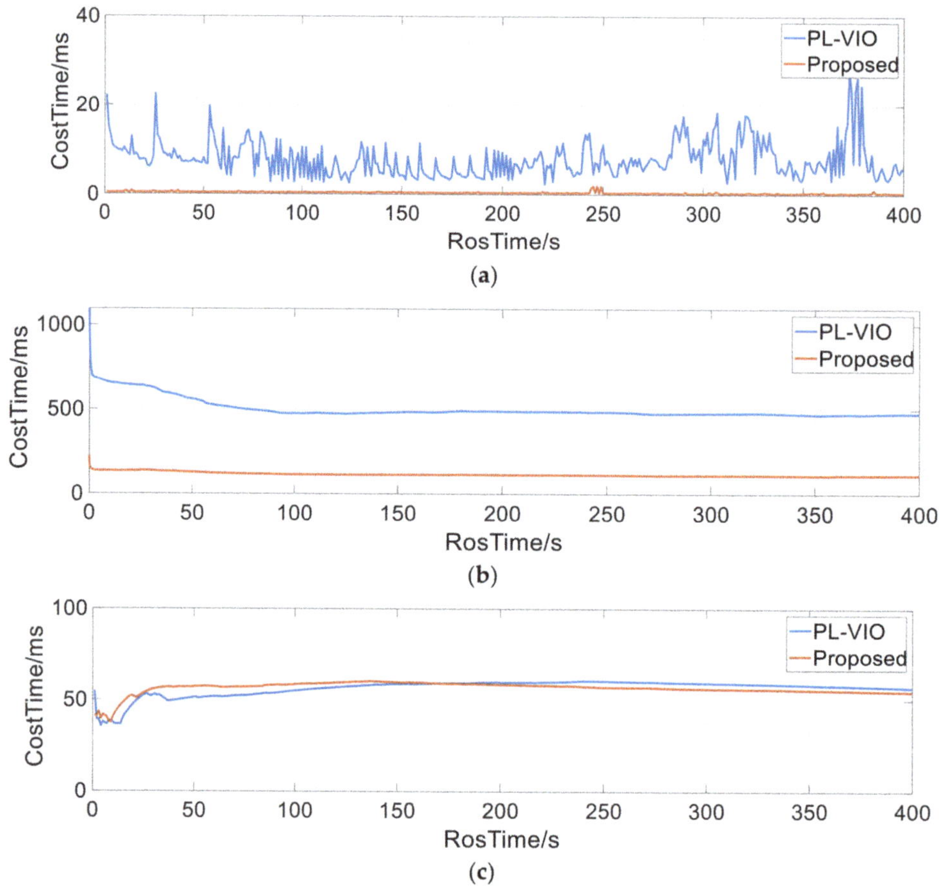

Figure 10. Real-time comparison experiment of Hong Kong 0428 dataset. (**a**) Line feature extraction and matching process. (**b**) Line feature tracking process. (**c**) Line feature residual optimization process.

In addition, as shown in Figure 10c, the time-consuming cost of the line feature residual optimization process is not much different from that of PL-VIO. Combining the above three time-consuming threads, it can be proved that our algorithm can achieve better real-time performance in different environments.

5.2. Positioning Accuracy

5.2.1. Indoor Environment

In this study, the EuROC dataset was used to compare and verify the positioning accuracy of each algorithm in the indoor environment. The experimental environment was in a factory with complex signal refraction and reflection conditions. LiDAR frequently fails in the experimental environment, so no comparison was made. The comparison of the point-line feature results extracted by PL-VIO and our algorithm in the experimental environment is shown in Figure 11.

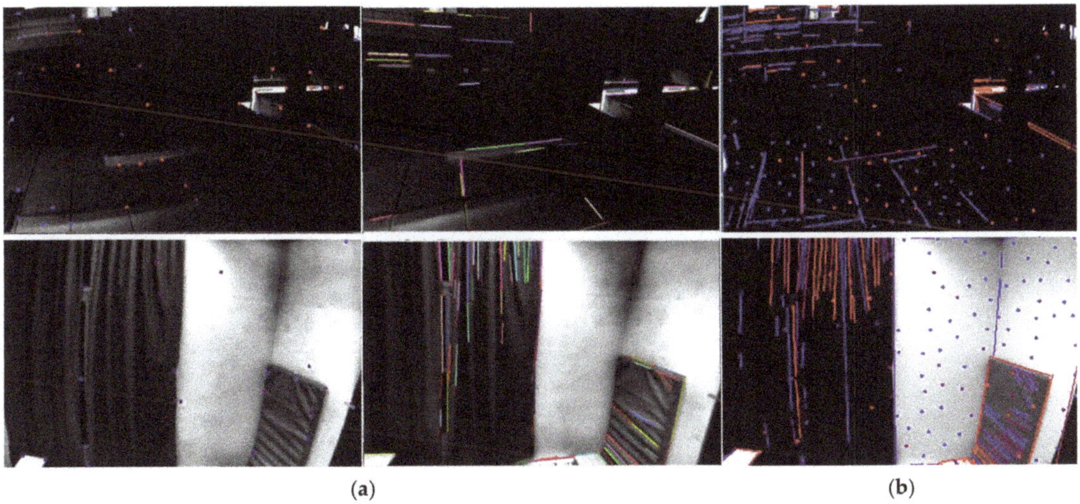

Figure 11. Comparison of point-line feature extraction results in poor lighting conditions and weak texture environment. (a) Point-line feature extraction results of PL-VIO. (b) Point-line feature extraction results of our algorithm.

As seen in Figures 12 and 13 and Table 1, the introducing line features in the image frames to add additional feature constraints can reduce the positioning error of the system to some extent, especially in areas with dim light and poor textures. For example, during the (160 s, 240 s) time, the UAV flight area is nearly full of darkness. Thus it is difficult for Harris corner point detection method to extract the corner points with large grayscale difference from the surrounding pixel blocks. The reduction in the number of effective feature points directly leads to poor feature tracking accuracy. Therefore, the absolute trajectory error of VINS-Mono based on point features is larger in this interval (as shown in Figure 13a). In contrast, PL-VIO based on point-line features and the present algorithm are less negatively affected by illumination, and the absolute trajectory error remains within 0.6 m. In a longitudinal comparison of similar algorithms based on point and line features, the accuracy of our algorithm is significantly improved over PL-VIO. These results are attributed to the high quality of matching by the geometric constraint strategy, which avoids the missegmentation of long-line features and then misclassification as invalid matches. The experimental results demonstrate the robustness and accuracy of this algorithm in the case of single system failure, which is important for localization in complex indoor environments.

Table 1. Motion estimation errors of each algorithm in indoor dataset.

Sequence	Vins_Mono (w/o loop)	Vins_Mono (w/ loop)	PL-VIO	LVI-SAM	Purposed
	ATE_RMSE(m)/Mean Error(m)				
MH_01_easy	0.213/0.189	0.188/0.158	0.093/0.081	0.181/0.147	0.073/0.062
MH_02_easy	0.235/0.193	0.188/0.157	0.072/0.062	0.182/0.167	0.045/0.039
MH_03_medium	0.399/0.321	0.402/0.315	0.260/0.234	0.400/0.308	0.056/0.050
MH_04_difficult	0.476/0.423	0.422/0.348	0.364/0.349	0.398/0.399	0.079/0.075
MH_05_difficult	0.426/0.384	0.370/0.309	0.251/0.238	0.380/0.287	0.139/0.127
V1_01_easy	0.157/0.137	0.145/0.121	0.078/0.067	0.142/0.119	0.040/0.037
V1_03_difficult	0.314/0.275	0.329/0.289	0.205/0.179	0.322/0.283	0.077/0.069
V2_01_easy	0.133/0.115	0.120/0.108	0.086/0.072	0.121/0.110	0.056/0.048
V2_02_medium	0.287/0.244	0.293/0.255	0.150/0.097	0.291/0.250	0.089/0.078
V2_03_difficult	0.343/0.299	0.351/0.315	0.273/0.249	0.351/0.308	0.098/0.092

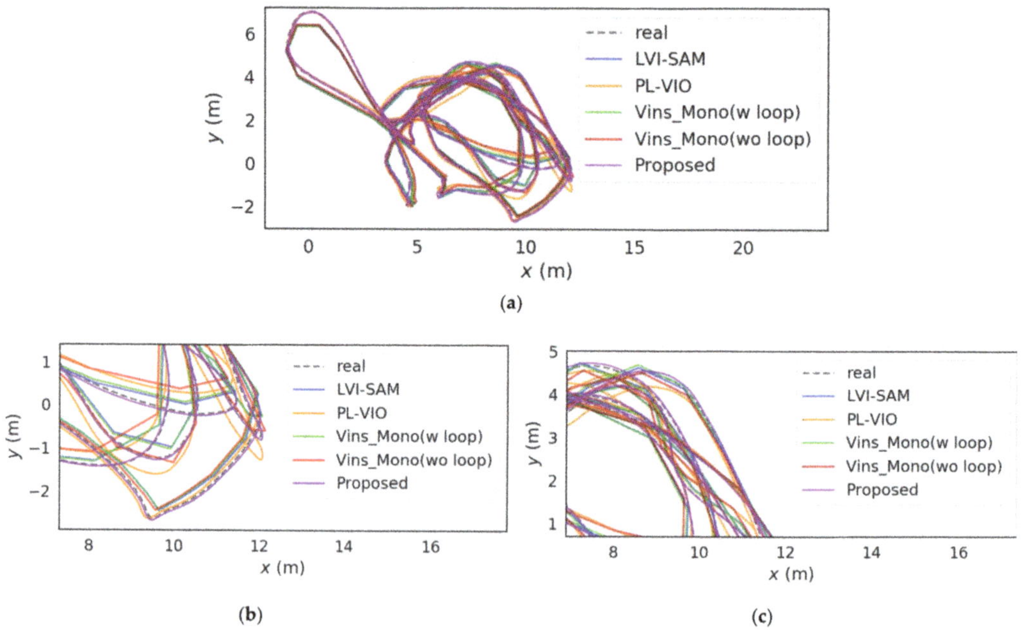

Figure 12. Comparison of trajectory fitting curve of each algorithm in the indoor dataset. (**a**) Global trajectory fitting curve. (**b**) Details of local trajectory. (**c**) Details of local trajectory.

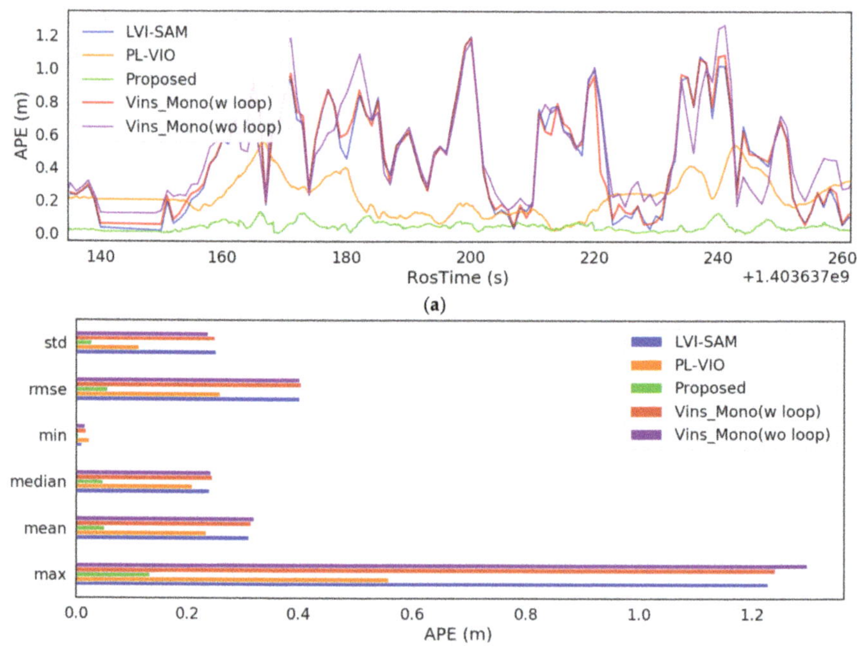

Figure 13. Comparison of positioning results of each algorithm in the indoor dataset. (**a**) APE_RMSE error fitting curve. (**b**) Comparison of index of absolute trajectory error.

5.2.2. Outdoor Environment

To evaluate the performance of the algorithm we conducted in the outdoor environment, the Hong Kong dataset was used for performance evaluation and it was compared with other similar advanced algorithms. The experimental equipment and environment are shown in Figure 14. The sensor models are as follows: the camera is BFLY-U3-23S6C-C, the LiDAR is HDL 32E Velodyne, IMU is Xsens Mti 10, and the GNSS receiver is u-blox M8T. In addition, we utilized the high-grade RTK GNSS/INS integrated navigation system, NovAtel SPAN-CPT, as the ground truth.

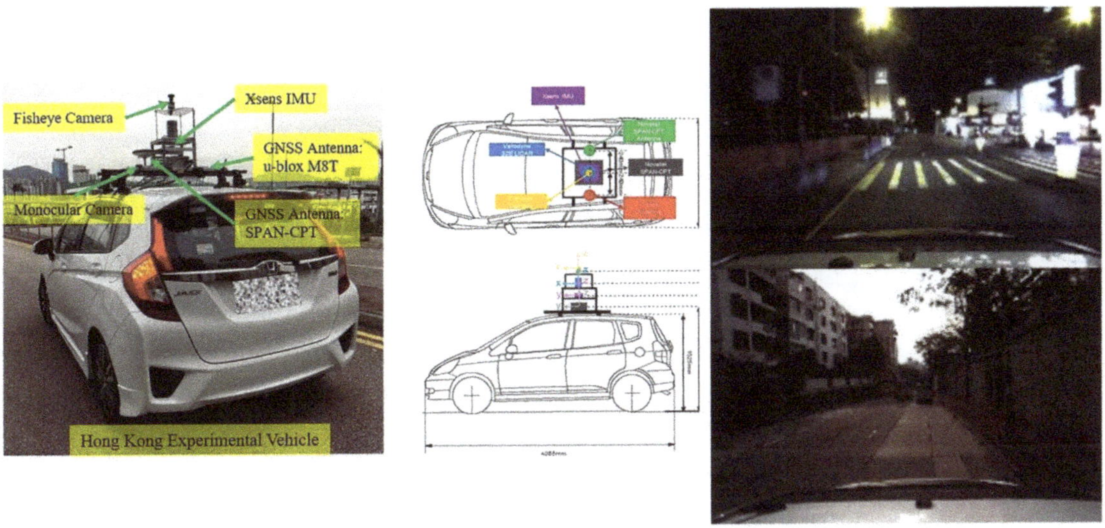

(a) (b)

Figure 14. Experimental equipment and environment. (**a**) The experimental vehicle and sensors setup. (**b**) Image of experimental environment.

To verify the superior performance of each aspect of our system, we performed ablation experiments, constructed without GNSS global correction (*), without visual line features (#), and our complete system (proposed), respectively. The experimental results are shown in Figures 15 and 16 and Table 2.

Table 2. Motion estimation errors of each algorithm on outdoor dataset.

Sequence	Hong Kong 0428	Hong Kong 0314
	ATE_RMSE(m)/Mean Error(m)	
Vins_Mono (w/o loop)	101.735/89.470	40.651/35.035
Vins_Mono (w/ loop)	76.179/67.535	19.191/15.617
LIO-SAM	7.181/6.787	41.933/39.672
LVI-SAM	9.764/9.061	3.065/2.557
Purposed(*)	9.475/8.884	2.842/2.456
Purposed(#)	5.808/5.436	2.595/2.041
Purposed	5.299/4.955	2.249/1.880

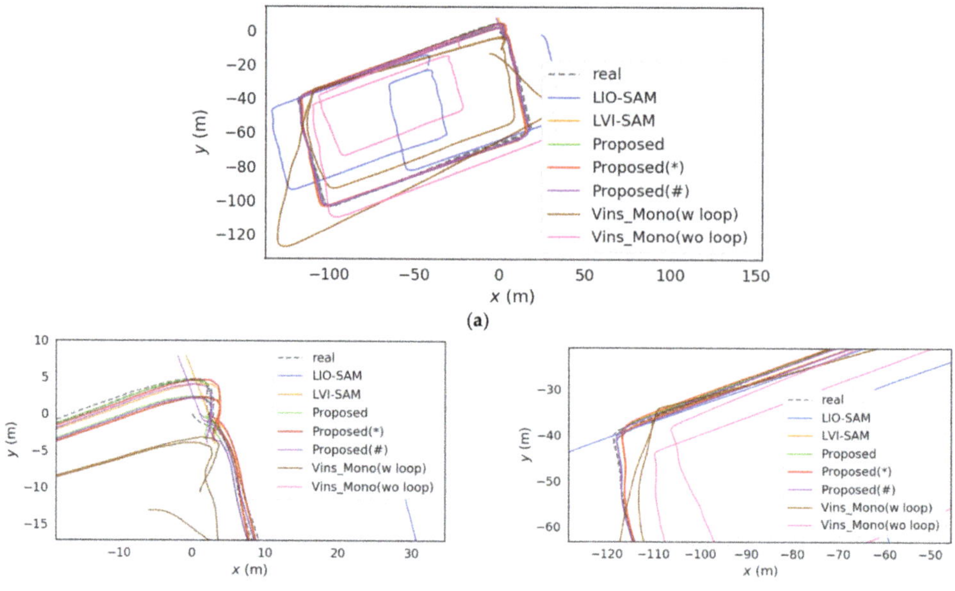

Figure 15. Comparison of trajectory fitting curve of each algorithm in the indoor dataset. (**a**) Global trajectory fitting curve. (**b**) Details of local trajectory. (**c**) Details of local trajectory.

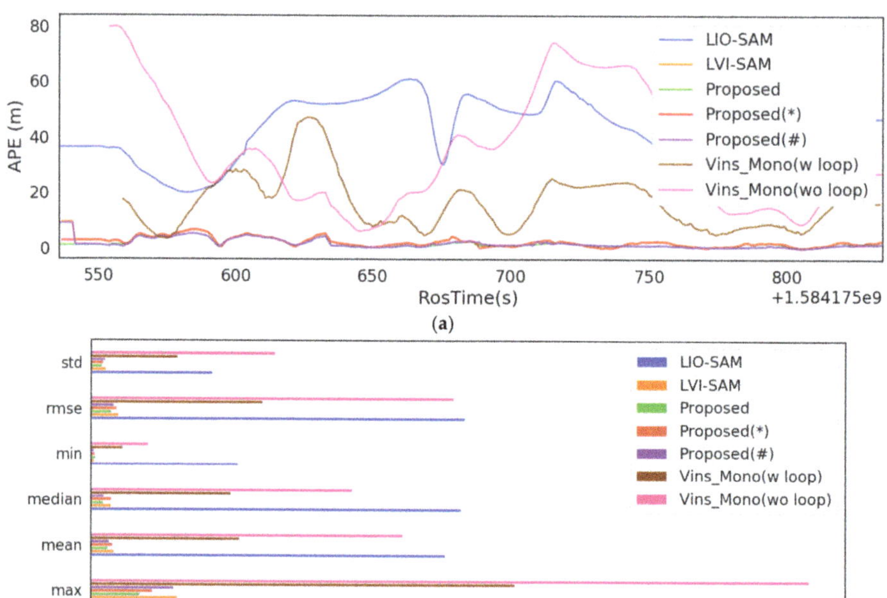

Figure 16. Comparison of positioning results of each algorithm on outdoor dataset. (**a**) APE_RMSE error fitting curve. (**b**) Comparison of index of absolute trajectory error.

From Figure 15, it can be seen that VIO and LIO, which are mainly based on a single sensor, each have different defects. First of all, VIO(VINS-Mono) is introduced. Before starting the movement, the moving carrier stopped at the roadside parking position for about 10 seconds. VIO was not given a large motion excitation during this period, which led to the VIO not being initialized properly. Secondly, the cumulative error caused by the scale uncertainty of the monocular camera increased significantly over time, and a large-scale estimation error was already generated at the second lap. Although the scale drift of LIO (LIO-SAM) is not large, it will immediately fail and keep restarting in the complex area of signal fold reflection. After LiDAR resumes operation, the translation and rotation of the current frame will be accumulated based on the positional estimation at the last frame that did not fail, resulting in the misjudgment of stopping the motion at the carrier motion to (50 m,150 m). When the carrier moves to the corner, LIO re-estimates the position and attitude. It was misjudged that the carrier stopped at (50 m,150 m) for a while and then began to turn, so it lost the estimated position and attitude for a period of time, which led to a large positioning error.

In a longitudinal comparison with the other LVIO system (LVI-SAM), we can conclude that our complete algorithm maintains a lower drift rate and localization integrity, which benefits from the extra constraint of line features and the global correction of GNSS. In conclusion, even in complex outdoor environments, our algorithm still outperforms other advanced algorithms.

5.3. Mapping Performance

As a demonstration of the superiority of our algorithm in building maps, we compared the building results with other advanced algorithms on different datasets. The visual line feature extraction and map building results are shown in Figure 17. Compared with PL-VIO, our algorithm has a great improvement in the number of visual line features extracted, which is attributed to the improved line feature extraction strategy. In a factory environment with complex lighting conditions, the line features in the actual environment will look minutely curved due to the refraction of light. Due to the proper value of the threshold value D of the density of homogeneous points, the angle tolerance of fitting pixels to approximate rectangles in this environment can be improved, thus increasing the number of line feature extraction. Further, the accuracy of the bit pose estimation is also substantially improved by the combination of the improved line feature extraction and tracking optimization strategies.

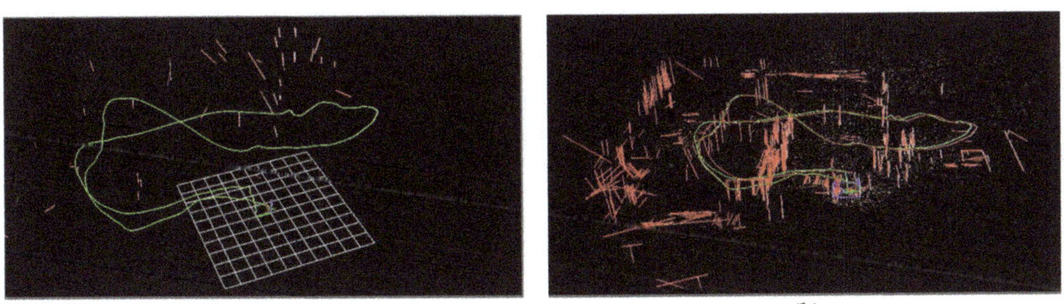

(a) **(b)**

Figure 17. Comparison of visual line feature extraction mapping. (**a**) show the mapping of each subsystem before improvement, and (**b**) show our algorithm mapping.

Further, comparison of the LiDAR point cloud detail views is shown in Figure 18. The more accurate VIO pose estimation after the line features are added provides a more accurate initial value for LiDAR scan matching, and reduces a large number of point cloud mismatching. Comparison of global point cloud trajectories is shown in Figure 19.

The area marked by circles demonstrates that the data drift caused by cumulative errors is significantly reduced by adding a GNSS factor and loop factor to our algorithm.

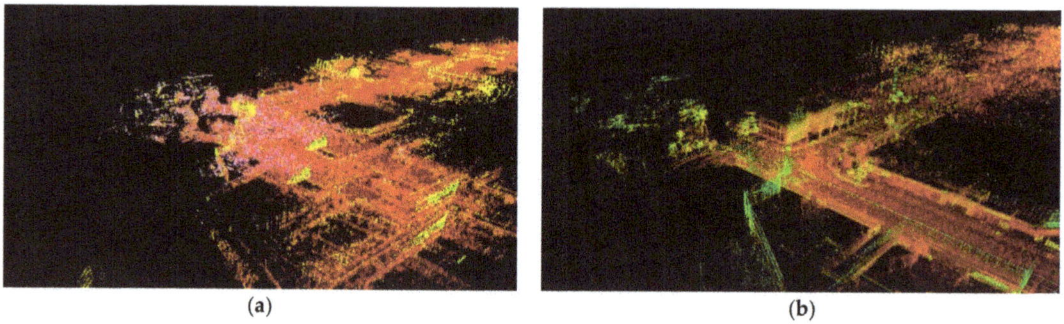

Figure 18. Comparison of LiDAR point cloud map details. (**a**) show the mapping of each subsystem before improvement, and (**b**) show our algorithm mapping.

Figure 19. Comparison of global point cloud trajectory. (**a**) show the mapping of each subsystem before improvement, and (**b**) show our algorithm mapping.

6. Discussion

Multi-sensor fusion positioning technology based on SLAM provides new opportunities for the high-precision positioning of mobile carriers. In this study, two problems that need to be further explored in the existing LVIO fusion system are proposed. The first problem is that LVIO system needs enough environmental feature information. According to the previous studies of Pumarola et al. [11] and Fu et al. [12], theoretically, the accuracy of the fusion system can be improved by increasing the constraint of visual line features. Huang et al. also proved that the average positioning error of the fusion system based on point-line feature can generally be improved, from the traditional 2.16% to 0.93% [18]. In this study, the steps of increasing visual line feature constraints are further optimized. The Monte Carlo method was used to select the appropriate scaling ratio and the density threshold of homogeneous points, which improves the angle tolerance of pixel fitting to line features. To a certain extent, it reduces the probability that short segments are wrongly judged as invalid features. According to the experiment of parameter selection in Section 3.1, compared with the indoor environment where the angle and translation of line features change little, the movement of line features between consecutive frames is more complicated in the outdoor environment. Therefore, the density threshold of homogeneous points needs to be lowered to reduce the probability that the valid line feature pairs are misjudged as invalid matches when turning sections. The results of outdoor real-time analysis shown in Section 5.1.2. show that the traditional method based on point-line features is difficult to match and track the visual line features in the outdoor environment, which takes a long time. However, the time consumption of this algorithm maintained a low level, which is beneficial to leave more time for back-end fusion optimization.

To solve the problem of line feature mismatching during the movement of carriers, Zhou et al. established a constraint equation using the 6-DOF Plücker coordinates of line features to perform matching optimization [19]. However, this increases the computational complexity of the fusion system, which is inconsistent with the lightweight requirements of autonomous driving positioning. In this study, the link of line feature constraint matching is simplified, and the original 6-DOF parameters are replaced by 4-DOF which represents the movement of line features for optimization. Thus, it can reduce the computational complexity of the system and effectively improve the inter-frame matching accuracy of line features. To explore the superiority of the proposed algorithm in real-time and positioning accuracy, we compared the precision and the time-consuming of three processes related to line features of this algorithm with several similar advanced algorithms in different environments. The experimental results show that our optimization strategy based on front-end point-line features effectively achieves the positive balance between reducing time consumption and improving accuracy.

The second problem to be solved is the global optimization of LVIO local pose estimation results by introducing global constraints. To further improve the positioning accuracy of local sensors, Qin et al. proposed a GNSS and local sensor fusion method to construct GNSS residual factors to correct the cumulative error of VIO [25]. Further, we propose a factor graph based on Bayesian network, in which GNSS observations are added as global constraint factors. The accumulated errors of LVIO are corrected by using GNSS observations within 0.1 s interval from LVIO keyframes as global constraint. In this study, it is proved that GNSS global constraint factor can effectively correct LVIO positioning error in the outdoor environment. It should be noted that since the coordinate of the current frame of LVIO is calculated from the coordinate of the previous frame, long-term observation or long moving distance will lead to more serious data drift. However, the GNSS observations are in the global coordinate, so long-term observation is not related to the data drift. Therefore, we can reasonably speculate that the longer the algorithm runs, the more obvious the correction effect of GNSS on LVIO will be. More comprehensively, LVIO will continue local positioning in GNSS rejection environment, so the positioning continuity of mobile carriers in different environments can be effectively guaranteed.

7. Conclusions

In this study, a LiDAR-Visual-Inertial Odometry based on optimized visual point-line features is proposed, taking advantage of the heterogeneous complementary characteristics of multiple sensors. First, a visual line feature extraction and matching optimization method is proposed. By improving the line feature extraction in the scale space and selecting the appropriate scaling ratio and same-sex point density threshold, the number of line features extracted in the light complex environment is largely improved to provide richer feature information for the front-end. Meanwhile, the original 6-DOF parameter optimization problem is further improved to a 4-DOF parameter optimization problem by using a least squares-based line feature constrained matching strategy. The complexity of the fusion system is reduced, and more accurate visual pose estimation is effectively accomplished. Second, the LiDAR point cloud is projected into the visual coordinates for depth correlation. Meanwhile, the initial pose estimation provided by the optimized VIO is used to help LiDAR scan matching. Finally, a factor graph method based on Bayesian networks is established. Two global constraint factors are added to the factor graph framework to constrain LVIO globally, which are the global constraint of GNSS factors from external sensors and the loop factor constraint of local sensors. The experimental results show that the algorithm can achieve real-time attitude estimation with good localization and mapping accuracy in different environments.

In the future, we will further improve and refine our work in the following aspects. First, the point cloud alignment algorithm of the loop factor in this study utilizes the traditional ICP algorithm, which is time-consuming to perform the nearest domain search using the KD tree. Thus, we will consider the improvement of the point cloud alignment algorithm next. Second, the inclusion of the GNSS factor in this study only utilizes the GNSS pseudo range single point positioning result. Although it is relatively simple and feasible on the vehicle platform with only one GNSS receiver, there is still room for improvement in the positioning accuracy of GNSS. A more accurate correction of LVIO by using higher accuracy RTK positioning results will be considered in the next step. Finally, since our proposed fusion system consists of two subsystems with high runtime computational resource requirements, we will work on reducing the resource occupation rate of the algorithm. Further, we will evaluate the positioning accuracy of the algorithm on vehicles with limited computing resources.

Author Contributions: Conceptualization, Z.Z., C.S., H.Z. and X.L.; methodology, X.H.; software, X.H.; validation, S.P. and L.D.; formal analysis, X.H.; investigation, W.G. and X.H.; resources, S.P., W.G. and X.H.; writing—original draft preparation, X.H.; writing—review and editing, W.G. and X.H.; supervision, S.P. and W.G.; project administration, W.G.; funding acquisition, S.P. All authors have read and agreed to the published version of the manuscript.

Funding: This research study was funded by the Fundamental Research Funds for the Central Universities (2242021R41134) and the Research Fund of the Ministry of Education of China and China Mobile (MCM20200J01).

Institutional Review Board Statement: Not applicable.

Informed Consent Statement: Not applicable.

Data Availability Statement: Not applicable.

Conflicts of Interest: The authors declare no conflict of interest.

References

1. Qian, Q.; Bai, T.M.; Bi, Y.F.; Qiao, C.Y.; Xiang, Z.Y. Monocular Simultaneous Localization and Mapping Initialization Method Based on Point and Line Features. *Acta Opt. Sin.* **2021**, *41*, 1215002.
2. Shan, T.; Englot, B. LeGO-LOAM: Lightweight and Ground-Optimized Lidar Odometry and Mapping on Variable Terrain. In Proceedings of the 2018 IEEE/RSJ International Conference on Intelligent Robots and Systems (IROS), Madrid, Spain, 1–5 October 2018; pp. 4758–4765.
3. Zuo, X.; Xie, X.; Liu, Y.; Huang, G. Robust visual SLAM with point and line features. In Proceedings of the 2017 IEEE/RSJ International Conference on Intelligent Robots and Systems (IROS), Vancouver, BC, Canada, 24–28 September 2017; pp. 1775–1782.
4. Zhang, J.; Singh, S. Laser-Visual-Inertial Odometry and Mapping with High Robustness and Low Drift. *J. Field Robot.* **2018**, *35*, 1242–1264. [CrossRef]
5. Qin, T.; Li, P.; Shen, S. Vins-mono: A robust and versatile monocular visual-inertial state estimator. *IEEE Trans. Robot.* **2018**, *34*, 1004–1020. [CrossRef]
6. Mur-Artal, R.; Tardós, J.D. ORB-SLAM2: An open-source slam system for monocular, stereo, and rgb-d cameras. *IEEE Trans. Robot.* **2018**, *33*, 1255–1262. [CrossRef]
7. Forster, C.; Carlone, L.; Dellaert, F.; Scaramuzza, D. On-manifold preintegration for real-time visual–inertial odometry. *IEEE Trans. Robot.* **2017**, *33*, 1–21. [CrossRef]
8. Forster, C.; Zhang, Z.; Gassner, M.; Werlberger, M.; Scaramuzza, D. SVO: Semi-direct visual odometry for monocular and multicamera systems. *IEEE Trans. Robot.* **2016**, *33*, 249–265. [CrossRef]
9. He, Y.; Zhao, J.; Guo, Y.; He, W.H.; Yuan, K. Pl-vio: Tightly-coupled monocular visual–inertial odometry using point and line features. *Sensors* **2018**, *18*, 1159. [CrossRef] [PubMed]
10. Wen, H.; Tian, J.; Li, D. PLS-VIO: Stereo Vision-inertial Odometry Based on Point and Line Features. In Proceedings of the 2020 International Conference on High Performance Big Data and Intelligent Systems (HPBD&IS), Shenzhen, China, 23 May 2020.
11. Pumarola, A.; Vakhitov, A.; Agudo, A.; Sanfeliu, A.; Moreno-Noguer, F. PL-SLAM: Real-time monocular visual SLAM with points and lines. In Proceedings of the 2017 IEEE International Conference on Robotics and Automation (ICRA), Singapore, 29 May–3 June 2017; pp. 4503–4508.
12. Fu, Q.; Wang, J.; Yu, H.; Ali, I.; Zhang, H. PL-VINS: Real-Time Monocular Visual-Inertial SLAM with Point and Line. [DB/OL]. Available online: https://arxiv.org/abs/2009.07462v1 (accessed on 27 November 2021).
13. Zhang, J.; Singh, S. LOAM: Lidar Odometry and Mapping in Real-time. In Proceedings of the 2014 Robotics: Science and Systems, Berkeley, CA, USA, 12–16 July 2014; pp. 9–17.

14. Shan, T.; Englot, B.; Meyers, D.; Wang, W.; Rus, D. LIO-SAM: Tightly-Coupled Lidar Inertial Odometry via Smoothing and Mapping. [DB/OL]. Available online: https://arxiv.org/abs/2007.00258v3 (accessed on 27 November 2021).
15. Xiang, Z.; Yu, J.; Li, J.; Su, J. ViLiVO: Virtual LiDAR-Visual Odometry for an Autonomous Vehicle with a Multi-Camera System. In Proceedings of the 2019 IEEE/RSJ International Conference on Intelligent Robots and Systems (IROS), Macao, China, 3–8 November 2019; pp. 2486–2492.
16. Chen, S.; Zhou, B.; Jiang, C.; Xue, W.; Li, Q. A LiDAR/Visual SLAM Backend with Loop Closure Detection and Graph Optimization. *Remote Sens.* **2021**, *13*, 2720. [CrossRef]
17. Lin, J.; Zheng, C.; Xu, W.; Zhang, F. R2LIVE: A Robust, Real-Time, LiDAR-Inertial-Visual Tightly-Coupled State Estimator and Mapping. [DB/OL]. Available online: https://arxiv.org/abs/2102.12400 (accessed on 27 November 2021).
18. Huang, S.; Ma, Z.; Mu, T.; Fu, H.; Hu, S. Lidar-Monocular Visual Odometry using Point and Line Features. In Proceedings of the 2020 IEEE International Conference on Robotics and Automation(ICRA), Online, 1–15 June 2020; pp. 1091–1097.
19. Zhou, L.; Wang, S.; Kaess, M. DPLVO: Direct Point-Line Monocular Visual Odometry. *IEEE Robot. Autom. Lett.* **2021**, *6*, 7113–7120. [CrossRef]
20. He, X.; Pan, S.G.; Tan, Y.; Gao, W.; Zhang, H. VIO-GNSS Location Algorithm Based on Point-Line Feature in Outdoor Scene. *Laser Optoelectron. Prog.* **2022**, *56*, 1815002.
21. Silva, V.D.; Roche, J.; Kondoz, A. Fusion of LiDAR and camera sensor data for environment sensing in driverless vehicles. *arXiv* **2018**, arXiv:1710.06230.
22. Liu, X.; Li, D.; Shi, J.; Li, A.; Jiang, L. A framework for low-cost Fusion Positioning with Single Frequency RTK/MEMS-IMU/VIO. *J. Phys. Conf. Ser.* **2021**, *1738*, 012007. [CrossRef]
23. Mascaro, R.; Teixeira, L.; Hinzmann, T.; Siegwart, R.; Gomsf, M.C. Graph-optimization based multi-sensor fusion for robust uav pose estimation. In Proceedings of the 2018 IEEE International Conference on Robotics and Automation(ICRA), Brisbane, Australia, 21–24 May 2018; pp. 1421–1428.
24. Woosik, L.; Eckenhoff, K.; Geneva, P.; Huang, G.Q. Intermittent GPS-aided VIO: Online Initialization and Calibration. In Proceedings of the 2018 IEEE International Conference on Robotics and Automation(ICRA), Paris, France, 31 May–6 June 2020; pp. 5724–5731.
25. Qin, T.; Cao, S.; Pan, J.; Shen, S. A General Optimization-Based Framework for Global Pose Estimation with Multiple Sensors. [DB/OL]. Available online: https://arxiv.org/abs/1901.03642 (accessed on 27 November 2021).
26. Fernandes, L.A.F.; Oliveira, M.M. Real-time line detection through an improved Hough transform voting scheme. *Pattern Recognit.* **2008**, *41*, 299–314. [CrossRef]
27. Nieto, M.; Cuevas, C.; Salgado, L.; Narciso, G. Line segment detection using weighted mean shift procedures on a 2D slice sampling strategy. *Pattern Anal. Appl.* **2011**, *14*, 149–163. [CrossRef]
28. Akinlar, C.; Topal, C. EDLines: A real-time line segment detector with a false detection control. *Pattern Recognit. Lett.* **2011**, *32*, 1633–1642. [CrossRef]
29. Gioi, R.; Jakubowicz, J.; Morel, J.M.; Randall, G. LSD: A fast line segment detector with a false detection control. *IEEE Trans. Pattern Anal. Mach. Intell.* **2010**, *32*, 722–732. [CrossRef] [PubMed]
30. Shan, T.; Englot, B.; Ratti, C.; Rus, D. LVI-SAM: Tightly-Coupled Lidar-Visual-Inertial Odometry via Smoothing and Mapping. [DB/OL]. Available online: https://arxiv.org/abs/2104.10831 (accessed on 27 November 2021).

remote sensing MDPI

Article

A Comparison of the Performances of Unmanned-Aerial-Vehicle (UAV) and Terrestrial Laser Scanning for Forest Plot Canopy Cover Estimation in *Pinus massoniana* Forests

Wenxia Dai [1], Qingfeng Guan [1,*], Shangshu Cai [2], Rundong Liu [3], Ruibo Chen [3], Qing Liu [3], Chao Chen [4] and Zhen Dong [2]

[1] School of Geography and Information Engineering, China University of Geosciences (Wuhan), Wuhan 430074, China; daiwenxia@cug.edu.cn
[2] State Key Laboratory of Information Engineering in Surveying, Mapping and Remote Sensing, Wuhan University, Wuhan 430079, China; shangshu_cai@whu.edu.cn (S.C.); dongzhenwhu@whu.edu.cn (Z.D.)
[3] Guangxi Zhuang Autonomous Region Institute of Natural Resources Remote Sensing, Nanning 530219, China; lrd_2021@163.com (R.L.); ruiboil@163.com (R.C.); liuqr369@163.com (Q.L.)
[4] Institute of Geophysics and Geomatics, China University of Geosciences (Wuhan), Wuhan 430074, China; lgxtz@cug.edu.cn
* Correspondence: guanqf@cug.edu.cn

Citation: Dai, W.; Guan, Q.; Cai, S.; Liu, R.; Chen, R.; Liu, Q.; Chen, C.; Dong, Z. A Comparison of the Performances of Unmanned-Aerial-Vehicle (UAV) and Terrestrial Laser Scanning for Forest Plot Canopy Cover Estimation in *Pinus massoniana* Forests. *Remote Sens.* **2022**, *14*, 1188. https://doi.org/10.3390/rs14051188

Academic Editors: Changhui Jiang, Yuwei Chen, Qian Meng, Panlong Wu, Bing Xu, Lianwu Guan, Wang Gao and Zeyu Li

Received: 25 January 2022
Accepted: 25 February 2022
Published: 28 February 2022

Abstract: Canopy cover is an important indicator and commonly used in forest management applications. Unmanned-Aerial-Vehicle (UAV)—Borne Laser Scanning (ULS) has drawn increasing attention as a new alternative source for forest field inventory due to its spatial resolution comparable to that of Terrestrial Laser Scanning (TLS). In this study, the performance of plot canopy cover estimations from ULS and TLS is investigated. The experiment was conducted in 16 plots from two *Pinus massoniana* forests with different stand conditions in Guangxi, China. Both the Canopy Height Model (CHM)-based and Individual Tree Delineation (ITD)-based methods were used to estimate the canopy cover. The influence of CHM pixel sizes on the estimations was also analyzed. Our results demonstrated that the accuracies of ULS (R^2: 0.992–0.996, $RMSE$: 0.591–0.820%) were better than those of TLS (R^2: 0.541–0.846, $RMSE$: 3.642–6.297%) when compared against the reference. The average difference between the ULS and TLS estimations was 6.91%, and the disagreement increased as the forest complexity increased. The reasonable CHM pixel sizes for the canopy cover estimations were 0.07–1.2 m for ULS and 0.07–1.5 m for TLS. This study can provide useful information for the selection of data sources and estimation methods in plot canopy cover mapping.

Keywords: forest point cloud; Unmanned Aerial Vehicle (UAV); Terrestrial Laser Scanning (TLS); canopy cover

1. Introduction

Forest canopy cover, which is defined as the proportion of the forest floor covered by the vertical projection of tree crowns [1,2], is directly related to the forest floor microclimate and light conditions [3–5] and is commonly used for biophysical and natural resource management applications. The spatially accurate mapping of canopy cover plays a critical role in forest stand structure classification [6], biomass production [7], wildfire behavior simulation [8], and wildlife habitat assessment [9,10].

Traditionally, canopy cover has been obtained from field measurements using sighting tubes [11], line intersect sampling [12], canopy photography [13], and the portable station field-map [14], which are laborious and time consuming. In addition, the field measurements obtained via these methods may be inaccurate because the crown boundaries can be difficult to distinguish in practice, and some subjectivity remains in the field measurement [2]. Compared with field measurements, remote sensing techniques can provide spatially continuous observations with a higher efficiency and at a lower cost. Light detection and ranging (LiDAR) is a promising tool for quantifying forest structural parameters because of its ability to assess 3D information with high precision [15–17]. LiDAR has the potential to replace field measurements or even be used to assess the quality of field measurements [5,18,19].

The potential of airborne laser scanning (ALS) from manned aircraft for estimating canopy cover has been investigated since commercial systems have become available. The simplest way to estimate canopy cover from ALS is to calculate the proportion of canopy hits above a specified height threshold [2]. However, this method is slightly biased because the ALS pulses are characterized by oblique observations and are not precisely vertical. Moreover, the sides of crowns are also observed at a certain scan angle, and this bias increases with the scan angle, becoming significant at approximately 40° (20° from the zenith) [13]. To eliminate the effect of oblique pulses on canopy cover estimations, several studies have utilized the rasterized canopy height model (CHM)-based method to estimate the canopy cover from ALS point clouds [2,20–22]. In these studies, canopy cover was calculated as the ratio of the number of canopy pixels to the total number of pixels. Allocating the canopy echoes to a grid based on XY coordinates was assumed to reduce this effect.

Several studies have also used mathematical models to estimate canopy cover from ALS. Mathematical models were constructed by investigating the correlations between airborne laser metrics and field-measured canopy cover [19,23,24]. For example, Holmgren et al. [23] utilized proportions of laser returns at certain height intervals derived from ALS data as explanatory variables in simple linear regression models for crown coverage estimation in southern Sweden. They reported a root-mean-square error ($RMSE$) of 4.9% for the tree crown coverage estimation. Melin et al. [24] compared with different remote sensing materials for predicting boreal forest canopy cover and observed a high correlation between the field-measured canopy cover and the selected LiDAR predictor.

The potential of terrestrial laser scanning (TLS) for mapping canopy gaps and structures has also been investigated [25–27]. However, compared with ALS canopy cover estimation studies, these studies are relatively fewer. The estimation of vertical canopy cover using TLS has not been studied as intensively. One Study [28] created a raster map of the canopy from TLS, and canopy cover was estimated as the proportion of canopy pixels with $RMSE$ of 8.0–17.9% and bias of 6.8–13.1%.

In recent years, improvements in the convenience and miniaturization of unmanned aerial vehicles (UAVs) have made it a powerful platform for forestry mapping. Combined with LiDAR, UAV laser scanning (ULS) provides higher data acquisition efficiency and flexibility at a lower cost than ALS [29,30] and provides detailed data comparable to that of TLS [31–34]. By providing a distinct combination of high spatial and temporal resolution, ULS narrows the gap between ALS and TLS systems and provides a new type of high-quality point cloud for forest investigations [33].

Previous studies have examined ULS as an alternative technology to ALS, which digitizes forests in a similar manner with higher altitude [30]. Previous ULS studies have predominantly focused on replicating existing forest attributes from ALS point clouds, such as forest height [35], tree crown diameter [29], stem volume [36], and aboveground biomass [30]. More recently, ULS has gained interest for its potential as an alternative technology to TLS owning to its near-ground perspective and dense point cloud characteristics. The first strict evaluation of ULS on in situ observations was provided by [33] and compared the DBH, tree height, and tree position results with those from TLS in a boreal forest. Similarly, [31] compared the ULS and TLS systems for canopy height and DBH estimation in a forest in the Netherlands, while [32] compared the performance of ULS and TLS with respect to explicit tree modeling and tree volume estimation in a Dutch temperate forest.

There is a growing tendency for foresters to use ULS instead of TLS in forest plot inventories, and this development will dramatically improve the efficiency and reduce costs for plot inventories. Accurate forest plot canopy cover estimations are crucial since mapping the distribution of canopy cover over large areas relies on the plot canopy cover estimations for model calibration and validation. It is necessary to study the performances of ULS and TLS in forest plot canopy cover estimations and how different methods may influence canopy cover estimations.

The main objectives of the present study were, therefore, threefold: (1) to investigate the performances of the recent rapidly developed ULS and the current widely used TLS techniques for plot canopy cover estimation under different forest stand conditions with respect to the manual references; (2) quantify the agreement and disagreement in the canopy cover estimations from ULS and TLS with respect to the CHM-based method and individual tree delineation (ITD)-based method; (3) clarify the influence of the pixel size on canopy cover estimation in the CHM-based method from ULS and TLS. The results from this study can provide practical guidance for the selection of data sources and estimation methods in plot canopy cover mapping.

2. Study Area and Materials

2.1. Study Area

The study was conducted in two coniferous forest sites in southern Guangxi Province, China: Guigang (23°7′N, 109°28′E) and Qinzhou (22°2′N, 108°34′E), as illustrated in Figure 1. The two sites are characterized by a subtropical monsoon climate with an annual rainfall of approximately 1600 mm and an annual average temperature of 21–23 °C. The Guigang site was established in a *Pinus massoniana* plantation with a very open understory and few evergreen microphanerophytes (Figure 2a). The *Pinus massoniana* plantation consists predominately of mature forest and had a low tree density of 133 trees/ha. The Qinzhou site was established in a natural forest of *Pinus massoniana*, which is a young forest with various understory species (Figure 2b). It comprises *Pinus massoniana* of mixed ages and has a tree density of 311 trees/ha, which is relatively higher than the Guigang site. The Guigang and Qinzhou sites had 11 and 5 plots, respectively, which were scanned by both the ULS and TLS. A total of sixteen plots were used in this study. The Guigang and Qinzhou plots are referred to as the GG plots and QZ plots, respectively, in the following sections.

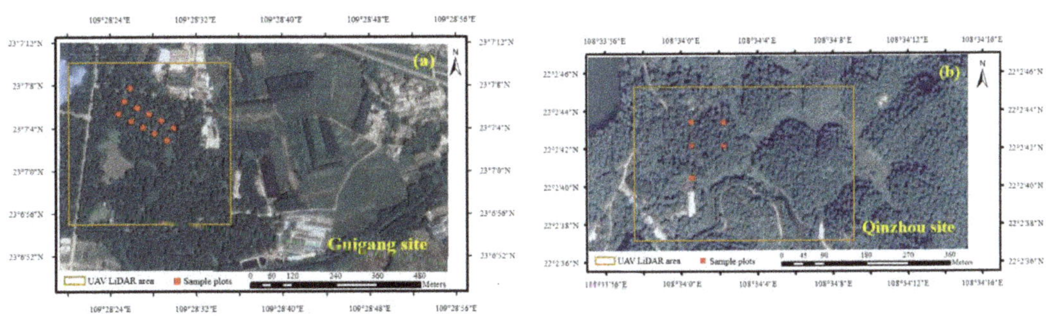

Figure 1. Map of the two test sites and sixteen sample plots in Guangxi, China: (**a**) Guigang site containing eleven plots and (**b**) Qinzhou site containing five plots analyzed in this study.

(a) **(b)**

Figure 2. Examples plots in Guigang and Qinzhou sites representing different forest structures: (**a**) GG plot was in mature forest and had a sparse tree density with less understory vegetation, and (**b**) QZ plot was in young forest and had a relatively high tree density representing various understory species.

2.2. *LiDAR Data Collection*

2.2.1. UAV Laser Scanning Data

The UAV-LiDAR data of the 16 plots were captured on 4 and 5 June 2020, with a Genius UAV LiDAR developed by SureStar, Beijing. The UAV LiDAR area coverage and plot distribution are shown in Figure 2. Each plot at the two sites had a fixed size of 30 m × 30 m. The UAV-LiDAR system consisted of a DJI M200 platform, an R-Fans-16 laser scanner, and an advanced navigation spatial dual-coupled GNSS and IMU sensor. The R-Fans-16 scanner features 16 scan layers with a 30° vertical field of view (FOV). It operates at a wavelength of 905 nm and supports 320,000 points per second. The measurement range of R-Fans-16 is 200 m. The flying height for data acquisition was 40 m above ground level with a speed of 5 m/s, resulting in an average point density of approximately 430 pts/m^2.

2.2.2. Terrestrial Laser Scanning Data

TLS data acquisition was completed on 28 and 29 May 2020, using the RIEGL VZ-400 laser scanner system. The RIEGL VZ-400 laser scanner system has a 360° field of view in the horizontal direction and 100° in the vertical direction, a maximum range of 600 m, and a measurement rate of 120,000 points per second. At the Guigang site, five scans were performed with an average scanning spacing of 35 m to complete a full coverage of the plots, since the plots in the Guigang site were characterized by clear visibility, little understory vegetation, and sparse tree density. Five scans were performed at the Qingzhou site with an average scanning spacing of 15 m to accommodate the high tree density and abundant understory vegetation in this region. Additional retro-reflective targets were set up during the data collection campaign for the co-registration of scans. The average point density of registered TLS data was 29,471 points per m^2, with a point spacing of approximately 6 mm.

2.3. Establishment of Reference Data

To evaluate the performance of the canopy cover estimation from TLS and ULS, the reference data for the 16 plots were generated by manual measurements from the co-registered TLS and ULS point clouds. The co-registered TLS and ULS point clouds combined the observations from under and above canopy view perspectives with high spatial resolution and were assumed to provide a more reliable canopy cover measurement compared to the use of either TLS or ULS point clouds alone. The ULS point clouds of the plots were manually co-registered to the multi-scan TLS point clouds by selecting common points from the crown boundary and stems. Manual fine-tuning was implemented when a discrepancy between the point clouds remained observable from the top view and two side views. Then, the digital terrain models (DTMs) were created at a 0.2 m resolution from the fused TLS and ULS point clouds using the cloth simulation filter (CSF) method [37]. The fused point cloud height was normalized by subtracting the ground surface height from the DTMs. Finally, the canopy boundaries in each plot were manually delineated from the fused and normalized point clouds, and the canopy cover for each plot was calculated as the ratio of the area of tree crown to the area of the plot. The canopy cover of all 16 plots ranged from 63.37% to 96.28%. Details of the canopy cover distribution are shown in Figure 3.

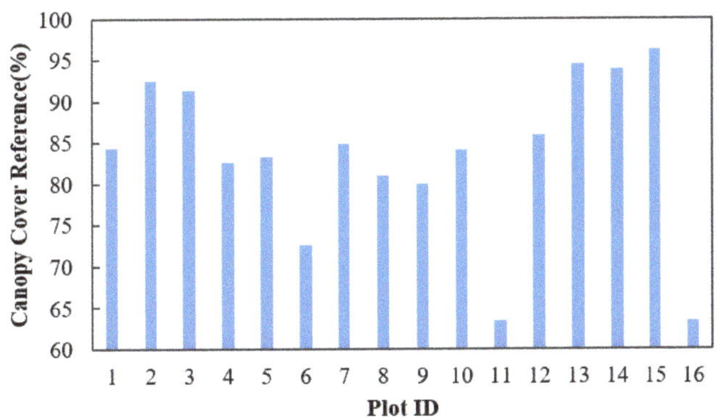

Figure 3. Canopy cover reference for the 16 plots analyzed in the study.

3. Methods

3.1. Canopy Cover Estimation Using CHM-Based Method

LiDAR-derived CHMs have been widely used to estimate canopy cover [2,21]. To guarantee independence of the canopy cover estimates from different data sources, CHM creations and canopy cover estimations were carried out separately in ULS and TLS point clouds. First, for each sample plot, the ground points and off-ground points were classified from the ULS and TLS point clouds separately, and their DTMs were generated using the CSF method [37] with a 0.2 m × 0.2 m resolution. The TLS and ULS data were normalized with respect to their corresponding DTMs. Then, the normalized point clouds were gridded, and the highest point in each grid was selected to construct the CHMs. To simultaneously describe the tree crown in as much detail as possible and reduce data redundancy, a raw pixel size that was slightly larger than the mean point spacing of the point cloud was used for CHM construction, as reported by [38,39]. In this study, 7 cm and 1 cm were used for the ULS and TLS point clouds, respectively, for CHM construction.

The CHM-based canopy cover was calculated as the percentage of pixels with a CHM value larger than a specified height threshold (canopy pixels):

$$\text{CanopyCover} = \sum CHM_{canopy} / \sum CHM_{total} \quad (1)$$

where CHM_{canopy} represents the number of canopy pixels (above a specific height threshold) in CHM, and CHM_{total} represents the total number of CHM pixels.

Considering the different understory vegetation arrangements in the plots of the two sites (as illustrated in Figure 2), different height thresholds were used to separate the crowns from the background for canopy cover estimation. In the GG plots, since these plots represented clear visibility with sparse understory vegetation, a distance of 2 m was used to separate the crown pixels. In the QZ plots, a distance of 5 m was used to extract crown pixels because of the relatively dense and high shrubs. The within-crown gaps would lead to the underestimation of canopy cover because their CHM values were relatively small (smaller than the height threshold) and were likely to be classified as non-canopy pixels. This situation often occurs in the crown where laser pulses penetrate the gap and reach the ground surface, resulting in a small height value in data collection. Therefore, we utilized the pit-free CHM method proposed by [38] to fill the within-crown gaps of the CHM before canopy cover estimation. This method works by simulating cloth sticking to the CHM surface and filling the within-crown gaps using the hardness of the simulated cloth. The pit-free CHM-based method was capable of filling within-crown gaps while keeping the original CHM pixel values unchanged. The CHM-based canopy cover estimation results from the ULS and TLS data are denoted as ULS_CHM and TLS_CHM, respectively, in the following sections.

We also explored the sensitivity of canopy cover estimation to different CHM pixel sizes for the ULS and TLS point clouds. The original CHMs of ULS and TLS were created under the raw pixel size, which was assumed to describe the crown structure in the most detail. Then, the pixel sizes were increased, from 0.07 to 4.8 m for the ULS and from 0.01 to 2.5 m for the TLS for CHM construction. The canopy cover estimations were then calculated using these CHMs. The canopy cover accuracy was evaluated by comparison with the reference data.

3.2. Canopy Cover Estimation Using ITD-Based Method

To guarantee the independence of the canopy cover estimates from different data sources and to minimize the influences of laser scanning data processing methods, individual tree delineations were manually conducted in the ULS and TLS point clouds separately. Automatic tree detection and modeling methods were not used in this study. Thus, the canopy cover evaluation results revealed the capacities of the applied laser scanning data while excluding the influence of the data processing approach, such as the individual tree detection. Then, each individual tree crown in a plot was accumulated and subtracted from the overlap area to calculate the total crown area. The ITD-based canopy cover was calculated as the percentage of the total crown area to the plot area. The ITD-based canopy cover estimation results from the ULS and TLS data are denoted as ULS_ITD and TLS_ITD, respectively, in the following sections.

3.3. Comparison Scheme and Accuracy Assessment

First, the four canopy cover estimates (ULS_CHM, TLS_CHM, ULS_ITD, and TLS_ITD) were compared with the reference data. Then, the agreement and disagreement in the canopy cover estimations from ULS and TLS were quantified in detail with respect to the CHM-based and ITD-based methods. Finally, the influence of the pixel size on canopy cover estimation in the CHM-based method from ULS and TLS was analyzed.

In this study, the accuracy of the estimated canopy cover was evaluated using the coefficient of determination (R^2) and root mean squared error ($RMSE$), which were calculated using the following equations:

$$R^2 = 1 - \frac{\sum_{i=1}^{n}(x_i - y_i)^2}{\sum_{i=1}^{n}(x_i - \overline{x})^2} \quad (2)$$

$$RMSE = \sqrt{\frac{\sum_{i=1}^{n}(x_i - y_i)^2}{n}} \quad (3)$$

where x_i and y_i are the values from the ith reference and estimated canopy cover values, $\overline{x}$ is the mean of the reference canopy cover, and n is the number of plots.

4. Results

4.1. Comparison of LiDAR Estimations and Reference

A comparison between the canopy cover estimations from the four methods (i.e., ULS_CHM, ULS_ITD, TLS_CHM, and TLS_ITD) and the reference data are shown in Figure 4 (for all the plots) and Figure 5 (for the GG plots alone). As illustrated in Figure 4, there was an overall moderate to high agreement between the LiDAR-estimated canopy cover and the reference data for all the plots, with R^2 values of 0.541–0.996, and $RMSE$ values of 0.591–6.297%. Among the four methods, the ULS_CHM method showed the highest accuracy, with an R^2 of 0.996 and an $RMSE$ of 0.591% for canopy cover estimation, while the ULS_ITD method had the second highest accuracy, with an R^2 of 0.992 and an $RMSE$ of 0.820%, followed by the TLS_ITD method (R^2 = 0.846, $RMSE$ = 3.642%) and the TLS_CHM method (R^2 = 0.541, $RMSE$ = 6.297%).

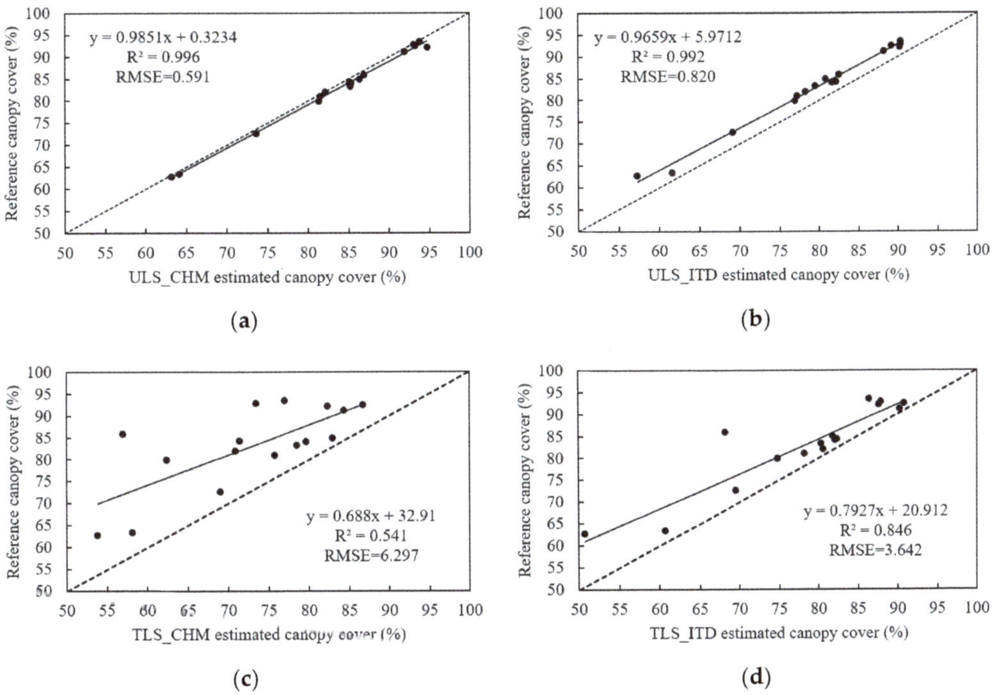

Figure 4. Comparison of canopy cover estimations derived from the reference data and LiDAR estimations for all the plots. Scatter plots with R^2, $RMSE$, and regression equations between reference data (y) and LiDAR-based estimations (x) are indicated for the (**a**) ULS_CHM, (**b**) ULS_ITD, (**c**) TLS_CHM, and (**d**) TLS_ITD estimations.

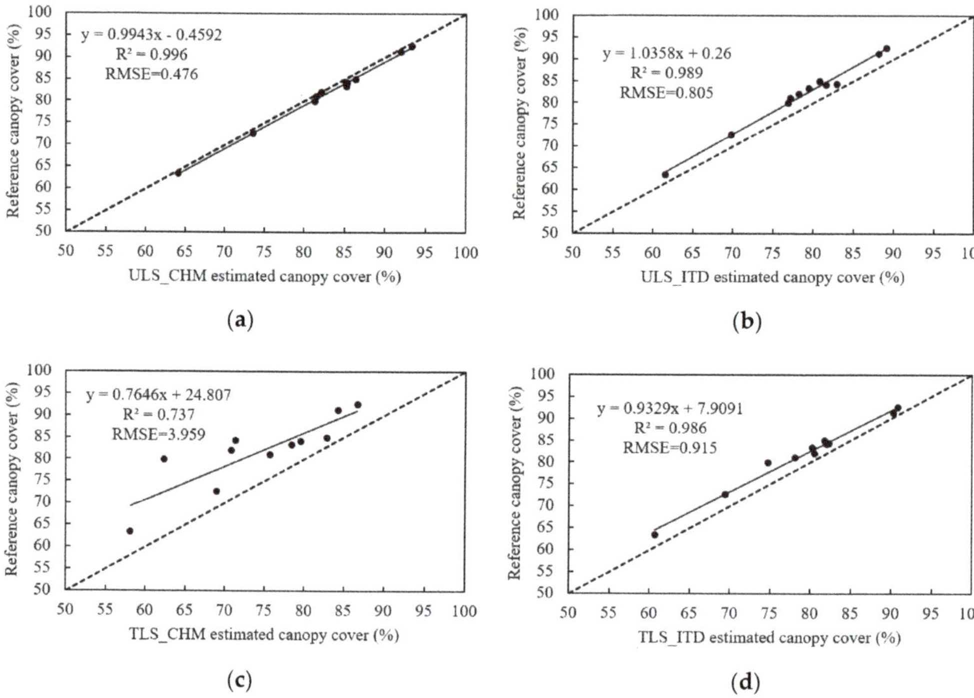

Figure 5. Comparison of canopy cover estimations derived from the reference data and LiDAR estimations for the GG plots alone. Scatter plots with R^2, $RMSE$, and regression equations between reference data (y) and LiDAR-based estimations (x) are shown for the (**a**) ULS_CHM, (**b**) ULS_ITD, (**c**) TLS_CHM, and (**d**) TLS_ITD estimations.

As illustrated in Figure 5, the results from the GG plots alone showed a similar tendency, with the ULS_CHM method showing the highest accuracy (R^2 of 0.996, $RMSE$ of 0.476%), followed by the ULS_ITD method (R^2 = 0.992, $RMSE$ = 0.685%) and the TLS_ITD method (R^2 = 0.986, $RMSE$ = 0.915%). The TLS_CHM method had the lowest accuracy (R^2 = 0.737, $RMSE$ = 3.959%). For different forest conditions, the GG plots had a better performance (R^2 = 0.737–0.996, $RMSE$ = 0.476–3.959%) than all other plots (R^2 = 0.541–0.996, $RMSE$ = 0.591–6.297%).

The overestimation and underestimation of the four methods against the reference data are summarized in Table 1, which shows the difference between the ULS/TLS estimated canopy cover and the reference data for all the plots, the GG plots only, and the QZ plots only. Overall, the ULS produced smaller deviations than TLS. The mean deviation for the ULS was 2.1% for all the plots and 7.46% for TLS. The ULS_CHM had the smallest deviations, followed by the ULS_ITD, TLS_ITD, and TLS_CHM. In addition, the ULS canopy cover estimations were more robust across different stand conditions and different methods than the TLS estimations. The ULS produced similar deviations between the GG and QZ plots (2.08% vs. 2.13%), whereas significant differences were observed between the corresponding deviations of TLS (4.93% and 13.03%).

Table 1. Canopy cover differences (%) calculated as ULS/TLS estimations minus reference data for all the plots, the GG plots only, and QZ plots only, where |mean| represents the mean of the absolute values of the difference.

		All Plots			GG Plots			QZ Plots		
		Min	Max	|Mean|	Min	Max	|Mean|	Min	Max	|Mean|
ULS	CHM-based	0.07	2.50	0.93	0.07	1.90	0.93	0.33	2.50	0.92
	ITD-based	−5.47	−1.75	3.26	−4.15	−1.75	3.22	−5.47	−2.13	3.34
	Mean			2.10			2.08			2.13
TLS	CHM-based	−28.94	−2.03	10.22	−17.58	−2.03	7.27	−28.94	−8.96	16.73
	ITD-based	−17.75	−1.06	4.69	−5.18	−1.06	2.59	−17.75	−4.70	9.32
	Mean			7.46			4.93			13.03

4.2. The Agreement and Disagreement in the Estimations from ULS and TLS

A more detailed comparison was conducted directly between the ULS and TLS estimations with respect to different forest conditions and estimation methods. The R^2, $RMSE$ and differences between the ULS and TLS estimations were summarized. The disagreement between the ULS and TLS estimations increased with increasing complexity of the forest stand with respect to these metrics.

A moderate agreement was observed between the ULS and TLS estimations when the CHM-based method was used. As illustrated in Figure 6a, the R^2 and $RMSE$ between the ULS_CHM and TLS_CHM estimations were 0.554 and 6.288% for all the plots. In the case of the ITD method, the R^2 and $RMSE$ between the ULS_ITD and TLS_ITD estimations were R^2 0.859 and 3.600% for all the plots (Figure 6b). For different forest conditions, the GG plots had a higher agreement between the ULS and TLS estimations than all the plots. Figure 6c,d illustrates the comparison on the GG plots, where the ULS and TLS produced an R^2 of 0.745 and an $RMSE$ of 3.913% for the CHM-based method, and R^2 of 0.985 and an $RMSE$ of 0.919% for the ITD method.

Table 2 summarizes the difference between the ULS and TLS estimations for all the plots, the GG plots only, and the QZ plots only. The number of all plots, GG plots, and QZ plots were 16, 11 and 5 respectively. The ULS estimations were larger overall than the TLS estimations for the CHM-based method, with an averaged difference of 11.15% for all the plots. The overestimations in the GG plots were lower than those in the QZ plots. The average overestimation was 8.19% for the GG plots and 17.65% for the QZ plots (Table 2). For the ITD-based method, the TLS estimations tended to be larger in the GG plots and lower in the QZ plots than the ULS estimations (Figure 6b). Only two plots had lower TLS_ITD estimations than the ULS_ITD estimations (Figure 6b). In the QZ plots, the average TLS_ITD estimations were 5.97% lower than ULS_ITD estimations.

Table 2. Canopy cover differences (%) calculated as ULS estimations minus TLS estimations for all the plots, GG plots only, and QZ plots only, where |mean| represents the mean of the absolute values of the difference.

	All Plots			GG Plots			QZ Plots		
	Min	Max	|Mean|	Min	Max	|Mean|	Min	Max	|Mean|
ULS_CHM-TLS_CHM	3.45	29.91	11.15	3.45	18.89	8.19	9.36	29.91	17.65
ULS_ITD-TLS_ITD	−2.33	14.34	2.67	−2.33	2.07	1.17	2.39	14.34	5.97
mean			6.91			4.68			11.81

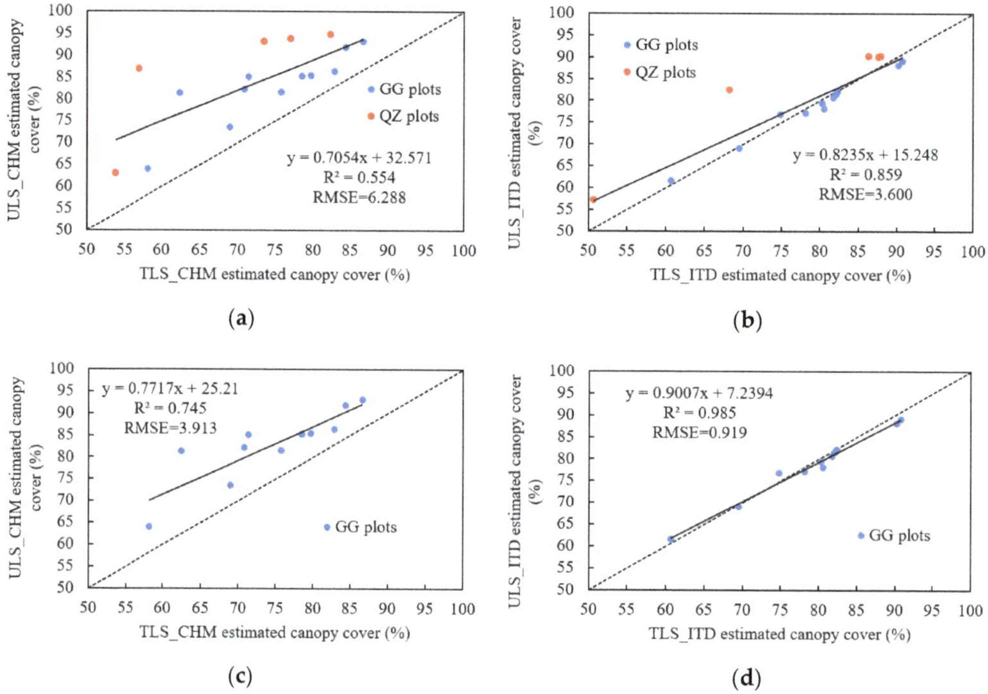

Figure 6. Comparisons of canopy cover estimations derived from ULS and TLS: (**a**,**b**) for all the plots, and (**c**,**d**) for only the GG plots.

4.3. Estimation Results of CHM-Based Canopy Cover with Different Pixel Size

Changes in R^2 and $RMSE$ between the CHM-based estimations using different pixel sizes and the reference data are presented in Figure 7. For the ULS_CHM method, the R^2 between the ULS_CHM estimations and the reference data decreased with an increase in the pixel size (Figure 7a). The R^2 decreased slowly from 0.996 to 0.959 with an increase in the pixel size range from 0.07 m (raw pixel size) to 1.2 m. The decrease rate of R^2 was significantly larger after the pixel size exceeded 1.2 m. Conversely, the $RMSE$ values increased as the pixel size increased.

For the TLS_CHM method, R^2 initially increased and then decreased with increasing pixel size (Figure 7b). The R^2 between the TLS_CHM estimations and reference data increased from 0.541 to 0.871 when the pixel size range increased from 0.01 m (raw pixel size) to 1.0 m. After the pixel size surpassed 1.0 m, R^2 rapidly decreased. The $RMSE$ values first decreased and then increased as the pixel size increased.

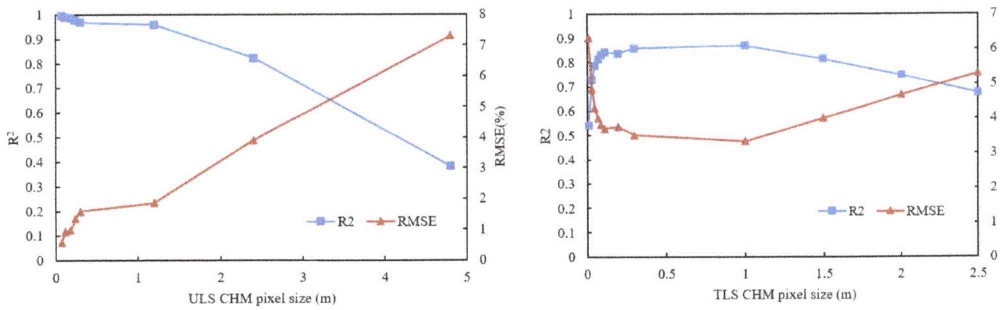

Figure 7. Changes in R^2 and *RMSE* between two CHM-based canopy cover estimations using different pixel sizes and reference data. (**a**) R^2 and *RMSE* between ULS_CHM estimations and reference data and (**b**) R^2 and *RMSE* between TLS_CHM estimations and reference data.

5. Discussion

5.1. Differences between LiDAR-Derived Canopy Cover and Reference Data

In this study, we compared four LiDAR-estimated canopy covers (ULS_CHM, ULS_ITD, TLS_CHM, and TLS_ITD) with the reference data. The ULS_CHM produced the highest accuracy, followed by ULS_ITD, TLS_ITD, and TLS_CHM. The results demonstrated that the canopy covers obtained using the ULS_CHM method were slightly higher than the reference data, and the canopy covers obtained from the other three methods were lower than the reference data. The higher canopy cover estimations obtained from the ULS_CHM method could be partly attributed to the following aspects: (i) some small between-crown gaps with similar size of within-crown gaps were also filled as canopy pixels, and (ii) the crown boundaries that adjoined the open ground in the horizontal plane tended to expand after the interpolation procedure of the pit-free method.

Since the original CHM method would underestimate the canopy cover owing to the existence of within-crown gaps, we utilized the pit-free method proposed by [38] to fill the within-crown gaps to mitigate the underestimation. Our results showed that the pit-free method could effectively remove the within-crown gaps (c1 and c2 in Figure 8a,b). However, the small between-crown gaps with similar within-crown gap size were also interpolated as canopy pixels (d1 and d2 in Figure 8a,b), and the crown boundaries adjacent to the open ground expanded after the pit-free method was applied (e1 and e2 in Figure 8a,b). It was difficult to distinguish between the within-crown gaps and the between-crown gaps with similar sizes and fill the within-crown gaps while ensuring that the between-crown gaps remained unchanged in the CHM smoothing process. The crown boundaries adjacent to the open ground expanded because there were height jumps between the crown boundaries and the adjoining ground. The pit-free method interpolated the pixel values of the adjoining ground and increased their height to reduce the height difference. Thus, several ground pixels were classified as canopy, and the canopy cover was slightly magnified.

In the TLS_CHM method, the canopy cover estimation was lower than the reference, which could be attributed to the incomplete tree crown structure generated from the TLS point clouds. Although the multi-scan mode was used in the TLS data collection, the crowns further away remained occluded when the laser beam was interrupted by stems or branches, and the upper crown of the higher trees was incomplete due to the limited field of view in the vertical direction (−40°–60°). These situations produced within-crown gaps, and these gaps could not be completely removed by the pit-free method, resulting in the underestimation of the TLS_CHM method.

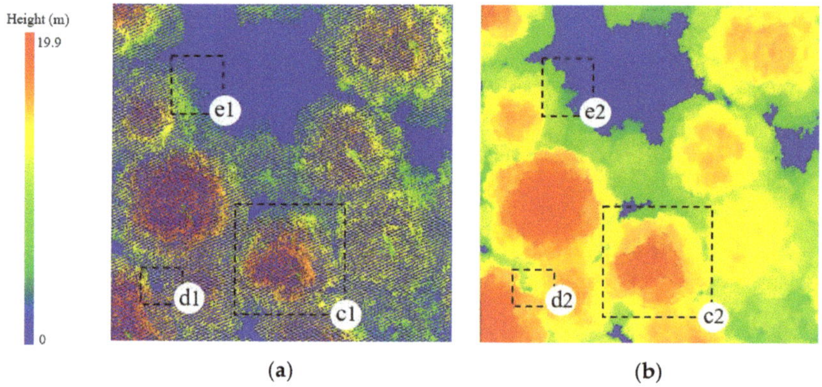

Figure 8. The pit-free method used in the ULS_CHM canopy cover estimation, where (**a**,**b**) represent the original CHM and the pit-free CHM, respectively, while (c1) (c2), (d1) (d2), and (e1) (e2) represent the within-crown gaps, between-crown gaps, and crown boundaries adjacent to the open ground in the original CHM and the pit-free CHM.

Both the ULS_ITD and TLS_ITD canopy cover estimations were smaller than the reference data, because the reference data were produced from the fused ULS and TLS point clouds. The TLS_ITD canopy cover estimations were lower when the crown boundaries were incomplete in the TLS point clouds. The slightly lower ULS_ITD estimations could be partly attributed to the lower point density of the ULS point clouds when compared with the fused point clouds.

5.2. Difference between ULS-Derived and TLS-Derived Canopy Cover Estimations

In the CHM method, our results demonstrated that the ULS estimations were larger than the TLS estimations for all the plots. The differences between the ULS_CHM and TLS_CHM estimations increased with the increased forest complexity. In the ITD method, the ULS estimations were smaller than the TLS estimations in the simple plots with little understory vegetation and low stem density (GG plots), and the ULS estimations were larger than the TLS estimations in the relatively complex plots with abundant understory growth and higher stem density (QZ plots).

Overall, the ULS tree crowns were more comprehensive than the TLS tree crowns, even when the multi-scan mode was used. Similar results were reported by [28], where ALS produced slight overestimation of canopy cover and TLS underestimated the canopy cover. TLS was vulnerable to the radial occlusion due to the side view perspective and produced gaps within the crowns. These gaps were large and difficult to fill by the CHM smooth method (pit-free), resulting in lower TLS estimations in the CHM-based method. This underestimation grew with increased forest complexity due to the increased occlusion. The QZ plots had denser understory vegetation and higher stem density than the GG plots, resulting in more occlusions in TLS than the GG plots. Therefore, the difference between the ULS_CHM and TLS_CHM estimations for the GG plots (Figure 9) was smaller than that of the QZ plots (Figure 10).

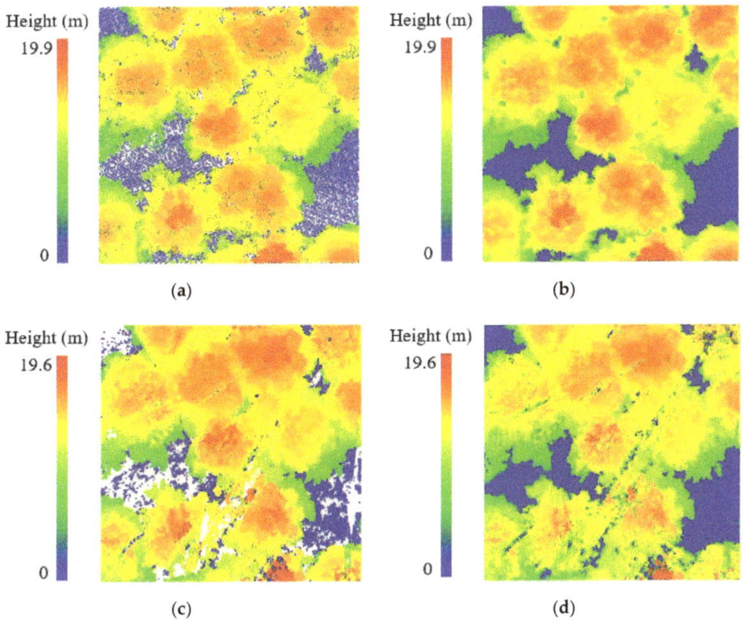

Figure 9. Difference between ULS_CHM and TLS_CHM in GG plot: (**a**,**c**) normalized ULS and TLS point clouds, and (**b**,**d**) corresponding pit-free CHMs.

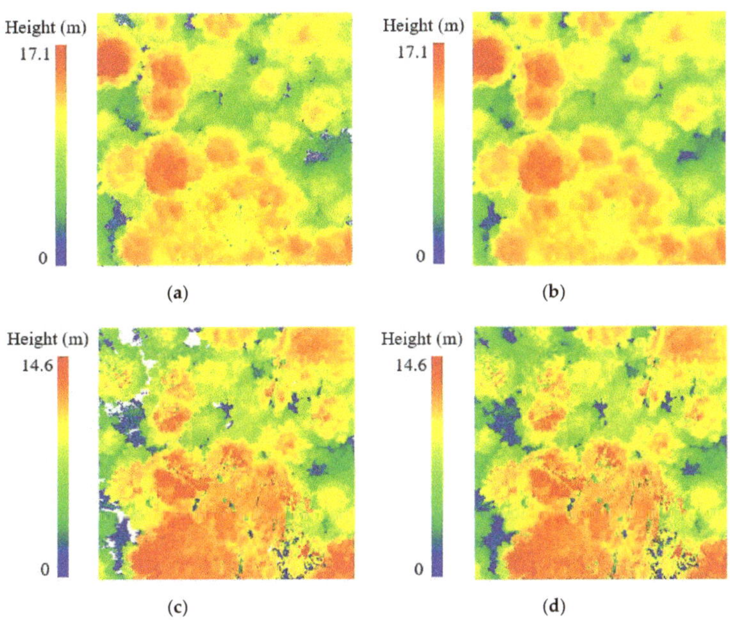

Figure 10. Difference between ULS_CHM and TLS_CHM in QZ plot: (**a**,**c**) normalized ULS and TLS point clouds and (**b**,**d**) corresponding pit-free CHMs.

In the ITD method, the TLS estimations were larger than the ULS estimations in simple plots, which can be partly attributed to the denser point density of the TLS point clouds and the fact that the incomplete tree crowns could be recovered as long as crown boundaries existed in the ITD method. The crown boundaries in the simple plots with little understory vegetation and low stem density were more likely to be collected than those in the relatively complex plots with abundant understory growth and higher stem density. Moreover, compared with the ULS crowns, the TLS crowns represented more details and larger areas (as illustrated in Figure 11e). The TLS crown boundaries were more compact, and their between-crown gaps were smaller than those of the ULS (Figure 11b,d). Therefore, the TLS_ITD produced slightly higher estimations than ULS in the simple plots. However, there were two GG plots that produced lower TLS_ITD estimations because their tree crown boundaries were incomplete and the ITD method could not recover the correct crowns areas.

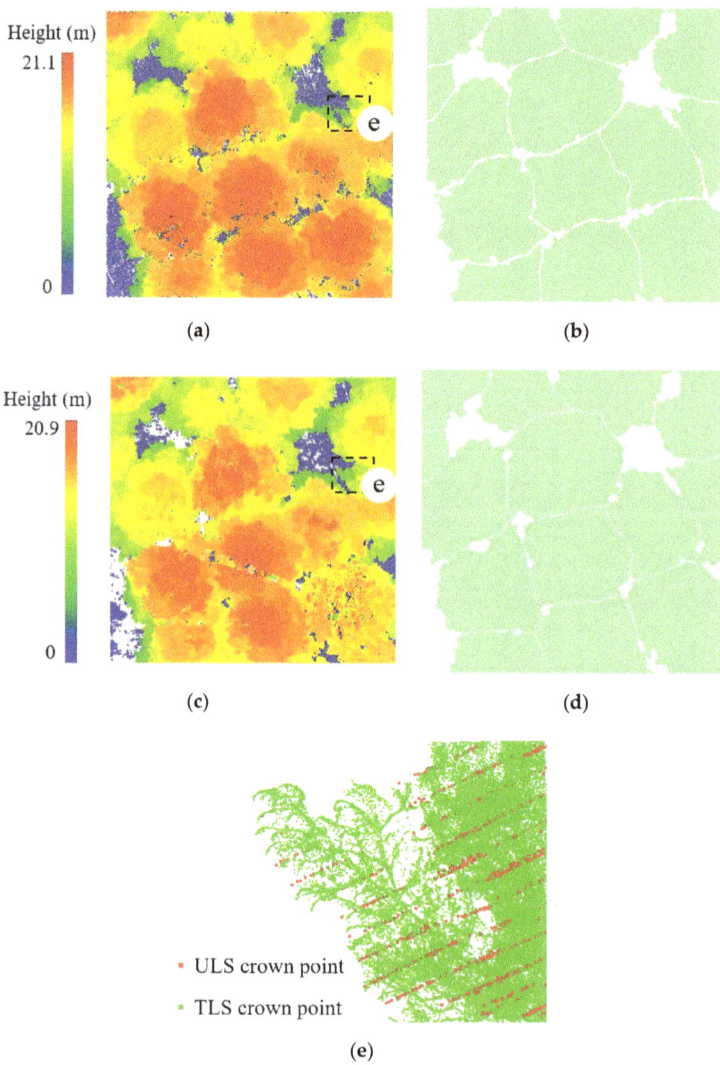

Figure 11. Difference between ULS_ITD and TLS_ITD in GG plot: (**a**,**c**) normalized ULS and TLS point clouds, (**b**,**d**) corresponding individual tree crown boundaries, and (**e**) the local detail for the overlapped ULS and TLS crown points.

For the QZ plots, the TLS estimations were lower than the ULS estimations in the ITD method. As illustrated in Figure 12b,d, the QZ plots had denser understory vegetation and higher stem density than the GG plots, which led to more severe occlusion and incomplete tree crowns in the TLS point clouds. The crown boundaries were incomplete and the ITD method cannot recover the correct crowns areas, resulting in the underestimation.

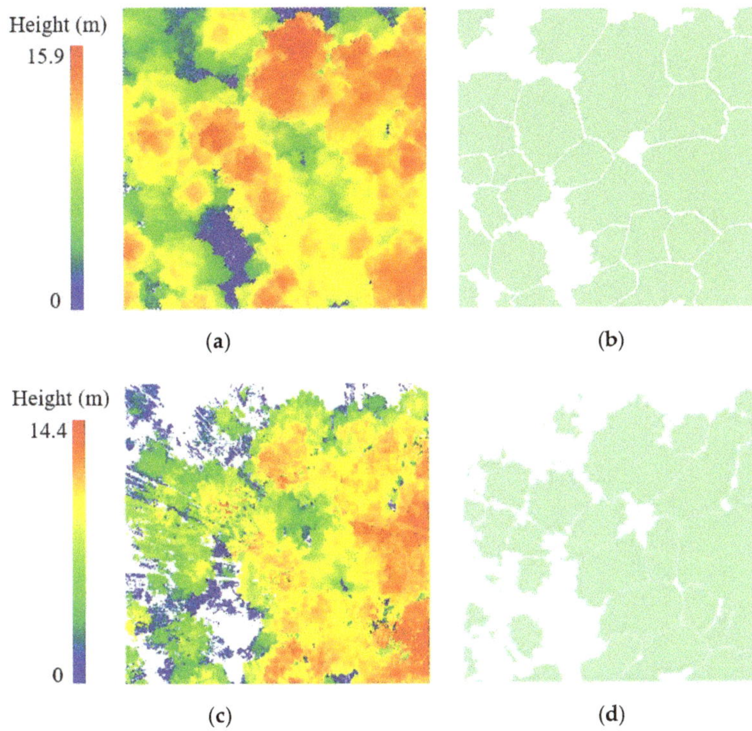

Figure 12. Difference between ULS_ITD and TLS_ITD in QZ plot: (**a**,**c**) the normalized ULS and TLS point clouds and (**b**,**d**) corresponding individual tree crown boundaries.

5.3. Effect of Pixel Size on the CHM-Based Canopy Cover Estimation Accuracy

The canopy cover estimation increased as the pixel size increased for both the ULS and TLS. This aligned with the results of [38]. When a larger pixel size was used, each within-crown gap was more likely to be assigned to a mixed pixel with crown points, and the gaps were neglected because only the highest point was selected. In addition, the crown boundaries expanded as the pixel size increased, which contributed to an increase in canopy cover values.

For the ULS point clouds, the raw pixel size of the CHM produced slightly overestimated estimations. With increasing pixel size, the overestimation increased when compared with the reference data. Therefore, R^2 decreased as the pixel size increased. The R^2 decreased slowly and remained relatively constant (0.959–0.996) when the pixel size increased from 0.07 m to 1.2 m, and then decreased significantly again when the pixel size exceeded 1.2 m. The increasing pixel size caused more gaps to be filled as crowns, and the tree crown edges were gradually expanded, resulting in an increase in canopy cover estimation. Reasonable and similar canopy cover estimations could be achieved at pixel sizes ranging from 0.07 m to 1.2 m, which could explain over 95% of the variations in the reference data.

For the TLS point clouds, the raw pixel size of the CHM was underestimated owing to the incomplete tree crown structure. The differences between the CHM estimations and reference data were gradually narrowed at first and then gradually increased with increasing pixel size owing to the increase in canopy cover values. The best agreement was achieved at a pixel size of 1.0 m with an R^2 of 0.871 and an $RMSE$ of 3.333%. Reasonable and similar canopy cover estimations could be achieved at a pixel size range of 0.07 m to 1.5 m, which could explain over 80% of the variations in the reference data.

6. Conclusions

Forest canopy cover plays a fundamental role in forest assessment and management. The Sample plot inventories are currently experiencing changes driven by the rapid development of UAV. This study provided a comprehensive cross-comparison of plot canopy cover from the recent rapidly developed ULS and current widely used TLS point clouds over 16 plots in *Pinus massoniana* forests with different stand conditions in Guangxi, China. Both the CHM- and ITD-based methods were used to estimate the canopy cover for both the ULS and TLS point clouds. Our results illustrated that, compared with the reference data, the ULS_CHM method was the most accurate, with an R^2 of 0.996 and $RMSE$ of 0.591%, followed by the ULS_ITD method (R^2 = 0.992, $RMSE$ = 0.820%), TLS_ITD method (R^2 = 0.846, $RMSE$ = 3.642%), and TLS_CHM method (R^2 = 0.541, $RMSE$ = 6.297%). When the ULS estimations were directly compared against the TLS estimations, most ULS estimations were larger than the TLS estimations, with an average difference of 6.91%, and the disagreement increased as the forest complexity increased. The ULS estimations were lower than the TLS estimations; this occurred when the crown boundaries were complete in the ITD method in the simple plots due to the more detailed crowns in the intermediate and suppressed layer than the ULS. In the CHM-based method, the reasonable CHM pixel sizes for the canopy cover estimations were 0.07–1.2 m for ULS and 0.07–1.5 m for TLS. In these ranges, the estimations were marginally influenced by the pixel size. Further work should investigate the estimation performance of canopy cover over large areas from different sources and extend the forest types.

Author Contributions: Conceptualization, W.D. and Z.D.; methodology, W.D. and S.C.; software, W.D. and S.C.; validation, Q.G., R.L., R.C. and Q.L.; formal analysis, W.D., S.C. and C.C.; writing, W.D. All authors have read and agreed to the published version of the manuscript.

Funding: This research was funded by the National Natural Science Foundation of China (Grant 42101456).

Data Availability Statement: The data that support the findings of this study are available on request from the author, (R.C.). The data are not publicly available because they contain information that could compromise the privacy of research participants.

Acknowledgments: The authors would like to acknowledge Guangxi Zhuang Autonomous Region Institute of Natural Resources Remote Sensing for providing the datasets for this study.

Conflicts of Interest: The authors declare no conflict of interest.

References

1. Jennings, S.B.; Brown, N.D.; Sheil, D. Assessing forest canopies and understorey illumination: Canopy closure, canopy cover and other measures. *Forestry* **1999**, *72*, 59–74. [CrossRef]
2. Korhonen, L.; Korpela, I.; Heiskanen, J.; Maltamo, M. Airborne discrete-return LIDAR data in the estimation of vertical canopy cover, angular canopy closure and leaf area index. *Remote Sens. Environ.* **2011**, *115*, 1065–1080. [CrossRef]
3. FAO. *On Definitions of Forest and Forest Change. Forest Resources Assessment Programme*; Working Paper; FAO: Rome, Italy, 2000; Volume 33, p. 15.
4. Gill, S.J.; Biging, G.S.; Murphy, E.C. 2000. Modeling conifer tree crown radius and estimating canopy cover. *For. Ecol. Manag.* **2000**, *126*, 405–416. [CrossRef]
5. Korhonen, L.; Korhonen, K.T.; Rautiainen, M.; Stenberg, P. Estimation of forest canopy cover: A comparison of field measurement techniques. *Silva Fenn.* **2006**, *40*, 577–588. [CrossRef]
6. Fiala, A.C.; Garman, S.L.; Gray, A.N. Comparison of five canopy cover estimation techniques in the western Oregon Cascades. *For. Ecol. Manag.* **2006**, *232*, 188–197. [CrossRef]

7. Chopping, M.; Moisen, G.G.; Su, L.; Lalibrete, A.; Rango, A.; Martonchik, J.; Peter, D.P.C. Large area mapping of southwestern forest crown cover, canopy height, and biomass using the NASA Multiangle Imaging Spectro-Radiometer. *Remote Sens. Environ.* **2008**, *112*, 2051–2063. [CrossRef]
8. Miller, J.D.; Knapp, E.E.; Key, C.H.; Skinner, C.N.; Isbell, C.J.; Creasy, R.M.; Sherlock, J.W. Calibration and validation of the relative differenced Normalized Burn Ratio (RdNBR) to three measures of fire severity in the Sierra Nevada and Klamath Mountains, California, USA. *Remote Sens. Environ.* **2009**, *113*, 645–656. [CrossRef]
9. Ganey, J.L.; Cassidy, R.H.; Block, W.M. *Estimating Canopy Cover in Forest Stands Used by Mexican Spotted owls: Do Stand-Exam Routines Provide Estimates Comparable to Field-Based Techniques? Res. Pap. RMRS-RP-72WWW*; US Department of Agriculture, Forest Service, Rocky Mountain Research Station: Fort Collins, CO, USA, 2008; Volume 8, p. 72.
10. Smart, L.; Swenson, J.; Christensen, N.; Sexton, J. Three-dimensional characterization of pine forest type and red-cockaded woodpecker habitat by small-footprint, discrete-return lidar. *For. Ecol. Manag.* **2012**, *281*, 100–110. [CrossRef]
11. Rautiainen, M.; Stenberg, P.; Nilson, T. Estimating canopy cover in Scots pine stands. *Silva Fenn.* **2005**, *39*, 137–142. [CrossRef]
12. Williams, M.S.; Patterson, P.L.; Mowrer, H.T. Comparison of ground sampling methods for estimating canopy cover. *For. Sci.* **2003**, *49*, 235–246.
13. Korhonen, L.; Heikkinen, J. Automated analysis of in situ canopy images for the estimation of forest canopy cover. *For. Sci.* **2009**, *55*, 323–334.
14. Santopuoli, G.; di Cristofaro, M.; Kraus, D.; Schuck, A.; Lasserre, B.; Marchetti, M. Biodiversity conservation and wood production in a Natura 2000 Mediterranean forest. A trade-off evaluation focused on the occurrence of microhabitats. *IForest* **2019**, *12*, 76. [CrossRef]
15. Liang, X.; Wang, Y.; Jaakkola, A.; Kukko, A.; Kaartinen, H.; Hyyppä, J.; Honkavaara, E.; Liu, J. Forest data collection using terrestrial image-based point clouds from a handheld camera compared to terrestrial and personal laser scanning. *IEEE Trans. Geosci. Remote Sens.* **2015**, *53*, 5117–5132. [CrossRef]
16. Sankey, T.; Donager, J.; McVay, J.; Sankey, J.B. UAV lidar and hyperspectral fusion for forest monitoring in the southwestern USA. *Remote Sens. Environ.* **2017**, *195*, 30–43. [CrossRef]
17. Schneider, F.D.; Kükenbrink, D.; Schaepman, M.E.; Schimel, D.S.; Morsdorf, F. Quantifying 3D structure and occlusion in dense tropical and temperate forests using close-range LiDAR. *Agric. For. Meteorol.* **2019**, *268*, 249–257. [CrossRef]
18. Smith, A.M.S.; Falkowski, M.J.; Hudak, A.T.; Evans, J.S.; Robinson, A.P.; Steele, C.M. A cross-comparison of field, spectral and lidar estimates of forest canopy cover. *Can. J. Remote Sens.* **2009**, *35*, 447–459. [CrossRef]
19. Wu, X.; Shen, X.; Cao, L.; Wang, G.; Cao, F. Assessment of individual tree detection and canopy cover estimation using unmanned aerial vehicle based light detection and ranging (UAV-LiDAR) data in planted forests. *Remote Sens.* **2019**, *11*, 908. [CrossRef]
20. Lee, A.C.; Lucas, R.M. A LiDAR-derived canopy density model for tree stem and crown mapping in Australian forests. *Remote Sens. Environ.* **2007**, *111*, 493–518. [CrossRef]
21. Ma, Q.; Su, Y.; Guo, Q. Comparison of canopy cover estimations from airborne LiDAR, aerial imagery, and satellite imagery. *IEEE J. Sel. Top. Appl. Earth Obs. Remote Sens.* **2017**, *10*, 4225–4236.
22. Liu, Q.; Fu, L.; Wang, G.; Li, S.; Li, Z.; Chen, E.; Pang, Y.; Hu, K. Improving estimation of forest canopy cover by introducing loss ratio of laser pulses using airborne LiDAR. *IEEE Trans. Geosci. Remote Sens.* **2019**, *58*, 567–585. [CrossRef]
23. Holmgren, J.; Johansson, F.; Olofsson, K.; Olsson, H.; Glimskär, A. *Estimation of Crown Coverage Using Airborne Laser Scanning*, 8th ed.; SilviLaser 2008 Organizing Committee: Bournemouth, UK, 2008.
24. Melin, M.; Korhonen, L.; Kukkonen, M.; Packalen, P. Assessing the performance of aerial image point cloud and spectral metrics in predicting boreal forest canopy cover. *ISPRS J. Photogramm. Remote Sens.* **2017**, *129*, 77–85. [CrossRef]
25. Danson, F.M.; Hetherington, D.; Morsdorf, F.; Koetz, B.; Allgower, B. Forest canopy gap fraction from terrestrial laser scanning. *IEEE Geosci. Remote Sens. Lett.* **2007**, *4*, 157–160. [CrossRef]
26. Seidel, D.; Fleck, S.; Leuschner, C. Analyzing forest canopies with ground-based laser scanning: A comparison with hemispherical photography. *Agric. For. Meteorol.* **2012**, *154*, 1–8. [CrossRef]
27. Cifuentes, R.; Van der Zande, D.; Farifteh, J.; Salas, C.; Coppin, P. Effects of voxel size and sampling setup on the estimation of forest canopy gap fraction from terrestrial laser scanning data. *Agric. For. Meteorol.* **2014**, *194*, 230–240. [CrossRef]
28. Korhonen, L.; Kaartinen, H.; Kukko, A.; Solberg, S.; Astrup, R. Estimating vertical canopy cover with terrestrial and airborne laser scanning. In Proceedings of the 10th international conference on LiDAR applications for assessing forest ecosystems (Silvilaser 2010), Freiburg, Germany, 14–17 September 2010.
29. Li, J.; Yang, B.; Cong, Y.; Cao, L.; Fu, X.; Dong, Z. 3D forest mapping using a low-cost UAV laser scanning system: Investigation and comparison. *Remote Sens.* **2019**, *11*, 717. [CrossRef]
30. Liu, K.; Shen, X.; Cao, L.; Wang, G.; Cao, F. Estimating forest structural attributes using uav-lidar data in ginkgo plantations. *ISPRS J. Photogramm. Remote Sens.* **2018**, *146*, 465–482. [CrossRef]
31. Brede, B.; Lau, A.; Bartholomeus, H.M.; Kooistra, L. Comparing RIEGL RiCOPTER UAV LiDAR derived canopy height and DBH with terrestrial LiDAR. *Sensors* **2017**, *17*, 2371. [CrossRef] [PubMed]
32. Brede, B.; Calders, K.; Lau, A.; Raumonen, P.; Bartholomeus, H.M.; Herold, M.; Kooistra, L. Non-destructive tree volume estimation through quantitative structure modelling: Comparing uav laser scanning with terrestrial lidar. *Remote Sens. Environ.* **2019**, *233*, 111355. [CrossRef]

33. Liang, X.; Wang, Y.; Pyörälä, J.; Lehtomäki, M.; Yu, X.; Kaartinen, H.; Kukko, E.H.; Issaoui, A.E.I.; Nevalainen, O.; Vaaja, M.; et al. Forest in situ observations using unmanned aerial vehicle as an alternative of terrestrial measurements. *For. Ecosyst.* **2019**, *6*, 20. [CrossRef]
34. Hillman, S.; Wallace, L.; Lucieer, A.; Reinke, K.; Turner, D.; Jones, S. A comparison of terrestrial and UAS sensors for measuring fuel hazard in a dry sclerophyll forest. *Int. J. Appl. Earth Obs. Geoinf.* **2021**, *95*, 102261. [CrossRef]
35. Jaakkola, A.; Hyyppä, J.; Kukko, A.; Yu, X.; Kaartinen, H.; Lehtomäki, M.; Lin, Y. A low-cost multi-sensoral mobile mapping system and its feasibility for tree measurements. *ISPRS J. Photogramm. Remote Sens.* **2010**, *65*, 514–522. [CrossRef]
36. Puliti, S.; Dash, J.P.; Watt, M.S.; Breidenbach, J.; Pearse, G.D. A comparison of UAV laser scanning, photogrammetry and airborne laser scanning for precision inventory of small-forest properties. *Forestry* **2020**, *93*, 150–162. [CrossRef]
37. Cai, S.; Zhang, W.; Liang, X.; Wan, P.; Qi, J.; Yu, S.; Shao, J. Filtering airborne LiDAR data through complementary cloth simulation and progressive TIN densification filters. *Remote Sens.* **2019**, *11*, 1037. [CrossRef]
38. Cai, S.; Zhang, W.; Jin, S.; Shao, J.; Li, L.; Yu, S.; Yan, G. Improving the estimation of canopy cover from UAV-LiDAR data using a pit-free CHM-based method. *Int. J. Digit. Earth* **2021**, *14*, 1477–1492. [CrossRef]
39. Zhao, D.; Pang, Y.; Li, Z.; Sun, G. Filling invalid values in a LiDAR-derived canopy height model with morphological crown control. *Int. J. Remote Sens.* **2013**, *34*, 4636–4654. [CrossRef]

remote sensing

Article

Intensity/Inertial Integration-Aided Feature Tracking on Event Cameras

Zeyu Li [1], Yong Liu [2,*], Feng Zhou [1] and Xiaowan Li [3]

1 College of Geodesy and Geomatics, Shandong University of Science and Technology, Qingdao 266590, China; lizeyu@sdust.edu.cn (Z.L.); zhouforme@sdust.edu.cn (F.Z.)
2 Institute of Cyber-Systems and Control, Zhejiang University, Hangzhou 310027, China
3 National Time Service Center, Chinese Academy of Sciences, Xi'an 710600, China; lixiaowan21@mails.ucas.ac.cn
* Correspondence: yongliu@iipc.zju.edu.cn

Abstract: Achieving efficient and accurate feature tracking on event cameras is a fundamental step for practical high-level applications, such as simultaneous localization and mapping (SLAM) and structure from motion (SfM) and visual odometry (VO) in GNSS (Global Navigation Satellite System)-denied environments. Although many asynchronous tracking methods purely using event flow have been proposed, they suffer from high computation demand and drift problems. In this paper, event information is still processed in the form of synthetic event frames to better adapt to the practical demands. Weighted fusion of multiple hypothesis testing with batch processing (WF-MHT-BP) is proposed based on loose integration of event, intensity, and inertial information. More specifically, with inertial information acting as priors, multiple hypothesis testing with batch processing (MHT-BP) produces coarse feature-tracking solutions on event frames in a batch processing way. With a time-related stochastic model, a weighted fusion mechanism fuses feature-tracking solutions from event and intensity frames compared with other state-of-the-art feature-tracking methods on event cameras. Evaluation on public datasets shows significant improvements on accuracy and efficiency and comparable performances in terms of feature-tracking length.

Keywords: event camera; feature tracking; intensity/inertial integration

Citation: Li, Z.; Liu, Y.; Zhou, F.; Li, X. Intensity/Inertial Integration-Aided Feature Tracking on Event Cameras. *Remote Sens.* **2022**, *14*, 1773. https://doi.org/10.3390/rs14081773

Academic Editor: Gemine Vivone

Received: 29 January 2022
Accepted: 3 April 2022
Published: 7 April 2022

1. Introduction

Event cameras, as a kind of bio-inspired sensor, trigger events on each pixel independently and asynchronously according to the changes of scene brightness. Compared with standard cameras, they output event flow, which is formed by coordinates on the image plane and the time when an event happens, namely x,y and t (μs or ns level). Another characteristic of event flow is polarity, which indicates the increase or decrease in brightness. The presence or absence of polarity information depends on manufacturers. For example, DAVIS 240C from Inivation provides polarity information, while IMX636 from Sony/Prophesee does not have polarity information.

Event cameras have the characteristics of low latency, high dynamic range, and low power consumption. Due to their different characteristics compared to traditional cameras, event cameras open up a new paradigm for a series of tasks such as VO (Visual Odometry), SLAM (Simultaneous Localization and Mapping), and SfM (Structure from Motion). Feature tracking on event cameras is one fundamental step toward the maturity of these practical applications, which have aroused the interest of a wide range of researchers [1,2].

Although many asynchronous feature-tracking methods have been proposed, template matching on event frames or event patches is still a major way to process event information, especially for high-level tasks [3]. By accumulating a certain number of events or calculating the significance of the incoming event, the contours of objects in the scene are formed in an image-like frame, which is closely related to the movement of the carrier.

Practical feature tracking on event frames still faces challenges from efficiency and accuracy problems. Efficiency is related to high event rates, which depends on carrier motions, scene (e.g., dynamic objects), texture, etc. Event rates may vary significantly, and thus, the frequency of event frames may be very high, increasing the computational difficulty of keeping tracking on a sufficient number of features.

Event flow is sparse, asynchronous, noisy, and only represents brightness changes; thus, event-to-frame transformation has the problem of low signal-to-noise ratio and "texture" loss. Therefore, feature tracking purely relying on event flow may have drift problems, deteriorating the accuracy and robustness of high-level tasks.

Purely relying on event flow information will lead to one problem: when the carrier is moving slowly, the time interval between event frames may be larger than that between images. The frequency of positioning or mapping solutions would not meet the requirements of users. For example, when a drone is performing autonomous exploration slowly in an unknown environment, timely positioning results are still needed for motion and path planning. Therefore, intensity images and inertial information can still provide timely updates as the basic support for high-level tasks.

From the view of bionics, the research on animal processing mechanisms for external information shows that different parts of the brain handle different senses with different attention, which jointly supports the decision of action and judgment. Visual information is not an exception, which is highly related to events, as events only happen when brightness changes. Both global views of the scenes and inertial information are still sensed and processed by the brain latently. Therefore, the fusion of event, intensity, and inertial information has support from bionic research [4,5].

Now, several event cameras, such as DAVIS 346 and CeleX-4, provide normal intensity images, angular velocity, and acceleration from embedded IMU (Inertial Measurement Unit), supporting the feasibility of using complementary information for feature tracking.

The advantages of multiple sensor fusion bring potentials to overcome accuracy and efficiency problems from purely event-information-based methods. Therefore, a new feature-tracking method on event frames is proposed in this paper. Its novelty can be summarized as follows:

1. An event-frame-based feature tracker by using multiple hypothesis testing with batch processing (MHT-BP) is proposed to provide initial tracking solution. In MHT-BP, four-parameter affine transformation is proposed to improve motion coverage of template matching, and batch processing is proposed to improve tracking efficiency.
2. Together with inertial information prediction, a time-related stochastic model and a constant-velocity model are proposed to loosely integrate the solutions of tracking solution from intensity image and initial tracking solutions, which improves tracking accuracy.
3. A comparison with other state-of-the-art methods is conducted on publicly available event-camera datasets in terms of tracking efficiency, accuracy, and length. The results show that the proposed method achieved significantly higher accuracy and efficiency and comparable feature-tracking lengths.

The rest of paper is constructed as follows: Section 2 reviews feature-detection and -tracking methods on event camera. Section 3 firstly illustrates the data stream that the proposed method deals with and gives a brief description of WF-MHT-BP and then presents inertial based rotation prediction, which acts as the priors for the next feature-tracking steps in Section 3.1. The method to generate event frame is introduced in Section 3.2. After that, MHT-BP, which is the tracking method purely relying on event frames, is introduced in Section 3.3. Section 3.4 presents the weighted integration of feature-tracking solutions from event frame and intensity frame. In Section 4, WF-MHT-BP is compared with two methods implemented in MATLAB and EKLT (Event-based Lucas–Kanade Tracker), implemented in C++ in terms of accuracy and efficiency, and EKLT in terms of feature-tracking length. Section 5 concludes the paper and gives the direction for future work.

2. Related Work

Feature tracking is an active research field, where a number of algorithms have been proposed. Traditional feature tracking on intensity images can be divided into feature-matching-based methods and template-based tracking methods [6]. Two representative methods are SIFT (Scale-Invariant Feature Transform) [7] and KLT [8,9] respectively. Recently, many deep-learning-based algorithms have been proposed to improve the available number and robustness of feature matching, such as SuperPoint [10] and D2Net [11]. However, the efficiency problem of deep-learning-based methods is an obstacle for practical applications, especially for mobile devices.

Due to the different characteristics of event cameras, feature tracking on event flow follows different paradigms. A practical way is to convert event flow to event frames. Usually, events are collected in a temporal window to form event frames, and then, traditional feature tracking paradigms can be applied [12,13]. To improve efficiency and accuracy, different event-to-frame transformations and feature-tracking methods are proposed [14].

Event-to-frame transformation is the first step for event-frame-based feature tracking. Time-surface (TS) is a kind of global 2D surface using exponential decay kernel [15] to emphasize events happening recently. Another global method is Event Map, proposed by Zhu et al. [12], to project the events in a selected spatio-temporal window on frames directly. Surface of active events (SAE) is a local form of processing 3D spatio-temporal domain that pays attention to the most recent event at each pixel [16]. Normally, feature detection on TS or SAE is more accurate than direct methods, as the response of events happening recently is larger. However, computational complexities of direct methods are much lower than that of TS or SAE. Besides, TS or SAE needs more memory, as at least floats are needed in event frames.

A number of event-camera-based feature-detection and tracking algorithms focusing on improving accuracy and efficiency have been proposed. Li et al. [17] proposed SAE-based FA-Harris corner detection algorithm directly on asynchronous events instead of event frames. Alzugaray et al. [18] proposed Arc* detector based on modified SAE filtering and subsequently proposed HASTE (multi-Hypothesis Asynchronous Speeded-up Tracking of Events), which purely tracked feature on an asynchronous patch using multi-hypothesis [19]. Tedaldi et al. [13] detected Harris features in intensity images and then used ICP (Iterative Closest Point) method to establish correspondences. Zhu et al. [12] proposed an affine transformation based Expectation-Maximization (EM) algorithm to align two patches in the consequent event frames.

The fusion of event frames and intensity images provides benefits for feature tracking. Gehrig et al. [20] proposed an event-camera-based tracker, which optimized brightness increment differences from intensity images and event flows. Dong et al. [21] proposed a template-based feature-tracking method to improve the robustness. They predicted feature-tracking solutions with events and used intensity to correct them. The calculation burdens of these methods cannot be ignored due to their high algorithm complexities. It is observed that only a few features are set to be tracked in these algorithms to ensure real-time performances, affecting the applications of high-level tasks.

Some high-level applications potentially achieve feature tracking by reconstructing 3D geometry. Usually, 3D coordinates of features act as prior information for feature tracking, as they can be projected to the image plane with predicted poses. Zhou et al. [3,22] tracked the pose of a stereo event camera and reconstructed 3D environments by minimizing spatio-temporal energy. Liu et al. [23] proposed spatial-temporal registration algorithm as a part of event-camera-based pose-estimation method. However, the performance of feature tracking and quality of high-level tasks (e.g., pose estimation, scene reconstruction) are closely related. Multiple factors affect feature-tracking accuracy. Moreover, computational burdens will increase with higher algorithm complexity.

In summary, feature tracking on event frames has efficiency and accuracy problems: (1) Tracking efficiency is low due to characteristics of event cameras and designed algorithms. The number of trackable feature points is small, which affects the stability of high-level tasks. (2) Purely tracking features on events easily cause accuracy problems. Although multiple sensor-fusion-based feature tracking has been proposed, the efficiency and accuracy problems still need to be further explored with all available information.

3. Methodology

The incoming data stream for an event-camera-based localization or mapping system is illustrated in Figure 1 as the basic input assumption of the proposed method. Event flow, intensity images, and inertial information will be received asynchronously. Note that polarity information of event flow is not a must for the proposed method. Normally, intensity images and inertial information have equal time intervals. If no dynamic objects are in the scene, event flow potentially represents carrier motion, and the "frequency" of event frames is not even if a constant number of events are collected. Therefore, the proposed method cannot predict the resource of the next input.

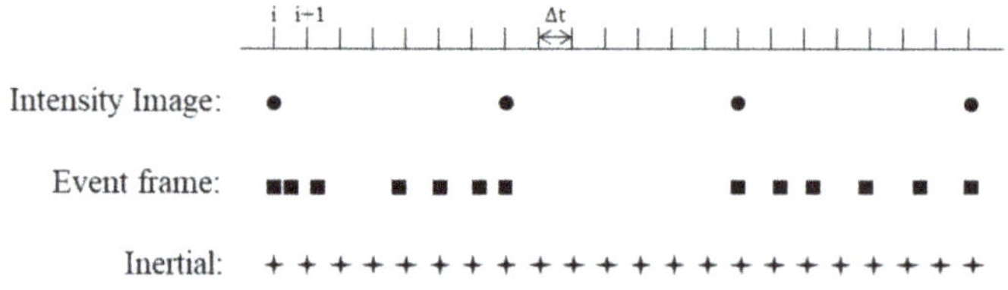

Figure 1. Different frequency of intensity images, event frames, and inertial updates.

Another characteristic of the data stream is that features are only detected on intensity images. Since each intensity frame will be attached with an event frame (see Section 3.2), the features to be tracked are directly projected to event frames with the same position.

The implementation of the proposed method can be illustrated in Figure 2. Firstly, IMU provides angular velocity w_t for feature rotation prediction, which acts as priors for KLT tracking and MHT-BP. Secondly, event flow is accumulated to generate event frame F_t for practical applications. Feature-detection module provides the positions of features for initialization, and features will be also re-detected if the number or the distribution score [24] of features are less than the threshold. Thirdly, with the assistance of rotation prediction from IMU, the method assigns MHT-BP for event frame F_t and KLT tracking for intensity image I_t. The weighted fusion mechanism will fuse their solutions. The tracking solution is outputted by MHT-BP or weighted fusion module depending on the applications. After the feature-tracking process is over, the method will automatically add detected features if the number or the distribution score [24] of features are not large enough. Shi-Tomasi corner feature point [25] was chosen as the feature to be tracked. Note that the proposed method is not limited to feature types and can be extended by other features, such as FAST [26].

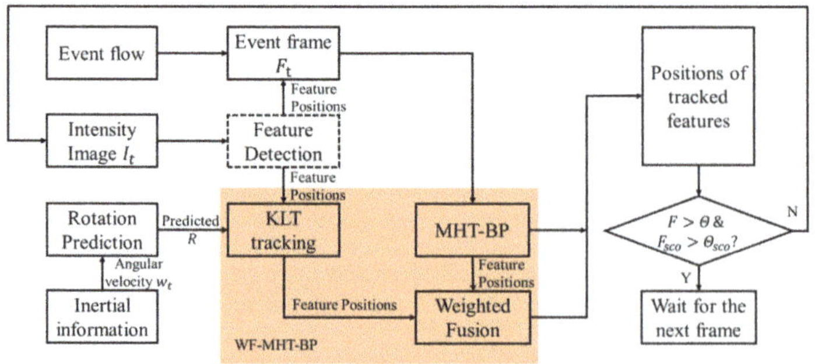

Figure 2. Flowchart of data stream and the proposed method WF-MHT-BP.

3.1. Inertial-Aided Rotation Prediction

According to the work of [27,28], the relationship between tracked features in two consecutive images can be approximated by Equation (1) considering the small translation relative to depth.

$$\mathrm{u}_{pred} = \mathrm{KRK}^{-1}\mathrm{u}_{last} \quad (1)$$

where K is the intrinsic matrix, and u_{pred} and u_{last} are 2D positions of the tracked feature and the predicted feature, respectively. The rotation matrix R can be integrated from angular velocity information w_t [27] as shown in Figure 2. u_{pred} will acts as the predicted feature position for the tracking.

3.2. Event-to-Frame Transformation

In this method, the generation of event frames follows that of Event Map instead of TS or SAE. Since TS- or SAE-based event-frame-generation method involves exponential computing, it is thus not used to reduce computing complexity.

Event-to-frame transformation follows two modes as shown in Equation (2): One is collecting a constant number τ_{FixNum} events. The events are projected on the image plane to form a binary image. Absence and presence of events within the temporal window are expressed by 0 and 1, respectively.

Another way is by the timestamps of intensity images. If one intensity image has been received, the events between the timestamp of last frame (intensity frame or event frame) and current timestamp, which is represented by $\tau_{FixTime}$, will be collected. That is, each intensity image has an attached event frame. This will cause one potential problem: If the number of events is too small, the generated event frame will be too sparse for reliable feature tracking. However, this situation will be alleviated by the following integration from intensity image:

$$\Pi(\mathrm{x},\mathrm{y}) = \begin{cases} 1 & if\ event(x,y,t) \in \tau_{FixNum} \text{ or } \tau_{FixTime} \\ 0 & others \end{cases} \quad (2)$$

3.3. Affine Transformation-Based Multiple Hypothesis Testing for Batch Processing (MHT-BP)

The work of [19] proposed five hypotheses purely based on event flow, which are Null, East, North, West, and South hypotheses. Firstly, a template and a model are generated from a time window on event flow. Then the alignment score that quantifies their differences is calculated to guide the selection of the above-mentioned five hypotheses.

Inspired by the work of [19], a batch processing-based multiple hypothesis testing using four-parameter affine transformation model (MHT-BP) is proposed to improve feature-tracking efficiency. Compared with the five hypotheses in [19], the four-parameter affine transformation model explores more hypotheses to improve matching accuracy. Moreover, batch processing is conducted to improve efficiency.

Since with the above-mentioned data stream, the possible minimal "frequency: of event frames is equal to that of intensity frame, the difference between consecutive frames is small enough for patch comparison. The small feature motions bring three benefits to improve efficiency and accuracy: (1) neighboring areas near features can provide supporting regions to make multiple hypotheses; (2) a set of features are able to share the same affine transformation, and therefore, batch processing can be conducted to improve efficiency; and (3) the search range of four parameters in the affine transformation model can be small.

The four-parameter affine transformation model is applied to generate multiple hypotheses. α, θ, $\Delta\mathrm{x}$, and $\Delta\mathrm{y}$ represent the variation on scale, rotation, and translation on X and Y, respectively. The four parameters between affine transformation are shown in Equation (3).

$$\mathrm{T}_{Aff} = \begin{pmatrix} \alpha\cos\theta & -\alpha\sin\theta & \Delta\mathrm{x} \\ \alpha\sin\theta & \alpha\cos\theta & \Delta\mathrm{y} \\ 0 & 0 & 1 \end{pmatrix} \quad (3)$$

The illustrative example of MHT-BP is illustrated in Figure 3 given two event frames F_t and F_{t+1}. A patch on F_t containing a set of features are selected with the four-parameter affine transformation model. The patch is transferred to generate multiple hypotheses; that is, each hypothesis corresponds to an affine transformation. After Gaussian blur, their differences are indicated by sum of absolute differences (SAD). The affine transformation model with minimum distance is chosen to establish correspondences between the two patches. This process will go on until all the features are involved.

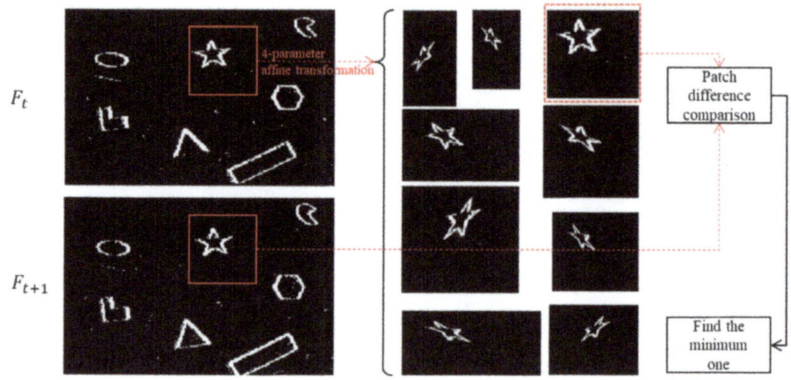

Figure 3. Illustrative example for multiple hypothesis testing with four-parameter affine transformation model.

3.4. Weighted Fusion Using Event/Intensity Information

MHT-BP is essentially a template-matching method purely using event information, which suffers from drift problems. Due to the high "frequency" of event frames, the mature KLT tracking results between intensity frames can guide the correspondence establishment between event frames. A constant-velocity model is used to re-predict the tracking solutions. A weighted fusion method is proposed to correct the drift and thus improve the accuracy. This provides two options for real-time processing and post processing. For real-time processing, features can be obtained from MHT-BP. For the post-processing tasks, such as bundle adjustment (BA) and SfM, poses and structure calculated from tracked features of event frames can still be obtained in a delayed manner.

As illustrated in Figure 4, the first row is one event frame sequence, and the second one is one intensity image sequence. The black dots and blue dots are detected features that share the same position. MHT-BP provides tracking solutions on each event frame. Due to the blurring effects of event frames, tracking solutions from MHT-BP suffer from drift problems. Drifts represented by uncertainties (shown in Figure 4) increase with time. However, since KLT tracking on the intensity sequence uses texture information, the increasing rate of drift is lower than that of MHT-BP through our tests. Therefore, the weighted fusion mechanism can reduce the drift error.

Firstly, the stochastic model of MHT-BP and KLT is expressed by σ^F and σ^I, respectively, as Equations (4) and (5). The uncertainty is related to time and tracking quality.

$$\sigma_F^2 = \beta\Delta t \tag{4}$$

$$\sigma_I^2 = \delta\Delta t \tag{5}$$

where Δt is time period from the timestamp of first intensity image. β and δ are drift in the unit of pixel per second, which can adjust the weights for fusing the solutions. The parameters need to be reasonably and empirically defined.

Secondly, the velocity of optical flow from KLT tracking is assumed to be constant due to the small-motion assumption. If event frames exist between two intensity images, feature-tracking solution can be linearly estimated on virtual intensity frame, such as $I_{t-\Delta t_1}$ and $I_{t-\Delta t_2}$ (see the dotted border of the second row in Figure 4) using Equations (6) and (7).

$$x_{KLT}^{t_i} = \Delta x \cdot \frac{t_i}{\Delta t} \tag{6}$$

$$y_{KLT}^{t_i} = \Delta y \cdot \frac{t_i}{\Delta t} \tag{7}$$

where Δx and Δy represent feature displacements between I_{t-1} and I_t on the x and y axis, respectively. Δt represents the time period between I_{t-1}, and I_t. t_i represents consumed time period from $t-1$.

Thirdly, with above-mentioned stochastic model and constant-velocity model, the weighted fusion can be conducted by using Equations (8) and (9). As shown in Figure 4, the red dots on last row show the result of weighted fusion.

$$x_{Fuse}^{t_i} = \frac{\sigma_F^2 \cdot x_{Event}^{t_i} + \sigma_I^2 \cdot x_{KLT}^{t_i}}{\sigma_I^2 + \sigma_F^2} \tag{8}$$

$$y_{Fuse}^{t_i} = \frac{\sigma_F^2 \cdot y_{Event}^{t_i} + \sigma_I^2 \cdot y_{KLT}^{t_i}}{\sigma_I^2 + \sigma_F^2} \tag{9}$$

The weighted fusion of MHT-BP and KLT solutions is named as WF-MHT-BP. Like the normal feature-tracking algorithm, it also detects new features after processing one event frame when the number of tracked feature or the distribution score is less than the threshold.

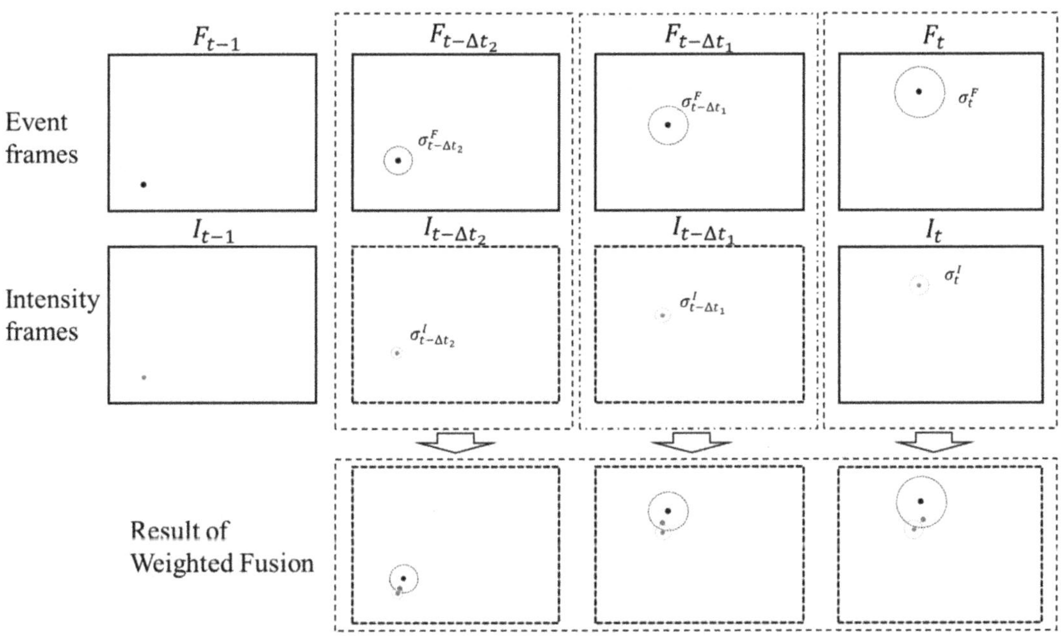

Figure 4. Illustration of the weighted fusion method for MHT-BP and KLT.

The proposed WF-MHT-BP can be summarized as Algorithm 1. Facing the structure of data stream shown in Figure 1, firstly, the inertial information is used to predict feature-tracking rotation. Then, if an event frame arrives, MHT-BP will purely use event information to generate tracking solutions. If an intensity image arrives, the proposed algorithm (WF-MHT-BP) can provide tracking solutions by fusing tracking solutions of KLT and MHT-BP. It should be noted that both MHT-BP and the proposed algorithm are able to output tracking solutions depending on when tracking solutions are needed.

Algorithm 1: Feature-tracking method based on integration of event, intensity and inertial information (WF-MHT-BP)

Input: Event $=\{x,y,t\}$, τ_{FixNum}, $\tau_{FixTime}$, search range of α, θ, Δx, Δy, β, δ. Intensity image I, angular velocity w, and threshold for feature detection Θ and Θ_{sco}
Output: Tracking solutions on the current intensity or event frame

Predict feature rotation using Equation (1) with angular velocity w
Generate event frame F_t by using Equation (2) with τ_{FixNum} and $\tau_{FixTime}$
If an event frame F_t is received, **then**
    Track features using MHT-BP with search range α, θ, Δx, Δy
If tracking solutions are needed, **then**
    Ouput tracking solutions on event frame F_t
If an intensity frame I_t is received, **then**
    Perform KLT tracking between I_t and I_{t-1}
    Perform weighted fusion of tracking solution from KLT and MHT-BP
    If NumOfTrackedFeature $\leq \Theta$ **or** DistributionScore $\leq \Theta_{sco}$, **then**
        Perform Shi-Tomasi detection on I_t
    If tracking solution is needed, **then**
        Output tracking solutions on intensity frame I_t

4. Experiments

The experiments chose 16 datasets from an event camera dataset publicly provided by University of Zurich [29], which uses a DAVIS 240C from Inivation, Zurich, Switzerland. It provides events flow, intensity images, and IMU measurements. The resolution of intensity images is 240 × 180. WF-MHT-BP runs on MATLAB platform in a computer with I5-10400F and 16 GB memory.

The input parameters are summarized in Table 1. τ_{FixNum} is set as 3000, which means when 3000 events arrive, an event frame will be generated. That is, an event frame will be formed once the number of events reaches 3000. $\alpha, \theta, \Delta x,$ and Δy define the search range of affine transformation as illustrated in Equation (3). β and δ are drift in the unit of pixel per second used in Equations (4) and (5). Θ and Θ_{sco} are set as 40 and 0.15, respectively, which means if the number or distribution score of tracked features is less than 40 or 0.15, Shi-Tomasi detection will be conducted to improve the number of newly detected features.

Table 1. Input parameters for the proposed method.

Input Parameters	Value
τ_{FixNum}	3000
α	(0.95, 1.05)
θ	$(-5^\circ, 5^\circ)$
Δx (pixel)	(−5, 5)
Δy (pixel)	(−5, 5)
β (pixel/s)	5
δ (pixel/s)	2
Θ	40
Θ_{sco}	0.15

The state-of-the-art methods are chosen from open-source event-frame-based feature tracking methods. One is probabilistic data association-based tracking (PDAT) method proposed by Zhu et al. [12]. The publicly available implementation is used. Another is ICP-based feature-tracking algorithm (High Temporal Resolution Tracking algorithm, HTRT) based on the work of Tedaldi et al. [13]. Since there are no original implementation provided by the authors, an implementation by a third-party is used (https://github.com/thomasjlew/davis_tracker, accessed on 29 February 2022). The modifications are made to provide better performances for the comparison. ICP maximum iterations is changed to 3. The feature is changed to Shi-Tomasi feature, which is the same with WF-MHT-BP.

Compared with purely using event information in the above-mentioned methods, EKLT in C++ version, which integrates event and intensity information, is also compared in accuracy, efficiency, and feature-tracking length. Since the proposed algorithm is implemented in MATLAB version, which is normally slower than C++, the performance of EKLT is listed as a reference.

Firstly, MHT-BP, which is the internal parts of WF-MHT-BP, is compared with an open-source template-matching method to show its improved efficiency and comparable accuracy. Then WF-MHT-BP is compared with PDAT, HTRT, and EKLT in tracking accuracy and efficiency. Finally, the feature-tracking length is compared between EKLT and WF-MHT-BP.

4.1. Feature Tracking Accuracy and Comparison between MHT-BP and FasT-Match

The goal of feature matching on event frames is to find the affine transformation in small ranges between consecutive frames. FasT-Match [30] was chosen as the baseline to compare the efficiency and patch matching error, as it achieves the same goal with MHT-BP. Moreover, it is a template-patch-matching-based method with similar control flow as MHT-BP. The difference of MHT-BP and FasT-Match are: (1) MHT-BP uses a simplified affine transformation model, but Fast-Match uses more complex transformation model with six parameters. (2) Batch processing is used in MHT-BP, but FasT-Match does not have the mechanism. (3) FasT-Match has a branch-and-bound search strategy to find the parameters in the transformation model, but MHT-BP does not use strict termination conditions to improve efficiency. FasT-Match are implemented in MATLAB, which is the same with MHT-BP. "shapes_rotation" was chosen for efficiency and template-matching error comparison.

Small, average normalized patch errors (NPE) mean higher similarities between two patches. Figure 5 shows the cumulative distribution function(CDF) of NPE and consumed time. The curve of MHT-BP and FasT-Match is very similar. Average NPEof FasT-Match and MHT-BP are 0.049 and 0.041, respectively, which means both of their matching errors are very small, and their differences can be ignored compared with the error ranges (0, 1). The mean consumed time of MHT-BP and FasT-Match is 3.53 ms and 80.40 ms, respectively. The consumed time of MHT-BP is much lower than FasT-Match. The reason for the acceptable error and reduced computational complexity is MHT-BP has the mechanism of batch processing and four-parameter affine transformation to improve accuracy and efficiency. It can be concluded that the majority of normalized patch errors by FasT-Match are slightly lower than that of MHT-BP method. However, the consumed time for MHT-BP is much lower than that of FasT-Match, which shows around a 10×–20× increase in speed. Although matching error of MHT-BP increased by around 19% compared with FasT-Match (0.049 vs. 0.041), the matching error is corrected in a timely manner by KLT tracking in WF-MHT-BP in the next step. Therefore, MHT-BP is chosen as the internal part of WF-MHT-BP.

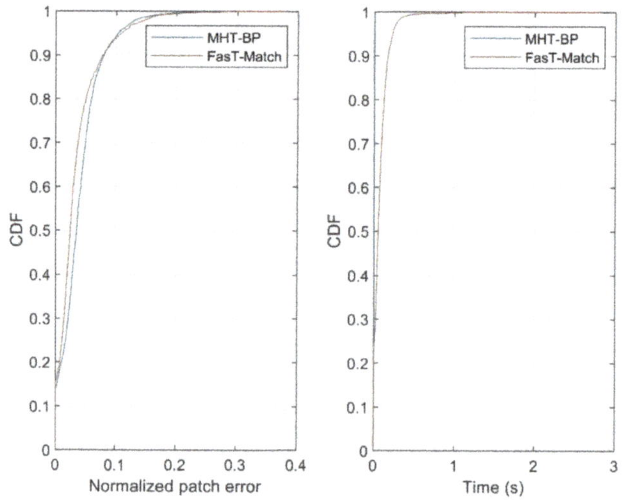

Figure 5. CDF of normalized patch error and consumed time for MHT-BP and FasT-Match.

4.2. Feature-Tracking Accuracy Comparison

Figure 6 shows feature-tracking solutions from the proposed method. The traces of tracked features are projected on the first intensity image to show the matching results. The 16 datasets with different scenarios of lighting, objects, and motion are compared in feature-tracking accuracy.

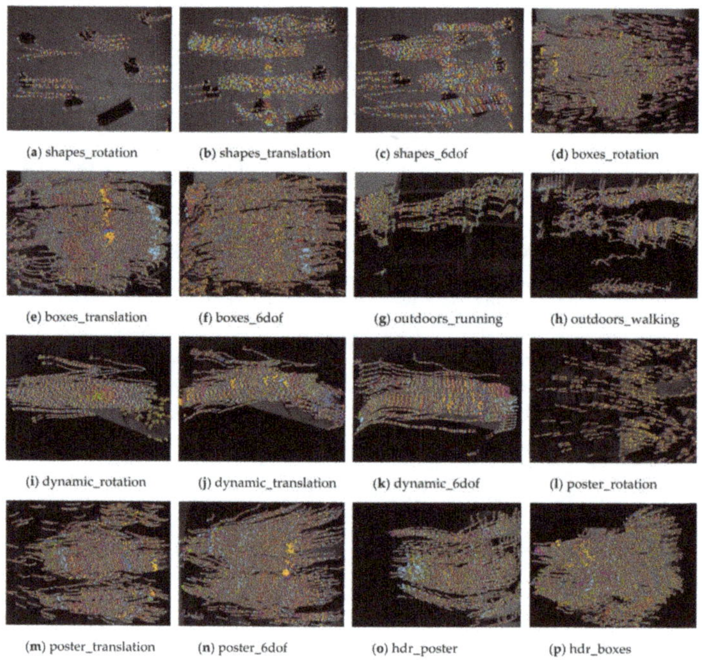

Figure 6. Illustration of feature tracking solutions from WF-MHT-BP (Each subgraph show the trajectories of tracked features projected on the first intensity image).

Both MHT-BP and WF-MHT-BP will generate tracking solutions. The solutions between frames are filtered by fundamental matrix based RANSAC (RANdom SAmple Consensus). For the parameters for RANSAC-based fundamental matrix, Sampson distance threshold is set as 0.1. The inlier ratio is used as the indicator for tracking accuracy. PDAT and HTRT follow in the same way. However, EKLT corrects tracking solutions from event information when an intensity frame arrives, and the inlier ratio between intensity frame is used.

For each dataset, the average inlier ratio is calculated as shown in Table 2. Note that the inliers from RANSAC are not used for the next feature-tracking process. For each method, different scenes with different light, object, and motion settings have different average inlier ratios. It is interesting to find that inlier ratio of HTRT for all the datasets is around 50%. The inlier ratio of PDAT ranges from 38.19% in "hdr_poster" scenario to 80.80% in "shapes_transaltion" scenario. In the scenario of "boxes"-, "hdr"-, and "poster"-related scenarios, inlier ratio of PDAT decreases rapidly, showing the difficulty from environmental factors.

Table 2. Inlier ratios of PDAT/HTRT/EKLT/WF-MHT-BP methods.

Dataset	PDAT	HTRT	EKLT	WF-MHT-BP
shapes_rotation	80.11%	52.94%	85.64%	**97.16%**
shapes_translation	80.80%	52.38%	89.83%	**93.54%**
shapes_6dof	80.36%	52.17%	84.74%	**94.41%**
boxes_rotation	41.97%	51.61%	85.99%	**99.97%**
boxes_translation	43.94%	51.52%	85.27%	**99.98%**
boxes_6dof	42.49%	51.85%	84.96%	**99.85%**
outdoors_running	55.06%	51.72%	98.86%	**98.04%**
outdoors_walking	77.92%	53.57%	95.54%	**99.18%**
dynamic_rotation	65.32%	50.84%	87.53%	**99.95%**
dynamic_translation	63.57%	50.62%	84.14%	**99.93%**
dynamic_6dof	70.66%	50.42%	84.74%	**99.78%**
poster_rotation	43.73%	51.22%	91.35%	**99.38%**
poster_translation	49.34%	50.84%	91.04%	**99.96%**
poster_6dof	41.65%	50.74%	87.20%	**99.87%**
hdr_poster	38.19%	50.57%	84.43%	**99.94%**
hdr_boxes	46.31%	50.70%	90.95%	**99.86%**
Average	57.59%	51.48%	88.26%	**98.80%**

The inlier ratio of EKLT is much better than PDAT and HTRT, which reaches 88.26% on average. WF-MHT-BP achieves the highest inlier ratio for all datasets. The main differences between EKLT and WF-MHT-BP are: (1) EKLT uses gradients to track features on event frame without batch processing, but MHT-BP in the proposed method uses four-parameter affine transformation for feature tracking in the way of batch processing. (2) WF-MHT-BP uses inertial information to predict the positions of features, while EKLT does not use inertial information. (3) KLT only correct drifts on arrived intensity images. WF-MHT-BP uses a simple and efficient fusion mechanism to correct tracking solutions from MHT-BP and current positions of tracked features. Therefore, the main reason is that WF-MHT-BP integrated all available factors for feature tracking, making tracking solution more accurate.

4.3. Feature-Tracking Efficiency Comparison

The efficiency of WF-MHT-BP is compared with PDAT, HTRT, and EKLT since they have different numbers of initialized features, which is meaningful for real applications, as lower time complexity will lead to more abundant time for frame processing. Since different methods have a different number of tracked features, the efficiency is quantified by the consumed time per tracked feature T_{PerFea}, which is calculated as:

$$T_{PerFea} = \frac{T_{PerFrm}}{N_{NumFea}} \tag{10}$$

where T_{PerFrm} is the consumed time on each frame, and N_{NumFea} is the number of tracked features on each frame. Note that all the visualization parts of all algorithms are closed to ensure accurate consumed-time statistics.

As shown in Table 3, HTRT has the highest computational complexity, which reaches 1062 ms for tracking one feature in "dynamic_rotation" scenario, which is not practical for real-time high-level tasks. EKLT still needs tens of milliseconds to track one feature between two intensity images. PDAT consumes less time to track one feature point than EKLT, but in the scenario of "dynamic_translation" and "outdoors_walking", it reaches 56 ms and 52 ms, respectively. WF-MHT-BP achieves the best efficiency among the four feature-tracking methods, which generally improves the efficiency by approximately three orders of magnitude compared with PDAT and EKLT and four orders compared with HTRT.

Table 3. The consumed time per tracked features of PDAT/HTRT/EKLT/WF-MHT-BP methods (unit: ms).

Dataset	PDAT	HTRT	EKLT	WF-MHT-BP
shapes_rotation	47	469	28	**1.00**
shapes_translation	48	417	29	**1.20**
shapes_6dof	56	356	25	**0.62**
boxes_rotation	14	405	62	**0.22**
boxes_translation	13	390	68	**0.15**
boxes_6dof	14	637	66	**0.36**
outdoors_running	23	473	52	**0.45**
outdoors_walking	52	151	15	**0.47**
dynamic_rotation	24	1062	75	**0.25**
dynamic_translation	56	759	54	**0.33**
dynamic_6dof	23	566	66	**0.19**
poster_rotation	16	440	47	**0.59**
poster_translation	18	290	58	**0.28**
poster_6dof	12	777	69	**0.13**
hdr_poster	12	367	85	**0.42**
hdr_boxes	12	214	62	**0.18**
Average	27.8	485.8	53.8	**0.43**

The main reason for high computational complexities of PDAT and HTRT is that the registration between two patches is done one by one. Besides, EM (Expectation Maximization) and ICP algorithms used in these methods are not suitable for real-time processing. Another reason for the high time consumption of HTRT is that the related event information to be processed needs to be searched from external memory for every event frame. The time consumption is at approximately the same level with EKLT at the beginning. As feature tracking continues processing, the time consumption becomes larger, resulting in the larger overall time consumption.

For EKLT algorithm, it involves the complex optimization of object function for tracking error and also does not have a batch-processing mechanism. Therefore, their consumed time is much larger than WF-MHT-BP. The efficiency problem is optimized in WF-MHT-BP by the batch-processing mechanism and simple loose integration with KLT tracking solution, which enables the lowest computational complexity of WF-MHT-BP.

4.4. Feature-Tracking Length Comparison

Feature-tracking length is another important indicator for feature tracking especially in VO or SLAM since the improved continuity of features can give more constraints for estimating poses or constructing maps. Without re-detecting features, feature-tracking length is quantified by the time period between the timestamp of first frame and the time when the number of tracked features decreases to 10% of the initial one. Due to the high computational complexity, HTRT is not further compared in its feature-tracking length.

The result is shown in Table 4: PDAT shows limited feature-tracking ability on tracking length, which means 90% of initialized feature can only last for 0.35 s. Its main difference with the other two methods is it purely uses event information for feature tracking, which can easily lose tracked features.

Table 4. Average duration of one track for PDAT/EKLT/WF-MHT-BP methods (unit: s).

Dataset	PDAT	EKLT	WF-MHT-BP
shapes_rotation	0.69	**3.30**	1.92
shapes_translation	1.58	6.82	**6.93**
shapes_6dof	1.21	**6.73**	5.70
boxes_rotation	0.01	**3.02**	2.70
boxes_translation	0.22	3.65	**3.93**
boxes_6dof	0.14	4.05	**4.41**
outdoors_running	0.16	1.58	**4.52**
outdoors_walking	0.46	3.45	**4.70**
dynamic_rotation	0.55	2.39	**3.24**
dynamic_translation	0.15	2.91	**4.48**
dynamic_6dof	0.31	2.75	**4.96**
poster_rotation	0.01	**2.76**	0.69
poster_translation	0.03	**3.38**	2.62
poster_6dof	0.03	**5.89**	3.89
hdr_poster	0.05	**4.84**	2.57
hdr_boxes	0.05	1.89	**3.35**
Average	0.35	3.71	**3.78**

WF-MHT-BP and EKLT show similar feature-tracking length statistics, showing the superiority in integrating the measurement from two sensors. The commonality between the two methods is that intensity information is used to correct the feature-tracking drifts from event information, which is helpful to continue tracking features.

5. Concluding Remarks

This paper proposes a loosely integrated feature tracking method on event frames using event, intensity, and inertial information to improve the accuracy and efficiency problem. MHT-BP, which involves four-parameter affine transformation and batch processing, is proposed to achieve fast and short-term feature matching. Then, a weighted fusion algorithm involving the constant velocity model and the stochastic model for drifts is proposed to reduce drifts. Next, it corrects the drift by weighted fusion in the way of post-processing, which is still meaningful for event-camera-based applications, such as SfM and SLAM.

MHT-BP is compared with FasT-Match, showing better efficiency at the expense of slight accuracy decline. In comparison with three state-of-the-art methods, including both event-information-based methods (PDAT and HTRT) and one multiple-sensor fusion-based method (EKLT), WF-MHT-BP shows the significant superiority on accuracy and efficiency and comparable feature-tracking lengths with EKLT.

In the future, the work can be refined and extended in the following aspects: First, feature detection is still conducted on intensity frames in the initialization and re-detection stage. If more features are needed, but intensity images have not arrived, the accuracy of high-level tasks, such as VO and SLAM, may be affected. Therefore, feature detection on event frames still needs to be further explored. Second, the final goal of event-frame-based feature tracking is to improve the robustness in challenging environments or motions. Thus, future work will focus on the application of WF-MHT-BP on high-level tasks, such as event-camera-based localization and mapping applications.

Author Contributions: Conceptualization, X.L. and Z.L.; methodology, Z.L.; software and validation, Z.L.; writing—original draft preparation, Z.L., F.Z., and Y.L.; writing—review and editing, F.Z. and Y.L.; supervision, Y.L.; project administration, Y.L.; funding acquisition Z.L. and Y.L. All authors have read and agreed to the published version of the manuscript.

Funding: This research was funded by project ZR2021QD148 supported by Shandong Provincial Natural Science Foundation and the Open Research Project (grant number ICT2021B17) of the State Key Laboratory of Industrial Control Technology, Zhejiang University, China.

Institutional Review Board Statement: Not applicable.

Informed Consent Statement: Not applicable.

Data Availability Statement: The data are available at URL: https://rpg.ifi.uzh.ch/davis_data.html (accessed on 29 January 2022).

Conflicts of Interest: The authors declare no conflict of interest.

References

1. Ma, Y.-Y.; Ye, Z.-H.; Liu, K.-H.; Chen, L. Event-based visual localization and mapping algorithms: A survey. *Acta Autom. Sin.* **2021**, *47*, 1484–1494.
2. Vidal, A.R.; Rebecq, H.; Horstschaefer, T.; Scaramuzza, D. Ultimate SLAM? Combining events, images, and IMU for robust visual SLAM in HDR and high-speed scenarios. *IEEE Robot. Autom. Lett.* **2018**, *3*, 994–1001. [CrossRef]
3. Zhou, Y.; Gallego, G.; Shen, S. Event-based stereo visual odometry. *IEEE Trans. Robot.* **2021**, *37*, 1433–1450. [CrossRef]
4. Franceschini, N. Small brains, smart machines: From fly vision to robot vision and back again. *Proc. IEEE* **2014**, *102*, 751–781. [CrossRef]
5. Shen, C. Brain-Like Navigation Technology Based on Inertial/Vision System. In *Intelligent Information Processing for Inertial-Based Navigation Systems*; Springer: Berlin/Heidelberg, Germany, 2021; pp. 95–114.
6. Zhai, M.; Xiang, X.; Lv, N.; Kong, X. Optical flow and scene flow estimation: A survey. *Pattern Recognit.* **2021**, *114*, 107861. [CrossRef]
7. Lowe, D.G. Object recognition from local scale-invariant features. In Proceedings of the Seventh IEEE International Conference on Computer Vision, Corfu, Greece, 20–27 September 1999; pp. 1150–1157.
8. Lucas, B.D.; Kanade, T. An iterative image registration technique with an application to stereo vision. In Proceedings of the 7th International Joint Conference on Artificial Intelligence, Vancouver, BC, Canada, 24–28 August 1981; pp. 674–679.
9. Tomasi, C.; Detection, T.K. Tracking of point features. *Int. J. Comput. Vis.* **1991**, *9*, 137–154. [CrossRef]
10. DeTone, D.; Malisiewicz, T.; Rabinovich, A. Superpoint: Self-supervised interest point detection and description. In Proceedings of the IEEE Conference on Computer Vision and Pattern Recognition Workshops, Salt Lake City, UT, USA, 18–23 June 2018; pp. 224–236.
11. Dusmanu, M.; Rocco, I.; Pajdla, T.; Pollefeys, M.; Sivic, J.; Torii, A.; Sattler, T. D2-net: A trainable CNN for joint description and detection of local features. In Proceedings of the IEEE/CVF Conference on Computer Vision and Pattern Recognition, Long Beach, CA, USA, 16–17 June 2019; pp. 8092–8101.
12. Zhu, A.Z.; Atanasov, N.; Daniilidis, K. Event-based feature tracking with probabilistic data association. In Proceedings of the 2017 IEEE International Conference on Robotics and Automation (ICRA), Singapore, 29 May–3 June 2017; pp. 4465–4470.
13. Tedaldi, D.; Gallego, G.; Mueggler, E.; Scaramuzza, D. Feature detection and tracking with the dynamic and active-pixel vision sensor (DAVIS). In Proceedings of the 2016 Second International Conference on Event-Based Control, Communication, and Signal Processing (EBCCSP), Krakow, Poland, 13–15 June 2016; pp. 1–7.
14. Gallego, G.; Delbruck, T.; Orchard, G.M.; Bartolozzi, C.; Taba, B.; Censi, A.; Leutenegger, S.; Davison, A.; Conradt, J.; Daniilidis, K.; et al. Event-based Vision: A Survey. *IEEE Trans. Pattern Anal. Mach. Intell.* **2020**. [CrossRef] [PubMed]
15. Delbruck, T. Frame-free dynamic digital vision. In Proceedings of the International Symposium on Secure-Life Electronics, Advanced Electronics for Quality Life and Society, Tokyo, Japan, 6–7 March 2008; pp. 21–26.
16. Mueggler, E.; Forster, C.; Baumli, N.; Gallego, G.; Scaramuzza, D. Lifetime estimation of events from Dynamic Vision Sensors. In Proceedings of the 2015 IEEE International Conference on Robotics and Automation (ICRA), Istanbul, Turkey, 26–30 May 2015; pp. 4874–4881.
17. Li, R.; Shi, D.; Zhang, Y.; Li, K.; Li, R. FA-Harris: A fast and asynchronous corner detector for event cameras. In Proceedings of the 2019 IEEE/RSJ International Conference on Intelligent Robots and Systems (IROS), Venetian Macao, Macau, 3–8 November 2019; pp. 6223–6229.
18. Alzugaray, I.; Chli, M. Asynchronous corner detection and tracking for event cameras in real time. *IEEE Robot. Autom. Lett.* **2018**, *3*, 3177–3184. [CrossRef]
19. Alzugaray Lopez, I.; Chli, M. HASTE: Multi-Hypothesis Asynchronous Speeded-up Tracking of Events. In Proceedings of the 31st British Machine Vision Virtual Conference (BMVC 2020), Virtual Event, 7–10 September 2020; p. 744.

20. Gehrig, D.; Rebecq, H.; Gallego, G.; Scaramuzza, D. EKLT: Asynchronous photometric feature tracking using events and frames. *Int. J. Comput. Vis.* **2020**, *128*, 601–618. [CrossRef]
21. Dong, Y.; Zhang, T. Standard and Event Cameras Fusion for Feature Tracking. In Proceedings of the 2021 International Conference on Machine Vision and Applications, Singapore, 25–27 July 2021; pp. 55–60.
22. Zhou, Y.; Gallego, G.; Rebecq, H.; Kneip, L.; Li, H.; Scaramuzza, D. Semi-dense 3D reconstruction with a stereo event camera. In Proceedings of the European Conference on Computer Vision (ECCV), Munich, Germany, 8–14 September 2018; pp. 235–251.
23. Liu, D.; Parra, A.; Chin, T.-J. Spatiotemporal Registration for Event-based Visual Odometry. In Proceedings of the IEEE/CVF Conference on Computer Vision and Pattern Recognition, Nashville, TN, USA, 19–25 June 2021; pp. 4937–4946.
24. Mur-Artal, R.; Tardós, J.D. ORB-SLAM2: An open-source slam system for monocular, stereo, and rgb-d cameras. *IEEE Trans. Robot.* **2017**, *33*, 1255–1262. [CrossRef]
25. Shi, J. Good features to track. In Proceedings of the 1994 IEEE Conference on Computer Vision and Pattern Recognition, Seattle, WA, USA, 21–23 June 1994; pp. 593–600.
26. Rosten, E.; Drummond, T. Machine learning for high-speed corner detection. In Proceedings of the European Conference on Computer Vision, Graz, Austria, 7–13 May 2006; pp. 430–443.
27. Hwangbo, M.; Kim, J.-S.; Kanade, T. Inertial-aided KLT feature tracking for a moving camera. In Proceedings of the 2009 IEEE/RSJ International Conference on Intelligent Robots and Systems, St. Louis, MO, USA, 10–15 October 2009; pp. 1909–1916.
28. Wei, H.; Tang, F.; Zhang, C.; Wu, Y. Highly Efficient Line Segment Tracking with an IMU-KLT Prediction and a Convex Geometric Distance Minimization. In Proceedings of the 2021 IEEE International Conference on Robotics and Automation (ICRA), Xi'an, China, 30 May–5 June 2021.
29. Mueggler, E.; Rebecq, H.; Gallego, G.; Delbruck, T.; Scaramuzza, D. The event-camera dataset and simulator: Event-based data for pose estimation, visual odometry, and SLAM. *Int. J. Robot. Res.* **2017**, *36*, 142–149. [CrossRef]
30. Korman, S.; Reichman, D.; Tsur, G.; Avidan, S. FasT-Match: Fast Affine Template Matching. In Proceedings of the 2013 IEEE Conference on Computer Vision and Pattern Recognition, 23–28 June 2013; pp. 2331–2338.

remote sensing

MDPI

Article

A Shrink-Branch-Bound Algorithm for eLoran Pseudorange Positioning Initialization

Kaiqi Liu [1,2,3,*], Jiangbin Yuan [1,3], Wenhe Yan [1,3], Chaozhong Yang [1,3], Wei Guo [1,3], Shifeng Li [1,2,3] and Yu Hua [1,2,3]

1 National Time Service Center, Chinese Academy of Sciences, Xi'an 710600, China; yuanjiangbin@ntsc.ac.cn (J.Y.); ywh@ntsc.ac.cn (W.Y.); ycz@ntsc.ac.cn (C.Y.); guowei@ntsc.ac.cn (W.G.); lishifeng@ntsc.ac.cn (S.L.); hy@ntsc.ac.cn (Y.H.)
2 University of Chinese Academy of Sciences, Beijing 100049, China
3 Key Laboratory of Precise Positioning and Timing Technology, Chinese Academy of Sciences, Xi'an 710600, China
* Correspondence: liukaiqi@ntsc.ac.cn

Abstract: Currently, eLoran is the ideal backup and supplement for global navigation satellite systems. The time synchronization accuracy between stations in the eLoran system has improved, providing conditions for eLoran pseudorange positioning. The pseudorange positioning of eLoran is a nonlinear least-squares problem and the location of the eLoran transmitting stations may cause the above problem to be non-convex. This makes the conventional pseudorange positioning algorithm strongly depend on the initial value when solving the eLoran pseudorange positioning. We propose a shrink-branch-bound (SBB) algorithm to solve the eLoran pseudorange positioning initialization problem. The algorithm first uses a shrink method to reduce the search space of the position estimator. Then, optimization is performed using a branch and bound algorithm within the shrunk region, where a trust region reflective algorithm is used for the lower bound process. The algorithm can help the receiver to complete the initial positioning without any initial value information. Simulation experiments verify that the algorithm has a success rate of more than 99.5% in solving the initialization problem of eLoran pseudorange positioning, and can be used as an initialization algorithm for pseudorange positioning problems for eLoran or other long-range terrestrial-based radio navigation system.

Keywords: pseudorange positioning; branch and bound; nonlinear least squares; eLoran; trust region reflective algorithm; initialization

Citation: Liu, K.; Yuan, J.; Yan, W.; Yang, C.; Guo, W.; Li, S.; Hua, Y. A Shrink-Branch-Bound Algorithm for eLoran Pseudorange Positioning Initialization. *Remote Sens.* **2022**, *14*, 1781. https://doi.org/10.3390/rs14081781

Academic Editors: Yuwei Chen, Changhui Jiang, Qian Meng, Bing Xu, Wang Gao, Panlong Wu, Lianwu Guan and Zeyu Li

Received: 4 March 2022
Accepted: 6 April 2022
Published: 7 April 2022

1. Introduction

Global navigation satellite system (GNSS) provides all-weather, all-day positioning, navigation, and timing (PNT) services in most outdoor environments. However, in cities or canyons, GNSS performance can degrade due to multipath or poor visibility [1–3]. In addition, the high vulnerability of GNSS to interference also seriously affects the security of PNT services [4–6]. Many algorithms have been developed to mitigate the performance degradation of GNSS receivers in dynamic multipath environments [7–10]. However, these algorithms can only improve receiver performance under certain conditions, and it is still difficult for GNSS receivers to work properly in scenarios with fewer visible satellites, such as cities or canyons. Geomagnetic, Wifi, Doppler, and pseudolite-based positioning technologies have been developed for GNSS denial scenarios [11–14], but these technologies can only provide positioning services in small areas, which cannot meet the positioning requirements of large cities or canyon scenes. In recent years, the eLoran system has regained attention due to its unique system performance, which is expected to solve the existing problems of GNSS [15,16]. The eLoran system is a terrestrial-based radio navigation system that transmits navigation information through a pulse signal with a carrier frequency of 100 kHz. The signal frequency band transmitted by the eLoran system is low and the transmission power is high. Therefore, the eLoran system has the advantages of wide coverage and good anti-interference performance, making it a good backup for GNSS [17–19].The traditional Loran navigation system uses a hyperbolic positioning method based on the time difference of arrival (TDOA) [20]. The receiver can only use the stations in a single chain for positioning. Therefore, it has the disadvantage of poor geometric dilution of precision (GDOP), limiting its positioning accuracy. In addition, the TDOA observations include delay errors along the two propagation paths, which makes it difficult to measure and remove abnormal propagation delays. This positioning method cannot directly solve the clock deviation between the receiver and the transmitting station. The eLoran positioning method is based on pseudorange measurement and uses a circular positioning method based on the time of arrival (*TOA*). This method has the following advantages. First, the receiver uses the signals of multiple chains and multiple stations for positioning, which significantly improves the GDOP factor. Second, the receiver can directly complete the clock error calculation. Third, it can be easily integrated with the wireless positioning system, which helps build an integrated world-ground PNT system [21,22]. Due to limited conditions, the eLoran positioning failed to attract attention in the past. With the transformation and upgrading of eLoran stations, the time between stations in different chains has been synchronized to Universal Time Coordinated (UTC) through technologies such as optical fiber, and the time synchronization accuracy reaches the nanosecond level, providing the basis for the use of eLoran positioning technology. In addition, the application of digital technology in eLoran receivers has improved their sensitivity, which allows them to receive signals from multiple chains and stations simultaneously. Owing to this technical background, the Loran positioning method has regained attention in recent years.

Groves briefly introduced the Loran pseudorange positioning method and pointed out that it was processed by analogy with GNSS-related methods [23]. Yan analyzed the feasibility of Loran pseudorange positioning and the influence of additional secondary factor (*ASF*) errors on various errors in pseudorange positioning [24]. Kim used the eLoran pseudorange measurements from multiple chains for positioning and performed real-world testing [25]. Peterson and Fang studied the integrated positioning of eLoran and GNSS and pointed out that eLoran pseudorange positioning is a necessary condition for integrated positioning [22,26]. In the above-mentioned literature, eLoran pseudorange positioning is regarded as a nonlinear least squares problem, and local optimization algorithms such as Newton-Raphson algorithm (NR) are used to solve it. However, the eLoran system is not specifically designed for pseudorange positioning, and the location of the eLoran transmitting station may make the problem non-convex. In addition, the nonlinear term in the eLoran pseudorange function is a complex nonlinear function with trigonometric functions, which may cause an ill-condition problem when using the first-order or second-order derivation information to optimize the objective function. Therefore, for many existing nonlinear least squares algorithms, when the selected initial values are inaccurate, convergence problems to local solutions or erroneous convergence results occur. This initial value dependence affects the ability of the receiver to locate autonomously and causes the receiver to experience localization errors under cold start. At present, there is no literature on the problem of eLoran pseudorange positioning under insufficient initial value information.

This study proposes a shrink-brand-bound (SBB) algorithm to solve the eLoran pseudorange positioning problem. The algorithm first obtains the shrunk region of the estimator through the shrink algorithm. The positioning problem is then solved within this compressed feasible region using a branch-and-bound algorithm, where a trust region reflective algorithm is used for each bound process [27,28]. The SBB algorithm has a global optimization capability and can achieve accurate positioning solutions without initial value information. The algorithm avoids the problem faced by the traditional nonlinear least-squares method by relying on the initial value when solving the Loran pseudorange positioning, which further improves the Loran positioning technology based on pseudorange measurement.

The rest of the paper is organized as follows. In Section 2, first, we describe the eLoran pseudorange measurement method and the error in the pseudorange. Then, we build a mathematical model of the eLoran pseudorange positioning and analyze the shortcomings of the NR algorithm in solving it. The principle of the SBB algorithm and the details of each part of the algorithm are introduced. In Section 3, we evaluate the performance of the SBB algorithm and other nonlinear least squares algorithms in solving the eLoran pseudorange positioning problem without initial value information through simulation experiments. Finally, we present the main conclusions of this paper.

2. Materials and Methods

In this section, we first introduce the pseudorange measurement technology and the error in the pseudorange. Secondly, we construct the mathematical model of eLoran pseudorange positioning and analyze the advantages and disadvantages of the traditional NR algorithm. Finally, we give the principle of the SBB algorithm and the details of each part of the algorithm.

2.1. Principle of eLoran's Pseudorange Measurement and Error Analysis

The eLoran positioning technology based on pseudo-range measurement includes two parts: pseudorange measurement technology and positioning algorithm. This section briefly describes the basic principles of pseudorange measurement technology and the error analysis in pseudorange measurement.

The basic principle of eLoran pseudorange measurement is shown in Figure 1. The receiver obtains the signal propagation delay or time of flight (*TOF*) by measuring the difference between the signal time of arrival (*TOA*) and the signal time of transmission (*TOT*). Usually, a certain characteristic point on the eLoran signal is selected as the *TOT*, such as the initial point or the zero-crossing point in the third circle. The *TOA* is obtained through the process of a group repetition period, carrier synchronization, and cycle identification. More details can be found in the references [24,29,30].

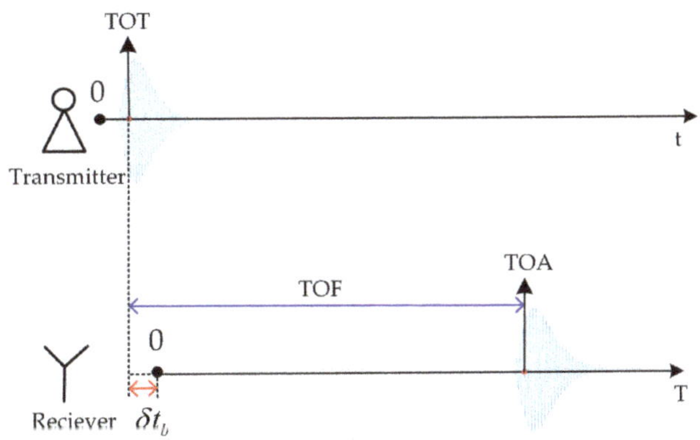

Figure 1. Schematic of eLoran pseudorange measurement principle.

The eLoran signal is mainly propagated by ground waves and its propagation process is affected by terrain, weather, and other conditions. Interference and noise also affect the *TOA* measurement during the receiver measurement process, so the *TOF*, which contains various additional time delay items, is not the true distance [31,32], as shown in Equation (1):

$$TOF = TOA - TOT = T_P + \Delta ASF(t) + \delta t_b + t_a + \eta(t), \quad (1)$$

where δt_b is the clock deviation between the receiver and the transmitting station, t_a is the receiver delay, $\eta(t)$ is the delay deviation caused by interference and noise in the TOF measurement process, and $\Delta ASF(t)$ is the time-related delay due to the ground wave propagation process time-varying factors such as weather. T_P is the delay term related to the propagation path as in Equation (2)

$$T_P = PF + SF + ASF, \tag{2}$$

where PF is the propagation delay of the signal through the atmosphere and is represented by Equation (3)

$$PF = \frac{s \cdot n_s}{c}, \tag{3}$$

where c is the speed of light in vacuum, s is the distance between the signal from the transmitter to the receiver; n_s is the refractive index of the atmosphere, which represents the ratio of the signal propagation speed in the atmosphere lower than the propagation speed in a vacuum. SF is the propagation delay of the signal through the entire seawater path, which is mainly related to the conductivity of the propagation path. ASF represents the propagation delay of eLoran signal caused by passing through a heterogeneous path of non-full seawater, which is mainly affected by parameters such as distance, surface impedance of the propagation path, and topography. ASF is an important factor affecting the positioning accuracy of eLoran, and it is often calibrated by eLoran differential station or ASF map [33–36].

In Equation (1), $\eta(t)$ and $\Delta ASF(t)$ are time-related delay items, which are difficult to calibrate. Figure 2 shows the statistical graph of the raw TOF value obtained by the receiver over time. The signal in the picture was transmitted from the Pucheng transmitting station (109.5438° E, 34.95043° N) and received in Lintong (109.2221° E, 34.3686° N). The fluctuation of the blue line in Figure 1 represents the TOF, which is affected by noise interference and its standard deviation is approximately 9 ns. The red line is the fitted curve of the data shown in blue, representing the fluctuation value with a standard deviation of approximately 10 ns. In order to present these time delays more clearly, we use the Fourier transform to analyze the spectrum of Figure 2a [37], and the obtained spectrum amplitude is shown in Figure 2b. In Figure 2b, we omit the spectrum after 0.001 Hz because its amplitude is too small. Among them, the amplitude at the lowest frequency is about 12 ns, which represents the deviation of the fitted curve in Figure 2a, that is, the delay introduced by $\Delta ASF(t)$. Other amplitudes due to measurement noise or interference are around 6 ns. As regards the delay error caused by measurement noise and interference $\eta(t)$, it is difficult to correct, so we uniformly regard it as noise. The error caused by $\Delta ASF(t)$ is often as high as more than 10 ns, so in high-precision eLoran positioning applications, the ASF prediction model is often used for calibration.

The propagation delay error calibration technology is essential for achieving high-precision positioning. There has been considerable research on this aspect [38–40]. Now consider the situation after the delay value is calibrated:

$$TOF_c = \tau + \delta t_b + \eta(t), \tag{4}$$

where TOF_c is the calibrated TOF, τ is the time delay value of the signal from the transmitting station to the receiver, δt_b is the clock deviation between the receiver and the transmitting station, $\eta(t)$ is the observation error introduced by the receiver due to time-varying factors such as interference, noise and $\Delta ASF(t)$. The t_a, SF and ASF in Equation (1) were calibrated. Multiplying both sides by the speed of light is the following pseudorange observation equation:

$$\rho = R_d + \rho_b + \eta, \tag{5}$$

where ρ is the pseudorange observation value of the station received by the receiver, R_d is the distance between the transmitting station and the receiver, ρ_b is the distance error caused by the clock deviation between the receiver and the transmitting station, and η is the distance error representing all other errors that are difficult to calibrate.

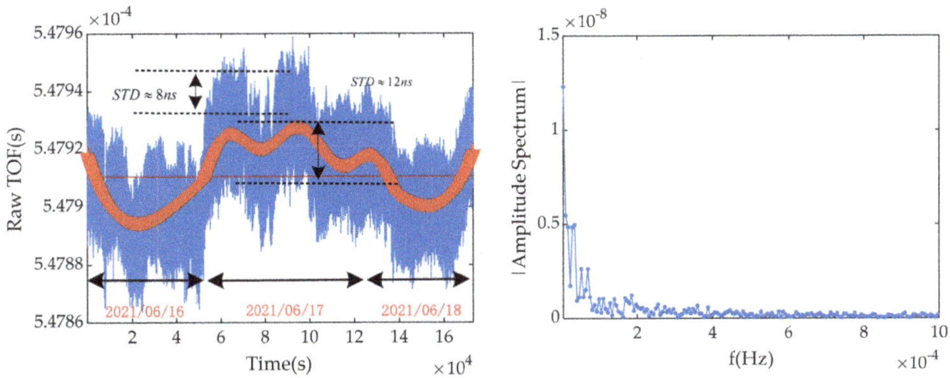

(**a**) (**b**)

Figure 2. The schematic diagram of measured propagation delay. (**a**) time delay in time domain (**b**) the amplitude spectrum of delays in the frequency domain.

It is worth noting that the eLoran signals mainly propagate through ground waves, and the transmitter and receiver are usually not within the line-of-sight range, so R_d cannot be calculated directly using the Euclidean distance formula but needs to be calculated using the great circle distance. The great circle refers to the shortest distance between two points on the surface of a sphere or ellipsoid. The Andoyer–Lambert formula is commonly used in the navigation field to calculate the distance between two points on the earth [41,42]. Suppose the position of the i-th station of eLoran is (λ^i, φ^i), and the position of the receiver is (λ, φ). Andoyer–Lambert's great circle distance formula is:

$$R_d^{(i)} = a\psi^{(i)} + \Delta S^{(i)}, \tag{6}$$

$$\begin{cases} \cos\psi^{(i)} = \sin\varphi^{(i)}\sin\varphi + \cos\varphi^{(i)}\cos\varphi\cos(\lambda - \lambda^{(i)}) \\ \Delta S = a\frac{f}{4}\left[\frac{\sin\psi^{(i)} - \psi^{(i)}}{1+\cos\psi^{(i)}}(\sin\varphi + \sin\varphi^{(i)})^2 - \frac{\sin\psi^{(i)} + \psi^{(i)}}{1-\cos\psi^{(i)}}\left(\sin\varphi - \sin\varphi^{(i)}\right)^2\right] \end{cases}, \tag{7}$$

Among them, λ^i, φ^i and λ, φ are the longitude and latitude of the transmitting station and the receiver, respectively, and ψ^i is the geocentric angle between the i-th eLoran station and the receiver. f and a are the basic geodetic parameters based on WGS-84; the former is the flattening of the ellipsoid, and the latter is the major axis radius of the reference ellipsoid.

2.2. eLoran Pseudorange Positioning Model and Conventional Positioning Algorithm

The eLoran pseudorange positioning is solving the estimator $\mathbf{x} = \begin{bmatrix} \varphi & \lambda & \delta t \end{bmatrix}^T$. Since the eLoran positioning is a plane positioning system, we only estimate the longitude λ and latitude φ. The principle of eLoran pseudorange positioning is shown in Figure 3. Each circle takes the transmitting station as the center and the calibrated pseudorange observation between point A and each transmitting station as the radius. The circles represent all possible solutions to the pseudorange observation of Equation (5). Since $\mathbf{x}$ contains three unknowns, the pseudorange observation equations of at least three stations are required to determine $\mathbf{x}$.

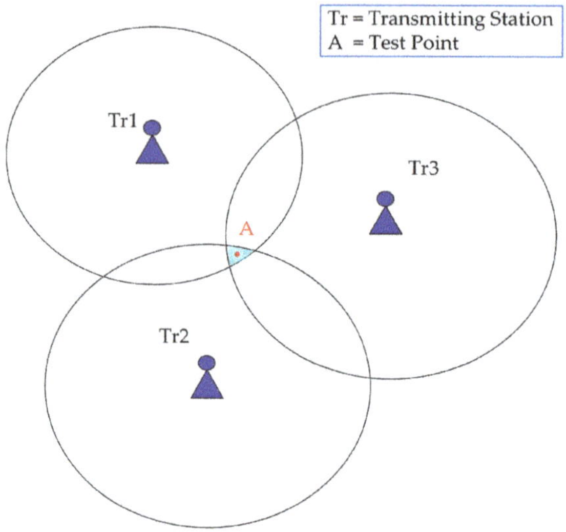

Figure 3. eLoran pseudorange positioning principle.

When we have no less than three pseudorange observation equations, we obtain **x** by solving the following equation set:

$$\begin{cases} \rho^{(1)} - R_d^{(1)}(\varphi,\lambda) - \rho_b(\delta t) = 0 \\ \rho^{(2)} - R_d^{(2)}(\varphi,\lambda) - \rho_b(\delta t) = 0 \\ \cdots \\ \rho^{(n)} - R_d^{(n)}(\varphi,\lambda) - \rho_b(\delta t) = 0 \end{cases} \tag{8}$$

The superscript of Equation (7) represents the eLoran station number. Owing to the existence of noise in the pseudorange observations, Equation (7) is often transformed into the following least-squares problem:

$$\min\{F(\mathbf{x})\} = \min\left\{\sum_{i=1}^{N}\left[\rho^{(i)} - R_d^{(i)}(\varphi,\lambda) - \rho_b(\delta t)\right]^2\right\}. \tag{9}$$

Equation (9) is the basic mathematical model of eLoran pseudorange positioning. The NR algorithm is widely used to solve the above problems. The algorithm linearizes Equation (9) through Taylor's formula and transforms it into a linear least-squares problem. The basic process is as follows:

First, we perform Taylor's first-order expansion of Equation (9) at $\mathbf{x}_{k-1}$, and obtain:

$$\boldsymbol{A} \cdot \Delta\mathbf{x} = \boldsymbol{B}, \tag{10}$$

where

$$\mathbf{H} = \begin{bmatrix} \frac{\partial R^1_{d,k-1}}{\partial \varphi} & \frac{\partial R^1_{d,k-1}}{\partial \lambda} & 1 \\ \frac{\partial R^2_{d,k-1}}{\partial \varphi} & \frac{\partial R^2_{d,k-1}}{\partial \lambda} & 1 \\ \vdots & \vdots & \vdots \\ \frac{\partial R^n_{d,k-1}}{\partial \varphi} & \frac{\partial R^n_{d,k-1}}{\partial \lambda} & 1 \end{bmatrix} \tag{11}$$

$$\Delta\mathbf{x} = \begin{bmatrix} \varphi_k - \varphi_{k-1} \\ \lambda_k - \lambda_{k-1} \\ \delta t_k - \delta t_{k-1} \end{bmatrix}, \tag{12}$$

$$\boldsymbol{B} = \begin{bmatrix} \rho^1 - \left(R^1_{d,k-1} + \rho_{b,k-1}\right) \\ \rho^2 - \left(R^2_{d,k-1} + \rho_{b,k-1}\right) \\ \vdots \\ \rho^n - \left(R^n_{d,k-1} + \rho_{b,k-1}\right) \end{bmatrix}. \tag{13}$$

Then, using the linear least-squares algorithm, the result is:

$$\Delta\mathbf{x} = \left(\mathbf{H}^T\mathbf{H}\right)^{-1}\mathbf{H}^T\boldsymbol{B}. \tag{14}$$

Finally, the state estimator is:

$$\widehat{\mathbf{x}_\mathbf{k}} = \mathbf{x}_{\mathbf{k}-1} + \Delta\mathbf{x}. \tag{15}$$

The advantage of this method is that it is simple, and if a suitable initial value $\mathbf{x}_0$ is selected, the convergence speed is fast and the solution is accurate. However, $F(\mathbf{x})$ is affected by the geometry of eLoran stations and may have local minima. Consider a special case, as shown in Figure 4, in which Tr represents the transmitting station, A is the test point, and the four stations are in linear distribution; a common feature as stations are often built along the coastline. It can be seen from the contour line of the function $F(\lambda, \varphi)$ that there is a local minimum value W in $F(\mathbf{x})$. This means that when using local optimization algorithms such as the NR algorithm [43] or the Levenberg–Marquardt (LM) algorithm [44] to solve the above problem, an inappropriate initial point will cause the algorithm to converge to a local minimum. We will confirm this with a simulation in Section 3. In addition, since the great-circle distance function contained in the eLoran pseudorange equation is a nonlinear term with trigonometric functions, which means that the optimization using the first-order and second-order derivation information of the objective function may face the problem of ill-condition, thereby converging to an erroraneous result. In view of this, it is necessary to design a global optimization algorithm to satisfy the positioning solution in the case of eLoran receiver cold-start.

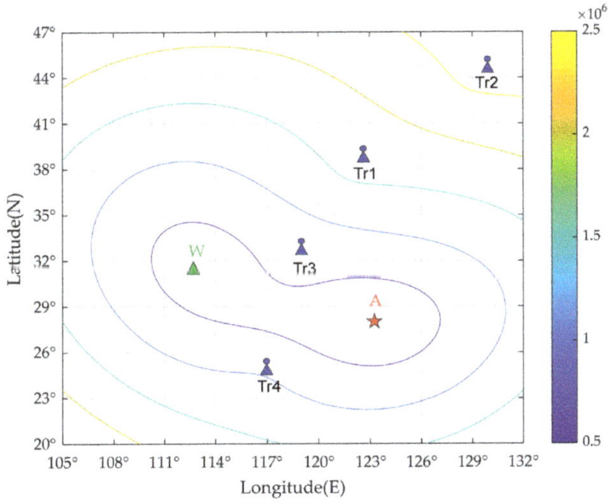

Figure 4. Contour map of $F(\varphi, \lambda)$ when the transmitting stations are linearly distributed.

2.3. The Shrink-Branch-Bound Algorithm

We define the eLoran positioning solution as the following optimization problem:

$$\{\mathbf{x}^*|F(\mathbf{x}^*) = F_{min}(\mathbf{x}), \mathbf{x} \in D\}, \tag{16}$$

where $F: D \to R$ is the objective function, and F is defined in Equation (9). D is the feasible region of $\mathbf{x}$, or search space. λ and φ in $\mathbf{x}$ have the following constraints

$$\mathbf{x} \in D = \begin{cases} -\pi \le \lambda \le \pi, \\ -\pi/2 \le \varphi \le \pi/2 \end{cases}, \tag{17}$$

The above boundary constraints represent the range of latitude and longitude coordinates of the earth. Since $\rho_b(\delta t)$ and δt have a linear relationship, the selection of the initial value of δt has no effect on the optimization process, so there is no need to consider the range of δt. From now on, we will refer D only to the feasible regions of λ and φ.

The SBB algorithm is a modification of the BB algorithm for the eLoran positioning problem. Before introducing the SBB algorithm, the BB algorithm needs to be described first. To solve the problem P, the BB algorithm first obtains a feasible solution as the optimal solution $\widehat{\mathbf{x}} \in D$ through a certain algorithm, and then iteratively divides the search space D into smaller subsets $D_{s1}, D_{s2}, \ldots, D_{sn}$. In each iteration process, when a solution $\mathbf{x}_1$ with a better objective function value can be found in a subset D_{si}, the current solution is updated to $\widehat{\mathbf{x}} = \mathbf{x}_1$, and the subset is divided into smaller subsets; the above process is repeated. If no solution in the subset is better than $\widehat{\mathbf{x}}$, the subset is pruned. When no subset can be pruned, $\widehat{\mathbf{x}}$ is the optimal value of P, and the iteration stops. The pseudocode for the generic BB algorithm is given in Algorithm 1 [28,45].

Algorithm 1 Generic Branch-and-Bound

1. Set $L = \{D\}$, initial $\mathbf{x}^* = \widehat{\mathbf{x}}$
2. While $L \neq \emptyset$
3. Select a subproblem D_s from L to explore
4. if a solution $\mathbf{x}_1 \in \left\{\mathbf{x} \in D_s \middle| F(\mathbf{x}) < F\left(\widehat{\mathbf{x}}\right)\right\}$ can be found, then $\widehat{\mathbf{x}} = \mathbf{x}_1$
5. if D_s cannot be pruned:
6. Partition D_s into $D_{s1}, D_{s2}, \ldots, D_{sn}$
7. Insert $D_{s1}, D_{s2}, \ldots, D_{sn}$ into L
8. Remove D_s from L
9. Return $\widehat{\mathbf{x}}$

The proposed SBB algorithm adds the process of shrinking the feasible region based on the BB algorithm and designs the corresponding branching strategy, bounding method, and pruned strategy according to the eLoran positioning problem. The basic flow chart of the SBB algorithm is shown in Figure 5. We introduce the SBB algorithm from the shrink method and the BB algorithm.

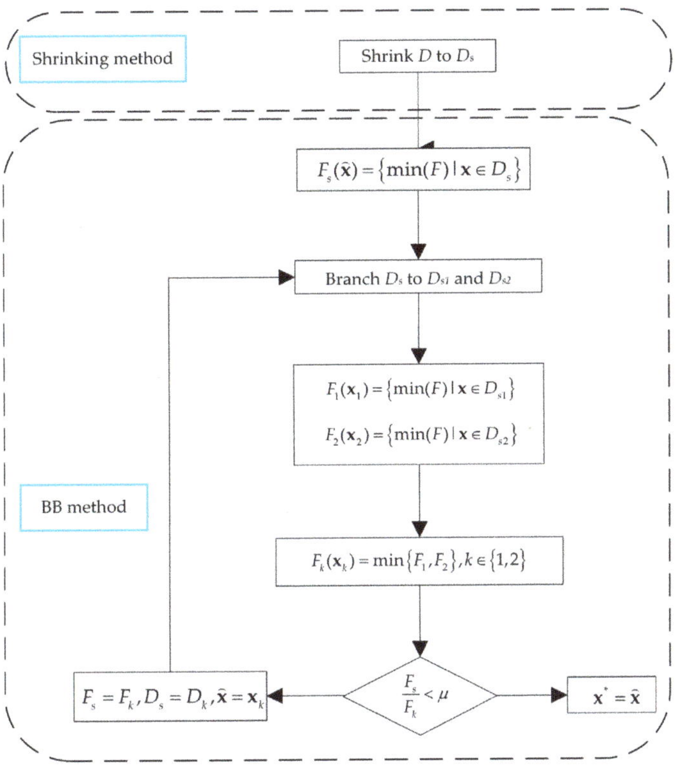

Figure 5. A diagram of the shrink-branch-bound algorithm.

2.3.1. The Shrink Method

From the basic principle of the BB algorithm, the search space D affects the amount of computation of the algorithm. If D can be shrunk, the subsequent BB algorithm can be significantly simplified. The range of D given by Equation (9) is derived from the range of latitude and longitude of the earth. Due to the limited coverage of the eLoran station, we can reduce D according to this feature.

The transmitting power of the eLoran transmitting station is usually fixed, and the eLoran receiver can receive signals from 800 km to 2500 km away from the transmitting station owing to the difference in the propagation path. When the receiver receives signals from multiple stations, it must be within the intersection of the coverage areas of these transmitters. Setting the range of this intersection as D_S, Figure 6 shows the basic schematic for determining D_S. The observable stations are TR1, TR2, TR3, and TR4. The prime vertical arc length between TR2 and TR4 is W, which can be estimated by Equation (18); the meridian arc length between TR1 and TR3 is L. It is estimated by Equation (19).

$$L = R_e(\varphi_{max} - \varphi_{min}), \quad (18)$$

$$W = R_e \cos(\overline{\varphi})(\lambda_{max} - \lambda_{min}), \quad (19)$$

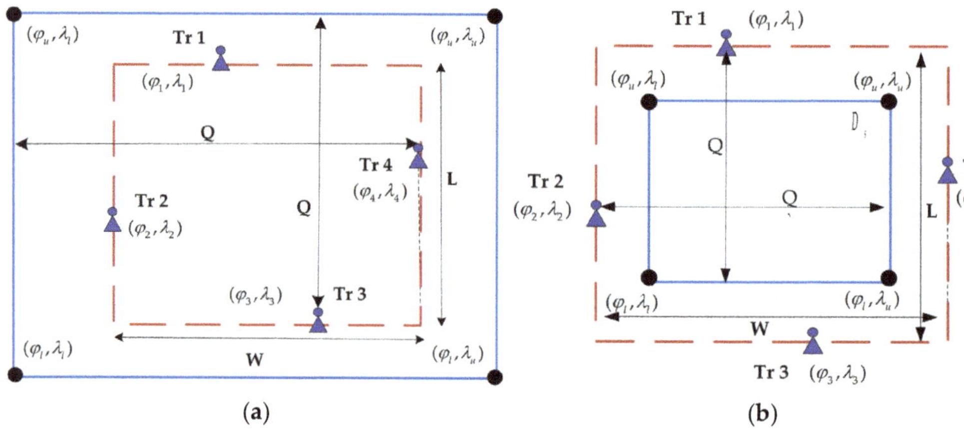

Figure 6. Shrinking diagram, (**a**) When $Q > W,\ Q > L$, the range of D_s. (**b**) When $Q < W,\ Q < L$, the range of D_s.

Assuming that the maximum working distance between the receiver and the transmitting station is Q, the range in the blue box of Figure 6 is D_s. Figure 6a,b show D_s under different conditions. The value of D_s can be calculated by Equation (20).

$$\begin{cases} \lambda \in [\lambda_l, \lambda_u] = \left[\ \min(\lambda_i) - \frac{(Q-W)}{R_e\cos(\overline{\varphi})}, \quad max(\lambda_i) - \frac{(W-Q)}{R_e\cos(\overline{\varphi})} \ \right] \\ \varphi \in [\varphi_l, \varphi_u] = \left[\ \min(\varphi_i) - \frac{(Q-L)}{R_e}, \quad max(\varphi_i) - \frac{(L-Q)}{R_e} \ \right] \end{cases}, \tag{20}$$

where R_e is the equivalent radius of the earth under the WGS-84 model. $\overline{\varphi}$ is the average latitude of the four stations. In practical applications, the setting of Q does not need to be precise but can be set as the maximum propagation distance according to the receiver performance and actual environment. In addition, Equation (20) is a general equation not limited to the two cases shown in Figure 6a,b. Therefore, once the receiver has identified the station information, Equation (20) can be used to calculate D_s.

2.3.2. The Branch and Bound Method in SBB Algorithm

The proposed branch-and-bound algorithm is as follows: First, a feasible solution $\widehat{\mathbf{x}}$ of F on D_s is obtained through a shrink algorithm, and $\widehat{\mathbf{x}}$ is assumed to be the global optimal solution. Then, we divide D_s into $D_{s1}$and D_{s2} and calculate the lower bounds $F_1(\mathbf{x}_1)$ and $F_2(\mathbf{x}_2)$ of function F on feasible domains D_{s1} and D_{s1}. We compare F_1 and F_2 and retain the subset D_{si} that has a lower bound F_i, where i = 1,2. Thereafter, we compare the order of F_s and F_i. If the order of F_i is smaller than F_s, we update the solution $\widehat{\mathbf{x}} = \mathbf{x}_1$, and divide D_{si} again and repeat the above steps. If F_s and F_i are of the same order, or the order of F_s is less than F_i, then $\widehat{\mathbf{x}}$ is the global optimal solution. The pseudocode of the SBB algorithm is shown in Algorithm 2.

Line 3 of the pseudocode is the branch strategy and we adopt the binary branch scheme as shown in Figure 7. The basic division principle is to make a vertical line at the midpoint of the broadest side of D_s to bisect D_s. Because the number of local minima on the F function is small, there is no need to divide D_s too much, and this binary branch strategy can effectively reduce the amount of calculation without losing the accuracy of the algorithm.

Algorithm 2 SBB Algorithm

1. Shrinking D to D_s, using Equation (20)
2. Take the initial value $\mathbf{x}_0 \in D$, use TRR algorithm to calculate $F_s(\widehat{\mathbf{x}}) = \{\min(F)|\mathbf{x} \in D_s, \mathbf{x}_0\}$
3. Branch D_s into D_{s1} and D_{s2}.
4. Calculate $F_1 = \{\min(F)|\mathbf{x} \in D_{s1}, \mathbf{x}_0\}$ and $F_2 = \{\min(F)|\mathbf{x} \in D_{s2}, \mathbf{x}_0\}$ and their corresponding solutions $\mathbf{x}_1$ and $\mathbf{x}_2$.
5. $F_k = \min\{F_1, F_2\}$ and F_s, where $k \in \{1,2\}$,
6. If $\frac{F_s}{F_k} < \mu$,
   Then $\mathbf{x}^* = \widehat{\mathbf{x}}$, the iteration ends;
7. If $\frac{F_s}{F_k} > \mu$,
   Then $F_s = F_k, D_s = D_{sk}, \widehat{\mathbf{x}} = \mathbf{x}_k$, and repeat steps 3–5.

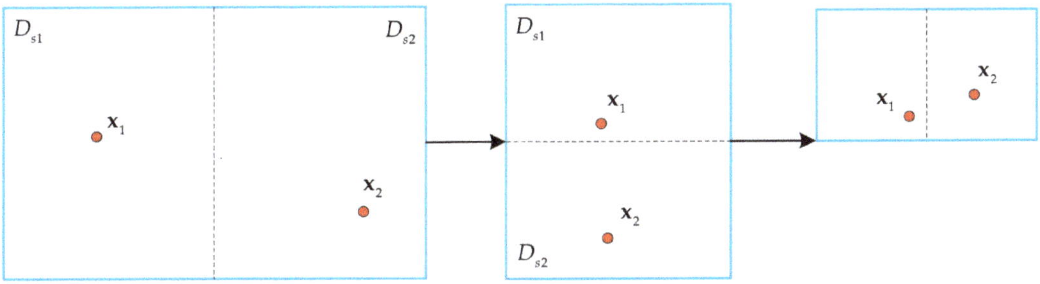

Figure 7. Diagram of the binary branching strategy.

In lines 2 and 4 of the pseudocode, it is necessary to calculate the lower bound of the objective function F in the specified feasible region, that is, to solve the following mathematical equation:

$$\{\mathbf{x}_s|F(\mathbf{x}_s) = F_{min}, \mathbf{x} \in D_s\} \tag{21}$$

D_s is determined by Equation (20). Equation (21) is a nonlinear least-squares problem with box constraints, which can be solved by the trust region reflective (TRR) algorithm. Based on the trust region algorithm, the trust region reflective method transforms the boundary-constrained optimization problem into an unconstrained optimization problem through reflection transformation so that each iteration result satisfies the boundary constraints [27]. The TRR algorithm uses the function $q(s)$ to fully approximate the behavior of the function $F(\mathbf{x})$ in the neighborhood N of $\mathbf{x}_k$, and find the tentative step s in this neighborhood. The pseudocode of the TRR algorithm is shown in Algorithm 3. In lines 3 and 4 of the pseudocode of Algorithm 3, the trust region model to be solved is as follows:

$$\min\{\ q(s) = \tfrac{1}{2}s^T Hs + s^T g,\quad \|s\| \le N\ \}, \tag{22}$$

where g is the gradient of the current $F(x_k)$, H is the Hessian matrix or the approximation of the Hessian matrix of $F(x_k)$, N is the trust region, and $\|\ \|$ is the 2-norm. For the solution of Equation (22), please refer to the literature [46,47]. Details of the reflection transformation method in line 5 can be found in the literature [27]. The approximation factor ρ_k of $q(s_k)$ to $F(s_k)$ in line 6 can be given by Equation (23):

$$\rho_k = \frac{F(\mathbf{x}_k) - F(\mathbf{x}_k + s_k)}{F(\mathbf{x}_k) - q(s_k)}, \tag{23}$$

When ρ_k is greater than the set value μ, it means that the current approximation effect of $q(s_k)$ to $F(\mathbf{x}_k)$ is good and the update step is $\mathbf{x}_{k+1} = \mathbf{x}_k + N_k$. Otherwise, the trust region radius N_k needs to be adjusted, the trust region sub-problem solved again, and the above process repeated.

Algorithm 3 TRR Algorithm

1. Initial $\mathbf{x}_0$, N_0 and μ
2. While $g(x_k) > \mu$
3. Build a trust region model $q(s)$
4. Solve the trust region subproblem, and get s_k
5. If $s_k \notin D_s$
6. Perform a reflection transform on s_k
7. Calculate the approximation ρ_k of $q(s_k)$ to $F(s_k)$ and update s_k or $\mathbf{x}_k$
8. Return $\widehat{\mathbf{x}}$

The TRR algorithm can make full use of the feature that the BB algorithm divides the feasible region. When the feasible region is divided, the box constraints will continue to shrink, and the probability of the trust region algorithm converging to the global optimal value will continue to increase. Using the TRR algorithm to obtain the lower bound of F under different feasible regions, the following inequalities must be satisfied.

$$\begin{cases} F_1 \leq F_s \\ F_2 \leq F_s \end{cases}, \tag{24}$$

where

$$F_s = \{F_{min}|\mathbf{x} \in D_s, \mathbf{x}_0\}, \tag{25}$$

$$F_1 = \{F_{min}|\mathbf{x} \in D_{s1}, \mathbf{x}_0\}, \tag{26}$$

$$F_2 = \{F_{min}|\mathbf{x} \in D_{s2}, \mathbf{x}_0\}, \tag{27}$$

Lines 5 and 6 of Algorithm 2 are the verification phase. We use μ to verify the convergence process, and μ can be a constant less than 5. When $F_s/F_k < \mu$, it means that F_s and F_k are of the same order, and the current iteration value is close to converging to the global optimal value, and the iteration ends. Otherwise, the above branch and bound process needs to be repeated.

2.3.3. Complexity Analysis

The main computational complexity of the proposed SBB algorithm is related to the number of branch iterations N and the convergence accuracy ε. In each iteration, the main computational complexity is related to the update of the bounding process of $F(\mathbf{x})$. More specifically, when we set the norm of the gradient of the solution to be $\|\nabla F\| \leq \varepsilon$, the upper bounds of the complexity required to solve steps (2) and (4) are $O(\varepsilon^{-2})$ and $O(2\varepsilon^{-2})$, respectively [48]. Considering the number of branch iterations N, the upper bound of the complexity of the SBB algorithm is $O((2N+1)\varepsilon^{-2})$. The upper bounds of the complexity of the following algorithms are shown in the Table 1.

Table 1. Algorithms Computational Complexity Comparison.

Algorithms	Computional Complexity
NR [49]	$O(kn^2m)$
LM [50]	$O(\varepsilon^{-2})$
Dogleg/TTR [48]	$O(\varepsilon^{-2})$
SBB	$O((2N+1)\varepsilon^{-2})$

The above table shows the upper bound of the computational complexity of different algorithms. Among them, the LM algorithm, the Dogleg algorithm, and the TTR algorithm are all Cauchy-related algorithms or Newton-like algorithms, and the upper bound of their complexity is $O(\varepsilon^{-2})$. The complexity of the NR algorithm is related to the number of iterations and the matrix calculation, where k is the number of iterations required, and m and n represent the dimensions of the estimator and the number of equations, respectively. It can be found that the complexity of the SBB algorithm compared with other algorithms mainly lies in N. Since we have shrunk D to D_s, this makes the number of branches N usually small, and we will confirm this in simulation experiments.

3. Results

The SBB algorithm is used to solve the initialization problem of eLoran pseudorange positioning. Therefore, the evaluation of the algorithm is mainly from two aspects. First, the algorithm should still be able to solve the position correctly when no initial value is available, which means that given a random initial value, the algorithm should be able to solve the position accurately. Secondly, the computational complexity of the algorithm should be at a reasonable level so that it can be implemented in the receiver. Based on the above evaluation criteria, this section is organized as follows: we first set the simulation parameters according to the actual station distribution. Then, the performance of various algorithms in solving the eLoran pseudorange positioning problem is compared. Finally, the reliability of the SBB algorithm was verified through simulation.

3.1. Simulation Parameter Settings

Assuming that the receiver at point A receives the signals from the four eLoran transmitting stations shown in Table 2, the calibrated pseudorange observations and geodesic distance values between point A and eLoran stations are shown in Table 3, and the atmospheric refractive index n_s is 1.000315. Where the calibrated pseudorange observations ρ are as described by Equation (5), they only include the clock deviation δt and the observation error η caused by time-varying delay factor. We set the clock error δt to be 5 μs and η follows a normal distribution, that is, $\eta \sim \mathbb{N}(0, 50)$.

Table 2. Transmitting station location and coordinates.

Transmitting Station Mark	Position	Longitude (E)	Latitude (N)
M	Rongcheng	122.3228	37.0644
T	Helong	129.1075	42.7199
Y	Xuancheng	118.886	31.0689
Z	Raoping	116.8958	23.7239

Table 3. Distance information from point A to each station.

Test Point (φ^*,λ^*) (N,E)	Transmitting Station	Distance (R_d/m)	PF (μs)	Pseudorange Observations (ρ/m)
A (27, 124)	M	1,128,758	3766.316	1,130,278
	T	1,806,302	6027.074	1,807,799
	Y	672,027	2242.348	673,547
	Z	801,620	2674.758	803,112

3.2. Analysis and Comparison of Simulation Results

Figure 8 shows the location of the transmitter station and receiver on the map. To clearly show the influence of (φ, λ) on $F(\mathbf{x})$, the contour of $F(\varphi, \lambda)$ is shown in Figure 9, where δt is set to a known value. The four black contours in Figure 9 are, respectively, surrounded by the solution sets of the four observation equations. The contour shape shows the non-convexity of $F(\varphi, \lambda)$, which is mainly related to the topology of the transmitting station. Take A as the test point and select the four positions shown in Table 3 as the initial value points. Since δt_0 has no effect on the optimization process, it will always be set to 0 in subsequent simulations.

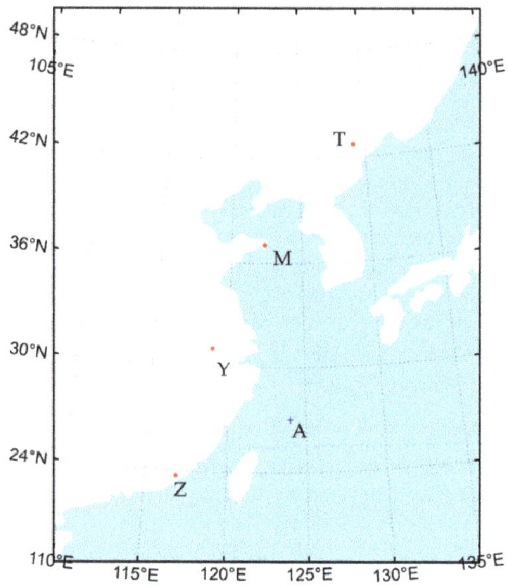

Figure 8. Stations location distribution.

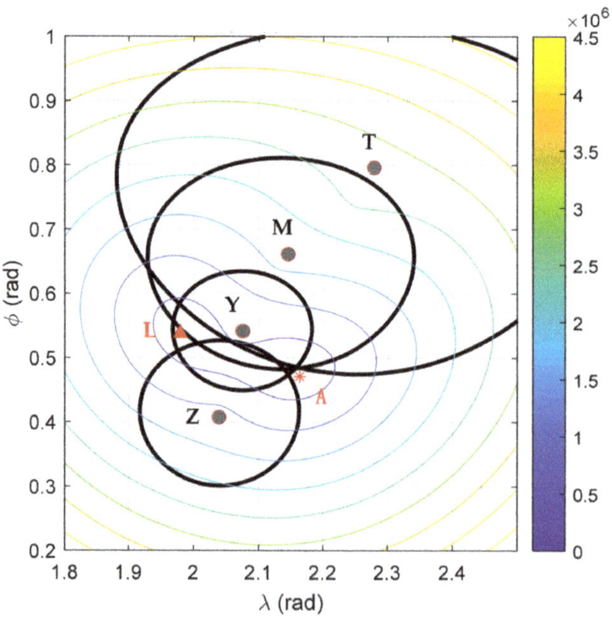

Figure 9. Contour plot of $F(\varphi, \lambda)$.

Since the initialization problem of eLoran has not been studied in the literature, there is a lack of competing algorithms for performance comparison. To this end, we select four commonly used nonlinear least squares methods, namely, the NR algorithm, the Levenberg–Marquardt (LM) algorithm, and the trust region Dogleg algorithm to compare with the SBB algorithm. The NR algorithm is a commonly used algorithm in positioning and is widely used in various pseudorange positioning scenarios. Its advantage is that the calculation is simple, and if the initial is suitable, it will converge quickly. Currently, only this algorithm is mentioned in the existing papers to solve the eLoran localization problem. The LM algorithm is an algorithm that combines the steepest descent method and Newton's method, and is currently widely used in nonlinear least squares. It is characterized by considering the stability of the steepest descent method and the fast convergence characteristics of Newton's method. This algorithm is a benchmark algorithm for solving nonlinear least squares problems based on the derivation algorithm, and is widely used in various scenarios. The LM algorithm can represent a series of scenarios based on the derivation algorithm to demonstrate the problem of solving the eLoran localization problem based on the first-order derivation and the second-order derivation algorithm.

The trust region Dogleg algorithm is representative of another large class of algorithms for solving nonlinear least squares algorithms. It is different from the line search algorithm; the algorithm first sets the step size, and then determines the search direction. The advantage of this algorithm is that it does not require a line search process when solving complex nonlinear least squares problems. Furthermore, even if the condition number of the objective function is poor, it is easy to introduce second-order information of the function. The above three algorithms represent the three most commonly used ideas for solving nonlinear least squares problems. The results are shown in Table 4.

Table 4. Convergence results of conventional algorithms under different initial points.

Point	Initial Points	NR Results	LM Results	Dogleg Results	SBB Results
(φ^*, λ^*) (N, E)	(φ_0, λ_0) (N, E)	(φ, λ) (N, E)	(φ, λ) (N, E)	(φ, λ) (N, E)	(φ, λ) (N, E)
A (27, 124)	0, 0	23.7162, 148.7832	31.2167, 103.7164	31.2167, 103.7164	27.0001, 124.0001
	28, 125	27.0001, 124.0001	27.0001, 124.0001	27.0001, 124.0001	27.0001, 124.0001
	40.1, 97.4	31.7164, 103.2167	31.21671, 03.7164	31.2167, 103.7164	27.0001, 124.0001
	32, 148.8	30.2167, 10.7162	27.0001, 124.0001	27.0001, 124.0001	27.0001, 124.0001
	28, 100	31.2167, 103.7164	31.2167, 103.7164	31.2167, 103.7164	27.0000, 123.9998
	15, 128	31.2195, 16.7159	26.9991, 124.0011	26.9991, 124.0011	26.9991, 124.0011

The data in red are the incorrect results, and the data in black are the correct results.

In Table 4, the data in red are the incorrect results, and the data in black are the correct results. The results of all algorithms may be incorrect due to the selection of initial values, except for the SBB algorithm. Among them, both the LM and Dogleg algorithms converge to (31.2167, 103.7164), which is the local minimum L shown in Figure 9. In addition, when the initial value point is close to Point A, both the LM and Dogleg algorithms converge correctly; when the initial value point is close to the local minimum point L, all the results of the above two are incorrect. The erroneous results of the NR algorithm may go beyond the feasible region D, mainly because the convergence of the NR algorithm may be out of control due to the lack of line search. Results from the above table verify that we need a global optimization algorithm to solve the eLoran pseudorange positioning problem when the initial value is not available.

We analyze how the SBB algorithm can always converge to the correct result, regardless of the change in the initial value.

Consider the shrink method of the SBB algorithm. Without loss of generality, we set Q in Equation (20) to 3000, and the reduced feasible region D_s is shown as the red box in Figure 10. It can be seen that D_s has been significantly reduced compared to D, which reduces the subsequent computation of the SBB algorithm.

To observe the global optimization performance of the SBB algorithm more clearly, Tables 5 and 6 show the iterative process of branch and bound under some initial value points.

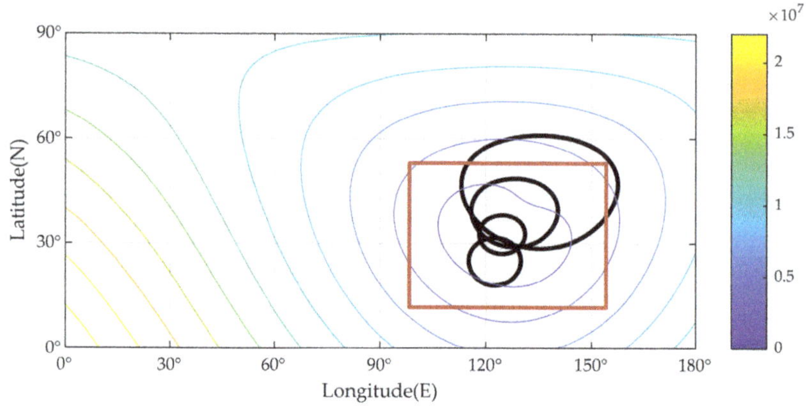

Figure 10. Contour line of F under feasible region D and feasible region D_s.

Table 5. Iterative process with initial value (0, 0, 0).

Feasible Region	F_{min}	x^* (φ, λ, δt) (deg, deg, μs)
D	3.5×10^9	31.2174, 103.7183, −2480
D_s	63.7	27.0001, 124.0001, 5.091
D_{s1}	3.5×10^9	31.2174, 103.7183, −2480
D_{s2}	63.7	27.0001, 124.0001, 5.091

Table 6. Iterative process with initial value (97.4, 40.1, 0).

Feasible Region	F_{min}	x^* (φ, λ, δt) (deg, deg, μs)
D	3.5×10^9	31.2147, 103.7225, −2480
D_s	3.5×10^9	31.2147, 103.7225, −2480
D_{s1}	3.5×10^9	31.2147, 103.7225, −2480
D_{s2}	99.8	26.9998, 124.0009, 4.933
D_{s1}	99.8	31.2147, 103.7225, −2480
D_{s2}	1.5×10^9	26.9998, 124.0009, 4.933

It can be seen from Tables 5 and 6 that, as the feasible region is continuously shrunk and divided, the SBB algorithm gradually converges to close to the global minimum.

To further verify the performance of the SBB algorithm, we designed the following simulation experiments: we randomly selected 1000 locations within D_s as test points and used the above mentioned algorithms to solve for these locations. Note that these locations were chosen to keep the GDOP as consistent as possible to avoid the impact of GDOP on location accuracy. For each algorithm, the initial value was randomly selected in D and D_s. When the positioning error was lower than the set threshold, the solution was successful. The statistical results of the success rate of these algorithms in solving these 1000 positions are shown in Figure 11.

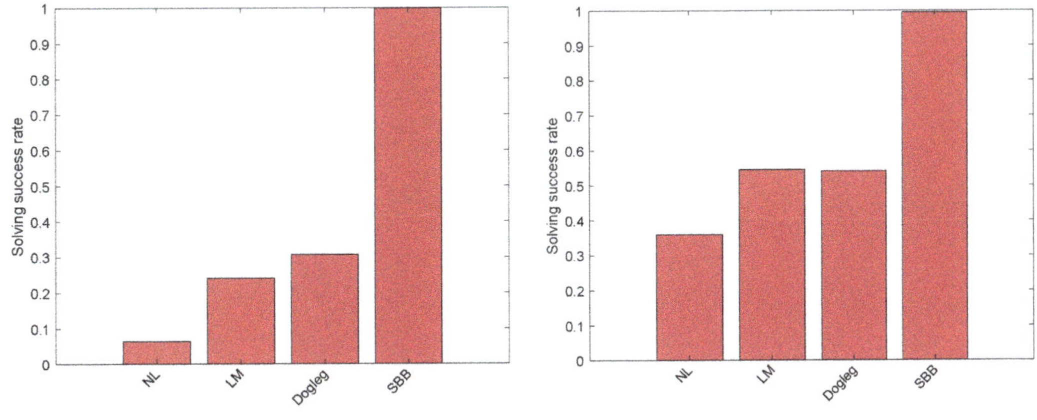

(**a**) (**b**)

Figure 11. Statistical chart of success rate of different algorithms. (**a**) $x_0 \in D$; (**b**) $x_0 \in D_s$.

As shown in Figure 11, the LM and Dogleg algorithms have a success rate of 55% in Figure 11b, while in Figure 11a, the success rates of the two are only 25% and 30%, respectively. This shows that the two algorithms depend strongly on the selection of initial values. The NL algorithm has the lowest success rate, and its solution probabilities are 5% and 35%, respectively, under the two initial value selection schemes. The main reason for the poor performance of the NL algorithm is that it lacks a line search process compared to the LM and Dogleg. The solution success rates of the SBB algorithm under the two initial value selection schemes are 99.9% and 99.5%, respectively, showing good global optimization performance. The possible reason for the failure of the SBB algorithm is that the algorithm will converge to the local minimum value when x_0 is selected very close to the local minimum value. Thus, when x_0 is selected in D, there is a smaller probability of selecting points close to the local minimum. Therefore, the success rate of the algorithm will be improved under $x_0 \in D$ compared to under $x_0 \in D_s$. To avoid choosing a point near the local minimum as the initial value when using the SBB algorithm, we can choose a point far away from all possible solutions as the initial value point, such as (0, 0).

Computational complexity affects the performance of an algorithm. The previous analysis of the complexity of the SBB algorithm showed that the number of branches, N, has an important impact on the complexity of the SBB algorithm. The figure shows the statistical graph of the number of branch iterations, N, required by the SBB algorithm to complete the positioning solution each time in 1000 positioning simulation experiments. Figure 12 shows that the SBB algorithm needs at most two branch iterations to complete the solution, and even only one branch is required in most cases. Comparing Figure 12a,b, it can be found that the probability that the latter requires two branches to solve is 46%, which is much higher than the 24% of the former. This is because when the initial value is randomly selected in D_s, there will be a higher probability of selecting the point close to the local minimum, which makes it converge to the global optimal value after two branches.

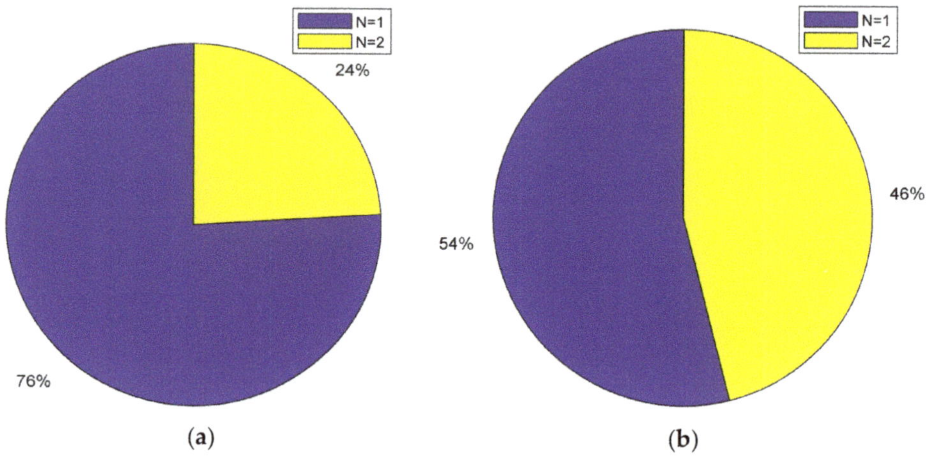

Figure 12. Pie chart of number of branch iterations N. (**a**) $x_0 \in D$; (**b**) $x_0 \in D_s$.

4. Discussion

Using pseudorange measurements for positioning in the eLoran system can make full use of the available eLoran stations, thereby expanding the coverage of the eLoran system and improving the positioning accuracy of the system. An important problem with eLoran pseudorange positioning, however, is that the geometric distribution of available eLoran transmitting stations may cause the positioning problem to be non-convex. This makes the existing pseudorange positioning algorithms such as NR algorithms extremely dependent on the selection of initial value. In practical positioning applications, it is difficult for the receiver to obtain reliable initial values in many cases. Therefore, conventional positioning algorithms may converge to wrong solutions due to lack of reliable initial values. At present, there is no literature to study the eLoran pseudorange localization initialization problem.

We transformed the eLoran pseudorange positioning into a nonlinear least squares problem with box constraints and proposed the shrink-branch-bound algorithm (SBB), a global optimization algorithm that can achieve accurate positioning without any initial value. The SBB algorithm first obtains the shrunk region of the estimator through the shrink method. The positioning problem is then solved within this shrunk feasible region using a branch-and-bound algorithm, where a trust region reflective algorithm is used for each bound process. We verified the performance of this method through simulation experiments. The results show that the success rate of the SBB algorithm to solve the position is more than 99.5%, when no initial value is available. However, the success rate of other conventional nonlinear least squares algorithms (such as LM algorithm, Dogleg algorithm) in this case is only around 50%. These results confirm that our proposed SBB algorithm can help the receiver to obtain correct positioning results when no initial value is available.

For the eLoran receiver, both the accuracy of the positioning algorithm and the computational complexity need to be considered. The computational complexity of the SBB algorithm is comparable to traditional Newton-based methods or Cauchy-related methods, which means that it can be implemented in the receiver.

5. Conclusions

eLoran is the ideal backup and supplement to GNSS systems. The improved accuracy of time synchronization between eLoran stations provides conditions for eLoran pseudorange positioning. We proposed a shrink-branch-bound (SBB) algorithm to solve the eLoran pseudorange positioning problem when the receiver has no initial value available. We verified the performance of the SBB algorithm through simulation experiments. The results show that the success rate of SBB algorithm in converging to the correct result without initial value is over 99.5%, which is more than 40% higher than that of conventional nonlinear least squares algorithms such as LM algorithm and Dogleg algorithm.

The proposed SBB algorithm is expected to make up for the defect that the existing eLoran pseudorange localization algorithm may converge to wrong results when no initial value is available, so it can be used as a cold-start algorithm for eLoran receivers. Therefore, the focus of follow-up research is to combine the SBB algorithm with the existing high-precision positioning algorithms, which is expected to further improve the positioning accuracy and reliability of the eLoran system under high dynamic conditions

Author Contributions: Conceptualization, K.L. and J.Y.; methodology, K.L.; software, K.L. and W.Y.; validation, K.L., J.Y. and W.Y.; formal analysis, S.L. and C.Y.; investigation, K.L.; resources, Y.H.; data curation, W.G.; writing—original draft preparation, K.L.; writing—review and editing, K.L.; visualization, W.Y.; supervision, J.Y.; project administration, S.L.; funding acquisition, C.Y., J.Y. All authors have read and agreed to the published version of the manuscript.

Funding: This work was supported by Chinese Academy of Sciences "Light of West China" Program (Grant No. E017YR1R10) and "Youth Innovation Promotion Association CAS" (GrantNo. 1188000YCZ).

Data Availability Statement: Restrictions apply to the availability of these data. The ownership of data belongs to the National Time Service Center (NTSC). These data can be available from the corresponding author with the permission of NTSC.

Acknowledgments: The authors would like to thank their colleagues for testing of the data provided in this manuscript. We are also very grateful to our reviewers who provided insight and expertise that greatly assisted the research.

Conflicts of Interest: The authors declare no conflict of interest.

References

1. Hussain, A.; Akhtar, F.; Khand, Z.H.; Rajput, A.; Shaukat, Z. Complexity and Limitations of GNSS Signal Reception in Highly Obstructed Enviroments. *Eng. Technol. Appl. Sci. Res.* **2021**, *11*, 6864–6868. [CrossRef]
2. Lu, D.; Jiang, S.; Cai, B.; Shangguan, W.; Liu, X.; Luan, J. Quantitative analysis of GNSS performance under railway obstruction environment. In Proceedings of the 2018 IEEE/ION Position, Location and Navigation Symposium (PLANS), Monterey, CA, USA, 23–26 April 2018; pp. 1074–1080. [CrossRef]
3. Wang, L.; Groves, P.D.; Ziebart, M.K. Multi-Constellation GNSS Performance Evaluation for Urban Canyons Using Large Virtual Reality City Models. *J. Navig.* **2012**, *65*, 459–476. [CrossRef]
4. Zhao, X.; Zhan, X.-Q.; Liu, X.; Li, S.-J. GNSS Vulnerability Analysis and Assessment. *J. Aeronaut. Astronaut. Aviat.* **2014**, *46*, 11–19. [CrossRef]
5. Dinesh, S. Globle Navigation Satellite System (GNSS) Spoofing: A Review of Growing Risks and Mitigation Steps. *Def. ST Tech. Bull.* **2013**, *6*, 42–61.
6. Panagiotis, P.; Jovanovic, A. Protection and Fundamental Vulnerability of GNSS. In Proceedings of the the 2008 IEEE International Workshop on Satellite and Space Communications 2008, Toulouse, France, 1–3 October 2008. [CrossRef]
7. Fascista, A.; Coluccia, A.; Ricci, G. A Pseudo Maximum likelihood approach to position estimation in dynamic multipath environments. *Signal Process.* **2020**, *181*, 107907. [CrossRef]
8. Lesouple, J.; Robert, T.; Sahmoudi, M.; Tourneret, J.-Y.; Vigneau, W. Multipath Mitigation for GNSS Positioning in an Urban Environment Using Sparse Estimation. *IEEE Trans. Intell. Transp. Syst.* **2018**, *20*, 1316–1328. [CrossRef]
9. Cheng, L.; Wang, K.; Ren, M.; Yan, G. Adaptive Filter Approach for Gps Multipath Estimation under Correntropy Criterion in Dynamic Multipath Environment. *IEEE Trans. Signal Process.* **2019**, *67*, 5798–5810. [CrossRef]
10. Cheng, C.; Tourneret, J.-Y. An EM-based multipath interference mitigation in GNSS receivers. *Signal Process.* **2019**, *162*, 141–152. [CrossRef]
11. Ferrigno, L.; Laracca, M.; Milano, F.; Cerro, G.; Bellitti, P.; Serpelloni, M.; Piedrafita, O.C. Magnetic Localization System for Short-Range Positioning: A Ready-to-Use Design Tool. *IEEE Trans. Instrum. Meas.* **2020**, *70*, 1–9. [CrossRef]
12. Wang, T.; Sui, T.; Liu, X.; Yuan, M.; Sun, G.; Gao, Z. WiFi positioning algorithm in tunnel based on Fuzzy C-means clustering and KNN algorithm. In Proceedings of the the 2019 Chinese Automation Congress (CAC), Hangzhou, China, 22–24 November 2019; pp. 567–571. [CrossRef]
13. Halili, R.; BniLam, N.; Yusuf, M.; Tanghe, E.; Joseph, W.; Weyn, M.; Berkvens, R. Vehicle Localization Using Doppler Shift and Time of Arrival Measurements in a Tunnel Environment. *Sensors* **2022**, *22*, 847. [CrossRef]

14. Sheng, C.; Gan, X.; Yu, B.; Zhang, J. Precise Point Positioning Algorithm for Pseudolite Combined with GNSS in a Constrained Observation Environment. *Sensors* **2020**, *20*, 1120. [CrossRef] [PubMed]
15. Griffioen, J.W.; Oonincx, P.J. Suitability of Low-Frequency Navigation Systems for Artillery Positioning in a GNSS Denied Environment. *J. Navig.* **2012**, *66*, 35–48. [CrossRef]
16. Johnson, G.W.; Swaszek, P.F.; Hartnett, R.J.; Shalaev, R.; Wiggins, M. An Evaluation of Eloran as a Backup to GPS. In Proceedings of the 2007 IEEE Conference on Technologies for Homeland Security, Woburn, MA, USA, 16–17 May 2007. [CrossRef]
17. Offermans, G.; Bartlett, S.; Schue, C. Providing a Resilient Timing and UTC Service Using eLoran in the United States. *Navigation* **2017**, *64*, 339–349. [CrossRef]
18. Son, P.-W.; Park, S.G.; Han, Y.; Seo, K. eLoran: Resilient Positioning, Navigation, and Timing Infrastructure in Maritime Areas. *IEEE Access* **2020**, *8*, 193708–193716. [CrossRef]
19. Narkus-Kramer, M.; Scales, W.; Calle, E. Evaluating Eloran as a Backup for Surveillance and Navigation: A Comparative Cost Analysis. In Proceedings of the the 2009 IEEE/AIAA 28th Digital Avionics Systems Conference, Orlando, FL, USA, 23–29 October 2009. [CrossRef]
20. Pierce, J.A. An Introduction to Loran. *IEEE Aerosp. Electron. Syst. Mag.* **1990**, *5*, 16–33. [CrossRef]
21. Peterson, B.; Hartnett, R.; Bruckner, D.; Heatherington, R.; Fiedler, R. Integrated GPS/LORAN: Structures and Issues. *Navigation* **1998**, *45*, 183–193. [CrossRef]
22. Peterson, B.B.; Lo, S.; Enge, P. Integrating Loran and Gnss for Safety of Life Applications. In Proceedings of the Institute of Navigation GNSS Conference, Savannah, GA, USA, 16–19 September 2008.
23. Groves, P.D. *Principles of GNSS, Inertial, and Multisensor Integrated Navigation Systems*; Boston Artech House: Norwood, MA, USA, 2013.
24. Yan, B.; Li, Y.; Guo, W.; Hua, Y. High-Accuracy Positioning Based on Pseudo-Ranges: Integrated Difference and Performance Analysis of the Loran System. *Sensors* **2020**, *20*, 4436. [CrossRef]
25. Kim, Y.; Fang, T.H.; Kim, D.; Seo, K.; Park, S.H. Loran-C Multiple Chain Positioning Using Toa Measurements. *J. Navig. Port Res.* **2019**, *43*, 23–32. [CrossRef]
26. Fang, T.H.; Kim, Y.; Park, S.G.; Seo, K.; Park, S.H. GPS and eLoran Integrated Navigation for Marine Applications Using Augmented Measurement Equation Based on Range Domain. *Int. J. Control. Autom. Syst.* **2020**, *18*, 2349–2359. [CrossRef]
27. Coleman, T.F.; Li, Y. A Reflective Newton Method for Minimizing a Quadratic Function Subject to Bounds on Some of the Variables. *SIAM J. Optim.* **1996**, *6*, 1040–1058. [CrossRef]
28. Morrison, D.R.; Jacobson, S.; Sauppe, J.J.; Sewell, E.C. Branch-and-bound algorithms: A survey of recent advances in searching, branching, and pruning. *Discret. Optim.* **2016**, *19*, 79–102. [CrossRef]
29. Shapiro, L.D. Time synchronization from Loran-C. *IEEE Spectr.* **1968**, *5*, 46–55. [CrossRef]
30. Yan, W.; Dong, M.; Li, S.; Yang, C.; Yuan, J.; Hu, Z.; Hua, Y. An eLoran Signal Cycle Identification Method Based on Joint Time–Frequency Domain. *Remote Sens.* **2022**, *14*, 250. [CrossRef]
31. Johler, J. Propagation of the Low-Frequency Radio Signal. *Proc. IRE* **1962**, *50*, 404–427. [CrossRef]
32. Samaddar, S.N. The Theory of Loran-C Ground Wave Propagation—A Review. *Navigation* **1979**, *26*, 173–187. [CrossRef]
33. Yang, S.H.; Lee, C.B.; Lee, Y.K.; Lee, J.K.; Kim, Y.J.; Lee, S.J. Accuracy Improvement Technique for Timing Application of Loran-C Signal. *IEEE Trans. Instrum. Meas.* **2011**, *60*, 2648–2654. [CrossRef]
34. Li, Y.; Hua, Y.; Yan, B.; Guo, W. Research on the eLoran Differential Timing Method. *Sensors* **2020**, *20*, 6518. [CrossRef]
35. Son, P.-W.; Rhee, J.H.; Hwang, J.; Seo, J. Universal Kriging for Loran Asf Map Generation. *IEEE Trans. Aerosp. Electron. Syst.* **2018**, *55*, 1828–1842. [CrossRef]
36. Kim, Y.; Park, S.H.; Fang, T.H.; Seo, K. A Test of a Loran-C Multi-Chain Positioning Method Using Asf Estimates. In Proceedings of the the 2019 European Navigation Conference (ENC), Warsaw, Poland, 9–12 April 2019. [CrossRef]
37. Felski, A.; Jaskólski, K.; Zwolak, K.; Piskur, P. Analysis of Satellite Compass Error's Spectrum. *Sensors* **2020**, *20*, 4067. [CrossRef]
38. Wang, L.-L.; Liang, Z.-C.; Pu, Y.-R.; Xi, X.-L. Method for Loran-C Additional Secondary Factor Correction Based on Neural Network and Transfer Learning. *IEEE Antennas Wirel. Propag. Lett.* **2021**, *21*, 332–336. [CrossRef]
39. Pu, Y.; Zheng, X.; Wang, D.-D.; Xi, X. Accuracy Improvement Model for Predicting Propagation Delay of Loran-C Signal Over a Long Distance. *IEEE Antennas Wirel. Propag. Lett.* **2021**, *20*, 582–586. [CrossRef]
40. Li, Y.; Hua, Y.; Yan, B.; Guo, W. Experimental Study on a Modif I Ed Method for Propagation Delay of Long Wave Signal. *IEEE Antennas Wirel. Propag. Lett.* **2019**, *18*, 1716–1720. [CrossRef]
41. Fisikopoulos, V. Geodesic Algorithms: An Experimental Study. *Int. Arch. Photogramm. Remote Sens. Spat. Inf. Sci.* **2019**, *42*, 45–47. [CrossRef]
42. Tseng, W.-K.; Guo, J.-L.; Liu, C.-P. A Comparison of Great Circle, Great Ellipse, and Geodesic Sailing. *J. Mar. Sci. Technol.* **2013**, *21*, 7. [CrossRef]
43. Kornerup, P.; Muller, J.-M. Choosing starting values for certain Newton–Raphson iterations. *Theor. Comput. Sci.* **2005**, *351*, 101–110. [CrossRef]
44. Moré, J.J. The Levenberg-Marquardt algorithm: Implementation and theory. In *Numerical Analysis*; Springer: Berlin, Germany, 1978; pp. 105–116.
45. Land, A.H.; Doig, A.G. An Automatic Method of Solving Discrete Programming Problems. *Econometrica* **1960**, *28*, 497. [CrossRef]

46. Steihaug, T. The Conjugate Gradient Method and Trust Regions in Large Scale Optimization. *SIAM J. Numer. Anal.* **1983**, *20*, 626–637. [CrossRef]
47. Byrd, R.H.; Schnabel, R.B.; Shultz, G.A. Approximate Solution of the Trust Region Problem by Minimization over Two-Dimensional Subspaces. *Math. Program.* **1988**, *40*, 247–263. [CrossRef]
48. Gratton, S.; Sartenaer, A.; Toint, P.L. Recursive Trust-Region Methods for Multiscale Nonlinear Optimization. *SIAM J. Optim.* **2008**, *19*, 414–444. [CrossRef]
49. Battiti, R. First- and Second-Order Methods for Learning: Between Steepest Descent and Newton's Method. *Neural Comput.* **1992**, *4*, 141–166. [CrossRef]
50. Ueda, K.; Yamashita, N. On a Global Complexity Bound of the Levenberg-Marquardt Method. *J. Optim. Theory Appl.* **2010**, *147*, 443–453. [CrossRef]

remote sensing

MDPI

Article

A Robot Pose Estimation Optimized Visual SLAM Algorithm Based on CO-HDC Instance Segmentation Network for Dynamic Scenes

Jinjie Chen [1], Fei Xie [1,*], Lei Huang [2], Jiquan Yang [1], Xixiang Liu [3] and Jianjun Shi [4]

[1] School of Electrical and Automation Engineering, Nanjing Normal University, Nanjing 210023, China; 21190420@njnu.edu.cn (J.C.); 63047@njnu.edu.cn (J.Y.)
[2] School of Mechanical & Electronic Engineering, Nanjing Forestry University, Nanjing 210037, China; huanglei@njfu.edu.cn
[3] College of Instrument Science & Engineering, Southeast University, Nanjing 210096, China; 101010902@seu.edu.cn
[4] Nanjing Zhongke Raycham Laser Technology Co., Ltd., Nanjing 210042, China; shijianjun@raycham.com
* Correspondence: xiefei@njnu.edu.cn

Citation: Chen, J.; Xie, F.; Huang, L.; Yang, J.; Liu, X.; Shi, J. A Robot Pose Estimation Optimized Visual SLAM Algorithm Based on CO-HDC Instance Segmentation Network for Dynamic Scenes. *Remote Sens.* **2022**, *14*, 2114. https://doi.org/10.3390/rs14092114

Academic Editors: Yuwei Chen, Changhui Jiang, Qian Meng, Bing Xu, Wang Gao, Panlong Wu, Lianwu Guan and Zeyu Li

Received: 14 March 2022
Accepted: 26 April 2022
Published: 28 April 2022

Abstract: In order to improve the accuracy of visual SLAM algorithms in a dynamic scene, instance segmentation is widely used to eliminate dynamic feature points. However, the existing segmentation technology has low accuracy, especially for the contour of the object, and the amount of calculation of instance segmentation is large, limiting the speed of visual SLAM based on instance segmentation. Therefore, this paper proposes a contour optimization hybrid dilated convolutional neural network (CO-HDC) algorithm, which can perform a lightweight calculation on the basis of improving the accuracy of contour segmentation. Firstly, a hybrid dilated convolutional neural network (HDC) is used to increase the receptive field, which is defined as the size of the region in the input that produces the feature. Secondly, the contour quality evaluation (CQE) algorithm is proposed to enhance the contour, retaining the highest quality contour and solving the problem of distinguishing dynamic feature points from static feature points at the contour. Finally, in order to match the mapping speed of visual SLAM, the Beetle Antennae Search Douglas–Peucker (BAS-DP) algorithm is proposed to lighten the contour extraction. The experimental results have demonstrated that the proposed visual SLAM based on the CO-HDC algorithm performs well in the field of pose estimation and map construction on the TUM dataset. Compared with ORB-SLAM2, the Root Mean Squared Error (Rmse) of the proposed method in absolute trajectory error is about 30 times smaller and is only 0.02 m.

Keywords: visual SLAM; instance segmentation; neural network; pose estimation

1. Introduction

Simultaneous localization and mapping (SLAM) is when a robot builds a map of the unknown environment during movement using vision, lidar, odometer and other sensors. At the same time, it carries out its own positioning [1,2]. SLAM can be used in various industries, and it will have wider applications in the future. In the driverless field, SLAM can be used to sense surrounding vehicles and scenes, creating a dynamic 3D map, which will make autonomous driving safer and more reliable [3,4]. In the 3D printing industry, by adding a camera to the printer, the SLAM algorithm can be used to determine whether the walking speed and the running path conform to the system setting [5]. In the medical field, the use of the SLAM algorithm can accurately perceive the patient's movement data during rehabilitation, which will help to assess the patient's physical condition [6].

SLAM consists of inferring the states of the robot and the environment. On the premise that the robot state is known, the target environment can be built through tracking algorithms, and the estimation problem of SLAM is proposed. The estimation problem is usually discussed in a Bayesian framework, focusing on reducing the cumulative error. The cumulative error can be estimated and adjusted through a closed-loop detection, returning to a mapped area [7], but this requires the system to match feature points or static landmarks accurately.

Different sensors affect the above errors and matching. At present, the main sensors used in SLAM include cameras, lidars, millimeter wave (mmWave) radar and the fusion of various sensors [8–10]. Examples of visual SLAM development in recent years include applying an echo state network (ESN) to a model image sequence [11,12], combining a neural network with visual SLAM [13], CPL-SLAM [14], using compact second-order statistics [15], a combination of points and lines to extract features [16], and others. It should be noted that the main purpose of the above methods is to improve the robustness and accuracy of feature point matching of visual SLAM. Lidar SLAM has been developing for a long time and now has widespread application. Paper [17] presents a 2D lidar-based SLAM algorithm, which is combined with a new structural unit encoding scheme (SEUS) algorithm, while the 2D lidar graph SLAM proposed in paper [18] is based on 3D "directional endpoint" features, performing better in robot mapping and exploration tasks. The cooperation of multiple robots can also improve the accuracy and efficiency of lidar SLAM [19–22]. Due to the advantages of mmWave in the spectrum and propagation characteristics [23], the application of mmWave in SLAM technology has become a new trend in recent years [24], and sub-centimeter SLAM can be achieved [25]. For instance, paper [26] proposed a maximum likelihood (ML) algorithm, which can achieve accurate SLAM in the challenging case of multiple-input single-output (MISO). Multi-sensor fusion can make up for the defects of single sensor and have more perfect perception [27]. For example, in the paper [28–30], the vision sensor and IMU are fused. Paper [28] proposes hybrid indoor localization systems using an IMU sensor and a smartphone camera, and adopts a UcoSLAM algorithm [31]. In addition, mainstream sensor fusion also includes lidar and vision [32,33], lidar and IMU [19,34], etc.

In order to show the advantages and disadvantages of the above different sensors more clearly, we have summarized them in four aspects: robustness, accuracy, cost and information provided, as shown in Table 1.

Table 1. The advantages and disadvantages of different sensors.

Sensor	Robustness	Accuracy	Cost	Information Provided
visual	susceptible to light	high	cheap	rich semantic information
lidar	high	higher	expensive	only depth and position
mmWave	higher	high in long distance, low in short distance	expensive	only distance and position
visual + IMU	susceptible to light	high	normal	rich semantic information
lidar + IMU	high	higher	expensive	only distance and position
visual + lidar	high	higher	more expensive	rich semantic information

It can be seen that visual sensors are the cheapest sensors [7] and can provide rich, high-dimensional semantic information [35], which can complete more intelligent tasks, although they have low robustness under current technological means. However, the traditional visual SLAM assumes a static environment. For an environment with dynamic objects, its accuracy decreases [36–38]. With the development of deep learning in computer vision and the increasing maturity of instance segmentation technology, the combination of visual SLAM and deep learning can identify and extract moving objects in the environment [39–41]. Through instance segmentation, dynamic objects in the environment are removed, and only static feature points are retained, which can significantly improve the accuracy of visual SLAM, such as You Only Look At CoefficienTs (YOLACT) [42]. Therefore, visual SLAM is no longer limited to static scenes. More and more researchers have begun to research the use of visual SLAM in dynamic scenes [43]. At present, the main SLAM algorithms based on dynamic feature point segmentation include DS-SLAM [44,45], DynaSLAM [46,47], LSD-SLAM + Deeplabv2 [48], SOF-SALM [49], ElasticFusion [50], RS-SLAM [51], DOT +

ORB-SLAM2 [52], etc. We evaluate the existing algorithms from five aspects: frontend, mapping, whether the segmentation network is independent, the accuracy of contour segmentation and the efficiency in dynamic environment. Among them, the frontend influences feature selection, extraction, matching and local map construction. Mapping affects the details of map construction, but the more details, the more calculation. An independent segmentation network reduces calculation time. The segmentation accuracy of contour will affect the elimination of dynamic feature points. We refer to papers [53,54] for the accuracy of contour segmentation and the efficiency in a dynamic environment. The details are shown in Table 2.

Table 2. The evaluation of existing visual SLAM based on dynamic feature point segmentation.

Algorithm	Frontend	Mapping	Whether Segmentation Network Is Independent	Accuracy of Contour Segmentation	Efficiency in Dynamic Environment
DS-SLAM	feature based	sparse	yes	low	higher
DynaSLAM	feature based	sparse	no	normal	high
LSD-SLAM + Deeplab V2	direct	semi dense	no	normal	low
SOF-SLAM	feature based	sparse	no	low	normal
ElasticFusion	ICP	dense	no	higher	low
RS-SLAM	feature based	dense	no	high	low
DOT + ORB-SLAM2	feature based	sparse	no	low	normal

As can be seen from the table, deep and high-dimension frontend processing can increase the accuracy of contour segmentation but also reduce the operation efficiency. Meanwhile, only DS-SLAM splits the segmentation network independently, which is beneficial to the operation efficiency of visual SLAM. In conclusion, current algorithms are difficult to achieve accurate contour segmentation and high operation efficiency at the same time. Once the contour segmentation is not accurate enough, it is easy to eliminate the static feature points from the contour by mistaking them for dynamic feature points, and it is also easy to retain the dynamic feature points by mistaking them for static feature points, which will reduce the accuracy of SLAM mapping in the later stage. At the same time, huge data adversely affects the real-time performance of visual SLAM. Therefore, aiming at the above problems, this paper proposes a visual SLAM based on the CO-HDC algorithm, which is an instance segmentation algorithm of contour optimization, including the CQE contour enhancement algorithm and Beetle Antennae Search Douglas–Peucker (BAS-DP) lightweight contour extraction algorithm. The main contributions of this paper are summarized as follows:

- To solve the problem of the imprecise segmentation of the object's contour, a hybrid dilated CNN is used as backbone network to increase the receptive field. In the empty convolution operation, the expansion rate of each layer can be designed as [1–3], and the top layer can obtain broader pixel information to improve the information utilization rate;
- CQE algorithm is proposed, which can enhance the contour of the object. CQE is composed of 4 convolution layers and 3 full connection layers. It is fused with hybrid dilated CNN to form an end-to-end contour enhancement network. This can significantly improve the elimination ability of dynamic feature points, especially the feature points falling on the contour;
- Although high-precision contour can be obtained through the CQE model, it needs a large amount of calculation, which adversely affects the real-time performance of visual SLAM based on instance segmentation. Therefore, the BAS-DP lightweight contour extraction algorithm is proposed. The BAS-DP algorithm converts the contour information surrounding the target into the best polygon surrounding the target,

which can greatly reduce the data file and make the calculation speed faster on the basis of preserving the contour accuracy.

The rest of the paper is organized as follows: In Section 2, the CO-HDC algorithm proposed in this paper is analyzed in detail, including hybrid dilated CNN, CQE, BAS-DP, global optimization module and mapping module. The test and results analysis are provided in Section 3. In Section 4, we further discuss our method and existing methods. The conclusions and future work are summarized in Section 5.

2. The Pose Estimation Optimized Visual SLAM Algorithm Based on CO-HDC Instance Segmentation Network

The instance SLAM is divided into three modules, as represented in Figure 1: tracking, global optimization and mapping module.

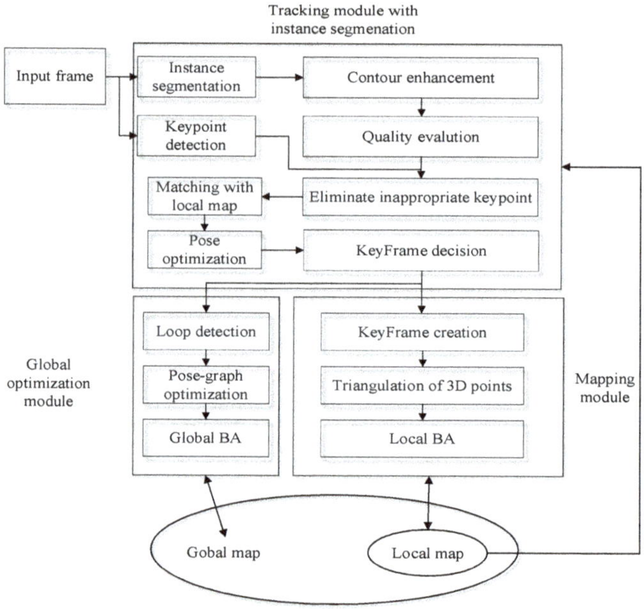

Figure 1. Visual SLAM algorithm with pose estimation optimized by instance segmentation architecture.

We add CO-HDC instance segmentation to the tracking module, which includes the CQE contour enhancement algorithm and BAS-DP lightweight contour extraction algorithm, and use the hybrid dilated convolutional neural network as the backbone network. CO-HDC can effectively improve the accuracy of dynamic feature point segmentation, especially the contour of the target. Tracking Module with instance segmentation minimizes the impact of dynamic objects. It means that pose estimation is more accurate and keyframe decisions are better. The global optimization module and mapping module can benefit from instance segmentation, which provides high-quality feather points. Loop detection makes the global optimization module able to work well. Therefore, a more accurate map can be built.

2.1. *Tracking Module with CO-HDC Instance Segmentation*

According to the input RGB image and the depth image, the algorithm front end performs feature point detection and feature descriptor calculation on the RGB image. Tracking Module is divided into the following steps:

Firstly, feature matching of two adjacent frames is performed according to the feature descriptor. A 2D-2D feature matching point set is obtained. Using CO-HDC instance segmentation to remove dynamic pixels can help feature point matching greatly. The working framework of CO-HDC is shown in Figure 2, which takes the vehicle detection commonly used in the industry as an example. Among them, the backbone network adopts a hybrid dilated CNN network, which can increase the ability of network feature extraction. Then, the contour of the detected target is strengthened to improve the accuracy of instance segmentation further. At the same time, BAS-DP is used to lighten the calculation of contour, which can speed up the visual SLAM.

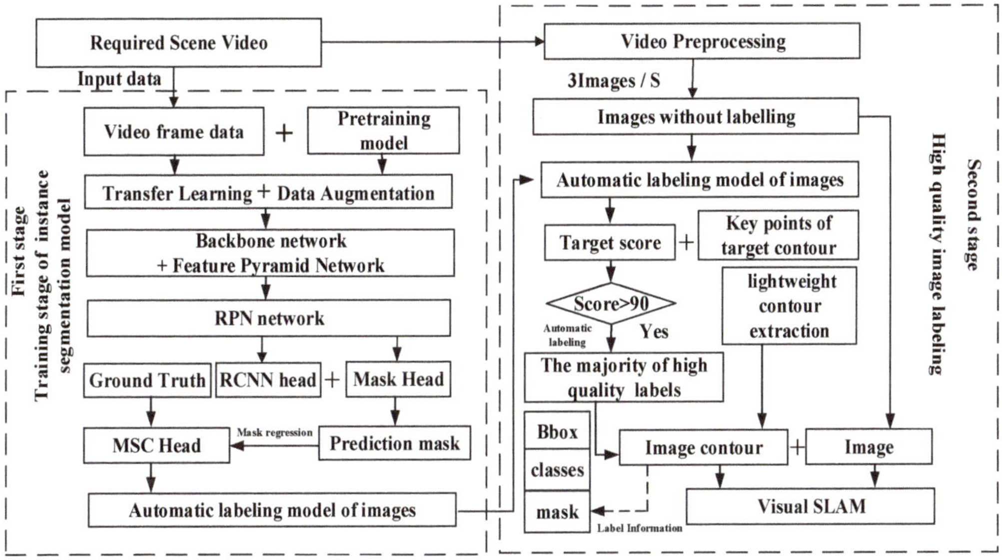

Figure 2. Framework of CO-HDC instance segmentation.

Secondly, according to the depth information of the image, the spatial three-dimensional coordinates of the 2D-2D feature matching point pairs are calculated to obtain a 3D-3D matching point set. The rotation and translation matrix between two adjacent frames of images can be calculated from the matched 3D-3D points.

Finally, the motion estimation error is optimized to obtain the pose estimation result with the smallest error. In this way, according to the input video stream, the incremental change of the camera pose can be continuously obtained. Therefore, the front end of the algorithm constructs a visual odometer [55].

2.1.1. Complex Feature Extraction Based on Hybrid Dilated CNN

Accurate instance segmentation will be conducive to the accuracy of SLAM composition and pose estimation. In order to improve the feature extraction ability of the backbone detector in the instance segmentation model, a dilated convolutional neural network is introduced into the network. With an increase in the number of insertion holes of the dilated convolutional neural network, the size of the receptive field will increase [56], but it also leads to the loss of continuous information, which is easy to cause the problem of meshing. In order to solve the problem of continuous information loss in grid sampling, the hybrid dilated convolutional neural network can be used to replace the dilated convolutional neural network.

Suppose an n-layer convolutional neural network, and the size of the convolution kernel of each layer is $K \times K$. The expansion rate is $[r_1, \cdots r_i, \cdots r_n]$. The purpose of constructing hybrid dilated convolutional neural network is that when a series of operations of dilated convolutions are completed, the extracted feature map can cover all pixels. The maximum distance between two non-zero pixels can be calculated by the following formula:

$$M_i = max[M_{i+1} - 2r_i,\ M_{i+1} - 2(M_{i+1} - r_i), r_i] \quad (1)$$

where r_i is the expansion rate of layer i,M_i is the maximum expansion rate of layer i. In order to make the final receptive field cover the whole region without any holes, an effective hybrid dilated convolutional neural network must meet $M_2 \leq K$. As shown in Figure 3, when the size of the convolution kernel $k = 3$, the expansion rate of each layer $r = [1,2,3]$, $M_2 = 2 \leq 3$ of all pixels can be covered.

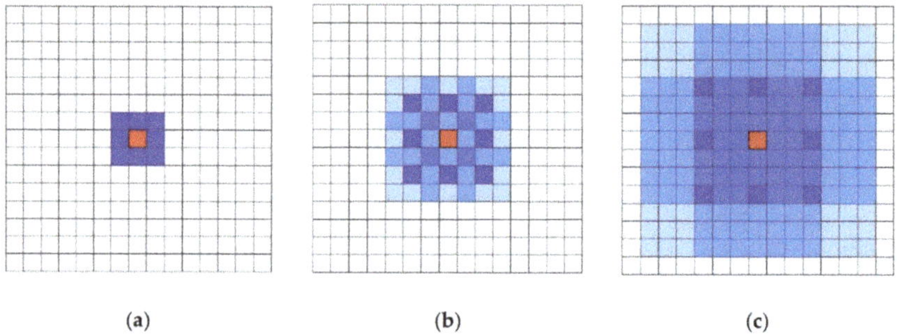

Figure 3. Diagram of hybrid dilated convolutional neural network with different expansion rates: (**a**) the diagram of HDC with expansion rate 1; (**b**) the diagram of HDC with expansion rate 2; (**c**) the diagram of HDC with expansion rate 3.

In order to highlight the improvement of the performance of the instance segmentation model by the hybrid dilated convolutional neural network, the traditional convolution core is replaced by the hybrid dilated convolution core. The backbone detector structure based on the hybrid dilated convolutional neural network is shown in Figure 4.

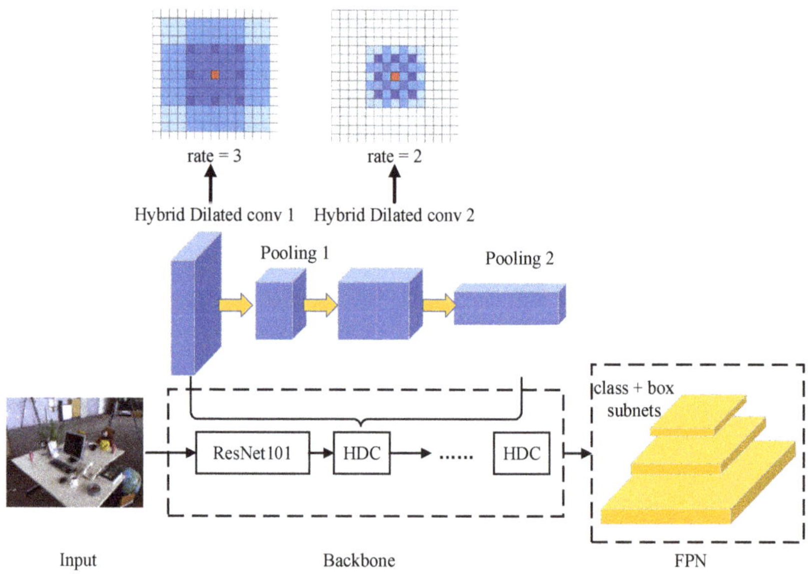

Figure 4. The diagram of backbone detector based on hybrid dilated convolutional neural network.

2.1.2. Contour Enhancement Based on CQE

However, only the hybrid dilated convolutional neural network in the first part is not enough. Any data generation network will produce some low-quality data, especially in the contour part. If the generated data is not judged and processed, a large number of low-quality data will be mixed, and the accuracy of feature points in the later stage will be seriously affected. Therefore, CQE, a discriminator, is proposed to judge the quality of image contour. It can remove the low-quality contour information and retain the high-quality contour to enhance the contour of the object.

In most instance segmentation networks, mean intersection over union (Miou) is calculated by the ratio of their cross area to their cumulative area, and the quality of predicted contour is measured by Miou, but it is necessary to ensure that they have the same height and width. However, the Miou calculated by this method is not linear with the quality of the predicted contour, so this method is inaccurate.

Therefore, the CQE algorithm is designed, working as a discriminator to evaluate the quality of contour. The evaluation mainly includes the accuracy of the surrounding target contour and target classification accuracy. Then, by setting the quality threshold, the contour with quality lower than the threshold is discarded, and the contour with quality higher than the threshold is retained. Finally, the contour above the threshold and the corresponding image data are combined to form the instance segmentation result.

The first is to evaluate the accuracy of the target contour. Due to the irregular shape surrounding the target contour, using the regression principle in the convolutional neural network, a CQE head is designed to regress the accuracy of the target contour in the generated data, which is supervised in the process of network training, and the irregular contour is well solved. The convolutional neural network can not only extract the features in the image but can also be used to regress the similarity between the two images. The CQE head is used to regress the true contour and the predicted contour. Calculate the complete intersection over the union (Ciou) value of the difference between the real contour and the predicted contour of each target, and normalize the Ciou to obtain *S*iou, which is the evaluation quality of the contour. Its range is between 0 and 1. By setting different *S*iou thresholds, different quality target contours can be obtained. The closer the value of *S*iou is to 1, the better the target contour prediction effect is.

The structural design of the CQE head is composed of four convolution layers and three full connection layers. For four convolution layers, the core size and the number of filters of all convolution layers are set to 3 and 256, respectively. For three fully connected layers, set the output of the first two FC layers to 1024 to connect all neurons. The C of the last FC layer is the number of categories to be classified. Finally, the CQE head outputs the contour quality Siou of each target.

Truth-contour and Predict-contour work together as the input of the CQE head. The Truth-contour exists in the characteristic graph, and the Predict-contour is the contour output by the CQE head. Because the output result of the CQE head is different from the size of the ROI characteristic diagram, two input structures are designed. Figure 5 shows two kinds of input structures of the CQE head.

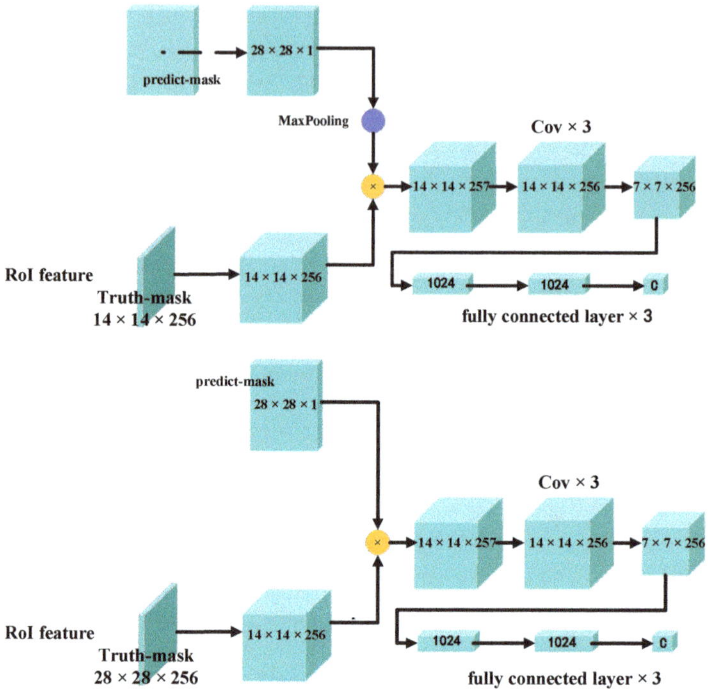

Figure 5. Two input structures of CQE head.

Among them, the input structure designed in the first figure in Figure 5 is to maximize the pool of the feature layer output by the CQE head through a convolution kernel with a size of 2 and a step size of 2 and then multiply it with the ROI feature map with a smaller size. The input structure designed in the second figure in Figure 5 is the CQE head, which is directly added to the larger ROI characteristic diagram without maximum pooling. Both structures can be used as inputs of the CQE head. The set CQE threshold is 0.9. When the CQE of each contour in the target is higher than 0.9, the generated contour quality is higher. When the CQE of the tag contour is lower than the threshold, the generated contour quality is low. The recognition process of contour enhancement using the CQE is shown in Figure 6.

2.1.3. The Lightweight Contour Extraction Algorithm Based on BAS-DP

A large number of high-precision instance segmentation can be obtained through the contour enhancement network. If all points on the target contour segmented by the instance are retained, the file will be too large, which will lead to slow SLAM operation time in the later stage and make it difficult to achieve the real-time effect. Therefore, a lightweight contour extraction algorithm based on BAS-DP is proposed. The algorithm converts the contour information surrounding the target into the best polygon surrounding the target. The number of coordinate points contained in the polygon is small, which can lighten the segmentation file while ensuring the accuracy of instance segmentation.

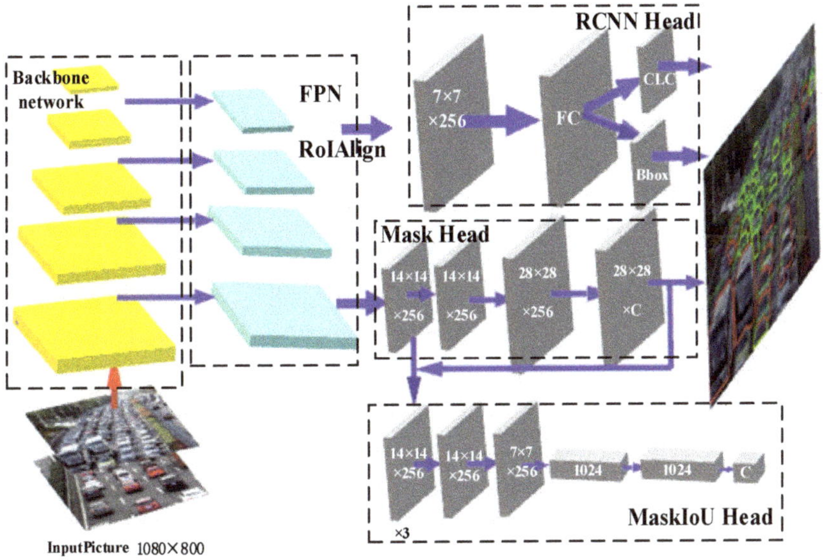

Figure 6. The recognition structure with contour enhancement using CQE.

Using the best polygon surrounding the target to replace the contour curve surrounding the target is the most direct and commonly used method. Therefore, it is necessary to convert the contour of the target into each turning point on the polygon surrounding the target. So, it is necessary to use a polygon approximation algorithm to convert the contour curve of the target into a polygon surrounding the target and then record the coordinates of key points on the polygon in the segmentation file.

Douglas–Peucker algorithm (DP algorithm) is a classical polygon approximation algorithm that can approximate the closed curve as a polygon and reduce the number of points as much as possible. It has the advantages of translation and rotation invariance. However, it needs to solve other points on the curve that do not belong to key points exhaustively, which requires a lot of calculation time. The Beetle antennae search algorithm (BA algorithm) is another classic polygon approximation algorithm that realizes efficient optimization by simulating longicorn beetle foraging. Beetle Antennae Search algorithm can realize optimization without knowing the specific form of function and gradient information. However, its accuracy is relatively low.

This paper proposes lightweight contour extraction algorithm based on BAS-DP, combining the advantages of the above two algorithms. The calculation steps are shown in Figure 7.

In the BAS-DP algorithm, parameter initialization includes the initial trial step attenuation factor H, step S, the ratio of step and whisker C, the number of iterations n and the number of parameters to be optimized k. Among them, the distance optimization function $f(x)$ is shown in Formula (2). According to this formula, the function values f_l and f_r corresponding to the left whisker position x_l and the right whisker position x_r of the longicorn beetle can be calculated, and the next position x of the longicorn beetle can be calculated at the same time. Perform calculating function $f(x)$ n times in total. The optimal function value corresponding to the last position x of the longicorn beetle is obtained as the optimal solution.

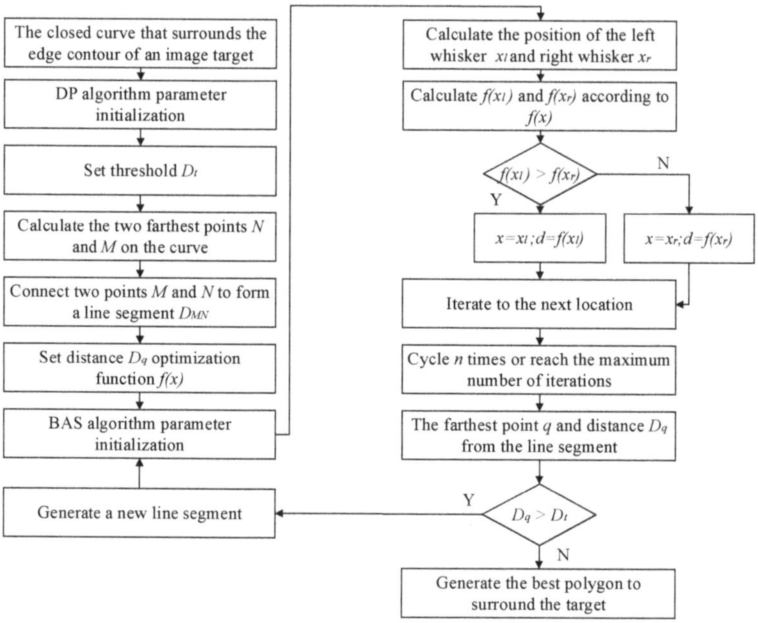

Figure 7. The calculation steps of BAS-DP lightweight contour extraction algorithm.

$$\begin{cases} d_{ir} = rand(k,1);\ d_0 = step/c \\ x_l = x + d_0 * dir/2;\ x_r = x - step * d_{ir}/2 \\ f_l = f(x_l);\ f_r = f(x_r) \\ x = x - s_{tep} * d_{ir} * s_{ign}(f_l - f_r) \end{cases} \tag{2}$$

The BAS-DP algorithm can reduce the size of the segmented file while maintaining the contour accuracy and improving the real-time performance of the later visual SLAM. Finally, the BAS-DP algorithm is combined with the hybrid dilated convolutional neural network and the CQE algorithm proposed in the previous two sections, forming the CO-HDC. Through this algorithm, a large number of high-quality instance segmentation images can be generated, and the data enhancement network needs only a small amount of data to record better accuracy, especially to solve the segmentation problem of the object contour.

2.2. Pose Optimization

Through the CO-HDC algorithm, we can accurately separate the object, especially the contour of the object, removing the feature points on the dynamic object and retaining the static feature points so as to achieve good feature point matching and complete pose estimation well. In visual SLAM, posture refers to the robot in spatial position and posture of the entire environment map. Both spatial position and robot posture position need to be accurately located in the three-dimensional space.

Figure 8 shows the principle of spatial measurement. It is assumed that in two adjacent frames, the camera has no distortion, and the two projection planes are parallel and coplanar. In the figure, P is an object, Z is its depth, f is the focal length of the camera, T is the center distance of two adjacent frames, O_l and O_r are the optical centers of two adjacent frames of the camera, respectively, and x_l and x_r are the horizontal axis coordinates of the projection of object P in two adjacent frames, respectively. The depth calculation formula of object P can be obtained from the relationship of similar triangles:

$$\frac{T - (x_l - x_r)}{Z - f} = \frac{T}{Z} \Rightarrow Z = \frac{fT}{x_l - x_r} \tag{3}$$

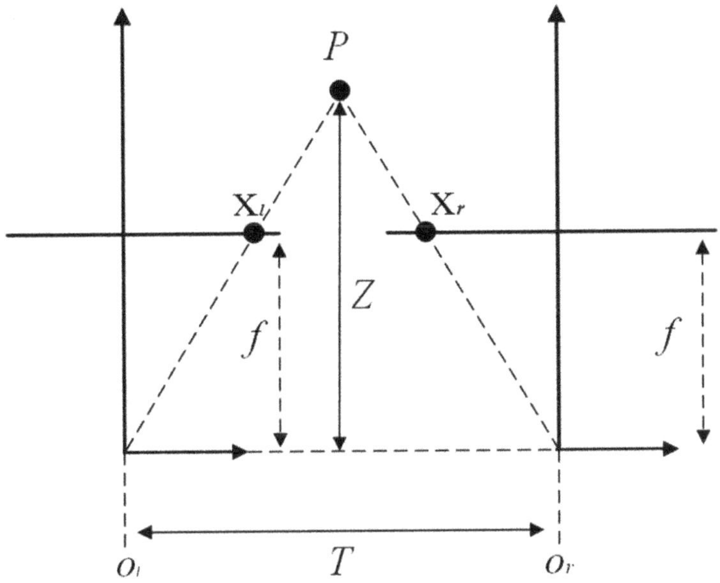

Figure 8. The principle of spatial measurement.

$d = x_l - x_r$ is defined as parallax, so that the depth information of the target point can be obtained through the parallax and f, T of the target. After obtaining the parallax map, the coordinates of the target point in the world coordinate system can be obtained through the re-projection matrix. The re-projection matrix is:

$$Q = \begin{bmatrix} 1 & 0 & 0 & -c_x \\ 0 & 1 & 0 & -c_y \\ 0 & 0 & 0 & f \\ 0 & 0 & \frac{-1}{T} & \frac{(c_x - c_y)}{T} \end{bmatrix} \tag{4}$$

In the above formula, c_x is the x coordinate value of the main point of the first frame, and c_y is the y coordinate value of the main point of the second frame. Assuming that the identified coordinate of the target point is (x, y), and the parallax in the two adjacent frames is D, its coordinate value in the world coordinate system can be recovered through Formula (5):

$$Q \begin{bmatrix} x \\ y \\ d \\ 1 \end{bmatrix} = \begin{bmatrix} x - c_x \\ y - c_y \\ f \\ \frac{-[d - (c_x - c_y)]}{T} \end{bmatrix} = \begin{bmatrix} X \\ Y \\ Z \\ W \end{bmatrix} \tag{5}$$

In proposed SLAM, the robot's posture is calculated through the translation vector and rotation quaternion number representation of seven paraments, as shown in the following type (6):

$$T = [x,\ y,\ z,\ qx,\ qy,\ qz,\ qw] \tag{6}$$

The first three are translation vectors. The last quaternion is the quaternion for rotation.

The task of the tracking thread is to calculate the posture of two adjacent frames according to the image change. This means not only the distance moved in the next frame should be calculated, but also the angle of rotation should be calculated. The results are then handed over to the back end, which accumulates and optimizes the relative positions between the two frames.

The images obtained by the pre-recognition before and after are $I1$ and $I2$. After feature extraction, the feature point $p1$ is obtained in $I1$. The feature point $p2$ is obtained in $I2$. Assuming the result of feature matching is that $p1$ is obtained and $p2$ is the closest point pair, it means that $p1$ and $p2$ is the projection of the same 3D point P on two frames of images.

$$p1 = KP,\ p2 = T(KP) \tag{7}$$

where, T is the camera's internal parameter matrix. When the camera is in different positions, point P obtains different pixel coordinates through the transformation of the internal parameter matrix. They are projection $p1$ and $p2$. K is the pose of $I1$ relative to $I2$. Assuming that multiple sets of point pairs can be matched between the two frames, the equation can be constructed by these point pairs to solve the relative pose. Specifically, it can be solved by solving the basis matrix and the homology matrix.

However, T must be calculated in the space P, where the whole environment's stationary conditions are valid. If the points in the pose estimation are in the process of moving, type (4) is set up. The error would arise. The worst-case scenario is to use the camera to participate in the pose estimation of all pixels for the same shipment. Then the pose estimation will always be 0.

2.3. Global Optimization Module and Mapping Module

The tracking module estimates the camera poses through keypoint matching and pose optimization. An instance segmentation function is added to the tracking thread, and the original image is segmented at the same time as the feature extraction. Then, the pixel coordinates of the human and the animal are obtained. Finally, some feature points distributed on the human or animal are removed from the original feature point.

After culling feature points, the feature matching and pose estimation are performed. After getting rid of the interference of the pixel points, the instance SLAM shows better anti-interference ability under dynamic scenes. The accuracy is greatly improved. This module also determines whether to insert a new keyframe. When a frame is considered suitable for a new keyframe, it is sent to the mapping module and global optimization module.

In the mapping module, to eliminate mismatches or inaccurate matches, a new 3D point is triangulated by inserting a keyframe, optimizing the projected points and lines and adding a projection matrix. This process is equivalent to minimizing the photometric difference between blocks of projected pixels u_i and the blocks corresponding to the 3D point on the current frame u_r. The model expression is:

$$\hat{u}_i = \underset{\hat{u}_i}{\operatorname{argmin}} \frac{1}{2} \sum_i || I_c(\hat{u}_i) - I_r[A(u_i)] ||_2^2 \tag{8}$$

where, I_c and I_r are the first and second frames, respectively, and A is the projection matrix. The projection matrix formula is as follows:

$$\begin{bmatrix} x' \\ y' \end{bmatrix} = R \begin{bmatrix} x \\ y \end{bmatrix} + \begin{bmatrix} t_x \\ t_y \end{bmatrix} \tag{9}$$

where, R is the matrix representing rotation and scaling, x and y are the coordinates before projection, and T_x and T_y represent translation distance.

In the process of global optimization, it is necessary to eliminate the accumulated errors caused by the odometer. The matching algorithm we use is a kind of image matching based on pixel value. Its purpose is to find a strict geometric transformation to make each pixel in the local map and the global map equal as much as possible. The inverse compositional algorithm can solve the problem of image matching, which is completed in three steps. The specific steps are given in the following formulations:

The first step is to calculate the Hessian matrix H:

$$H = \sum_x \left[\nabla I_{PM}(x)\frac{\partial W}{\partial P}\right]^T \left[\nabla I_{PM}(x)\frac{\partial W}{\partial p}\right] \tag{10}$$

where, I_{PM} is the global map image, x is the coordinates of pixels in the image, $P = [(\Delta x,\ \cdot y,\ \theta)]T$ represents translation and rotation vectors, and $I\ (W\ (x;\ P))$ represents the Euclidean transformation of vector P on image $I(x)$.

The second step is to calculate the new vector $\Delta\, p$:

$$\Delta p = H^{-1}\sum_x \left[\nabla I_{PM}(x)\frac{\partial W}{\partial p}\right] [I_{LM}(W(x;p)) - I_{PM}(x)]^2 \tag{11}$$

where, I_{LM} is the image of a local subgraph.

Step 3: Update vector p:

$$p = p + \Delta p \tag{12}$$

The final output p of the algorithm represents the translation and rotation between maps, which can eliminate the accumulated errors in global map construction, and also solves the problem of trajectory drift that often occurs in visual SLAM.

3. Tests and Results Analysis

In order to demonstrate the advantages of the CO-HDC instance segmentation algorithm proposed in this paper and test the actual effect of visual SLAM based on CO-HDC instance segmentation, our experiment will be divided into two parts. Firstly, we will experiment with the performance of the CO-HDC instance segmentation algorithm. Secondly, we will test the performance of the visual SLAM based on the CO-HDC instance segmentation algorithm proposed in this paper and judge the effect of feature point matching and real-time modeling.

3.1. Experiment of CO-HDC Instance Segmentation Algorithm

In order to test the accuracy and efficiency of the proposed contour enhancement instance segmentation algorithm, the following experiments are carried out:

- the selection of network hyperparameters to achieve the precise and fast segmentation;
- comparison of different backbone networks.

3.1.1. The Network Hyperparameters Selection and Controlled Experiment

Instance segmentation can remove the dynamic object, which increases the accuracy of visual SLAM. In order to integrate with visual SLAM better, the instance segmentation network model needs to be optimized. Therefore, ten comparative experiments were conducted under hybrid dilated CNN to select appropriate network parameters and observe the effect of transfer learning on training time, accuracy and training data volume. The hyperparameters selection and the corresponding results are shown in Table 3. *mAP* is the average precision, and *mIoU* is the average intersection ratio. In this paper, *mAP* and *mIoU* are used to evaluate the quality of network training structure. In order to strictly evaluate the performance of the method, the thresholds of *mAP* are set to 0.5 and 0.7, respectively. Those greater than or equal to the threshold are true positive, while those less than the threshold are false positive. The *mIoU* and *mAP* indicators for each experiment are shown in the last three rows of the table for detailed analysis of the experiment contents and results.

Table 3. Hyperparameters selection comparison experiments.

Hyperparameters	Test 1	Test 2	Test 3	Test 4	Test 5
Train obj.	2081	2081	2520	3005	3005
Val obj.	537	537	632	826	826
Train imag.	680	680	820	1014	1014
Val imag.	120	120	140	180	180
Epochs	100	200	200	400	400
Mini-mask Shape	56 × 56	56 × 56	56 × 56	56 × 56	56 × 56
Img. size	1024 × 800	1024 × 800	1024 × 800	1024 × 800	1920 × 1080
RPN Anchor Scales	(32, 64, 128, 256)	(32, 64, 128, 256)	(32, 64, 128, 256)	(32, 64, 128, 256)	(32, 64, 128, 256)
Pre-train Model	NO	NO	NO	NO	NO
mIoU	0.485	0.492	0.535	0.498	0.294
mAP(*IoU* > 0.5)	0.569	0.586	0.495	0.565	0.395
mAP(*IoU* > 0.7)	0.472	0.488	0.406	0.485	0.289
Hyperparameters	**Test 6**	**Test 7**	**Test 8**	**Test 9**	**Test 10**
Train obj.	3005	3005	3005	1573	1573
Val obj.	826	826	826	537	537
Train imag.	1014	1014	1014	480	480
Val imag.	180	180	180	120	120
Epochs	100	100	100	100	100
Mini-mask Shape	28 × 28	28 × 28	28 × 28	28 × 28	28 × 28
Img. size	1024 × 800	1920 × 1080	1920 × 1080	1920 × 1080	1920 × 1080
RPN Anchor Scales	(32, 64, 128, 256)	(16, 32, 64, 128)	(8, 16, 32, 64)	(8, 16, 32,64)	(8, 16, 32,64)
Pre-train Model	NO	NO	NO	NO	Yes
mIoU	0.545	0.565	0.652	0.429	0.684
mAP(*IoU* > 0.5)	0.558	0.573	0.716	0.345	0.725
mAP(*IoU* > 0.7)	0.489	0.493	0.575	0.294	0.585

Train obj. and Val obj. correspond to the total number of training objectives and verification objectives of the training, respectively. Train imag. and Val imag. are the number of training images and verification images. Epochs is the number of iterations of all training sets, and the Mini-mask Shape is the minimum mask size. Img. Size is the size of the input image, and RPN Anchor Scales is the proportion Size of the Anchor. The Pretrain Model is the 80 classification pre-training model of coco data sets.

Test 1 and Test 2 use the same Non-Maximum Suppression (NMS) threshold, the basic learning rate, and other hyperparameters but use different amounts of epochs. Feeding all data into the network for iteration is called an epoch, and the number of epochs is set to 100 and 200, respectively. With the increase of epochs, the value of *mAP* (*IoU* > 0. 5) in test 1 increased from 0.569 to 0.586 with a low volatility effect. So, on a low number of iterations, it was still easy to converge, indicating that the convergence effect of the algorithm in this paper was great.

In Test 3 and Test 6, we used images of more data for training and testing, and epochs were the same as before. The results showed a decrease in detection rate, which was later improved in test 4 by increasing the number of epochs, resulting in an *mAP* (*IoU* > 0. 5) of 0. 565.

In Test 5, we evaluate the effects of the image width and height, the size of the training images from 1024 × 800 to 1920 × 1080, learning rate from the default of 0. 001 to 0. 02, the rest of the parameters like Test 4. We get a poor performance of the algorithm (*mAP* (*IoU* > 0. 5) = 0.395). It indicates that the accuracy of images of high resolution is low under the current parameters.

In Test 6, we reduced the size of the mini-mask from 56 × 56 to 28 × 28, and compared with Test 4; we found some improvement in network performance.

Therefore, in Test 7, we reduced the Scales of RPN Anchor and improved the input image resolution to 1920 × 1080 and the small mask to 28 × 28. It was found that the performance of the network was greatly improved, which was close to the network performance in Test 6.

In Test 8, we used the same configuration as Test 7 and further reduced the RPN Anchor Scales. It was found that the performance of the network with reduced RPN Anchor Scales was greatly improved, and (8, 16, 32, 64) was considered the best RPN Anchor Scales of the network.

In Test 9, in order to improve the training accuracy, reduce the training time and prevent network overfitting, we reduced the amount of training data on the basis of Test 8 and found that the network performance decreased significantly.

In Test 10, we substantially recompressed the training data on the basis of Test 9, other parameters remained unchanged, and we used 80 classification models of the pre-trained COCO data sets for transfer learning. The results showed that the network performance was basically the same as that of Test 8, and the network performance reached a higher level, but the training time was half that of Test 8. Network performance can accurately detect and segment vehicle images.

Through 10 comparative experiments, it can be seen that the more training data, the higher the image resolution, the smaller the mask and the smaller the scale of RPN anchor will lead to better network performance. The results show that 100 epochs are enough to achieve convergence for target detection. At the same time, an increasing pre-training model can reduce the training data. In conclusion, Test 10 achieves the most perfect balance among training data, image resolution, mask size, epochs, scale of the RPN anchor and other parameters. Appropriate data volume and resolution ensure not only high speed but also high precision. At the same time, the transfer learning method can reduce the training data, training time and improve the detection accuracy. Therefore, we set the parameters of Test 10 as our optimal network parameters and carried out subsequent experiments and studies with the parameters of Test 10.

3.1.2. Comparison of Different Backbone Networks

Under the network configuration parameters of Test 10, a comparative test was conducted for different backbone networks to demonstrate the advantages of HDC-Net. The neural networks of HDC-Net, ResNet50, Res-Net101 and MobileNetV1 were all composed of residual blocks, which simplified their architectures with residual learning, reduced their computational overhead and well solved the gradient vanishing problem.

Its performance was compared in four aspects. Network training time, image detection time per second, network model weight and accuracy (S > 90 means that SMask is greater than 90). Accuracy is the ratio of high-quality labels to all labels. It can be seen from the Table 4 that when HDC-Net is used as the backbone network, the training time is 13.21 h, which is quite similar to ResNet50; the speeds of these four networks are 6.65 sheets per second, 6.25 sheets per second, 4.6 sheets per second and 5.2 sheets per second respectively, and HDC-Net has the fastest speed for calibrating the image. In the model size comparison test, when HDC-Net is used as the backbone network, the label model size is the smallest. When HDC-Net, ResNe50, ResNet101 and MobileNet V1 are used as the backbone network, the accuracy of the vehicle image label is 95.1%, 93.4%, 93.8% and 84.5%, respectively. It can be seen that although HDC-Net has a slight increase in training time compared with ResNet50, it is far ahead of other backbone networks in terms of speed, model weight and accuracy. Therefore, HDC-Net has the best performance.

Table 4. Performance comparison of four backbone networks.

Backbone Network	Train Time/h	Speed/FPS	Model Weight /MB	Accuracy S > 90
HDCNet	13.21	6.65	163.21	95.1%
ResNet50	12.65	6.25	186.75	93.4%
ResNet101	20.73	4.60	268.86	93.8%
MobileNet V1	14. 61	5.27	207.82	84.5%

3.2. Experiment of Visual SLAM Based on CO-HDC

In this paper, two sets of tests are carried out to evaluate the visual SLAM based on CO-HDC. The first set of tests is that dynamic feature points for single-frame pictures in motion and intermediate results are shown. The second set of tests is that the instance visual SLAM based on CO-HDC proposed in this paper and ORB-SLAM2 algorithms are run on the TUM RGBD public dataset. Other than this, experimental results are compared with each other.

The dataset used in this paper are rgbd_dataset_freiburg3_walking_xyz (dataset one), rgbd_dataset_freiburg3_walking_halfsphere (dataset two) and rgbd_dataset_freiburg3_walking_static (dataset three) in the TUM dataset Dynamic Objects. This dataset contains moving people, and the camera is also in motion to evaluate the robustness of the SLAM system or motion calculations in scenes with fast-moving dynamic objects. In the dataset, the video frame rate is 30 Hz, and the sequence contains a full sensor resolution is 640 × 480. The ground real trajectory is obtained from a motion capture system of eight high speed tracking cameras.

3.2.1. Feature Point Extraction and Matching after CO-HDC Instance Segmentation

A comparison between ORB-SLAM2 and the proposed visual SLAM based on CO-HDC instance segmentation is carried out. ORB-SLAM2 assumes that feature points in the scenes are static, and feature points matching is performed directly after feature points extraction. However, this may lead to pose estimation errors and map relative drifts under dynamic environments. At the same time, the proposed visual SLAM segments the dynamic objects and retains static feature points. Moreover, it performs feature points matching using static point only.

Firstly, the feature point extraction and matching in the ORB-SLAM2 algorithm are performed. The two adjacent frames in the video sequence of the dataset are randomly selected, as shown in Figure 9a,b. Figure 9c,d show the feature extraction in the ORB-SLAM2 algorithm, where some feature points fall on the human body. Then, the feature matching is shown in Figure 9e.

In the BAS-DP algorithm, parameter initialization includes the initial trial step attenuation factor H, step S, the ratio of step and whisker C, the number of iterations n and the number of parameters to be optimized k. Among them, the distance optimization function $f(x)$ is shown in Formula (2). According to this formula, the function values f_l and f_r corresponding to the left whisker position x_l and the right whisker position x_r of the longicorn beetle can be calculated, and the next position x of the longicorn beetle can be calculated at the same time. Perform calculating function $f(x)$ n times in total. The optimal function value corresponding to the last position x of the longicorn beetle is obtained as the optimal solution.

Figure 9. Results of feature extraction and matching based on ORB-SLAM2: (**a**) Original Figure 1; (**b**) Original Figure 2; (**c**) Feature points extracted before screening of original Figure 1; (**d**) Feature points extracted before screening of original Figure 2; (**e**) The ORB matching results of original Figures 1 and 2.

3.2.2. Using Datasets to Test the Preference of ORB-SLAM2 and Instance Visual SLAM Based on CO-HDC Algorithm

The dataset provides an automated assessment tool for visual odometer system drift and global attitude error for SLAM systems, which is divided into absolute trajectory errors (ATE) and relative pose errors (RPE). The ATE difference is used to calculate the difference between the actual values and estimated values of the camera pose of the SLAM system. The RPE is used to calculate the difference between the pose changes on the same two timestamps. Firstly, the estimated value is aligned with the real value according to the timestamp of the pose. The drift of the system is also evaluated. From Figures 11–13, the RPE of instance SLAM based on CO-HDC is much smaller than ORB-SLAM2. The amount of change in pose is calculated at the same time. From Figures 14–16, it can be concluded that the proposed SLAM performs better than ORB-SLAM2, as the ATE of the proposed SLAM is also smaller than ORB-SLAM2. In Table 5, compared with ORB-SLAM2, the Rmse of the proposed method in absolute trajectory error is about 30 times smaller and is only 0.02 m. The comparison in Tables 6 and 7 also confirms the advantages of the proposed SLAM.

Figure 10. Results of feature extraction and matching based on proposed SLAM: (**a**) Figure for dynamic dot culling of first frame; (**b**) Figure for dynamic dot culling of second frame; (**c**) Feature points extracted after screening of first frame; (**d**) Feature points extracted after screening of second frame; (**e**) The ORB matching results of original Figures 1 and 2 after screening.

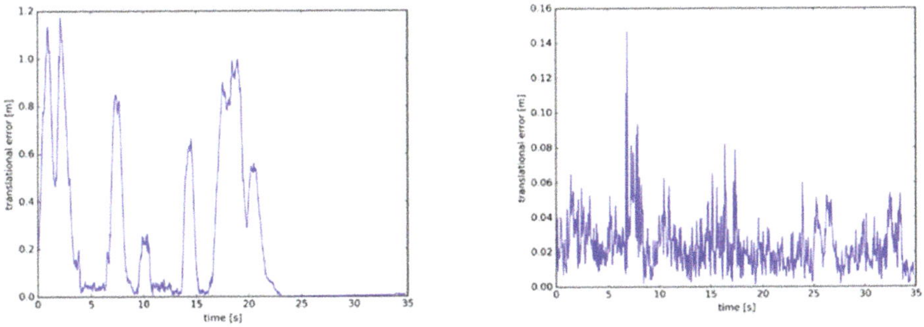

(**a**) (**b**)

Figure 11. Relative pose error of dataset one: (**a**) The relative pose error of dataset one using ORB-SLAM2; (**b**) The relative pose error of dataset one using instance SLAM based on CO-HDC.

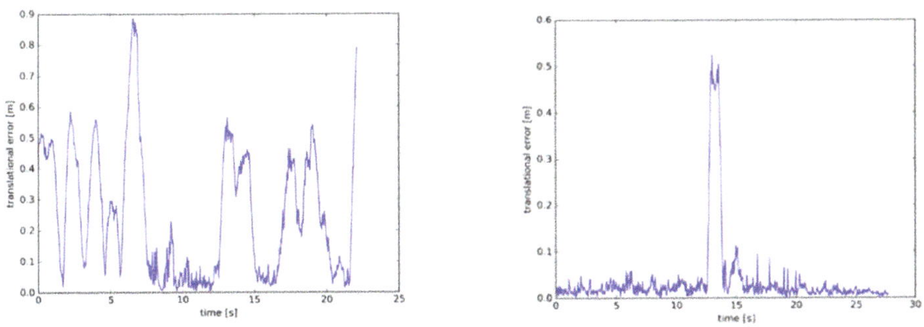

(**a**) (**b**)

Figure 12. Relative pose error of dataset two: (**a**) The relative pose error of dataset two using ORB-SLAM2; (**b**) The relative pose error of dataset two using instance SLAM based on CO-HDC.

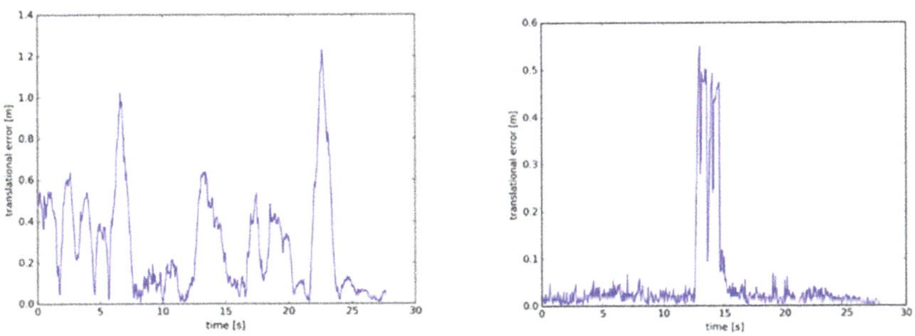

(**a**) (**b**)

Figure 13. Relative pose error of dataset three: (**a**) The relative pose error of dataset two using ORB-SLAM2; (**b**) The relative pose error of dataset two using instance SLAM based on CO-HDC.

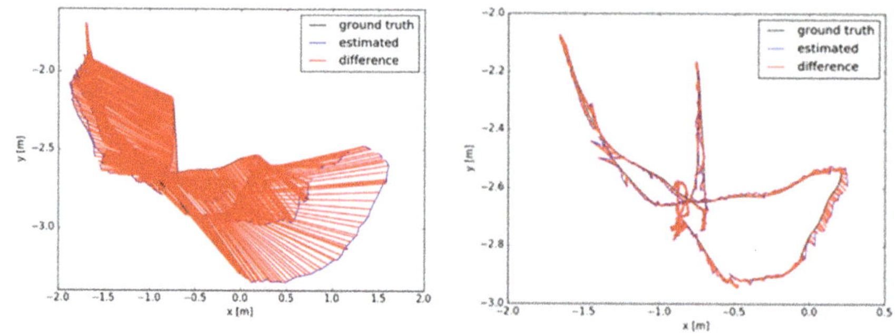

(**a**) (**b**)

Figure 14. Absolute trajectory error of dataset one: (**a**) The absolute trajectory error of dataset one using ORB-SLAM2; (**b**) The absolute trajectory error of dataset one using instance SLAM based on CO-HDC.

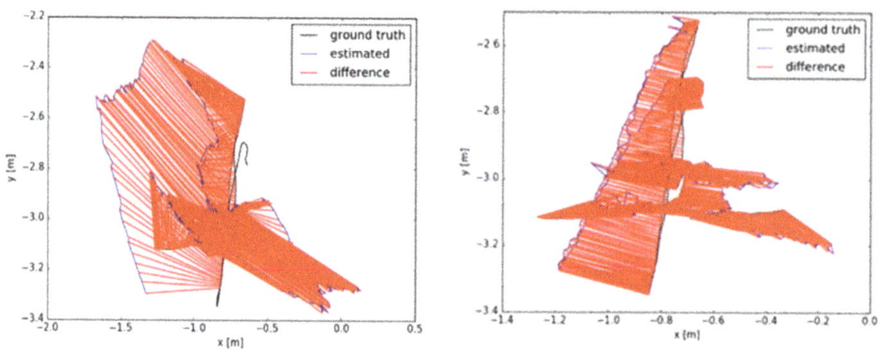

(**a**) (**b**)

Figure 15. Absolute trajectory error of dataset two: (**a**) The absolute trajectory error of dataset two using ORB-SLAM2; (**b**) The absolute trajectory error of dataset two using instance SLAM based on CO-HDC.

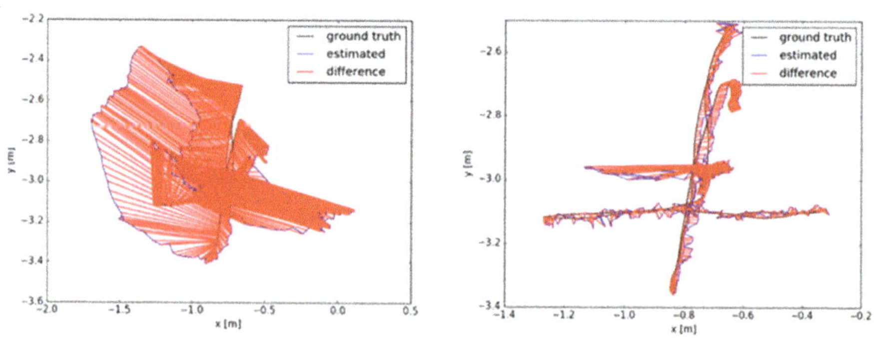

(**a**) (**b**)

Figure 16. Absolute trajectory error of dataset three: (**a**) The absolute trajectory error of dataset two using ORB-SLAM2; (**b**) The absolute trajectory error of dataset two using instance SLAM based on CO-HDC.

Table 5. Pose error representative value of dataset one.

Evaluation	Methods	Rmse (m)	Mean (m)	Median (m)	Std (m)	Min (m)	Max (m)
Absolute trajectory error	ORB-SLAM2	0.760252	0.690474	0.639742	0.318165	0.022187	1.715618
	Proposed SLAM	0.027541	0.023047	0.018764	0.015077	0.001505	0.141699
Relative pose error	ORB-SLAM2	1.134662	0.922296	0.845839	0.660930	0.000000	3.203089
	Proposed SLAM	0.038877	0.033508	0.030002	0.019715	0.000000	0.186828

Table 6. Pose error representative value of dataset two.

Evaluation	Methods	Rmse (m)	Mean (m)	Median (m)	Std (m)	Min (m)	Max (m)
Absolute trajectory error	ORB-SLAM2	0.638354	0.560560	0.635890	0.305399	0.050749	1.246406
	Proposed SLAM	0.209539	0.195746	0.203446	0.074766	0.029710	0.364841
Relative pose error	ORB-SLAM2	0.957366	0.763961	0.734479	0.526331	0.000000	2.128197
	Proposed SLAM	0.326175	0.240677	0.103095	0.220147	0.000000	0.584625

Table 7. Pose error representative value of dataset three.

Evaluation	Methods	Rmse (m)	Mean (m)	Median (m)	Std (m)	Min (m)	Max (m)
Absolute trajectory error	ORB-SLAM2	0.597385	0.503305	0.461168	0.321796	0.033516	1.243515
	Proposed SLAM	0.071849	0.195746	0.030831	057592	0.003704	0.428562
Relative pose error	ORB-SLAM2	0.927718	0.763961	0.734479	0.526331	0.000000	2.128197
	Proposed SLAM	0.117698	0.052240	0.023353	0.105470	0.000000	0.606306

The platform of this experiment is a personal laptop configured as CPU I7 7700HQ, GPU 1050TI and 16G memory. The evaluation tool is used to compare the errors of the two systems running the above two datasets.

Through the above experiments, comparing ORB-SLAM2 and instance SLAM based on CO-HDC, we can see that the performance of instance SLAM based on CO-HDC is better than traditional SLAM.

4. Discussion

Visual SLAM based on instance segmentation has been widely used due to its high accuracy in dynamic environments. At present, eliminating dynamic feature points to improve the accuracy of visual SLAM is a widely recognized method in academic circles [57,58]. Alejo Concha et al. use this technology to prolong the time of world-locked mobile AR experiences, letting users have a more satisfying experience [59]. Fessl [60] and Sanchez-Lopez [61] et al. have applied them in the field of aircraft. In addition, it has been widely used in location-aware communication [62], medical [6], 3D printing [5] and other fields [63]. However, this method has two major problems: the accuracy of dynamic point elimination is not high, and the elimination speed is slow. To solve these two problems, we propose a CO-HDC instance segmentation model, which consists of a CQE contour enhancement algorithm and a BAS-DP lightweight contour extraction algorithm.

Firstly, the main reason for the low accuracy of dynamic feature point elimination is the low accuracy of object contour segmentation, which makes it difficult to distinguish whether the feature points at the object contour are dynamic feature points or static feature points. To solve this problem, we propose a CQE contour enhancement algorithm. By evaluating the contour of the object, the optimal contour is selected as the output. In order to solve this problem, Chang et al. introduced the optical flow method to detect moving objects [64]. The optical flow method obtains the motion information of the object by calculating the change of pixels between adjacent frames. This method can not only work when the camera is in motion but also get the three-dimensional structure of the object. However, the optical flow method is too sensitive to the change of illumination intensity, and it needs to assume that the brightness of object pixels is constant. This is difficult to achieve in most cases. In addition, the optical flow method is difficult to recognize fast-moving objects. Therefore, in contrast, the method proposed in this paper has stronger robustness and can better adapt to a complex environment.

Secondly, in order to match the mapping speed of visual SLAM based on instance segmentation, instance segmentation needs to have a faster segmentation speed. The BAS-DP lightweight contour extraction algorithm proposed in this paper can effectively reduce the amount of calculation while ensuring accuracy by using the most similar polygon contour. In order to solve the same problem, Xiong et al. optimized the backbone network and accelerated the segmentation speed by designing a semantic segmentation head based on deformable convolution [65]. However, this method depends on the selection of keyframes in the video sequence. Therefore, compared with it, the method proposed in this paper is more practical.

5. Conclusions

This paper has presented a pose estimation optimized visual SLAM algorithm based on the CO-HDC instance segmentation network for dynamic scenes. CO-HDC instance segmentation includes the CQE contour enhancement algorithm and the BAS-DP lightweight contour extraction algorithm. The CQE contour enhancement algorithm improves the segmentation accuracy at the contour of dynamic objects. The problem of excessive calculation of instance segmentation is overcome by the BAS-DP algorithm. As the test results show, the proposed algorithm can reduce pose estimation errors and map relative drifts under dynamic environments compared to ORB-SLAM2.

In the future, visual SLAM based on instance segmentation has broad development space, including the driverless field, 3D printing industry, location-aware communication, aircraft and other fields. Instance segmentation can not only improve the accuracy of visual SLAM but also provide rich object information in the scene. In future work, the proposed algorithm would be further implemented and demonstrated in the embedded system to fit more robots under complex environments.

Author Contributions: Conceptualization, J.C. and F.X.; methodology, J.C., F.X. and L.H.; software, F.X. and X.L.; validation, J.C., F.X., J.Y. and J.S.; formal analysis, F.X.; investigation, X.L. and J.S.; resources, L.H. and J.Y.; data curation, X.L.; writing—original draft preparation, J.C. and F.X.; writing—review and editing, J.C., F.X. and X.L.; visualization, J.S.; supervision, J.Y.; project administration, F.X.; funding acquisition, J.Y. and J.S. All authors have read and agreed to the published version of the manuscript.

Funding: This work was partially supported by the National Natural Science Foundation of China (Grant No. 41974033), the Scientific and Technological Achievements Program of Jiangsu Province (BA2020004), and 2020 Industrial Transformation and Upgrading Project of Industry and Information Technology Department of Jiangsu Province (JITC-2000AX0676-71), Postgraduate Research & Practice Innovation Program of Jiangsu Province.

Data Availability Statement: Publicly available datasets were analyzed in this study, 2 April 2022. The dataset can be found here: https://vision.in.tum.de/data/datasets/rgbd-dataset/download (accessed on 12 March 2022).

Conflicts of Interest: The authors declare no conflict of interest.

References

1. García-Fernández, Á.F.; Hostettler, R.; Särkkä, S. Rao-Blackwellized Posterior Linearization Blackward SLAM. *IEEE Trans. Veh. Technol.* **2019**, *68*, 4734–4747. [CrossRef]
2. Evers, C.; Naylor, P.A. Optimized Self-localization for SLAM in Dynamic Scenes Using Probability Hypothesis Density filters. *IEEE Trans. Signal Process.* **2018**, *66*, 863–878. [CrossRef]
3. Lee, J.; Hwang, S.; Kim, W.J.; Lee, S. SAM-Net: LiDAR Depth Inpainting for 3D Static Map Generation. *IEEE Trans. Intell. Transp. Syst.* **2021**, *22*, 1–16. [CrossRef]
4. Cattaneo, D.; Vaghi, M.; Valada, A. LCDNET: Deep Loop Closure Detection and Point Cloud Registration for Lidar SLAM. *IEEE Trans. Robot.* **2022**, *38*, 1–20. [CrossRef]
5. Li, J.; Aubin-Fournier, P.L.; Skonieczny, K. SLAAM: Simultaneous Localization and Additive Manufacturing. *IEEE Trans. Robot.* **2021**, *37*, 334–349. [CrossRef]

6. Hussain, A.; Memon, A.R.; Wang, H.; Wang, Y.; Miao, Y.; Zhang, X. S-VIT: Stereo Visual-Inertial Tracking of Lower Limb for Physio therapy Rehabilitation in Context of Comprehensive Evaluation of SLAM Systems. *IEEE Trans. Autom. Sci. Eng.* **2021**, *19*, 1550–1562. [CrossRef]
7. Mur-Artal, R.; Tardós, J.D. ORB-SLAM2: An Open-Source SLAM System for Monocular, Stereo, and RGB-D Cameras. *IEEE Trans. Robot.* **2017**, *33*, 1255–1262. [CrossRef]
8. Shi, Q.; Zhao, S.; Cui, X.; Lu, M.; Jia, M. Anchor self-localization algorithm based on UWB ranging and inertial measurements. *Tsinghua Sci. Technol.* **2019**, *24*, 728–737. [CrossRef]
9. Yang, S.; Scherer, S. Monocular Object and Plane SLAM in Structured Environments. *IEEE Robot. Autom. Lett.* **2019**, *4*, 3145–3152. [CrossRef]
10. Han, F.; Wang, H.; Huang, G.; Zhang, H. Sequence-based sparse optimization methods for long-term loop closure detection in visual SLAM. *Auton. Robot.* **2018**, *42*, 1323–1335. [CrossRef]
11. Yuan, J.; Zhu, W.; Dong, X.; Sun, F.; Zhang, X.; Sun, Q.; Huang, Y. A Novel Approach to Inage-Sequence-Based Mobile Robot Place Recognition. *IEEE Trans. Syst.* **2021**, *51*, 5377–5391.
12. Ntalampiras, S. Moving Vehicle Classification Using Wireless Acoustic Sensor Networks. *IEEE Trans. Eng. Top. Comput. Intell.* **2018**, *2*, 129–138. [CrossRef]
13. Zhu, J.; Jia, Y.; Li, M.; Shen, W. A New System to Construct Dense Map with Pyramid Stereo Matching Network and ORB-SLAM2. In Proceedings of the International Conference on Computer and Communications, Chengdu, China, 10–13 December 2021; pp. 430–435.
14. Fan, T.; Wang, H.; Rubenstein, M.; Murphey, T. CPL-SLAM: Efficient and Certifiably Correct Planar Graph-Based SLAM Using the Complex Number Representation. *IEEE Trans. Robot.* **2020**, *36*, 1719–1737. [CrossRef]
15. Han, L.; Xu, L.; Bobkov, D.; Steinbach, E.; Fang, L. Real-Time Global Registration for Globally Consistent RGB-D SLAM. *IEEE Trans. Robot.* **2019**, *35*, 498–508. [CrossRef]
16. Gomez-Ojeda, R.; Moreno, F.A.; Zuniga-Noël, D.; Scaramuzza, D.; Gonzalez-Jimenez, J. PL-SLAM: A Stereo SLAM System through the Combination of Points and Line Segments. *IEEE Trans. Robot.* **2019**, *35*, 734–746. [CrossRef]
17. Gao, H.; Zhang, X.; Yuan, J.; Song, J.; Fang, Y. A Novel Global Localization Approach Based on Structual Unit Encoding and Multiple Hypothesis Tracking. *IEEE Trans. Instrum. Meas.* **2019**, *68*, 4427–4442. [CrossRef]
18. Gao, H.; Zhang, X.; Wen, J.; Yuan, J.; Fang, Y. Autonomous Indoor Exploration Via Polygon Map Construction and Graph-Based SLAM Using Directional Endpoint Features. *IEEE Trans. Autom. Sci. Eng.* **2019**, *16*, 1531–1542. [CrossRef]
19. Xie, Y.; Zhang, Y.; Chen, L.; Cheng, H.; Tu, W.; Cao, D.; Li, Q. RDC-SLAM: A Real-Time Distributed Cooperative SLAM System Based on 3D LiDAR. *IEEE Trans. Intell. Transp. Syst.* **2021**, *22*, 1–10. [CrossRef]
20. Guo, C.X.; Sartipi, K.; DuToit, R.C.; Georgiou, G.A.; Li, R.; O'Leary, J.; Nerurkar, E.D.; Hesch, J.A.; Roumeliotis, S.I. Resource-Aware Large- Scale Cooperative Three -Dimensional Mapping Using Multiple Mobile Devices. *IEEE Trans. Robot.* **2018**, *34*, 1349–1369. [CrossRef]
21. Zhang, Y.; Chen, L.; Zhe, X.; Tian, W. Three-Dimensional Cooperative Mapping for Connected and Automated Vehicles. *IEEE Trans. Ind. Elecronice* **2020**, *67*, 6649–6657. [CrossRef]
22. Chu, X.; Lu, Z.; Gesbert, D.; Wang, L.; Wen, X. Vehicle Localization via Cooperative Channel Mapping. *IEEE Trans. Veh. Technol.* **2021**, *70*, 5719–5733. [CrossRef]
23. Yassin, A.; Nasser, Y.; Al-Dubai, A.Y.; Awad, M. MOSAIC: Simultaneous Localization and Environment Mapping Using nnWave Without A-Priori Knowledge. *IEEE Access* **2018**, *6*, 68932–68947. [CrossRef]
24. De Lima, C.; Belot, D.; Berkvens, R.; Bourdoux, A.; Dardari, D.; Guillaud, M.; Isomursu, M.; Lohan, E.-S.; Miao, Y.; Barreto, A.N.; et al. Convergent Communication, Sensing and Localization in 6G Systems: An Overview of Technologies, Opportunities Challenges. *IEEE Access* **2021**, *9*, 26902–26925. [CrossRef]
25. Aladsani, M.; Alkhateeb, A.; Trichopoulos, G.C. Leveraging mmWAVE Imaging and Communications for Simultaneous Localization and Mapping. In Proceedings of the 2019 IEEE International Conference on Acoustics, Speech and Signal Processing (ICASSP), Brighton, UK, 12–17 May 2019; pp. 4539–4543.
26. Fascista, A.; Coluccia, A.; Wymeersch, H.; Seco-Granados, G. Downlink Single-Snapshot Localization and Mapping with a Single-Antenna Receiver. *IEEE Trans. Wirel. Commun.* **2021**, *20*, 4672–4684. [CrossRef]
27. Saputra, M.R.U.; Lu, C.X.; de Gusmao, P.P.B.; Wang, B.; Markham, A.; Trigoni, N. Graph-Based Thermal-Inertial SLAM With Probabilistic Neural Networks. *IEEE Trans. Robot.* **2021**, *37*, 1–19. [CrossRef]
28. Poulose, A.; Han, D.S. Hybrid Indoor Localization Using IMU Sensors and Smartphone Camera. *Sensors* **2019**, *19*, 5084. [CrossRef] [PubMed]
29. Jung, J.H.; Choe, Y.; Park, C.G. Photometric Visual-Inertial Navigation with Uncertainty-Aware Ensembles. *IEEE Trans. Robot.* **2021**, *37*, 1–14. [CrossRef]
30. Campos, C.; Elvira, R.; Rodríguez, J.J.G.; Montiel, J.M.; Tardós, J.D. ORB-SLAM3: An Accurate Open-Source Library for Visual, Visual-Inertial, and Multimap SLAM. *IEEE Trans. Robot.* **2021**, *37*, 1874–1890. [CrossRef]
31. Muñoz-Salinas, R.; Medina-Carnicer, R. UcoSLAM: Simultaneous localization and mapping by fusion of keypoints and squared planar markers. *Pattern Recognit.* **2020**, *101*, 107193. [CrossRef]
32. Ding, X.; Wang, Y.; Xiong, R.; Li, D.; Tang, L.; Yin, H.; Zhao, L. Persistent Stereo Visual Localization on Cross-Modal Invariant Map. *IEEE Trans. Intell. Transp. Syst.* **2020**, *21*, 4646–4658. [CrossRef]

33. Chou, C.C.; Chou, C.F. Efficient and Accurate Tightly-Coupled Visual-Lidar SLAM. *IEEE Trans. Intell. Transp. Syst.* **2021**, *22*, 1–15. [CrossRef]
34. Wu, Y.; Li, Y.; Li, W.; Li, H.; Lu, R. Robust Lidar-Based Localization Scheme for Unmanned Ground Vehicle via Multisensor Fusion. *IEEE Trans. Neural Netw. Learn. Syst.* **2021**, *32*, 5633–5643. [CrossRef] [PubMed]
35. Li, J.; Hu, S.; Li, Q.; Chen, J.; Leung, V.C.; Song, H. Global Visual and Semantic Observations for Outdoor Robot Localization. *IEEE Trans. Netw. Sci. Eng.* **2021**, *8*, 2909–2921. [CrossRef]
36. Han, S.; Xi, Z. Dynamic scene semantics SLAM based on semantic segmentation. *IEEE Access* **2020**, *8*, 43563–43570. [CrossRef]
37. Li, F.; Chen, W.; Xu, W.; Huang, L.; Li, D.; Cai, S.; Yang, M.; Xiong, X.; Liu, Y.; Li, W. A mobile robot visual SLAM system with enhanced semantics segmentation. *IEEE Access* **2020**, *8*, 25442–25458. [CrossRef]
38. Zhang, Z.; Zhang, J.; Tang, Q. Mask R-CNN based on semantic RGB-D SLAM for dynamics scenes. In Proceedings of the IEEE International Conference on Advanced Intelligent Mechatronics, Hong Kong, China, 8–12 July 2019; pp. 1151–1156.
39. Ai, Y.; Rui, T.; Lud, M.; Fu, F.; Liu, S.; Wang, S. DDL-SLAM: A robust RGB-D SLAM in dynamic environments combined with deep learning. *IEEE Access* **2020**, *8*, 162335–162342. [CrossRef]
40. Javed, Z.; Kim, G.-W. A comparative study of recent real time semantic segmentation algorithms for visual semantic SLAM. In Proceedings of the IEEE International Conference on Big Date and Smart Computing, Online, 10–13 December; pp. 474–476.
41. Qian, H.; Ding, P. An improved ORB-SLAM2 in dynamic scene with instance segmentation. In Proceedings of the International Workshop on Research, Education and Development on Unmanned Aerial Systems, Cranfield, UK, 25–27 November 2019; pp. 185–191.
42. Bolya, D.; Zhou, C.; Xiao, F.; Lee, Y.J. YOLACT Real-time Instance Segmentation. In Proceedings of the IEEE International Conference on Computer Vision, Seoul, Korea, 27 October–2 November 2019; pp. 9157–9166.
43. Bista, S.R.; Hall, D.; Talbot, B.; Zhang, H.; Dayoub, F.; Sünderhauf, N. Evaluating the impact of semantic segmentation and pose estimation on dense semantic SLAM. In Proceedings of the IEEE International Conference on Intelligent Robots and Systems, Prague, Czech Republic, 27 September–1 October 2021; pp. 5328–5335.
44. Yu, C.; Liu, Z.; Liu, X.J.; Xie, F.; Yang, Y.; Wei, Q.; Fei, Q. DS-SLAM: A Semantic Visual SLAM towards Dynamic Environments. In Proceedings of the IEEE International Conference on Intelligent Robots and Systems (IROS), Madrid, Spain, 1–5 October 2018; pp. 1168–1174.
45. Wu, Y.; Luo, L.; Yin, S.; Yu, M.; Qiao, F.; Huang, H.; Shi, X.; Wei, Q.; Liu, X. An FPGA Based Energy Efficient DS-SLAM Accelerator for Mobile Robots in Dynamic Environment. *Appl. Sci.* **2021**, *11*, 1828. [CrossRef]
46. Bescos, B.; Fácil, J.M.; Civera, J.; Neira, J. DynaSLAM: Tracking, Mapping, and Inpainting in Dynamic Scenes. *IEEE Robot. Autom. Lett.* **2018**, *3*, 4076–4083. [CrossRef]
47. Bescos, B.; Campos, C.; Tardós, J.D.; Neira, J. DynaSLAM II: Tightly-Coupled Multi-Object Tracking and SLAM. *IEEE Robot. Autom. Lett.* **2021**, *6*, 5191–5198. [CrossRef]
48. Endo, Y.; Sato, K.; Yamashita, A.; Matsubayashi, K. Indoor Positioning and Obstacle Detection for Visually Impaired Navigetion System based on LSD-SLAM. In Proceedings of the International Conference on Biometrics and Kansei Engineering, Kyoto, Japan, 15–17 September 2017; pp. 158–162.
49. Cui, L.; Ma, C. SOF-SLAM: A Semantic Visual SLAM for Dynamic Environments. *IEEE Access* **2019**, *7*, 166528–166539. [CrossRef]
50. Whelan, T.; Leutenegger, S.; Salas-Moreno, R.; Ben, G.; Davison, A. ElasticFusion: Dense SLAM without A Pose Graph. In Proceedings of the Conference on Robotics—Science and Systems, Rome, Italy, 13–17 July; pp. 1–23.
51. Ran, T.; Yuan, L.; Zhang, J.; Tang, D.; He, L. RS-SLAM: A robust semantic SLAM in dynamic environment based on RGB-D sensor. *IEEE Sens. J.* **2021**, *21*, 20657–20664. [CrossRef]
52. Ballester, I.; Fontan, A.; Civera, J.; Strobl, K.H.; Triebel, R. DOT: Dynamic object tracking for visual SLAM. In Proceedings of the IEEE International Conference on Robotics and Automation, Xi'an, China, 31 May–4 June 2021; pp. 11705–11711.
53. Mingachev, E.; Lavrenov, R.; Tsoy, T.; Matsuno, F.; Svinin, M.; Suthakorn, J.; Magid, E. Comparison of ROS-Based Monocular Visual SLAM Methods: DSO, LDSO, ORB-SLAM2 and DynaSLAM. In Proceedings of the Interactive Collaborative Robotics, St Petersburg, Russia, 7–9 October 2020; pp. 222–233.
54. Xia, L.; Cui, J.; Shen, R.; Xu, X.; Gao, Y.; Li, X. A survey of image semantics-based visual simultaneous localization and mapping: Application-oriented solutions to autonomous navigation of mobile robots. *Int. J. Adv. Robot. Syst.* **2020**, *17*, 1–17. [CrossRef]
55. Sun, Y.; Liu, M.; Meng, M.Q.-H. Improving RGB-D SLAM in dynamic environments: A motion removal approach. *Robot. Auton. Syst.* **2017**, *89*, 110–122. [CrossRef]
56. Xie, X.; Li, C.; Yang, X.; Xi, J.; Chen, T. Dynamic Receptive Field-Based Object Detection in Aerial Imaging. *Acta Opt. Sin.* **2020**, *40*, 0415001.
57. Dong, X.; Ouyang, Z.; Guo, Z.; Niu, J. Polarmask-tracker: Lightweight multi-object tracking and segmentation model for edge device. In Proceedings of the IEEE International Conference on Parallel, New York, NY, USA, 30 September–3 October 2021; pp. 689–696.
58. Fu, J.; Liu, J.; Tian, H.; Li, Y.; Bao, Y.; Fang, Z.; Lu, H. Dual attention network for scene segmentation. In Proceedings of the IEEE Conference on Computer Vision and Pattern Recognition, Long Beach, CA, USA, 15 June 2019; pp. 3141–3149.
59. Concha, A.; Burri, M.; Briales, J.; Forster, C.; Oth, L. Instant Visual Odometry Initialization for Mobile AR. *IEEE Trans. Vis. Comput. Graph.* **2021**, *27*, 4226–4235. [CrossRef] [PubMed]

60. Faessler, M.; Fontana, F.; Forster, C.; Scaramuzza, D. Automatic Re-Initialization and Failure Recovery for Aggressive Flight with a Monocular Vision-Based Quadrotor. In Proceedings of the 2015 IEEE International Conference on Robotics and Automation (ICRA), Seattle, WA, USA, 26–30 May 2015; pp. 1722–1729.
61. Sanchez-Lopez, J.L.; Arellano-Quintana, V.; Tognon, M.; Campoy, P.; Franchi, A. Visual Marker based Multi-Sensor Fusion State Estimaion. In Proceedings of the 2017 IFAC, Toulouse, France, 9–14 July 2017; pp. 16003–16008.
62. Leitinger, E.; Meyer, F.; Hlawatsch, F.; Witrisal, K.; Tufvesson, F.; Win, M.Z. A Belief Propagation Algorithm for Multipath-Based SLAM. *IEEE Trans. Wirel. Commun.* **2019**, *18*, 5613–5629. [CrossRef]
63. Xiang, Z.; Bao, A.; Su, J. Hybrid bird's-eye edge based semantic visual SLAM for automated valet parking. In Proceedings of the IEEE International Conference on Robotics and Automation, Xi'an, China, 31 May–4 June 2021; pp. 11546–11552.
64. Chang, J.; Dong, N.; Li, D. A real-time dynamics object segmentation framework for SLAM system in dynamic scenes. *IEEE Trans. Instrum. Meas.* **2021**, *70*, 2513708–2513716. [CrossRef]
65. Xiong, Y.; Liao, R.; Zhao, H.; Hu, R.; Bai, M.; Yumer, E.; Urtasun, R. UPSNet: A unified panoptic segmentation network. In Proceedings of the IEEE Conference on Computer Vision and Pattern Recognition, Long Beach, CA, USA, 15 June 2019; pp. 8810–8818.

remote sensing

MDPI

Article

LiDAR-Inertial-GNSS Fusion Positioning System in Urban Environment: Local Accurate Registration and Global Drift-Free

Xuan He [1,2], Shuguo Pan [1,2,*], Wang Gao [1,2] and Xinyu Lu [1,2]

1 School of Instrument Science and Engineering, Southeast University, Nanjing 210096, China; hexuan@seu.edu.cn (X.H.); gaow@seu.edu.cn (W.G.); 220213597@seu.edu.cn (X.L.)
2 Key Laboratory of Micro-Inertial Instrument and Advanced Navigation Technology, Southeast University, Nanjing 210096, China
* Correspondence: psg@seu.edu.cn

Abstract: Aiming at the insufficient accuracy and accumulated error of the point cloud registration of LiDAR-inertial odometry (LIO) in an urban environment, we propose a LiDAR-inertial-GNSS fusion positioning algorithm based on voxelized accurate registration. Firstly, a voxelized point cloud downsampling method based on curvature segmentation is proposed. Rough classification is carried out by the curvature threshold, and the voxelized point cloud downsampling is performed using HashMap instead of a random sample consensus algorithm. Secondly, a point cloud registration model based on the nearest neighbors of the point and neighborhood point sets is constructed. Furthermore, an iterative termination threshold is set to reduce the probability of the local optimal solution. The registration time of a single frame point cloud is increased by an order of magnitude. Finally, we propose a LIO-GNSS fusion positioning model based on graph optimization that uses GNSS observations weighted by confidence to globally correct local drift. The experimental results show that the average root mean square error of the absolute trajectory error of our algorithm is 1.58m on average in a large-scale outdoor environment, which is approximately 83.5% higher than that of similar algorithms. It is fully proved that our algorithm can realize a more continuous and accurate position and attitude estimation and map reconstruction in urban environments.

Keywords: LiDAR-inertial odometry; point cloud registration; multi-sensor fusion

Citation: He, X.; Pan, S.; Gao, W.; Lu, X. LiDAR-Inertial-GNSS Fusion Positioning System in Urban Environment: Local Accurate Registration and Global Drift-Free. *Remote Sens.* **2022**, *14*, 2104. https://doi.org/10.3390/rs14092104

Academic Editor: Francesco Nex

Received: 14 March 2022
Accepted: 25 April 2022
Published: 27 April 2022

1. Introduction

For any autonomous robot system, such as unmanned aerial vehicles and autonomous vehicles, the accurate and robust localization of a mobile carrier is one of the fundamental technologies [1]. Traditionally, the integrated navigation and positioning technology based on the global navigation satellite system (GNSS) and inertial navigation system (INS) is usually regarded as a reliable method to achieve high-accuracy positioning [2]. However, in complex urban environments, there are a large number of GNSS multipath or rejection areas due to the blockage of GNSS signals by urban objects such as tall buildings, tunnels and street trees. As a result, the integrated positioning method based on GNSS/INS is not effective in achieving a continuous and robust positioning of targets in large urban environments. In summary, there is an urgent need to upgrade and expand the traditional positioning techniques by introducing heterogeneous and complementary measurement information from other sensors.

In recent years, the multi-sensor fusion positioning technology based on simultaneous localization and mapping (SLAM) has received extensive attention from related enterprises and researchers [3]. It can not only make use of the excellent characteristics of cameras, LiDAR and other sensors, including the independence from environmental occlusion and signal refraction in complex areas, but can also effectively make up for the signal lock-out defect of GNSS signals in the parking lot or tunnel area. Moreover, incremental map reconstruction can be achieved by sensing the external environment. Depending on the primary sensor, SLAM-based multi-sensor fusion positioning solutions can be divided into vision-based SLAM and LiDAR-based SLAM [4]. Due to the superiority of the sensors, the solution of LiDAR-based SLAM allows for a higher frequency and more accurate acquisition of spatial fingerprint information, thus achieving a more accurate positioning than vision-based SLAM [5–7]. Secondly, analyzed at the algorithm level, LiDAR odometry is more lightweight in processing environmental features than visual odometry and more suitable for vehicle-mounted platforms with limited computational resources [8,9]. Therefore, the LiDAR-inertial odometry (LIO)-based SLAM scheme is widely used to obtain 3D geographic information of a complex environment, as well as carrier positioning and map reconstruction.

Throughout the development of the LiDAR-based SLAM, it can be seen that the registration of the point cloud of LiDAR is a key step in the pose estimation of a mobile carrier. It strictly affects the pose estimation and the map reconstruction results. The commonly used point cloud registration methods include normal distribution transform (NDT) [10], iterative closest point (ICP) [11], generalized iterative closest point (GICP) [12] and other improved algorithms [13–16]. The core of NDT algorithm is used to take the probability density function of the source point cloud and the target point cloud as the objective function; then, it uses a nonlinear optimization method to minimize the probability density between them to obtain the optimal solution. Andreasson et al. [17] avoids an explicit nearest neighbor search by establishing segmented continuous and differentiable probability distributions, and the registration speed is effectively improved. Although the real-time performance is better, the covariance matrix needs to be constructed at multiple points, which has a low robustness in the sparse area of the point cloud. Caballero et al. [18] proposed an improved NDT algorithm that was used to model the alignment problem as a distance field. The optimization equation is constructed by using the distance between the feature points of the current frame and the prior map, which improves the speed by an order of magnitude. However, the robustness of the localization algorithm is not guaranteed for unknown sections where the priori map is missing or unreliable [19].

As another method of point cloud registration, the ICP algorithm has a higher positioning accuracy than NDT, but it needs to search for the nearest neighbor again and obtain the transformation matrix in each iteration process, so the calculation efficiency needs to be improved. Koide et al. [20] proposed a generalized iterative nearest point algorithm that used a Gaussian probability model to fit the distribution of the point cloud to reduce the computational complexity. However, its accuracy is still limited by the maximum number of iterations. In addition, the algorithm is heavily influenced by the observation noise and the accuracy of the initial positional transformation matrix, and there is a risk of the algorithm falling into local minima. In order to break out of the logical limitation of being limited to local optimal solutions, Yang et al. [21] proposed Go-ICP, a branch-and-bound scheme to impose domain restrictions on the objective function of rigid alignment. This processing reduced the abnormal influence of the local minimum, and made the registration result of the point cloud approach to the global optimal solution. In 2021, Pan et al. [22] proposed MULLS-ICP, which uses an improved ICP algorithm based on double-threshold filtering and multi-scale linear least squares to realize the registration between the current frame and local sub-map, but the high computational cost of multiple filtering is difficult to adapt to the vehicle platform with limited computational resources. To sum up, on the basis of reducing the calculation cost, a high-precision real-time point cloud registration algorithm suitable for a vehicle platform still needs to be investigated.

In addition, as a local sensor integrator, the LIO has a cumulative offset between its local map and the global map when it performs a positional estimation of the current frame, which largely limits the positioning accuracy of the LIO position building scheme in large outdoor environments. Fortunately, the global observation information from GNSS can provide a credible global constraint correction for LIO [23]. Conversely, LIO systems can also compensate for the limitations of GNSS in terms of continuous precise positioning due to multipath effects and non-line-of-sight (NLOS) problems. Therefore, LIO-GNSS fusion positioning technology provides a feasible technical scheme for realizing globally weak drift and locally accurate positioning and mapping targets.

The mainstream LIO-GNSS fusion algorithms can be divided into two categories, filter-based methods and optimization-based methods, based on the method of sensor measurement data fusion. Li et al. [24] used the filter-based method as the integration strategy. They use the extended Kalman filter to realize LIO-GNSS tight coupling, but did not set up an anomaly detection mechanism, so it was prone to the dispersion of the positional estimates in GNSS multipath regions or point cloud degradation regions. To resolve this issue, Li et al. [25] uses an edge fault-tolerant mechanism to improve the robustness of the algorithm in case of single-sensor failure. However, it weakens the linearization error at the cost of increasing the amount of computation, which is contrary to the lightweight principle of large outdoor scenes. As another fusion method, the optimization-based method uses multiple iterations to approach the optimal solution, which can effectively handle such non-linear heterogeneous data fusion problems. Soloviev et al. [26] proposed an optimization-based LIO-GNSS scheme, but only the horizontal components of GNSS measurements were used to optimize the LIO pose estimation results, with low utilization of the measurement information. Shan et al. [27] puts forward an optimization framework that introduces 3D GNSS measurement factors to assist LIO, but the measurement information of a single key frame is redundant, and the reliability of GNSS factors added when driving to the GNSS multipath area is poor. Sun et al. [28] proposed a GNSS corner factor to constrain the local pose, but it does not consider the shortage of corners on straight road sections, so its application in a large-scale complex outdoor environment is limited.

From the above analysis, it can be seen that the research points of the LIO-GNSS fusion scheme are as follows:

1. Realizing real-time and high-precision point cloud alignment based on compressed computational costs.
2. On the basis of making full use of GNSS measurement information, global cumulative error correction of LIO is carried out by GNSS.

To address the above issues, in this contribution, we propose a LiDAR-inertial-GNSS fusion positioning system based on voxelized accurate registration. Firstly, a voxelized point cloud downsampling method based on curvature segmentation is proposed. Rough classification is carried out by a curvature threshold, and the voxelized point cloud downsampling is performed using HashMap instead of the random sample consensus algorithm. Therefore, the spatial distribution attributes of the source point cloud are retained to a greater extent. Secondly, a point cloud registration model based on the nearest neighbors of the point and neighborhood point sets is constructed. Thirdly, an optimization-based method is used to build a higher-order Markov model based on sliding windows, and a GNSS factor and loop factor are introduced into the factor graph to constrain LIO globally. Finally, on this basis, a GNSS residual construction method based on the GNSS reliability weight is proposed to make full use of GNSS measurement information. Therefore, the goal of positioning and mapping with a light weight, high precision and high applicability in a complex urban environment can be achieved.

2. System Overview

The proposed algorithm framework is shown in Figure 1. The main functions of each module are as follows.

The front-end of the system is mainly used to preprocess IMU observations and LiDAR original point cloud sequences, and to optimize the generation of local maps by inter-frame matching. The LiDAR raw point cloud sequence is clustered and segmented by a breadth-first-search combined with the Euclidean angle threshold, and then edge and plane feature point clouds are extracted. These two types of feature clouds are downsampled for point cloud alignment, and the local inter-frame matching is optimized using the IMU pre-integration as the initial pose estimate. Finally, the LIO local pose estimates are used to pre-process the GNSS global observations, including the temporal interpolation alignment of GNSS and LIO local observations and coordinate system alignment, so as to achieve the space–time synchronization among sensors.

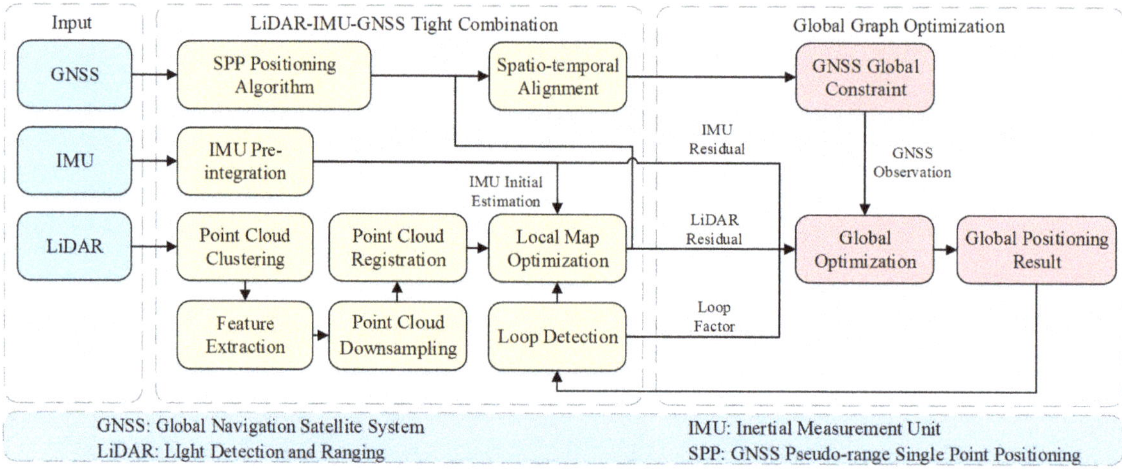

Figure 1. General framework of the algorithm. LIO's pose estimation results are used as local optimization factors, and GNSS pseudo-range single point positioning (SPP) results are used as global optimization factors for global constraint.

The back-end mainly uses the residuals of pose estimates of each sensor to optimize the map. The residual factors from local sensors include the IMU pre-integration and LiDAR observation residual, whereas the global residual factors include the GNSS observation residual and loop residual. It should be noted that the global residual factors are added only when their existence is detected, and, when there is no global residual factor, the system only performs local position, such as when the carrier is travelling in a flat and straight tunnel environment. When global corrections are available, the obtained global positioning results are used to update the local pose estimates in the sliding window to obtain the best pose estimates with local accurate registration and global drift-free.

3. Point Cloud Voxelization Downsampling and Alignment

The accuracy of the registration of the environmental point cloud extracted by LiDAR strictly affects the result of the subsequent local pose estimation. Therefore, the processing steps of the front-end point cloud of the system need to be described in detail. This paper mainly involves the improved point cloud downsampling method and registration method.

3.1. Voxelized Downsampling Based on Curvature Segmentation

This paper presents a voxelized downsampling method based on curvature segmentation. Given a set of raw point cloud sequences collected by LiDAR, all points in the raw point cloud sequences are traversed and coarse clustering is performed using a breadth-first algorithm. Furthermore, the geometric angle threshold based on Euclidean distance is used to finely segment the point cloud clusters with similar depth. Let the scanning center of LiDAR be O and the two adjacent edge points p_a and p_b in the point cloud cluster with depths d_a and d_b, respectively ($d_a > d_b$). Let the number of point clouds in the point cloud cluster where point p_i is located be M. Then, the roughness of point cloud p_i is:

$$c = \frac{1}{|M| \cdot ||d_i||} \left|\left| \sum_{j \neq i} (d_i - d_j) \right|\right|, \ i,j \in M \tag{1}$$

Set the roughness threshold as $\bar{c}$, then traverse M. We classify the points of $c < \bar{c}$ as the set of edge feature points, classify the points of $c > \bar{c}$ as the set of plane feature points and perform downsampling operations on them, respectively.

This method is mainly used in the feature extraction step of LiDAR odometry [4]; we extend it to the downsampling step. This means that, for any application where downsampling of point clouds is required, such as artefact inspection, the method can better restore the spatial distribution properties of point clouds by downsampling in clusters.

Next, this paper proposes a point cloud downsampling strategy based on HashMap, instead of the random sample consensus (RANSAC), so that the downsampling result of the point cloud is closer to the approximate center of gravity of voxels. Let the coordinate of a feature point in a set of point cloud sequences in the voxel space be $p(x,y,z)$. If the voxel grid size is r, the dimension of the voxel grid in the x direction is $D_x = (x_{\max} - x_{\min})/r$, and the index of p in the x direction within the voxel grid is $h_x = (x - x_{\min})/r$. The same applies to the y and z directions.

After obtaining the 3D index of feature points in the voxel space, if the random sorting strategy of [11] is adopted, the sorting complexity will be $O((m+n) * \log(m+n))$, which has a negative impact on the down-sampling time. Therefore, this paper uses the hash function to sort the index of feature points quickly and map them to N containers $(N = 80)$. The hash function is:

$$hash(h_x, h_y, h_z) = (h_x + h_y \cdot D_x + h_z \cdot D_x \cdot D_y)\ \%N \ \ \mathbb{R}^3 \to \mathbb{R} \tag{2}$$

To avoid hash conflicts, set the conflict detection conditions as follows:

$$hash(h_x, h_y, h_z) = hash(h'_x, h'_y, h'_z)\ \left(h_x \neq h'_x \middle| h_y \neq h'_y \middle| h_z \neq h'_z\right) \tag{3}$$

Once the hash conflict is detected, the index value in the current container is output and the container is emptied, and the new index value is put into the container.

To sum up, the main improvement of this section lies in extending the curvature segmentation step originally used for feature extraction to the downsampling step, and using hash mapping instead of the random sampling method for point cloud sampling. For the LiDAR odometer, using the clustering line and surface features again after the feature extraction step can improve the accuracy of downsampling single-frame or discontinuous point clouds at a weak time cost, thus providing more accurate point cloud distribution results for the pose estimation step between consecutive frames. In addition, using HashMap to downsample can further improve the sampling efficiency, and the time consumption of quadratic curvature segmentation is almost negligible. For other applications that need to downsample point clouds, the point cloud clustering method based on curvature segmentation can restore the spatial distribution of point clouds more accurately, and the benefits of this method are extensive and obvious.

The results and time consumption of the improved point cloud downsampling process are shown in Figure 2 and Table 1. Cloud number $M = 112624$, the line feature extraction threshold is 1, the surface feature extraction threshold is 0.1 and r = 0.3. It can be seen from Figure 2c that the present method has a clearer reduction in the spatial distribution of diagonal lines within a rectangular point cloud. Therefore, it can be proved that our method can retain the texture feature information of the source point cloud to a greater extent, and the accuracy and real-time performance of the downsampling results can be improved.

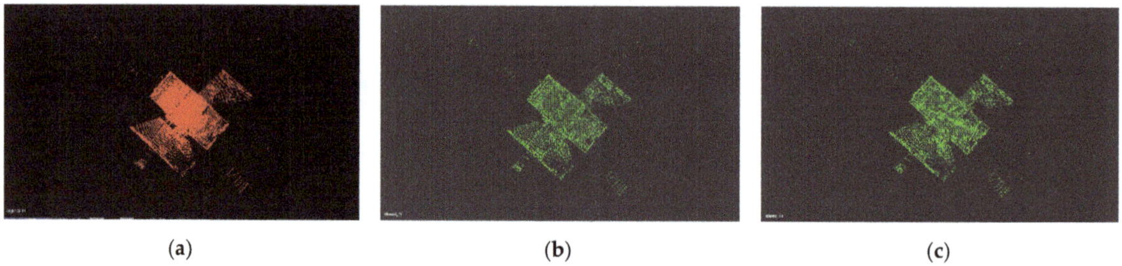

Figure 2. Comparison of point cloud downsampling results. (**a**) Source point cloud; (**b**) downsampling results before improvement; (**c**) downsampling results after improvement.

Table 1. Comparison of the number of point clouds and time consumption after downsampling.

Point Cloud Type	Number of Point Clouds	Time Consumption
Source point cloud	112,624	-
Before improvement	4100	0.004 s
After improvement	5929	0.002 s

3.2. Voxelized Point Cloud Registration

The purpose of point cloud registration is to update the rigid frame transformation of a moving carrier by comparing two consecutive frames of point clouds or similar point clouds detected by a loopback to solve for the carrier's pose. Traditional LIO usually uses ICP to realize the precise registration of point clouds. The ICP can be briefly described as follows: given a set of source point cloud $A = \{a_1, a_2, \ldots, a_n\}$ and target point cloud $B = \{b_1, b_2, \ldots, b_n\}$, the nearest neighbor search of KDTree is used to obtain the inter-frame pose transformation relationship $b_i = Ta_i$, and the optimal solution is achieved through multiple iterations. However, an unreasonable initial position selection will make ICP fall into the misunderstanding of the local optimal solution, and the calculation resource consumption of the single-point nearest neighbor search is large. In view of the defects of the ICP algorithm, this paper utilizes a method based on the distribution of feature points in voxels, as shown in Figure 3.

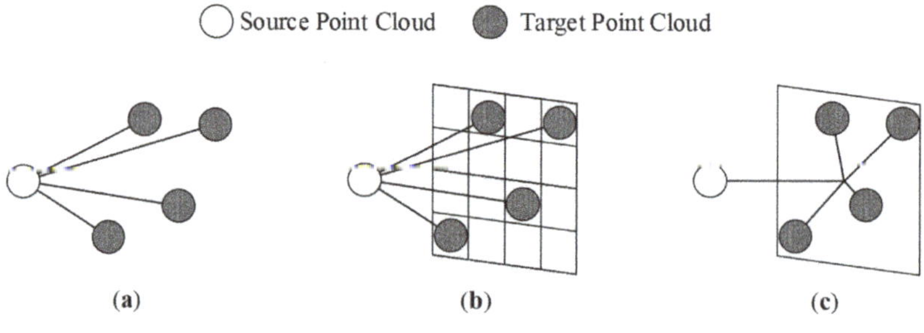

Figure 3. Comparison of point cloud registration strategies. (**a**) ICP/GICP; (**b**) NDT; (**c**) our algorithm.

As shown in Figure 3. the problem of constructing the nearest neighbor model of a point pair by using a tree diagram is transformed into constructing the nearest neighbor model of a point and a neighborhood point set. Firstly, the two sets of point cloud sequences are approximated as Gaussian distributions, i.e., $a_i \sim N(\hat{a}_i, \Sigma_i^A)$ and $b_i \sim N(\hat{b}_i, \Sigma_i^B)$, where $i \in (1, n)$. Σ_i^A and Σ_i^B are covariance matrices of two sets of point cloud sequences, respectively. Let the distance between a pair of corresponding points between the target point cloud and the source point cloud be:

$$d_i = b_i - Ta_i \tag{4}$$

Let the neighborhood point set of a_i be $B_{a_i} = \{b_j | \|a_i - b_j\| < \lambda\}$, where λ is the neighborhood judgment threshold. Thus, the distance between the extended point and the neighborhood point set is:

$$\hat{d}_i = \sum_j \left(\hat{b}_j - T\hat{a}_i\right) \tag{5}$$

As a result of $a_i \sim N(\hat{a}_i, \Sigma_i^A)$ and $b_i \sim N(\hat{b}_i, \Sigma_i^B)$, the rigid body transformation error e_i is calculated as:

$$e_i \sim (\sum_j \left(\hat{b}_j - T\hat{a}_i\right), \sum_j \left(\Sigma_j^B - T\Sigma_i^A T^T\right)) \tag{6}$$

In this way, the smoothing of all neighboring point clouds in the neighborhood of a_i is achieved. Let $\mu = N(\sum_j \left(\hat{b}_j - T\hat{a}_i\right)$ and $\Sigma = \sum_j \left(\Sigma_j^B - T\Sigma_i^A T^T\right)$; because e_i is a high-dimensional Gaussian distribution, its probability density function expansion form is:

$$P(e_i) = \frac{1}{\sqrt{(2\pi)^N \det(\Sigma)}} \exp\left\{-\frac{1}{2}(e_i - \mu)^T \Sigma^{-1}(e_i - \mu)\right\} \tag{7}$$

The negative logarithmic form of Equation (7) is:

$$-\ln(P(e_i)) = \frac{1}{2}\ln\left[(2\pi)^N \det(\Sigma)\right] + \frac{1}{2}(e_i - \mu)^T (\Sigma)^{-1}(e_i - \mu) \tag{8}$$

Solving the inter-frame pose transformation matrix T by maximum likelihood method:

$$T = \underset{T}{\operatorname{argmax}} \prod_i P(e_i) = \underset{T}{\operatorname{argmin}} \sum_i e_i{}^T (\Sigma_j^B - T\Sigma_i^A T^T) e_i{}^T \tag{9}$$

Furthermore, after introducing the number N_i of point clouds in the neighborhood a_i, Equation (9) can be written as:

$$\begin{cases} T = \underset{T}{\operatorname{argmin}} \sum_i (N_i \hat{e}_i^T \Sigma_i^{-1} \hat{e}_i) \\ \hat{e}_i = \frac{\sum_j b_j}{N_i} - Ta_i \\ \hat{\Sigma}_i = \frac{\sum_j \Sigma_j^B}{N_i} + T\Sigma_i^A T^T \end{cases} \tag{10}$$

In addition to the smoothing of all neighboring point clouds in the neighborhood of a_i, an iteration termination threshold ε was established to avoid falling into a blind region of local optima after multiple iterations as follows:

$$|RMSE_{k+1} - RMSE_k| > \varepsilon \tag{11}$$

where $RMSE_{k+1}$ and $RMSE_k$ are the root mean square error of the previous $k+1$ iterations and the previous k iterations, respectively. The iteration is completed when the absolute value of the change in the root mean square error $|RMSE_{k+1} - RMSE_k| \leq \varepsilon$, or the maximum number of iterations, is reached.

4. Graph Optimization Framework

4.1. Local Pose Map Structure

Authors should discuss the results and how they can be interpreted from the perspective of previous studies and of the working hypotheses. The findings and their implications should be discussed in the broadest context possible. Future research directions may also be highlighted.

The local state vectors in the local coordinate system in which the LiDAR and IMU are located are given as follows:

$$\begin{aligned} \mathcal{X}^L &= \left[x_{b1}, x_{b2}, \ldots, x_{bi}, d_1^e, d_2^e, \ldots, d_k^e, d_1^p, d_2^p, \ldots, d_k^p\right] \\ x_{bi} &= \left[p_{bi}^L, q_{bi}^L, v_{bi}^L, b_a, b_g\right] \end{aligned} \tag{12}$$

where x_{bi} denotes the state quantity after pre-integration of the ith IMU at t_k, including position p_{bi}^L, rotation q_{bi}^L, speed v_{bi}^L and IMU bias b_a, b_g. d_k^e is the distance from the LiDAR feature point at t_{k-1} to the matching edge feature at t_k, and d_k^p is the distance from the feature point at t_{k-1} to the matching planar feature at t_k.

From this, the Gauss–Newton method can be used instead of the fastest gradient descent method used in [27] to minimize all cost functions so as to reduce the number of iterations for rapid convergence to a locally optimal estimate. The local optimization function is constructed as follows:

$$\min_{\mathcal{X}} \left\{ \sum d_k^e + \sum d_k^p + \sum_{k \in B} \left\| r_B\left(\hat{z}_{k+1}^k, \mathcal{X}\right) \right\|_{\Sigma_b}^2 \right\} \tag{13}$$

where $\sum d_k^e + \sum d_k^p$ is used to solve the carrier pose x_{tk}^{LiDAR} in the local coordinate system of LiDAR at time t_k. $r_B\left(\hat{z}_i^{i-1}, \mathcal{X}\right)$ and Σ_b are IMU measurement residuals and covariance matrices, respectively. The meanings of the terms are described below.

4.1.1. IMU Pre-integration Factor

Let $\left[\alpha_i^{i+1}, \theta_i^{i+1}, \beta_i^{i+1}\right]^T$ be the IMU pre-integration calculation value between the ith and $i+1$th LiDAR key frames. Details of the derivation of the IMU pre-integration are presented in Appendix A. Δt_i is the time interval between the two LiDAR key frames, and the spatial transformation matrix from the IMU coordinate system to the LiDAR coordinate system in ith frame is represented by $R_L^{b_i}$. The IMU residual can be obtained as follows:

$$r_B\left(\hat{z}_i^{i+1}, \mathcal{X}\right) = \begin{bmatrix} \delta\alpha_i^{i+1} \\ \delta\theta_i^{i+1} \\ \delta\beta_i^{i+1} \\ \delta b_a \\ \delta b_g \end{bmatrix} = \begin{bmatrix} R_L^{b_i}\left(p_{b_{i+1}}^L - v_{b_i}^L \Delta t_i - p_{b_i}^L + \frac{1}{2} g \Delta t_i^2\right) - \hat{\alpha}_i^{i+1} \\ 2\left[q_{b_i}^{L^{-1}} \otimes q_{b_{i+1}}^L \otimes \hat{\theta}_i^{i+1^{-1}}\right]_{xyz} \\ R_L^{b_i}\left(v_{i+1}^L - v_i^L + g\Delta t_i\right) - \hat{\beta}_i^{i+1} \\ b_{ai+1} - b_{ai} \\ b_{\omega i+1} - b_{\omega i} \end{bmatrix} \tag{14}$$

where the symbol $[\cdot]_{xyz}$ represents extracting the real part of the quaternion used to calculate the rotation state error, and $\otimes$ represents the quaternion multiplication.

After the pose estimation of the previous key frame is completed, the IMU acceleration bias and gyroscope bigotry will be updated, the update amounts are set as $\delta \breve{b}_a^{bi}$ and $\delta \breve{b}_g^{bi}$ and the pre-integration calculation value at this time is updated as follows:

$$\begin{aligned} \alpha_i^{i+1} &= \hat{\alpha}_i^{i+1} + \frac{\delta\hat{\alpha}_i^{i+1}}{\delta b_a} \delta \breve{b}_a + \frac{\delta\hat{\alpha}_i^{i+1}}{\delta b_\omega} \delta \breve{b}_\omega \\ \theta_i^{i+1} &= \hat{\theta}_i^{i+1} \cdot Exp\left(\frac{\delta\hat{\theta}_i^{i+1}}{\delta b_\omega} \delta \breve{b}_\omega\right) \\ \beta_i^{i+1} &= \hat{\beta}_i^{i+1} + \frac{\delta\hat{\beta}_i^{i+1}}{\delta b_a} \delta \breve{b}_a + \frac{\delta\hat{\beta}_i^{i+1}}{\delta b_\omega} \delta \breve{b}_\omega \end{aligned} \tag{15}$$

4.1.2. LiDAR Factor

The feature point cloud extracted by LiDAR can be divided into two types: line features and surface features. The LiDAR residuals of the two types need to be constructed separately and then summed to obtain the total LiDAR residuals. Details of the specific derivation of LiDAR residuals are presented in Appendix B. Figure 4 shows the schematic diagram of LiDAR residual construction.

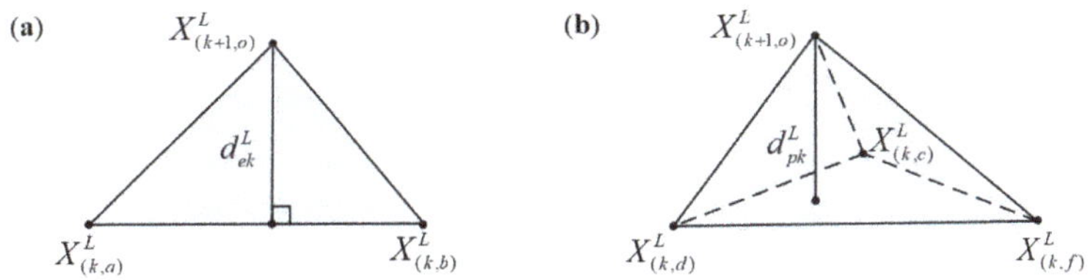

Figure 4. Schematic diagram of LiDAR residual construction. (**a**) Line characteristic residual construction; (**b**) surface characteristic residual construction.

As shown in Figure 4, let a feature point obtained in the $k+1$th scan have the coordinates of $X^L_{(k+1,o)}$ in the LiDAR coordinate system, and the coordinates of two end points of the line features matched with it in the kth scan are $X^L_{(k,a)}$ and $X^L_{(k,b)}$. The residual error of the line features can be expressed by the point-to-line distance:

$$d^L_{ek} = \frac{\left|\left(X^L_{(k+1,o)} - X^L_{(k,a)}\right) \times \left(X^L_{(k+1,o)} - X^L_{(k,b)}\right)\right|}{\left|X^L_{(k,a)} - X^L_{(k,b)}\right|} \tag{16}$$

Similarly, if the surface features that match it in the kth scan are represented as $X^L_{(k,c)}$, $X^L_{(k,d)}$ and $X^L_{(k,f)}$, then the surface feature residual can be represented by the point-to-surface distance:

$$d^L_{pk} = \frac{\left|\left(X^L_{(k+1,o)} - X^L_{(k,d)}\right) \cdot \left(\left(X^L_{(k,c)} - X^L_{(k,d)}\right) \times \left(X^L_{(k,c)} - X^L_{(k,f)}\right)\right)\right|}{\left|\left(X^L_{(k,c)} - X^L_{(k,d)}\right) \times \left(X^L_{(k,c)} - X^L_{(k,f)}\right)\right|} \tag{17}$$

4.2. *Spatial Unification of Multi-Sensor Poses*

Constructing the time—space correlation of each sensor is a fundamental task in multi-sensor fusion optimization. For this system, it is necessary to spatially unify the positional estimation results of LiDAR and IMU in the local map with the GNSS measurements in the global map. Therefore, the spatial unification strategy of the multi-sensor pose involved in this paper is shown in Figure 5.

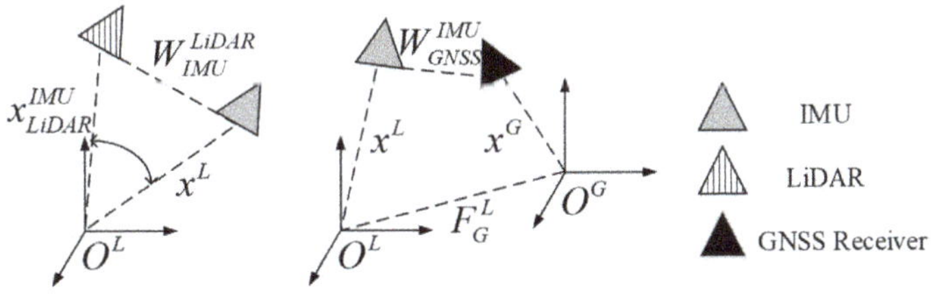

Figure 5. Schematic diagram of the spatial unification of multi-sensor poses. The spatial association of the poses of LiDAR and IMU in the local coordinate system is on the left, and the spatial association of the poses of IMU and GNSS receivers in the global coordinate system is on the right.

As shown in Figure 5, W_{GNSS}^{IMU} is the external parameter conversion matrix from IMU to GNSS and W_{IMU}^{LiDAR} is the external parameter conversion matrix from LiDAR to IMU. Since the hardware is fixed to the mobile carrier, both are calibrated to a constant value. The left figure shows the positional conversion between LiDAR and IMU in the local coordinate system, whereas the right figure shows the multi-sensor positional spatial unification from the local to the global coordinate system. Here, it is necessary to introduce the spatial transformation parameters F_G^L (including translation p_G^L and rotation q_G^L) to correlate the two positional spaces. The spatial unification of the multi-sensor pose at the moment of t can be expressed as:

$$\min_{q_L^G, p_L^G} \frac{1}{2} \sum_{i=1}^{j} \| P_t^G - \left(\left(q_G^L \right)^T \cdot \left(q_t^L \cdot W_{GNSS}^{IMU} + p_t^L \right) + p_L^G \right) \|_2^2 \tag{18}$$

where the initial value of F_G^L is set as the unit matrix. Every time the GNSS factor is added to solve the global optimum, the value of F_G^L at the next moment will be updated, thus correcting the cumulative offset between the local and global coordinate systems.

4.3. Global Pose Map Structure

Global pose map construction can be regarded as a nonlinear optimization problem; that is, the nonlinear optimization of the state vector in the sliding window. Different from the factor graph method adopted in [27], this paper adopts the graph optimization method to directly construct the residual block in the original pose graph structure for nonlinear optimization, and only optimizes the key frames in the sliding window. However, the factor graph based on GTSAM [29] needs to construct the optimization problem into a new graph corresponding to the original pose graph, with the optimization variables as the vertices and error terms as the edges. The complicated constraint relationship among the vertices is more favorable toward the optimization accuracy. However, once a new key frame is detected, all of its associated constraint nodes will be updated, which is complicated and takes too long in the engineering field. Therefore, in order to meet the requirements of the lightweight and real-time performance of the vehicle platform, we choose not to build a new constraint-related Bayesian network, but to construct the residual error and nonlinear optimization in the original pose map structure. The global pose optimization framework proposed in this paper is shown in Figure 6.

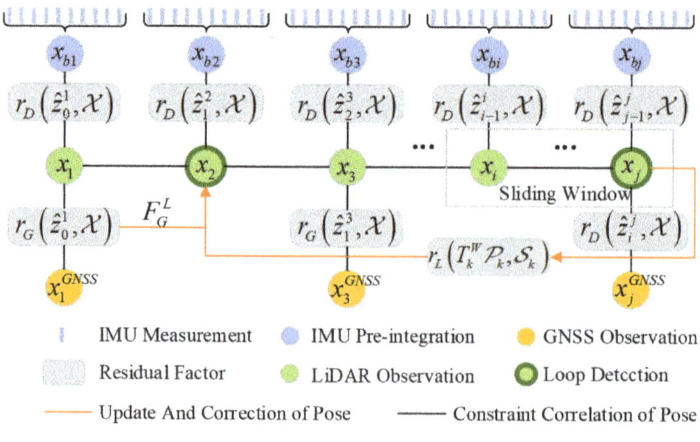

Figure 6. LiDAR-IMU-GNSS fusion framework based on graph optimization.

The global optimization function is constructed as follows:

$$\mathcal{X} = \underset{\mathcal{X}}{\operatorname{argmin}} \sum_{t=0}^{n} \left(\| z_t^L - h_t^L(\mathcal{X}) \|_{\Sigma_t^k}^2 + \rho \| z_t^G - h_t^G(\mathcal{X}) \|_{\Sigma_t^k}^2 \right) + r_{loop}\left(T_k^W \mathcal{P}_k, \mathcal{S}_k \right) \tag{19}$$

where ρ is the GNSS confidence level expressed by the covariance of the error in the GNSS observations obtained by the pseudo-range single point positioning (SPP) algorithm solution. T_k^W is the pose transformation matrix between the current global point cloud $\mathcal{P}_k$ and the local point cloud $\mathcal{S}_k$ derived from the inter-frame local matching. The specific meaning of each sensor cost function in the formula are as follows.

4.3.1. LIO Factor

According to Section 4.1, the position p_t^L and rotation q_t^L of the carrier in the local coordinate system at the moment t can be obtained. Therefore, the LIO local residual factor can be constructed as follows:

$$r_D\left(\hat{z}_t^{t-1}, \mathcal{X} \right) = \begin{bmatrix} {q_{t-1}^L}^{-1}(p_t^L - p_{t-1}^L) \\ {q_{t-1}^L}^{-1} q_t^L \end{bmatrix} \ominus \begin{bmatrix} {q_{t-1}^G}^{-1}(p_t^G - p_{t-1}^G) \\ {q_{t-1}^G}^{-1} q_t^G \end{bmatrix} \tag{20}$$

where the symbol $\ominus$ represents the quaternion subtraction.

4.3.2. GNSS Factor

Set the time interval between two frames of GNSS observations as Δt and realize the time alignment with LIO pose estimation by interpolation. Cubic spline interpolation is used for position interpolation and spherical linear interpolation is used for quaternion interpolation. Now, given the GNSS measurement p_t^{GNSS} in the ENU coordinate system and the LIO positional observations p_t^G in the global coordinate system, the GNSS residual factor is expressed as follows:

$$r_G\left(\hat{z}_t^{t-1}, \mathcal{X} \right) = p_t^G - p_t^{GNSS} \tag{21}$$

When the carrier moves to the GNSS signal confidence region, in order to fully and reliably utilize the GNSS observations, the GNSS factor is added with the GNSS confidence as the weight. The GNSS confidence is determined by the number of visible and effective GNSS satellites. After GNSS participates in the global pose estimation, it will update the pose conversion parameter F_G^L between the local coordinate system and the global coordinate system. This ensures that, even if the mobile carrier enters a GNSS-rejected environment (e.g., indoor car parks and tunnels), our algorithm can provide a more accurate initial observation after GNSS correction.

4.3.3. Loop Factor

Considering the possible overlap of the moving vehicle driving areas, it is necessary to add a loop detection link to establish possible loop constraints between non-adjacent frames. According to Equation (5), the loop factors can be constructed as follows:

$$r_L\left(T_k^W \mathcal{P}_k, \mathcal{S}_k \right) = \begin{cases} T_k^W = \underset{k}{\operatorname{argmin}} \sum \left(N_k \hat{e}_k^T \Sigma_k^{-1} \hat{e}_k \right) \\ \hat{e}_k = \frac{\sum_k \mathcal{P}_k}{N_k} - T_k^W \mathcal{S}_k \\ \hat{\Sigma}_k = \frac{\sum_k \Sigma_k^{\mathcal{P}}}{N_k} + T_k^W \Sigma_k^{\mathcal{S}} T_k^{WT} \end{cases} \tag{22}$$

Using the optimized point cloud registration method in Section 3.2, we optimize and correct the historical trajectory through the registration between the point cloud of the prior local map and the current global point cloud. This method ensures that the positional estimates converge to the global optimum result.

5. Experimental Setup and Results

5.1. Point Cloud Registration Results

To verify the superiority of the point cloud registration algorithm used in this paper, we compared the registration results of the ICP algorithm used in traditional LiDAR odometry with our algorithm. The comparison results are shown in Figures 7–9.

(a) (b) (c)

Figure 7. Point cloud registration results. (**a**) Original source and target point clouds. (**b**) Alignment results using ICP algorithm. (**c**) Alignment results using our algorithm.

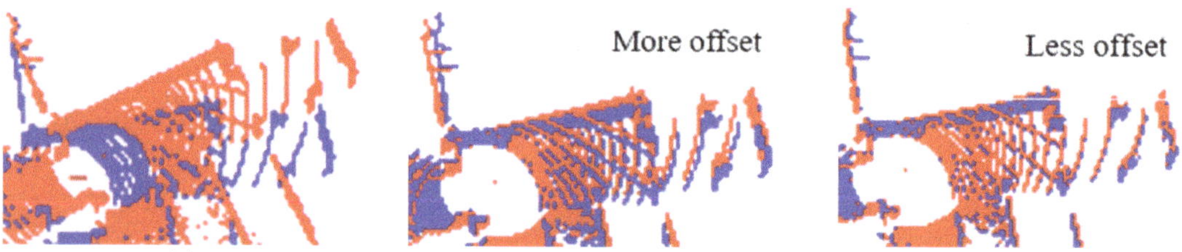

(a) (b) (c)

Figure 8. Detail diagram of point registration results of the point cloud registration results. (**a**) Original source and target point clouds. (**b**) Alignment results using ICP algorithm. (**c**) Alignment results using our algorithm.

(a) (b) (c)

Figure 9. A circular expansion of the point cloud alignment results. (**a**) Original source and target point clouds. (**b**) Alignment results using ICP algorithm. (**c**) Alignment results using our algorithm.

Compared with the typical indoor environment, for mobile carriers in complex urban environments, the angles and translations between the source and target point clouds during the continuous frame and the loopback detection may be larger. As shown in Figure 7, when the initial position is unreasonable, the registration results of the ICP algorithm cannot fully approach the global optimal solution, which is detrimental to both the local pose estimation and loopback correction of the mobile carrier. However, the registration accuracy of our algorithm is not affected by the large positional transformation of the vehicle platform. Compared with the traditional ICP algorithm, the registration result of our algorithm suited the needs of the vehicle platform better.

Furthermore, in order to avoid the contingency of the registered objects, we made a comparative experiment on the source point clouds with different rotation angles and translation distances, and quantitatively compared the registration accuracy and time consumption of each algorithm. The test results are shown in Tables 2 and 3.

Table 2. Registration results of different algorithms for point clouds with different angles.

Rotation Angles	4°	15°	30°	60°	90°
	Time Consumption (ms)/Root Mean Square Error (m)				
PCL_ICP	212.79/0.00055	244.08/0.00138	314.76/0.15597	393.49/6.01775	350.92/8.04554
PCL_GICP	244.45/0.00055	319.93/0.00056	861.81/0.00057	1763.89/24.3384	1375.59/25.1845
PCL_NDT	267.46/0.00350	563.40/0.01668	332.04/6.11404	202.97/17.9595	346.27/29.0983
Proposed	31.71/0.00023	37.25/0.00037	46.72/0.00050	51.80/1.72794	40.80/2.67116

Table 3. Registration results of different algorithms for point clouds with different translation distances.

Translation Distances	0.5 m	1 m	2 m	4 m	5 m
	Time Consumption (ms)/Root Mean Square Error (m)				
PCL_ICP	213.11/0.02398	221.52/0.02398	254.83/0.02409	312.12/0.04216	303.797/0.86212
PCL_GICP	215.33/0.02415	222.46/0.24152	254.359/0.02415	303.31/0.02415	354.088/0.02415
PCL_NDT	456.43/0.03024	723.031/0.03375	1186.57/0.02511	577.64/2.28457	279.047/2.68836
Proposed	31.9215/0.00106	33.44/0.01716	41.26/0.01265	31.19/0.03011	33.88/0.03047

From the vertical comparison between Tables 2 and 3, it can be seen that the point cloud registration algorithm is more sensitive to rotation, which means that, if the vehicle rotates at a large angle in the city, the point cloud registration between consecutive frames is perhaps less reliable. It may even lead to a failure of the pose estimation, as has been demonstrated in [5]. However, as the rotation angle increases, it can be seen from Table 2 that the registration accuracy of our algorithm decreases the least, and, at the extreme 90° rotation angle, the accuracy is still more than four times better than the other algorithms, with a root mean square error of approximately 2.67116 m. On the other hand, for large translations (5 m) between the source and target point clouds, our algorithm also shows an excellent registration accuracy, with a root mean square error of approximately 0.03047 m. It is worth noting that, in the aspect of single-point cloud registration, the registration time of our algorithm is increased by approximately one order of magnitude compared with others. To sum up, it is sufficient to verify the superiority of the proposed registration algorithm in terms of compressing time and improving the registration accuracy.

5.2. Positioning Accuracy

In this paper, the absolute trajectory error (ATE) was selected as the evaluation index of SLAM system positioning accuracy so as to directly reflect the difference between the global position estimation of the moving carrier and the ground truth. The absolute trajectory error is calculated as follows.

$$A_i := g_i^{-1} S p_i \tag{23}$$

where A_i is the absolute trajectory error of the SLAM system in the ith frame, g_i and p_i are the ground truth and the estimated pose, respectively, and S is the transformation matrix between the ground truth and the estimated pose. In this paper, the mean error (ATE_ME) and root mean square error (ATE_RMSE) of the absolute trajectory error were selected as evaluation criterion.

5.2.1. Public Dataset

To verify the positioning accuracy of the fusion algorithm in different outdoor environments, the KITTI_RAW dataset [30], which includes a variety of outdoor scenes, was used to evaluate the localization accuracy of the fusion algorithm and to compare it with other similar advanced algorithms. The experimental data acquisition platform is as follows: LiDAR point cloud data are acquired by Velodyne HDL~64 line LiDAR, with horizontal field angle of view of 360°, vertical field angle range of $(-24.8°, +2°)$, horizontal resolution range of $(0.08°, 0.35°)$, vertical angle resolution of 0.4° and scanning frequency of 10 Hz, which can meet the requirements of in-vehicle point cloud data acquisition. The GPS/IMU integrated system adopts OXTS RT3003, with a GPS output frequency of 1 Hz/s and an IMU output frequency of 100 Hz. The ground truth is provided by a high-precision integrated navigation system.

Four different outdoor scenarios were used to validate the performance of the fusion algorithm, including urban environments, open area, highway and forest road. The voxel grid size of the fusion algorithm was set to $0.3 \times 0.2 \times 0.3$, the maximum iteration threshold was set to 30 and the iteration termination tolerance threshold was set to 1×10^{-8}, so as to meet the real-time requirements and ensure the stable number of feature point clouds participating in the matching in sparse areas of outdoor environments. The comparison of the experimental results is shown in Figures 10 and 11 and Table 4.

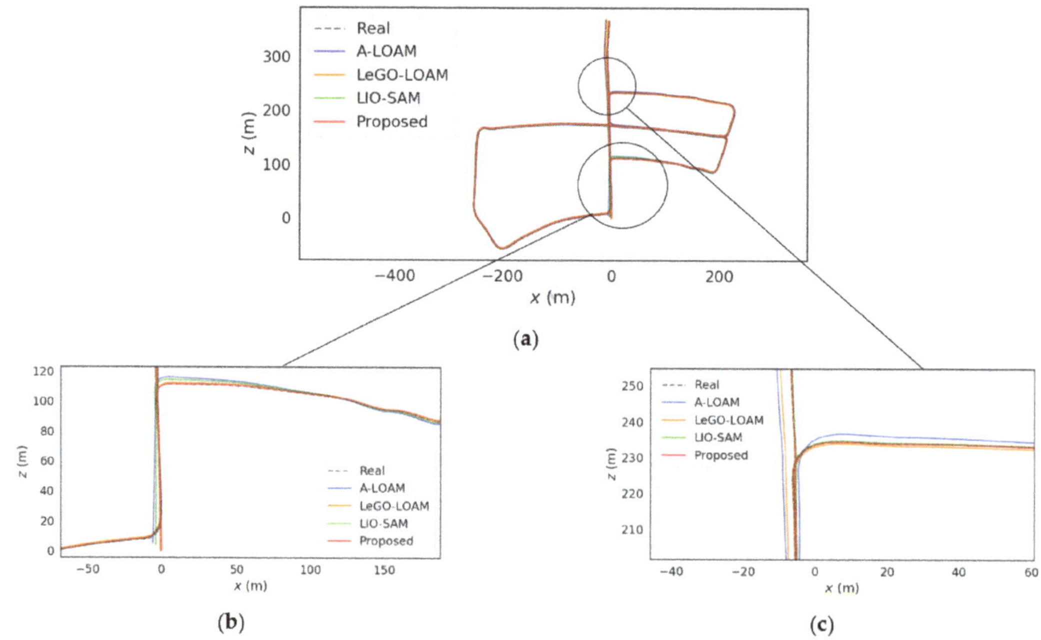

Figure 10. Comparison of the estimated trajectories. (**a**) Global positioning trajectory. (**b**) Local details of the trajectory. (**c**) Local details of the trajectory.

Table 4. Comparison of ATE_RMSE(m) of each algorithm in KITTI_RAW dataset.

Sequence	Urban Environments		Open Area	Highway	Forest Road
	09_30_0018	09_30_0027	09_30_0016	10_03_0042	09_30_0033
A-LOAM	3.545	1.181	0.475	21.058	8.693
LeGO-LOAM	2.253	2.060	0.515	186.608	4.585
LIO-SAM	1.881	0.905	0.423	19.745	10.109
Proposed	1.056	0.400	0.222	6.239	3.669

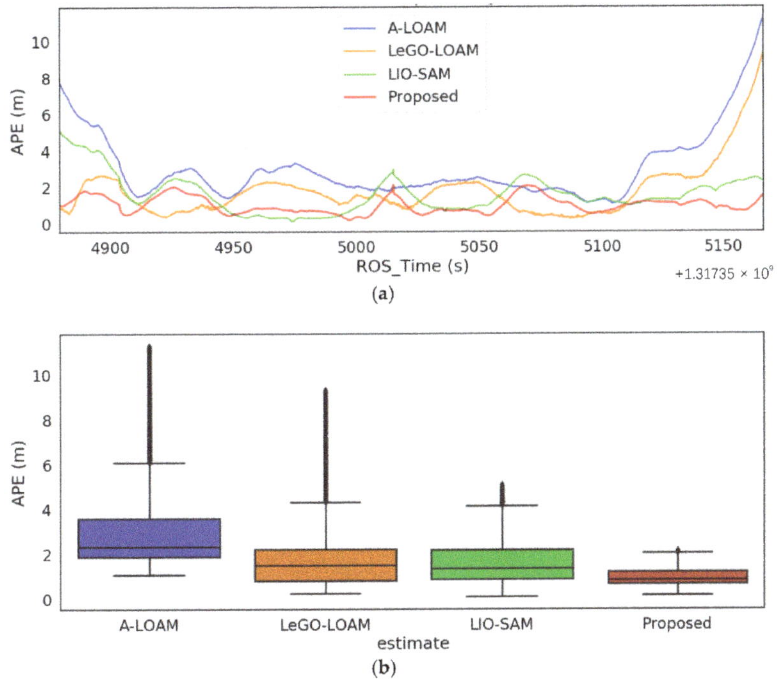

Figure 11. Comparison of the positioning error of each algorithm. (**a**) APE fitting curve. (**b**) The box diagram of APE.

Figure 10 shows the comparison of the positioning results of each algorithm in the 09_30_0018 dataset representing the urban environment. As shown in Figure 11b, both LiDAR odometry (LO), represented by A-LOAM, and LIO, represented by LeGO-LOAM, show significant degradation in the position estimation results in the first 50 s and the last 50 s. The reason is that LO and LIO systems only rely mainly on LiDAR to extract spatial geometric feature information. Once LiDAR feature constraints are sparse or fail, the carrier state estimation degradation will occur in this feature direction, and additional constraints need to be added. The first 50 s and last 50 s are both flat, open roads with sparse point cloud features, which are susceptible to the degradation of the LiDAR positional optimization results. However, the number of GNSS visible satellites in the flat and open road is enough, and using GNSS observations as global constraints can greatly improve the positioning accuracy and robustness in sparse areas of point clouds. As can be seen from Figure 11b, the ATE_RMSE of both the LIO-SAM with GNSS global constraints and the present algorithm is stable between (0 m, 2 m), and the positioning accuracy remains stable in the sparse region of the point cloud features in the latter 50 s without large data drift. In addition, from the box diagram shown in Figure 9c, it can be seen that the positional outliers estimated by LIO-SAM are reduced by approximately 80% compared with the LIO system. Furthermore, the positional estimation errors of our algorithm are concentrated between (0.68 m, 1.23 m) with very few outliers, which fully demonstrates the superiority of the proposed algorithm in its positioning accuracy in urban environments.

5.2.2. Urban Dataset

To further investigate the extent to which improvements in both the front-end and back-end components of the fusion algorithm improve its positioning accuracy, we conducted ablation experiments in a complex environment of GNSS signals. We constructed a system without GNSS global correction (-), a system without smoothed voxelized point cloud registration and loopback correction (#) and a complete system (Proposed), respectively. The experimental environment is the complex reflection area of GNSS in the urban environment. The experimental platform includes: the ground truth, which is provided by NovAtel SPAN-CPT positioning results; the LiDAR point cloud, which is acquired by HDL 32E Velodyne LiDAR, where the horizontal field of view angle is 360°, the vertical field of view angle range is $(-30°, +10°)$ and the scanning frequency is 10 Hz, which is suitable for in-vehicle point cloud data acquisition; IMU, which is Xsens Mti 10, and the update frequency of the pose is 100 Hz; the GNSS receiver, which is u-blox M8T, and the update frequency is 1 Hz.

Different from the KITTI_RAW dataset, the GNSS confidence parameter in this experiment is not fixed. After solving the raw observation data collected by u-blox with the SPP algorithm, we obtained the GNSS confidence covariance as the GNSS factor weight parameter. This is more in line with the real urban environment, where GNSS reflected and refracted signals interfere with the direct signal superimposed, thus causing the pseudo-range and carrier phase observations to deviate from the true value of the direct signal. The experimental results are shown in Figures 12 and 13 and Table 5.

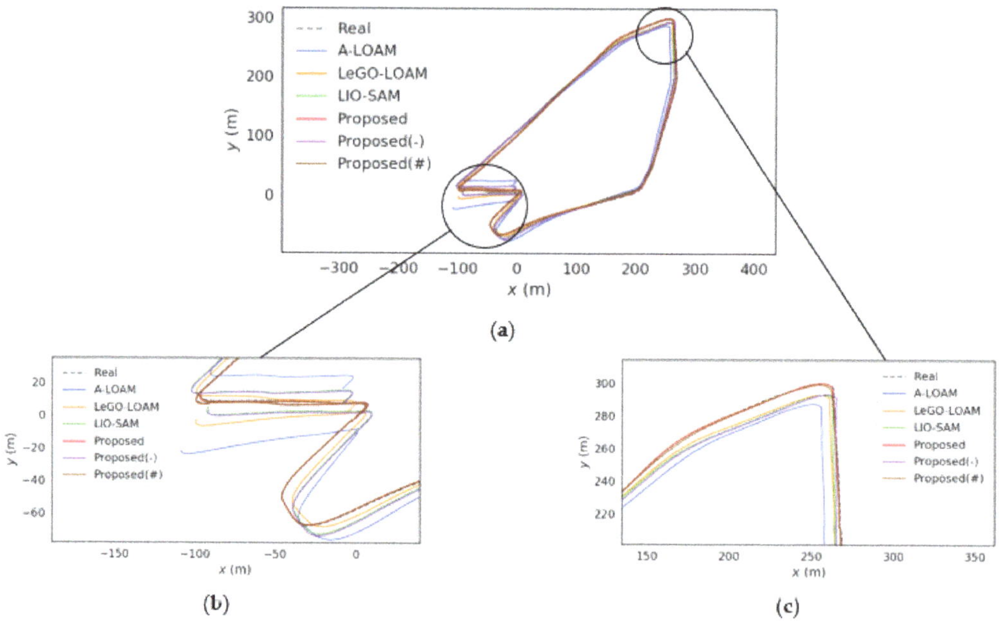

Figure 12. Comparison of the estimated trajectories. (**a**) Global positioning trajectory. (**b**) Local details of the trajectory. (**c**) Local details of the trajectory.

Table 5. Motion estimation errors of each algorithm on outdoor dataset.

Sequence	**Hong Kong 0428**	**Hong Kong 0314**
	ATE_RMSE(m)/ATE_ME(m)	
A-LOAM	41.933/37.672	23.220/21.143
LeGO-LOAM	43.441/40.515	10.250/9.644
LIO-SAM	19.042/12.464	7.181/6.786
Proposed(-)	16.456/14.119	6.785/6.198
Proposed(#)	4.671/3.847	1.816/1.382
Proposed	2.265/1.901	1.573/1.260

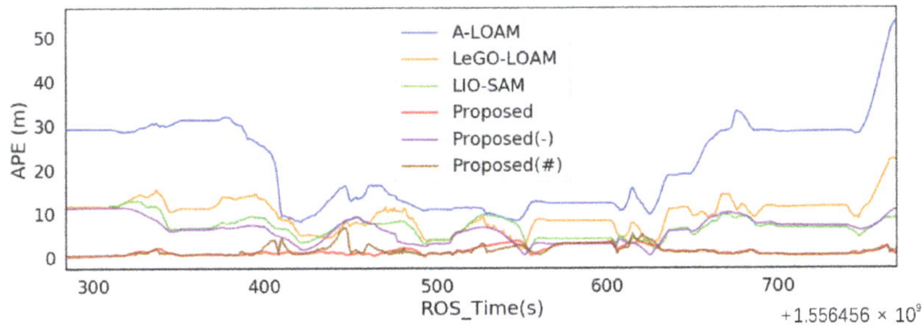

(**a**)

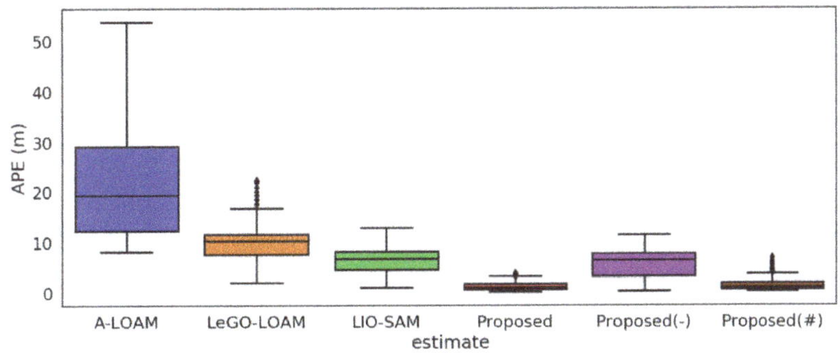

(**b**)

Figure 13. Comparison of the positioning error of each algorithm. (**a**) APE fitting curve. (**b**) The box diagram of APE.

As can be seen from Figure 13a, firstly, due to the accurate registration of the front-end point cloud, the absolute trajectory error of Proposed(-) decreases slightly compared to the pre-improved system. In the initial parking section of the dataset, the traditional ICP algorithm suffers from the problem of over-iterations, and the result is not the global optimal solution. However, the data smoothing processing and the setting of the iteration termination threshold of our algorithm can solve this problem well, providing a better initial value for the positional matching. The absolute trajectory error within the first 25 s drops by approximately 8m compared to LIO-SAM. Secondly, compared to LIO-SAM, which uses GNSS observations directly as global constraints without filtering, we introduced GNSS confidence into the optimization equation. It allows our algorithm to remain unaffected by poor-quality GNSS observations and to maintain a better positioning accuracy in the latter 50 s in areas with dense tall buildings and poor-quality point cloud distribution. In contrast, due to the poor quality of GNSS observations involved in optimization (LIO-SAM) and the low accuracy of LiDAR loop detection as a global constraint (A-LOAM and LeGO-LOAM), all other similar algorithms have a cumulative increase in the absolute trajectory error with steeper slopes. This is extremely detrimental for vehicle-mounted platforms driving in realistic large outdoor environments. From the experimental results, it can be seen that the global optimization link in our complete algorithm can well suppress the local cumulative drift and make the pose estimation result move more towards the global optimal solution.

In summary, driven by a combination of respective front-end and back-end improvements, our complete algorithm achieves a higher positioning accuracy than other comparable algorithms within real urban environments. Furthermore, as a result of the inherent advantage of local sensors not being subject to signal refraction environmental interference, it compensates for positioning outliers arising from multipath effects in traditional GNSS positioning in urban environments. It fully ensures the integrity and reliability of the fusion system's positioning.

5.3. Time-Consuming Performance

In this paper, the KITTI 09_30_0033 sequence is randomly selected to verify the real-time performance of our algorithm and the similar algorithms. In this paper, the above-mentioned three similar advanced algorithms are selected as control algorithms to compare with our algorithm, so as to verify the superior real-time performance of this algorithm in three stages: downsampling, point cloud registration and optimization. The experimental results are shown in Figure 14.

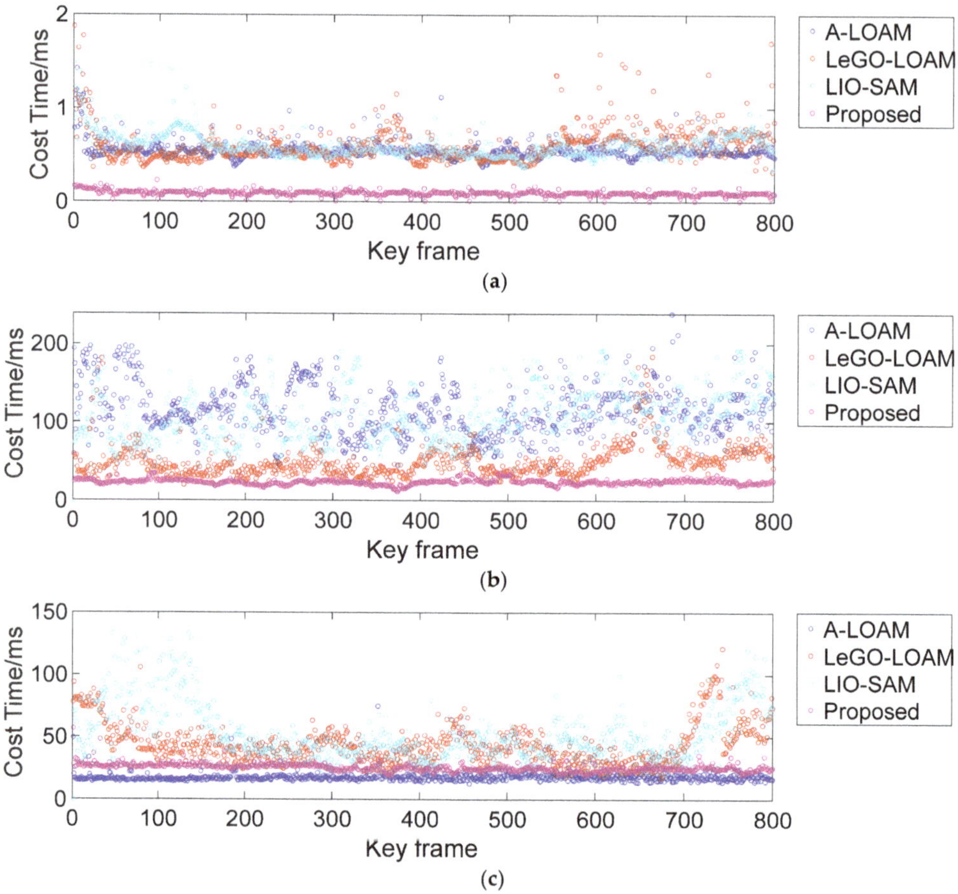

Figure 14. Time-consuming comparison of three processes. (**a**) Point cloud downsampling process. (**b**) Point cloud registration process. (**c**) Position global optimization process.

The time consumption of the point cloud downsampling is shown in Figure 14a. The downsampling process of the three other algorithms uses RANSAC as the core algorithm, but its iterative approximation speed is slow, at approximately (0.5 ms, 1 ms), and the filtering and fitting quality of the depth information is not good. In contrast, our proposed algorithm uses HashMap instead of random sampling, which improves the speed of filtering out similar points in voxels to a certain extent and reduces the time consumption by two to five times compared with the traditional downsampling method. Although the time-consuming ratio of this process is relatively small in typical indoor environments or short-term positioning processes, for vehicles driving in large outdoor environments with complex point cloud environments for long periods of time, the accumulation of tiny instances of time consumption will lead to a cumulative increase in positioning time consumption. Therefore, the time-consuming compression in point cloud downsampling in this paper is beneficial for ensuring real-time vehicle positioning.

The time consumption of the point cloud registration is shown in Figure 14b, which shows that the time taken for the LIO-SAM and A-LOAM point cloud registration step is in the range of (50 ms, 200 ms). By extracting and separating the ground point clouds, LeGO-LOAM can inhibit the time-consuming increase caused by outlier registration due to the interference between non-identical cluster point clouds to a certain extent. The point cloud registration step takes approximately (50 ms, 100 ms). There are two possible factors for the obvious fluctuation of the above algorithm. The first is the fluctuation in the registration time due to the changing distribution of the ambient point cloud. The second is that, according to the experiments in Section 5.1, the rotation angle and displacement between the source and the target point cloud also have a certain influence on time consumption, which is obvious in the urban driving environment, where the speed and driving direction change irregularly. However, the time consumption of our algorithm is stable between (20 ms, 30 ms). The smoothing of the single-to-many distribution of the point cloud sequence greatly reduces the effect of the sparsity of the point cloud distribution on the alignment time, ensuring that the point cloud alignment step is both time-efficient and stable.

The time consumption for the positional optimization step is shown in Figure 14c. The time taken to match the local map to the global map for LIO-SAM and LeGO-LOAM is around (30 ms, 130 ms). Due to the addition of GNSS sensors and the interference of some GNSS observations with low confidence, the total time consumption of LIO-SAM is even higher than that of LeGO-LOAM. However, A-LOAM takes (20 ms, 25 ms). The main reason is that the observation of only one sensor in LiDAR needs to be optimized, and the residual block is directly constructed in the original pose map structure, which reduces the computational burden of the multidimensional factor map. In this algorithm, the optimization method of A-LOAM is used for reference. It can be seen that, although the GNSS sensor is added, the time consumption is still stable at approximately 30 ms. The reason is that the global constraint of GNSS provides a more accurate transformation matrix from the local map to the global map for the fusion system, which makes it easier for the objective function of map matching to converge to the optimal solution. Secondly, we use the Gauss–Newton method instead of the steepest gradient descent method used in [27] to minimize all of the cost functions, so as to reduce the number of iterations to converge quickly to the locally optimal estimate, which is one of the main reasons for the decrease in time consumption. The average total time of each algorithm in a single frame is shown in Table 6.

Table 6. Comparison of average total time consumption of each algorithm in single frame.

	A-LOAM	LeGO-LOAM	LIO-SAM	Proposed
Time Consumption/ms	138.07	126.29	223.15	108.76

In summary, thanks to the double improvement of this algorithm in the front-end and back-end of our system, the time consumption of all three steps involved is compressed. Although the optimization vector of the GNSS sensor is newly added, it has a better real-time performance than other similar algorithms.

5.4. Mapping Results in the Real Urban Environment

As shown in Figure 15, this section shows the comparison of mapping results between our algorithm and similar advanced algorithms. Figure 15a shows the ground truth in the real outdoor environment, which is obtained by NovAtel SPAN-CPT. The vehicle travels for one week from the starting point in the lower right corner and then returns, and the trajectory is almost closed. Figure 15b shows the mapping result of LIO-SAM, and it is clear that the section near the end of the journey deviates significantly from the actual path travelled. The reason can be attributed to the fast displacement of the carrier, which leads to an increased difference in the point cloud clusters captured between the front and back frames, including rotation and displacement, resulting in the point cloud registration in the loopback detection of LIO-SAM being prone to failure and the loopback constraint results not being ideal. In addition, the GNSS constraint strategy adopted by LIO-SAM has a poor global correction effect on local sensors. However, compared with LIO-SAM, our algorithm has an obvious loop detection accuracy and GNSS global constraint effectiveness.

(a) (b) (c)

Figure 15. Comparison of mapping effects of various algorithms. (**a**) The ground truth. (**b**) The mapping results of LIO-SAM. (**c**) The mapping results of the proposed algorithm.

As shown in Figure 15c, the mapping trajectory of the algorithm proposed in this paper is basically fitted with the true value of the driving trajectory, and, in the loop road section in the lower right corner, the trajectories passing through the same landmark twice are basically coincident. The reasons are as follows: firstly, smoothing the point cloud clusters can improve the fault tolerance of the point cloud registration between the front and back frames; secondly, the weighted GNSS global constraint can eliminate the GNSS measurements with large observation gross errors, thus achieving superior mapping results.

6. Discussion

SLAM-based multi-sensor fusion positioning technology expands the application field of traditional GNSS-based mapping techniques and makes continuous and reliable positioning in complex urban environments with good/intermittent/rejection GNSS a reality. The superior sensing capability of LIO for natural sources has been demonstrated in the literature [5,6] and others. However, the accuracy limitation of point cloud registration and the local pose drift of LIO limit the application of LIO in large outdoor environments to some extent. This paper focuses on the above two technical bottlenecks faced by the existing LiDAR SLAM and proposes simplified but effective improvement solutions.

First of all, the primary technical bottleneck is how to improve the accuracy of point cloud registration. Koideet al. [20] demonstrated that smoothing the point cloud cluster can improve the fault tolerance of point cloud registration and improve the accuracy of the point cloud from 1.20m to 0.89m. However, from our practical tests, it has been shown that, in long and large turning or translating sections in real environments, if the number of iterations is not limited, it may still fall into the local optimum problem. Therefore, on the basis of constructing a registration equation based on smooth voxelization filtering, we further use the judgment condition of iteration termination for the secondary constraint, so as to reduce the over-fitting problem of local point clouds in a typical environment. According to the registration experiment in Section 5.1 and the positional estimation accuracy experiment in Section 5.2, it can be seen that our registration method has a higher accuracy than the conventional methods. According to Figure 11, it can be seen that accurate point cloud alignment leads to an accurate initial value estimation of the positional attitude, which has a considerable positive effect on the positional estimation of the vehicle platform with a complex moving state.

Secondly, there is the challenge of how to effectively use GNSS measurements and loopback detection mechanisms to converge the local positioning results of LIO to a globally optimal solution. Firstly, the main step of loop detection dependency is point cloud registration. The previous paragraph has specifically analyzed the superiority of our method. Thanks to the positive point cloud registration results, we can reason that our global constraint using loopback detection is superior, as demonstrated in Section 5.2.2. Next, some related research has been carried out on the research of GNSS and LIO fusion positioning. Optimization-based methods have been applied in [31,32] and so on. However, taking [27] for example, in the traditional graph optimization model, the covariance effect of measurements is only used to determine whether to add factors or not, which is crude for the screening of the measurements, especially for GNSS, a measurement signal that is greatly influenced by the environmental catadioptric surface, where its observation information has not been fully applied. Therefore, in this paper, on the basis of the rough screening of GNSS observations, the covariance of the quantitative measurements is added as a weight to the graph optimization model. It achieves a more adequate and accurate global constraint on the LIO local poses using GNSS observations. From the ablation experiment results in Section 4.3.2, it is evident that the method of the weighted GNSS residual added in this paper achieves a satisfactory positioning accuracy. According to the above results, we can draw a reasonable inference: even when entering an indoor parking lot or tunnel, the accumulated error of LIO in this algorithm will start to accumulate from a lower initial value of drift. In contrast, for the LIO algorithms without GNSS constraints, they already have a large deviation between the local map and the global map before entering the denial environment. Therefore, the cumulative error range of this algorithm is acceptable. Once the GNSS signal is restored, the local error will be corrected within the time alignment interval of 0.1s as set in Section 4.3.2. This provides an accurate and continuous positional estimation for the in-vehicle platform during travel in complex urban environments.

7. Conclusions

In this paper, a LiDAR-IMU-GNSS fusion positioning algorithm with accurate local alignment and weak global drift is proposed for the high-precision continuous positioning of mobile carriers in complex urban environments.

Firstly, a voxelized point cloud downsampling method based on curvature segmentation is proposed. Rough classification is carried out by a curvature threshold, and the voxelized point cloud downsampling is performed using HashMap instead of RANSAC, so that the spatial feature distribution attributes of the source point cloud, including texture feature information, such as surfaces and curves, are retained to a greater extent.

Secondly, a point cloud registration model based on the nearest neighbors of the point and neighborhood point sets is constructed. Furthermore, an iterative termination threshold is set to reduce the probability of the local optimal solution. This greatly improves the real-time performance of the point cloud registration and can also play a large role in aligning the point cloud between the front and back frames of a fast-moving carrier with large displacement.

Finally, we propose a LIO-GNSS fusion positioning model based on graph optimization that uses GNSS observations weighted by confidence to globally correct local drift. In addition, the loop detection mechanism using the above-mentioned point cloud registration algorithm is also added into the fusion system, resulting in further global constraints of the driving areas with prior maps. Experimental results show that our algorithm can realize a more continuous and accurate pose estimation and map reconstruction in complex urban environments than similar state-of-the-art algorithms.

In the future work, there are still several issues in our work that deserve further exploration. Firstly, we plan to build a deeper constraint relationship between LIO and GNSS and make use of the rich planar features perceived by LiDAR in the urban environments to compensate for GNSS occlusion or the presence of multipath areas in the direction of the constraint. It will reduce the probability of the unreliability of SPP positioning results in the urban environments. Secondly, considering the environments with multi-sensor failures, such as tunnels with a high environmental texture similarity, where both GNSS rejection and point cloud degradation failures exist. We consider a more accurate degradation direction detection using the degeneracy factor (DF) algorithm proposed in [5], and make a non-linear optimization correction for the positional attitude in that direction. Finally, the perception ability of 3D environmental features by using only LiDAR, IMU and GNSS sensors is still relatively limited. We plan to use the observation residuals from other sensors to add multi-dimensional feature constraints to the fusion positioning algorithm, such as cameras, wheel odometers and so on, so as to make full use of environmental features and realize accurate and real-time navigation and positioning targets with high environmental universality.

Author Contributions: Conceptualization, X.L.; methodology, X.H.; software, X.H.; validation, S.P.; formal analysis, X.H.; investigation, X.L.; resources, W.G.; data curation, X.H.; writing—original draft preparation, X.L.; writing—review and editing, X.H.; visualization, X.H.; supervision, S.P. and W.G.; project administration, W.G.; funding acquisition, S.P. All authors have read and agreed to the published version of the manuscript.

Funding: This research study was funded by the National Key Research and Development Program of China (2021YFB3900804) and the Research Fund of the Ministry of Education of China and China Mobile (MCM20200J01).

Data Availability Statement: Not applicable.

Conflicts of Interest: The authors declare no conflict of interest.

Appendix A

Mathematical Derivation of the Equations (14) and (15).

First, given the original accelerometer and gyroscope measurement values of IMU as follows:

$$\begin{aligned} \hat{a}_t &= a_t + b_{at} + R_W^L g_w + n_a \\ \hat{\omega}_t &= \omega_t + b_{\omega t} + n_{\omega} \end{aligned} \tag{A1}$$

where $n_a \sim N(0,\sigma_a^2)$ and $n_{\omega} \sim N(0,\sigma_{\omega}^2)$ are white Gaussian noise of accelerometer and gyroscope, respectively. The following mathematical derivations are all completed in the IMU body coordinate system.

Therefore, the position, rotation and velocity between the ith IMU frame and the $i+1$th IMU frame can be obtained:

$$\begin{aligned} p_{i+1} &= p_i + v_i\Delta t_i + \iint_{t\in[t_i,t_{i+1}]}\left(R_W^b(\hat{a}_t - b_t - n_a) - g_w\right)dt^2 \\ q_{i+1} &= q_i Exp(\hat{\omega}_t - b_t - n_{\omega})\Delta t_i \\ v_{i+1} &= v_i + \int_{t\in[t_i,t_{i+1}]}\left(R_W^b(\hat{a}_t - b_t - n_a) - g_w\right)dt \end{aligned} \tag{A2}$$

To avoid repeated calculation of IMU parameters during pose estimation, pre-integration is introduced to simplify calculation, namely:

$$\begin{aligned} p_{i+1} &= p_i + v_i\Delta t_i - \tfrac{1}{2}g_w\Delta t_i^2 + R_b^w\alpha_{i+1}^i \\ q_{i+1} &= q_i \otimes \theta_{i+1}^i \\ v_{i+1} &= v_i - g_w\Delta t_i + R_b^w\beta_{i+1}^i \end{aligned} \tag{A3}$$

where $\left[\alpha_i^{i+1},\theta_i^{i+1},\beta_i^{i+1}\right]^T$ is the IMU pre-integration value. It can be inferred from [28] that the IMU pre-integration value is only related to the IMU bias at different times. Since the

IMU bias change is very small, we assume that the pre-integration change is linear with the IMU bias, and then $\left[\alpha_i^{i+1}, \theta_i^{i+1}, \beta_i^{i+1}\right]^T$ after each pose estimation can be recorded as:

$$\begin{aligned}
\alpha_i^{i+1} &= \hat{\alpha}_i^{i+1} + \frac{\delta\hat{\alpha}_i^{i+1}}{\delta b_a}\delta \breve{b}_a + \frac{\delta\hat{\alpha}_i^{i+1}}{\delta b_\omega}\delta \breve{b}_\omega \\
\theta_i^{i+1} &= \hat{\theta}_i^{i+1} \otimes \begin{bmatrix} 1 \\ \frac{1}{2}\frac{\delta\hat{\theta}_i^{i+1}}{\delta b_\omega}\delta \breve{b}_\omega \end{bmatrix} \\
\beta_i^{i+1} &= \hat{\beta}_i^{i+1} + \frac{\delta\hat{\beta}_i^{i+1}}{\delta b_a}\delta \breve{b}_a + \frac{\delta\hat{\beta}_i^{i+1}}{\delta b_\omega}\delta \breve{b}_\omega
\end{aligned} \tag{A4}$$

Equation (A4) is the pre-integration form in the continuous time between the two IMU frames, and the actual IMU pre-integration is the incremental in the discrete time. Therefore, the mid-point integration is used for discretization, and the matrix form of the discrete IMU state error transfer equation is obtained:

$$\begin{bmatrix} \delta\alpha_{i+1} \\ \delta\theta_{i+1} \\ \delta\beta_{i+1} \\ \delta b_{ai+1} \\ \delta b_{\omega i+1} \end{bmatrix} = F_i \begin{bmatrix} \delta\alpha_i \\ \delta\theta_i \\ \delta\beta_i \\ \delta b_{ai} \\ \delta b_{\omega i} \end{bmatrix} + V_i \begin{bmatrix} n_{\alpha_i} \\ n_{\omega_i} \\ n_{\alpha_{i+1}} \\ n_{\omega_{i+1}} \\ n_{b_a} \\ n_{b_\omega} \end{bmatrix} \tag{A5}$$

where F_i and V_i are matrix abbreviations, with specific values as follows:

$$F_i \begin{bmatrix} \delta\alpha_i \\ \delta\theta_i \\ \delta\beta_i \\ \delta b_{ai} \\ \delta b_{\omega i} \end{bmatrix} = \begin{bmatrix} I & f_{01} & \delta t & f_{03} & f_{04} \\ 0 & f_{11} & 0 & 0 & -\delta t \\ 0 & f_{21} & I & f_{23} & f_{24} \\ 0 & 0 & 0 & I & 0 \\ 0 & 0 & 0 & 0 & I \end{bmatrix} \tag{A6}$$

$$\begin{aligned}
f_{01} &= \frac{\delta t}{2} f_{21} = -\frac{1}{4} q_k(\hat{a}_i - b_i) \times \delta t^2 - \frac{1}{4} q_{i+1}(\hat{a}_{i+1} - b_i) \times \left[I - \left(\frac{\hat{\omega}_i + \hat{\omega}_{i+1}}{2} - b_i\right) \times \delta t\right]\delta t^2 \\
f_{03} &= -\frac{1}{4}(q_i + q_{i+1})\delta t^2 \\
f_{04} &= \frac{\delta t}{2} f_{24} = \frac{1}{4} q_{i+1}(\hat{a}_{i+1} - b_i) \times \delta t^3 \\
f_{11} &= I - \left(\frac{\hat{\omega}_i + \hat{\omega}_{i+1}}{2} - b_k\right) \times \delta t \\
f_{21} &= -\frac{1}{2} q_i(\hat{a}_i - b_i) \times \delta t - \frac{1}{2} q_{i+1}(\hat{a}_{i+1} - b_i) \times \left[I - \left(\frac{\hat{\omega}_i + \hat{\omega}_{i+1}}{2} - b_i\right) \times \delta t\right]\delta t \\
f_{23} &= -\frac{1}{2}(q_i + q_{i+1})\delta t^2 \\
f_{24} &= \frac{1}{2} q_{i+1}(\hat{a}_{i+1} - b_i) \times \delta t^2
\end{aligned} \tag{A7}$$

$$V_i = \begin{bmatrix} v_{00} & v_{01} & v_{02} & v_{03} & 0 & 0 \\ 0 & \frac{\delta t}{2} & 0 & \frac{\delta t}{2} & 0 & 0 \\ \frac{q_i \delta t}{2} & v_{21} & \frac{q_{i+1}\delta t}{2} & v_{23} & 0 & 0 \\ 0 & 0 & 0 & 0 & \delta t & 0 \\ 0 & 0 & 0 & 0 & 0 & \delta t \end{bmatrix} \tag{A8}$$

$$\begin{aligned}
v_{00} &= -\frac{1}{4} q_i \delta t^2 \\
v_{01} &= v_{03} = \frac{\delta t}{2} v_{21} = \frac{1}{4} q_{i+1}(\hat{a}_{i+1} - b_i) \times \delta t^2 \frac{\delta t}{2} \\
v_{02} &= -\frac{1}{4} q_{i+1} \delta t^2 \\
v_{21} &= v_{23} = \frac{1}{4} q_{i+1}(\hat{a}_{i+1} - b_i) \times \delta t^2
\end{aligned} \tag{A9}$$

Let $z_i^{15\times1} = [\delta\alpha_i, \delta\theta_i, \delta\beta_i, \delta b_{ai}, \delta b_{\omega i}]^T$ and $z_{i+1}^{15\times1} = [\delta\alpha_{i+1}, \delta\theta_{i+1}, \delta\beta_{i+1}, \delta b_{ai+1}, \delta b_{\omega i+1}]^T$ be the error state vector of the i th frame and the i+1 th frame, respectively, and $n = \left[n_{\alpha_i}, n_{\omega_i}, n_{\alpha_{i+1}}, n_{\omega_{i+1}}, n_{b_a}, n_{b_\omega}\right]^T$ is the noise vector; then, Equation (A5) can be written as:

$$\delta z_{i+1}^{15\times1} = F^{15\times15}\delta z_i^{15\times1} + V^{15\times18}n^{18\times1} \tag{A10}$$

where the initial Jacobian value is $J_i = I$ and the Jacobian iteration formula in the process of nonlinear optimization is:

$$J_{i+1}^{15\times15} = F^{15\times15}J_i^{15\times15} \tag{A11}$$

The iterative formula of the covariance of pre-integration in the nonlinear optimization process is:

$$\sum\nolimits_{i+1}^{15\times15} = F\sum\nolimits_i^{15\times15}F^T + Vn_iV^T \tag{A12}$$

After the pre-integration derivation, Equation (14) is the variable quantity of position, rotation, velocity and IMU bias between two frames.

Appendix B

Mathematical Derivation of the Equations (16) and (17).

Equation (16) can be explained using the plane vector method. Let $\left|\left(X_{(k+1,o)}^L - X_{(k,a)}^L\right) \times \left(X_{(k+1,o)}^L - X_{(k,b)}^L\right)\right|$ be the area of the parallelogram formed by three points $X_{(k+1,o)}^L$, $X_{(k,a)}^L$ and $X_{(k,b)}^L$. Let the spatial coordinates of the three points be $X_{(k+1,o)}^L(x_0, y_0, z_0)$, $X_{(k,a)}^L(x_1, y_1, z_1)$ and $X_{(k,b)}^L(x_2, y_2, z_2)$, from which, the three vectors are constructed as:

$$\begin{cases} \overrightarrow{X_{(k+1,o)}^L X_{(k,a)}^L} = X_{(k+1,o)}^L - X_{(k,a)}^L = (x_1 - x_0, y_1 - y_0, z_1 - z_0) \\ \overrightarrow{X_{(k+1,o)}^L X_{(k,b)}^L} = X_{(k+1,o)}^L - X_{(k,b)}^L = (x_2 - x_0, y_2 - y_0, z_2 - z_0) \\ \overrightarrow{X_{(k,a)}^L X_{(k,b)}^L} = X_{(k,o)}^L - X_{(k,b)}^L = (x_2 - x_1, y_2 - y_1, z_2 - z_1) \end{cases} \tag{A13}$$

The molecules of Equation (11) can be obtained as follows:

$$\left|\left(X_{(k+1,o)}^L - X_{(k,a)}^L\right) \times \left(X_{(k+1,o)}^L - X_{(k,b)}^L\right)\right| = \begin{vmatrix} X_{(k+1,o)}^L & X_{(k,a)}^L & X_{(k,b)}^L \\ (x_1 - x_0) & (y_1 - y_0) & (z_1 - z_0) \\ (x_2 - x_0) & (y_2 - y_0) & (z_2 - z_0) \end{vmatrix} \tag{A14}$$

The distance between the point $X_{(k+1,o)}^L$ and the line $X_{(k,a)}^L X_{(k,b)}^L$ represented by Equation (11) is:

$$\begin{aligned} d_{ek}^L &= \frac{\left|\left(X_{(k+1,o)}^L - X_{(k,a)}^L\right) \times \left(X_{(k+1,o)}^L - X_{(k,b)}^L\right)\right|}{\left|X_{(k,a)}^L - X_{(k,b)}^L\right|} \\ &= sqrt\{[(y_1 - y_0) * (z_2 - z_0) - (y_2 - y_0) * (z_1 - z_0)] \\ &* [(y_1 - y_0) * (z_2 - z_0) - (y_2 - y_0) * (z_1 - z_0)] \\ &+ [(x_2 - x_0) * (z_1 - z_0) - (x_1 - x_0) * (z_2 - z_0)] \\ &* [(x_2 - x_0) * (z_1 - z_0) - (x_1 - x_0) * (z_2 - z_0)] \\ &\left.\begin{array}{l} +[(x_1 - x_0) * (y_2 - y_0) - (x_2 - x_0) * (y_1 - y_0)] \\ *[(x_1 - x_0) * (y_2 - y_0) - (x_2 - x_0) * (y_1 - y_0)] \end{array}\right\}] \\ &/ sqrt[(x_2 - x_1) * (x_2 - x_1) + (y_2 - y_1) * (y_2 - y_1) + (z_2 - z_1) * (z_2 - z_1)] \end{aligned} \tag{A15}$$

Similarly, the molecule of Equation (17) can be expressed as the volume of a triangular pyramid composed of four points: $X^L_{(k+1,\rho)}$, $X^L_{(k,c)}$, $X^L_{(k,d)}$ and $X^L_{(k,f)}$ in a geometric sense. It can be known that $\left|\left(X^L_{(k,c)} - X^L_{(k,d)}\right) \times \left(X^L_{(k,c)} - X^L_{(k,f)}\right)\right|$ is twice the area of the base. Let the spatial coordinates of the four points be $X^L_{(k+1,\rho)}(x_3, y_3, z_3)$, $X^L_{(k,c)}(x_4, y_4, z_4)$, $X^L_{(k,d)}(x_5, y_5, z_5)$ and $X^L_{(k,f)}(x_6, y_6, z_6)$, so the three vectors required for constructing are:

$$\begin{cases} \overrightarrow{X^L_{(k+1,\rho)}X^L_{(k,d)}} = X^L_{(k+1,\rho)} - X^L_{(k,d)} = (x_5 - x_3, y_5 - y_3, z_5 - z_3) \\ \overrightarrow{X^L_{(k,c)}X^L_{(k,d)}} = X^L_{(k,c)} - X^L_{(k,d)} = (x_5 - x_4, y_5 - y_4, z_5 - z_4) \\ \overrightarrow{X^L_{(k,c)}X^L_{(k,f)}} = X^L_{(k,c)} - X^L_{(k,f)} = (x_6 - x_4, y_6 - y_4, z_6 - z_4) \end{cases} \tag{A16}$$

The molecules of Equation (17) can be obtained as follows:

$$\begin{aligned} \left|\left(X^L_{(k,c)} - X^L_{(k,d)}\right) \times \left(X^L_{(k,c)} - X^L_{(k,f)}\right)\right| &= \begin{vmatrix} X^L_{(k,c)} & X^L_{(k,d)} & X^L_{(k,f)} \\ (x_5 - x_4) & (y_5 - y_4) & (z_5 - z_4) \\ (x_6 - x_4) & (y_6 - y_4) & (z_6 - z_4) \end{vmatrix} \\ &= sqrt(s_a * s_a, s_b * s_b, s_c * s_c) \end{aligned} \tag{A17}$$

where s_a, s_b and s_c represent the component vectors of x, y and z axes, respectively:

$$\begin{cases} s_a = (y_5 - y_4) * (z_6 - z_4) - (y_6 - y_4) * (z_5 - z_4) \\ s_b = (z_5 - z_4) * (x_6 - x_4) - (z_6 - z_4) * (x_5 - x_4) \\ s_c = (x_5 - x_4) * (y_6 - y_4) - (x_6 - x_4) * (y_5 - y_4) \end{cases} \tag{A18}$$

Therefore, the point-to-surface distance can be obtained as follows:

$$\begin{aligned} d^L_{pk} &= \frac{\left|\left(X^L_{(k+1,\rho)} - X^L_{(k,d)}\right) \cdot \left(\left(X^L_{(k,c)} - X^L_{(k,d)}\right) \times \left(X^L_{(k,c)} - X^L_{(k,f)}\right)\right)\right|}{\left|\left(X^L_{(k,c)} - X^L_{(k,d)}\right) \times \left(X^L_{(k,c)} - X^L_{(k,f)}\right)\right|} \\ &= \frac{(x_4 - x_3) * s_a + (y_4 - y_3) * s_b + (z_4 - z_3) * s_c}{= sqrt(s_a * s_a, s_b * s_b, s_c * s_c)} \end{aligned} \tag{A19}$$

References

1. Mascaro, R.; Teixeira, L.; Hinzmann, T.; Siegwart, R.; Chli, M. GOMSF: Graph-Optimization based Multi-Sensor Fusion for robust UAV pose estimation. In Proceedings of the 2018 IEEE International Conference on Robotics and Automation (ICRA), Brisbane, Australia, 21–25 May 2018; pp. 1421–1428.
2. Lee, W.; Eckenhoff, K.; Geneva, P.; Huang, G.Q. Intermittent GPS-aided VIO: Online Initialization and Calibration. In Proceedings of the 2020 IEEE International Conference on Robotics and Automation (ICRA), Paris, France, 31 May–31 August 2020; pp. 5724–5731.
3. Zhang, J.; Khoshelham, K.; Khodabandeh, A. Seamless Vehicle Positioning by Lidar-GNSS Integration: Standalone and Multi-Epoch Scenarios. *Remote Sens.* **2021**, *13*, 4525. [CrossRef]
4. Forster, C.; Carione, L.; Dellaert, F.; Scaramuzza, D. On-Manifold Preintegration for Real-Time Visual–Inertial Odometry. *IEEE Trans. Robot.* **2017**, *33*, 1–21. [CrossRef]
5. Shan, T.; Englot, B. LeGO-LOAM: Lightweight and Ground-Optimized Lidar Odometry and Mapping on Variable Terrain. In Proceedings of the 2018 IEEE/RSJ International Conference on Intelligent Robots and Systems (IROS), Madrid, Spain, 1–5 October 2018; pp. 4758–4765.
6. Li, S.; Li, J.; Tian, B.; Chen, L.; Wang, L.; Li, G. A laser SLAM method for unmanned vehicles in point cloud degenerated tunnel environments. *Acta Geod. Cartogr. Sin.* **2021**, *50*, 1487–1499.
7. Gong, Z.; Liu, P.; Wen, F.; Ying, R.D.; Ji, X.W.; Miao, R.H.; Xue, W.Y. Graph-Based Adaptive Fusion of GNSS and VIO Under Intermittent GNSS-Degraded Environment. *IEEE Trans. Instrum. Meas.* **2021**, *70*, 9268091. [CrossRef]
8. Chou, C.C.; Chou, C.F. Efficient and Accurate Tightly-Coupled Visual-Lidar SLAM. *IEEE Trans. Intell. Transp. Syst.* **2021**, 1–15. [CrossRef]

9. He, X.; Gao, W.; Sheng, C.Z.; Zhang, Z.T.; Pan, S.G.; Duan, L.J.; Zhang, H.; Lu, X.Y. LiDAR-Visual-Inertial Odometry Based on Optimized Visual Point-Line Features. *Remote Sens.* **2022**, *14*, 622. [CrossRef]
10. Biber, P.; Strasser, W. The normal distributions transform: A new approach to laser scan matching. In Proceedings of the 2003 IEEE/RSJ International Conference on Intelligent Robots and Systems (IROS), Las Vegas, NV, USA, 27–31 October 2003; pp. 2743–2748.
11. Besl, P.J.; Mckay, N.D. A method for registration of 3-D shapes. *IEEE Trans. Pattern Anal. Mach. Intell.* **1992**, *14*, 239–256. [CrossRef]
12. Servos, J.; Waslander, S.L. Multi-Channel Generalized-ICP. In Proceedings of the 2014 IEEE International Conference on Robotics and Automation (ICRA), Hong Kong, China, 31 May–7 June 2014; pp. 3644–3649.
13. Du, S.Y.; Liu, J.; Bi, B.; Zhu, J.H.; Xue, J.R. New iterative closest point algorithm for isotropic scaling registration of point sets with noise. *J. Vis. Commun. Image Represent.* **2016**, *38*, 207–216. [CrossRef]
14. Wu, Z.; Chen, H.; Du, S. Robust Affine Iterative Closest Point Algorithm Based on Correntropy for 2D Point Set Registration. In Proceedings of the IEEE International Joint Conference on Neural Networks (IJCNN), Vancouver, BC, Canada, 24–29 July 2016; pp. 1415–1419.
15. Wu, L.Y.; Xiong, L.; Bi, D.Y.; Fang, T.; Du, S.Y.; Cui, W.T. Robust Affine Registration Based on Corner Point Guided ICP Algorithm. In Proceedings of the IEEE International Conference on Systems, Man, and Cybernetics (SMC), Banff, AB, Canada, 5–8 October 2017; pp. 537–541.
16. Grisetti, G.; Stachniss, C.; Burgard, W. Improved techniques for grid mapping with Rao-Blackwellized particle filters. *IEEE Trans. Robot.* **2007**, *23*, 34–46. [CrossRef]
17. Andreasson, H.; Stoyanov, T. Real-Time Registration of RGB-D Data Using Local Visual Features and 3D-NDT Registration. [DB/OL]. Available online: https://www.researchgate.net/publication/267688026_Real_Time_Registration_of_RGB-D_Data_using_Local_Visual_Features_and_3D-NDT_Registration (accessed on 14 March 2022).
18. Caballero, F.; Merino, L. DLL: Direct LIDAR Localization. A map-based localization approach for aerial robots. In Proceedings of the 2021 IEEE/RSJ International Conference on Intelligent Robots and Systems (IROS), Prague, Czech Republic, 27 September–1 October 2021; pp. 5491–5498.
19. Chen, C.; Yang, B.; Tian, M.; Li, J.; Zou, X.; Wu, W.; Song, Y. Automatic registration of vehicle-borne mobile mapping laser point cloud and sequent panoramas. *Acta Geo Daetica Cartogr. Sin.* **2018**, *47*, 215–224.
20. Koide, K.; Yokozukam, M.; Oishi, S.; Banno, A. Voxelized GICP for Fast and Accurate 3D Point Cloud Registration. In Proceedings of the 2021 IEEE International Conference on Robotics and Automation (ICRA), Xi'an, China, 30 May–5 June 2021; pp. 11054–11059.
21. Yang, J.; Li, H.D.; Campbell, D.; Jia, Y. Go-ICP: A Globally Optimal Solution to 3D ICP Point-Set Registration. *IEEE Trans. Pattern Anal. Mach. Intell.* **2016**, *38*, 2241–2254. [CrossRef] [PubMed]
22. Pan, Y.; Xiao, P.; He, Y.; Shao, Z.; Li, Z. MULLS: Versatile LiDAR SLAM via Multi-metric Linear Least Square. In Proceedings of the 2021 IEEE International Conference on Robotics and Automation (ICRA), Xi'an, China, 30 May–5 June 2021; pp. 11633–11640.
23. Qin, C.; Ye, H.; Pranata, C.E.; Han, J.; Zhang, S.; Liu, M. LINS: A Lidar-Inertial State Estimator for Robust and Efficient Navigation. In Proceedings of the 2020 IEEE International Conference on Robotics and Automation (ICRA), Paris, France, 31 May–31 August 2020; pp. 8899–8906.
24. Li, W.; Liu, G.; Cui, X.; Lu, M. Feature-Aided RTK/LiDAR/INS Integrated Positioning System with Parallel Filters in the Ambiguity-Position-Joint Domain for Urban Environments. *Remote Sens.* **2021**, *13*, 2013. [CrossRef]
25. Li, X.X.; Wang, H.D.; Li, S.Y.; Feng, S.Q.; Wang, X.B.; Liao, J.C. GIL: A tightly coupled GNSS PPP/INS/LiDAR method for precise vehicle navigation. *Satell. Navig.* **2021**, *26*, 2. [CrossRef]
26. Soloviev, A. Tight Coupling of GPS, INS, and Laser for Urban Navigation. *IEEE Trans. Aerosp. Electron. Syst.* **2010**, *46*, 1731–1746. [CrossRef]
27. Shan, T.; Englot, B.; Meyers, D.; Wang, W.; Ratti, C.; Rus, D. LIO-SAM: Tightly-coupled Lidar Inertial Odometry via Smoothing and Mapping. In Proceedings of the 2020 IEEE/RSJ International Conference on Intelligent Robots and Systems (IROS), Las Vegas, NV, USA, 24 October–24 January 2021; pp. 5136–5142.
28. Sun, X.; Guan, H.; Su, Y.; Xu, G.; Guo, Q. A tightly coupled SLAM method for precise urban mapping. *Acta Geod. Cartogr. Sin.* **2021**, *50*, 1585–1593.
29. GTSAM. Available online: https://gtsam.org/tutorials/intro.html (accessed on 14 April 2022).
30. Geiger, A.; Lenz, P.; Stiller, C.; Urtasun, R. Vision meets robotics: The KITTI dataset. *Int. J. Robot. Res.* **2013**, *32*, 1231–1237. [CrossRef]
31. Chen, S.; Zhou, B.; Jiang, C.; Xue, W.; Li, Q. A LiDAR/Visual SLAM Backend with Loop Closure Detection and Graph Optimization. *Remote Sens.* **2021**, *13*, 2720. [CrossRef]
32. Qin, T.; Li, P.; Shen, S. Vins-mono: A robust and versatile monocular visual-inertial state estimator. *IEEE Trans. Robot.* **2018**, *34*, 1004–1020. [CrossRef]

remote sensing

MDPI

Article

A Dual w-Test Based Quality Control Algorithm for Integrated IMU/GNSS Navigation in Urban Areas

Rui Sun [1,2], Ming Qiu [1], Fei Liu [2,3,*], Zhi Wang [2,3] and Washington Yotto Ochieng [1,4]

1 College of Civil Aviation, Nanjing University of Aeronautics and Astronautics, Nanjing 211106, China; rui.sun@nuaa.edu.cn (R.S.); qm1211@nuaa.edu.cn (M.Q.); w.ochieng@imperial.ac.uk (W.Y.O.)
2 Zhe Jiang Key Laboratory of General Aviation Operation Technology, General Aviation Institute of Zhejiang Jiande, Hangzhou 311612, China; wangzhi@camic.cn
3 CAAC Key Laboratory of General Aviation Operation, Civil Aviation Management Institute of China, Beijing 100102, China
4 Department of Civil and Environmental Engineering, Imperial College London, London SW7 2AZ, UK
* Correspondence: liufei@camic.cn

Abstract: Integration of the Global Navigation Satellite System (GNSS), with Inertial Measurement Unit (IMU) sensors to improve navigation performance, is widely used in many land-based applications. However, further application, especially in urban areas, is limited by the quality (due mainly to multipath effects) and availability of GNSS measurements, with a significant impact on performance, especially from low grade integration. To maximize the potential of GNSS measurements, this paper proposes a dual w-test-based quality control algorithm for integrated IMU/GNSS navigation in urban areas. Quality control is achieved through fault detection and exclusion (FDE) with the capability to detect simultaneous multiple faults in measurements from different satellites. The remaining fault-free GNSS measurements are fused with IMU sensor measurements to obtain the final improved state solution. The effectiveness of the algorithm is validated in a deep urban field test. Compared to the cases without fault exclusion, the results show improvements of about 24% and 30% in horizontal and vertical positioning components, respectively.

Keywords: GNSS; IMU; urban positioning; fault detection and exclusion

Citation: Sun, R.; Qiu, M.; Liu, F.; Wang, Z.; Ochieng, W.Y. A Dual w-Test Based Quality Control Algorithm for Integrated IMU/GNSS Navigation in Urban Areas. *Remote Sens.* **2022**, *14*, 2132. https://doi.org/10.3390/rs14092132

Academic Editors: Yuwei Chen, Changhui Jiang, Qian Meng, Bing Xu, Wang Gao, Panlong Wu, Lianwu Guan and Zeyu Li

Received: 17 March 2022
Accepted: 20 April 2022
Published: 29 April 2022

1. Introduction

The emerging mission-critical applications in urban areas are placing more stringent requirements on the underpinning positioning, navigation, and timing (PNT) systems [1]. Due to complementary characteristics, GNSS and Inertial Measurement Unit (IMU) sensors are commonly used in an integrated architecture to support location-based services. However, in urban areas, GNSS signals are susceptible to attenuation and blockage in the built environment, resulting in multipath effects and non-line of sight (NLOS) reception. The satellite faults, defined in this paper, describe corresponding measurements that have acceptable errors, irrespective of the source and type of failure. These errors in the measurements will affect the accuracy and reliability of positioning from integrated IMU/GNSS systems. Therefore, it is particularly important to develop an effective fault detection scheme that can be applied to GNSS measurements so as to ensure quality control of integrated IMU/GNSS systems.

Fault Detection and Exclusion (FDE)-based GNSS measurements quality control has been investigated for many years. The basic FDE methods include: (1) range and position comparison [2]; (2) minimum least squares residuals [3]; (3) parity space [4]; (4) maximum slope (MS) [5]. The four methods have been shown to be largely equivalent.

The performance of FDE algorithms is related to GNSS signal quality and the number of visible satellites. With the increase in constellations beyond GPS, there are more visible satellites and better signal design, greatly improving positioning quality, and promoting the development of FDE algorithms. Some new FDE algorithms have appeared, such as: GPS Integrity Channel (GIC), which is a hybrid between the GIC approach and the maximum solution separation RAIM technique [6]; Novel Integrity Optimized RAIM [7]; Optimally Weighted Average Solution [8]. Given that the probability of multiple faults in a single constellation is relatively small, the above FDE algorithms assume a single fault at a time.

In medium to high density built environments coupled with the increase in the number of constellations, the probability of simultaneous multiple faults increases. Therefore, increasing research effort is dedicated to developing algorithms for simultaneous multiple FDE. These methods include the use of statistics, calculated based on the w-test, to detect and identify outlier faults [9]. A theoretical analysis of the principle of double satellites faults in 2009, as well as their successful elimination through experiments, is presented in [10]. The Solution Separation (SS) algorithm was also applied to Advanced RAIM (ARAIM) research [11]. A point to note is that, for 4-D positioning and geometry permitting, there must be at least five visible satellites for fault detection and at least six visible satellites for fault exclusion in a single constellation. When the number of satellites is insufficient, these FDE algorithms are unavailable, thus affecting the quality of GNSS positioning with potential safety risks.

To address the problem of GNSS measurement quality, additional sensors are also used to aid GNSS FDE by considering the various error characteristics of each sensor [12].

Comparison of FDE performance, based on loosely-coupled and tightly-coupled IMU/GPS integration modes, is also analyzed in some literature [13,14]. A multiple fault detection and elimination algorithm, based on pseudorange comparison, is proposed and used for vehicle GNSS/IMU integrated navigation and positioning [15], but it needs initial database generation. In real situations, multipath effects and poor user-satellite geometry result in excessive positioning errors in urban areas, and the methods above cannot verify the correctness or reliability of the FDE algorithms. In addition, the a priori parameters of the measurement covariance matrix cannot be determined in these urban areas. This increases the probability of incorrect fault detection resulting in excessive final positioning errors. A series of adaptive Kalman filters (AKF) have been developed to overcome the limitation of using a priori statistics to model errors that have time-varying characteristics [16–18]. The adaptive indicators may take on a range of roles, including an adjustment of the covariance matrix of the state estimation vector, the covariance matrix of the process vector, and the covariance matrix of measurement vector [19–21]. None of the adaptive indicators in the above fusion methods, however, have been adjusted specifically for the errors caused by multipath signals and NLOS that are common in urban areas.

In recent years, with the continuous emergence of multi-sensors, the integrated navigation system of multi-source fusion has also ushered in a vigorous development. Altimeter, wheel odometer, magnetometer, etc., improve the accuracy and reliability of navigation information by providing additional information such as position, speed, and altitude to the GNSS/IMU integration system. From the perspective of technology integration, the research on the integration of GNSS, INS, and emerging visual navigation technology is extremely hot. Li developed a semi-tightly coupled GNSS PPP/S-VINS integration framework for better navigation performance in urban environments [22]. On this basis, Li further studied GNSS/LiDAR/INS tightly coupled integrated navigation [23]. However, the above method is in the theoretical research stage, and the high cost of the sensor is not conducive to popularization.

Another idea for quality control is to assign appropriate weights to the GNSS measurements to mitigate the effects of multipath/NLOS signals. The commonly used method is to determine weight based on the quality of GNSS signals. This usually involves the use of one or more characteristics of GNSS signals (e.g., satellite elevation angle, C/N_0, or a combination of the two) to assign corresponding weights to GNSS measurements. Other weighting-based quality control methods include Huber [24], Bifactor reduction model [25], Robust estimation based on M-estimation principle [26], Robust Bayesian estimation [27], and Danish [28]. However, application of appropriate weighting, in different scenarios in urban environments, is difficult. Given the limitations of the state-of-the-art methods above, this paper proposes a dual w-test-based quality control algorithm for integrated IMU/GNSS navigation in urban environments. The contributions are summarized below.

(1) A dual w-test is proposed, which achieves multiple fault detection from the observation domain, thus solving the problem of false alarms in the traditional w-test.

(2) A range detection is proposed to detect the subsets generated after dual w-test, and a scoring strategy is proposed to select the optimal subset. Starting from the location domain, the proposed algorithm is able to reduce the miss detection rate and, therefore, ensure the quality of the output position.

2. Algorithm Design

The proposed tightly-coupled algorithm is illustrated in Figure 1 and comprises two parts. In the first part, a dual w-test-based FDE model is designed for multiple failure detection in urban areas. In the second part, a scoring strategy is used to exclude faulty measurements. The remaining satellites are then fused with IMU sensor measurements to compute the final state.

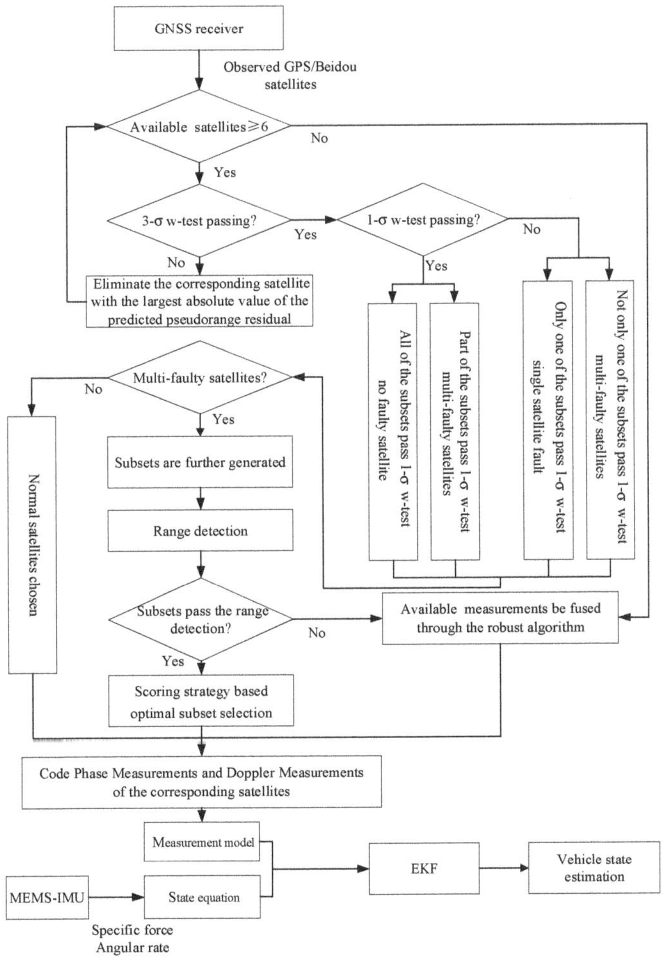

Figure 1. System Framework.

2.1. Dual w-Test

2.1.1. Traditional w-Test

Due to the non-linear relationship between the GNSS pseudorange observation and state variables [29], the linearized pseudorange observation equation can be written as (1).

$$Y = HX + \varepsilon \tag{1}$$

Here, Y is the difference between the observed pseudorange and computed pseudorange from the initial state, H is the measurement matrix, X is the user's state vector, and ε is the observation error vector. The weighted least squares solution for the state vector X is (2).

$$X_{\text{WLS}} = \left(H^T W H\right)^{-1} H^T W Y \tag{2}$$

where, W is the weighting matrix. With $W = (cov(\varepsilon))^{-1}$, based on Equation (2), the residual vector r is derived as:

$$r = Y - HX_{WLS} = \left(I - H\left(H^T W H\right)^{-1} H^T W\right)\varepsilon = S\varepsilon \tag{3}$$

After obtaining the residual vector r, the sum of the squares of the residual or error (SSE) vector is used as the statistics for GNSS fault detection, which is defined as (4)

$$SSE = r^T W r \tag{4}$$

Based on weighted least squares residuals, GNSS pseudorange measurements with significant errors are detected and eliminated by overall and local inspection methods. The overall test assumes that when observations do not contain gross errors, the observation errors follow the Gaussian distribution. Hence, the statistic SSE follows the chi-square distribution with degrees of freedom $(n - m)$, where n is the total number of satellites observed, and m is the dimension of the state. When the test statistic exceeds the global threshold, there is at least one faulty satellite. The global test threshold T_G is:

$$T_G = \chi^2_{1-P_{FA},(m-n)} \tag{5}$$

where P_{FA} is the probability of false alarm, which is selected according to specific application scenarios, and χ^2 denotes the probability density of the chi-square distribution. When the statistic exceeds the global threshold, it is necessary to find the failing measurement or gross error in observations, using the traditional w-test. The test normalizes the residual as a new statistic. The specific expression is (6):

$$w_i = \frac{e_i^T r}{\sqrt{e_i^T S e_i}}, i = 1 : m \tag{6}$$

where e_i is the unit vector whose i-th element is 1. When the i-th observation has no error, the variance of the corresponding observation noise ε_i is σ^2, with w_i following the normal distribution $N(0, \sigma^2)$. $|w_i|_{\max}$ is then compared with the w-test threshold T_L. If $|w_i|_{\max}$ exceeds the threshold, it is considered that the corresponding observation contains gross error. Then, the traditional w-test eliminates the corresponding satellite. The expression of the w-test threshold is:

$$T_L = N_{1-P_{FA}/2}\left(0, \sigma^2\right) \tag{7}$$

The traditional w-test only identifies one faulty satellite at a time, and the $|w_i|_{\max}$ corresponding satellite is eliminated. At the same time, in order to confirm whether there are any faulty satellites in the remaining satellites, all the remaining satellites after w-test are regarded as a new corpus again, and a new round of fault detection is performed.

Therefore, the w-test method is suitable for the case of highly redundant observation data, and it is assumed that only one failure occurs at a time. In urban environments, however, this condition may not be met. Therefore, this paper adopts $3-\sigma$ and $1-\sigma$ w-test double w-test, as shown in the following subsection.

2.1.2. $3-\sigma$ and $1-\sigma$ Dual w-Test

Different from the traditional w-test, this paper firstly adopts $3-\sigma$ w-test. The first $3-\sigma$ w-test is to prevent the observation noise variance from being too small, as well as too strict, for the corresponding w-test threshold. At the same time, in the $3-\sigma$ w-test, the method of excluding satellites is not to eliminate the $|w_i|_{\max}$ corresponding satellite, but it is to eliminate the corresponding satellite with the largest absolute value of the predicted pseudorange residual when the $|w_i|_{\max}$ exceeds the threshold. The predicted pseudorange residual is calculated as (8):

$$\Delta\rho = \rho_{IMU} - \rho_{GNSS} = r_{IMU} + c\left(dt_R - dt^S\right) + I_\rho + T_\rho - \rho_{GNSS} \tag{8}$$

where, ρ_{GNSS} is the observed pseudorange, ρ_{IMU} is the pseudorange predicted by *IMU*, r_{IMU} is the geometric range between the observed satellite and the user position estimated by *IMU*. dt_R and dt^S are the receiver and satellite clock errors, respectively, I_ρ and T_ρ are tropospheric and ionospheric corrections, respectively. The r_{IMU} can be calculated as (9):

$$r_{IMU} = \sqrt{\left(X_k^G - X_k^{IMU}\right)^2 + \left(Y_k^G - Y_k^{IMU}\right)^2 + \left(Z_k^G - Z_k^{IMU}\right)^2} \tag{9}$$

$\left(X_k^G, Y_k^G, Z_k^G\right)$ is the satellite position at epoch k, $\left(X_k^{IMU}, Y_k^{IMU}, Z_k^{IMU}\right)$ is user positions estimated using IMU data at epoch k. However, due to the complexity of urban environments, it is impossible to ensure correct detection using the $3-\sigma$ w-test. Therefore, the positions calculated before and after each $3-\sigma$ w-test are saved until either no faulty satellite measurements are detected or the number of remaining observed satellites is insufficient. Then, in order to ensure that multiple faults can be detected, this paper takes each subset obtained after the $3-\sigma$ w-test, removing a satellite each time, and performing the $1-\sigma$ w-test on C_m^1 each subset. The results can be one of four cases:

1. The universal set and all subsets pass the $1-\sigma$ w-test.
2. The universal set and some subsets pass the $1-\sigma$ w-test.
3. The universal set does not pass the $1-\sigma$ w-test, and only one of the subsets passes the $1-\sigma$ w-test.
4. The universal set does not pass the $1-\sigma$ w-test, with more than one subset passing the $1-\sigma$ w-test.

The fault conditions at a given epoch can be determined by considering the test results in these four cases. In case 1, we consider that there is no faulty satellite at this epoch, as the universal set and all subsets have passed the w-test. In case 2, the high correlation of each satellite will result in the universal set passing the test, while the low correlation of the faulty satellite in the subset, after one satellite exclusion, can result in the subset not passing the test. Therefore, in this case, we consider that there are multiple faults. In case 3, as a single satellite fault can lead to the universal set not passing the w-test, the subset can only pass the w-test in the case that the faulty satellite is excluded. Therefore, a single fault case is considered in this case. In case 4, faulty satellites in the universal set and subsets can lead to the failure to pass the w-test for a part or all of the subsets. Therefore, the existence of multiple faults is considered in this case. Satellite selection is then made according to the fault conditions. In case 1, all of the satellites at this epoch are selected for a further GNSS/IMU integration. In case 3, the satellites in the subset, which passed the w-test, are selected for further fusion. Considering cases 2 and 4 with multiple faults, the C_m^2 subsets are further generated, which are then subjected to range detection. The range is calculated by the difference between the predicted position estimated by the IMU data and the position calculated by the selected subset in the proposed algorithm. The expression for range detection is:

$$\begin{bmatrix} e \\ n \\ u \end{bmatrix} = \begin{bmatrix} -\sin\lambda_0 & \cos\lambda_0 & 0 \\ -\sin\varphi_0\cos\lambda_0 & -\sin\varphi_0\sin\lambda_0 & \cos\varphi_0 \\ \cos\varphi_0\cos\lambda_0 & \cos\varphi_0 sin\lambda_0 & \sin\varphi_0 \end{bmatrix} \begin{bmatrix} x_s - x_0 \\ y_s - y_0 \\ z_s - z_0 \end{bmatrix} \tag{10}$$

where λ_0 and φ_0 are, respectively, the latitude and longitude corresponding to the predicted position. x_s, y_s, z_s and x_0, y_0, z_0, respectively, are the coordinates of the calculated position and the predicted position in the WGS-84 coordinate system. Then,$|e|$,$|n|$, and $|u|$ are compared with the range threshold. Here, the threshold of the range value is set as 17 m, as the city speed limit is around 60 km/h (i.e., 17 m/s). Only the subsets that pass the range detection test are used further for the optimal subset selection.

2.2. Scoring Strategy Based Optimal Subset Selection

After range detection, the subsets that pass the test are selected. The optimal subset within these selected subsets is chosen, and the corresponding measurements in the optimal subset are used to integrate with the IMU data to calculate position. The strategy uses a scoring mechanism to subtract the positions calculated using the selected subsets from the predicted position at the current epoch. The predicted position can be obtained from that of the previous epoch combined with inertial navigation information. The difference in position is then scored according to the following formula, based on a weighting method, in which the smaller the *JointCost* the higher the score. Finally, the satellites corresponding to the position difference with the highest score are selected to be combined with the inertial navigation. The *JointCost* is calculated as:

$$JointCost = \frac{\boldsymbol{Cost}(1) - Cost1min}{Cost1max - Cost1min} + \frac{\boldsymbol{Cost}(2) - Cost2min}{Cost2max - Cost2min} + \frac{\boldsymbol{Cost}(3) - Cost3min}{Cost3max - Cost3min} \tag{11}$$

Here, $Cost1max, Cost2max, Cost3max$ are the maximum values of longitude, latitude, and height difference among all the position differences. $Cost1min$, $Cost2min$, and $Cost3min$ are the minimum values of longitude, latitude, and height difference among all the position differences. $\boldsymbol{Cost}(1), \boldsymbol{Cost}(2), \boldsymbol{Cost}(3)$ are all the longitude, latitude, and height difference among all the position differences.

2.3. IMU/GNSS Integration

In this section, an Extended Kalman Filter (EKF), based on linearization of nonlinear models, is used as the data fusion algorithm [30]. The state vector for the EKF is:

$$\boldsymbol{X} = \begin{bmatrix} (\delta \boldsymbol{r}^e_{INS})^T & (\delta \boldsymbol{v}^e_{INS})^T & (\boldsymbol{\phi}^e_{INS})^T & \boldsymbol{b}_g{}^T & \boldsymbol{b}_a{}^T & \boldsymbol{s}_g{}^T & \boldsymbol{s}_a{}^T & \boldsymbol{t}_{GPS} & \boldsymbol{\delta t}_{GPS} & \boldsymbol{t}_{BDS} & \boldsymbol{\delta t}_{BDS} \end{bmatrix} \tag{12}$$

where, $\delta \boldsymbol{r}^e_{INS}$, $\delta \boldsymbol{v}^e_{INS}$, and $\boldsymbol{\phi}^e_{INS}$ are the three-axis error vectors of IMU position, velocity, altitude in the ECEF framework e; $\boldsymbol{b}_g$, $\boldsymbol{b}_a$, $\boldsymbol{s}_g$, and $\boldsymbol{s}_a$ are the three-axis acceleration and gyroscope bias and scale factor error; $\boldsymbol{t}_{GPS}$ and $\boldsymbol{\delta t}_{GPS}$ are the receiver clock error and drift rate of GPS satellite; $\boldsymbol{t}_{BDS}$ and $\boldsymbol{\delta t}_{BDS}$ are the clock error and drift rate of Beidou satellite. The system model is then formed as a first-order state equation in (13):

$$\dot{\boldsymbol{X}} = \boldsymbol{F}\boldsymbol{X} + \boldsymbol{G}\boldsymbol{W} \tag{13}$$

where $\dot{\boldsymbol{X}}$ is the first derivative of $\boldsymbol{X}$. $\boldsymbol{F}$ is the dynamic transition matrix, $\boldsymbol{G}$ is the noise driven matrix, and $\boldsymbol{W}$ is the system noise. The measurement model is given by:

$$\boldsymbol{Z} = \boldsymbol{H}\boldsymbol{X} + \boldsymbol{V} \tag{14}$$

where $\boldsymbol{Z}$ is the measurements vector, $\boldsymbol{H}$ is the measurement mapping matrix, and $\boldsymbol{V}$ represents the measurement noise. In this paper, if the number of visible satellites is n, the pseudorange error and the Doppler measurement error are used to form measurement vector $\boldsymbol{Z}$ as:

$$\boldsymbol{Z} = \begin{bmatrix} \rho_{1,GPS}^{IMU} - \rho_{1}^{GPS} \\ \vdots \\ \rho_{l,GPS}^{IMU} - \rho_{l}^{GPS} \\ \rho_{1,BDS}^{IMU} - \rho_{1}^{BDS} \\ \vdots \\ \rho_{m,BDS}^{IMU} - \rho_{m}^{BDS} \\ f_{1,GPS}^{IMU} - f_{1}^{GPS} \\ \vdots \\ f_{l,GPS}^{IMU} - f_{l}^{GPS} \\ f_{1,BDS}^{IMU} - f_{1}^{BDS} \\ \vdots \\ f_{m,BDS}^{IMU} - f_{m}^{BDS} \end{bmatrix}_{2n\times 1} \tag{15}$$

where ρ_{GNSS}^{IMU} and f_{GNSS}^{IMU} denote IMU-derived GNSS pseudorange and Doppler measurements respectively. Based on the derivations in [30], ρ^{GNSS} and f^{GNSS} refer to pseudorange and Doppler measurements decoded from GNSS observation data, respectively. l and m refer to the number of GPS and BDS visible satellites. After discretization of (13) and (14), the discrete form of the Kalman filtering procedure can be split into two stages, as follows:

Prediction stage:

$$\hat{\boldsymbol{X}}_{k,k-1} = \boldsymbol{\Phi}_{k,k-1}\hat{\boldsymbol{X}}_{k-1} \tag{16}$$

$$\boldsymbol{P}_{k,k-1} = \boldsymbol{\Phi}_{k,k-1}\boldsymbol{P}_{k-1}\boldsymbol{\Phi}_{k,k-1}^{T} + \boldsymbol{Q}_{k-1} \tag{17}$$

Update stage:

$$\boldsymbol{K}_k = \boldsymbol{P}_{k,k-1}\boldsymbol{H}_k{}^T\left(\boldsymbol{H}_k\boldsymbol{P}_{k,k-1}\boldsymbol{H}_k{}^T + \boldsymbol{R}_k\right)^{-1} \tag{18}$$

$$\hat{\boldsymbol{X}}_k = \hat{\boldsymbol{X}}_{k,k-1} + \boldsymbol{K}_k(\boldsymbol{Z}_k - \boldsymbol{H}_k\hat{\boldsymbol{X}}_{k,k-1}) \tag{19}$$

$$\boldsymbol{P}_k = (\boldsymbol{I} - \boldsymbol{K}_k\boldsymbol{H}_k)\boldsymbol{P}_{k,k-1}(\boldsymbol{I} - \boldsymbol{K}_k\boldsymbol{H}_k)^T + \boldsymbol{K}_k\boldsymbol{R}_k\boldsymbol{K}_k{}^T \tag{20}$$

where,$\hat{\boldsymbol{X}}_k$ is the system state vector estimates at time epoch k; $\boldsymbol{\Phi}_k$ is the system transition matrix at time epoch k; $\boldsymbol{P}_k$ is the error covariance matrix at time epoch k; $\boldsymbol{Q}_k$ is the system noise covariance matrix at time epoch k; $\boldsymbol{R}_k$ is the measurement noise covariance matrix at time epoch k; $\boldsymbol{H}_k$ is the measurement matrix at time epoch k; $\boldsymbol{K}_k$ is the Kalman gain matrix at time epoch k; $\Theta_{k,k-1}$ is the matrix/vector Θ propagation from time epoch $k-1$ to k.Table 1 has illustrated the parameters and their value or initial value used for the EKF. The setting of the system noise covariance matrix $\boldsymbol{Q}$ is based on experience. The ***diag*** means that the matrix is a diagonal matrix and the values in the bracket are the diagonal elements. The initial value of error covariance matrix of the state vector $\boldsymbol{P}$, noted as $\boldsymbol{P}_0$, is calculated by the historical data collected from the IMU and GNSS receiver. The covariance matrix of the measurement noise $\boldsymbol{R}$ is set based on the statistical data collected from GNSS receiver.

Table 1. The parameters used for the EKF.

Parameter	Initial Value
Q	$diag(I_{3\times3}*0.0004^2\ \ I_{3\times3}*0.0005^2\ \ I_{3\times3}*0.000035^2\ \ I_{3\times3}*0.00000032^2\ \ I_{3\times3}*0.0001^2\ \ I_{3\times3}*0.00001^2\ \ I_{3\times3}*0.00001^2\ \ 0.001^2\ \ 0.002^2\ \ 0.001^2\ \ 0.002^2)$
R	$I_{n\times n}*1.5^2$
P_0	$diag(I_{3\times3}*0.025^2\ \ I_{3\times3}*0.075^2\ \ I_{3\times3}*0.000035^2\ \ I_{3\times3}*0.0000097^2\ \ I_{3\times3}*0.003^2\ \ I_{3\times3}*0.0025^2\ \ I_{3\times3}*0.003^2\ \ 0.02^2\ \ 0.03^2\ \ 0.02^2\ \ 0.03^2)$

If positions calculated by all subsets do not pass the range detection test or the number of satellites cannot meet the condition of the w-test, then all the satellite measurements and the inertial navigation output information are fused through the robust algorithm. The robust algorithm introduces a fault detection factor D to scale $\boldsymbol{R}$. D is given as:

$$D_{ii} = \begin{cases} 1, & \left|\tilde{\mathfrak{R}}_{k,i}\right| \leq Tm \\ \frac{\left|\tilde{\mathfrak{R}}_{k,i}\right|}{Tm}, & \left|\tilde{\mathfrak{R}}_{k,i}\right| > \mathrm{Tm} \end{cases} \tag{21}$$

$$\tilde{\mathfrak{R}}_{k,i} = \frac{\mathfrak{R}_{k,i}}{\sqrt{\mathbb{C}_{k,ii}}} \tag{22}$$

$\mathfrak{R}_k = \boldsymbol{Z}_k - \boldsymbol{H}_k\hat{\boldsymbol{X}}_{k,k-1}$ is the innovation sequence, which exhibits a white Gaussian sequence of mean zero and covariance $\mathbb{C}_k$ where $\mathbb{C}_k = \boldsymbol{H}_k\boldsymbol{P}_{k,k-1}\boldsymbol{H}_k{}^T + \boldsymbol{R}_k$. Tm is a constant value, which is valued according to the specific scenario. Then, the elements in $\boldsymbol{R}$ are given as:

$$\overline{\boldsymbol{R}}_{k,ii} = D_{ii} \cdot \boldsymbol{R}_{k,ii} \tag{23}$$

3. Test and Validation

3.1. Simulation

Faults are simulated and added to data from UAV flight tests to test the proposed quality control algorithm. The UAV flight data were collected in Nantou City, Taiwan, shown in Figure 2. The UAV used in the test is AXH-E230 from AVIX Technology (Toronto, ON, Canada), and it was flown semi-automatically with a smart power control module to perform autonomous intelligent navigation flight mission. The onboard equipment setup included: (1) a dual-frequency GNSS receiver, Trimble BD 982 (Sunnyvale, CA, USA), with a sampling rate of 10 Hz for the raw pseudorange measurements collection; (2) a STIM-300 IMU (Sensonor, Horten, Norway), with a sampling rate of 100 Hz for UAV acceleration and angular rate collection; (3) an on-board VLP-16 Velodyne Lidar (San Jose, CA, USA) to provide centimeter-level positioning accuracy for the reference trajectory generation in the experiment. The speed of UAV was less than 10 m/s during the flight, and the height was about 60 m AGL (with the ground elevation around 120 m). The fault scenarios in Table 2 were specified in order to compare the proposed algorithm with the traditional IMU/GNSS tightly-coupled (TC) without fault exclusion, the TC with traditional w-test quality control (FDE TC), and the TC with Robust filter (AKF TC) in [31].

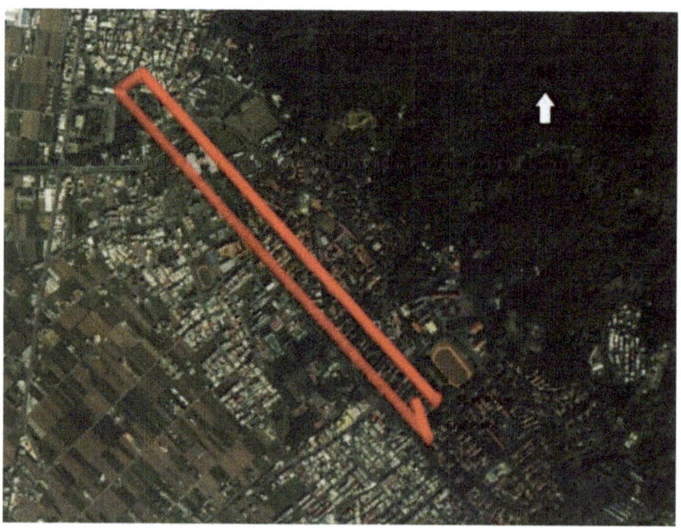

Figure 2. Unmanned aerial vehicle (UAV) flight trajectory.

Table 2. The defined scenarios.

Scenarios	Time Interval of Faults (s)	Error Sources
1	30	10 m, 30 m step errors added to the pseudoranges of two satellites
2	30	10 m, 50 m step errors added to the pseudoranges of two satellites
3	30	30 m, 30 m step errors added to the pseudoranges of two satellites
4	30	30 m, 50 m step errors added to the pseudoranges of two satellites

In the different scenarios above, for each selected satellite, an error of 10 m, 30 m, or 50 m was injected into the pseudo-range observation of the satellite during the corresponding fault duration. Based on the derivations in [32], UAV flight in the urban environment is subjected to multipath interference to produce similar errors, with error magnitudes less than 10 m having little impact on the satellite navigation and positioning results, and is hence ignored as constituting failure. At the same time, considering the characteristics of UAV in urban low-altitude areas, fault duration is selected as 30 s. In order to verify the validity of the algorithm, in terms of accuracy, this paper uses the Root Mean Square Error (RMSE) metric to compare the performance of the TC, FDE TC, AKF TC, and the proposed methods. The errors of the position, calculated from the candidate algorithms, are shown in Figure 3. The RMSE of the positions for the candidate algorithms are represented in Table 3.

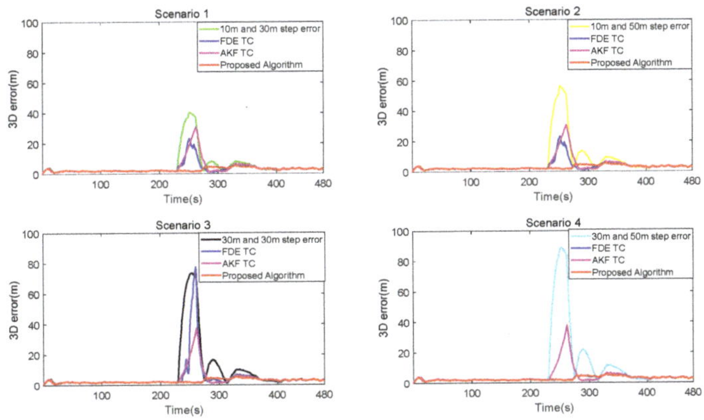

Figure 3. The positioning error of TC, FDE TC, AKF TC, and the proposed algorithm in the four fault scenarios.

Table 3. Comparison of algorithm performance between TC, FDE TC, AKF TC, and the proposed algorithm in the different fault scenarios.

Scenarios	TC	FDE TC		AKF TC		Proposed Algorithm	
	RMSE (m)	RMSE (m)	Improvement (%)	RMSE (m)	Improvement (%)	RMSE (m)	Improvement (%)
1	9.62	4.92	48.89	6.26	34.97	2.98	69.07
2	13.04	4.92	62.28	6.27	51.89	2.98	77.17
3	17.36	11.28	35	6.78	60.92	2.98	82.85
4	20.73	2.98	85.64	6.82	67.10	2.98	85.64

It can be seen, in Figure 3, that TC position error increases rapidly after pseudorange errors are introduced in the four scenarios. This indicates that, without FDE, the IMU/GNSS integrated navigation positioning quality is seriously degraded and results in divergence in the filter estimated results. Therefore, quality control of the GNSS measurement is essential. Meanwhile, by observing the position errors of the FDE TC in different scenarios, it can be seen that, in most cases, when two satellites simultaneously fail, the performance of FDE TC is poor. Only when one satellite has a 30 m step error, and one satellite has a 50 m step error, does FDE TC correctly identify the two faulty satellites in all epochs and eliminate them.

In the other three scenarios, however, the corresponding faulty satellites could not be correctly detected and excluded in all epochs by FDE TC, resulting in a large positioning error. In scenario 3, the maximum positioning error of the FDE TC method even exceeds that of the traditional TC. This is mainly because, in scenario 3, the two satellites add the same step error. As a result, the test statistics of other satellites are strongly correlated with the two faulty satellites, resulting in the maximum test statistics exceeding the traditional w-test threshold. When the satellite with the maximum test statistics exceeding the threshold is eliminated based on a traditional w-test, the redundancy of the observation data is further reduced, so the remaining faulty satellite cannot be detected in the subsequent traditional w-test. The satellite faults still exist in the GNSS measurements, so the positioning performance of the FDE TC is comparable to that of the traditional TC without FDE. It can be seen from Table 3 that the FDE TC, in the above four different scenarios, has similar accuracy to the traditional TC in some cases. However, in scenarios 1 and 2, the FDE TC can still eliminate all faulty satellites in some epochs, but the faulty satellites cannot be correctly eliminated all the time by FDE TC. As a result, the positioning performance of FDE TC is improved by 49% and 62% compared with the traditional TC, respectively. On the other hand, although AKF TC cannot eliminate faults, it reduces the weight of fault observations, thus ensuring the navigation performance to a certain extent. The positioning performance of AKF TC is improved by 35%, 52%, 61%, and 67% compared with the traditional TC, respectively.

However, compared with the above algorithm, the proposed algorithm significantly improves positioning accuracy. This also shows that the proposed algorithm can correctly detect the satellites with the step errors in the above four different cases. The 3D positioning RMSE of the algorithm proposed in this paper, in four different fault scenarios, is 2.98 m. Compared with 9.62 m, 13.04 m, 17.36 m, and 20.73 m of the traditional TC, the accuracy is improved by 69.07%, 77.17%, 82.85%, and 85.64%, respectively. In summary, the above results show that the algorithm proposed in this paper can correctly detect the faulty satellites in the real-data field scenarios with the simulated step errors. Compared with the traditional TC, FDE TC, and AKF TC, it is able to provide a significant improvement in the position solutions.

3.2. Field Test

In order to further validate the performance of the proposed algorithm in an urban environment, a field test was carried out in a deep urban environment in Taipei. The experimental data acquisition equipment contained a low-cost IMU Stim-300 and a GNSS receiver Trimble BD 982, with a sampling rate of 250 Hz and 1 Hz, respectively. The reference trajectory was obtained by an integrated high-grade GNSS receiver and iNAV-RQH IMU with the commercial software NovAtel Inertial Explorer. The experimental test environment is shown in Figure 4, and the reference trajectory is shown in Figure 5. PDOP values during the test are always very high, with the highest value above 16, exhibiting the characteristics of the deep urban environment, as seen in Figure 6. The number of visible satellites is shown in Figure 7.

In order to evaluate the performance of the proposed algorithm, the results of the proposed algorithm are compared with those of the traditional TC, FDE TC, and AKF TC. The errors in position, velocity, and altitude, calculated from the algorithms, are shown in Figures 8–10. The accuracies (RMSE) of the position, velocity, and altitude for the algorithms are given in Tables 3–6.

Figure 4. Environments of field test.

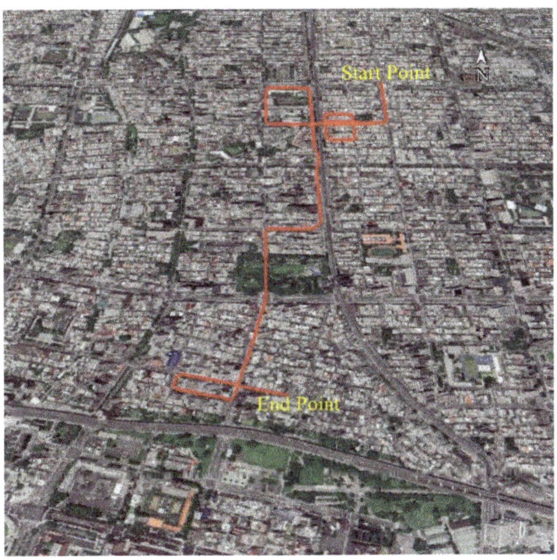

Figure 5. Vehicle trajectory in field test.

From Figure 8 and Table 3, the AKF TC position RMSE is 4.40 m in the horizontal direction and 8.94 m in the vertical direction (Down), which is an improvement of 11.65% and 17.15% compared to the 4.98 m and 10.79 m of the TC. The FDE TC vertical position RMSE is 9.66 m, whose performance is not as good as AKF TC, but the performance is better in the horizontal direction. However, neither is as much improved as the algorithm proposed in this paper. The position RMSE of the algorithm proposed is 3.79 m and 7.51 m in the horizontal and vertical directions. The results represent improvements of 23.90% and 30.40% compared to TC without FDE, 7.79% and 22.26% over FDE TC, as well as 13.86% and 15.88% over AKF TC, respectively. As shown in Figure 11, the algorithm proposed in this paper has a better performance in urban environments in the horizontal directions.

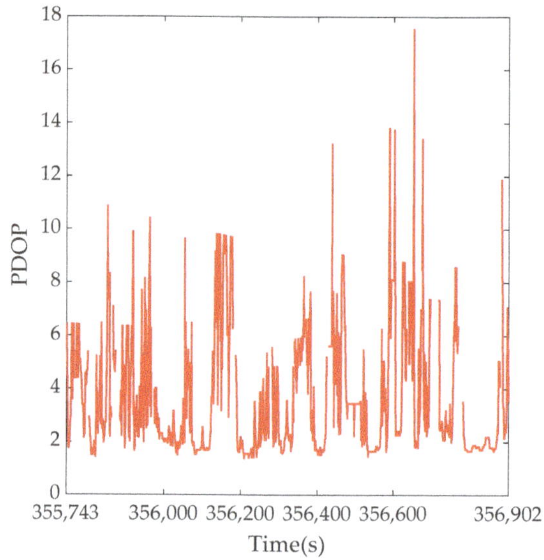

Figure 6. PDOP in field test.

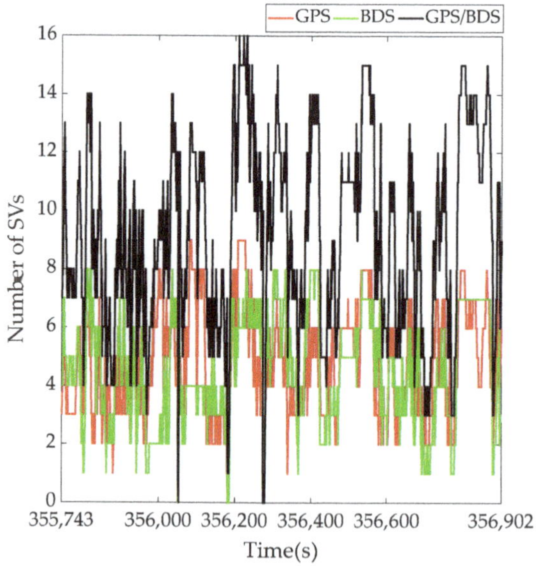

Figure 7. Visible satellite number in field test.

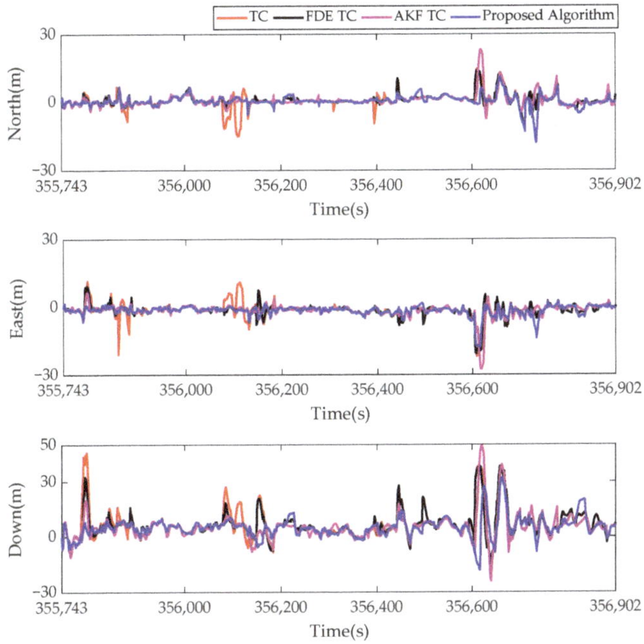

Figure 8. The position error of TC, FDE TC, AKF TC, and the proposed algorithm in field test.

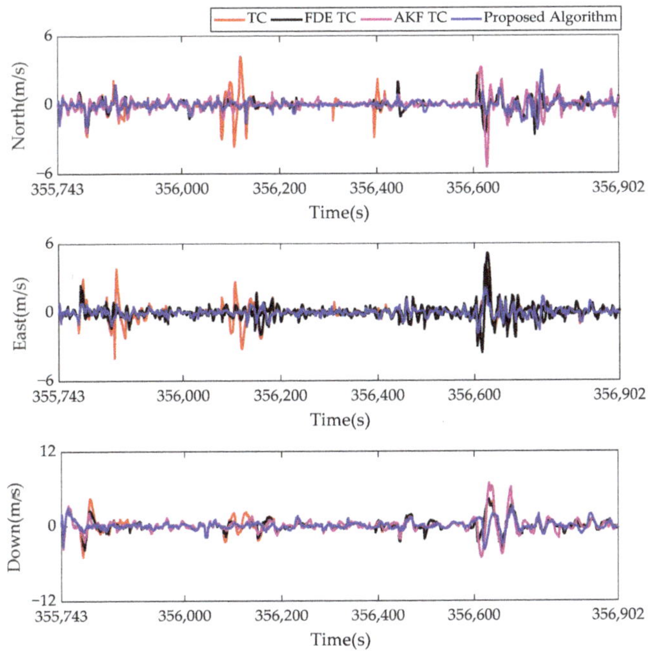

Figure 9. The velocity error of TC, FDE TC, AKF TC, and the proposed algorithm in field test.

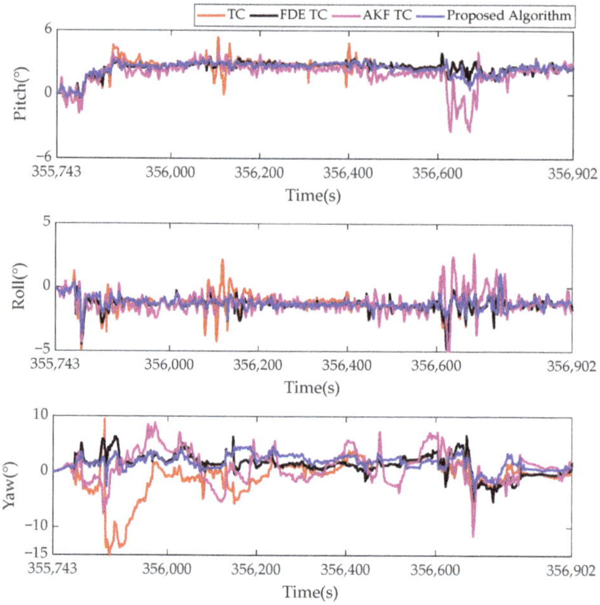

Figure 10. The altitude error of TC, FDE TC, AKF TC, and the proposed algorithm in field test.

Table 4. The position RMSE of TC, FDE TC, AKF TC, and the proposed algorithm in field test.

Algorithm	RMSE (m)			
	North	East	2D	Down
TC	3.31	3.72	4.98	10.79
FDE TC	2.65	3.15	4.11	9.66
Improvement over TC (%)	19.94	15.32	17.47	10.47
AKF TC	2.94	3.27	4.40	8.94
Improvement over TC (%)	11.18	12.10	11.65	17.15
Proposed algorithm	2.55	2.80	3.79	7.51
Improvement over TC (%)	22.96	24.73	23.90	30.40
Improvement over FDE TC (%)	3.77	11.11	7.79	22.26
Improvement over AKF TC (%)	13.27	14.37	13.86	15.88

Table 5. The velocity RMSE of TC, FDE TC, AKF TC, and the proposed algorithm in field test.

Algorithm	RMSE (m/s)			
	North	East	2D	Down
TC	0.68	0.71	0.98	1.07
FDE TC	0.48	0.55	0.73	0.93
Improvement over TC (%)	29.41	22.54	25.51	13.08
AKF TC	0.63	0.63	0.89	1.21
Improvement over TC (%)	7.35	11.27	9.18	−13.08
Proposed algorithm	0.45	0.38	0.59	0.72
Improvement over TC (%)	33.82	46.48	39.80	32.71
Improvement over FDE TC (%)	6.25	30.91	19.18	22.58
Improvement over AKF TC (%)	28.57	39.68	33.71	40.50

Table 6. The altitude RMSE of TC, FDE TC, AKF TC, and the proposed algorithm in field test.

Algorithm	RMSE (°)		
	Pitch	Roll	Yaw
TC	2.70	1.39	3.43
FDE TC	2.62	1.38	2.28
Improvement over TC (%)	2.96	0.72	33.53
AKF TC	2.33	1.43	3.08
Improvement over TC (%)	13.70	−2.88	10.20
Proposed algorithm	2.58	1.27	2.25
Improvement over TC (%)	4.44	8.63	34.4
Improvement over FDE TC (%)	1.53	7.97	1.32
Improvement over AKF TC (%)	−10.73	11.19	27.60

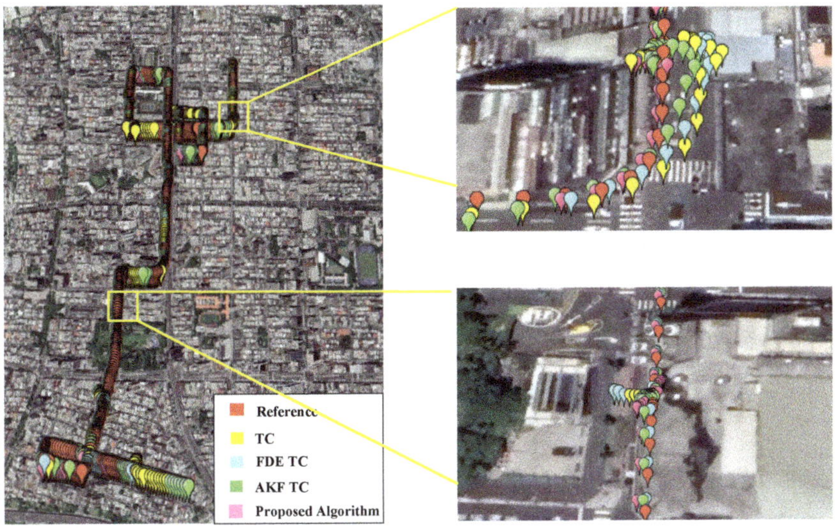

Figure 11. Trajectory comparison for TC, FDE LC, AKF TC, and the proposed algorithm in field test.

It can be seen from Figure 9 and Table 5 that the horizontal and vertical velocity RMSE of the traditional TC scheme without FDE are 0.98 m/s and 1.07 m/s, with the corresponding values, from the proposed algorithm, of 0.59 m/s and 0.72 m/s. These correspond to improvements of 40% and 33%, respectively. While the AKF TC gives an RMSE for horizontal velocity of 0.89 m/s, the performance in the vertical direction deteriorates by 13.08% due to its inability to be accurately adjusted, specifically, for the errors caused by multipath signals and NLOS that are common in urban areas. Compared with the 0.73 m/s and 0.93 m/s of FDE TC, the proposed algorithm in this paper improves by 19% and 23%. This shows that correct fault detection and elimination is effective for quality control.

For the performance of altitude determination in Figure 10 and Table 6, pitch, roll, and yaw RMSE of the traditional TC scheme without FDE are 2.70°, 1.39°, and 3.43°, with the corresponding values from the FDE TC of 2.62°, 1.38°, and 2.28°. These correspond to improvements of 2.96%, 0.72%, and 33.53%, respectively. It is worth noting that the correction of yaw information has always been a difficult problem in the GNSS/IMU integrated navigation algorithm, and the yaw RMSE of FDE TC has dropped by 27.6%. This further illustrates the importance of quality control. While the AKF TC gives an RMSE for pitch angle of 2.33°, the performance in the roll angle deteriorates by 2.88%, and there is less improvement in the yaw angle. The proposed algorithm has improved the estimation results of pitch angle, roll angle, and yaw angle by 2%, 8%, and 1% compared with FDE TC, respectively. Although the performance of the proposed algorithm in this paper is not good in the pitch angle, compared with AKF TC, the overall performance of the proposed algorithm in this paper is better, which improves by 11.19% and 27.60% in roll and yaw angles.

4. Conclusions

This paper has developed a dual w-test-based quality control algorithm for IMU/GNSS integrated navigation in urban areas. Simulation and field test results show that the proposed algorithm is capable of achieving quality control for integrated IMU/GNSS navigation. The experimental results in deep urban environments show that the proposed integration algorithm can improve positioning accuracy compared to the cases without fault exclusion by about 24% and 30%, compared to FDE TC by about 8% and 22%, and compared to AKF TC by about 14% and 16% in the horizontal and vertical directions, respectively. However, the current work does not suit for the case of insufficient visible satellites, as the dual w-test cannot be carried out without enough of a degree of freedom in the statistic SSE. In future work, we will continue to develop more advanced quality control methods, including seeking a better robust algorithm when the number of satellites is insufficient and designing a corresponding failure detection algorithm according to the failure mechanisms of different sensors, such as inertial sensors, vision sensors, and lidar.

Author Contributions: Conceptualization, M.Q. and R.S.; Data curation, F.L.; Formal analysis, M.Q.; Methodology, M.Q. and R.S.; Software, F.L.; Supervision, Z.W. and W.Y.O.; Writing—original draft, M.Q.; Writing—review & editing, R.S. and W.Y.O. All authors have read and agreed to the published version of the manuscript.

Funding: This work was supported in part by the sponsorship of the National Natural Science Foundation of China under Grant 71731001, U1933130, 41974033, 42174025, in part by Natural Science Foundation of Jiangsu Province under Grant BK20212569 and in part by Zhe Jiang Key laboratory of General Aviation Operation technology (General Aviation Institute of Zhejiang JianDe) under Grant JDGA2020-11.

Conflicts of Interest: The authors declare no conflict of interest.

References

1. Du, Y.; Wang, J.; Rizos, C.; El-Mowafy, A. Vulnerabilities and integrity of precise point positioning for intelligent transport systems: Overview and analysis. *Satell. Navig.* **2021**, *2*, 3. [CrossRef]
2. Lee, Y.C. Analysis of range and position comparison methods as a means to provide GPS integrity in the user receiver. In Proceedings of the The User Recver Us Institute of Navigation Annual Meeting, Seattle, WA, USA, 24–26 June 1986; Volume 42, pp. 1–4.
3. Parkinson, B.W.; Axelrad, P. A basis for the development of operational algorithms for simplified GPS integrity checking. In Proceedings of the Satellite Division's First Technical Meeting, Colorado Spring, CO, USA, 21–25 September 1987.
4. Sturza, M.A. Navigation system integrity monitoring using redundant measurements. *Navigation* **1988**, *35*, 483–501. [CrossRef]
5. Brown, R.G.; Mcburney, P.W. Self-Contained GPS integrity check using maximum solution separation. *Navigation* **1988**, *35*, 41–53. [CrossRef]
6. Virball, V.G.; Michalson, W.R. A GPS integrity channel based fault detection and exclusion algorithm using maximum solution separation. In Proceedings of the Position Location and Navigation Symposium, IEEE, Las Vegas, NV, USA, 11–15 April 1994.
7. Hwang, P.Y.; Brown, R.G. RAIM-FDE revisited: A new breakthrough in availability performance with NIORAIM (novel integrity-optimized RAIM). *Navigation* **2006**, *53*, 41–52. [CrossRef]
8. Lee, Y.C.; Fernow, J.P.; McLaughlin, M.P. GPS and Galileo with RAIM or WAAS for vertically guided approaches. In Proceedings of the 18th International Technical Meeting of the Satellite Division of the Institute of Navigation, Long Beach, CA, USA, 13–16 September 2005; pp. 302–326.
9. Hewitson, S.; Wang, J. GNSS Receiver Autonomous Integrity Monitoring (RAIM) for multiple outliers. *Eur. J. Navig.* **2006**, *4*, 47–57.
10. Knight, N.L.; Wang, J.; Rizos, C. GNSS integrity monitoring for two satellite faults. In Proceedings of the International Global Navigation Satellite Systems Society IGNSS Symposium 2009, Surfers Paradise, Australia, 1–3 December 2009.

11. Joerger, M.; Stevanovic, S.; Chan, F.C.; Langel, S.; Pervan, B. Integrity risk and continuity risk for fault detection and exclusion using solution separation ARAIM. In Proceedings of the International Technical Meeting of the Satellite Division of the Institute of Navigation, Nashville, TN, USA, 16–20 September 2013.
12. Bruggemann, T.S. GPS fault detection with IMU and aircraft dynamics. *IEEE Trans. Aerosp. Electron. Syst.* **2011**, *47*, 305–316. [CrossRef]
13. Petovello, M.G. Real-Time Integration of a Tactical-Grade IMU and GPS for High-Accuracy Positioning and Navigation. Ph.D. Thesis, University of Calgary, Calgary, AB, Canada, 2003.
14. Hewitson, S.; Wang, J. GNSS receiver autonomous integrity monitoring with a dynamic model. *J. Navig.* **2007**, *60*, 247–263. [CrossRef]
15. Sun, R.; Wang, J.; Cheng, Q.; Mao, Y.; Ochieng, W.Y. A new IMU-aided multiple GNSS fault detection and exclusion algorithm for integrated navigation in urban environments. *GPS Solut.* **2021**, *25*, 147. [CrossRef]
16. Mohamed, A.H.; Schwarz, K.P. Adaptive kalman filtering for INS/GPS. *J. Geodesy.* **1999**, *73*, 193–203. [CrossRef]
17. Hide, C.; Moore, T.; Smith, M. Adaptive kalman filtering for lowc ost INS/GPS. *J. Navig.* **2003**, *56*, 143–152. [CrossRef]
18. Xiao, Z.; Zhao, P.; Li, S. Adaptive fuzzy Kalman filter based on INS/GPS integrated navigation system. *J. Chin. Inert. Technol.* **2010**, *18*, 203.
19. Xian, Z.; Hu, X.; Lian, J. Robust innovation-based adaptive Kalman filter for INS/GPS land navigation. In Proceedings of the Chinese Automation Congress, Changsha, China, 7–8 November 2013; pp. 374–379.
20. Liu, Y.; Fan, X.; Lv, C.; Wu, J.; Li, L.; Ding, D. An innovative information fusion method with adaptive Kalman filter for integrated INS/GPS navigation of autonomous vehicles. *Mech. Syst. Signal Process* **2018**, *100*, 605–616. [CrossRef]
21. Yan, F.; Li, S.; Zhang, E.; Chen, Q. An intelligent adaptive Kalman filter for integrated navigation systems. *IEEE Access* **2020**, *8*, 213306–213317. [CrossRef]
22. Li, X.; Wang, X.; Liao, J.; Li, X.; Lyu, H. Semi-tightly coupled integration of multi-GNSS PPP and S-VINS for precise positioning in GNSS-challenged environments. *Satell. Navig.* **2021**, *2*, 1. [CrossRef]
23. Li, X.; Wang, H.; Li, S.; Feng, S.; Wang, X.; Liao, J. GIL:a tightly coupled GNSS PPP/INS/LiDAR method for precise vehicle navigation. *Satell. Navig.* **2021**, *2*, 17. [CrossRef]
24. Wang, H.; Li, H.; Zhang, W.; Zuo, J.; Wang, H. Derivative-free huber-kalman smoothing based on alternating minimization. *Signal Process* **2019**, *163*, 115–122. [CrossRef]
25. Yang, Y.; Song, L.; Xu, T. Robust estimator for correlated observations based on bifactor equivalent weights. *J. Geodesy* **2002**, *76*, 353–358. [CrossRef]
26. Yang, Y.; Cheng, M.K.; Shum, C.K.; Tapley, B.D. Robust estimation of systematic errors of satellite laser range. *J. Geodesy* **1999**, *73*, 345–349. [CrossRef]
27. Yang, Y. Robust bayesian estimation. *J. Geodesy* **1991**, *65*, 145–150.
28. Kuusniemi, H.; Wieser, A.; Lachapelle, G.; Takala, J. User-level reliability monitoring in urban personal satellite-navigation. *IEEE Trans. Aerosp. Electron. Syst.* **2007**, *43*, 1305–1318. [CrossRef]
29. Teunissen, P. Quality control and GPS, chapter 7. In *GPS for Geodesy*, 2nd ed.; Springer: Berlin/Heidelberg, Germany, 1998.
30. Sun, R.; Zhang, W.; Zheng, J.; Ochieng, W.Y. GNSS/INS integration with integrity monitoring for UAV No-fly zone management. *Remote Sens.* **2020**, *12*, 524. [CrossRef]
31. Wu, F.; Nie, J.; He, Z. Classified adaptive filtering to GPS/INS integrated navigation based on predicted residuals and selecting weight filtering. *Geomat. Inf. Sci. Wuhan Univ.* **2012**, *37*, 261–264.
32. Cheng, Q.; Chen, P.; Sun, R.; Wang, J.; Mao, Y.; Ochieng, W.Y. A new faulty GNSS measurement detection and exclusion algorithm for urban vehicle positioning. *Remote Sens.* **2021**, *13*, 2117. [CrossRef]

remote sensing

MDPI

Article

Smartphone-Based Unconstrained Step Detection Fusing a Variable Sliding Window and an Adaptive Threshold

Ying Xu [1,*], Guofeng Li [1], Zeyu Li [1], Hao Yu [1], Jianhui Cui [1], Jin Wang [1] and Yu Chen [2]

1 College of Geodesy and Geomatics, Shandong University of Science and Technology, Qingdao 299590, China; guofengli@sdust.edu.cn (G.L.); lizeyu@sdust.edu.cn (Z.L.); yuhao@sdust.edu.cn (H.Y.); jhcui@sdust.edu.cn (J.C.); wangjin@sdust.edu.cn (J.W.)
2 Shandong Provincial Institute of Land Surveying and Mapping, Jinan 250013, China; chenyugtchy@shandong.cn
* Correspondence: yingxu@sdust.edu.cn

Abstract: Step detection for smartphones plays an important role in the pedestrian dead reckoning (PDR) for indoor positioning. Aiming at the problem of low step detection accuracy of smartphones in complex unconstrained states in PDR, smartphone-based unconstrained step detection method fusing a variable sliding window and an adaptive threshold is proposed. In this method, the dynamic updating algorithm of a peak threshold is developed, and the minimum peak value filtered after a sliding window filter is used as the adaptive peak threshold, which solves the problem that the peak threshold of different motion states is difficult to update adaptively. Then, a variable sliding window collaborative time threshold method is proposed, which solves the problem that the adjacent windows cannot be contacted, and the initial peak and the end peak are difficult to accurately identify. To evaluate the performance of the proposed unconstrained step detection algorithm, 50 experiments in constrained and unconstrained states are conducted by 25 volunteers holding 21 different types of smartphones. Experimental results show: The average step counting accuracy of the proposed unconstrained step detection algorithm is over 98%. Compared with the open source program Stepcount, the average step counting accuracy of the proposed algorithm is improved by 10.0%. The smartphone-based unconstrained step detection fusing a variable sliding window and an adaptive threshold has a strong ability to adapt to complex unconstrained states, and the average step counting accuracy rate is only 0.6% lower than that of constrained states. This algorithm has a wide audience and is friendly for different genders and smartphones with different prices.

Keywords: step detection; indoor positioning; unconstrained state; peak detectors; adaptive threshold; variable sliding window

Citation: Xu, Y.; Li, G.; Li, Z.; Yu, H.; Cui, J.; Wang, J.; Chen, Y. Smartphone-Based Unconstrained Step Detection Fusing a Variable Sliding Window and an Adaptive Threshold. *Remote Sens.* **2022**, *14*, 2926. https://doi.org/10.3390/rs14122926

Academic Editor: Liang Chen

Received: 27 April 2022
Accepted: 17 June 2022
Published: 19 June 2022

1. Introduction

The Global Navigation Satellite System (GNSS) is dominant in providing outdoor positioning service due to its coverage and high accuracy. However, people spend about 80% of their time indoors every day according to incomplete statistics. During the epidemic, medical staff or volunteers needed to grasp the dynamic position of personnel in isolated hotels and isolated wards in real time. When an indoor fire occurs, rescuers need to know the exact location of trapped people in time. In the construction of a smart city [1–3], and the tracking of pandemics [4], indoor positioning is the basic technology. In short, indoor positioning has broad application prospects. However, GNSS cannot provide services indoor. Researchers have proposed to use pedestrian dead reckoning (PDR) or PDR and wireless sensor fusion to achieve indoor positioning [5]. Smartphone-based step detection is necessary for PDR to determine pedestrian trajectory information [6,7]. In addition, smartphones bring many conveniences to people's lives with their rich functions and applications. Among these functions, step detection plays a role in health care for obese patients, has become a physical therapy to control chronic low back pain [8], monitor the fall of the elderly [9], and can also be applied to daily fitness training [10]. MI band, Huawei band and other commercial products record users' daily steps, and then give health tips. The commonly used step detection algorithms are zero velocity update (ZUPT) [11,12], autocorrelation analysis [13], peak detection [14], etc.

ZUPT refers to the lower limb being in a static state at a certain time during the walking process, and the walking speed is zero at this time, or the output values of acceleration sensor and angular velocity sensor will be approximately zero when the foot makes contact with the ground. The ZUPT method generally requires the sensor to be fixed in a specific position of the lower limbs, such as calves, feet, etc. [11,12]. Obviously, the smartphone does not have the basic conditions for this method.

When pedestrians walk continuously, there is a high correlation between the front and back gait cycles. The auto correlation analysis detects the number of steps by judging the correlation coefficient of the two cycles. Pan [13] uses auto correlation analysis to calculate steps. The experimental results show that the average step counting accuracy can reach 97.8% when the pedestrians dynamically switch the position carried by the smartphone, but this experiment failed to consider the important factor of changing the movement state of pedestrians at any time on the step-counting accuracy. Additionally, the calculation of correlation coefficient is large, which affects the timeliness of the algorithm.

Peak detection also detected the number of steps according to the periodicity of the pedestrian's continuous walking. Unlike the auto correlation analysis method, the peak detection takes the number of peaks (valleys) generated by the acceleration sensor or gyroscope sensor as steps. However, the pseudo-peak restricts the accuracy of the detection steps of the peak detection method. At present, the pseudo-peak is mainly eliminated by setting the threshold. Xu et al. [7] used a fixed threshold method to remove pseudo-peak, which has high accuracy when pedestrian motion and smartphone carrying mode are constrained. Cho et al. [14] used the sign-of-slope method and average threshold method to realize peak detection; Zhang et al. [15] used the mean value of the acceleration amplitude of the previous window to dynamically update the acceleration threshold; Wang [16] adaptively selects the threshold of acceleration according to the average value of the difference between peaks and valleys in unit time. Ryu et al. [17] proposed an adaptive threshold method, which uses the average value of the first five consecutive peaks as the adaptive threshold. However, these algorithms are easy to cause misjudgment in multiple motion modes. Dirican et al. [18] proposed a threshold-based unconstrained step counting algorithm. Unlike other ways of thinking about setting update thresholds based on acceleration data, this method sets the real and imaginary parts of the data transformed by fast Fourier transform (FFT) to different thresholds and achieves the update of thresholds by averaging the current and previous thresholds. This method achieves an adaptive update of acceleration thresholds and can adapt to a variety of unconstrained states, but the accuracy of the pedometer for running states is only 41.7%.

In addition to the pseudo-peak affecting the accuracy of step counting, selecting the appropriate sliding window is also helpful to indirectly improve the accuracy of step counting. Currently, the sliding window size is mainly based on the time required for a single step, so that the pedestrian can continuously output the step counting results without delay for each step; however, this requires that the size of the sliding window should essentially match the time used for a single step. Ning et al. [19] sets the size of three sliding windows according to the three states of motion of going up and down, walking and running, but the size of these windows are still fixed, so it is difficult to adjust to different users. Kang et al. [20] proposes a method of changing the sliding window, which mainly determines the size of the sliding window adaptively according to the walking frequency, but the accuracy of the walking frequency will affect the determination of the sliding window. It can be seen that it is difficult to ascertain the size of the sliding window when realizing continuous single-step counting. As long as the influence of sliding window size on real-time performance is within an acceptable range, looking for a sliding window which is not based on single-step detection, and solving the problem that a large sliding window separates the connection between neighboring windows, is also an ideal solution to the problem.

In summary, in the motion environment and motion state, the random transformation of smartphone carrying mode and the complex unconstrained state with interference factors, the step detection algorithm based on peak detection has the problem that the adaptive threshold is difficult to accurately update dynamically and the sliding window is difficult to accurately determine, which restricts the accuracy of peak detection. In view of this, smartphone-based unconstrained step detection fusing a variable sliding windows and an adaptive threshold is proposed in this paper. The algorithm uses the minimum peak filtered by the sliding window as the adaptive threshold to solve the problem that the peak threshold is difficult to update adaptively when the pedestrian state changes. The algorithm is a step detection algorithm of a variable sliding window. It is not based on the commonly used single-step detection of a sliding window, but realizes the variable sliding window on the basis of a fixed sliding window of 1 s, which ensures the close connection between the windows. At the same time, the cooperative time threshold solves the problem that the initial peak and the final peak make it difficult to distinguish the authenticity in the fixed sliding window. Using this algorithm, the accuracy of step counting can be guaranteed under complex unconstrained conditions. There are three aspects of contributions for the smartphone-based unconstrained step detection method proposed in this paper. First, it allows users to carry smartphones at multi points for indoor positioning. Second, users can freely switch the way of carrying and the state of motion for indoor positioning. Third, smartphones with different valences have high step detection accuracy.

The Section 1 introduces the research background and existing problems of the step detection algorithm. The preprocessing process and motion state recognition process of the step detection algorithm are described in the Section 2. Smartphone-based unconstrained step detection fusing a variable sliding window and an adaptive threshold is proposed in the Section 3. The Section 4 evaluates the step counting performance of smartphone-based unconstrained step detection fusing a variable sliding window and an adaptive threshold in constrained and unconstrained states through 50 groups of experiments. The Section 5 summarizes the work of this paper.

2. Step Detection Preprocessing and Motion State Recognition

As the sensors in the smartphone can detect the periodic changes of pedestrians, smartphones can detect the step numbers. Both acceleration sensors and gyroscope sensors can detect the periodic change when walking. However, the sensitivity of the gyroscope sensor is depressed, and the acceleration sensor is mostly used for step detection [21].

With each step forward, the pedestrian will produce a vertical motion and forward motion. The vertical axis of the three-axis accelerometer will produce an approximate sinusoidal wave, and the number of peaks (valleys) detected can be used as the number of steps of the pedestrian. However, the location of the smartphone carried by pedestrians is changeable, and it is difficult to accurately identify which single axis is in the vertical state, but the influence of sensor attitude can be reduced by calculating the overall acceleration. Formula (1) is the formula for calculating the overall acceleration.

$$a_c(t) = \sqrt{a_x^2 + a_y^2 + a_z^2} \tag{1}$$

In the formula, a_x, a_y, a_z represent the accelerometer output values of the X-axis, Y-axis and Z-axis at t time, and $a_c(t)$ represents the overall acceleration.

The signal characteristics of triaxial acceleration and overall acceleration were compared by experiment. Figure 1 shows the triaxial acceleration signal and the overall acceleration signal when walking at will. In the experiment, there are three states: walking and the smartphone is flat, normal walking and putting the smartphone next to the ear, running and swing hand. It can be seen from Figure 1 that the most sensitive axis has undergone three transformations, namely the Z-axis, Y-axis and X-axis, whereas the overall acceleration signal shows significant periodical changes. Therefore, the overall acceleration is adopted in the step detection algorithm in this paper.

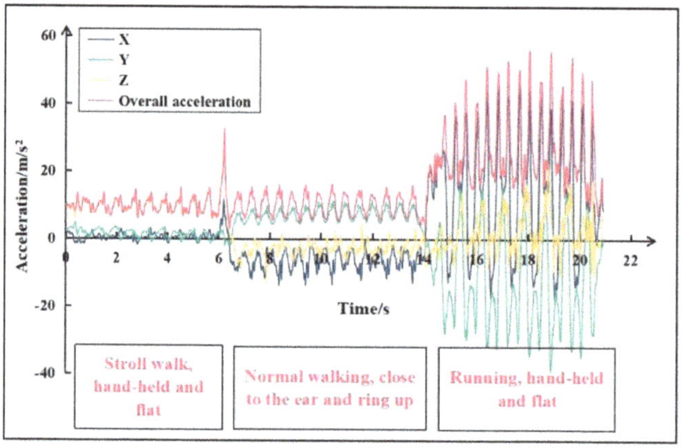

Figure 1. Triaxial acceleration and overall acceleration when walking at will.

The accuracy of the acceleration sensor of the smartphone is low, which leads to too many burr points of the original overall acceleration signal. It needs to be filtered to reduce the interference of more burrs before step detection. In addition, in the process of movement, pedestrians will be accompanied by multiple motion states. Different motion states often have different peak threshold. Therefore, it is necessary to identify motion state before step detection. The following focuses on the methods of step detection data preprocessing and motion state recognition.

2.1. Data Preprocessing of Step Detection

In order to remove the white Gaussian noise, Guo et al. [22] uses the weighted moving average method and the Kalman filter to preprocess the original resultant acceleration data to remove the influence of the Gaussian white noise, and then uses the Butterworth filter to refine the denoising. But the excessive and complicated filtering methods increased the data processing time. Zhang et al. [15] adopted the sliding window filter method to weaken the multi-peak phenomenon, which is a more common and better smoothing method for filter data, but the method loses the characteristics values of the data. Alabadleh et al. [23] used the Kalman filter and high-pass filter to smooth data, which removed gravity and outliers, but the algorithm had some complexity. Liu et al. [24] adopted a low-pass filter to eliminate signal noise, and the low-pass filter can keep the characteristics of the data very well. In this paper, a Finite Impulse Response (FIR) low-pass filter based on a Hamming window [25] is used to preprocess the ensemble acceleration signal, where the order of the filter is 10 and the length of the Hamming window is 11. As the actual output frequency of some smartphones does not match the sampling frequency, the sampling frequency used in the paper is the actual output frequency and the pass-band frequency is 5 Hz. Figure 2 shows the original overall acceleration signal, and Figure 3 compares a sliding window filter (the window size is 15 samples) and a FIR low-pass filter. The FIR low-pass filter retains the large and small peaks of the original overall acceleration signal (shown by the black arrows in Figure 3), reflecting the different characteristics of the left and right footsteps of the human body when walking.

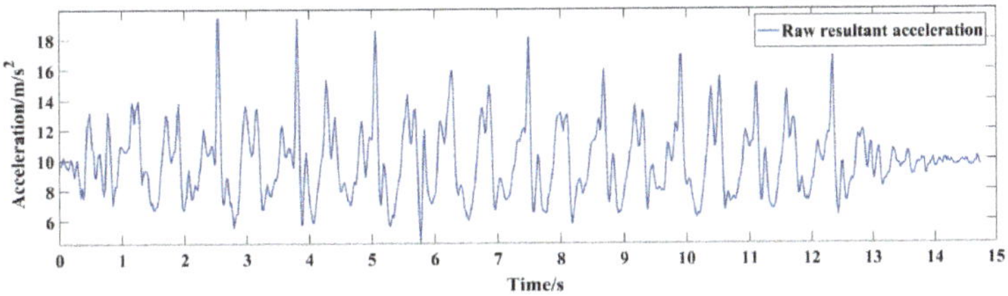

Figure 2. Raw overall acceleration signal.

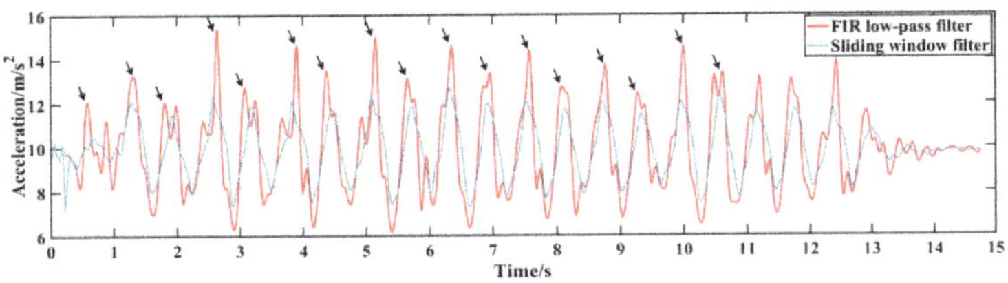

Figure 3. FIR low-pass filter and sliding window filter.

2.2. Motion State Recognition

Pedestrians lift one foot off the ground, move to a new position and after that, put it back on the ground, which is known as a single step [17]. Figure 4 shows the decomposition diagram of pedestrian single-step action. Since the step detection algorithm in this paper is mainly applied to PDR, stroll walking, normal walking and running in the previous progress are considered, which are general division methods.

Figure 4. Single step action decomposition.

When pedestrians are in different states of motion, the time and the peak acceleration of a single step are different. If a single fixed time threshold and a peak threshold are used to realize step counting, it is difficult to ensure the accuracy of steps. Identifying different states of motion and setting or updating different thresholds according to different states of motion can effectively improve the step counting accuracy. Zhang [26] uses the finite state machine method to distinguish whether the pedestrian is at rest or in motion, but does not make a further division of the motion state. In the study of Chen et al. [27], based on the inherent correlation between the state of motion and the maximum acceleration, the maximum acceleration threshold is set to identify the state of motion of pedestrians. With this method, it is easy to misjudge the state of motion under more complex unconstrained conditions. For example, the acceleration caused by the arm swing of the hand-held smartphone during stroll walking is similar to that of normal walking and when the smartphone is flat. Generally speaking, the step frequency of stroll walking, normal walking and running increases in turn, so the state of motion can be identified based on the step frequency. Using FFT, the time domain information can be converted into frequency domain information, and with the exception of the first DC point, the point with the largest amplitude is taken as the step frequency.

In this paper, through 50 experimental tests of 25 people, it is found that the walking frequency of continuous walking should be less than 1.6 Hz, the walking frequency of continuous normal walking should be less than 2 Hz, and the walking frequency of continuous running is 2–3.5 Hz. Limited by navigation factors and non-competitive state, the frequency above 3.5 Hz is mostly caused by interference factors such as body shaking, typing, video brushing, etc. Therefore, the frequency above 3.5 Hz is regarded as an interference state in this paper. When the state of motion is stable, the frequency calculated by FFT will be similar to the real step frequency. To accurately identify the step frequency, FFT requires at least 256 samples. In this paper, when the number of samples is less than 256, FFT judgment is not enabled. When the number of samples is more than 256, the window size of FFT is set to 256 samples, and the sliding time is 1 s (the actual samples in 1 s). Because the FFT used in this paper requires 256 samples (about 5 s), if there is multiple switching of state of motion in the window, it will reduce the accuracy of FFT to judge the step frequency. In order to prevent large step counting errors caused by misjudgment, the frequency threshold should be as small as possible. Theoretically, the time threshold should correspond to the step frequency, but it is found that in the walking state (stroll walking, normal walking), the time required for a single step is similar to that of running. So in order to avoid misjudging some true peaks because of the time threshold, in this paper, the time thresholds of all exercise states are set to smaller values. According to many experimental tests and references, the spectrum, the step frequency threshold, the time threshold and the peak threshold of walking, normal walking and running are shown in Figure 5 and Table 1.

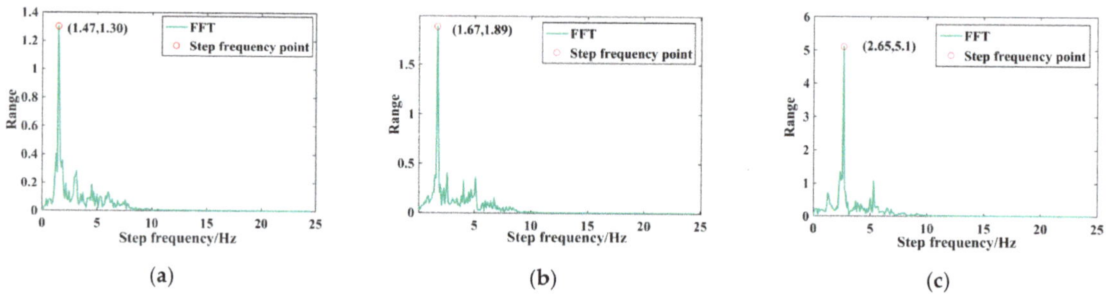

Figure 5. Spectrum of pedestrian stroll walking, normal walking and running: (**a**) Stroll walking (**b**) Normal walking (**c**) Running.

Table 1. Empirical values of step frequency threshold, time threshold and peak threshold.

Motion State	Step Frequency Threshold (Hz)	Time Threshold (ms)	Peak Threshold (m/s^2)
stroll walking	[0,1.6]	>333	>10.6
normal walking	[1.6,1.8]	>333	>11.0
running	[1.8,3.5]	>286	>11.6
interference factors	>3.5	—	—

3. Smartphone-Based Unconstrained Step Detection Fusing Variable Sliding Window and Adaptive Threshold

In order to address the problem of the low accuracy of step detection algorithms in a complex unconstrained state, an unconstrained step detection algorithm for smartphones is proposed in this paper. In this algorithm, the minimum peak filtered by the sliding window is used as the adaptive threshold. The algorithm is a step detection algorithm of a variable sliding window, which ensures the close connection between the windows. At the same time, the cooperative time threshold solves the problem that the initial peak and the final peak are difficult to distinguish the authenticity in the fixed sliding window. The flow chart of the algorithm is illustrated in Figure 6, and the steps of the algorithm are as following:

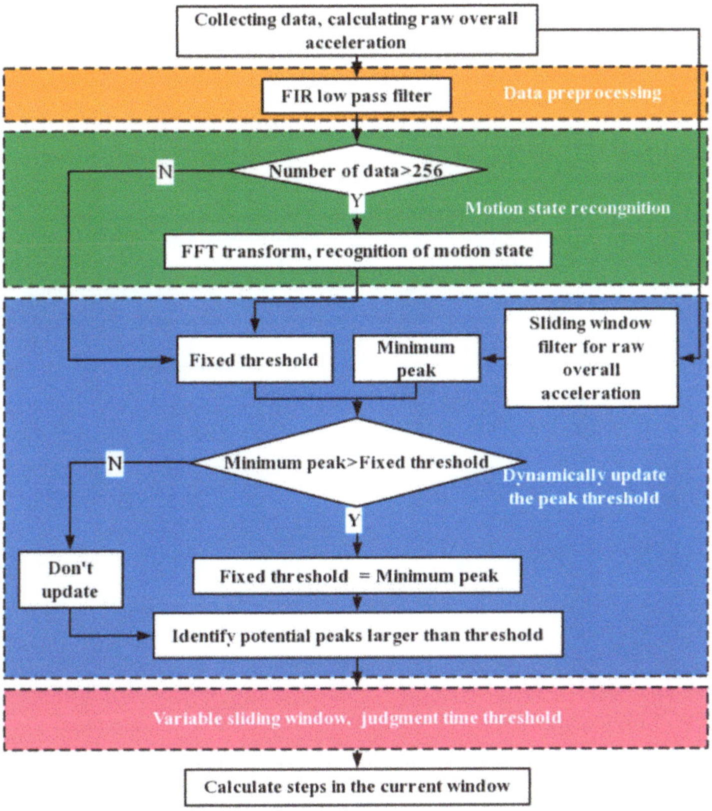

Figure 6. Flow chart of the unconstrained step detection algorithm.

The first step is data preprocessing. The FIR low-pass filter is used to denoise the original overall acceleration signal.

The second step is to identify the motion state and gain the fixed threshold for the third step. The motion state is identified by FFT. The initial peak threshold and time threshold are matched according to different motion states.

The third step is to dynamically update the peak threshold and eliminate the pseudo-peak. It is found that the waveform filtered by a sliding window is smaller than that filtered by a FIR low-pass filter. Based on these characteristics, the minimum peak value detected by sliding window filtering in the window is used in this paper as an adaptive threshold, and the adaptive threshold is used to replace the fixed threshold for step detection based on the waveform after the FIR low-pass filter. This peak threshold dynamic updating algorithm is proposed in detail in Section 3.1.

The fourth step is the variable sliding window cooperative time threshold to eliminate the pseudo-peak. Since the fixed sliding window cuts off the connection between the adjacent windows, the discrimination of the initial peak and the final peak of each window is easily missed. To solve this problem, the variable sliding window cooperative time threshold pseudo-peak elimination method is proposed to eliminate the pseudo-peak, and the structure and process of this method will be introduced in detail in Section 3.2.

The fifth step is to calculate the steps for the current window.

3.1. Dynamically Update the Peak Threshold

When affected by height, weight, health status, walking habits and other factors, it is difficult to adapt to different users and different unconstrained states only by a fixed threshold. It is found that the waveform filtered by a sliding window is smaller than that filtered by the FIR low-pass filter. To dynamically update the peak threshold, this paper takes the minimum peak detected after a sliding window filter as the adaptive threshold based on this feature. If the adaptive threshold is greater than the empirical threshold, the empirical threshold is used as the adaptive threshold.

Figure 7 is a group of experiments running on flat ground. In the experiment, adaptive peak threshold and fixed peak threshold (11.6 m/s^2) are used to eliminate pseudo-peak and count steps, respectively. The experimenter ran 29 steps, and the step number results of adaptive peak threshold and fixed peak threshold were 29 steps and 40 steps, respectively.

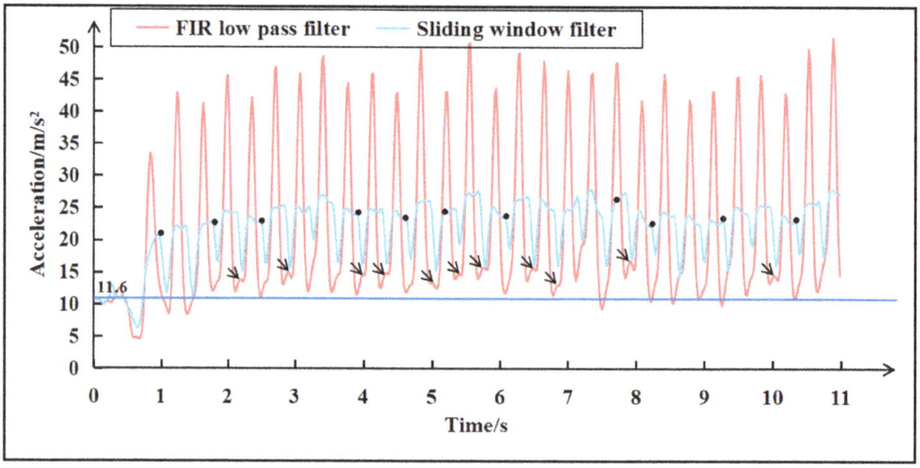

Figure 7. Pseudo-peak elimination by adaptive threshold in running state.

In Figure 7, the arrow marks the pseudo-peak, and the blue horizontal line is the fixed threshold line. The black dots are adaptive thresholds obtained by a sliding window filter. It can be seen that the fixed threshold is difficult to eliminate pseudo-peaks marked by arrows, whereas adaptive thresholds can eliminate them.

3.2. Pseudo-Peak Elimination Method with a Variable Sliding Window Cooperative Time Threshold

Because the size of the sliding window determined in this paper is larger than that of the single-step sliding window, there may be several true and pseudo-peaks in the window. The positions of true and pseudo-peaks can be divided into three categories: the initial peak in front of the window, the adjacent peak in the window and the peak at the end of the window.

As shown in Figure 8, the initial peak of the second window and the peak at the end of the ninth window are pseudo-peaks in fact. The fixed sliding window separates the connection between the adjacent windows. If the fixed sliding window of 1 s is adopted, the two pseudo-peaks pointed by the arrow will be misjudged as true peaks. In order to solve this problem, this paper proposes a pseudo-peak elimination method with a variable sliding window cooperative time threshold, which ensures the connection between the adjacent windows, and can judge the starting peak and the end peak. Figure 9 is a flowchart of this method.

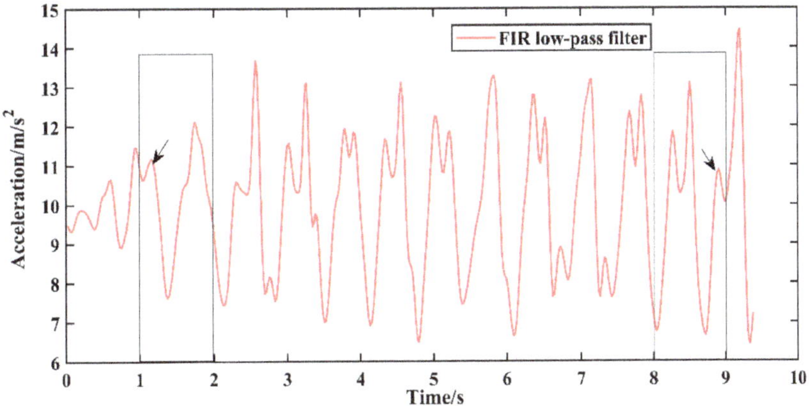

Figure 8. Pseudo-peaks in front of the window and at the end of the window.

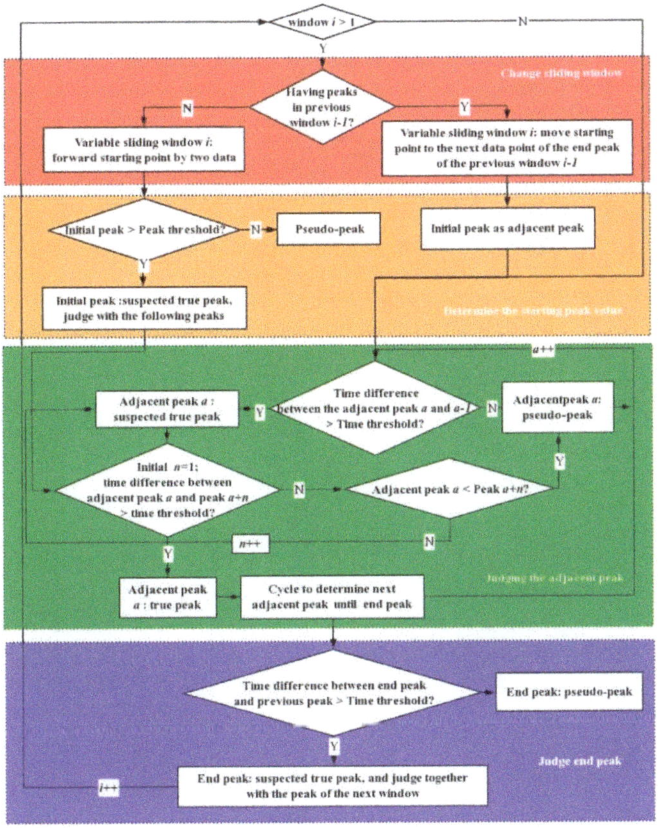

Figure 9. Flow chart of pseudo-peak elimination method with variable sliding window cooperative time threshold.

According to the flow chart, the specific implementation steps of this method are as follows:

Step 1: judging adjacent peaks. If there are two or more peaks in the i ($i \geq 1$) window, there are neighboring peaks in the window. The time threshold is used to judge the adjacent peaks. If the time difference between peak a and peak a-1 is less than the time threshold, then peak a is a pseudo-peak. Otherwise, peak a is suspected to be a true peak, which needs to be judged cyclically with the subsequent peak $a + n$ (n = 1,2,3 …). If the time difference between peak a and peak $a + n$ is less than the time threshold, then peak value is compared. If peak a is greater than peak $a + n$, peak a is a suspected true peak, and the program executes n++ to continue the loop judgment. If peak a is less than peak $a + n$, peak a is a pseudo-peak. If the time difference between peak a and peak $a + n$ is greater than the time threshold, peak a is determined to be the true peak. After achieving the judgment of peak a, the program ends this cycle and moves to judge the next peak a++.

Step 2: judging the end peak. If the time difference between the end peak and the previous peak is greater than the time threshold, the end peak is suspected to be the true peak, which needs to be judged together with the starting peak of the next window. Otherwise, the end peak is a pseudo-peak.

Step 3: variable sliding window. If the i ($i \geq 1$) window contains a peak, the starting point of the i + 1 window is set to the next data point of the end peak of the i window (Figure 10a). If the i window does not contain a true peak, the starting point of the i + 1 window is moved forward two data points(Figure 10b).

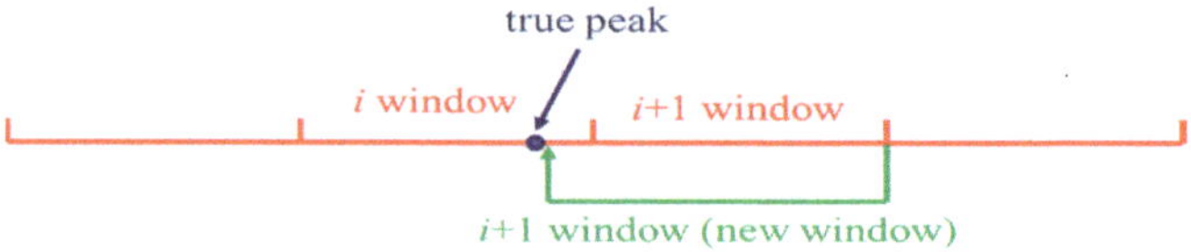

(**a**)

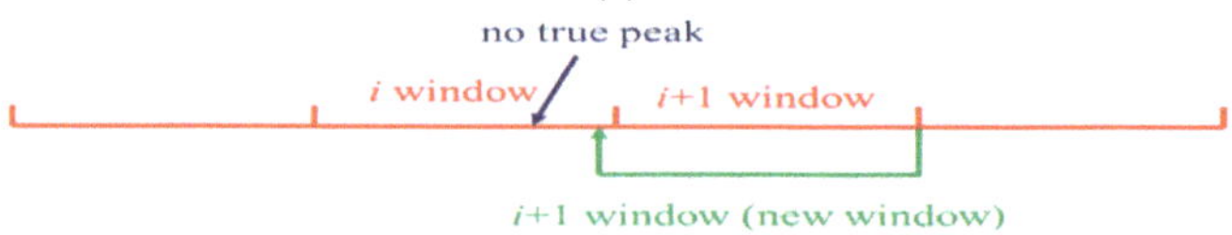

(**b**)

Figure 10. Schematic diagram of the variable sliding window: (**a**) The i ($i \geq 1$) window contains a peak (**b**) The i window does not contain a true peak. Note: the green arrows indicate the starting point of new windows.

Step 4: judging the initial peak. If the previous window does not contain a peak, and the initial peak is greater than the peak threshold, the initial peak is suspected to be a true peak, which needs to be judged with the next peak. The judgment method is the same as that of peak a and the peak $a + n$ in step 1. If the initial peak is less than the peak threshold, the initial peak is a pseudo-peak. If the previous window contains a peak, the initial peak is equivalent to the adjacent peak, which can be judged by the method in step 1.

Step 5: circularly executing step 1, step 2, step 3 and step 4 until the window ends.

4. Experiment and Analysis

In order to evaluate the accuracy of smartphone-based unconstrained step detection fusing a variable sliding window and an adaptive threshold, 25 volunteers (9 females and 16 males) were recruited. The volunteers' height ranged from 155 cm to 185 cm and the weight ranged from 41 kg to 93 kg. The experimental smartphones were all volunteers' own phones, including 21 models of Huawei, Apple, Samsung, Honor, Xiaomi, Oneplus, Oppo and Vivo. Figure 11 shows 25 smartphone brands and models of the volunteers in the experiment.

Figure 11. Brands and models of 25 smartphones.

4.1. Experimental Setup

The experiment was carried out on the campus of Shandong University of Science and Technology. The four kinds of states of volunteers' motion state, motion environment, smartphone carrying mode and interference factors are shown in Table 2. In total, 25 volunteers were divided into 5 groups to carry out the step detection experiments in a constrained state (smartphones keep flat) and an unconstrained state. The experimental process in a constrained state is shown in Table 3, and the experimental process in an unconstrained state is shown in Table 4. Then, 5 volunteers in each group collected data according to the experimental process of Tables 3 and 4, and accurately counted the actual walking steps. Figure 12 is the schematic diagram of data collected by volunteers in a constrained state, and Figure 13 is the schematic diagram of an unconstrained state. The data collected by the volunteers, as well as attribute information such as height, weight and smartphone models have been uploaded to GitHub (https://github.com/jackleenotjackma/StepCountingData.git (accessed on 25 December 2021).

Table 2. Experimental motion environment, motion state, smartphone carrying mode and interference factor.

Test Scenario	Motion State	Smartphone Carrying Mode	Interference Factor
Flat ground(A1)	Standing(B1)	Trouser pocket(C1)	Typing(D1)
Go upstairs(A2)	Stroll walking(B2)	Hold and swing hand(C2)	Brush video(D2)
Go downstairs(A3)	Normal walking(B3)	Hand-held and flat(C3)	—
—	Running(B4)	Close to the ear and ringer up(C4)	—

Table 3. Experiment process of 5 constrained state groups.

Experimental Group	Experimental Process
Constrained: first group	A1,B2,C3
Constrained: second group	A1,B3,C3
Constrained: third group	A1,B4,C3
Constrained: fourth group	A2,B3,C3
Constrained: fifth group	A3,B3,C3

Table 4. Five groups of unconstrained state experiment process.

Experimental Group	Experimental Process
Unconstrained: first group	A1,B1,D2–A1,B2,C2–A1,B3,C3–A1,B4,C1
Unconstrained: second group	A1,B2,C4–A1,B1,D1–A1,B3,C2–A1,B4,C2
Unconstrained: third group	A1,B4,C2–A1,B3,C2–A1,B3,C3–A1,B2,D2–A1,B1,D1
Unconstrained: fourth group	A1,B2,C3–A2,B3,C2–A1,B3,C2–A3,B3,C1–A1,B2,D2–A1,B1,D1
Unconstrained: fifth group	A1,B2,C3–A2,B3,C1–A1,B3,C2–A3,B3,C4–A1,B2,D2–A1,B1,D1

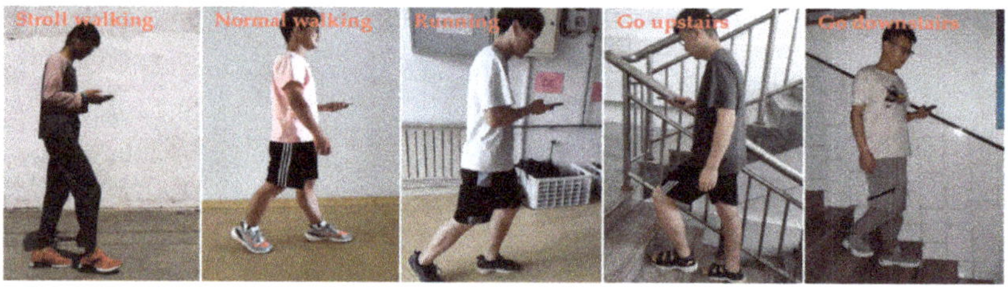

Figure 12. Constrained experiment process.

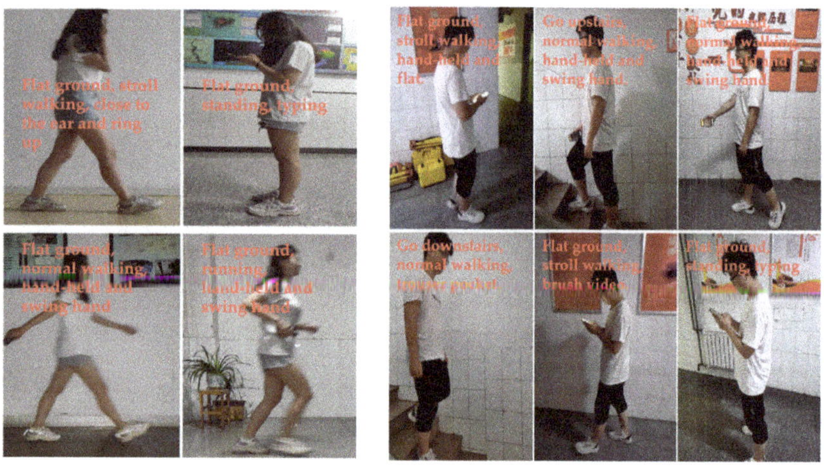

(a) (b)

Figure 13. Unconstrained experiment process: (**a**) Schematic diagram of the second group of experimental process (**b**) Schematic diagram of the fourth group of experimental process.

4.2. Experimental Results and Analysis

A total of 50 sets of data were obtained in the experiment, and the steps were between 72 steps and 343 steps. Unconstrained step count detection algorithm for smartphones is compared with the open source program Stepcount [28], which is well-known on GitHub.

Stepcount is a peak detection algorithm based on an adaptive threshold. The actual step number was taken as the reference value.

The step counting results of 25 volunteers in a constrained state and an unconstrained state are shown in Table 5. Figures 14 and 15 show the step counting accuracy of 25 volunteers in a constrained state and an unconstrained state, respectively. It can be seen from Table 5 and Figures 14 and 15 that under constraint conditions, the lowest step counting accuracy of the proposed smartphone-based unconstrained step detection fusing a variable sliding window and an adaptive threshold is 96.3%, whereas the lowest correct rate of step counting by the Stepcount is only 45.4%. In the unconstrained state, the lowest correct rate of the proposed smartphone-based unconstrained step detection fusing variable sliding window and adaptive threshold is 95.3%, whereas it is only 57.5% for Stepcount. Regardless of the constrained state or the unconstrained state, the step counting accuracy of the proposed smartphone-based unconstrained step detection fusing a variable sliding window and an adaptive threshold is stable, whereas the step counting accuracy of the Stepcount fluctuates greatly. In the constraint condition, the average step counting accuracy of the unconstrained step detection algorithm and the Stepcount are about 99.0% and 90.1%, respectively. In the unconstrained state, the average step accuracy of the unconstrained step detection algorithm and Stepcount are about 98.4% and 87.4%, respectively. In the constrained state and unconstrained state, the step accuracy of the proposed smartphone-based unconstrained step detection fusing a variable sliding window and an adaptive threshold is about 8.9% and 11% higher than that of the Stepcount.

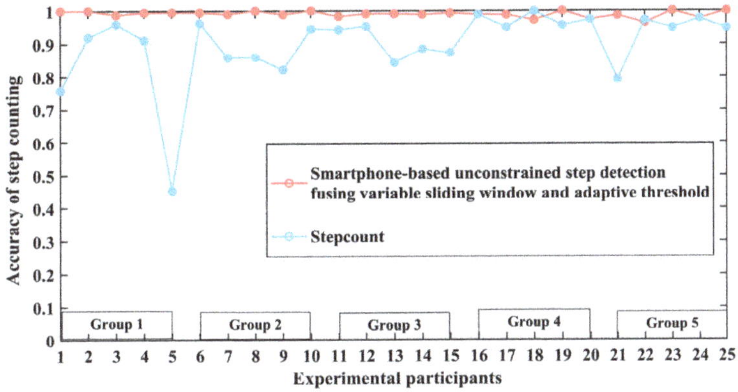

Figure 14. Step counting accuracy in constrained state.

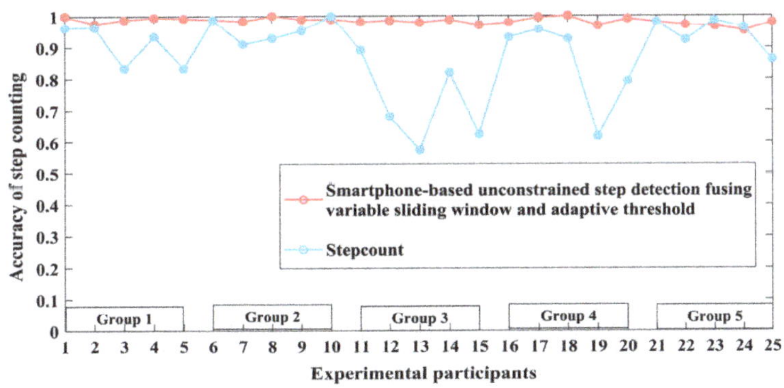

Figure 15. Step counting accuracy in unconstrained state.

Table 5. Constrained state and unconstrained state step counting result.

Experimental Group	Actual Steps	Stepcount Correct Rate (%)	Smartphone-Based Unconstrained Step Detection Fusing a Variable Sliding Window and an Adaptive Threshold Correct Rate (%)	Actual Steps	Stepcount Correct Rate (%)	Smartphone-Based Unconstrained Step Detection Fusing a Variable Sliding Window and an Adaptive Threshold Correct Rate (%)
		Constrained State			Unconstrained State	
Group 1	186	75.8	100.0	241	96.3	100.0
	150	92.0	100.0	240	96.3	97.5
	170	95.9	98.8	240	83.3	98.8
	191	91.1	99.5	184	93.5	99.5
	196	45.4	99.5	240	83.3	99.2
Group 2	210	96.2	99.5	160	98.8	98.8
	199	85.9	99.0	180	91.1	98.3
	150	86.0	100.0	171	93.0	100.0
	180	82.2	98.9	195	95.4	98.9
	235	94.5	100.0	182	100.0	98.9
Group 3	240	94.2	98.3	260	89.2	98.1
	343	95.3	99.1	210	68.1	98.6
	230	84.3	99.1	240	57.5	97.9
	183	88.5	98.9	240	82.1	98.8
	284	87.3	99.3	240	62.5	97.1
Group 4	94	98.9	98.9	194	93.3	97.9
	80	95.0	98.8	212	95.8	99.5
	107	100.0	97.2	235	92.8	100.0
	114	95.6	100.0	260	61.9	96.9
	114	97.4	97.4	209	79.4	99.0
Group 5	72	79.2	98.6	161	98.1	98.1
	108	97.2	96.3	170	92.4	97.1
	96	94.8	100.0	214	98.6	96.7
	132	97.7	97.7	192	96.4	95.3
	131	94.7	100.0	237	86.1	97.9

In order to explore the adaptability of the proposed smartphone-based unconstrained step detection fusing a variable sliding window and an adaptive threshold, the performance of this algorithm is analyzed from the perspective of smartphone price and gender of volunteers.

The smartphones of the 25 volunteers were divided into three types according to prices: mid-low-level (1200–2599 RMB), mid-level (2600–3599 RMB) and high-level (over 3600 RMB). Figure 16 shows the step counting accuracy of the smartphone-based unconstrained step detection fusing a variable sliding window and an adaptive threshold with different prices in the unconstrained state. It can be seen from the figure that the smartphone-based unconstrained step detection fusing a variable sliding window and an adaptive threshold has the best step counting accuracy for high-level smartphones, and there is no obvious difference between mid-level smartphones and mid-low-level smartphones. In the unconstrained state, the average step counting accuracy of the smartphone-based unconstrained step detection fusing variable sliding window and adaptive threshold is 99.1%, 98.0% and 98.0% for high-level, mid-level and mid-low-level smartphones respectively.

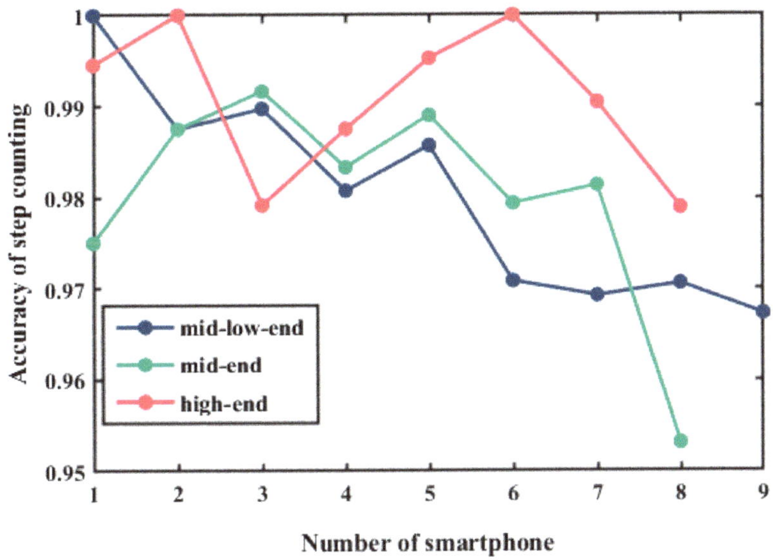

Figure 16. Step counting accuracy rate of smartphones with different prices in unconstrained state.

From a gender perspective, in the unconstrained state, the average step counting accuracy rate of the smartphone-based unconstrained step detection fusing a variable sliding window and an adaptive threshold is 98.2% for male and 98.7% for female, respectively. The difference between them is only 0.5%. It can be seen that the smartphone-based unconstrained step detection fusing a variable sliding window and an adaptive threshold has good adaptability to different genders.

5. Conclusions

Aiming at the problem of low step detection accuracy of PDR in an unconstrained state, this paper proposes a step detection algorithm for smartphones. In this algorithm, the pseudo-peaks are eliminated by preprocessing acceleration using a FIR low-pass filter, FFT recognition gait, dynamic updating peak threshold and a variable sliding window cooperative time threshold; and, finally, the step counting is realized.

The FIR low-pass filter is used to denoise the overall acceleration signal. FFT is used to identify stroll walking, normal walking, running and interference state. The minimum peak value after a sliding window filter is used to dynamically update the peak value threshold, which solves for the problem that the fixed peak value threshold has low adaptability in an unconstrained state. A method of a variable sliding window cooperative time threshold is proposed, which ensures the connection between adjacent windows and makes up for the problem that the fixed window cannot judge the initial peak and the end peak. In order to evaluate the step counting performance of the proposed algorithm, 50 experiments in constrained and unconstrained states are conducted by 25 volunteers holding 21 different types of smartphones. The experimental results show that, the average step counting accuracy of the proposed algorithm is 98.4% in an unconstrained state, which is 10.0% higher than that of the result from the open source program Stepcount. This proposed algorithm has a strong ability to adapt to complex unconstrained states and it is friendly for different genders and mobile phones with different prices. In the future, we plan to carry out indoor positioning research with this step detection method and use it in the construction of a smart city and the tracking of pandemics.

Author Contributions: Conceptualization, Y.X., G.L. and Z.L.; methodology, Y.X. and G.L.; software, G.L. and Z.L.; validation, Y.X., Z.L., H.Y. and J.C.; formal analysis, G.L., H.Y. and J.C.; investigation, G.L., H.Y., J.C. and J.W.; data curation, G.L.; writing—original draft preparation, G.L.; writing—review and editing, Y.X., H.Y., J.C., J.W. and Y.C.; supervision, Y.X. and Z.L.; project administration, Y.X.; funding acquisition, Y.X. All authors have read and agreed to the published version of the manuscript.

Funding: This research was funded by National Natural Science Foundation of China (42174035), Talent introduction plan for Youth Innovation Team in universities of Shandong Province (innovation team of satellite position and navigation) and Shandong University of Science and Technology school-level scientific research team (2019TDJH103).

Data Availability Statement: Not applicable.

Conflicts of Interest: The authors declare no conflict of interest.

References

1. Sobnath, D.; Rehman, I.U.; Nasralla, M.M. Smart cities to improve mobility and quality of life of the visually impaired. In *Technological Trends in Improved Mobility of the Visually Impaired*; Springer: Berlin/Heidelberg, Germany, 2019; pp. 3–28.
2. Nasralla, M.M. Sustainable virtual reality patient rehabilitation systems with IoT sensors using virtual smart cities. *Sustainability* **2021**, *13*, 4716. [CrossRef]
3. Hazarika, A.; Poddar, S.; Nasralla, M.M.; Rahaman, H. Area and energy efficient shift and accumulator unit for object detection in IoT applications. *Alex. Eng. J.* **2022**, *61*, 795–809. [CrossRef]
4. Kumar, S.A.; Nasralla, M.M.; García-Magariño, I.; Kumar, H. A machine-learning scraping tool for data fusion in the analysis of sentiments about pandemics for supporting business decisions with human-centric AI explanations. *PeerJ Comput. Sci.* **2021**, *7*, e713. [CrossRef] [PubMed]
5. Kuang, J.; Niu, X.; Chen, X. Robust Pedestrian Dead Reckoning Based on MEMS-IMU for Smartphones. *Sensors* **2018**, *18*, 1391. [CrossRef] [PubMed]
6. Huang, Y.; Bi, L.; Fu, W.; Yang, R.; Bi, X. A Precise Step Counting Algorithm Based on Acceleration Correlation Analysis. In Proceedings of the 2019 IEEE 13th International Conference on Anti-counterfeiting, Security, and Identification (ASID), Xiamen, China, 25–27 October 2019; pp. 299–302. [CrossRef]
7. Xu, L.; Xiong, Z.; Liu, J.; Wang, Z.; Ding, Y. A Novel Pedestrian Dead Reckoning Algorithm for Multi-Mode Recognition Based on Smartphones. *Remote Sens.* **2019**, *11*, 294. [CrossRef]
8. Lang, A.E.; Hendrick, P.A.; Clay, L.; Mondal, P.; Trask, C.M.; Bath, B.; Penz, E.D.; Stewart, S.A.; Baxter, G.D.; Hurley, D.A.; et al. A randomized controlled trial investigating effects of an individualized pedometer driven walking program on chronic low back pain. *BMC Musculoskel. Dis.* **2021**, *22*, 206. [CrossRef] [PubMed]
9. Kang, H.; Lee, B. Comparison of Gait variables and Relative Risk of Falls According to Walking Speed During Flat and Obstacles Walking of Fallers and Non-Fallers in Korean Elderly Women. *Exerc. Sci.* **2022**, *31*, 80–87. [CrossRef]
10. Du, C. Step-Counting Function of Adolescent Physical Training APP Based on Artificial Intelligence. *Math. Probl. Eng.* **2021**, *2021*, 5582598. [CrossRef]
11. Xie, D.; Jiang, J.; Wu, J.; Yan, P.; Tang, Y.; Zhang, C.; Liu, J. A Robust GNSS/PDR Integration Scheme with GRU-Based Zero-Velocity Detection for Mass-Pedestrians. *Remote Sens.* **2022**, *14*, 300. [CrossRef]
12. Wang, Z.; Zhao, H.; Qiu, S.; Gao, Q. Stance-Phase Detection for ZUPT-Aided Foot-Mounted Pedestrian Navigation System. *IEEE/ASME Trans. Mechatron.* **2015**, *20*, 3170–3181. [CrossRef]
13. Pan, M.; Lin, H. A Step Counting Algorithm for Smartphone Users: Design and Implementation. *IEEE Sens. J.* **2015**, *15*, 2296–2305. [CrossRef]
14. Cho, Y.; Cho, H.; Kyung, C.M. Design and Implementation of Practical Step Detection Algorithm for Wrist-Worn Devices. *IEEE Sens. J.* **2016**, *16*, 7720–7730. [CrossRef]
15. Zhang, H.; Duan, Q.; Pan, D.; Bei, H. Integrated iBeacon/PDR Indoor Positioning System Using Extended Kalman Filter. In *Advances in Materials, Machinery, Electrical Engineering (AMMEE 2017)*; Atlantis Press: Amsterdam, The Netherlands, 2017; pp. 9–16. [CrossRef]
16. Wang, P. Research on Indoor Positioning Technology Based on WiFi and Inertial Sensor. Master's Thesis, Lanzhou Jiaotong University, Lanzhou, China, 2019. [CrossRef]
17. Ryu, U.; Ahn, K.; Kim, E.; Kim, M.; Kim, B.; Woo, S.; Chang, Y. Adaptive Step Detection Algorithm for Wireless Smart Step Counter. In Proceedings of the 2013 International Conference on Information Science and Applications (ICISA), Pattaya, Thailand, 24–26 June 2013; pp. 1–4. [CrossRef]
18. Dirican, A.C.; Aksoy, S. Step counting using smartphone accelerometer and fast Fourier transform. *Sigma J. Eng. Nat. Sci.* **2017**, *8*, 175–182.
19. Ning, Y.; Liang, J.; Wang, J. Algorithm of adaptive threshold gait detection in multiple motion states. *J. Chin. Inert. Technol.* **2020**, *28*, 172–178. [CrossRef]

20. Kang, X.; Huang, B.; Yang, R.; Qi, G. Accurately Counting Steps of the Pedestrian with Varying Walking Speeds. In Proceedings of the 2018 IEEE SmartWorld, Ubiquitous Intelligence & Computing, Advanced & Trusted Computing, Scalable Computing & Communications, Cloud & Big Data Computing, Internet of People and Smart City Innovation (SmartWorld/SCALCOM/UIC/ATC/CBDCom/IOP/SCI), Guangzhou, China, 8–12 October 2018; pp. 679–686.
21. Bassett, D.R.; Toth, L.P.; Lamunion, S.R.; Crouter, S.E. Step Counting: A Review of Measurement Considerations and Health-Related Applications. *Sports Med.* **2017**, *47*, 1303–1315. [CrossRef] [PubMed]
22. Guo, Y.; Liu, Q.; Ji, X.; Li, G.; Wang, S. Pedestrian Gait Analysis Based on Mobile Phone Accelerometer. *J. Chin. Inert. Technol.* **2017**, *25*, 708–712. [CrossRef]
23. Alabadleh, A. Step detection algorithm for accurate distance estimation using dynamic step length. In Proceedings of the IEEE International Conference on Mobile Data Management, Daejeon, Korea, 29 May–1 June 2017; pp. 324–327. [CrossRef]
24. Liu, Y.; Chen, Y.; Shi, L.; Tian, Z.; Zhou, M.; Li, L. Accelerometer Based Joint Step Detection and Adaptive Step Length Estimation Algorithm Using Handheld Devices. *J. Commun.* **2015**, *10*, 520–525. [CrossRef]
25. Hannah, A.; Agordzo, G.K. A Design of a low-pass FIR filter using Hamming Window Functions in Matlab. *Comput. Eng. Intell. Syst.* **2020**, *11*, 24–30. [CrossRef]
26. Zhang, H. Research on Bluetooth and Dead Reckoning Based Indoor Positoning Algorithm. Master's Thesis, Chongqing University, Chongqing, China, 2017.
27. Chen, G.; Li, F.; Zhang, Y. Pedometer method based on adaptive peak detection algorithm. *J. Chin. Inert. Technol.* **2016**, *23*, 315–321. [CrossRef]
28. STEPCOUNT. Available online: https://github.com/finnfu/stepcount (accessed on 25 December 2021).

remote sensing

MDPI

Article

Evaluation of Forest Features Determining GNSS Positioning Accuracy of a Novel Low-Cost, Mobile RTK System Using LiDAR and TreeNet

Omid Abdi [1,*], Jori Uusitalo [1], Julius Pietarinen [2] and Antti Lajunen [2]

[1] Department of Forest Sciences, University of Helsinki, Latokartanonkaari 7, 00014 Helsinki, Finland; jori.uusitalo@helsinki.fi
[2] Department of Agricultural Sciences, University of Helsinki, Koetilantie 5, 00790 Helsinki, Finland; julius.pietarinen@helsinki.fi (J.P.); antti.lajunen@helsinki.fi (A.L.)
* Correspondence: omid.abdi@helsinki.fi; Tel.: +358-2941-58466

Citation: Abdi, O.; Uusitalo, J.; Pietarinen, J.; Lajunen, A. Evaluation of Forest Features Determining GNSS Positioning Accuracy of a Novel Low-Cost, Mobile RTK System Using LiDAR and TreeNet. *Remote Sens.* **2022**, *14*, 2856. https://doi.org/10.3390/rs14122856

Academic Editors: Yuwei Chen, Changhui Jiang, Qian Meng, Bing Xu, Wang Gao, Panlong Wu, Lianwu Guan and Zeyu Li

Received: 28 April 2022
Accepted: 10 June 2022
Published: 15 June 2022

Abstract: Accurate positioning is one of the main components and challenges for precision forestry. This study was established to test the feasibility of a low-cost GNSS receiver, u-blox ZED-F9P, in movable RTK mode with features that determine its positioning accuracy following logging trails in the forest environment. The accuracy of the low-cost receiver was controlled via a geodetic-grade receiver and high-density LiDAR data. The features of nearby logging trails were extracted from the LiDAR data in three main categories: tree characteristics; ground-surface conditions; and crown-surface conditions. An object-based TreeNet approach was used to explore the influential features of the receiver's positioning accuracy. The results of the TreeNet model indicated that tree height, ground elevation, aspect, canopy-surface elevation, and tree density were the top influencing features. The partial dependence plots showed that tree height above 14 m, ground elevation above 134 m, western direction, canopy-surface elevation above 138 m, and tree density above 30% significantly increased positioning errors by the low-cost receiver over southern Finland. Overall, the low-cost receiver showed high performance in acquiring reliable and consistent positions, when integrated with LiDAR data. The system has a strong potential for navigating machinery in the pathway of precision harvesting in commercial forests.

Keywords: mobile RTK; low-cost GNSS receiver; positioning accuracy; LiDAR data; tree characteristics; terrain conditions; precision forestry; TreeNet; geographic object-based approach; commercial forests

1. Introduction

The combination of low-cost global navigation satellite system (GNSS) with real-time kinematic (RTK) has streamlined determining centimeter-level positioning accuracy of vehicles feasible for precise practices [1,2]. However, the feasibility of this system is little known in forestry, where precision forestry is growing due to its advantages in reducing operational costs and ecological impacts.

Accurate positioning is one of the main components of precision forestry, along with remote sensing data and geospatial information systems. As forest operations become more autonomous, the demand for highly accurate positioning increases [3,4]. Machine navigation and control rely on reliable and accurate positioning to perform forest operations. On the other hand, any inaccuracy in positioning increases the costs of operations and our carbon footprint. It also decreases machine operational robustness and safety, with huge implications on the quality of production and environment [5]. Furthermore, in commercial forests, harvesters collect a huge amount of data from the processed single trees over large areas and forest stands. The data are economically valuable and include the bucking information and positions, which can be used for mapping and predicting forest attributes [4,6]. However, any positioning errors of the data significantly degrade the performance of the models used to estimate the merchantable timber volume collected by harvesters [7,8].

Various metrics are introduced to measure the performance of GNSS positioning, such as the availability of sufficient signals, the continuity and integrity of the signals, and most importantly, the accuracy of positioning [9]. The environmental factors have an undeniable impact on the quality of signals. For example, signal blockage and multipath errors result from site-specific conditions or atmospheric factors. The poor visibility of satellites due to trees' occlusion or terrain conditions is one of the main sources of single blockage, which hampers the geometry of the GNSS and increases the optimal time for the initialization of the system [10].

Earlier studies have focused on the impact of forest type or forest cover density [11–14] in the positioning accuracy of GNSS receivers in forest environments. For example, Feng et al. (2021) [14] explored the effect of forest types and crown size on the accuracy of positioning for individual trees through GNSS receivers. They found that the error of positioning in broadleaved forests is higher than in coniferous forests, while the size of crown did show no significant impact on the increasing the error of positioning. Likewise, Murgaš et al. [13] tested the accuracy of a mapping-grade device for the positioning of inventory plots under open sky and forest canopy conditions. They reported the increase of positioning errors under canopy condition, while the coniferous forests and young stands showed lower impacts on the positioning errors of the GNSS receiver. Moreover, few studies have considered the influence of GNSS receivers' factors along with the forest-related factors. Ordóñez Galán et al. [15] tested the influence of various forest cover features and GPS-related factors on the positioning accuracy of a DGPS receiver. They reported that the influence of forest cover features on the positioning accuracy is significant in comparison with the GPS factors. However, they concluded that there is no priority between different forest variables on the accuracy of the positions. Piedallu and Gégout [16] evaluated the accuracy of GPS positioning based on the type of receiver, forest cover type, the components of GPS survey, and the season. They reported the impacts of all factors on the positioning accuracy of GPS, except the season of data recording. The influence of high density of forest cover on the accuracy of positioning is higher than other variables. However, the consideration of other factors, such as terrain variables, which may affect the positioning accuracy of GNSS receivers in a forest environment, has been somewhat diminished in earlier studies. Valbuena et al. [17] explored both terrain and forest variables. However, they excluded the terrain variables in the final modelling and concluded that the leaf area index, the relative spacing index between trees, and the wood volume can express the positioning accuracy of a GNSS receiver in a pine forest. Meanwhile, Pini et al. [1] concluded that terrain conditions are not only significant for the accuracy of positions, but are also effective for the accurate heading of vehicles. Kabir et al. [18] reported a significant decrease in in the accuracy of GNSS receivers in the mountainous areas relative to orchards or open fields. Many of these studies were developed to introduce an appropriate GNSS receiver for measuring the accurate locations of individual trees or inventory plots under forest canopy. Hence, the measurements were carried out as static, with several minutes to record an accurate position. Although positioning accuracy through mapping-grade or geodetic-grade GNSS receivers is reliable, their high cost, the difficulty in carrying them in forest conditions, and the complexity of using them have led to a degradation of their efficiency for cost-sensitive and small size applications, such as operations in precision agriculture or precision forestry. Therefore, the need for a new generation of cost-effective GNSS receivers with high positioning accuracy and simplicity of usage under forest conditions, such as u-blox modules, is inevitable.

In modern forestry, the use of high-density LiDAR data is growing for mapping the forest environment, such as individual tree characteristics [7,19,20], aboveground biomass estimation [21,22], forest disturbances [23], and logging trail detection [24,25]. Mapping forest features depends on reliable and accurate field measurements for both attributes and positions. For this purpose, we need receivers that are able to acquire positions at centimetre-level accuracy, such as geodetic-grade GNSS receivers, to be compromised with LiDAR-derived forest metrics. However, there has also been some research that introduced the relatively expensive approaches by integrating GNSS, IMU, and mobile laser scanning (MLS) to improve the positioning accuracy for solving the simultaneous localization and mapping (SLAM) problem under forest canopy [26,27].

Despite the high positioning accuracy, the cost of establishment a traditional RTK-related receiver, e.g., a geodetic-grade receiver, is approximately 10 times higher than a low-cost RTK receiver [28], which has made it inappropriate for cost-sensitive and small size applications, e.g., in forestry or agriculture applications. The reliable accuracy and continuity of low-cost GNSS receivers are reported for a variety of applications, mostly in non-forest environments, such as surveying and mapping [29,30], monitoring [28,31], Android smartphone positioning [32], precision agriculture [2], and urban environments [33]. Many of the earlier studies reported reliable positioning of the low-cos receivers, such as u-blox modules [28–30,34–36] in a desirable environment condition, for example, an open sky with a wide range in availability of satellites. However, their efficiency might be degraded in an obstructed environment or in the dynamic mode of positioning RTK. Jackson et al. [37] evaluated the positioning accuracy of five low-cost GNSS receivers for RTK positioning under different environments, in both static and dynamic conditions. The results indicated that the positioning errors of the low-cost receivers, in static tests, was less than 10 cm in less complex areas, such as rural environments. However, the error reached over 1 m in complex areas, such as urban and suburban environments. The positioning accuracy of the receivers was different in dynamic tests, and the optimal accuracy is reported 1.5 cm to 1.8 m for the suitable receiver. Likewise, Kadeřábek et al. [2] tested the performance of various types of RTK receivers in horizontal positioning under the modes of static or dynamic. They concluded that the accuracy of positioning is significantly lower in a dynamic mode rather than a static mode. They emphasised that accelerating the speed increases the error of positioning. Janos and Kuras 2021 [35] tested the positioning accuracy of a low-cost GNSS receiver, u-blox ZED-F9P, in the RTK mode under different environment conditions. They found that the type of antenna has a significant impact on the increase of positioning errors in a complex environment, such as urban canyons.

Although a variety of studies have explored the feasibility of traditional GNSS receivers in the forest environment, our understanding concerning the efficiency of newly low-cost receivers and the factors that affect their positioning accuracy is limited, particularly in commercial forests, where the monitoring of machines or recording of the position of processed single trees by harvesters [27,38,39] has become widespread in forest operations. Hence, this research was designed to test the feasibility of using low-cost GNSS receivers and RTK correction signal to determine precise positions in forests under the rotation forest management (RFM) system in southern Finland. Specifically, we want to evaluate the positioning accuracy of the u-blox ZED-F9P in combination with high-density LiDAR data. Moreover, we will explore features that affect the accuracy of a low-cost GNSS receiver in the forest using the TreeNet algorithm.

2. Materials and Methods

2.1. Description of the Study Area

We selected three forest stands for our experiment from Karpanmaa forests in southern Finland (Figure 1). One stand is young and had undergone its first commercial thinning. Two other stands are mature and are waiting for their second/third commercial thinning operations. The tree species compositions are pine, spruce, and birch, with a predominance of pine in the three stands. The age of stands is distributed from 34 to 72 years. The height of trees is between 5 and 29 m. The ground elevation of the forest ranges between 127 and 149 m.

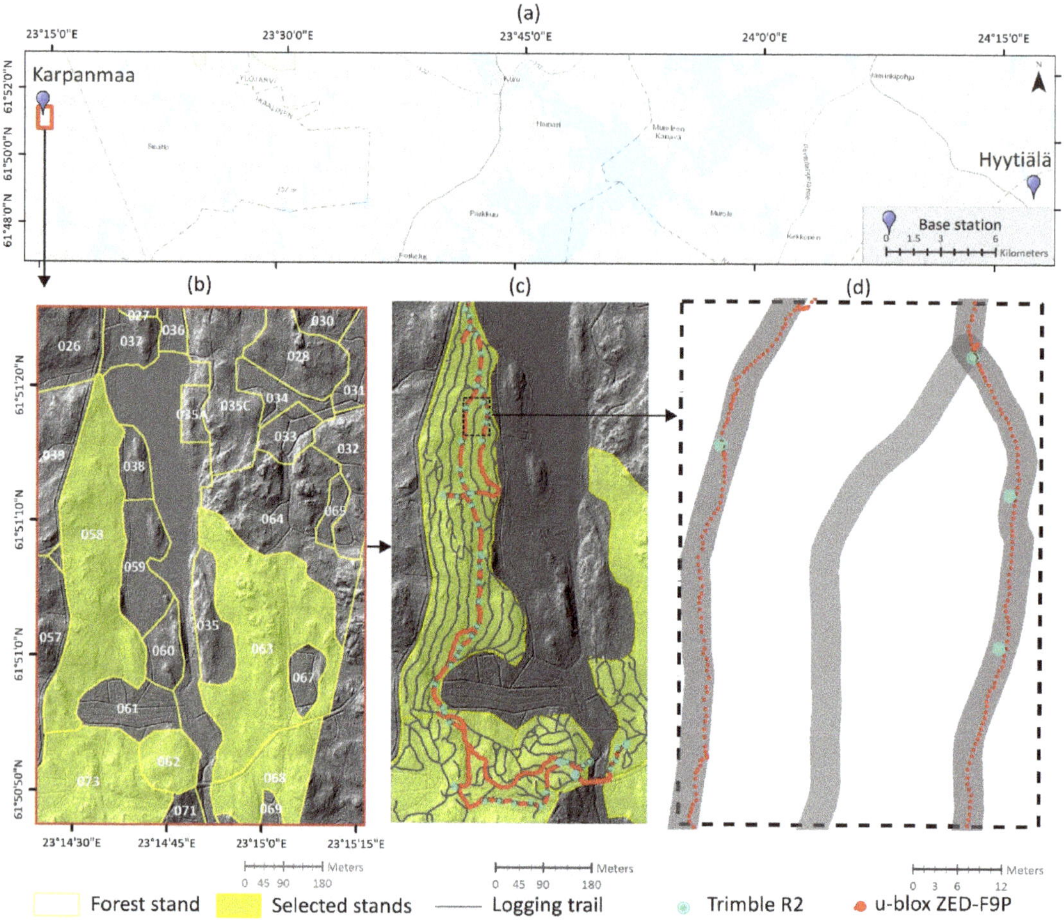

Figure 1. The study site and field measurements in southern Finland: (**a**) the locations of base stations and study area, (**b**) selected stands in Karpanmaa site, (**c**) logging trails, measured positions using the Trimble R2 and u-blox ZED-F9P and (**d**) the patterns of the recorded positions by u-blox ZED-F9P on the logging trails within an example of the study stands.

2.2. Data

We used the high-density LiDAR data, under the license of the National Land Survey of Finland (NLS), recorded in 2020 for the selected stands. The data have a density of greater than 5 points/m^2 along with horizontal and altimetric errors of less than 45 and 10 cm [40], respectively. The features affecting the positioning accuracy of the u-blox ZED-F9P were mapped from the LiDAR-derived metrics such as the digital terrain model (DTM), digital surface model (DSM), point density, and signal intensity. Logging trails were detected from the high-density LiDAR data based on the U-Net convolutional neural network approach developed by Abdi et al. [25]. In addition, we obtained orthophotos from the databases of NLS [41]. The attributes of the forest stands were collected from the databases of Finsilva Oy.

2.3. GNSS Devices

Three GNSS receivers were used, including an Oregon® 750t (Garmin Ltd., Olathe, KS, USA), a u-blox ZED-F9P (u-blox, Thalwil, Switzerland), and a Trimble R2 (Trimble Inc., Sunnyvale, CA, USA), for specialised applications during our field operations.

The Oregon® 750t receiver was used for navigating the approximate locations of the selected logging trails.

We used the u-blox ZED-F9P to identify features affecting the positioning accuracy of the low-cost GNSS receivers in the forest. The u-blox ZED-F9P is a multi-band GNSS receiver that can measure positions at centimetre-level accuracies in RTK mode. This receiver obtains signals from multiple bands (L1, L2/E5b/B21) of all four global GNSS constellations including GPS, Galileo, BeiDou, and GLONASS via a Multi band GNSS antenna ANN-MB-00 (SMA) (u-blox, Thalwil, Switzerland) [42]. The antenna is designed in a small, compact size. It can be easily mounted on different machinery due to its magnetic fixed installation base and a long cable of 5 m [43]. The GNSS and RTK integration has accelerated its convergence time (down to less than 10 sec) [42]. Moreover, performance with the application for conducting unmanned autonomous vehicles (UAV), automatic and semi-automatic machinery, and robotic machines has improved [44]. The u-blox F9P was equipped with advanced anti-spoofing and anti-jamming algorithms that guarantee highly accurate positioning and navigation information. The receiver and antenna are both waterproof and can also operate under extreme temperatures (−40 °C to +85 °C) [45].

We used the Trimble R2 receiver, paired with Trimble TSC7 (Trimble Inc., Sunnyvale, CA, USA), for collecting accurate control points during recording data through u-blox F9P. The Trimble R2 can acquire high positioning accuracy in RTK mode both horizontally (1 cm to 1 ppm RMS) and vertically (2 cm to 1 ppm RMS).

2.4. Research Sulky

We modified a sulky for transporting the u-blox and its compartment for recording positioning data in the forest. The sulky includes two bicycle wheels of 28″ in width. The length, width, and height of the main body of the sulky are 1.2 m, 65 cm, and 87 cm, respectively. It was equipped with a veneer plate (103 cm × 53 cm) for holding the devices. A thin plate (50 cm × 40 cm) was installed on the veneer plate for fixing the magnetic antenna. The sulky was controlled by a draught pole (adjustable up to 1.3 m), which includes a trapezoid-shaped handle to make pulling easier for the user (Figure 2).

2.5. Field Measurements

We established the local base stations to send corrected data to the rover receiver in the vicinity of our study site (Figure 1a). The base and movable stations were developed using a SparkFun GPS-RTK2 Board (SparkFun Electronics, Boulder, CO, USA) with ublox-ZED-F9P module and the SMA. The receivers of the base stations were configured based on an NTRIP-protocol via the internet server (rtk2go.com) to provide RTCM V3.2 standard correction signals.

We selected a number of logging trails with about 2 km for our experiment (Figure 1c). The shapefiles of the logging trails were converted into GPS Exchange Format (GPX) and imported into the Garmin device for spotting out the logging trails in the field.

We started our measurement from 8:32 a.m. and ended at 12:37 p.m. (GMT). The routes were so designed to pass through different species with diversity in age, height, density, canopy cover, and topographic conditions. Our speed was close to the normal speed of harvesters and forwarders (i.e., 44 to 56 m/min) in the forest.

The u-blox ZED-F9P receiver and its compartments, including the SAM, 4G TP-LINK M7200 modem (TP-Link Technologies Co., Shenzhen, China), and Laptop computer, were mounted on a research sulky that was designed for this purpose (Figure 2). The configuration period of the receiver was set to 1 Hz. All four GNSS constellations were selected to receive adequate and strong signals to acquire positions by the u-blox F9P in RTK mode. A Raspberry Pi minicomputer with RTK-LIB open-source program was used for operating the RTK-station. We used u-center evaluation software program to monitor and process all aspects of recorded data (e.g., horizontal and vertical positions, accuracies, time, tracking of GNSS constellations, etc.) during the operation of the u-blox GNSS receiver [46]. All recorded data were captured in ASCII format, imported into ArcGIS (Esri, Redlands, CA, USA), and integrated with the object features to be used for TreeNet analysis.

Figure 2. Compartments for field experiment: (**a**) research sulky, (**b**) u-blox ZED-F9P, (**c**) Multi band GNSS antenna ANN-MB-00 (SMA), (**d**) modem, and (**e**) computer system.

We recorded the positions of 64 control points on the logging trails with an accuracy of less than 1 cm by the Trimble R2, as references, in RTK mode. The absolute errors were calculated between the measured positions by the u-blox F9P and the positions of the control points. The one-sample t-test was used to determine whether the mean of errors exceeded the optimal directions of an image pixel (i.e., 50 cm) derived from the high-density LiDAR data at a significance level of 0.05.

2.6. High-Density Laser Scanning Features

We applied the binning interpolation method based on the maximum cell-assignment method to generate a digital surface model (DSM) from the high-density LiDAR points. The linear interpolation method was used to fill void areas in the DSM. The digital terrain model (DTM) was created based on the interpolation from ground points as well [47]. The features representing the surface topography and canopy conditions, such as slope gradient, aspect, topography/canopy position, plan curvature, profile curvature, and mean curvature [48,49], were derived from the DTM and DSM (Figure 3).

We subtracted the DTM from the DSM to reach the canopy height model (CHM) [50]. The canopy density was generated based on the ratio of the number of nonground points to the total number of points in an object [51]. The tree canopy was delineated from the CHM and used for measuring the canopy cover within an object [52]. The intensity image was created from the range of pulse-intensity values of the laser points [53]. It was applied in combination with orthophotos to determine species types within the objects.

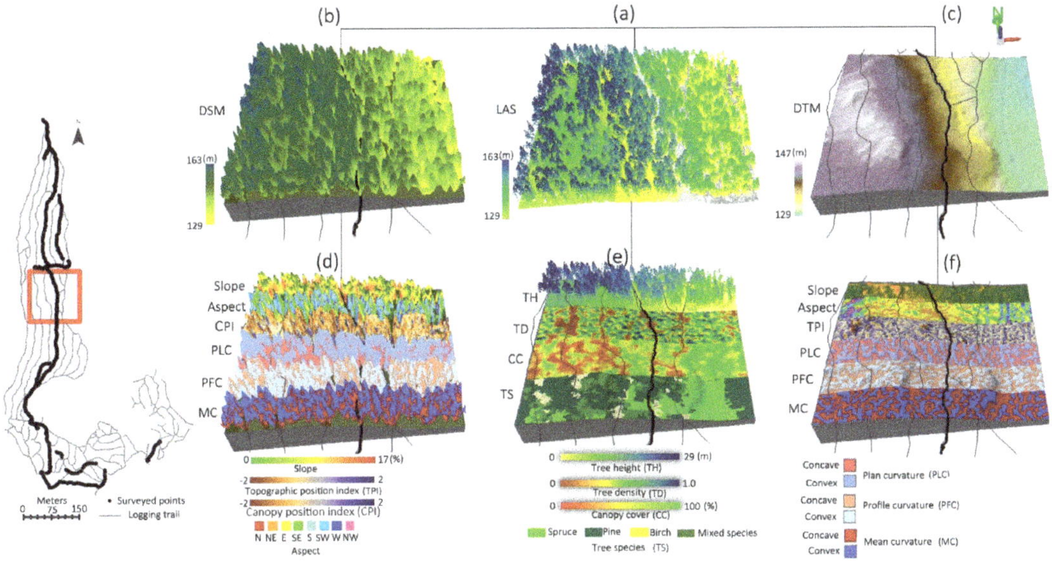

Figure 3. Object features in an example area of the study area. (**a**) 3D perspectives of high-density LiDAR point clouds, its metric derivatives including (**b**) the digital terrain model (DTM) and (**c**) digital surface model (DSM). (**d**) The ground-surface features were extracted from the DTM and (**e**) the canopy characteristics and (**f**) the canopy-surface features were extracted from the LiDAR points and the DSM, respectively.

2.7. Object Features

A buffer of 10 m was delineated around the logging trails, and then the area was segmented into the homogenous units (objects) based on the similarity in the spectral properties of the adjacent cells in the high-resolution orthophotos. The average size of the objects was obtained around 18 sq.m. The values of the derived features from the LiDAR data (Table 1) and the accuracy of positioning by u-blox ZED-F9P were summarized within the objects. This single database was used for analyzing the relationship between the target and the features using TreeNet.

2.8. TreeNet Regression

We used the TreeNet regression algorithm to determine influential features on the positioning accuracy of the u-blox ZED-F9P. Numerous advantages have been reported for TreeNet, in comparison with other machine learning based approaches. In addition to its highly accurate predictions, TreeNet is not sensitive to errors in data or missing data. No data pre-processing (e.g., transformation, normalisation, or reduction) or preselection of the variables is required. TreeNet is strong against overfitting, and the process of growing trees is extraordinarily fast [54,55].

TreeNet begins with an initial model, which consists of a very small tree. This simple model is deliberately weak. The residuals are computed for each data in the first model and are used to grow the second tree. The residuals of the second tree are then computed and used to grow the third tree. Likewise, this process repeats to generate a sequence of hundreds or thousands of trees, in order to achieve an optimal tree. All trees contribute to the optimal model. The final model prediction is based on the total contribution of the individual trees, which is known as score. The accuracy of the TreeNet score will improve steadily by increasing the number of trees until to reach an optimal number of trees [55].

We used a dataset including 2000 sample objects. Twenty percent of the sample objects were randomly assigned as the testing set and the remaining 80% as the learning set. The

TreeNet loss function was set on Huber-M. The mean square error (MSE) was chosen as the criterion for determining the number of optimal trees. We set the initial tree size at 10,000 and generated 12 TreeNet models based on the different learning rates and tree complexity levels. The optimal model was selected based on the one that recorded the minimum MSE, and its parameters were tuned for the final TreeNet model. The performance of the model was tested using the area under the receiver operating characteristic (ROC) curve. A value greater than 0.9 represents high performance, while values less than 0.7 indicate low performance [56] of the TreeNet model.

The influence of features on the positioning accuracy of the u-blox ZED-F9P was determined via relative importance [55]. The importance values of the features are ranged between 0 and 100. The most influential feature gains a value of 100 and the remaining features are rescaled to reflect their importance relative to this feature. We produced partial dependence (PD) plots for individual and pairs of features that contributed to the predicted positioning accuracy in the model. The PD plots represent the response of the target variable to individual or pairs of features, as all remaining features are taken into account [57].

Table 1. Object features affecting on the accuracy of positioning by u-blox ZED-F9P, derived from different metrics of high-density LiDAR data.

Category	Features	Description
Ground-surface/canopy-surface conditions	Elevation	The mean of ground/canopy elevation (m) [58] in an object.
	Slope (°)	The average of maximum changes in elevation value [59] within each object.
	Aspect	The direction of compass of downhill slope [59] in each object.
	Topographic position index (TPI)	TPI measures the difference between the elevation of the central point against the average elevation of the ground surface in an object. The positive values indicate the higher elevation of the central points and vice versa [60].
	Canopy position index (CPI)	CPI measures the difference between the elevation of the central point against the average elevation of the canopy surface in an object. The positive values indicate the higher elevation of the central points and vice versa.
	Plan curvature	The curvature of the surface (ground or canopy) perpendicular to the direction of slope. The positive values indicate the convex surface and negative values indicate the concave surface [48,49].
	Profile curvature	The curvature of the surface (ground or canopy) in the direction of the maximum slope in each object. The negative values indicate the convex position and positive values indicate concave surface [48,49].
	Mean curvature	The combination of the plan and profile curvatures within an object [48,49].
Tree characteristics	Canopy height	The difference between the elevation of canopy surface and ground surface in an object [50].
	Canopy density	The density of nonground returns of LiDAR points in an object [51].
	Canopy cover	The percentage of canopy cover within an object, delineated from the CHM [52].
	Species type	The type of species trees extracted from the intensity image, derived from the high-density LiDAR data and orthophoto images.

3. Results

3.1. The Accuracy of Positions

The mean of absolute errors of positions between those measured by u-blox ZED-F9P and control points was obtained as about 43 cm. The result of the one-sample t-test showed that the positioning accuracy of u-blox has no significant deviation from the area of logging trails (test value = 0.5 m, *p*-value > 0.05). However, the absolute errors are distributed between 1.5 cm to 1.8 m (Figure 4).

Figure 4. Distribution of absolute errors of positions recorded for u-blox ZED-F9P in control points.

3.2. TreeNet Performance

The optimal TreeNet model was obtained after building 490 trees with a learning rate of 0.011 and tree complexity of 7 (Figure 5). The ROC values were 0.977 for the training dataset and 0.745 for the testing dataset. It indicates that TreeNet demonstrated high performance for expressing the features that are determining the positioning accuracy of the GNSS receivers in forest.

3.3. Features' Importance

The results of evaluation the importance values of features indicated that all of the features have affected the positioning accuracy of u-blox ZED-F9P. The top influential feature was tree height. Then, the influential features were ranked by their importance to the tree height feature. The ground elevation and aspect have gained 79% and 62% importance of the tree height, respectively. The importance of the other features has steadily decreased from the canopy surface elevation to the surface aspect. Two out of five top influential features were classified in the category of tree characteristics (Figure 6).

3.4. Marginal Effect of Individual Features

The interpretation of univariate PD plots regarding the tree characteristics shows that when tree height increases to 14 m, the error in positioning by u-blox ZED-F9P increases (Figure 7a). By increasing tree density, the positioning error drastically increases, at a density over 0.25 (Figure 7b). Canopy cover of more than 30% shows a positive response to error in positions of the GNSS receiver (Figure 7c). The mixed species and pine show a positive response to the high number of errors among tree species (Figure 7d).

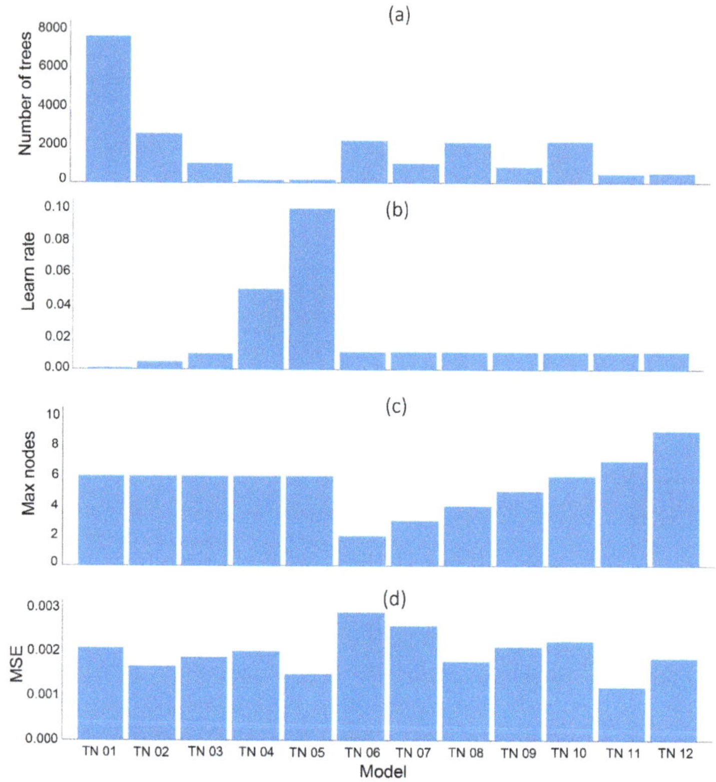

Figure 5. Comparison between different TreeNet (TN) models regarding (**a**) the number of trees, (**b**) learning rates, and (**c**) tree complexity for determining optimal TreeNet model based on (**d**) the mean square error (MSE).

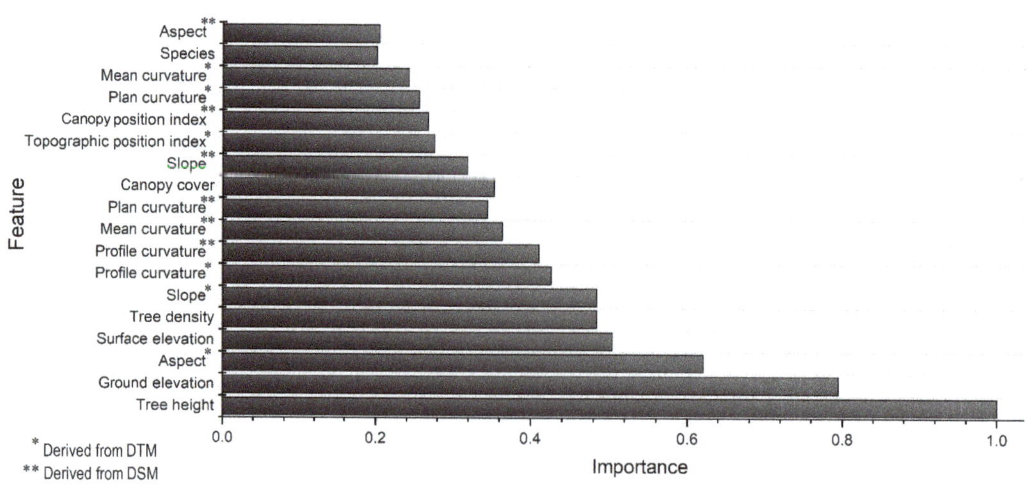

Figure 6. Importance scores of the features affecting the positioning accuracy of u-blox ZED-F9P. The top influential feature is tree height. Other features are ranked based on their importance to the tree height.

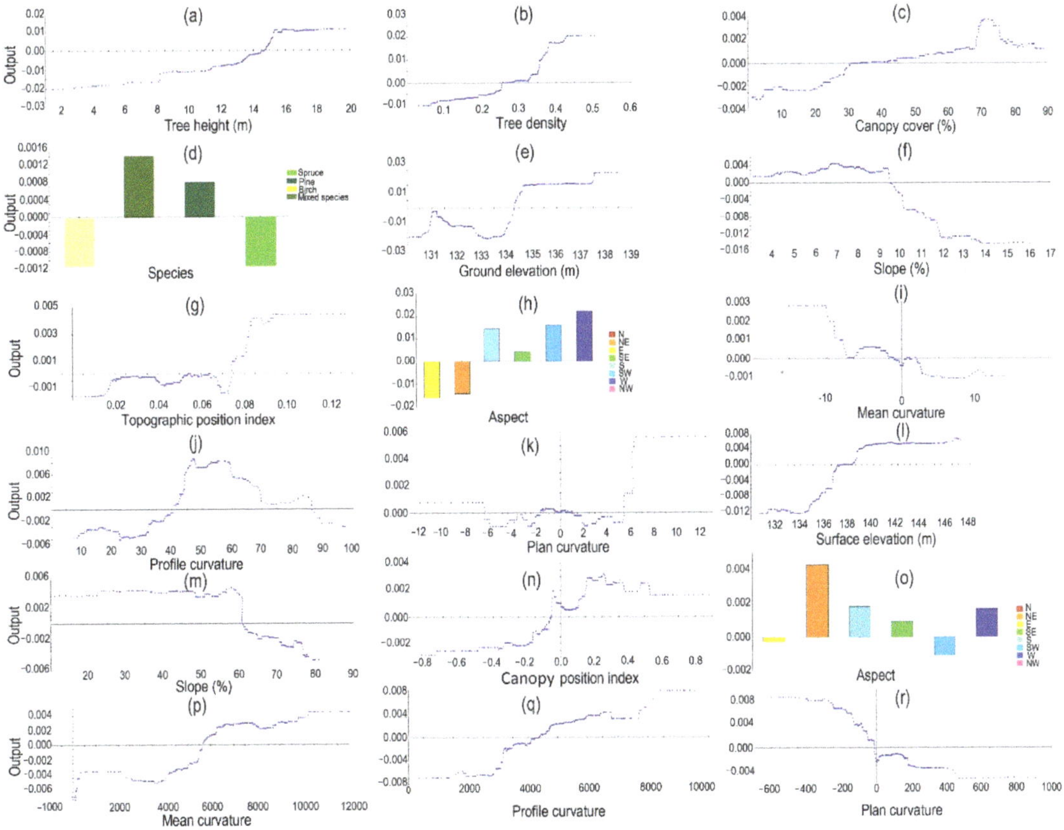

Figure 7. Univariate partial dependence plots for features affecting the positioning accuracy of the u-blox ZED-F9P. (**a–d**) tree characteristics, (**e–k**) topography conditions, and (**l–r**) canopy-surface conditions.

The univariate PD plots of the topographic features show that, when the ground elevation exceeds 134 m in the study area, the error of positioning by u-blox increases (Figure 7e). The areas with a slope of less than 10% show a positive response to the errors (Figure 7f). Increasing complexity in the topographic position increases the errors (Figure 7g). The western direction shows the greatest errors among the topographic aspects. The southern, south-western, and south-eastern directions show a positive response to the errors (Figure 7h). The concave curvatures show a positive response to errors in plan, profile, and mean curvatures (Figure 7i–k).

The PD plots show that the surface elevation positively responds to the high error in recorded positions by u-blox ZED-F9P after 137 m (Figure 7l). The areas with a surface slope of less than 60% show an increase in the errors (Figure 7m). High complexity in the canopy-surface position coincides with high errors in positioning (Figure 7n). Canopies with the domination of the north-eastern direction show higher errors (Figure 7o). The canopies with concave curvatures demonstrate mostly high errors totally, vertically, and horizontally (Figure 7p–r).

3.5. Marginal Effects of Pairs of Features

Figure 8 shows the interactions of five top pairs of features on the positioning accuracy of u-blox ZED-F9P. The increasing height of trees (Figure 8a), tree density (Figure 8c), surface elevation (Figure 8f), and ground elevation (Figure 8j) in the western and southern portions increased the probability of errors in the receiver. The interaction of large trees and high tree-density (Figure 8b), surface elevation (Figure 8d), and ground elevation (Figure 8g) increased the errors. The interaction of increasing the tree density and surface elevation (Figure 8e) and ground elevation (Figure 8i) led to errors in positions. Whenever both ground elevation and surface elevation increased, the errors increased (Figure 8h).

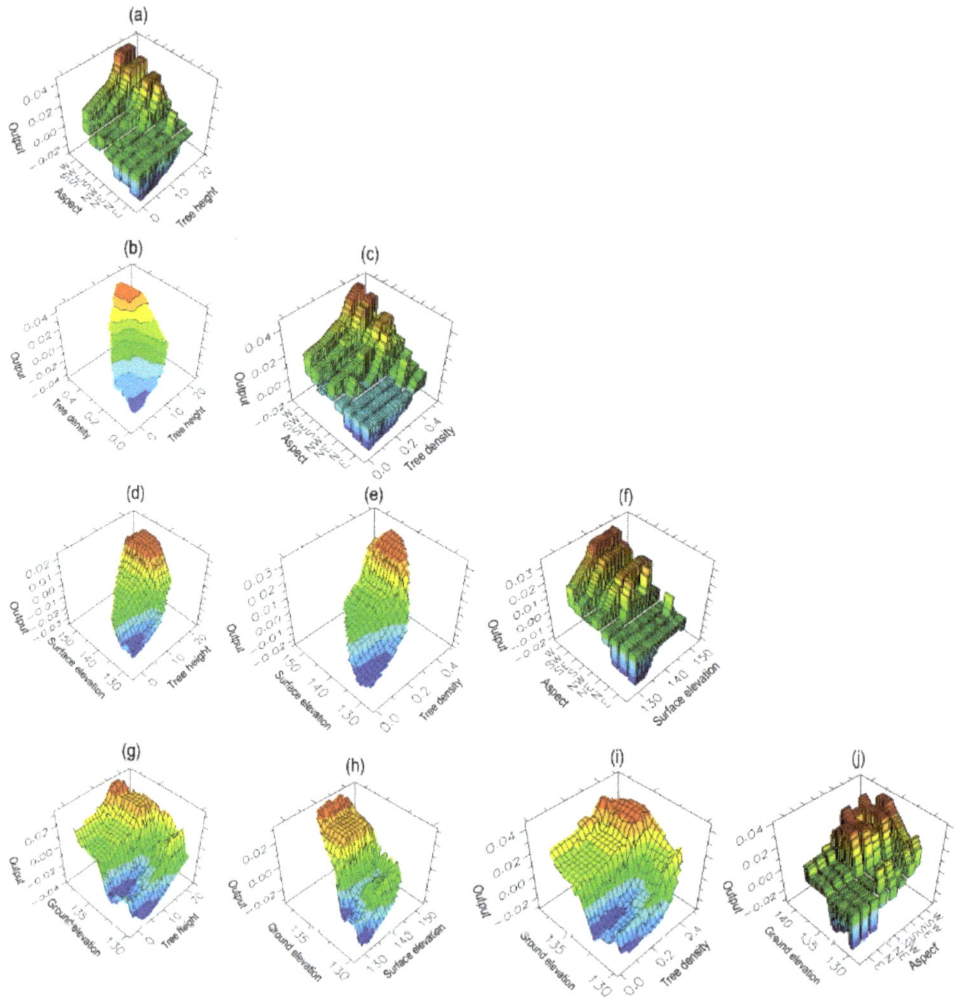

Figure 8. Bivariate partial dependence plots for five top features affecting the positioning accuracy of the u-blox ZED-F9P. (**a**) tree height and aspect, (**b**) tree height and tree density, (**c**) tree density and aspect, (**d**) tree height and surface elevation, (**e**) tree density and surface elevation, (**f**) surface elevation and aspect, (**g**) tree height and ground elevation, (**h**) surface elevation and ground elevation, (**i**) tree density and ground elevation, and (**j**) aspect and ground elevation.

4. Discussion

4.1. The Positioning Accuracy of the Low-Cost GNSS Receiver

We reach an absolute error of 0.43 m for positioning accuracy by the low-cost u-blox ZED-F9P GNSS receiver with its equipped standard patch antenna in movable RTK mode in forest environment. This level of positioning accuracy is promising for forest operations, particularly relative to the positioning accuracy of current GNSS receivers used by vehicles in forest. Alternatively, the development of LiDAR systems has provided the possibility of producing high precision maps of the forest environment and tree characteristics at centimetre-level accuracy, with a significant reduction in costs and improvement in time of processing. Our findings verify the trust in positioning by the low-cost receiver in RTK mode for integration with the forest features derived from high-density LiDAR data, such as logging trails. This may have wide implications for the improvement of the safety of crews, autonomous navigation, ergonomics, and reduction of environmental impacts and costs [61] during forest operations down the pathways of precision forestry. Although the positioning accuracy of the low-cost GNSS receivers in static mode was reported higher than the RTK mode in complex non-forest environments [2,37], their positioning accuracy was considerable in this mode [28,30] as well. We should stress that our results are only based on using the standard patch antenna. Further work is required to test the performance of the low-cost GNSS receivers when equipped with additional antennas in the forest environment. Previous experiments acknowledged significant improvement in the positioning accuracy of the low-cost receivers, for example, when using with a geodetic-grade antenna [29,34,35].

The u-blox ZED-F9P obtains signals in multiband from four global GNSS. Multi-GNSS contributes to increasing the continuity and integrity of positioning by the receivers, particularly in environments with obstacles [62,63]. The effect of receiver types on the accuracy of positioning was reported in earlier studies. The survey-grade devices recorded higher accuracy than consumer-grade [16,64], mapping-grade [13,65], or smartphone-grade [66–68] devices.

4.2. The Performance of TreeNet

This study applied high-density LiDAR data and a novel object-based TreeNet approach to determine influential features that degrade the positioning accuracy of the novel developed low-cost GNSS receivers in a forest context. Earlier studies mostly modelled the influential variables of positioning accuracy using traditional regression models [12,14,17,66], which are limited with few variables and data. Conversely, we adopted TreeNet, as one of the most powerful machine learning algorithms, with remarkable abilities in handling big data and numerous variables without any preselection, pre-processing, or reduction in dataset. It reveals that a combination of forest characteristics and terrain features express the positioning errors of GNSS receivers. However, the importance of features are different.

4.3. Influential Forest Features

Based on our analyses, LiDAR-derived tree height is the top feature that influence on the positioning accuracy of the low-cost GNSS receiver in movable RTK mode. Tree density is among top five influential features (Figure 6). The complexity of the forest structure causes multipath effects [69], which is one of the main sources of increasing positioning errors in forest. Tree's characteristics, such as height, volume, tree density and canopy may block or weaken the signals [70]. A closed canopy can cause cycle slips [63], which clogs the signals to reach the receivers. Our analyses demonstrated that the low-cost GNSS receiver continuously recorded the signals throughout the logging trails, and it was resistant against cycle slips effects. Conversely, the great majority of work has focused on the canopy cover [11,13,14,16,71–73] as the main factors that affect the positioning accuracy of the GNSS receivers. Nevertheless, no priority between the forest cover factors on the positioning accuracy of GNSS was reported in the research of Ordóñez Galán et al. [15], while our study shows that there is a distinct difference between the impacts of tree characteristic factors on the positioning accuracy of the GNSS receiver. Moreover, some earlier studies reported the higher impact of the broadleaved tree species on increasing the errors of positioning by the GNSS receiver [13,14]. Our research showed that the importance of the tree species is less than other tree characteristics. The influence of pine and mixture species is positive against the spruce and birch.

Using high-density LiDAR data enabled us to take into account precisely some features of tree characteristics that were less a focus of earlier studies, such as tree height or tree density, not in plot scale but over the entire surveyed logging trails. For example, due to limitations involved in using traditional methods to measure tree height, the preponderance of the studies focused instead on forest cover or forest type [13–16,64] as potential effective factors of tree characteristics to determine the positioning accuracy of GNSS receivers. We considered trees' characteristics inside an object, which is much more similar to the natural condition of a forest stand. Furthermore, we carried out the experiments in a season which the leaf of some tree species such as birch is almost off. Hence, canopy cover or tree species did show lower importance than tree height or tree density in the current research. Likewise, topographic conditions affect the signals and cause the multipath effects. Our study revealed that topography directions and elevation are among top five important features that determine the positioning accuracy of the low-cost GNSS receiver (Figure 6). Aspect and slope derived from the DTM showed higher importance than was shown by the corresponding features derived from the DSM. The high ability of DTM to visualise the morphology of the bare earth under the forest canopy [74], such as variations in slope directions and values, might be one reason for this difference. On the other hand, the high variation in the curvature of the forest canopy may lead to a higher influence of DSM-derived plan curvature and mean curvature when compared to the corresponding derivatives from the DTM. Although there is no holistic research about the impact of topographic conditions in forest environment, some studies verified significant impacts of terrain in non-forest environments on the positioning accuracy of the GNSS receivers [1,18]. Based on the PD plots, the positioning errors of the low-cost GNSS receiver increase in forest areas with tree height above 14 m, tree density above 30%, western topographic directions or high elevation (Figure 6). These thresholds are based on our results in southern Finland, and similar studies should be repeated to achieve fixed values over the Nordic region. The performance of the antenna, as a sky view, is presented for a specific measured position during our experiment in Figure 9. The number of satellites that are used in navigation with a valid fixed position in the eastern direction is higher than in the western direction. We can infer that the geometric location of the satellites and their signal qualities may cause that aspect to be one of the top features determining the accuracy of the positioning by the low-cost GNSS receiver. Lower fixing rates and position errors of GNSS on west aspects were reported by D'eon and Delparte [75]. However, they reported that the differences in these values between different directions were not significant. The effect of aspect and convex slopes on the odds of missing signals was reported in the forest as well. Zimmerman and Keefe [76] verified that the alert delay of GNSS in the west directions is higher than in the east directions, which increases the error of positioning under forest canopies.

In addition, our analyses indicated that the interaction of the conditioning features intensifies the positioning errors by the receiver (Figure 8). It seems that using extra antenna or geodetic antenna [29–31,35] may mitigate the impact of the forest structure or topographic conditions on the positioning errors of this type of low-cost GNSS receivers in forest environments.

4.4. The Application

The new generation of harvesters is equipped with sensors, computers, and GNSS receivers that store big data obtained from the processed trees and machine parameters as a standard format of StanForD (Standard for Forest Data and communication) [3]. Despite the high capacity of these data for modelling forest productivity, the errors in positions significantly degrade the efficiency of these data [6–8]. This type of low-cost GNSS receiver can improve the accuracy, integrity, and continuity of positions for the harvesters, with a significant impact on increasing the efficiency of forest productivity maps to Improve the sustainability of future rotations and precision forestry. Moreover, understanding the influential features that affect the positioning accuracy of the low-cost GNSS devices contributes to developing algorithms for the correction of positioning errors or for the selection of appropriate low-cost receivers or antennas to minimize the influence of environmental features on positioning accuracy during forest operations.

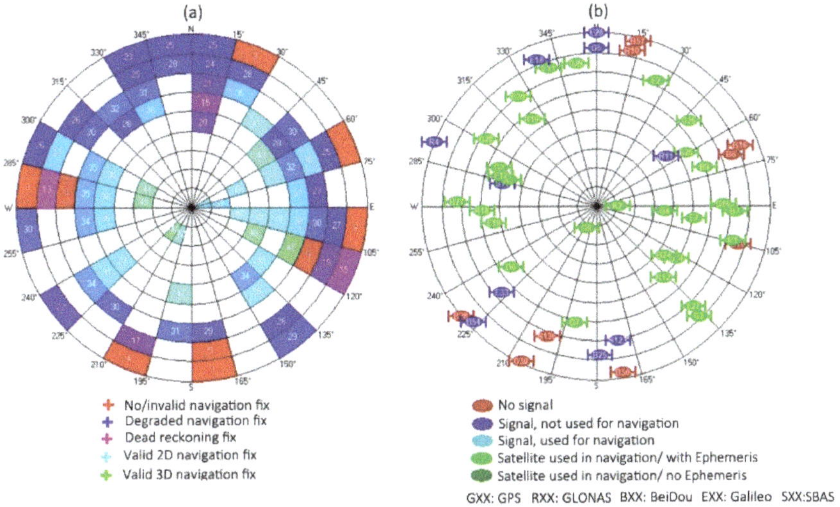

Figure 9. Sky view for a specific location for the antenna of u-blox ZED-F9P: (**a**) the quality of sky view and (**b**) the position of satellites that were used in navigation.

5. Conclusions

In this study, we presented a geographic object-based TreeNet approach to determine influential environmental features that affect the positioning accuracy of a newly developed, low-cost, high-precision GNSS receiver, u-blox ZED-F9P, in forests. The experiment concentrated on some logging trails in commercial forests in Southern Finland. The low-cost receiver showed reliable positioning accuracy when integrated with high-density LiDAR data in the forest. The TreeNet model showed a high performance for expressing features that determine the positioning accuracy of the low-cost receiver in the forest. A combination of features increased the positioning errors of the low-cost receiver, in which the most important feature was tree height and then the topographic features, such as elevation and slope direction over the study stands. In the current research, we merely used the standard patched antenna packed with the low-cost receiver. However, we suggest testing the efficiency of other types of antennas, e.g., geodetic-grade ones, or a combination of antennas with the low-cost receiver to improve the positioning accuracy in the forest environment.

Author Contributions: Conceptualization, O.A., J.U., A.L. and J.P.; methodology, O.A., J.U., A.L. and J.P.; data provision, J.U.; data preparation, O.A. and J.P.; software and programming, O.A. and J.P.; field investigation and sampling, O.A., J.U., A.L. and J.P.; visualization, O.A.; writing—original draft preparation, O.A.; writing—review and editing, J.U., A.L. and J.P.; supervision, J.U. and A.L.; project administration, J.U. All authors have read and agreed to the published version of the manuscript.

Funding: Open access funding provided by University of Helsinki. This work has been funded by the public-private partnership grant established for the professorship of forest operation and logistics at the University of Helsinki, grant number 7820148 and by the proof-of-concept-grant by the Faculty of Agriculture and Forestry, University of Helsinki, grant number 78004041.

Acknowledgments: We would like thank Pekka Huuhka for constructing the research sulky and Mikko Leinonen for assisting in the field operations. We would also like to express our gratitude to Finsilva Oyj for providing the access to their forest holdings and related forest inventory databases.

Conflicts of Interest: The authors declare no conflict of interest. The funders had no role in the design of the study; in the collection, analyses, or interpretation of data; in the writing of the manuscript, or in the decision to publish the results.

References

1. Pini, M.; Marucco, G.; Falco, G.; Nicola, M.; Wilde, W.D. Experimental Testbed and Methodology for the Assessment of RTK GNSS Receivers Used in Precision Agriculture. *IEEE Access* **2020**, *8*, 14690–14703. [CrossRef]
2. Kadeřábek, J.; Shapoval, V.; Matějka, P.; Kroulík, M.; Kumhála, F. Comparison of Four RTK Receivers Operating in the Static and Dynamic Modes Using Measurement Robotic Arm. *Sensors* **2021**, *21*, 7794. [CrossRef] [PubMed]
3. Möller, J.J.; Arlinger, J.; Hannrup, B.; Larsson, W.; Barth, A. Harvester data as a base for management of forest operations and feedback to forest owners. In Proceedings of the 4th Forest Engineering Conference: Innovation in Forest Engineering—Adapting to Structural Change, Stellenbosch, South Africa, 5–7 April 2011; Ackerman, P., Ham, H., Elizabeth, G., Eds.; Department of Forest and Wood Science: White River, South Africa, 2011. ISBN 978-0-7972-1284-8.
4. Olivera, A.; Visser, R. Using the harvester on-board computer capability to move towards precision forestry. *N. Z. J. For.* **2016**, *60*, 3–7. [CrossRef]
5. Auat Cheein, F.; Torres-Torriti, M.; Rosell-Polo, J.R. Usability analysis of scan matching techniques for localization of field machinery in avocado groves. *Comput. Electron. Agric.* **2019**, *162*, 941–950. [CrossRef]
6. Kemmerer, J.; Labelle, E.R. Using harvester data from on-board computers: A review of key findings, opportunities and challenges. *Eur. J. For. Res.* **2021**, *140*, 1–17. [CrossRef]
7. Saukkola, A.; Melkas, T.; Riekki, K.; Sirparanta, S.; Peuhkurinen, J.; Holopainen, M.; Hyyppä, J.; Vastaranta, M. Predicting Forest Inventory Attributes Using Airborne Laser Scanning, Aerial Imagery, and Harvester Data. *Remote Sens.* **2019**, *11*, 797. [CrossRef]
8. Noordermeer, L.; Næsset, E.; Gobakken, T. Effects of harvester positioning errors on merchantable timber volume predicted and estimated from airborne laser scanner data in mature Norway spruce forests. *Silva Fenn.* **2022**, *56*, 10608. [CrossRef]
9. Schwarzbach, P.; Michler, A.; Tauscher, P.; Michler, O. An Empirical Study on V2X Enhanced Low-Cost GNSS Cooperative Positioning in Urban Environments. *Sensors* **2019**, *19*, 5201. [CrossRef]
10. Donahue, B.; Wentzel, J.; Berg, R. *Guidelines for RTK/RTN GNSS Surveying in Canada*; Natural Resources Canada: Vancouver, BC, Canada, 2013.
11. Sigrist, P.; Coppin, P.; Hermy, M. Impact of forest canopy on quality and accuracy of GPS measurements. *Int. J. Remote Sens.* **1999**, *20*, 3595–3610. [CrossRef]
12. Naesset, E.; Jonmeister, T. Assessing Point Accuracy of DGPS Under Forest Canopy Before Data Acquisition, in the Field and After Postprocessing. *Scand. J. For. Res.* **2002**, *17*, 351–358. [CrossRef]
13. Murgaš, V.; Sačkov, I.; Sedliak, M.; Tunák, D.; Chudý, F. Assessing horizontal accuracy of inventory plots in forests with different mix of tree species composition and development stage. *J. For. Sci.* **2018**, *64*, 478–485. [CrossRef]
14. Feng, T.; Chen, S.; Feng, Z.; Shen, C.; Tian, Y. Effects of Canopy and Multi-Epoch Observations on Single-Point Positioning Errors of a GNSS in Coniferous and Broadleaved Forests. *Remote Sens.* **2021**, *13*, 2325. [CrossRef]
15. Ordóñez Galán, C.; Rodríguez-Pérez, J.R.; Martínez Torres, J.; García Nieto, P.J. Analysis of the influence of forest environments on the accuracy of GPS measurements by using genetic algorithms. *Math. Comput. Model.* **2011**, *54*, 1829–1834. [CrossRef]
16. Piedallu, C.; Gégout, J.-C. Effects of Forest Environment and Survey Protocol on GPS Accuracy. *Photogramm. Eng. Remote Sens.* **2005**, *71*, 1071–1078. [CrossRef]
17. Valbuena, R.; Mauro, F.; Rodríguez-Solano, R.; Manzanera, J.A. Partial Least Squares for Discriminating Variance Components in Global Navigation Satellite Systems Accuracy Obtained Under Scots Pine Canopies. *For. Sci.* **2012**, *58*, 139–153. [CrossRef]
18. Kabir, M.S.N.; Song, M.-Z.; Sung, N.-S.; Chung, S.-O.; Kim, Y.-J.; Noguchi, N.; Hong, S.-J. Performance comparison of single and multi-GNSS receivers under agricultural fields in Korea. *Eng. Agric. Environ. Food* **2016**, *9*, 27–35. [CrossRef]
19. White, J.C.; Wulder, M.A.; Varhola, A.; Vastaranta, M.; Coops, N.C.; Cook, B.D.; Pitt, D.; Woods, M. A best practices guide for generating forest inventory attributes from airborne laser scanning data using an area-based approach. *For. Chron.* **2013**, *89*, 722–723. [CrossRef]
20. Maltamo, M.; Næsset, E.; Vauhkonen, J. *Forestry Applications of Airborne Laser Scanning: Concepts and Case Studies*; Maltamo, M., Næsset, E., Vauhkonen, J., Eds.; Springer: Dordrecht, The Netherlands, 2014; ISBN 978-94-017-8662-1.
21. Su, Y. *The Use of LiDAR in Multi-Scale Forestry Applications*; University of California: Merced, CA, USA, 2017.
22. Ene, L.T.; Næsset, E.; Gobakken, T.; Mauya, E.W.; Bollandsås, O.M.; Gregoire, T.G.; Ståhl, G.; Zahabu, E. Large-scale estimation of aboveground biomass in miombo woodlands using airborne laser scanning and national forest inventory data. *Remote Sens. Environ.* **2016**, *186*, 626–636. [CrossRef]
23. Noordermeer, L.; Økseter, R.; Ørka, H.O.; Gobakken, T.; Næsset, E.; Bollandsås, O.M. Classifications of Forest Change by Using Bitemporal Airborne Laser Scanner Data. *Remote Sens.* **2019**, *11*, 2145. [CrossRef]
24. Sherba, J.; Blesius, L.; Davis, J. Object-Based Classification of Abandoned Logging Roads under Heavy Canopy Using LiDAR. *Remote Sens.* **2014**, *6*, 4043–4060. [CrossRef]
25. Abdi, O.; Uusitalo, J.; Kivinen, V.-P. Logging Trail Segmentation via a Novel U-Net Convolutional Neural Network and High-Density Laser Scanning Data. *Remote Sens.* **2022**, *14*, 349. [CrossRef]
26. Qian, C.; Liu, H.; Tang, J.; Chen, Y.; Kaartinen, H.; Kukko, A.; Zhu, L.; Liang, X.; Chen, L.; Hyyppä, J. An Integrated GNSS/INS/LiDAR-SLAM Positioning Method for Highly Accurate Forest Stem Mapping. *Remote Sens.* **2017**, *9*, 3. [CrossRef]
27. Kukko, A.; Kaijaluoto, R.; Kaartinen, H.; Lehtola, V.V.; Jaakkola, A.; Hyyppä, J. Graph SLAM correction for single scanner MLS forest data under boreal forest canopy. *ISPRS J. Photogramm. Remote Sens.* **2017**, *132*, 199–209. [CrossRef]

28. Poluzzi, L.; Tavasci, L.; Corsini, F.; Barbarella, M.; Gandolfi, S. Low-cost GNSS sensors for monitoring applications. *Appl. Geomat.* **2020**, *12*, 35–44. [CrossRef]
29. Tsakiri, M.; Sioulis, A.; Piniotis, G. Compliance of low-cost, single-frequency GNSS receivers to standards consistent with ISO for control surveying. *Int. J. Metrol. Qual. Eng.* **2017**, *8*, 11. [CrossRef]
30. Wielgocka, N.; Hadas, T.; Kaczmarek, A.; Marut, G. Feasibility of Using Low-Cost Dual-Frequency GNSS Receivers for Land Surveying. *Sensors* **2021**, *21*, 1956. [CrossRef]
31. Xue, C.; Psimoulis, P.; Zhang, Q.; Meng, X. Analysis of the performance of closely spaced low-cost multi-GNSS receivers. *Appl. Geomat.* **2021**, *13*, 415–435. [CrossRef]
32. Semler, Q.; Mangin, L.; Moussaoui, A.; Semin, E. Development of a Low-Cost Centimetric Gnss Positioning Solution for Android Applications. *Int. Arch. Photogramm. Remote Sens. Spat. Inf. Sci.* **2019**, *XLII-2/W17*, 309–314. [CrossRef]
33. Li, T.; Zhang, H.; Gao, Z.; Chen, Q.; Niu, X. High-Accuracy Positioning in Urban Environments Using Single-Frequency Multi-GNSS RTK/MEMS-IMU Integration. *Remote Sens.* **2018**, *10*, 205. [CrossRef]
34. Yuwono; Handoko, E.Y.; Cahyadi, M.N.; Rahmadiansah, A.; Yudha, I.S.; Sari, A. Assessment of the Single Frequency Low Cost GPS RTK Positioning. *IOP Conf. Ser. Earth Environ. Sci.* **2019**, *280*, 12025. [CrossRef]
35. Janos, D.; Kuras, P. Evaluation of Low-Cost GNSS Receiver under Demanding Conditions in RTK Network Mode. *Sensors* **2021**, *21*, 5552. [CrossRef] [PubMed]
36. Krietemeyer, A.; van der Marel, H.; van de Giesen, N.; Veldhuis, M.-C.T. High Quality Zenith Tropospheric Delay Estimation Using a Low-Cost Dual-Frequency Receiver and Relative Antenna Calibration. *Remote Sens.* **2020**, *12*, 1393. [CrossRef]
37. Jackson, J.; Saborio, R.; Ghazanfar, S.A.; Gebre-Egziabher, D.; Davis, B. *Evaluation of Low-Cost, Centimeter-Level Accuracy OEM GNSS Receivers*; MN/RC 2018-10; University of Minnesota: Minneapolis, MN, USA, 2018.
38. Hauglin, M.; Hansen, E.H.; Næsset, E.; Busterud, B.E.; Gjevestad, J.G.O.; Gobakken, T. Accurate single-tree positions from a harvester: A test of two global satellite-based positioning systems. *Scand. J. For. Res.* **2017**, *32*, 774–781. [CrossRef]
39. Kaartinen, H.; Hyyppä, J.; Vastaranta, M.; Kukko, A.; Jaakkola, A.; Yu, X.; Pyörälä, J.; Liang, X.; Liu, J.; Wang, Y.; et al. Accuracy of Kinematic Positioning Using Global Satellite Navigation Systems under Forest Canopies. *Forests* **2015**, *6*, 3218–3236. [CrossRef]
40. National Land Survey of Finland. Laser Scanning Data 5 p. Available online: https://www.maanmittauslaitos.fi/en/maps-and-spatial-data/expert-users/product-descriptions/laser-scanning-data-5-p (accessed on 6 May 2021).
41. National Land Survey of Finland. NLS Orthophotos. Available online: https://tiedostopalvelu.maanmittauslaitos.fi/tp/kartta?lang=en (accessed on 1 May 2021).
42. U-blox AG. ZED-F9P: U-blox F9 High Precision GNSS Module. Integration Manual. Available online: https://www.u-blox.com/en/docs/UBX-18010802 (accessed on 15 November 2021).
43. U-blox AG. ANN-MB Series: Multi-band, High Precision GNSS Antennas. Data Sheet. Available online: https://www.u-blox.com/sites/default/files/ANN-MB_DataSheet_%28UBX-18049862%29.pdf (accessed on 15 November 2021).
44. U-blox AG. ZED-F9P: Application Note. Available online: https://www.u-blox.com/en/docs/UBX-19009093 (accessed on 15 November 2021).
45. U-blox AG. ZED-F9P Module: U-blox F9 High Precision GNSS Module. Available online: https://www.u-blox.com/en/ubx-viewer/view/ZED-F9P_ProductSummary_UBX-17005151?url=https%3A%2F%2Fwww.u-blox.com%2Fsites%2Fdefault%2Ffiles%2FZED-F9P_ProductSummary_UBX-17005151.pdf (accessed on 1 December 2021).
46. U-blox AG. U-Center: GNSS Evaluation Software for Windows. User Guide. Available online: https://www.u-blox.com/en/docs/UBX-13005250 (accessed on 1 December 2021).
47. ESRI. Lidar Solutions in ArcGIS: Estimating Forest Canopy Density and Height. Available online: https://desktop.arcgis.com/en/arcmap/latest/manage-data/las-dataset/lidar-solutions-estimating-forest-density-and-height.htm (accessed on 10 December 2021).
48. Wilson, J.P. *Environmental Applications of Digital Terrain Modeling*, 1st ed.; Wiley-Blackwell: Hoboken, NJ, USA, 2018.
49. Florinsky, I.V. *Digital Terrain Analysis in Soil Science and Geology*; Academic Press: Oxford, UK; Amsterdam, The Netherlands, 2012; ISBN 978-0-12-385036-2.
50. Dong, P.; Chen, Q. *LiDAR Remote Sensing and Applications*, 1st ed.; CRC Press: Boca Raton, FL, USA, 2017; ISBN 9781482243017.
51. Evans, J.; Hudak, A.; Faux, R.; Smith, A.M. Discrete Return Lidar in Natural Resources: Recommendations for Project Planning, Data Processing, and Deliverables. *Remote Sens.* **2009**, *1*, 776–794. [CrossRef]
52. García, R.; Suárez, J.C.; Patenaude, G. Delineation of individual tree crowns for LiDAR tree and stand parameter estimation in Scottish woodlands. In *The European Information Society*; Cartwright, W., Gartner, G., Meng, L., Peterson, M.P., Fabrikant, S.I., Wachowicz, M., Eds.; Springer: Berlin/Heidelberg, Germany, 2007; pp. 55–85. ISBN 978-3-540-72384-4.
53. ESRI. Creating Intensity Images from Lidar in ArcGIS. Available online: https://desktop.arcgis.com/en/arcmap/latest/manage-data/las-dataset/lidar-solutions-creating-intensity-images-from-lidar.htm (accessed on 12 December 2021).
54. Friedman, J.H. Stochastic gradient boosting. *Comput. Stat. Data Anal.* **2002**, *38*, 367–378. [CrossRef]
55. Salford Systems. Salford Predictive Modeler: Introducing TreeNet® Gradient Boosting Machine. 2019. Available online: https://www.minitab.com/en-us/products/spm/user-guides/ (accessed on 11 December 2021).
56. Swets, J.A. Measuring the accuracy of diagnostic systems. *Science* **1988**, *240*, 1285–1293. [CrossRef]
57. Friedman, J.H. Greedy function approximation: A gradient boosting machine. *Ann. Statist.* **2001**, *29*, 1189–1232. [CrossRef]

58. Hutchinson, M.F. A new procedure for gridding elevation and stream line data with automatic removal of spurious pits. *J. Hydrol.* **1989**, *106*, 211–232. [CrossRef]
59. Burrough, P.A.; McDonnell, R.A. *Principles of Geographical Information Systems*; Oxford University Press: New York, NY, USA, 1998; ISBN 9780198748618.
60. Szypuła, B. Digital Elevation Models in Geomorphology. In *Hydro-Geomorphology—Models and Trends*; Shukla, D.P., Ed.; InTech: Rijeka, Croatia, 2017; ISBN 978-953-51-3573-9.
61. Holpp, M.; Kroulik, M.; Kviz, Z.; Anken, T.; Sauter, M.; Hensel, O. Large-scale field evaluation of driving performance and ergonomic effects of satellite-based guidance systems. *Biosyst. Eng.* **2013**, *116*, 190–197. [CrossRef]
62. Santra, A.; Mahato, S.; Mandal, S.; Dan, S.; Verma, P.; Banerjee, P.; Bose, A. Augmentation of GNSS utility by IRNSS/NavIC constellation over the Indian region. *Adv. Space Res.* **2019**, *63*, 2995–3008. [CrossRef]
63. Brach, M. Rapid Static Positioning Using a Four System GNSS Receivers in the Forest Environment. *Forests* **2022**, *13*, 45. [CrossRef]
64. Wing, M.G. Consumer-Grade GPS Receiver Measurement Accuracy in Varying Forest Conditions. *Res. J. For.* **2011**, *5*, 78–88. [CrossRef]
65. Joyce, M.; Moen, R. *Accuracy of a Modular GPS/GLONASS Receiver*; NRRI/TR-2018/28, 2018, Release 1.0; University of Minnesota Duluth: Duluth, MN, USA, 2018.
66. Konnestad, A.J. *On the Accuracy of GNSS in Forests: A Test of Consumer-Grade GNSS Equipment, Smartphones and Open-Source Postprocessing Software under Forest Canopies, for Mapping of Forest Species*; Norwegian University of Life Sciences: As, Norway, 2018.
67. Pesyna, K.M.; Heath, R.W.; Humphreys, T.E. Centimeter Positioning with a Smartphone-Quality GNSS Antenna. In Proceedings of the 27th International Technical Meeting of the Satellite Division of The Institute of Navigation (ION GNSS+ 2014), Tampa, FL, USA, 8–12 September 2014.
68. Merry, K.; Bettinger, P. Smartphone GPS accuracy study in an urban environment. *PLoS ONE* **2019**, *14*, e0219890. [CrossRef]
69. Ritchie, D.A. *Factors That Affect the Global Positioning System and Global Navigation Satellite System in an Urban and Forested Environment*; East Tennessee State University: Johnson City, TN, USA, 2007.
70. Brach, M.; Stereńczak, K.; Bolibok, L.; Kwaśny, Ł.; Krok, G.; Laszkowski, M. Impacts of forest spatial structure on variation of the multipath phenomenon of navigation satellite signals. *Folia For. Pol.* **2019**, *61*, 3–21. [CrossRef]
71. Pirti, A. Accuracy Analysis of GPS Positioning Near the Forest Environment. *Croat. J. For. Eng.* **2008**, *29*, 189–199.
72. Simwanda, M.; Wing, M.G.; Sessions, J. Evaluating Global Positioning System Accuracy for Forest Biomass Transportation Tracking within Varying Forest Canopy. *West. J. Appl. For.* **2011**, *26*, 165–173. [CrossRef]
73. Yan, F.; Hu, X.; Xu, L.; Wu, Y. Construction and Accuracy Analysis of a BDS/GPS-Integrated Positioning Algorithm for Forests. *Croat. J. For. Eng.* **2021**, *42*, 321–335. [CrossRef]
74. Kobal, M.; Bertoncelj, I.; Pirotti, F.; Dakskobler, I.; Kutnar, L. Using lidar data to analyse sinkhole characteristics relevant for understory vegetation under forest cover-case study of a high karst area in the dinaric mountains. *PLoS ONE* **2015**, *10*, e0122070. [CrossRef]
75. D'Eon, R.G.; Delparte, D. Effects of radio-collar position and orientation on GPS radio-collar performance, and the implications of PDOP in data screening. *J. Appl. Ecol.* **2005**, *42*, 383–388. [CrossRef]
76. Zimbelman, E.G.; Keefe, R.F. Real-time positioning in logging: Effects of forest stand characteristics, topography, and line-of-sight obstructions on GNSS-RF transponder accuracy and radio signal propagation. *PLoS ONE* **2018**, *13*, e0191017. [CrossRef] [PubMed]

remote sensing

Review

A Review of Multi-Sensor Fusion SLAM Systems Based on 3D LIDAR

Xiaobin Xu [1,2,*,†], Lei Zhang [1,2,†], Jian Yang [3], Chenfei Cao [1,2], Wen Wang [1,2], Yingying Ran [1,2], Zhiying Tan [1,2] and Minzhou Luo [1,2]

1 College of Mechanical and Electrical Engineering, Hohai University, Changzhou 213022, China; 201319010033@hhu.edu.cn (L.Z.); 211319010001@hhu.edu.cn (C.C.); 211619010135@hhu.edu.cn (W.W.); 200219030004@hhu.edu.cn (Y.R.); zytan@hhu.edu.cn (Z.T.); lmz@hhuc.edu.cn (M.L.)
2 Jiangsu Key Laboratory of Special Robot Technology, Hohai University, Changzhou 213022, China
3 College of Mechanical Engineering, Yangzhou University, Yangzhou 225127, China; jianyang@yzu.edu.cn
* Correspondence: xxbtc@hhu.edu.cn
† These authors contributed equal to this work.

Abstract: The ability of intelligent unmanned platforms to achieve autonomous navigation and positioning in a large-scale environment has become increasingly demanding, in which LIDAR-based Simultaneous Localization and Mapping (SLAM) is the mainstream of research schemes. However, the LIDAR-based SLAM system will degenerate and affect the localization and mapping effects in extreme environments with high dynamics or sparse features. In recent years, a large number of LIDAR-based multi-sensor fusion SLAM works have emerged in order to obtain a more stable and robust system. In this work, the development process of LIDAR-based multi-sensor fusion SLAM and the latest research work are highlighted. After summarizing the basic idea of SLAM and the necessity of multi-sensor fusion, this paper introduces the basic principles and recent work of multi-sensor fusion in detail from four aspects based on the types of fused sensors and data coupling methods. Meanwhile, we review some SLAM datasets and compare the performance of five open-source algorithms using the UrbanNav dataset. Finally, the development trend and popular research directions of SLAM based on 3D LIDAR multi-sensor fusion are discussed and summarized.

Keywords: SLAM; LIDAR; multi-sensor fusion; coupling methods

Citation: Xu, X.; Zhang, L.; Yang, J.; Cao, C.; Wang, W.; Ran, Y.; Tan, Z.; Luo, M. A Review of Multi-Sensor Fusion SLAM Systems Based on 3D LIDAR. *Remote Sens.* **2022**, *14*, 2835. https://doi.org/10.3390/rs14122835

Academic Editors: Yuwei Chen, Changhui Jiang, Qian Meng, Bing Xu, Wang Gao, Panlong Wu, Lianwu Guan and Zeyu Li

Received: 5 May 2022
Accepted: 10 June 2022
Published: 13 June 2022

1. Introduction

A mobile robot is a complex system integrating computer technology, sensor technology, information processing, electronic engineering, automation, and artificial intelligence. With the assistance of artificial intelligence technology, mobile robots with versatile functions are widely used in the fields of emergency rescue, industrial automation, and smart life. Precise positioning is one of the key technologies for mobile robots to complete tasks autonomously. With the rapid development of robot technology, a single sensor can no longer meet the increasingly rich functional requirements of robots. Therefore, the technology of information fusion of multi-source sensing has gradually attracted attention.

Mobile robots are widely used in indoor environments. 2D LIDAR has become the choice for indoor navigation and positioning due to the advantages of high-precision ranging and reduced data volume. However, with the increasing demand for outdoor scenes, robots are gradually moving towards increasingly complex open scenes. Driven by the DARPA (Defense Advanced Research Projects Agency Ground Challenge) [1,2], multi-line 3D LIDAR became known and began to be widely used in outdoor scenes. 3D LIDAR has stronger environmental awareness but at the cost of expensive price, high data volume, and processing difficulty. In recent years, with the popularization of 3D LIDAR and the enhancement of the computing power of embedded processors, the positioning technology based on 3D LIDAR has developed rapidly. 3D LIDAR provides high-density point clouds with richer matching methods and better robustness for the matching between frames. Furthermore, it can be fused with image and odometer information [3] to enhance the positioning accuracy, which has become the mainstream sensor used in many fields such as unmanned driving, robot autonomous navigation, and drone flight control.

For a SLAM system, accurate position and orientation estimation are essential. Scholars have conducted a lot of studies, including vision-based methods and LIDAR-based methods, to realize real-time high-precision 6-DOF state estimation for mobile robots. However, a single sensor system has limitations. On the one hand, the dependence of vision on initialization and the sensitivity to the sum of illumination leads to the instability of the system. On the other hand, the sparse information provided by LIDAR rapidly degenerates the positioning in unstructured scenes. In addition, the rapid motion mode and long-term error accumulation further invalidates the odometer. Therefore, many auxiliary sensors such as IMU, GPS, MEMS, and UWB are added to the positioning system to solve the above problems. In recent years, there have been many LIDAR SLAM review literatures. Most of them introduce the development process of the entire 3D LIDAR SLAM in simple terms, which includes huge but too complicated content.

Tee [4] presents a detailed analysis and comparison of several common open-source solutions for 2D SLAM. The advantages and disadvantages of each method are demonstrated by simulation and experiment. However, 3D SLAM has not been addressed. Bresson [5] reviews LIDAR SLAM related to the large-scale problem faced by autonomous driving. Similarly, reference [6] is an earlier SLAM review, which discusses in detail the basic issues of SLAM and many works in the development of SLAM, including long-distance SLAM, theoretical performance analysis, semantic association, and development directions. Both works summarize the classic theories and work in the field of SLAM, however, the related content of multi-sensor fusion is not involved.

Debeunne [7] divides SLAM into three parts: image-based, LIDAR-based, and image-LIDAR fusion. The integration of SLAM work, complicated integration methods, and the development process of data fusion have not been mentioned. Taheri [8] proposes a SLAM review showing the development of SLAM by reviewing important works. It summarizes and looks forward to SLAM work from multiple directions and stages. However, this work mainly summarizes the visual SLAM, and the reference value of LIDAR SLAM is limited. Zhou [9] summarizes the SLAM algorithm based on 3D LIDAR from the aspects of optimization framework, key SLAM modules, and future research hotspots. Subsequently, this work compares the performance of various SLAM algorithms in detail, which has high reference value.

It can be seen that most of the relevant reviews of SLAM are based on key modules such as front-end matching, closed-loop detection, back-end optimization, and mapping, focusing on the development history and latest works of SLAM. This paper will summarize the multi-sensor fusion SLAM algorithms based on 3D LIDAR from different perspectives. The contributions are:

- The multi-sensor fusion SLAM systems in recent years are categorized and summarized according to the types of fused sensors and the means of data coupling.
- This work fully demonstrates the development of multi-sensor fusion positioning and reviews the works of both loosely coupled and tightly coupled systems, so as to help readers better understand the development and latest progress of multi-sensor fusion SLAM.
- This paper reviews some SLAM datasets and compares the performance of five open-source algorithms using the UrbanNav dataset.

This paper provides a detailed overview of multi-sensor systems through five main sections. The first section details the necessity of multi-sensor fusion in localization systems. In Section 2, the basic problems to be solved by SLAM and the classical framework are presented. In Section 3, the related works of the loosely coupled system are reviewed in detail in two parts according to the sensor types. Similarly, the related works on tightly coupled systems are reviewed in Section 4. A comparison table is given at the end of each section. Finally, a summary of the full text and an outlook for the follow-up works are presented. Abbreviations used in this paper are summarized in Table 1.

Table 1. Abbreviations for terms.

Full Name	Abbreviation
Simultaneous Localization and Mapping	SLAM
Laser Detection and Ranging	LIDAR
Degrees of Freedom	DOF
Micro Electro-Mechanical System	MEMS
Ultra-Wide Band	UWB
Inertial Measurement Unit	IMU
Iterative Closest Point	ICP
Graphic Processing Unit	GPU
Robot Operating System	ROS
LIDAR Odometry and Mapping	LOAM
Lidar Odometry	LO
Visual Odometry	VO
Visual-Inertial Odometry	VIO
LIDAR-Inertial Odometry	LIO
LIDAR-Visual-Inertial	LVI
Extended Kalman Filter	EKF
Multi-State Constrained Kalman Filter	MSCKF
Unmanned Aerial Vehicles	UAV

2. Simultaneous Localization and Mapping System

Over the past few decades, SLAM techniques have come a long way. SLAM systems based on various sensors have been developed, such as LIDAR, cameras, millimeter-wave radar, ultrasonic sensors, etc. As early as in 1990, the feature-based fusion SLAM framework [10], as shown in Figure 1, was established and it is still in use today. The SLAM problem has evolved from two independent modules, localization and mapping, into a complete system that integrates the two. The two modules promote each other. The high-precision odometer composed of multiple sensors provides real-time pose estimation for the robot and the basis for the reconstruction and stitching of the 3D scene. Similarly, high-precision 3D reconstruction provides important data for pose estimation for feature-based odometry. Even a separate odometer system is also inseparable from the establishment or storage of temporary local maps to assist pose estimation.

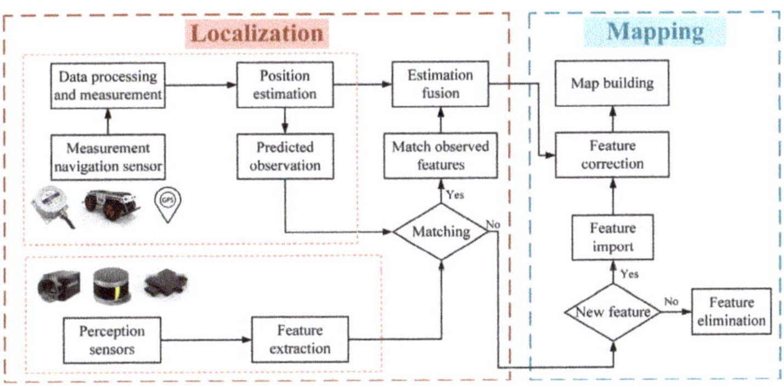

Figure 1. Feature-based fusion SLAM framework.

Most modern SLAM systems are divided into two parts: front end and back end (as shown in Figure 2). The front end is responsible for estimating the current frame pose in real time and storing the corresponding map information. The back end is responsible for large-scale pose and scene optimization. Loop closure detection is one of the key issues of

SLAM, which helps the robot identify visited scenes and trigger global-scale drift correction. Large-scale global optimization is also the main difference between SLAM and modern odometry. The two methods have many similarities in pose estimation. Most modern multi-sensor fusion technologies act in the front end to achieve high-precision and low drift of the odometer systems by the means of information complementation, local pose fusion, and multi-data source filtering.

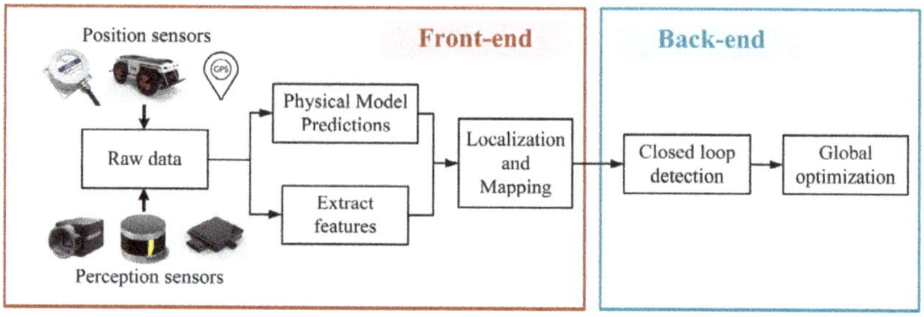

Figure 2. The main components of a SLAM system.

The development of single sensor system is relatively mature, among which LIDAR, camera, and IMU are the most common sensors in SLAM systems. 3D LIDAR can provide the system with rich structural information of the environment. However, the data are discrete and numerous. The camera can capture color and texture in the environment at high speed, but depth cannot be directly perceived by the camera and it tends to be perturbed by light. The IMU can sensitively perceive weak changes of the system in a very short period of time, but long-term drift is inevitable. The characteristics of the three are distinct, and their advantages and disadvantages are obvious. Single-sensor SLAM systems are fragile and full of uncertainty. They are not capable of dealing with multiple complex environments such as high-speed scenes, small spaces, and open and large scenes at the same time.

Therefore, multi-sensor fusion has become a new trend in the development of SLAM systems. Most of the SLAM and odometry systems for multimodal sensor fusion use a combination of LIDAR, camera, and IMU, which can be categorized as loosely coupled or tightly coupled modes. The loosely coupled system processes measurement data of each sensor separately and fuses them in a filter that marginalizes data of the current frame to achieve the latest state estimation results. The tightly coupled system jointly optimizes the measurement data of all sensors and combines the observation characteristics and physical models of each sensor to obtain a more robust pose estimation. The loosely coupled system has the advantages of small calculation amount, simple system structure, and easy implementation. However, its positioning accuracy usually has limitations. In contrast, a tightly coupled system is computationally intensive so that its implementation is difficult, but it gains a more accurate state estimation in complex and changeable environments.

Based on these three sensors, a number of multi-sensor fusion simultaneous localization and mapping works have emerged in recent years. In this paper, according to the coupling method of the system and the types of sensors to be fused, these works are divided into LIDAR-IMU loosely coupled system, Visual-LIDAR-IMU loosely coupled system, LIDAR-IMU tight coupled system, and Visual-LIDAR-IMU tight coupled system. The development of SLAM is a process of transition from loose coupling to tight coupling. The classification of some of the SLAM systems mentioned in this paper and the developmental relationship between them are shown in Figure 3.

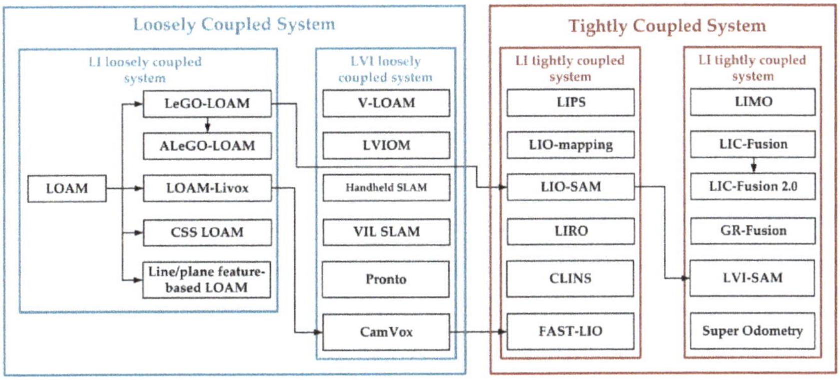

Figure 3. Classification of parts of the work and the relationship between them.

3. Multi-Sensor Loosely Coupled System Based on LIDAR

The emergence of loosely coupled systems has opened up a new stage in the development of multi-sensor fusion systems. Until now, its application in low-cost platforms with limited computation powers is still wide. Loose coupling is mostly applied in three aspects: multi-sensor-based stage pose estimation, raw data-based information complementarity, and sensor-assisted pose constraints.

3.1. LIDAR-IMU Loosely Coupled System

Most loosely coupled systems appear in earlier works. The popularity and wide attention of LIDAR-based 3D SLAM technology can be attributed to Zhang's original work, the LOAM algorithm [11]. One of his important contributions was to extract the information of effective edge and plane feature points from the complex point cloud. Furthermore, the point-to-line and plane distances are used to construct an error function and solve the nonlinear optimization problem of the pose, as shown in Figure 4. Even the earliest LIDAR SLAM systems already had the concept of sensor fusion. This work uses the integration operation of the gyroscope and accelerometer of the six-axis IMU to obtain the prior pose, which further improves the accuracy of the LIDAR odometer. However, LOAM does not have loop closure detection and global pose optimization at the back-end. After that, many LIDAR-IMU loosely coupled systems were improved and perfected on the basis of LOAM. The work presented in this section not only focuses on sensor data fusion, but also improving the point cloud registration of the front end and the overall optimization of the back end.

Shan [12] proposed the LeGO-LOAM algorithm on the basis of LOAM, which introduced point cloud clustering and ground segmentation into data preprocessing to speed up point cloud registration [13]. At the same time, a simple acceleration formula is used to process the IMU data for point cloud distortion correction and provide a priori pose. The IMU has played the same role in line/plane feature-based LOAM and two-stage LOAM. The feature extractions have drawn more attention. The normal vector of point is used to extend the feature types in both methods [14,15]. Different from both of the above methods, CSS-based LOAM and ALeGO-LOAM enhance the feature quality of LOAM [16,17]. Since the curvature scale space method and adaptive cloud sampling method are put forward, more accuracy edge points and plane points are extracted. However, this loose integration method does not effectively exclude the influence of the measurement bias of the IMU itself. Moreover, the IMU is merely a supplementary means.

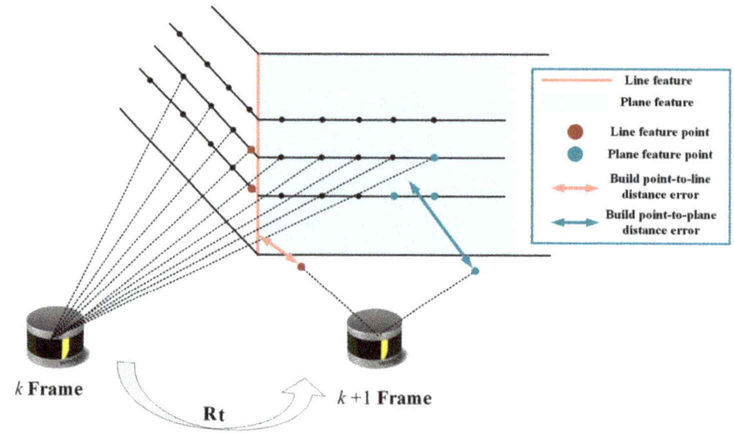

Figure 4. Inter-frame pose estimation in LOAM systems.

In addition to the improvement in feature extraction, improved ICP and GPU acceleration are also optimized. ICP is widely used in SLAM front end. Many works in recent years have proposed faster and more robust ICP variants in order to guarantee the real-time (10 Hz) performance of LIDAR odometry in real-life regularization scenarios. The latest work [18] introduces symmetric KL divergence in the front end ICP algorithm. Its optimization object includes not only the distance between points, but also the difference in distribution shape. In order to ensure the real-time performance and calculation accuracy of the front end, GPU acceleration is applied to point cloud computing including SuMa [19], Elastic-LiDAR Fusion [20], and Droeschel [21]. The shape of the distance data is approximated as a set of disks called Surfel [22], which is convenient for solving point-to-plane problem in GPU. Due to GPU acceleration, the dense point cloud is exhibited in front of people. The visual content is more understandable. To some extent, these point clouds have semantic meaning. Moreover, these methods can be used in virtual reality and augmented reality. However, portable devices and mobile robots still lack powerful GPUs.

It is essential to adopt overall optimization of the back end for the SLAM system. In the latest work [23], Yue enriched feature types. He used the Multi-metric Linear Least Squares Iterative Closest Point (MULLS ICP) algorithm based on categorical feature points to efficiently estimate self-motion and construct a submap-based PGO (Pose Graph Optimization) backend optimization. Effective loop closure detection is a significant procedure of SLAM. In [24,25], the geometric and intensity information of the point cloud are encoded, and a global point cloud descriptor is set to implement a rotation-invariant loop-closure matching algorithm, which clarifies the appropriate optimization timing for the SLAM back-end. In [26], a 2D histogram, converted by all key frames of point clouds, determines where the loop closure occurs by calculating the similarity between the current frame and historical key frames. The loop detection is added in LOAM and LeGO-LOAM.

In addition, there is a branch of 3D LIDAR—solid-state LIDAR. It has the advantages of stable performance and low cost compared to the traditional multi-line LDIAR such as VLP-16 and VLP-32. However, by using irregular scans, this kind of LIDAR with smaller field of view tends to result in motion blurring. Solid-state LIDAR-based SLAM is a new topic. LOAM-Livox [27] is the one of the most representative works. According to the unique scanning method and sensor characteristics of the Livox radar, the author designed a SLAM system suitable for Livox with LOAM as a reference. The system removes the unqualified point cloud and extracts the line and surface features. The pose is iteratively solved by constructing the residuals of the line and surface distances. However, IMU is not used in this method.

The loose coupling of the inertial system processes the IMU data for point cloud distortion correction and provide a priori pose. In this framework, the effect of sensor fusion is limited. Thus, most of the existing algorithms improve front end and back end. The related work on the LIDAR-IMU tightly coupled system is compared in Table 2.

Table 2. LI loosely coupled system.

Year	Method	Author	Strength	Problem
2014	LOAM [11]	J. Zhang et al.	Low-drift. Low-computational complexity.	Lack of closed loop and backend optimization.
2018	LeGO-LOAM [12]	T. Shan et al.	Ground segmentation. Two-stage optimization.	Closed loop detection accuracy is low. Complex terrain failure.
2018	SuMa [19]	J. Behley et al.	A surfel-based map can be used for pose estimation and loop closure detection.	High complexity. High computational cost.
2019	Line/plane feature-based LOAM [14]	X. Huo et al.	Explicit line/plane features.	Limited to an unstructured environment.
2019	ALeGO-LOAM [17]	S. Lee et al.	Adaptive cloud sampling method.	Limited to an unstructured environment.
2019	CSS-based LOAM [16]	C. Gonzalez et al.	Curvature scale space method.	High complexity.
2019	A loop closure for LOAM [26]	J. Lin et al.	2D histogram-based closed loop.	Ineffective in open large scenes.
2020	Two-stage feature-based LOAM [15]	S. Zhang et al.	Two-stage features. Surface normal vector estimation.	High complexity.
2020	Loam-livox [27]	J. Lin et al.	Solid State LIDAR. Intensity values assist in feature extraction. Interpolation to resolve motion distortion.	No backend. No inertial system
2021	MULLS [23]	Y. Pan et al.	Optimized point cloud features. Strong real-time.	Limited to an unstructured environment.
2021	LiTAMIN2 [18]	M. Yokozuka et al.	More accurate front-end registration. Faster point cloud registration.	Lack of backend optimization.

It can be seen from the table that most of the LI loosely coupled systems in recent years focus on the completion and optimization of the system. Finding accurate front end matching and efficient back end optimization methods are their major innovations. Although this part of the work did not make outstanding contributions to data fusion, it provided a stable platform and interface for the subsequent work and accelerated the development of SLAM technology. A summary of the methods in this section according to different strategies is shown in Figure 4.

3.2. LIDAR-Visual-IMU Loosely Coupled System

LIDAR odometer degradation occurs in unstructured and repetitive environments. Even for the positioning with the assistance of IMU, it still cannot work properly for long periods. In contrast, vision sensors do not require specific structural features such as edges and planes, which rely on sufficient texture and color information to accomplish localization. However, vision sensors cannot obtain depth information intuitively. Therefore, combining a camera with LIDAR provides a complementary solution. The LO and VO of LIDAR-Visual-IMU loosely coupled systems mostly operate independently but they share positioning information to each other for pose correction and smoother estimation.

The authors of LOAM extend their algorithms by combining feature tracking of monocular cameras with IMU in V-LOAM [28]. They correlated feature point depths with point clouds to produce a visual-inertial odometry for LIDAR scan matching. However, VO in this work only provides pose before LO and the final error solution is exactly the same as LOAM, which is without vision coupling. Subsequently, the authors of V-LOAM released its iterative version [29]. The improved method employs a sequential parallel processing flow to solve the motion estimation from coarse to fine. The system uses visual-inertial coupling methods to estimate motion and perform the scan matching to refine motion estimation and mapping further. The resulting system enables high-frequency, low-latency motion estimation, as well as dense, accurate 3D map registration.

Tim [30] took a different approach and chose to perform visual localization in a known point cloud map. The method utilizes the map acquired by the camera to track the pose of the monocular camera in a given 3D LIDAR map. Its local BA-based visual odometry system reconstructs sparse 3D points from image features and continuously matches them with the map to track the pose of the camera in an online manner. However, this approach requires the map to be obtained prior. This fusion method is obviously contrary to the original intention of SLAM and is more like an odometer. Zhang [31] combined LOAM-based LIDAR odometry with optical flow tracking-based visual odometry. On the premise of culling dynamic objects, the system weights and fuses the pose results of the two according to the number of valid features. However, this method cannot run in real-time. Pose optimization is still performed independently and without data association.

With the rapid development of VIO systems [32–34], visual-inertial odometry has gradually become a research hotspot in SLAM with its high performance to price ratio and high positioning accuracy. It has a profound impact on multi-sensor fusion in LO systems. Wang [35] proposed a LIDAR-Visual-Inertial SLAM system based on V-LOAM and VINS-MONO. He used a V-LOAM-based approach for mileage estimation and back-end global pose graph optimization by maintaining key frame database. In this approach, the pose estimation result of LO can correct the VIO system.

Shao [36] uses a binocular camera to form a LIDAR-Visual-Inertial system, which is divided into two parts: binocular-based VIO and LIDAR-based LO. The binocular VIO system employs stereo matching and IMU measurements to perform IMU pre-integration sum and tight coupling of pose graphs to marginalize lag frames, which provides LO with an accurate and reliable initial pose. Based on LOAM, LO adds vision-based closed-loop detection and pose graph-based back end optimization, and uses iSAM2 [37,38] to incrementally optimize the LIDAR odometry factor and closed-loop factor. This work has approached tightly coupled systems, but the VIO and LO of the system are still relatively independent. Efficient closed-loop detection and back-end optimization make up for these shortcomings and lay the foundation for a large number of tightly coupled systems that appeared later.

Khattak [39] proposed another loosely coupled method similar to V-LOAM, which uses the inertial prior results of visual and thermal imaging for LIDAR scan matching. To adapt to a variety of complex environments, the authors employed visual and thermal imaging inertial odometry to work in long tunnels without illumination. In [40], the authors combined the VO and LO systems with a leg odometer. In its core, an Extended Kalman Filter (EKF) fuses IMU and legged odometry measurements for attitude and velocity estimation. The system also integrates attitude corrections from VO and LO and corrects attitude drift in a loosely coupled manner. This method has a good localization effect on the legged robot platform.

Lowe [41] proposed a LIDAR-aided vision SLAM system, which employs a novel feature depth and depth uncertainty estimation method. The system uniformly parameterizes three different types of visual features using measurements from LIDAR, camera, and IMU, simultaneously. The system has good adaptability to handheld devices.

CamVox [42] is the first Livox LIDAR SLAM system for assisted vision. The system is built on ORB-SLAM2 [43] and uses Livox to provide a more accurate depth estimation for the camera. Unlike LOAM-Livox, IMU is used for distortion correction of non-repetitively scanned point clouds. In addition, the authors utilized the non-repeating scanning feature of Livox LIDAR to perform automatic calibrations between the camera and LIDAR at uncontrolled scenes. The system achieved better pose estimation results than VINS-MONO and ORB-SLAM2. Shin et al. [44] believe that the relatively sparse point cloud is not meaningful for the depth associated with visual features. They applied the direct method [45] to the combination of low-line LIDAR and camera to implement a loosely coupled SLAM system, which addresses the sparsity problem in data association.

The latest representative loosely coupled work [46] proposed a system composed of multiple odometry methods. The system takes point clouds and images as outputs. Pose estimation algorithms include GICP (Generalized Iterative Closest Point) [47], P2P-ICP (Point-to-Plane Iterative Closest Point) [48], NDT (Normal Distributions Transform) [49], ColorICP (Color Iterative Closest Point) [50], and Huang's method of combining LiDAR and camera data [51]. The system utilizes multiple odometers to improve integrity and robustness. Point cloud-based localization evaluation methods and scoring criteria are defined to generate the optimal pose results. However, the system does not have data association or sharing. Wang [52] proposed a LIDAR-assisted VIO system, which relies on the voxel map structure to efficiently assign the depth information of LIDAR to visual features. Moreover, this work innovatively introduced the vanishing point information in the image into the visual odometry to reduce the rotation drift further. The localization accuracy of this method is superior to the state-of-the-art VIO and LIO systems. Table 3 summarizes the related works on the LIDAR-Visual-IMU loosely coupled system.

Table 3. LVI loosely coupled system.

Year	Method	Author	Strength	Problem
2015	V-LOAM [28]	J. Zhang et al.	Visual feature fusion point cloud depth.	Weak correlation between vision and LIDAR.
2016	Monocular Camera Localization [30]	T. Caselitz et al.	Rely on a priori maps Local BA.	Unknown environment failure.
2018	LVIOM [29]	J. Zhang et al.	VIO preprocessing. Addresses sensor degradation issues. Staged pose estimation.	Inertial system state stops updating when vision fails.
2018	Handheld SLAM [41]	T. Lowe et al.	The incorporation of depth uncertainty. Unified parameterization of different features.	System failure when vision is unavailable.
2018	Direct Visual SLAM for Camera-LiDAR System [44]	Y. Shin et al.	Direct method. Sliding window-based pose graph optimization.	Not available in open areas. Poor closed-loop detection performance.
2019	VIL SLAM [35]	Z. Wang et al.	VIO and LO assist each other. Addresses sensor degradation issues.	Closed loop unavailable when vision fails.
2019	Stereo Visual Inertial LiDAR SLAM [36]	W. Shao et al.	Stereo VIO provides initial pose. Factor graph optimization.	No raw data association between VIO and LO. Sngle factors.
2020	Pronto [40]	M. Camurri et al.	EKF Fusion Leg Odometer and IMU. LO and VO corrected pose estimation.	Drifts seriously. Not bound by historical data.

Table 3. *Cont.*

Year	Method	Author	Strength	Problem
2020	CamVox [42]	Y. Zhu et al.	Livox LIDAR aids depth estimation. An automatic calibration method.	No inertial system.. No LO.
2021	Redundant Odometry [46]	A. Reinke et al.	Multiple algorithms in parallel. Filter the best results.	High computational cost. No data association.
2021	LiDAR-Visual-Inertial Estimator [52]	P. Wang et al.	LIDAR assists the VIO system Voxel map structures share depth. Vanishing Point optimizes rotation estimation.	System failure when vision is unavailable.

This part of the work becomes more flexible with the introduction of vision. The point cloud increases the stability of the depth acquisition of visual features. In addition, the robustness of system positioning is also stronger. However, loose coupling leads to the independence between vision and LIDAR. The constraints between the data are not strong enough.

4. Multi-Sensor Tightly Coupled System Based on LIDAR

Positioning and mapping techniques are applied to more complex and changeable scenes with the rapid development of robotics. The previous loosely coupled system has the advantages of real-time and low computational complexity. However, it is still difficult to guarantee the accuracy in high-speed motion or degradation scenarios. With its high-frequency motion response characteristics, IMU has always been an indispensable sensor for mobile robots. For tightly coupled systems, it is a key issue to effectively fuse the IMU with other odometers.

Tightly coupled systems based on IMU assistance have made a breakthrough in visual odometry [53,54]. In this work, the IMU pre-integration formula, error transfer model, and definition of residual are deduced, which have a profound impact on the subsequent development of LIO and VIO. Moreover, these equations and models become the theoretical basis for joint optimization of tightly coupled systems. The world coordinate system is defined as W, the binding of the robot coordinate system and the IMU coordinate system is defined as B; the state vector $\mathbf{x}$ of robot can be defined as:

$$\mathbf{x} = \left[\mathbf{R}^{\mathrm{T}}, \mathbf{p}^{\mathrm{T}}, \mathbf{v}^{\mathrm{T}}, \mathbf{b}^{\mathrm{T}}\right]^{\mathrm{T}} \tag{1}$$

where $\mathbf{R} \in SO(3)$ which is the rotation group, $\mathbf{p}$ and $\mathbf{v}$ are the position and velocity vectors of the robot, and $\mathbf{b}$ is the bias of the IMU. The measurement value of the IMU can be written as:

$$\hat{\omega}_t = \omega_t + \mathbf{b}_t^{\omega} + \mathbf{n}_t^{\omega} \tag{2}$$

$$\hat{\mathbf{a}}_t = \mathbf{R}_t^{\mathrm{BW}}(\mathbf{a}_t - \mathbf{g}) + \mathbf{b}_t^{a} + \mathbf{n}_t^{a} \tag{3}$$

where $\hat{\omega}$ and $\hat{\mathbf{a}}$ represent the angular velocity and angular velocity measurements of the IMU, $\mathbf{b}$ and $\mathbf{n}$ represent the bias and noise of the gyroscope and accelerometer, respectively, ω and $\mathbf{a}$ represent the true value, $\mathbf{g}$ represents the local gravitational acceleration, $\mathbf{R}^{\mathrm{BW}}$ represents the world coordinate system to the rotation matrix of the IMU coordinate system, and t represents the time. Using the discrete pre-integration method in [53], the relative motion can be obtained. $\Delta\mathbf{v}_{ij}$, $\Delta\mathbf{p}_{ij}$ and $\Delta\mathbf{R}_{ij}$ can be expressed as:

$$\Delta\mathbf{v}_{ij} = \mathbf{R}_i^{\mathrm{T}}(\mathbf{v}_j - \mathbf{v}_i - \mathbf{g}\Delta t_{ij}) \tag{4}$$

$$\Delta\mathbf{p}_{ij} = \mathbf{R}_i^{\mathrm{T}}\left(\mathbf{p}_j - \mathbf{p}_i - \mathbf{v}_i\Delta t_{ij} - \frac{1}{2}\mathbf{g}\Delta t_{ij}^2\right) \tag{5}$$

$$\Delta \mathbf{R}_{ij} = \mathbf{R}_i^{\mathrm{T}} \mathbf{R}_j \tag{6}$$

The residuals for the terms are expressed as:

$$\mathbf{r}_{ij} = \left[\mathbf{r}_{\Delta v_{ij}}, \mathbf{r}_{\Delta R_{ij}}, \mathbf{r}_{\Delta p_{ij}}, \mathbf{r}_{b_{ij}^{\omega}}, \mathbf{r}_{b_{ij}^{a}} \right] \tag{7}$$

For the specific derivation of residuals, please refer to the literature [53,54]. With the pre-integration formula and the definition of the error term, the coupling relationship between the IMU and the world coordinate system can be decoupled in the process of joint optimization. The system can update the biases of the IMU to ensure that the IMU data are added and optimized. Hence, a measured value closer to the true one is obtained. The definition of the residual makes it easier for the IMU to combine the residual terms of other sensor odometers to create a more complete error function. This is also the rationale of the tightly coupled optimization. The factor graph representation of the tightly coupled system is shown in Figure 5.

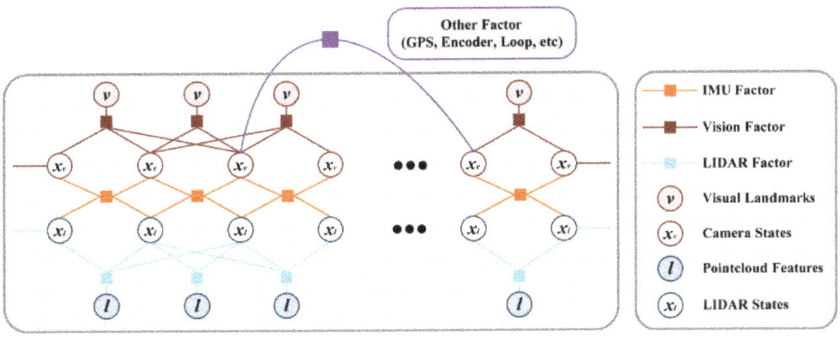

Figure 5. The factor graph for tightly coupled systems.

4.1. LIDAR-IMU Tightly Coupled System

On the theoretical basis of pre-integration, a large amount of LIO-related work has gradually emerged in recent years. One of the early approaches to tightly couple LIDAR and IMU was proposed in LIPS [55], which employs a graph-based optimization framework. In this framework, a planar representation of the closest point is proposed. A set of point clouds is parameterized as plane features, and then the residuals-function is converted into the differences between the plane parameters of two frames, which, together with the residual term of the IMU pre-integration, constitutes the final optimization function. This tightly coupled approach was deeply influenced by the VINS series and began to emerge in the field of LIDAR SLAM. This form lays a solid foundation for the subsequent LIO and LVI tightly coupled systems.

The pre-integration of the IMU was used for removing the distortion of the raw point cloud in Gentil's work [56]. It tightly integrates the IMU and LIDAR data into a batch manifold optimization formulation, which describes the motion in the LIDAR scan based on the extrinsic parameters of IMU. The system also considers the first-order form of the pre-integration error to the time difference and solves the problem of hardware time asynchrony. Ye [57] proposed a tightly coupled LIDAR inertial localization and mapping framework, LIOM (LIO-mapping), which jointly optimizes measurements from LIDAR and IMU. A sliding-window model was further used to maintain a certain scale of optimization data. The accuracy of LIOM is better than that of LOAM. However, real-time cannot be achieved since LIOM is designed to process the measurements by all sensors.

Inspired by Hess's work [58], Ding [59] introduced the subgraph representation of 2D SLAM into the 3D LIDAR odometry and added inertial data to establish motion prediction and constraints between frames. In this system, the 3D occupancy grid method is utilized to replace the 2D occupancy grid to realize the pose measurement of all 6 degrees of freedom. Finally, the iterative solution is performed in the solver Ceres [60]. The system innovatively joins the environmental change detection (ECD) module, which can detect whether the known surrounding environment has changed. However, this feature is not used to eliminate the influence of unknown dynamic environment on SLAM.

The authors of LeGO-LOAM released LIO's follow-up work LIO-SAM [61] in combination with IMU-related theories. The system builds the LIDAR-inertial odometry on a factor graph, and multiple relative and absolute measurements including closed loops are incorporated into the system as factors, as shown in Figure 6. The innovation of LIO-SAM is to marginalize old pose and point cloud data to replace matching scans to global maps. The system uses local map matching instead of global matching to significantly improve real-time performance. In addition, the system also adds the GPS absolute positioning factor [62], which is used to correct the long-term drift of the system. However, since feature extraction relies on geometric environments, this method still cannot work for a long time in open scenes.

The latest work of LIRO [63] proposed a sensor fusion scheme combining LIO with UWB ranging. The solution can be easily deployed with minimal cost and time. The system tightly couples IMU, LIDAR, and UWB data with timestamp-based robot states in a sliding window to construct a cost function consisting of UWB, LO, and IMU pre-integrations. Finally, a factor graph model is used to incrementally marginalize and update the data within the window. However, the usage scenarios of UWB have great limitations. The system will no longer have an advantage in a huge range of occlusion scenarios.

The tightly coupled LIO system proposed by Chen [64] further refines the front-end. He proposed an efficient algorithm to simultaneously extract the explicit mixed features of the original point cloud, including ground features, edge features, and plane features. The system also introduced a deep learning-based LPD-Net to generate global descriptors for point clouds. The loop closures detection can be accomplished in the key frame database. This method greatly improves the accuracy of closed loop detection. In order to ensure the real-time performance of the system, Li [65] proposed a quantitative evaluation method for point cloud feature constraints and a screening algorithm for key features. An effective compromise is traded off between accuracy and computational cost. Lv [66] proposed a high-accuracy continuous-time trajectory estimation framework for LIO systems to efficiently fuse high-frequency asynchronous sensor data. The system uses a non-rigid body registration method for continuous-time trajectory estimation. Dynamic and static control points are defined to further optimize trajectory estimation. At the same time, a two-stage closed-loop correction method is proposed to effectively update the closed-loop pose and control points, respectively. However, the computational cost of closing the loop is not reported, nor does it address the uncertainty in motion that might suffer from motion degradation.

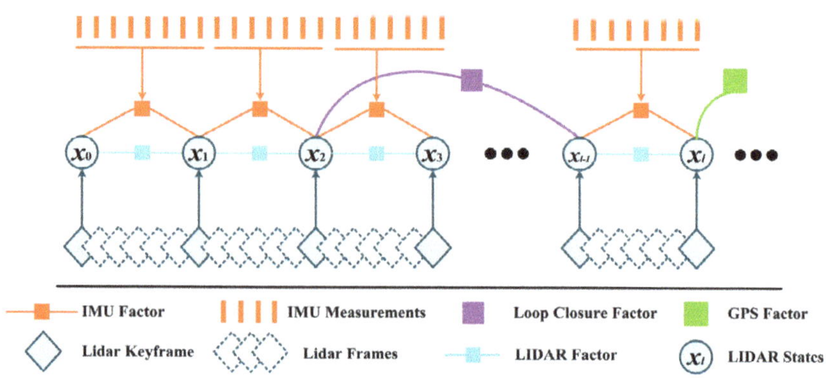

Figure 6. Factor graph structure of LIO-SAM.

RF-LIO [67] is a dynamic SLAM framework proposed on the basis of LIO-SAM. The system adaptively adds a multi-resolution range image composed of point clouds and removes moving objects using tightly coupled LIDAR inertial odometry. The LIDAR scans are then matched with the subgraphs. Therefore, it can obtain accurate pose estimation results even in highly dynamic environments. LIO tightly coupled systems based on solid-state LIDAR have also been gradually attracting attention. However, LIO degradation still occurs when moving in open scenes for a long time. FAST-LIO [68] proposed an efficient and robust LIO framework based on tightly coupled iterative Kalman filters for UAV systems. However, the system discards the impact of historical data on the current state. Global pose correction cannot be performed.

The tight coupling of the inertial system will undoubtedly increase the computational burden of the system while improving the accuracy. Most of the existing algorithms improve computing speed by marginalizing historical data or limiting local map capacity. The back end optimization generally only builds the pose graph of the LIDAR without adding the bias and speed measured by the IMU. These methods achieve excellent results in most scenarios. However, due to the dependence on geometric features, once the inertial system loses the LO constraint in the open unstructured scene, the SLAM will suffer serious drift and degradation. The related work on the LIDAR-IMU tightly coupled system is compared in Table 4.

Table 4. LI tightly coupled system.

Year	Method	Author	Strength	Problem
2018	LIPS [55]	P. Geneva et al.	The singularity free plane factor. Preintegration factor. Graph optimization.	High computational cost. No backend or local optimization.
2019	IN2LAMA [56]	C. Le Gentil et al.	Pre-integration to remove distortion. Unified representation of inertial data and point cloud.	The open outdoor scene fails.
2019	LIO-mapping [57]	H. Ye et al.	Sliding window. Local optimization.	High computational cost. Not real time.
2020	LiDAR Inertial Odometry [58]	W. Ding et al.	The occupancy grid based LO. Map updates in dynamic environments.	Degradation in unstructured scenes.
2020	LIO-SAM [61]	T. Shan et al.	Sliding window. Add GPS factor. Marginalize historical frames and generate local maps.	Poor closed loop detection. Degradation in open scenes.
2021	LIRO [63]	T.-M. Nguyen et al.	UWB constraints. Build fusion error.	UWB usage scenarios are limited.
2021	Inertial Aided 3D LiDAR SLAM [64]	W. Chen et al.	Refine point cloud feature classification. Closed Loop Detection Based on LPD-Net.	Degradation in unstructured scenes.
2021	KFS-LIO [65]	W. Li et al.	Point cloud feature filtering Efficient Computing.	Poor closed loop detection.
2021	CLINS [66]	J. Lv et al.	The two-state continuous-time trajectory correction method. Optimization based on dynamic and static control points.	High computational cost. Affected by sensor degradation.
2021	RF-LIO [67]	IEEE	Remove dynamic objects. Match scan to the submap.	Low dynamic object removal rate.
2021	FAST-LIO [68]	W. Xu et al.	Iterated Kalman Filter. Fast and efficient.	Cumulative error. No global optimization.

With the development and improvement of the IMU pre-integration theory, the LO system can establish a stronger constraint relationship with the IMU. The localization accuracy of the SLAM system has also been further improved. However, tight coupling involves a great amount of computation. Finding a balance between the speed and the precision is the difficulty of this stage of work.

4.2. LIDAR-Visual-IMU Tightly Coupled System

Although the study of visual SLAM started relatively late, it quickly became a research hotspot of SLAM technology due to its advantages of small size and low cost. Many VIO works have reported in recent years. Vision forms an excellent complementation with LIDAR because it is not constrained by the structure of the scene. Therefore, LVI systems have received increasing attention owing to their stronger robustness in sensor degradation scenarios.

The strong data association between point clouds and images enables the system to tightly combine multiple effective features in the preprocessing stage. They will play an important role in the matching and optimization process. The LVI tightly coupled system is divided into two coupling methods based on optimization and filtering. The optimization-based approach tightly integrates the error models of individual sensors and reduces the sensitivity of time synchronization by using local maps or sliding windows. This method simultaneously optimizes historical poses and achieves real-time performance with BA. In addition, this paper classifies the tight association of sensor data level as a special category of the tightly coupled. Although no joint optimization is performed during the pose solving, these works strongly correlate the data through preprocessing, and already contain the necessary key information of the sensor in single objective function. They all have a closed system, which is less scalable and compatible with other sensors. Filter-based approaches merely use the sensor data of the current frame and rely on the time synchronization of each data. Since the influence of historical data on the current pose is not considered, the amount of computation is relatively small and the scalability is relatively good.

LIMO [69] is one of the multi-sensor fusion positioning systems. The system performs strong correlation between point clouds and images through a variety of data preprocessing methods to achieve a stable and robust system. The system performs foreground segmentation, plane fitting, and ground fitting on point clouds for different scenes so as to obtain the best depth estimation of visual features. The system combines the 3D-2D PNP (Perspective-n-Point-Problem) [70] pose estimation method and the 2D-2D epipolar constraint [71] to achieve a good localization effect. Reference [72] is another example of strong data correlation, where point and line features are extracted in the image, and the position information of points and lines are obtained by a similar method to LIMO. Furthermore, a 3D-3D ICP model and reprojection error functions of points and lines are constructed, which achieves higher accuracy pose estimation. Recent work [73] fuses data from two sensors in a higher-level space using geometric features co-detected in LIDAR scan and image data. Correspondences between 3D lines extracted in the LIDAR scan and 2D lines detected in the image were determined. Wang [74] proposed a DSP-SLAM system by combining object detection with SLAM. This work uses a DeepSDF network to generate object shape vectors and 7D poses from keyframed point cloud and image data. Sparse point clouds and image segmentation results are used as observations to minimize surface loss and depth rendering loss functions. Object reconstruction and pose update are added to the ORB-SLAM2-based BA factor graph to simultaneously optimize camera pose, map points, and object pose. These works do a lot of meaningful work on data association, which makes the whole system more robust. However, these systems obviously lack the tight coupling of the optimization process. The combination of correlation between raw data and integration of errors will lead to a stronger system.

On the other hand, the very matured framework of the LIO system paves the way for the establishment of the LVI tightly coupled system, which has led to the emergence of a large number of tightly coupled systems based on the optimized LIDAR-Vision-IMU in the past two years. GR-Fusion [75] uses camera, IMU, LIDAR, GNSS, and encoder of motion chassis as main sensors to build a factor graph model in a sliding window. The LIDAR factor, the visual factor, the IMU factor, and the odometry factor are added as primary constraints to the factor graph. Meanwhile, local constraints are tightly coupled with GNSS constraints to optimize the global state of the robot. The system can detect the degradation of the sensor in real time and flexibly configure multi-working modes, which is suitable for a wide range of scenarios. LVIO-Fusion [76] adopts a similar system architecture. The difference is that the binocular camera is used as the visual sensor. This paper innovatively proposed different optimization strategies for straight and turning. Moreover, reinforcement learning networks are introduced to adaptively adjust the weights of sensors in different scenarios. LVI-SAM [77] is the latest work by the authors of LeGO-LOAM and LIO-SAM. The system consists of VIS (Visual-Inertial System) and LIS (LIDAR-Inertial System). VIS can provide pose prior for LIS, which can provide pose estimation and accurate feature point depth for VIS initialization. However, this system does not consider the marginalization of the LIO system and the problem of timestamp synchronization.

Some recent odometry systems also employ a tightly coupled approach to obtain low-drift pose estimates. Super Odometry [78] and MetroLoc [79] both use an LVI system with an IMU as the main sensors. The system consists of three parts: IMU odometer, VIO, and LIO. The observation data provided by VIO and LIO can constrain the bias of the IMU. On the other hand, the constrained IMU odometry provides predictions for VIO and LIO to achieve coarse-to-fine pose estimation, which is shown in Figure 7. The system can simultaneously extend GPS and wheel odometer with robustness to sensor degradation. Wisth [80] proposed a tightly coupled system based on LIDAR and a binocular camera. The factor graph optimization problem is composed of initial prior factor, visual factor, line factor, plane factor, and IMU pre-integration, and is solved by GTSAM [81]. The system shares a representation of vision-based point features and point cloud-based line and area features. The reprojection error function is then defined by parameterizing different features. In addition, based on the data timestamp of the camera, the system splits and merges the adjacent scan data of LIDAR to realize the time hard synchronization at the software level. There is no doubt that these works have good real-time positioning effect. However, a pure odometer system often discards historical data and only focuses on current or local observations, which leads to the loss of correlation and optimization of global data. Therefore, excellent front end odometer and reasonable back end optimization are necessary for SLAM system.

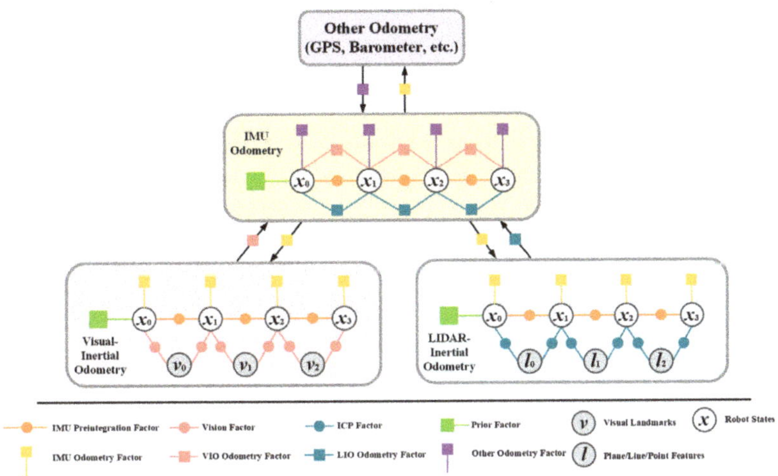

Figure 7. Factor graph structure of Super Odometry.

Filter-based methods also play an important role in the field of multi-sensor fusion. For joint state optimization, many methods use the EKF or the MSCKF framework [82]. Yang [83] used MSCKF to tightly couple planar features from RGB-D sensors, IMU measurements, and visual point features within 3.5 m. To limit the scale of the state vector, the system linearly marginalizes most of the point features and retains a few point features with plane-enhanced constraints in the state vector as SLAM features. The LIC-Fusion proposed by Zuo [84] adopts the MSCKF fusion framework, which tightly combines IMU measurements, extracted LIDAR edge features, and sparse visual features. Subsequently, in the latest follow-up work, LIC-Fusion 2.0 [85], the authors introduced a sliding-window-based planar feature tracking method to efficiently process 3D LIDAR point clouds in real time. R2LIVE [86] is a tightly coupled work based on solid-state LIDAR. The system combines an error-state-based iterative Kalman filtering front end and a new step of factor graph optimization-based sliding window optimization to refine the visual pose and landmark estimation. It achieves high accuracy and robustness in harsh scenarios such as indoors, outdoors, tunnels, and high-speed motion. These methods are fast and computationally inexpensive, but are sensitive to time synchronization. Measurements during filtering may degrade or fail. Therefore, a special sorting mechanism is required to guarantee the correct order of the measurement results of the different sensors.

In this paper, the LVI tightly coupled system is divided into three parts: strong data correlation, nonlinear optimization tight coupling, and state filter. Among them, the tightly coupled front end based on optimization is the main implementation. Table 5 compares the related works of the LIDAR-Vision-IMU tightly coupled system.

Table 5. LVI Tightly coupled system.

Year	Method	Author	Strength	Problem
2018	LIMO [69]	J. Graeter et al.	Point cloud scene segmentation to optimize depth estimation. Epipolar Constraint. Optimization PnP Solution	Unused LO. Sparse map.
2019	Tightly-coupled aided inertial navigation [83]	Y. Yang et al.	MSCKF. Point and plane features.	LIDAR is unnecessary.
2019	LIC-Fusion [84]	X. Zuo et al.	MSCKF. Point and Line Features.	High computational cost. Time synchronization is sensitive.. Unresolved sensor degradation.
2020	LIC-Fusion 2.0 [85]	X. Zuo et al.	MSCKF. Sliding window based plane feature tracking.	Time synchronization is sensitive.. Unresolved sensor degradation.
2020	LIDAR-Monocular Visual Odometry [72]	S.-S. Huang et al.	Reprojection error combined with ICP. Get depth of point and line features simultaneously.	Poor closed-loop detection performance. High computational cost.
2021	LIDAR-Monocular Surface Reconstruction [73]	V. Amblard et al.	Match line features of point clouds and images. Calculate reprojection error for points and lines.	Inertial measurement not used.
2021	GR-Fusion [75]	T. Wang et al.	Factor graph optimization. Address sensor degradation. GNSS global constraints.	No apparent problem.
2021	Lvio-Fusion [76]	Y. Jia et al.	Two-stage pose estimation. Factor graph optimization. Reinforcement learning adjusts factor weights.	High computational cost. Difficult to deploy.

Table 5. *Cont.*

Year	Method	Author	Strength	Problem
2021	LVI-SAM [77]	T. Shan et al.	Factor graph optimization. VIS and LIS complement each other. Optimize depth information.	Poor closed loop performance.
2021	Super Odometry [78]	S. Zhao et al.	IMU as the core. LIO and VIO operate independently. Jointly optimized pose results. Address sensor degradation.	High computational cost.
2021	Tightly Coupled LVI Odometry [80]	D. Wisth et al.	Factor graph optimization. Unified feature representation. Efficient time synchronization.	Unresolved sensor degradation.
2021	DSP-SLAM [74]	J. Wang et al.	Add object reconstruction to factor graph. The DeepSDF network extracts objects.	No coupled inertial system. Poor closed loop performance.
2021	R2LIVE [86]	J. Lin et al.	The error-state iterated Kalman filter. Factor graph optimization.	No closed loop detection and overall backend optimization.

The emergence of a complete system in which LIDAR, vision, and IMU cooperate and complement each other is a milestone for multi-sensor fusion SLAM. The integration is not limited to these three sensors. Wheel/leg odometer and GNSS have also been effectively integrated into the system. Similarly, the increase in computational complexity is one of the toughest problems. In addition, there are some details that need to be optimized, such as dynamic environments, unstructured environments, rain and snow weather.

5. Performance Evaluation

5.1. SLAM Datasets

Evaluating the performance of SLAM algorithms is often inseparable from the help of open-source datasets. The mobile carriers for research and application of 3D LIDAR SLAM include unmanned vehicles, unmanned ships, and unmanned aerial vehicles. However, the current LIDAR point cloud datasets are mainly for autonomous driving scenarios. Data collection in outdoor scenes is complex and cumbersome, involving time synchronization, coordinate calibration, and calibration among various sensors. Public datasets save the time for data preparation for algorithmic research. The sequences and benchmarking frameworks provided also facilitate algorithm development.

The current public LIDAR-based datasets in the field include: KITTI dataset [87], which is currently the largest international evaluation dataset for autonomous driving scenarios and is also the most commonly used dataset in academia. The Waymo dataset [88] is a data-open project of the autonomous driving company Waymo. The PandaSet dataset [89] is used to develop safe and controllable autonomous driving technology in complex environments and extreme weathers. Oxford Robotcar dataset [90] is a public dataset proposed by Oxford University Robotics Laboratory. The UrbanNav dataset [91] provides a challenging data source to the community to further accelerate the study of accurate and robust positioning in challenging urban canyons. The UrbanNav dataset includes complex and dynamic urban road environments and closed tunnel environments. It also provides the real pose of the GNSS system as a reference. Compared with the commonly used KITTI dataset, the collection environment of the UrbanNav dataset is closer to the complex environment of unmanned driving. At the same time, the related team of the UrbanNav dataset provides an overview [92] of LIDAR Odometry, which uses this dataset to evaluate open-source algorithms based on point clouds and features, respectively.

Obviously, the complex scene of the city can magnify the advantages and disadvantages of the SLAM algorithm. Therefore, this paper uses the UrbanNav dataset to evaluate some open-source algorithms in the following section.

5.2. *Performance Comparison*

The evaluation of the SLAM algorithm is mainly based on evaluating the positioning accuracy. The relative pose error (RPE) is used to describe the accuracy of the pose difference between two frames separated by a certain time difference. The changes of the real pose and the estimated pose are calculated at the same time interval. Then, the difference between the two is calculated to obtain the relative pose error. Afterwards, the relative pose error of each period of time can be counted by the root mean square error (RMSE) to obtain the overall value. The absolute trajectory error (ATE) describes the direct difference between the estimated pose and the real pose, which can intuitively reflect the accuracy of the algorithm and the global consistency of the trajectory. Many SLAM algorithms and review papers have analyzed the performance of open-source algorithms using datasets in the experimental part. Jonnavithula [93] provides an overview of existing LO systems for the application environment of autonomous driving. This paper uses the KITTI dataset to experimentally verify some of the reviewed algorithms. Huang [92] uses the UrbanNav dataset for comparison to demonstrate the pros and cons of point cloud-based and feature-based localization methods. Yokozuka [18] also conducted many comparative experiments using the KITTI dataset in the experimental part of his algorithm.

This paper selects five open-source 3D LIDAR SLAM algorithms for testing and evaluation. They are A-LOAM [94], LeGO-LOAM [12], SC-LeGO-LOAM [24], LIO-SAM [61], and F-LOAM [95]. A-LOAM is an open-source version that uses an optimizer for code simplification based on LOAM. SC-LeGO-LOAM uses scan context to optimize loop closure detection based on LeGO-LOAM. F-LOAM is an odometer system that only relies on LIDAR but has good performance. We apply them to ROS running on a laptop with an Intle i7-10875H CPU to achieve the functionality, and the platform has 2 × 8 GB of RAM memory and an RTX2060 GPU. All algorithms are evaluated and compared in experiments based on the UrbanNav public dataset benchmark. We collect experimental results under the same conditions and carry out performance metrics in order to evaluate the performance of the tested algorithms. This paper focuses on two competitive datasets, UrbanNav-HK-Medium-Urban-1 (Data1) and UrbanNav-HK-Tunnel-1 (Data2). Data1 are in an urban center area with heavy traffic and towering buildings. Data2 were collected while moving fast in a closed tunnel. The pictures corresponding to the two scenarios are shown in Figure 8.

(a) **Data1** (b) **Data2**

Figure 8. Demonstration of the scenarios in the three urban datasets. (**a**) Data1: Variety of dynamic vehicles and numerous high-rising buildings in UrbanNav-HK-Medium-Urban-1. (**b**) Data2: Closed tunnel in UrbanNav-HK-Tunnel-1.

First, we focus on the mapping effects of the five algorithms. Mapping is based on positioning, which can intuitively show the overall operation effect of SLAM. We take

Data1 as an example to show the official environment map and the different results of each algorithm (as shown in Figure 9).

The comparison chart intuitively shows the mapping results of the five algorithms. A-LOAM suffers severe drift in traffic-congested sections, which leads to poor localization and mapping. The mappings of LeGO-LOAM, SC-LeGO-LOAM, and LIO-SAM are good and the localizations are effectively constrained by the closed loop. However, it is worth emphasizing that the search radius of the closed loop detection of LeGO-LOAM and LIO-SAM is adjusted to 50 m before the closed loop can be accurately identified. F-LOAM still has high global consistency in the absence of loop closure detection.

Second, trajectories of different algorithms and ground truth are plotted together for comparison (as shown in Figure 10). In order to facilitate the observation, we take the trajectory generated by the first lap of data for comparison. The first loop closure occurs at the zoomed-in position on the left, which is the same as that marked in the point cloud map.

The trajectories generated by the five algorithms are clearly shown in Figure 10. The trajectory of A-LOAM has poor positioning accuracy when there are many traffic jams or dynamic vehicles. Therefore, it basically loses the positioning ability in the urban canyon environment. The other four algorithms have good global consistency. It is worth emphasizing that F-LOAM only deviates a small distance from the closed loop position without loop closure detection. The positioning accuracies of the five algorithms are listed in Table 6. We use the RMSE and mean of relative pose errors to describe the accuracy of the algorithm. The odometer's average processing time for per-frame (APTFP) is used to describe the algorithm efficiency.

Table 6. Performance comparison of five algorithms.

Methods	Relative Translation Error (m)		Relative Rotation Error (deg)		Odometry APTPF (s)
	RMSE	Mean	RMSE	Mean	
A-LOAM	1.532	0.963	1.467	1.054	0.013
LeGO-LOAM	0.475	0.322	1.263	0.674	0.009
SC-LeGO-LOAM	0.482	0.325	1.278	0.671	0.009
LIO-SAM	0.537	0.374	**0.836**	**0.428**	0.012
F-LOAM	**0.386**	**0.287**	1.125	0.604	**0.005**

From the data in the table, it can be seen that the overall performances of F-LOAM and LIO-SAM are better. The relative translation error of F-LOAM is the smallest, while the relative rotational error of LIO-SAM is the smallest. The addition of IMU pre-integration can effectively improve the positioning accuracy of rotation. The positioning accuracy and processing time of A-LOAM are relatively poor. We recommend using LIO-SAM when there is a closed loop in the environment dealt with. If no closed loop presents, the real-time performance of F-LOAM is better.

Finally, the operation on Data2 will be summarized and discussed. The tunnel environment in Data2 is more challenging than congested urban canyons. There are very few structural features and the light changes rapidly in the closed tunnel. The tunnel seems to be a long corridor so that the point cloud data generated by LIDAR is basically the same no matter where it is located. Unfortunately, none of the five algorithms can effectively complete the positioning process without using GPS data. However, they show different behaviors and have different odometer failure locations. The real point cloud map and the actual performance of the five algorithms are shown in Figure 11.

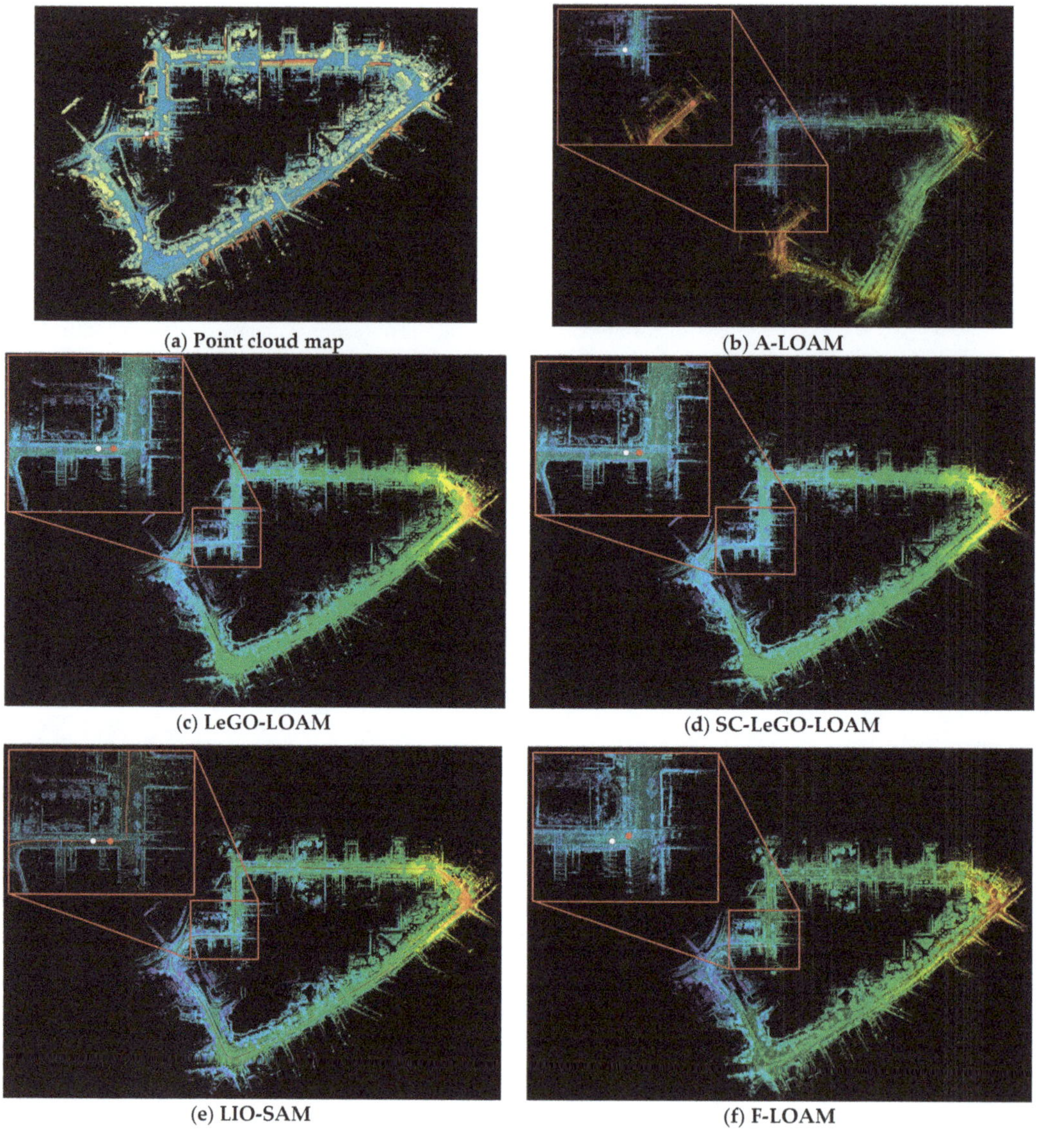

Figure 9. Point cloud maps generated by different algorithms. (**a**) The actual point cloud map given by the dataset. (**b**) Result of A-LOAM. (**c**) Result of LeGO-LOAM. (**d**) Result of SC-LeGO-LOAM. (**e**) Result of LIO-SAM. (**f**) Result of F-LOAM. We zoomed in where the loop was first generated to see the effect of generating the map. The white dots indicate the starting point and the red dots indicate the ending point.

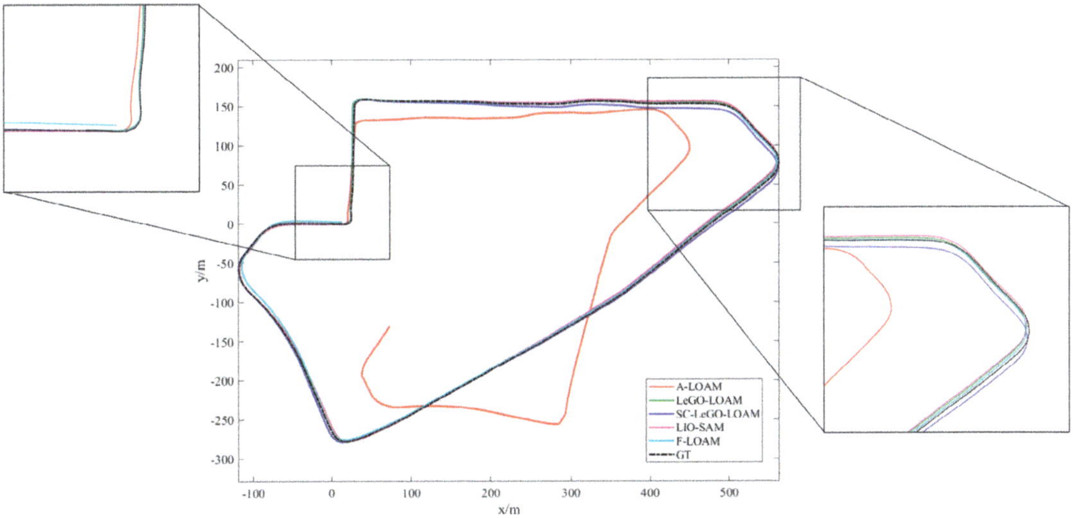

Figure 10. Trajectories generated by different algorithms. The first occurrence of the closed loop and the farthest distance from the closed loop are zoomed in for a closer comparison of their differences.

The A-LOAM was disabled before entering the tunnel due to the large number of dynamic vehicles congested around the LIDAR. The results of LeGO-LOAM and SC-LeGO-LOAM are similar. They both degenerate when they first enter the tunnel because their system compositions are basically the same. F-LOAM performs slightly better than the previous two thanks to the feature weights it assigns. Degradation of F-LOAM occurs after a certain distance in the tunnel. Finally, LIO-SAM performs the best. The front end of LIO-SAM is also feature-based. However, the system can still run robustly for some time as the LO degenerates due to the addition of IMU data constraints. This allows it to travel the longest distance effectively in the tunnel. Obviously, the tight coupling of the IMU cannot completely solve the long-distance tunneling problem.

The long corridor problem is a difficult problem often faced in practical applications. The above experiments show that assigning weights to features and adding IMU pre-integration can effectively alleviate the phenomenon of odometer degradation. Clearly, feature-based methods encounter a bottleneck in the tunnel environment. This problem will be better solved if the precise control model and kinematic model of the robot chassis are combined with a feature-based odometry system or aided by visual features [96]. At the same time, UWB is also suitable for such closed scenes within a certain range. This is also an effective way of carrying out multi-sensor fusion.

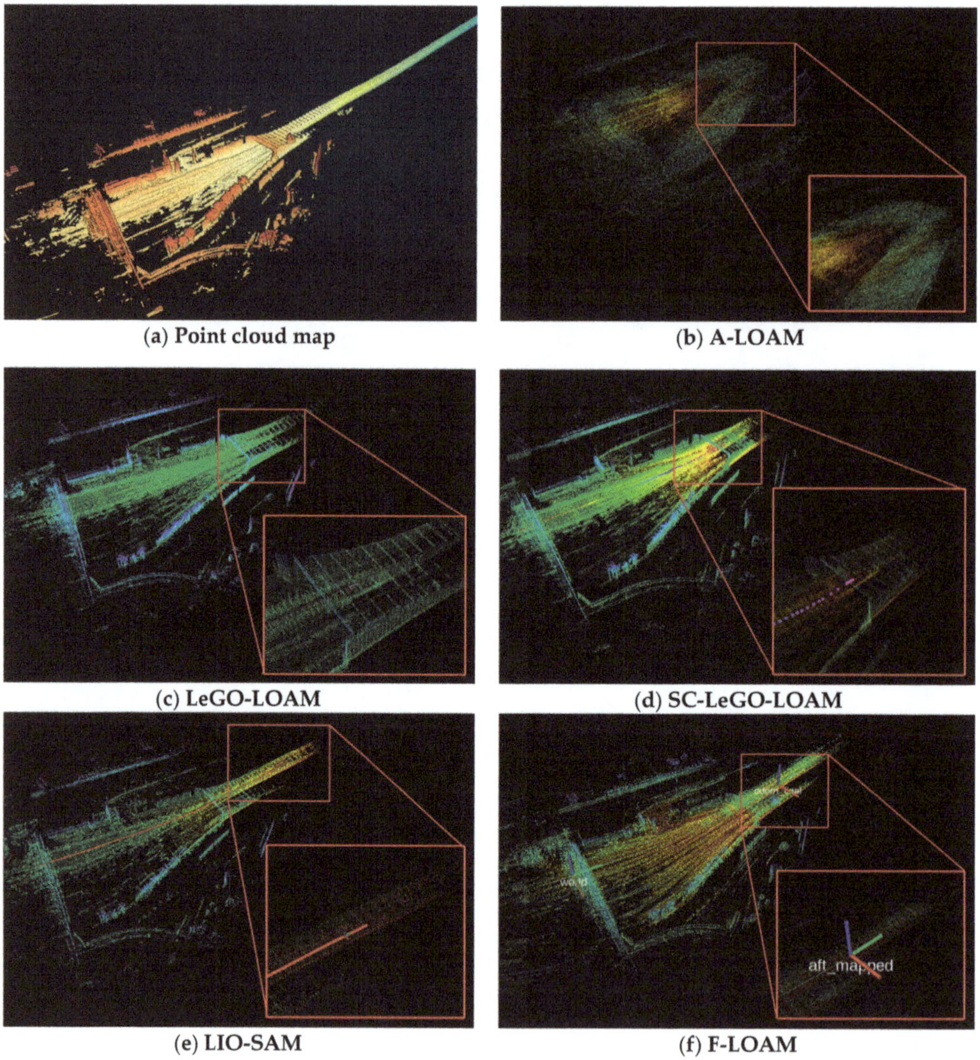

Figure 11. The mapping effect of different algorithms in the tunnel and where the degradation occurs. (**a**) The actual point cloud map given by the dataset. (**b**) Result of A-LOAM. (**c**) Result of LeGO-LOAM. (**d**) Result of SC-LeGO-LOAM. (**e**) Result of LIO-SAM. (**f**) Result of F-LOAM. The exact location where the degradation occurs is zoomed in so that the mapping effect and the effective distance of the odometer can be accurately compared.

6. Conclusions and Future Outlook

The development of SLAM technology based on 3D LIDAR in recent years has been rapid. Among them, excellent works of multi-sensor fusion have emerged in an endless stream. Throughout the development history of fusion SLAM, we have seen from filter-based probabilistic methods to information-based optimization methods; from raw data-assisted front end fusion to error-coupling-based back end optimization; from single sensor systems to complex systems with multiple subsystems coupled; from independent error models to tightly coupled complete graphical models. Various application scenarios and demands have promoted the diversity of SLAM technology. Continuous advancements in sensor technology provide the foundation and impetus for it.

This paper mainly classifies and summarizes the papers and works that have appeared in recent years based on the data coupling method of the SLAM system. The main innovations of the paper are mentioned together when describing the details of data association. The strengths and weaknesses of each work are based on the qualitative analysis of the system composition. However, our work is far from perfect. Obviously, this paper does not list all related works for readers' reference. Only a portion of representative works are shown. There are more works based on deep learning for multi-sensor fusions, mostly used in environment perception, object detection and semantic segmentation. They may play auxiliary roles in SLAM systems.

Multi-sensor fusion is a key to building robust systems. Complex systems based on multi-sensors need to be lightweight, accurate, scalable, and versatile for SLAM. From the experimental part, we know that dynamic environment, object occlusion, and long corridor environment are the key challenges for feature-based SLAM methods. Combining the sensor with the control model of the robot or vehicle can effectively alleviate the problem of odometer degradation in special cases. With the increase of the number of sensors, the amount of data, and the continuous expansion of application scenarios, it is difficult for SLAM systems to further improve the accuracy of positioning and mapping within a specified computing time. Therefore, SLAM has large development space in the applications of various scenes. Distributed multi-robot collaboration, land–air collaboration, and sea–air collaboration systems can effectively solve the problems faced in large scenes. In addition, hardware acceleration and parallel processing feature extraction and pose optimization can effectively relieve the computational pressure of the system due to the multi-sensor data fusion. On the other hand, deep learning is undoubtedly one of the hottest directions at present. There have been a lot of efforts towards combining deep learning with SLAM systems. The application of deep learning can be seen in almost all key steps such as feature extraction, depth estimation, environment perception, pose estimation, and semantic map. In the current works, deep learning only replaced limited parts of the SLAM system. For example, optimizing depth estimation of monocular camera to obtain landmark points, directly estimating pose without feature extraction, perceiving the environment to distinguish moving objects, and building high-precision semantic maps. These are research directions with great potential in the future. The application of deep learning will further improve and expand the performance and functions of SLAM. In future work, the combination of data fusion of multiple sensors and deep learning to optimize and improve the SLAM algorithm will receive more attention.

Author Contributions: Collection and organization of references, C.C. and W.W.; summarization and classification of references, X.X and L.Z.; writing—original draft preparation, L.Z.; writing—review and editing, L.Z., X.X. and J.Y.; supervision, Y.R., Z.T. and M.L. All authors have read and agreed to the published version of the manuscript.

Funding: This research was funded by Changzhou Sci&Tech Program (Grant No. CE20215041), the Fundamental Research Funds for the Central Universities (Grant No. B220202023), and Jiangsu Key R&D Program (Grant No. BE2020082-1).

Data Availability Statement: Not applicable.

Conflicts of Interest: The authors declare no conflict of interest.

References

1. Montemerlo, M.; Becker, J.; Bhat, S.; Dahlkamp, H.; Dolgov, D.; Ettinger, S.; Haehnel, D.; Hilden, T.; Hoffmann, G.; Huhnke, B.; et al. Junior: The Stanford Entry in the Urban Challenge. *J. Field Robot.* **2008**, *25*, 569–597. [CrossRef]
2. Levinson, J.; Askeland, J.; Becker, J.; Dolson, J.; Held, D.; Kammel, S.; Kolter, J.Z.; Langer, D.; Pink, O.; Pratt, V.; et al. Towards Fully Autonomous Driving: Systems and Algorithms. In Proceedings of the 2011 IEEE Intelligent Vehicles Symposium (IV), Baden-Baden, Germany, 5–9 June 2011; pp. 163–168.

3. He, X.; Gao, W.; Sheng, C.Z.; Zhang, Z.T.; Pan, S.G.; Duan, L.J.; Zhang, H.; Lu, X.Y. LiDAR-Visual-Inertial Odometry Based on Optimized Visual Point-Line Features. *Remote Sens.* **2022**, *14*, 662. [CrossRef]
4. Tee, Y.K.; Han, Y.C. Lidar-Based 2D SLAM for Mobile Robot in an Indoor Environment: A Review. In Proceedings of the 2021 International Conference on Green Energy, Computing and Sustainable Technology (GECOST), Miri, Malaysia, 7–9 July 2021; pp. 1–7.
5. Bresson, G.; Alsayed, Z.; Yu, L.; Glaser, S. Simultaneous Localization and Mapping: A Survey of Current Trends in Autonomous Driving. *IEEE Trans. Intell. Vehic.* **2017**, *2*, 194–220. [CrossRef]
6. Cadena, C.; Carlone, L.; Carrillo, H.; Latif, Y.; Scaramuzza, D.; Neira, J.; Reid, I.; Leonard, J.J. Past, Present, and Future of Simultaneous Localization and Mapping: Toward the Robust-Perception Age. *IEEE Trans. Robot.* **2016**, *32*, 1309–1332. [CrossRef]
7. Debeunne, C.; Vivet, D. A Review of Visual-LiDAR Fusion Based Simultaneous Localization and Mapping. *Sensors* **2020**, *20*, 2068. [CrossRef] [PubMed]
8. Taheri, H.; Xia, Z.C. SLAM; Definition and Evolution. *Eng. Appl. Artif. Intell.* **2021**, *97*, 104032. [CrossRef]
9. Zhiguo, Z.; Jiangwei, C.; Shunfan, D. Overview of 3D Lidar SLAM Algorithms. *Chin. J. Sci. Instrum.* **2021**, *42*, 13–27.
10. Leonard, J.J.; Durrant-Whyte, H.F. *Directed Sonar Sensing for Mobile Robot. Navigation*; Kluwer Academic Publishers: Norwood, MA, USA, 1992.
11. Ji, Z.; Singh, S. LOAM: Lidar Odometry and Mapping in Real-Time. In Proceedings of the Robotics: Science and Systems Conference (RSS), Berkeley, CA, USA, 12–14 July 2014.
12. Shan, T.; Englot, B. LeGO-LOAM: Lightweight and Ground-Optimized Lidar Odometry and Mapping on Variable Terrain. In Proceedings of the 2018 IEEE/RSJ International Conference on Intelligent Robots and Systems (IROS), Madrid, Spain, 1–5 October 2018; pp. 4758–4765.
13. Zermas, D.; Izzat, I.; Papanikolopoulos, N. Fast Segmentation of 3D Point Clouds: A Paradigm on LiDAR Data for Autonomous Vehicle Applications. In Proceedings of the 2017 IEEE International Conference on Robotics and Automation (ICRA), Singapore, 29 May–3 June 2017; pp. 5067–5073.
14. Huo, X.; Dou, L.; Lu, H.; Tian, B.; Du, M. A Line/Plane Feature-Based Lidar Inertial Odometry and Mapping. In Proceedings of the 2019 Chinese Control Conference (CCC), Guangzhou, China, 27–30 July 2019; pp. 4377–4382.
15. Zhang, S.; Xiao, L.; Nie, Y.; Dai, B.; Hu, C. Lidar Odometry and Mapping Based on Two-Stage Feature Extraction. In Proceedings of the 2020 39th Chinese Control Conference (CCC), Shenyang, China, 27–29 July 2020; pp. 3966–3971.
16. Gonzalez, C.; Adams, M. An Improved Feature Extractor for the Lidar Odometry and Mapping (LOAM) Algorithm. In Proceedings of the 2019 International Conference on Control, Automation and Information Sciences (ICCAIS), Chengdu, China, 23–26 October 2019; pp. 1–7.
17. Lee, S.-W.; Hsu, C.-M.; Lee, M.-C.; Fu, Y.-T.; Atas, F.; Tsai, A. Fast Point Cloud Feature Extraction for Real-Time SLAM. In Proceedings of the 2019 International Automatic Control Conference (CACS), Keelung, Taiwan, 13–16 November 2019; pp. 1–6.
18. Yokozuka, M.; Koide, K.; Oishi, S.; Banno, A. LiTAMIN2: Ultra Light LiDAR-Based SLAM Using Geometric Approximation Applied with KL-Divergence. In Proceedings of the 2021 IEEE International Conference on Robotics and Automation (ICRA), Xi'an, China, 30 May–5 June 2021; pp. 11619–11625.
19. Behley, J.; Stachniss, C. Efficient Surfel-Based SLAM Using 3D Laser Range Data in Urban Environments. In Proceedings of the 14th Conference on Robotics-Science and Systems (RSS), Pittsburgh, PA, USA, 26–30 June 2018.
20. Park, C.; Moghadam, P.; Kim, S.; Elfes, A.; Fookes, C.; Sridharan, S. Elastic LiDAR Fusion: Dense Map-Centric Continuous-Time SLAM. In Proceedings of the 2018 IEEE International Conference on Robotics and Automation (ICRA), Brisbane, QLD, Australia, 21–25 May 2018; pp. 1206–1213.
21. Droeschel, D.; Behnke, S. Efficient Continuous-Time SLAM for 3D Lidar-Based Online Mapping. In Proceedings of the 2018 IEEE International Conference on Robotics and Automation (ICRA), Brisbane, QLD, Australia, 21–25 May 2018; pp. 5000–5007.
22. Pfister, H.; Zwicker, M.; van Baar, J.; Gross, M. Surfels: Surface Elements as Rendering Primitives. In Proceedings of the Computer Graphics Annual Conference, New Orleans, LA, USA, 23–28 July 2000; pp. 335–342.
23. Pan, Y.; Xiao, P.C.A.; He, Y.J.; Shao, Z.L.; Li, Z.S. MULLS: Versatile LiDAR SLAM via Multi-Metric Linear Least Square. In Proceedings of the 2021 IEEE International Conference on Robotics and Automation (ICRA), Xi'an, China, 30 May–5 June 2021; pp. 11633–11640.
24. Kim, G.; Kim, A.; Kosecka, J. Scan Context: Egocentric Spatial Descriptor for Place Recognition within 3D Point Cloud Map. In Proceedings of the 2018 IEEE/RSJ International Conference on Intelligent Robots and Systems (IROS), Madrid, Spain, 1–5 October 2018; pp. 4802–4809.
25. Wang, H.; Wang, C.; Xie, L.H. Intensity Scan Context: Coding Intensity and Geometry Relations for Loop Closure Detection. In Proceedings of the 2020 IEEE International Conference on Robotics and Automation (ICRA), Paris, France, 31 May–31 August 2020; pp. 2095–2101.
26. Lin, J.; Zhang, F. A Fast, Complete, Point Cloud Based Loop Closure for LiDAR Odometry and Mapping. *arXiv* **2019**. [CrossRef]
27. Lin, J.; Zhang, F. Loam Livox: A Fast, Robust, High-Precision LiDAR Odometry and Mapping Package for LiDARs of Small FoV. In Proceedings of the 2020 IEEE International Conference on Robotics and Automation (ICRA), Paris, France, 31 May–31 August 2020; pp. 3126–3131.
28. Zhang, J.; Singh, S. Visual-Lidar Odometry and Mapping: Low-Drift, Robust, and Fast. In Proceedings of the 2015 IEEE International Conference on Robotics and Automation (ICRA), Seattle, WA, USA, 26–30 May 2015; pp. 2174–2181.

29. Zhang, J.; Singh, S. Laser-Visual-Inertial Odometry and Mapping with High Robustness and Low Drift. *J. Field Robot.* **2018**, *35*, 1242–1264. [CrossRef]
30. Caselitz, T.; Steder, B.; Ruhnke, M.; Burgard, W. Monocular Camera Localization in 3D LiDAR Maps. In Proceedings of the 2016 IEEE/RSJ International Conference on Intelligent Robots and Systems (IROS), Daejeon, Korea, 9–14 October 2016; pp. 1926–1931.
31. Zhang, L.; Xu, X.; Cao, C.; He, J.; Ran, Y.; Tan, Z.; Luo, M. Robot Pose Estimation Method Based on Image and Point Cloud Fusion with Dynamic Feature Elimination. *Chin. J. Lasers* **2022**, *49*, 0610001.
32. Qin, T.; Li, P.; Shen, S. VINS-Mono: A Robust and Versatile Monocular Visual-Inertial State Estimator. *IEEE Trans. Robot.* **2018**, *34*, 1004–1020. [CrossRef]
33. Zhang, M.; Han, S.; Wang, S.; Liu, X.; Hu, M.; Zhao, J. Stereo Visual Inertial Mapping Algorithm for Autonomous Mobile Robot. In Proceedings of the 2020 3rd International Conference on Intelligent Robotic and Control Engineering (IRCE), Oxford, UK, 10–12 August 2020; pp. 97–104.
34. Qin, T.; Cao, S.; Pan, J.; Shen, S. A General Optimization-Based Framework for Global Pose Estimation with Multiple Sensors. *arXiv* **2019**. [CrossRef]
35. Wang, Z.; Zhang, J.; Chen, S. Robust High Accuracy Visual-Inertial-Laser SLAM System. In Proceedings of the 2019 IEEE/RSJ International Conference on Intelligent Robots and Systems (IROS), Macau, China, 4–8 November 2019; pp. 6636–6641.
36. Shao, W.; Vijayarangan, S.; Li, C.; Kantor, G. Stereo Visual Inertial LiDAR Simultaneous Localization and Mapping. In Proceedings of the 2019 IEEE/RSJ International Conference on Intelligent Robots and Systems (IROS), Macau, China, 4–8 November 2019; pp. 370–377.
37. Kaess, M.; Ranganathan, A.; Dellaert, F. iSAM: Incremental Smoothing and Mapping. *IEEE Trans. Robot.* **2008**, *24*, 1365–1378. [CrossRef]
38. Kaess, M.; Johannsson, H.; Roberts, R. iSAM2: Incremental Smoothing and Mapping with Fluid Relinearization and Incremental Variable Reordering. In Proceedings of the 2011 IEEE International Conference on Robotics and Automation (ICRA), Shanghai, China, 9–13 May 2011; pp. 3281–3288.
39. Khattak, S.; Nguyen, H.D.; Mascarich, F. Complementary Multi-Modal Sensor Fusion for Resilient Robot Pose Estimation in Subterranean Environments. In Proceedings of the 2020 International Conference on Unmanned Aircraft Systems (ICUAS), Athens, Greece, 1–4 September 2020; pp. 1024–1029.
40. Camurri, M.; Ramezani, M.; Nobili, S.; Fallon, M. Pronto: A Multi-Sensor State Estimator for Legged Robots in Real-World Scenarios. *Front. Robot. AI* **2020**, *7*, 18. [CrossRef] [PubMed]
41. Lowe, T.; Kim, S.; Cox, M. Complementary Perception for Handheld SLAM. *IEEE Robot. Autom. Lett.* **2018**, *3*, 1104–1111. [CrossRef]
42. Zhu, Y.; Zheng, C.; Yuan, C. CamVox: A Low-cost and Accurate Lidar-assisted Visual SLAM System. In Proceedings of the 2021 IEEE International Conference on Robotics and Automation (ICRA), Xi'an, China, 30 May–5 June 2021; pp. 5049–5055.
43. Mur-Artal, R.; Tardós, J.D. ORB-SLAM2: An Open-Source SLAM System for Monocular, Stereo, and RGB-D Cameras. *IEEE Trans. Robot.* **2017**, *33*, 1255–1262. [CrossRef]
44. Shin, Y.; Park, Y.; Kim, A. Direct Visual SLAM Using Sparse Depth for Camera-LiDAR System. In Proceedings of the 2018 IEEE International Conference on Robotics and Automation (ICRA), Brisbane, QLD, Australia, 21–25 May 2018; pp. 5144–5151.
45. Engel, J.; Koltun, V.; Cremers, D. Direct Sparse Odometry. *IEEE Trans. Pattern Anal. Mach. Intell.* **2016**, *40*, 611–625. [CrossRef]
46. Reinke, A.; Chen, X.; Stachniss, C. Simple but Effective Redundant Odometry for Autonomous Vehicles. In Proceedings of the 2021 IEEE International Conference on Robotics and Automation (ICRA), Xi'an, China, 30 May–5 June 2021; pp. 9631–9637.
47. Segal, A.; Hhnel, D.; Thrun, S. Generalized-ICP. In Proceedings of the Robotics: Science and Systems V (RSS), Seattle, WA, USA, 28 June–1 July 2009.
48. Rusinkiewicz, S.; Levoy, M. Efficient variants of the ICP algorithm. In Proceedings of the Proceedings Third International Conference on 3-D Digital Imaging and Modeling, Quebec, QC, Canada, 28 May–1 June 2001; pp. 145–152.
49. Biber, P.; Strasser, W. The normal distributions transform: A new approach to laser scan matching. In Proceedings of the Proceedings 2003 IEEE/RSJ International Conference on Intelligent Robots and Systems (IROS), Las Vegas, NV, USA, 27–31 October 2003; pp. 2743–2748.
50. Park, J.; Zhou, Q.Y.; Koltun, V. Colored Point Cloud Registration Revisited. In Proceedings of the 2017 IEEE International Conference on Computer Vision (ICCV), Venice, Italy, 22–29 October 2017; pp. 143–152.
51. Huang, K.; Stachniss, C. Joint ego-motion estimation using a laser scanner and a monocular Camera through relative orientation estimation and 1-DoF ICP. In Proceedings of the 2018 IEEE/RSJ International Conference on Intelligent Robots and Systems (IROS), Madrid, Spain, 1–5 October 2018; pp. 671–677.
52. Wang, P.; Fang, Z.; Zhao, S. Vanishing Point Aided LiDAR-Visual-Inertial Estimator. In Proceedings of the 2021 IEEE International Conference on Robotics and Automation (ICRA), Xi'an, China, 30 May–5 June 2021; pp. 13120–13126.
53. Forster, C.; Carlone, L.; Dellaert, F.; Scaramuzza, D. IMU Preintegration on Manifold for Efficient Visual-Inertial Maximum-a-Posteriori Estimation. In Proceedings of the 2015 Robotics Science and Systems (RSS), Rome, Italy, 17 July 2015.
54. Forster, C.; Carlone, L.; Dellaert, F. On-Manifold Preintegration for Real-Time Visual–Inertial Odometry. *IEEE Trans. Robot.* **2017**, *33*, 1–21. [CrossRef]
55. Geneva, P.; Eckenhoff, K.; Yang, Y. LIPS: LiDAR-Inertial 3D Plane SLAM. In Proceedings of the 2018 IEEE/RSJ International Conference on Intelligent Robots and Systems (IROS), Madrid, Spain, 1–5 October 2018; pp. 123–130.

56. Gentil, C.L.; Vidal-Calleja, T.; Huang, S. IN2LAMA: INertial Lidar Localisation And Mapping. In Proceedings of the 2019 International Conference on Robotics and Automation (ICRA), Montreal, QC, Canada, 20–24 May 2019; pp. 6388–6394.
57. Ye, H.; Chen, Y.; Liu, M. Tightly Coupled 3D Lidar Inertial Odometry and Mapping. In Proceedings of the 2019 International Conference on Robotics and Automation (ICRA), Montreal, QC, Canada, 20–24 May 2019; pp. 3144–3150.
58. Hess, W.; Kohler, D.; Rapp, H. Real-time loop closure in 2D LIDAR SLAM. In Proceedings of the 2016 IEEE International Conference on Robotics and Automation (ICRA), Stockholm, Sweden, 16–21 May 2016; pp. 4322–4328.
59. Ding, W.; Hou, S.; Gao, H. LiDAR Inertial Odometry Aided Robust LiDAR Localization System in Changing City Scenes. In Proceedings of the 2020 IEEE International Conference on Robotics and Automation (ICRA), Paris, France, 31 May–31 August 2020; pp. 1271–1278.
60. Ceres Solver. Available online: http://ceres-solver.org (accessed on 4 May 2022).
61. Shan, T.; Englot, B.; Meyers, D.; Wang, W.; Ratti, C.; Rus, D. LIO-SAM: Tightly-Coupled Lidar Inertial Odometry via Smoothing and Mapping. In Proceedings of the 2020 IEEE/RSJ International Conference on Intelligent Robots and Systems (IROS), Las Vegas, NV, USA, 24 October–24 January 2021; pp. 5135–5142.
62. Moore, T.; Stouch, D. A Generalized Extended Kalman Filter Implementation for the Robot Operating System. In Proceedings of the 13th International Conference on Intelligent Autonomous Systems (IAS), Padova, Italy, 15–18 July 2016; pp. 335–348.
63. Nguyen, T.-M.; Cao, M.; Yuan, S.; Lyu, Y.; Nguyen, T.H.; Xie, L. LIRO: Tightly Coupled Lidar-Inertia-Ranging Odometry. In Proceedings of the 2021 IEEE International Conference on Robotics and Automation (ICRA), Xi'an, China, 30 May–5 June 2021; pp. 14484–14490.
64. Chen, W.; Zhao, H.; Shen, Q.; Xiong, C.; Zhou, S.; Liu, Y.-H. Inertial Aided 3D LiDAR SLAM with Hybrid Geometric Primitives in Large-Scale Environments. In Proceedings of the 2021 IEEE International Conference on Robotics and Automation (ICRA), Xi'an, China, 30 May–5 June 2021; pp. 11566–11572.
65. Li, W.; Hu, Y.; Han, Y.; Li, X. KFS-LIO: Key-Feature Selection for Lightweight Lidar Inertial Odometry. In Proceedings of the 2021 IEEE International Conference on Robotics and Automation (ICRA), Xi'an, China, 30 May–5 June 2021; pp. 5042–5048.
66. Lv, J.; Hu, K.; Xu, J.; Liu, Y.; Ma, X.; Zuo, X. CLINS: Continuous-Time Trajectory Estimation for LiDAR-Inertial System. In Proceedings of the 2021 IEEE/RSJ International Conference on Intelligent Robots and Systems (IROS), Prague, Czech Republic, 27 September–1 October 2021; pp. 6657–6663.
67. RF-LIO: Removal-First Tightly-Coupled Lidar Inertial Odometry in High Dynamic Environments. In Proceedings of the 2021 IEEE/RSJ International Conference on Intelligent Robots and Systems (IROS), Prague, Czech Republic, 27 September–1 October 2021; pp. 4421–4428.
68. Xu, W.; Zhang, F. FAST-LIO: A Fast, Robust LiDAR-Inertial Odometry Package by Tightly-Coupled Iterated Kalman Filter. *IEEE Robot. Autom. Lett.* **2021**, *6*, 3317–3324. [CrossRef]
69. Graeter, J.; Wilczynski, A.; Lauer, M. LIMO: Lidar-Monocular Visual Odometry. In Proceedings of the 2018 IEEE/RSJ International Conference on Intelligent Robots and Systems (IROS), Madrid, Spain, 1–5 October 2018; pp. 7872–7879.
70. Szeliski, R. *Computer Vision: Algorithms and Applications*; Springer-Verlag: London, UK, 2010.
71. Hartley, R.; Zisserman, A. *Multiple View Geometry in Computer Vision*, 2nd ed.; Cambridge University Press: Cambridge, UK, 2004.
72. Huang, S.-S.; Ma, Z.-Y.; Mu, T.-J.; Fu, H.; Hu, S.-M. Lidar-Monocular Visual Odometry Using Point and Line Features. In Proceedings of the 2020 IEEE International Conference on Robotics and Automation (ICRA), Paris, France, 31 May–31 August 2020; pp. 1091–1097.
73. Amblard, V.; Osedach, T.P.; Croux, A.; Speck, A.; Leonard, J.J. Lidar-Monocular Surface Reconstruction Using Line Segments. In Proceedings of the 2021 IEEE International Conference on Robotics and Automation (ICRA), Xi'an, China, 30 May–5 June 2021; pp. 5631–5637.
74. Wang, J.; Rünz, M.; Agapito, L. DSP-SLAM: Object Oriented SLAM with Deep Shape Priors. In Proceedings of the 2021 International Conference on 3D Vision (3DV), London, UK, 1–3 December 2021; pp. 1362–1371.
75. Wang, T.; Su, Y.; Shao, S.; Yao, C.; Wang, Z. GR-Fusion: Multi-Sensor Fusion SLAM for Ground Robots with High Robustness and Low Drift. In Proceedings of the 2021 IEEE/RSJ International Conference on Intelligent Robots and Systems (IROS), Prague, Czech Republic, 27 September–1 October 2021; pp. 5440–5447.
76. Jia, Y.; Luo, H.; Zhao, F.; Jiang, G.; Li, Y.; Yan, J.; Jiang, Z.; Wang, Z. Lvio-Fusion: A Self-Adaptive Multi-Sensor Fusion SLAM Framework Using Actor-Critic Method. In Proceedings of the 2021 IEEE/RSJ International Conference on Intelligent Robots and Systems (IROS), Prague, Czech Republic, 27 September–1 October 2021; pp. 286–293.
77. Shan, T.; Englot, B.; Ratti, C.; Rus, D. LVI-SAM: Tightly-Coupled Lidar-Visual-Inertial Odometry via Smoothing and Mapping. In Proceedings of the Proceedings-IEEE International Conference on Robotics and Automation, Xi'an, China, 30 May–5 June 2021; pp. 5692–5698.
78. Zhao, S.; Zhang, H.; Wang, P.; Nogueira, L.; Scherer, S. Super Odometry: IMU-Centric LiDAR-Visual-Inertial Estimator for Challenging Environments. In Proceedings of the 2021 IEEE/RSJ International Conference on Intelligent Robots and Systems (IROS), Prague, Czech Republic, 27 September–1 October 2021; pp. 8729–8736.
79. Wang, Y.; Song, W.; Zhang, Y.; Huang, F.; Tu, Z.; Lou, Y. MetroLoc: Metro Vehicle Mapping and Localization with LiDAR-Camera-Inertial Integration. *arXiv* **2021**. [CrossRef]
80. Wisth, D.; Camurri, M.; Das, S.; Fallon, M. Unified Multi-Modal Landmark Tracking for Tightly Coupled Lidar-Visual-Inertial Odometry. *IEEE Robot. Autom. Lett.* **2021**, *6*, 1004–1011. [CrossRef]

81. Dellaert, F.; Kaess, M. Factor graphs for robot perception. *Found. Trends Robot.* **2017**, *6*, 1–139. [CrossRef]
82. Mourikis, A.I.; Roumeliotis, S.I. A Multi-State Constraint Kalman Filter for Vision-Aided Inertial Navigation. In Proceedings of the Proceedings 2007 IEEE International Conference on Robotics and Automation, Rome, Italy, 10–14 April 2007; pp. 3565–3572.
83. Yang, Y.; Geneva, P.; Zuo, X.; Eckenhoff, K.; Liu, Y.; Huang, G. Tightly-Coupled Aided Inertial Navigation with Point and Plane Features. In Proceedings of the 2019 International Conference on Robotics and Automation (ICRA), Montreal, QC, Canada, 20–24 May 2019; pp. 6094–6100.
84. Zuo, X.; Geneva, P.; Lee, W.; Liu, Y.; Huang, G. LIC-Fusion: LiDAR-Inertial-Camera Odometry. In Proceedings of the 2019 IEEE/RSJ International Conference on Intelligent Robots and Systems (IROS), Macau, China, 3–8 November 2019; pp. 5848–5854.
85. Zuo, X.; Yang, Y.; Geneva, P.; Lv, J.; Liu, Y.; Huang, G.; Pollefeys, M. LIC-Fusion 2.0: LiDAR-Inertial-Camera Odometry with Sliding-Window Plane-Feature Tracking. In Proceedings of the 2020 IEEE/RSJ International Conference on Intelligent Robots and Systems (IROS), Las Vegas, NV, USA, 10 February 2020; pp. 5112–5119.
86. Lin, J.; Chunran, Z.; Xu, W.; Zhang, F. R2LIVE: A Robust, Real-Time, LiDAR-Inertial-Visual Tightly-Coupled State Estimator and Mapping. *IEEE Robot. Autom. Lett.* **2021**, *6*, 7469–7476. [CrossRef]
87. Geiger, A.; Lenz, P.; Urtasun, R. Are we ready for autonomous driving? the kitti vision benchmark suite. In Proceedings of the 2012 IEEE Conference on Computer Vision and Pattern Recognition, Providence, RI, USA, 16–21 June 2012; pp. 3354–3361.
88. Waymo Open Dataset. Available online: https://waymo.com/open/data (accessed on 4 May 2022).
89. PandaSet Open Datasets. Available online: https://scale.com/open-datasets/pandaset (accessed on 4 May 2022).
90. Maddern, W.; Pascoe, G.; Linegar, C.; Newman, P. 1 year, 1000 km: The Oxford RobotCar dataset. *Int. J. Robot. Res.* **2016**, *36*, 3–15. [CrossRef]
91. Hsu, L.T.; Kubo, N.; Wen, W.; Chen, W.; Liu, Z.; Suzuki, T.; Meguro, J. UrbanNav: An open-sourced multisensory dataset for benchmarking positioning algorithms designed for urban areas. In Proceedings of the 34th International Technical Meeting of the Satellite Division of The Institute of Navigation, St. Louis, MO, USA, 20–24 September 2021; pp. 226–256.
92. Huang, F.; Wen, W.; Zhang, J.; Hsu, L.T. Point wise or Feature wise? Benchmark Comparison of public Available LiDAR Odometry Algorithms in Urban Canyons. *arXiv* **2021**. [CrossRef]
93. Jonnavithula, N.; Lyu, Y.; Zhang, Z. LiDAR Odometry Methodologies for Autonomous Driving: A Survey. *arXiv* **2021**. [CrossRef]
94. LOAM. Available online: https://github.com/HKUST-Aerial-Robotics/A-LOAM (accessed on 4 May 2022).
95. Wang, H.; Wang, C.; Chen, C.-L.; Xie, L. F-loam: Fast lidar odometry and mapping. In Proceedings of the 2021 IEEE/RSJ International Conference on Intelligent Robots and Systems (IROS), Prague, Czech Republic, 27 September–1 October 2021; pp. 4390–4396.
96. Chen, S.; Zhou, B.; Jiang, C.; Xue, W.; Li, Q. A LiDAR/Visual SLAM Backend with Loop Closure Detection and Graph Optimization. *Remote Sens.* **2021**, *13*, 2720. [CrossRef]

remote sensing

Article

LIDAR-Inertial Real-Time State Estimator with Rod-Shaped and Planar Feature

Hong Liu [1,2], Shuguo Pan [1,2,*], Wang Gao [1,2], Chun Ma [1,2], Fengshuo Jia [1,2] and Xinyu Lu [1,2]

[1] School of Instrument Science and Engineering, Southeast University, Nanjing 210096, China
[2] Key Laboratory of Micro-Inertial Instrument and Advanced Navigation Technology, Southeast University, Nanjing 210096, China
* Correspondence: psg@seu.edu.cn

Abstract: State estimation and mapping based on Light Detection and Ranging (LIDAR) are important for autonomous systems. Point cloud registration is a crucial module affecting the accuracy and real-time performance of LIDAR simultaneous localization and mapping (SLAM). In this paper, a novel point cloud feature selection for LIDAR-inertial tightly coupled systems is proposed. In the front-end, a point cloud registration is carried out after marking rod-shaped and planar feature information which is different from the existing LIDAR and inertial measurement unit (IMU) integration scheme. This preprocessing method subsequently reduces the outliers. IMU pre-integration outputs high-frequency result and is used to provide the initial value for LIDAR solution. In the scan-to-map module, a computationally efficient graph optimization framework is applied. Moreover, the LIDAR odometry further constrains the IMU states. In the back-end, the optimization based on sliding-window incorporates the LIDAR-inertial measurement and loop closure global constraints to reduce the cumulative error. Combining the front-end and back-end, we propose the low drift and high real-time LIDAR-inertial positioning system. Furthermore, we conducted an exhaustive comparison in open data sequences and real-word experiments. The proposed system outperforms much higher positioning accuracy than the state-of-the-art methods in various scenarios. Compared with the LIO-SAM, the absolute trajectory error (ATE) average RMSE (Root Mean Square Error) in this study increases by 64.45% in M2DGR street dataset (street_01, 04, 07, 10) and 24.85% in our actual scene datasets. In the most time-consuming mapping module of each system, our system runtime can also be significantly reduced due to the front-end preprocessing and back-end graph model.

Keywords: tightly-coupled integration; LIDAR-inertial SLAM; rod-shaped and planar feature; sliding-window; graph optimization framework

Citation: Liu, H.; Pan, S.; Gao, W.; Ma, C.; Jia, F.; Lu, X. LIDAR-Inertial Real-Time State Estimator with Rod-Shaped and Planar Feature. *Remote Sens.* **2022**, *14*, 4031. https://doi.org/10.3390/rs14164031

Academic Editor: Andrzej Stateczny

Received: 8 June 2022
Accepted: 16 August 2022
Published: 18 August 2022

1. Introduction

Accurate and reliable state estimation is a fundamental requirement of mobile robot and automatic driving. In urban environments, indoor environments and other complex scenes, it is difficult to achieve a high precision of positioning requirements with the traditional GNSS/INS integrated.

In recent years, visual/LIDAR simultaneous localization and mapping have made certain developments. On the one hand, visual slam can achieve six degrees-of-freedom state estimation just by camera, but it is seriously affected by the illumination and low texture feature [1]. On the other hand, the laser sensor directly obtains depth information and has high resolution, which can also work at night and achieve accurate pose estimation. Therefore, this research mainly focuses on LIDAR simultaneous localization and mapping.

LIDAR odometry and mapping (LOAM) [2] is an earlier proposed LIDAR slam algorithm. Iterative ICP algorithm is a common method for point cloud matching, which is time-consuming for registration, and it is easy to fall into a local minimum [3]. LOAM replaces ICP with point-to-line and point-to-plane matching. It consists of two subsystems.

The odometry system performs point to line/plane feature matching to calculate pose between scans. The features of line and plane are judged according to point curvature. At the same time, LOAM effectively eliminates the unreliable parallel points and occlusion points. And it performs the distortion compensation by motion interpolation. The mapping system performs scan to map matching and runs at lower frequency, which can perform higher accuracy state estimation. By combining these two systems, LOAM achieves low drift and low-computational complexity, which has been ranked as the top in the LIDAR based method on the KITTI odometry benchmark site [4]. However, LOAM still has some flaws; its point cloud is stored in global voxel. Without key frame selection, it is difficult to integrate observation information of other sensors and perform global optimization.

F-LOAM adopts a two-stage distortion compensation method to reduce the computational cost and improve the real-time performance [5], but there are still no global optimization methods such as loop closure, resulting in large cumulative errors over a long period of time. Liu et al. propose a method based on deep learning for extracting feature points and obtaining the descriptors in LIDAR odometry. It also adopts the two-step state estimation for long distance experiment, which has a good performance for LIDAR of various resolutions [6]. V-LOAM introduces the visual odometry as the front-end of the laser odometry, further improving the accuracy of slam [7]. HDL_GRAPH_SLAM [8] is an algorithm that can fuse LIDAR, IMU and GNSS sensors, but the scan registration accuracy is low which is based on NDT [9]. It is also prone to drift in non-plane because of the flat ground constraint. LeGo-LOAM implements point cloud segmentation to reduce the number of features, and two-step registration provides the initial value for LIDAR mapping module. LeGo-LOAM firstly covers the key frame selection and loop detection [10]. However, there is obvious drift in the large scene testing experiment and the IMU is only used to remove distortion. LIO-mapping [11] is a joint state estimation problem based on the ideas of LOAM and VINS-Mono [12]. The front-end vision part is replaced by the LIDAR front-end for feature extraction and state estimation. However, the optimization problem is too large to be real-time, which makes it hard to apply in a mobile device. LINS is a tightly coupled LIDAR-inertial odometry (LIO) system based on the filter method [13]. The iterative error state Kalman Filter is used to correct the state estimation of the robot, but there is still a problem that the robot will drift when it runs for a long time without global constraints. LiLi-OM puts forward an adaptive keyframe selection for both solid-state and traditional LIDAR. It also introduces a metric weighting function during sensor fusion [14]. However, lacking a point cloud processing, the system stability is inadequate. LIO-SAM [15] is also a tightly-coupled LIO system, which is based on the incremental smoothing and mapping framework iSAM2 [16]. In addition, the loop closure factor and GPS factor can be added to the global optimization factor graph. In spite of this, its IMU constraints do not enter the factor graph optimization system, which may result in loss of constraint information between IMU and LIDAR measurements. In the actual scene test, LIO-SAM will appear at unstable states such as point cloud matching errors, especially when the carrier movement is in a large scene. Zhang et al. proposed the LIDAR-inertial odometry with an adaptive covariance estimation algorithm which is based on loosely-coupled method. It achieves better result compared to the tightly-coupled method [17].

In short, the existing LIDAR slam algorithms are mainly for small scenes. But for the complex scene or the great motion change, they are prone to cumulative errors and poor robustness.

Meanwhile, the processing of point cloud data affects the accuracy of point cloud registration for LIDAR slam. Douillard et al. introduced a method which jointly determines the ground and individual objects on the ground in three-dimensional space, including overhanging structures, but it requires a large amount of computation time, limiting online applications [17]. B. Douillard et al. proposed a priori ground extraction way. Segmentation of dense 3D data is optimized via a simple yet efficient voxel of the space. This approach provides near-real-time performance, but is not sufficient for real-time positioning scenarios [18]. M. Himmelsbach et al. proposed that 3D point clouds are projected onto 2D grids on the ground plane, and then point clouds were segmented on the occupied grids [19]. The algorithm has fast speed and is suitable for online segmentation. However, the method tends to result in weak segmentation. When two objects are relatively close to each other, it is prone to misrecognition, especially in the z-axis direction. In 2019, Seungcheol Park et al. proposed Curved-Voxel Clustering. The point cloud coordinates are converted from cartesian coordinates to spherical coordinates, and each point cloud is assigned to the voxel in the corresponding spherical coordinate system. Hash tables establish associations between indexes and points. When clustering, lookup is implemented using the hash tables [20]. Chen et al. use IMU to assist the point cloud registration and introduce the inertial error model for mobile laser scanning, which could effectively reduce the error with low time cost [21].

This paper mainly aims to improve the accuracy of LIDAR point cloud registration under the condition of real-time positioning, so as to ensure the robustness of the system. The contributions of this paper are summarized as follows:

1. A quick and effective feature extraction method is proposed. Due to the information of rod-shaped and planar feature, edge points and surface points are extracted reasonably to calculate curvature with the low computational cost.
2. IMU pre-integration is used to provide the initial value for LIDAR odometry, and the LIDAR odometry further constrains the pre-integrated IMU states.
3. A graph optimization model is used to solve the scan-to-map module, which greatly improves the speed of the traditional algorithm. Another graph optimization model is used to globally optimize the pre-integrated IMU measurements residuals, inter-frame matching residuals and loop residuals, which improves the accuracy and stability of LIO system effectively.

2. System Overview

The overall framework of this system is shown in the Figure 1. LIDAR and IMU measurements are the inputs for the system.

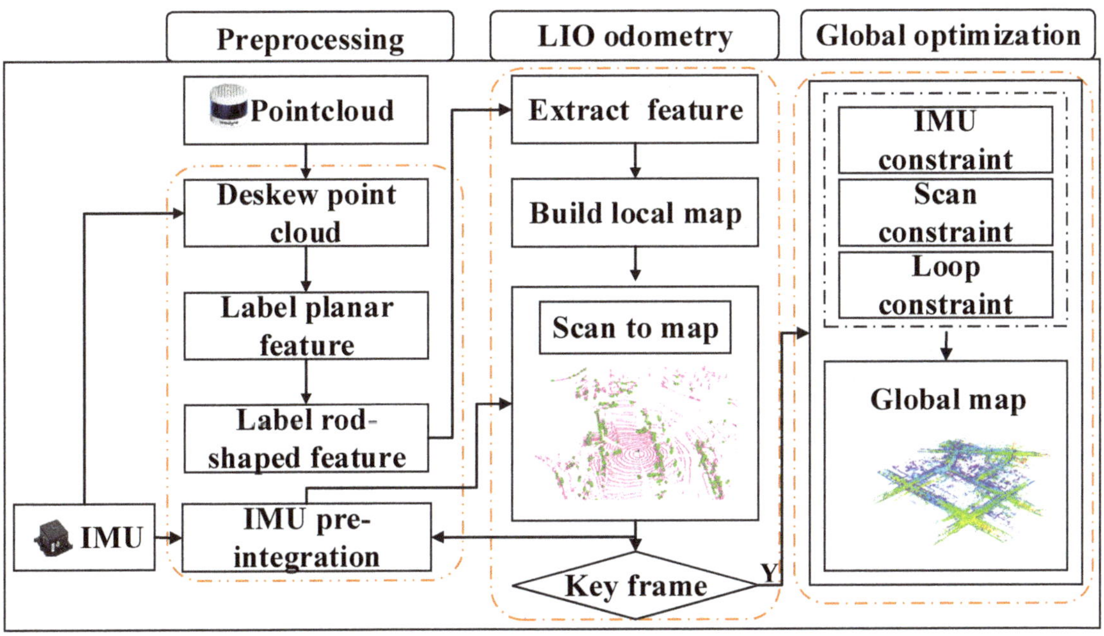

Figure 1. Overall framework of our LIO system.

The system can be divided into three parts.

First, *the preprocessing module*, the raw point clouds are de-skewed using gyroscope data and IMU pre-integration value. Current scan's point cloud is projected to the 2D image. The depth characteristic value is used to remove the outlier points. Image is used to segment the planar feature and cluster the rod-shaped feature information.

Then, *the LIO odometry module*, IMU pre-integration results are used to estimate motion pose. The scan-to-map between current frame and local map is performed. In the scan-to-map module, we introduce the graph optimization model which can enhance the speed and accuracy of the solution, and a sliding window-based way is applied to update and maintain the local map.

At last, *the global optimization module*, if the current frame is judged to be a keyframe, LIDAR scan-to-scan residuals, pre-integrated IMU residuals and loop residuals are optimized via the slide window optimization. Information of marginalization is used for prior constraints. Loop closure is detected and performed in an effective way, which is beneficial to reduce cumulative error.

According to this system, we get the 6-DOF pose estimation and a real-time updated global map. Exhaustive comparisons have been conducted to prove the superiority of our system.

We define notations and frame definitions throughout the article. $(\cdot)^W$ is considered as world frame. In the LIO system, the origin of the world coordinate is identified as the first LIDAR frame $(\cdot)^B$ is the body frame and $(\cdot)^L$ is the LIDAR frame. Rotation is represented by rotation matrices R and quaternions q. So R_W^B and q_W^B is the rotation from world frame to body frame, and p_W^B is the translation. $\otimes$ is defined as the multiplication between two quaternions.

3. The Preprocessing Module

In this study, the current LIDAR point cloud is projected onto the current 2D image grid, which is represented by a matrix. The horizontal index unit of moment frame is the horizontal resolution of each frame, and the vertical unit is the vertical resolution. For example, the size of the projected image matrix of 16-line LIDAR is 16 ×1800. The value of the image grid stores the depth of each point, and the points will be removed if there is an outlier value. The operating point cloud data on the basis of two-dimensional images can significantly improve the computing speed.

After this process, reliable estimation of LIDAR´s per scan is a necessary prerequisite. In this paper, IMU pre-integration is used to obtain the relative translational motion at the beginning and the end of each scan. Based on this method, point cloud distortion can be eliminated. In the meantime, the raw point clouds from each scan are rotationally de-skewed using gyroscope data.

3.1. Label Planar Feature Information

The sensor carrier is moving on the ground and the LIDAR is mounted horizontally. The ground is observed with the beams below. We can get a rough but fast estimate of the plane from the number of rows of the image matrix. In the estimation plane, accurate ground points can be marked by judging the angle of each point to the ground.

$$a = \tan^{-1}\frac{dz}{\sqrt{(dx)^2+(dy)^2}} = \frac{OP_2}{OP_1} \tag{1}$$

As shown in the Figure 2, P_1 and P_2 are two laser beams reflection points. The angle a corresponding to the points of adjacent laser beams should have a small value if there is no barrier. Points on the ground can be marked according to the size of the included angle value. Dx, dy and dz represent the differences of the two laser beams in the three directions, respectively.

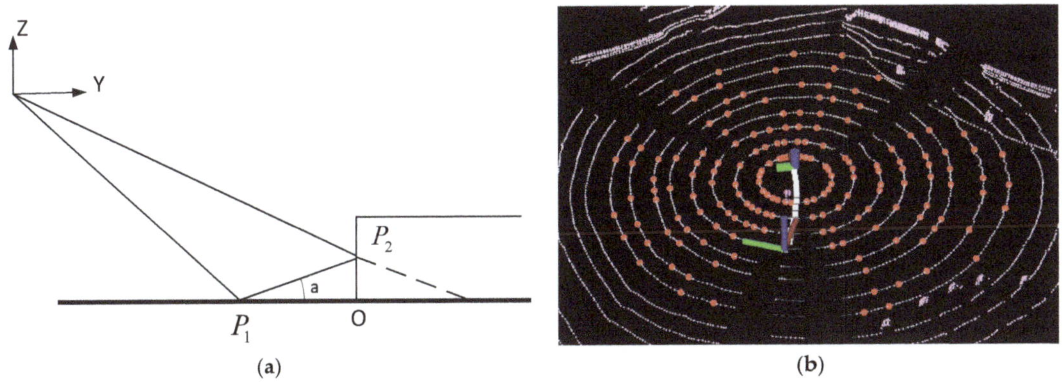

Figure 2. Illustration of the planar feature information (**a**) The angle between the lines for judging the surface; (**b**) Red points are the planar feature in the scan.

3.2. Label Rod-Shape Feature Information

Figure 3 is the top view; OC is the measurement of the first beam and OD is the second.

$$B = \tan^{-1}\frac{r_2 \sin\alpha}{r_1 - r_2 \cos\alpha} = \frac{MD}{OM} \tag{2}$$

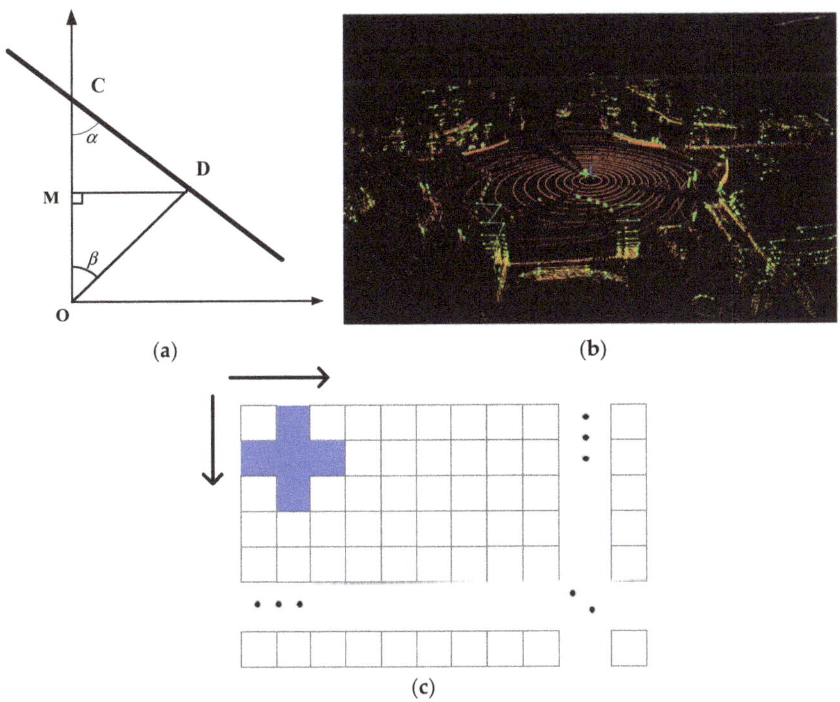

Figure 3. Illustration of the rod-shape feature cluster (**a**) The fast and accurate method for segmentation; (**b**) Green points are the edge feature in the scan; (**c**) The BFS search method in current scan.

In this formula, r_1 and r_2 are the depth values measured by the two beams. A is the angle between the two laser beams. For example, the angle is 0.2° in x direction and 2° in y direction in Velodyne16.

If C and D are on different objects, the angle β between OC and OD will be greater than a certain threshold. In this work, we set this threshold at 20°. According to this method, we can cluster the same cluster objects quickly and accurately.

The point cloud of our two-dimensional image model can be traversed quickly. We traversed each pixel in the 2D image and calculated the included angle for four points around each pixel. And each pixel is searched by BFS algorithm [22].

As shown in the Figure 3c, the surrounding points consist of the left, right, lower and top pixels. If the angle between the surrounding points is smaller than the threshold, we decided they are the same object. If a point is marked, then it will be skipped. So, the algorithm complexity is $\Theta(\mathrm{N})$, where N is the number of image pixels.

For the subsequent processing of point cloud, the influence of disorderly points and inaccurate points can be avoided. For example, when the carrier is driving, leaves, small objects, grass and weeds can be removed. These are difficult to observe through two consecutive frames of scanning, which are the main factors affecting the pose solution.

3.3. Feature Extraction

Through the segmentation and clustering of planar and rod-shaped information in the previous steps, the extraction of edge point and planar point are carried out in the rod-shape and planar feature information. Curvature is defined as follows:

$$c = \frac{1}{|S| \bullet ||\mathrm{Pr}_i^L||} || \sum_{j \in S, j \neq i} (\mathrm{Pr}_j^L - \mathrm{Pr}_i^L) || \tag{3}$$

In this formula, S is the set of continuous points from the same row of the 2D image. Pr is the point range. In this work, S is set to 10. And c is the curvature value.

Similar to LOAM, the depth information of each point is used to eliminate parallel points and occluded points, which have certain influence on the subsequent solution. In the feature extraction process, each frame is divided into 6 sub-images, which has a resolution of 16×300. Edge points and planar points are extracted from each sub-image, which are determined according to threshold c_{th} and p_{th}. In this work, c_{th} and p_{th} are chosen to be 1 and 0.1. The edge point set and planar point set extracted from k frame are ε_k and s_k.

After this process, we selectively obtain stable features and reduce the calculation pressure at the back-end. At the same time, this step improves the reliability of front-end scan registration.

4. The LIO Odometry Module

4.1. IMU Pre-Integration

LIDAR and IMU work in different frequencies. Usually, the LIDAR is 10 to 30 Hz and the IMU is 100 to 500 Hz. The pre-integration integrates the IMU measurement values between each adjacent frame of LIDAR, and adopts a value to express it. Through this step, we can get the output of the two sensors at the same frequency

The measurements of IMU include angular velocity $\widetilde{\omega_B}(t)$ and acceleration $\widetilde{a_B}(t)$. The measured values are all under the B coordinate system. And the measurement equation can be modeled as:

$$\widetilde{\omega_B}(t) = \omega_B(t) + b^{\omega}(t) + \eta^{\omega}(t) \tag{4}$$

$$\widetilde{a_B}(t) = R_W^B(t)(a_W(t) - g^W) + b^a(t) + \eta^a(t) \tag{5}$$

The measured values are affected by the slowly varying bias $b(t)$ and white noise $\eta(t)$. The acceleration of gravity in the world system. $g^W = [0, 0, g]^T$ is the gravity vector, which affects the measurement. So, it should be subtracted.

In our work, noise is ignored and the biases are considered the constant during the pre-integration period. The current state value can be obtained based on the derive of pre-integration. Assuming that j is the current frame and i is the last frame. The attitude rotation matrix R^j_{WB}, velocity v^j_{WB}, and position p^j_{WB} can be expressed as:

$$R^j_{WB} = R^i_{WB}\Delta R_{ij}Exp(J^g_{\Delta R_{ij}}\delta b^i_\omega) \tag{6}$$

$$v^j_{WB} = v^i_{WB} + g^W\Delta t_{ij} + R^i_{WB}\left(\Delta v_{ij} + J^\omega_{\Delta v_{ij}}\delta b^i_\omega + J^a_{\Delta v_{ij}}\delta b^i_a\right) \tag{7}$$

$$p^j_{WB} = p^i_{WB} + v^i_{WB}\Delta t_{ij} + \frac{1}{2}g^W\Delta t_{ij}{}^2 + R^i_{WB}\left(\Delta p_{ij} + J^\omega_{\Delta p_{ij}}\delta b^i_\omega + J^a_{\Delta p_{ij}}\delta b^i_a\right) \tag{8}$$

J is for Jacobians, and the details can be found in [23]. The $J^a_{(\cdot)}b_a$ and $J^\omega_{(\cdot)}b_\omega$ means a first-order approximation of the effect of changing the biases to avoid repeated integration. Meanwhile, the terms of pre-integration ΔR_{ij}, Δv_{ij} and Δp_{ij} can be computed between the frame i and j:

$$\Delta R_{ij} = \prod_{k=i}^{j-1} Exp((\omega^k_B - b^i_\omega)\Delta t) \tag{9}$$

$$\Delta v_{ij} = \sum_{k=i}^{j-1} \Delta R_{ik}(a^k_B - b^i_a)\Delta t \tag{10}$$

$$\Delta p_{ij} = \sum_{k=i}^{j-1}(\Delta v_{ik}\Delta t + \frac{1}{2}\Delta R_{ik}(a^k_B - b^i_a)\Delta t^2) \tag{11}$$

4.2. Build Local Map

In point cloud registration, the iterative closest point (ICP) algorithm is the most commonly scan registration method. However, as the urban scenes consists of lots of moving targets, ICP registration failure rate is high, which is directly based on raw point data. And ICP is improper for localization and mapping in real time due to its large amount of computation. In LOAM, pose estimation depends on scan-to-scan matching for quick estimation. However, this method is prone to cumulative error. In our work, current scan and local map are matched according to the predict value of IMU pre-integration. Meanwhile, the scan-to-map result is used to correct the IMU accumulative errors.

A local map associated with the current LIDAR frame is constructed. A fixed number of key frame maps within a certain range are constructed by sliding window method. The local map is converted to the W coordinate system. Edge points and planar points of a local map form the voxel map. And the points in local map are down-sampled to eliminate the duplicated features. In order to improve the point cloud matching speed, the feature information of the local map is stored in the data structure of KD-tree [24] for the convenience of subsequent search.

Therefore, this paper adopts the registration method based on feature points. After the feature points with the same type are obtained through preprocess, the graph optimization model is used to iteratively locate for current scan and local map.

4.3. Pose Estimation

For each edge point $p_\varepsilon \in \varepsilon_k$, we search for the nearest five points on the local map and calculate the mean and covariance matrices for the five points. When the distribution of points approximates a straight line, one eigenvalue of the covariance matrix will be significantly larger than the rest. In this study, the eigenvector corresponding to the eigenvalue u^{sm}_ε is the main direction of the line, and p^{sm}_ε is the geometric center of the five points in the Figure 4. If the line feature satisfies the condition, the distance between the current edge point and the line can be calculated, and the best pose estimation of the current point in the local map can be obtained by minimizing the distance. The distance calculation formula:

$$f_{\varepsilon}(p_{\varepsilon}) = p_n \bullet ((T_k p_{\varepsilon} - p_{\varepsilon}^{sm}) \times u_{\varepsilon}^{sm}) \tag{12}$$

where symbol • is the dot product and × is the cross product. p_n is the unit vector.

$$p_n = \frac{(T_k p_{\varepsilon} - p_{\varepsilon}^{sm}) \times u_{\varepsilon}^{sm}}{||(T_k p_{\varepsilon} - p_{\varepsilon}^{sm}) \times u_{\varepsilon}^{sm}||} \tag{13}$$

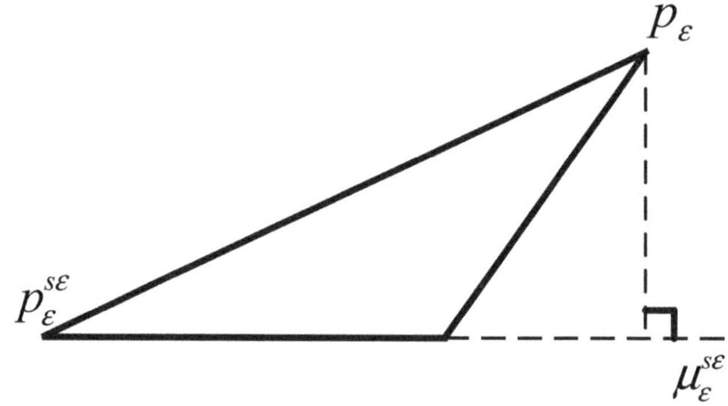

Figure 4. Illustration of the edge point-to-line residual. u_{ε}^{sm} s the main direction of the line, and p_s^{sm} is the geometric center of the five nearest points in the local map.

In the same way, each planar point in the current scan $p_s \in s_k$, we search for five points on the local map to form a plane. However, the difference is that the eigenvector corresponding to the minimum eigenvalue of the five-point covariance matrix is the normal vector corresponding to this plane. As shown in the Figure 5, u_s^{sm} is the main direction of the normal vector. p_s^{sm} is the geometric center of five planar points.

$$f_s(p_s) = (T_k p_s - p_s^{sm}) \bullet u_s^{sm} \tag{14}$$

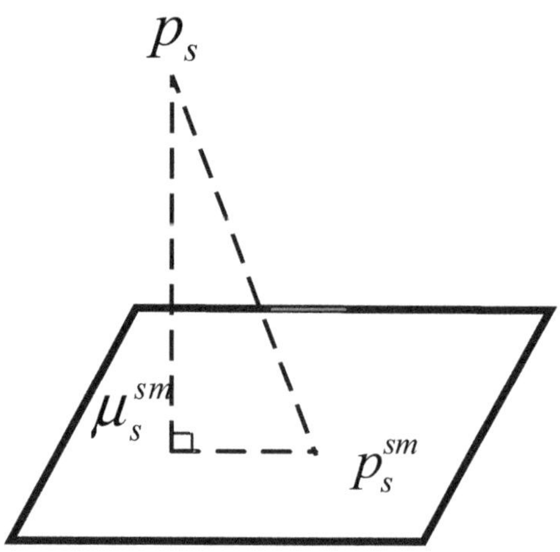

Figure 5. Illustration of the planar point-to-plane residual. p_s^{sm} is the geometric center of five planar points. u_s^{sm} is the main direction of the normal vector.

Therefore, this optimization problem can be constructed as:

$$\min\left\{\sum f_{\varepsilon}(p_{\varepsilon})+\sum f_{s}(p_{s})\right\} \tag{15}$$

The graph optimization algorithm is used to solve the nonlinear optimization problem. Thus, accurate state estimation can be obtained. Jacobian's derivation can be based on the mathematical model of left disturbance with $\delta\varphi\epsilon se(3)$ [25].

$$\begin{aligned} J_p &= \frac{\partial(TP)}{\partial\delta\varphi} \\ &= \lim_{\delta\varphi\to 0}\frac{(\exp(\delta\varphi)(TP)-(TP))}{\partial\varphi} \\ &= \begin{bmatrix} I_{3*3} & -[TP]_{\times} \\ 0_{1*3} & 0_{1*3} \end{bmatrix} \end{aligned} \tag{16}$$

where $[TP]_{\times}$ transforms 4D point expression $\{x,y,z,1\}$ into 3D point expression $\{x,y,z\}$ and calculates its skew symmetric matrix. Jacobian matrix with edge residual can be derived by:

$$J_{\varepsilon}=\frac{\partial f_{\varepsilon}(p_{\varepsilon})}{\partial(TP)}\frac{\partial(TP)}{\partial\delta\varphi}=p_n\bullet\left(u_{\varepsilon}^{sm}\times J_p\right) \tag{17}$$

In the same way, we also can derive:

$$J_s=\frac{\partial f_s(p_s)}{\partial(TP)}\frac{\partial(TP)}{\partial\delta\varphi}=u_{\varepsilon}^{sm}\bullet J_p \tag{18}$$

According to above formula, the estimation can be calculated by iterative optimization until it converges. In the work, the local map size is set to within 50 m radius. We propose the new optimization model, deduce the corresponding residual and Jacobian, and improve the solving speed significantly compared to other algorithms (see Section 6).

5. The Global Optimization Module

If the motion change in the current scan is greater than a certain threshold (10° in rotation and 0.5 m in translation) compared with that of the previous scan, the current frame will be judged as a key frame, and it will enter the global optimization which is based on sliding window.

In this paper, the state vector in the sliding window is defined as $\chi=[x_0,\ x_1,x_2,\ldots,\ x_n]$. And $\chi_i=\left[p_{b_i}^W,q_{b_i}^W,\ v_{b_i}^W,b_a,b_g\right]$. For the n keyframe window width, these states are obtained by minimizing

$$\underbrace{min}_{x_n}\left\{\|R_p(\widetilde{\chi})\|^2+\sum_{k=1}^{n}L\left(z_{b_j}^{b_i},\chi\right)+\sum_{k=1}^{n}\varkappa\left(z_{b_j}^{b_i},\chi\right)+\sum_{k=1}^{n}F\left(z_{b_j}^{b_{loop}},\chi\right)\right\} \tag{19}$$

In this formula, $R_p(\widetilde{x})$ means the prior residual according to the measurements which are marginalized out because of the sliding window. $L(x_k)$, $\varkappa(x_k)$ and $F(x_k)$ denote the LIDAR, IMU and loop closure error terms. Figure 6 shows the optimization process.

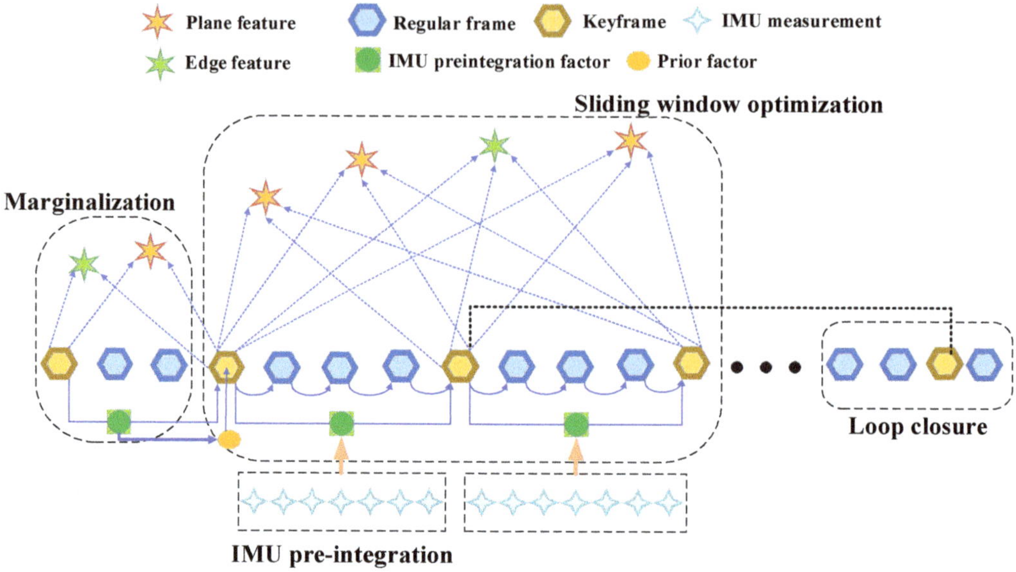

Figure 6. The optimization contains prior term, LIDAR term, IMU term and loop term. Prior term is generated by marginalization. Observations of LIDAR can provide scan-to-scan constraints and IMU pre-integration forms the constraints between keyframes. Loop closure is used for reduce the drift for the long-time running.

5.1. Prior Term

The purpose of marginalization is to bound the computational complexity. For the states out of the sliding window, they cannot be directly throwed away, because it will destroy the original constraint relationship and lose the constraint information. This work selectively marginalizes out x_i from the sliding window via Schur-complement [26], and convert measurements corresponding to marginalized states into the prior.

5.2. LIDAR Term

Through the previous scan-to-map calculation of each scan (see Section 4.3), The LIDAR state variation between two adjacent frames is added into the graph optimization model as scan-to-scan constraint.

$$L\left(z_{h_j}^{b_i}, \chi\right) = \Delta T_{ij} = T_i^T T_j \tag{20}$$

This work assumes that j and i are the current and previous frame, respectively. This term can inhibit the accumulation of cumulative errors over a long time.

5.3. IMU Term

When IMU measurements are available, the residual between two continuous frames can be calculated, and the residual is defined as:

$$\varkappa\left(\hat{Z}_{b_j}^{b_i}, \chi\right) = \begin{bmatrix} \delta\alpha_{b_j}^{b_i} \\ \delta\beta_{b_j}^{b_i} \\ \delta\theta_{b_j}^{b_i} \\ \delta b^a \\ \delta b^\omega \end{bmatrix} = \begin{bmatrix} R_W^{b_i}\left(p_{b_j}^W - p_{b_i}^W - v_i^W \Delta t + \frac{1}{2} g^W \Delta t^2\right) - \hat{\alpha}_{b_j}^{b_i} \\ R_W^{b_i}\left(v_j^W - v_i^W \Delta t + g^W \Delta t\right) - \hat{\beta}_{b_j}^{b_i} \\ 2\left[\left(\hat{q}_{b_j}^{b_i}\right)^{-1} \otimes \left(q_{b_i}^{\omega}\right)^{-1} \otimes q_{b_j}^{\omega}\right]_{xyz} \\ b_j^a - b_i^a \\ b_j^\omega - b_i^\omega \end{bmatrix} \tag{21}$$

where $[\bullet]_{xyz}$ is the imaginary part of a quaternion. $\hat{\alpha}^{b_i}_{b_j}$, $\hat{\beta}^{b_i}_{b_j}$ and $\hat{q}^{b_i}_{b_j}$ are the pre-integration of position, velocity, and rotation between j and i under the assumption that b^a and b^ω are stable.

5.4. Loop Term

Loop closure is an important step to correct the accumulated error in SLAM system. In this study, the function is realized by distance detection. In the current frame, we search for the distance coordinates of nearby key frames. And frames within the geometric radius 15m can be marked as the candidate loop closure frames.

We select the nearest frame from the candidate frame as the previous key frame T^W_{loop}. Then a certain number of point clouds are found near the previous key frame, which are used for a small local map. In this module, the number of points in optimization model is less, so ICP is used to calculate the relative transformation $T^W_{W'}$ of similar scenes. The residual between the previous key frame and current frame can be obtained:

$$F\left(\hat{z}^{b_{loop}}_{b_j}, \chi\right) = \Delta T_{loop,j} = \left(T^W_{loop}\right)^{-1} T^W_{W'} T^{W'}_j \tag{22}$$

In order not to affect the real-time performance, loop closure detection and mapping for ICP are in another thread.

6. Evaluation

In order to verify our algorithm, we have conducted public dataset experiments and real-word experiments. The proposed algorithm is operated on a laptop which consists of an Intel-i7 CPU and 16G of memory. The operating system is Ubuntu18.04 and ROS Melodic [27]. We use evo [28] to evaluate accuracy. The optimization library we used are GTSAM [29] and Ceres [30].

6.1. Validation of M2DGR Datasets

M2DGR [31] datasets were recorded using ground robots. As shown in Figure 7, A HDL 32E Velodyne LiDAR (labeled 3 in the figure) was used to scan the surrounding environment and obtain the 3D point cloud. The IMU device is Handsfree A9 (labeled 5 in the figure), which is a 9-axis sensor. In outdoors, the satellite visibility is good so that the GNSS-RTK suite (labeled 4 in the figure) outputs high-precision ground truth. For indoor environments, the ground truth trajectories are recorded with a motion-capture system which consists of twelve highspeed tracking cameras. The spatial relationship among different sensors have been calibrated.

In order to test the robustness of our algorithm, we adopt tests of different scenarios, and the data information is shown in the Table 1. Street_01, 04, 07, 10 are collected on the street. In the street dataset, there are buildings discontinuously. The structured environment has rich geometric feature information. However, the switching of unstructured scene has the unpredictable influence on LIDAR odometry. Various weeds, leaves and other environmental factors affect the positioning accuracy in Street_04 around the lawn. Loop is set for loop closure detection, which is important for the validation in back-end graph optimization, and the motion state of zigzag brings challenges for interframe motion estimation. When the running time is longer than 500 s, we think that it is long-term to test for stability and robustness. Gate_02 is collected around the large circular gate. It is easy to satisfy the loopback condition. The ground robot is always rotating in Circle_02 scene, which is difficult for the feature matching, and Hall_05 is collected for indoor environment. There is a large amount of overlap and structured feature during the experiment.

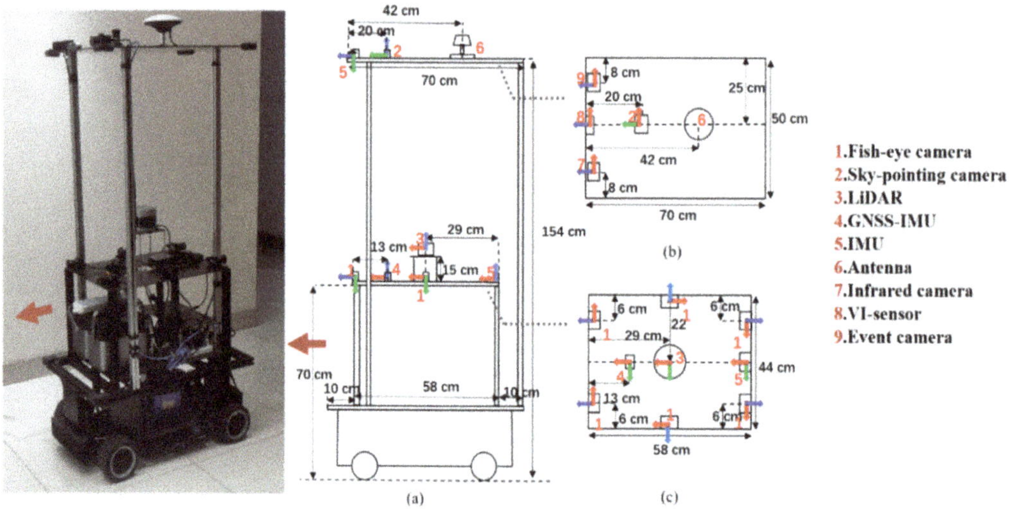

Figure 7. The sensor suite of M2DGR. Sensors are strictly calibrated and using the same time stamps.

Table 1. The dataset contains all kinds of scene.

	Street_01	Street_04	Street_07	Street_10	Gate_02	Circle_02	Hall_05
Durations/s	1028	858	929	910	327	244	402
Description of features	Street and buildings, zigzag, long-term	Around lawn, loop back, long-term	Zigzag, long-term	Zigzag, long-term	Loop back, around gate	Circle, rotation	Indoor, large overlap

We contrast our system with ALOAM, LeGo-LOAM and LIO-SAM. ALOAM only depends on LIDAR; the core of the algorithm is the same as LOAM, but it is achieved according to Ceres for the code readability. LeGo-LOAM uses IMU data to help remove motion distortions from point clouds. LIO-SAM is a tightly coupled LIDAR and IMU approach, but there is no front-end processing of point clouds and the traditional solution algorithm in scan-to-map is more time consuming.

We choose some typical scenarios such as zigzag, rotation and loop. In these cases, point cloud mismatching often occurs due to the violent motion of the carrier. The long-term run is to verify the elimination of the accumulated error of the LIO system.

In Table 2, the bold and italic values indicate the minimum error. Seven groups of experiments prove that our algorithm improves the accuracy in most scenarios. Especially in a scene such as a street. In the sequence "Circle_02", our LIO system has a higher error than LeGo-LOAM. That is because "Circle_02" is collected in a fixed scene and the ground robot is always rotating. This motion state has slightly bad effect on IMU pre-integration. Other than "Circle_02", our system benefits from the tightly couple of inertial and LIDAR information. In the sequence "Gate_02" and "Hall_05", the four algorithms perform equally well. These two scenes are simple and rich in structural features. However, in the street sequence, our algorithm can greatly improve the performance.

Table 2. Absolute trajectory error (ATE) RMSE (m) of the four algorithms in seven datasets.

	Street_01	Street_04	Street_07	Street_10	Gate_02	Circle_02	Hall_05
ALOAM	7.661	3.582	27.590	22.075	0.361	1.391	1.029
LeGo-LOAM	3.269	1.193	14.583	31.024	0.485	***0.288***	1.034
LIO-SAM	6.390	1.133	4.693	2.569	0.326	0.618	1.053
OURS	***1.362***	***0.836***	***1.579***	***1.479***	***0.313***	0.409	***0.980***

6.1.1. Positioning Performance Analysis

Street_01 is chosen for our analysis. Figure 8 shows the trajectories of the four algorithms in street_01 in X-Y plot. The accumulative error of four algorithms can be obtained from the detail diagram. Our system makes reasonable use of feature information, which effectively improve the accuracy of point cloud matching. Planar points are extracted from ground surface feature information, and edge points are extracted from rod feature information. We notice that ALOAM, LeGo-LOAM and LIO-SAM will drift and have a large deviation when it comes to turning. However, in this system, the point cloud registration has been greatly improved.

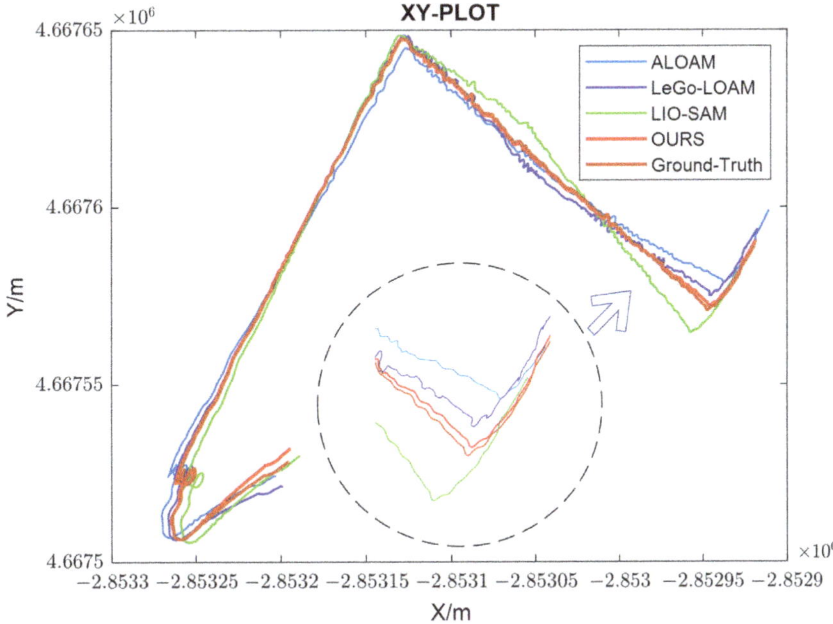

Figure 8. The trajectories of the four algorithms on street_01. Our trajectory is closest to the ground truth. The detail diagram is near the end of this test.

More detailed evaluations have been conducted. Figure 9 shows absolute trajectory error variation for 4 algorithms in street_01. ALOAM have largest error without IMU measurements, and its scan-to-scan method has a bad influence, which is easy for providing inaccurate information in the scan-to-map module. LeGo-LOAM applies a two-step scan-to-scan method, which is beneficial for improving efficiency, but it still introduces much error. Moreover, without IMU constraint, loose-coupled LIO system such as LeGo-LOAM cannot adequately make use of sensor observation information. In LIO-SAM, there is no point cloud preprocessing section. Lots of unstable observations also have bad effects on point cloud registration.

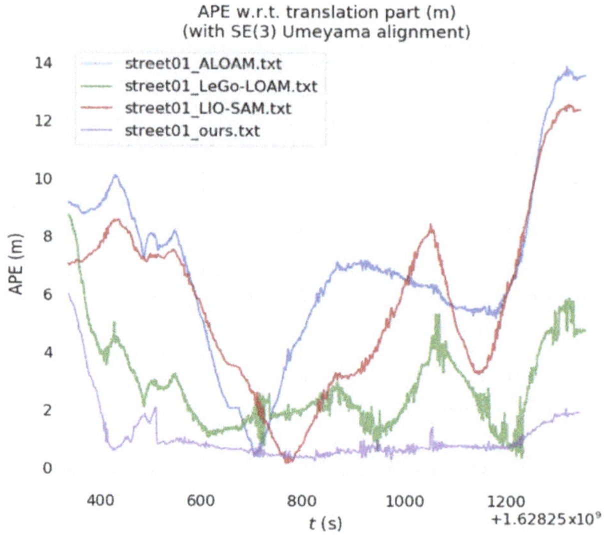

Figure 9. Illustration of the absolute trajectory error variation. Our system has been kept a low level.

This work not only tests RMSE in Table 2. Figure 10 displays each evaluation parameter. Our system has a good performance in different indicators. Figure 11 is the box diagram, which is used to display the dispersion of a set of data. The system also has the lowest deviation.

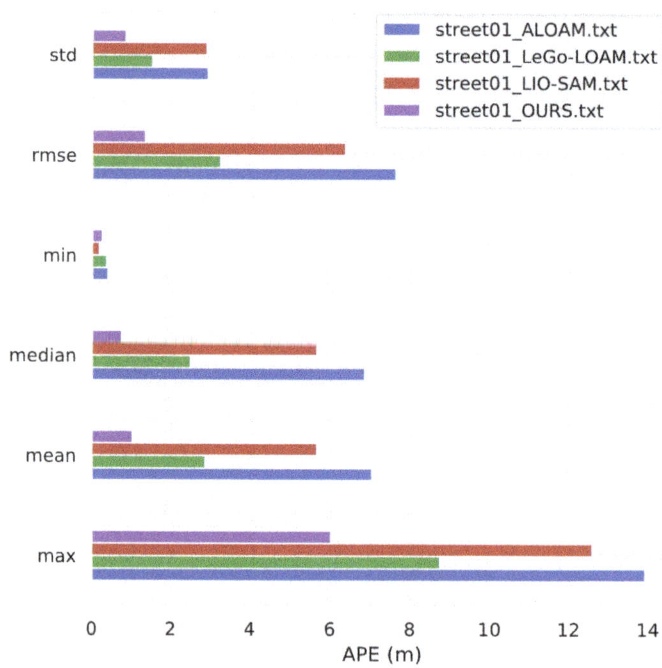

Figure 10. Illustration of A variety of indicators. The superiority of our algorithm can be concluded from the statistical graphs.

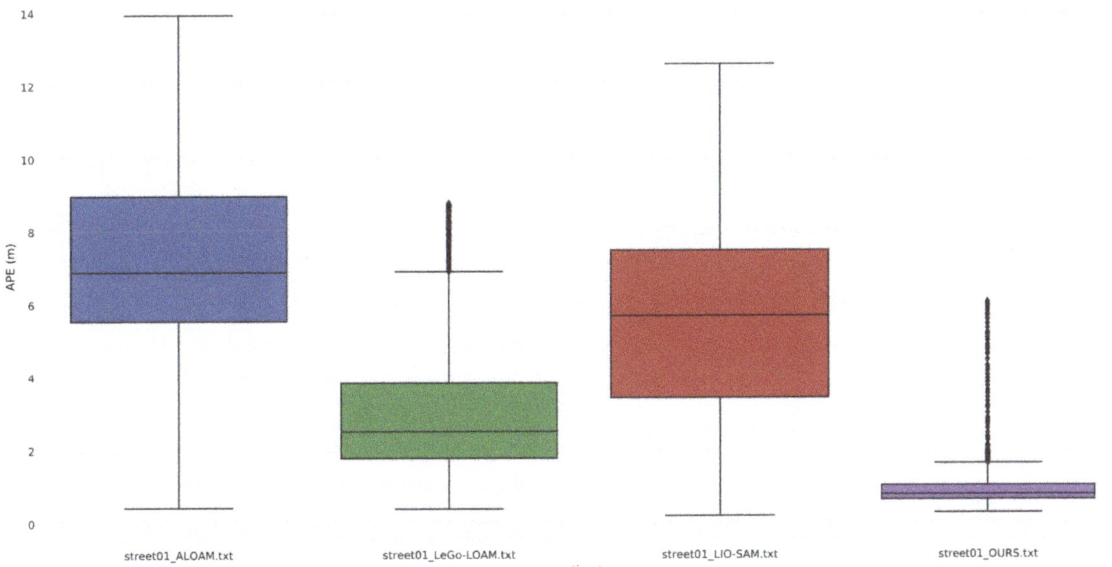

Figure 11. Box diagram which is used to reflect the characteristics of distribution of data.

Four algorithms have been compared. We now analyze the difference between our system and ground truth. Figure 12 is the display diagram of trajectory and truth value in the X-Y plane.

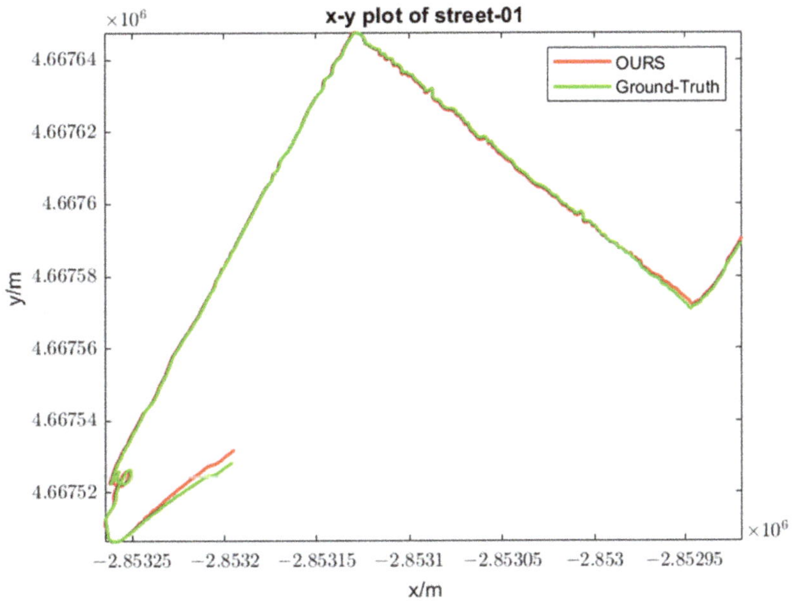

Figure 12. The detailed trajectories of ground truth and our system in street_01. They are aligned well.

Figure 13 shows error changes in three directions throughout the period. We can see that the system has an obvious deviation at the start time. That is because the optimization process takes time to converge and correct. At the same time, local map takes time to build.

After a long run, the error is remained low in our system. It verifies the robustness and high precision of the work.

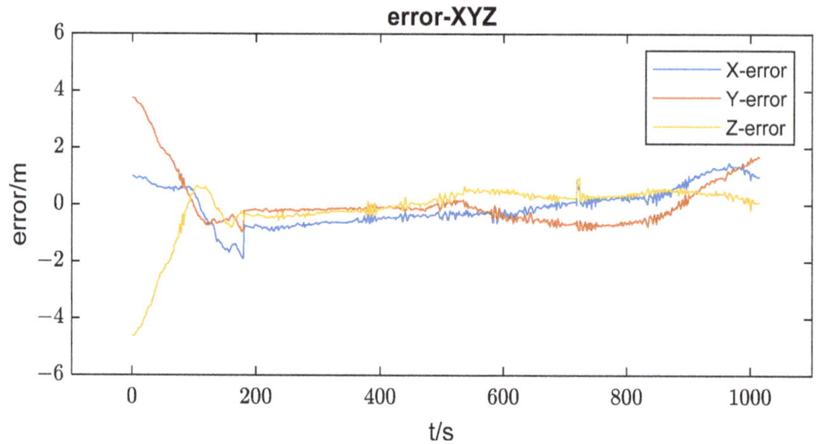

Figure 13. Error changes in three directions. The initial error is a little larger, after that the error is small.

6.1.2. Runtime Performance Analysis

Meanwhile, our experiments proved that the time consumption of our scan-to-map module is significantly reduced from Table 3. Four algorithms have this module which is the most time-consuming. So, we choose the cost time of this module for comparison. The bold and italic values indicate the minimum time consumption. We can see the obvious advantage of our algorithm.

Table 3. The time consumption (ms) in seven datasets. All are recorded in the same platform.

	Street_01	Street_04	Street_07	Street_10	Gate_02	Circle_02	Hall_05
ALOAM	295.875	250.093	309.164	251.751	230.516	294.688	106.374
LeGo-LOAM	132.986	93.184	149.694	124.700	112.126	122.932	57.938
LIO-SAM	61.900	43.461	83.957	76.541	57.518	90.450	21.308
OURS	***37.600***	***27.427***	***66.093***	***50.888***	***35.215***	***34.628***	***14.296***

LOAM and LeGo-LOAM use a scan-to-scan match to provide odometry, which means using the current scan and last scan to do the scan matching, and the result offers an initial guess for mapping. LIO-SAM and our system use IMU pre-integration, which is of high frequency, and we use the back-end result to suppress IMU drift. Even more, thanks to the edge points extracted from the rod-shaped information and the planar points extracted from the ground surface information, many outliers are not in the operation. Accuracy and speed are greatly improved.

We still choose street_01 for analysis. Table 4 shows the number of frames four systems processed.

Table 4. The scan-to-map frames in street_01.

	ALOAM	LeGo-LOAM	LIO-SAM	OURS
Scan-to-map frames	2788	2566	5128	5133

Figure 14 shows the processing time of each frame. We can clearly see the lowest cost time of our system. In ALOAM, the mapping module uses the global map and map maintenance is time-consuming. LeGo-LOAM and LIO-SAM are the same, which apply Levenberg–Marquardt algorithm [32] of 30 iterations for optimization. Our algorithm uses faster and more accurate graph optimization model to solve the scan-to-map module (see Sections 4.2 and 4.3).

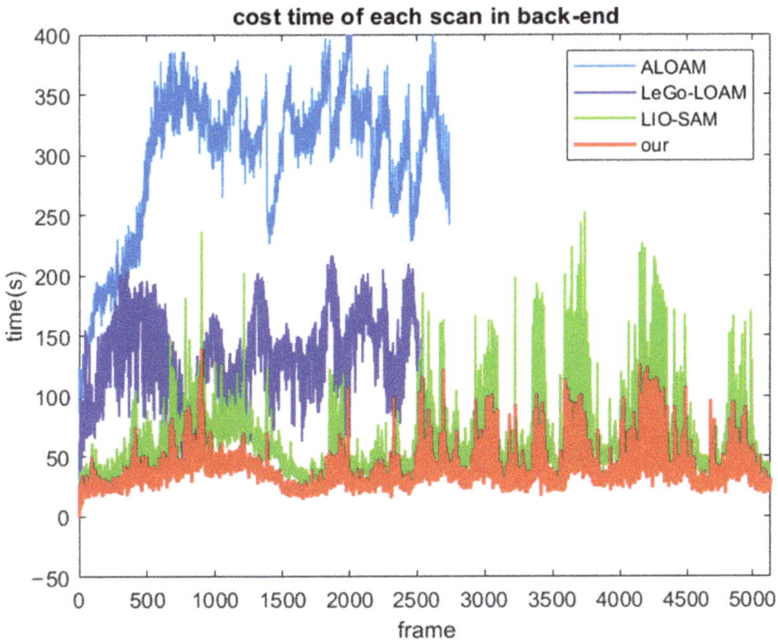

Figure 14. Processing time of each scan. The red is obviously lower than other three.

The computationally-efficient system is meaningful for mobile terminal and other platforms with limited computing resources.

6.2. Validation of Our Datasets

To further test our system, LIDAR has less beams and IMU is of different quality. We set up a sensor suite composed of a VLP16 Velodyne and an ADIS16488 IMU (see Figure 15). Sensors have hardware time synchronization because of GPS pulse per second (PPS). RTK/IMU combined navigation results are used as the truth value, which is after NovAtel Inertial Explorer software post-processing. Our aim is to prove the versatility of our algorithm. We pick two typical scenarios. One is in a campus (dataset_01) and the other is on a city road (datasetet_02).

In dataset_01, (see Figure 16) the speed of our car is about 6 m/s. There is rich feature, but there is accumulated error in long-term run. Various weeds, leaves and other environmental factors affect the positioning accuracy.

Dataset_02 was collected in the wide urban road (see Figure 17) and the speed of our car is about 14 m/s. It contains a large number of buildings. Dynamic objects will affect the point cloud matching accuracy. The density of point cloud in open space is small and it is difficult to have a good performance on localization.

(**a**) (**b**)

Figure 15. Our sensor suite. All have hardware time synchronization (**a**) IMU is in the LIDAR below. Integrated navigation device is used to gain ground truth. (**b**) The device is mounted on top of the vehicle.

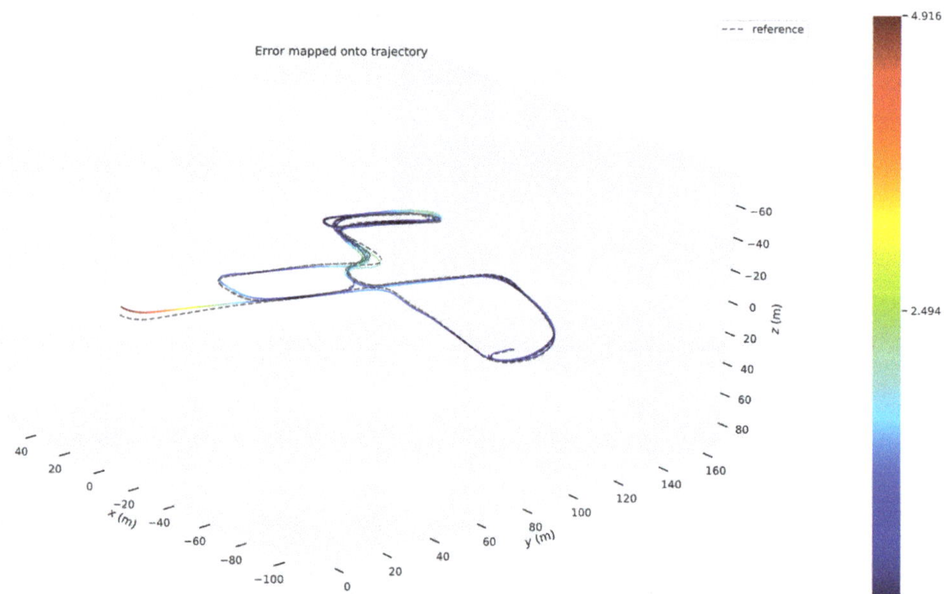

(**a**)

Figure 16. *Cont.*

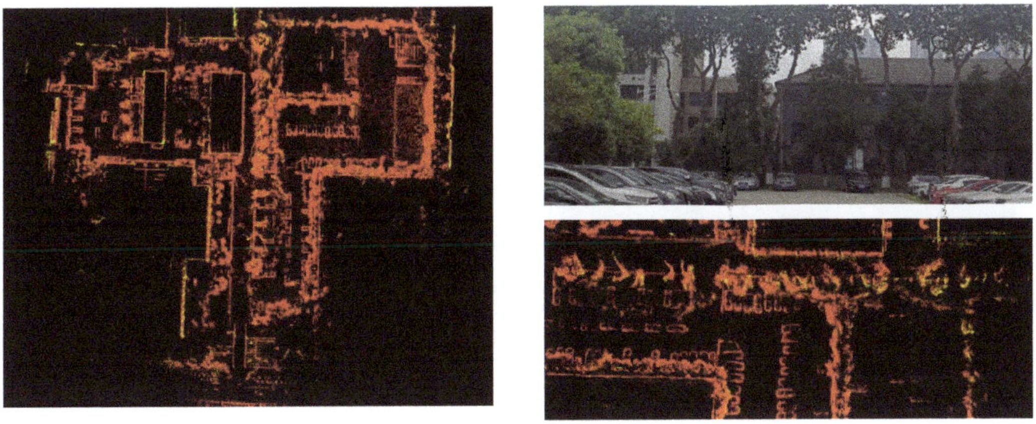

(**b**) (**c**)

Figure 16. Trajectory and mapping are generated by our system. (**a**) Our trajectory and ground truth. Different colors represent the error values. (**b**) The mapping result is rendered with LIDAR intensity value from the top view during the positioning process. (**c**) The top panel is the specific real-word environment picked out of the whole trajectory. The bottom panel shows the detail from LIDAR mapping correspond to the top panel.

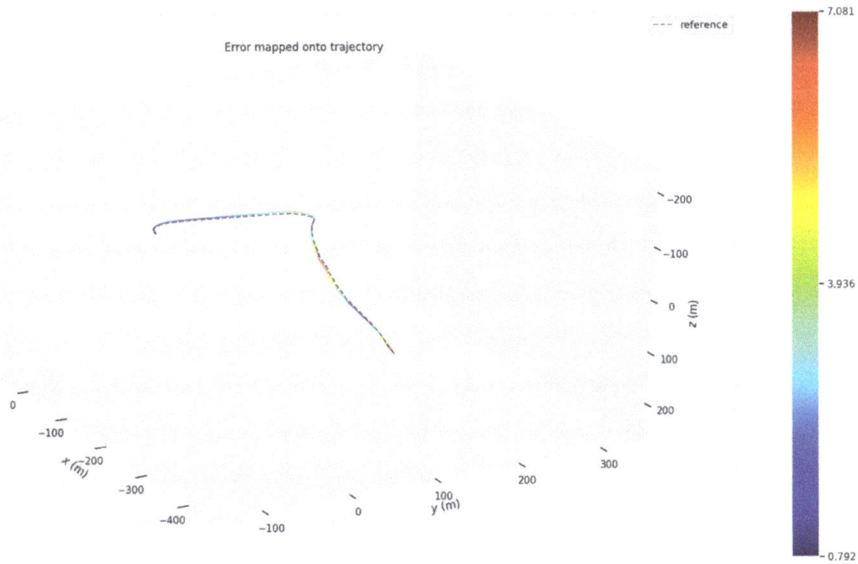

(**a**)

Figure 17. *Cont.*

(**b**) (**c**)

Figure 17. Trajectory and mapping in dataset_02. (**a**) Our trajectory and ground truth in the urban road. (**b**) The mapping result is rendered with LIDAR intensity value from bird-eye view. (**c**) The road scene.

In feature-rich areas(dataset_01), we can conclude that LIDAR will have good performance than the wide-open spaces(dataset_02) from the Table 5. The bold and italic values indicate the minimum error. Compared with the LIO-SAM, the RMSE in the study increases by 24.2% in dataset_01 and 25.0% in dataset_02.

Table 5. Absolute trajectory error (ATE) RMSE (m) in our real-word experiments.

	Durations/s	Description of Features	A-LOAM	LeGo-LOAM	LIO-SAM	OURS
dataset_01	678	campus	4.040	1.264	1.272	***0.964***
dataset_02	240	wide rode	6.391	5.933	4.205	***3.152***

In the Figures 16a and 17a, they show the corresponding trajectories of the two datasets and different colors represent the error values. In the Figures 16b and 17b, the global reconstruction of the two scenes is built. Due to the multiple constraints of the back-end optimization, we obtain a globally consistent point cloud map. According to Equation (19), the robustness and reliability of the map can be guaranteed. The map shows the structural details in the bottom panel of Figure 16c. We can clearly see the cars and the trunk of the tree in the dataset in bird-eye view. In the Figure 17c, the wide roads have great influence for LIDAR slam (see Table 5). However, our algorithm can also reduce the error.

The time performance is consistent with the M2DGR dataset analysis (see Section 6.1.2). In dataset_01, due to the features' richness in campus scenes, the feature information relationship in scan-to-map needs much time for calculation. However, the average time consumption is 64.379 ms, which can satisfy the real-time requirement (LIDAR is 10 HZ sampling frequency), and the average time consumption is 25.874 ms in the road test in the dataset_02, which is computationally efficient. Our system can achieve a good estimation result with less time cost.

7. Conclusions and Future Perspectives

According to the datasets and our own data experiments, compared to LIDAR only positioning (ALOAM), the positioning accuracy and robustness is significantly improved. Then only IMU data helps the point cloud to remove distortion (LeGo-LOAM), the tightly coupled LIO has lower drift, and compared to LIO-SAM, segmentation and clustering are used to mark feature information. The point cloud matching is more accurate and the runtime of scan-to-map module is much less.

In this paper, we propose an improved LIO system. Firstly, it makes reasonable use of the feature information of point cloud and effectively improves the accuracy of point cloud matching. Point cloud registration is carried out after marking rod-shaped and planar feature information which is different from the existing LIDAR-inertial integration scheme. The optimized edge points and planar points extraction modes reduce the computation of scan-to-map and improve the real-time performance. Secondly, prediction of IMU odometry and correction of LIDAR odometry improve the accuracy and frequency of the mapping module, which is inspired by LIO-SAM. Comparing this to the front-end odometry in traditional scan-to-scan mode, the tightly coupled mode of system greatly improves the performance of LIO. Thirdly, the scan-to-map based on the graph optimization model is of great significance to speed up the solution and decrease error. Therefore, the system does not apply the Levenberg–Marquardt algorithm, which is adopted in Lego-LOAM and LIO-SAM. Fourthly, the robust back-end optimization system including effective loop closure suppress the cumulative drift of LIO odometry, and IMU measurements residuals add more constraints information between IMU and LIDAR measurements compared to LIO-SAM. The optimization mode based on sliding window ensure full use of sensors information under real-time conditions. Experiments show that the real-time performance and accuracy of our algorithm exceed that of most state-of-the-art systems in various typical environments.

It can be seen from Table 2 that the positioning accuracy (RMSE) can be improved by 25–78% (the average increment is 64.45%) in the M2DGR street datasets compared to the current tightly coupled LIDAR SLAM algorithms (LIO-SAM). After optimizing the extraction mode of edge points and planar points, our system processes more frames and takes less time on average, effectively improving real-time performance. In our actual scene datasets, the RMSE in the study increases by 24.4% in dataset_01 and 25.0% in dataset_02.

We draw a conclusion that we propose the low drift and high real-time LIDAR-inertial positioning and mapping system, which is of great importance in indoor locating and other GNSS occlusion area. At the same time, it can provide high precision point cloud image for scene understanding in automatic system. For the back-end optimization framework, we can easily add other measurements such as GNSS for global restriction.

In the future, we noticed that it is necessary to improve the initialization process to reduce initial error. It is very important to judge the rod-shaped feature information and the planar feature information in the research process of this paper. This work gets thresholds according to experience temporarily. We will focus on online threshold estimation and adaptive threshold selection. Also, it is worth mentioning that LIO system is prone to Z direction drift in the large scene. Then more constraints will be introduced to suppress drift in our next step. Furthermore, according to the recent study [33–36], the positioning and mapping system based on solid state LIDAR can significantly reduce the hardware cost. Therefore, the research of solid-state LIDAR- inertial system is worth exploring.

Author Contributions: Conceptualization, H.L., S.P., W.G. and C.M.; methodology, H.L.; software, F.J. and C.M.; validation, S.P. and F.J.; formal analysis, H.L. and C.M.; investigation, W.G. and F.J.; resources, S.P., W.G. and H.L.; writing—original draft preparation, H.L. and X.L.; writing—review and editing, W.G. and X.L.; supervision, S.P. and W.G. All authors have read and agreed to the published version of the manuscript.

Funding: This research study was funded by the National Key Research and Development Program of China (No. 2021YFB3900804), the Research Fund of Ministry of Education of China and China Mobile, (No. MCM20200J01) and the Fundamental Research Funds for the Central Universities (No. 2242021R41134).

Data Availability Statement: Not applicable.

Conflicts of Interest: The authors declare no conflict of interest.

References

1. Cadena, C.; Carlone, L.; Carrillo, H.; Latif, Y.; Scaramuzza, D.; Neira, J.; Reid, I.; Leonard, J.J. Past, Present, and Future of Simultaneous Localization and Mapping: Toward the Robust-Perception Age. *IEEE Trans. Robot.* **2016**, *32*, 1309–1332. [CrossRef]
2. Zhang, J.; Singh, S. Low-drift and real-time lidar odometry and mapping. *Auton. Robot.* **2017**, *41*, 401–416. [CrossRef]
3. Besl, P.J.; McKay, N.D. A method for registration of 3-D shapes. *IEEE Trans. Pattern Anal. Mach. Intell.* **1992**, *14*, 239–256. [CrossRef]
4. Geiger, A.; Lenz, P.; Urtasun, R. Are We Ready for Autonomous Driving? The KITTI Vision Benchmark Suite. In Proceedings of the 2012 IEEE Conference on Computer Vision and Pattern Recognition, Providence, RI, USA, 16–21 June 2012.
5. Wang, H.; Wang, C.; Chen, C.L.; Xie, L. F-LOAM: Fast LiDAR Odometry and Mapping. In Proceedings of the 2021 IEEE/RSJ International Conference on Intelligent Robots and Systems (IROS), Prague, Czech Republic, 27 September–1 October 2021; pp. 4390–4396.
6. Liu, T.; Wang, Y.; Niu, X.; Chang, L.; Zhang, T.; Liu, J. LiDAR Odometry by Deep Learning-Based Feature Points with Two-Step Pose Estimation. *Remote Sens.* **2022**, *14*, 2764. [CrossRef]
7. Zhang, J.; Singh, S. Visual-Lidar Odometry and Mapping: Low-Drift, Robust, and Fast. In Proceedings of the 2015 IEEE International Conference on Robotics and Automation (ICRA), Seattle, WA, USA, 26–30 May 2015.
8. Koide, K.; Miura, J.; Menegatti, E. A portable three-dimensional LIDAR-based system for long-term and wide-area people behavior measurement. *Int. J. Adv. Robot. Syst.* **2019**, *16*, 1729881419841532. [CrossRef]
9. Biber, P.; Strasser, W. The normal distributions transform: A new approach to laser scan matching. In Proceedings of the Proceedings 2003 IEEE/RSJ International Conference on Intelligent Robots and Systems (IROS 2003) (Cat. No.03CH37453), Las Vegas, NV, USA, 27–31 October 2003.
10. Shan, T.; Englot, B. LeGO-LOAM: Lightweight and Ground-Optimized Lidar Odometry and Mapping on Variable Terrain. In Proceedings of the 2018 IEEE/RSJ International Conference on Intelligent Robots and Systems (IROS), Madrid, Spain, 1–5 October 2018.
11. Ye, H.; Chen, Y.; Liu, M. Tightly Coupled 3D Lidar Inertial Odometry and Mapping. In Proceedings of the 2019 International Conference on Robotics and Automation (ICRA), Montreal, QC, Canada, 20–24 May 2019.
12. Qin, T.; Li, P.; Shen, S. VINS-Mono: A Robust and Versatile Monocular Visual-Inertial State Estimator. *IEEE Trans. Robot.* **2018**, *34*, 1004–1020. [CrossRef]
13. Qin, C.; Ye, H.; Pranata, C.E.; Han, J.; Zhang, S.; Liu, M. LINS: A Lidar-Inertial State Estimator for Robust and Efficient Navigation. In Proceedings of the 2020 IEEE International Conference on Robotics and Automation (ICRA), Paris, France, 31 May–31 August 2020.
14. Li, K.; Li, M.; Hanebeck, U.D. Towards High-Performance Solid-State-LiDAR-Inertial Odometry and Mapping. *IEEE Robot. Autom. Lett.* **2021**, *6*, 5167–5174. [CrossRef]
15. Shan, T.; Englot, B.; Meyers, D.; Wang, W.; Ratti, C.; Rus, D. LIO-SAM: Tightly-coupled Lidar Inertial Odometry via Smoothing and Mapping. In Proceedings of the 2020 IEEE/RSJ International Conference on Intelligent Robots and Systems (IROS), Las Vegas, NV, USA, 24 October–24 January 2020.
16. Kaess, M.; Ranganathan, A.; Dellaert, F. iSAM: Incremental Smoothing and Mapping. *IEEE Trans. Robot.* **2008**, *24*, 1365–1378. [CrossRef]
17. Zhang, J.; Wen, W.; Huang, F.; Chen, X.; Hsu, L.-T. Coarse-to-Fine Loosely-Coupled LiDAR-Inertial Odometry for Urban Positioning and Mapping. *Remote Sens.* **2021**, *13*, 2371. [CrossRef]
18. Douillard, B.; Underwood, J.; Vlaskine, V.; Quadros, A.; Singh, S. A Pipeline for the Segmentation and Classification of 3D Point Clouds. In *Experimental Robotics: The 12th International Symposium on Experimental Robotics*; Khatib, O., Kumar, V., Sukhatme, G., Eds.; Springer: Berlin/Heidelberg, Germany, 2014; pp. 585–600.
19. Himmelsbach, M.; Hundelshausen, F.v.; Wuensche, H.-J. Wuensche Fast segmentation of 3D point clouds for ground vehicles. In Proceedings of the 2010 IEEE Intelligent Vehicles Symposium, La Jolla, CA, USA, 21–24 June 2010.
20. Park, S.; Wang, S.; Lim, H.; Kang, U. Curved-Voxel Clustering for Accurate Segmentation of 3D LiDAR Point Clouds with Real-Time Performance. In Proceedings of the 2019 IEEE/RSJ International Conference on Intelligent Robots and Systems (IROS), Macau, China, 3–8 November 2019; pp. 6459–6464.
21. Chen, Z.; Li, Q.; Li, J.; Zhang, D.; Yu, J.; Yin, Y.; Lv, S.; Liang, A. IMU-Aided Registration of MLS Point Clouds Using Inertial Trajectory Error Model and Least Squares Optimization. *Remote Sens.* **2022**, *14*, 1365. [CrossRef]
22. Bogoslavskyi, I.; Stachniss, C. Fast range image-based segmentation of sparse 3D laser scans for online operation. In Proceedings of the 2016 IEEE/RSJ International Conference on Intelligent Robots and Systems (IROS), Daejeon, Korea, 9–14 October 2016.
23. Forster, C.; Carlone, L.; Dellaert, F.; Scaramuzza, D. On-Manifold Preintegration for Real-Time Visual—Inertial Odometry. *IEEE Trans. Robot.* **2017**, *33*, 1–21. [CrossRef]
24. Bentley, J.L. Multidimensional binary search trees used for associative searching. Commun. *ACM* **1975**, *18*, 509–517. [CrossRef]
25. Barfoot, T.D. *State Estimation for Robotics: A matrix Lie Group Approach*; Cambridge University Press: Cambridge, UK, 2016.
26. Sibley, G.; Matthies, L.; Sukhatme, G. Sliding window filter with application to planetary landing. *J. Field Robot.* **2010**, *27*, 587–608. [CrossRef]

27. Będkowski, J.; Pełka, M.; Majek, K.; Fitri, T.; Naruniec, J. Open source robotic 3D mapping framework with ROS—Robot Operating System, PCL—Point Cloud Library and Cloud Compare. In Proceedings of the 2015 International Conference on Electrical Engineering and Informatics (ICEEI), Denpasar, Indonesia, 10–11 August 2015.
28. Sanfourche, M.; Vittori, V.; le Besnerais, G. evo: A realtime embedded stereo odometry for MAV applications. In Proceedings of the 2013 IEEE/RSJ International Conference on Intelligent Robots and Systems, Tokyo, Japan, 3–7 November 2013; Available online: https://github.com/MichaelGrupp/evo (accessed on 25 May 2022).
29. Kaess, M.; Johannsson, H.; Roberts, R.; Ila, V.; Leonard, J.; Dellaert, F. iSAM2: Incremental smoothing and mapping with fluid relinearization and incremental variable reordering. In Proceedings of the 2011 IEEE International Conference on Robotics and Automation, Shanghai, China, 9–13 May 2011.
30. Agarwal, S.; Mierle, K. Ceres Solver. Available online: http://ceres-solver.org (accessed on 25 May 2022).
31. Yin, J.; Li, A.; Li, T.; Yu, W.; Zou, D. M2DGR: A Multi-Sensor and Multi-Scenario SLAM Dataset for Ground Robots. *IEEE Robot. Autom. Lett.* **2022**, *7*, 2266–2273. Available online: https://github.com/SJTU-ViSYS/M2DGR (accessed on 25 May 2022). [CrossRef]
32. Moré, J.J. *The Levenberg-Marquardt Algorithm: Implementation and Theory*; Springer: Berlin/Heidelberg, Germany, 1977.
33. Bai, C.; Xiao, T.; Chen, Y.; Wang, H.; Zhang, F.; Gao, X. Faster-LIO: Lightweight Tightly Coupled Lidar-Inertial Odometry Using Parallel Sparse Incremental Voxels. *IEEE Robot. Autom. Lett.* **2022**, *7*, 4861–4868. [CrossRef]
34. Xu, W.; Zhang, F. FAST-LIO: A Fast, Robust LiDAR-Inertial Odometry Package by Tightly-Coupled Iterated Kalman Filter. *IEEE Robot. Autom. Lett.* **2021**, *6*, 3317–3324. [CrossRef]
35. Xu, W.; Cai, Y.; He, D.; Lin, J.; Zhang, F. FAST-LIO2: Fast Direct LiDAR-Inertial Odometry. *IEEE Trans. Robot.* **2022**, *38*, 2053–2073. [CrossRef]
36. Lin, J.; Zhang, F. Loam livox: A fast, robust, high-precision LiDAR odometry and mapping package for LiDARs of small FoV. In Proceedings of the 2020 IEEE International Conference on Robotics and Automation (ICRA), Paris, France, 31 May–31 August 2020.

remote sensing

Article

FPS: Fast Path Planner Algorithm Based on Sparse Visibility Graph and Bidirectional Breadth-First Search

Qunzhao Li [1], Fei Xie [1,*], Jing Zhao [2], Bing Xu [3], Jiquan Yang [1], Xixiang Liu [4] and Hongbo Suo [5]

[1] School of Electrical and Automation Engineering, Nanjing Normal University, Nanjing 210023, China; 201846063@njnu.edu.cn (Q.L.); 63047@njnu.edu.cn (J.Y.)
[2] College of Automation & College of Artificial Intelligence, Nanjing University of Posts and Telecommunications, Nanjing 210023, China; zhaojing@njupt.edu.cn
[3] Department of Aeronautical and Aviation Engineering, The Hong Kong Polytechnic University, Hong Kong, China; pbing.xu@polyu.edu.hk
[4] College of Instrument Science & Engineering, Southeast University, Nanjing 210096, China; 101010902@seu.edu.cn
[5] Nanjing Zhongke Raycham Laser Technology Co., Ltd., Nanjing 210023, China; suohongbo@raycham.com
* Correspondence: xiefei@njnu.edu.cn

Abstract: The majority of planning algorithms used are based on the occupancy grid maps, but in complicated situations, the occupancy grid maps have a significant search overhead. This paper proposed a path planner based on the visibility graph (v-graph) for the mobile robot that uses sparse methods to speed up and simplify the construction of the v-graph. Firstly, the complementary grid framework is designed to reduce graph updating iteration costs during the data collection process in each data frame. Secondly, a filter approach based on the edge length and the number of vertices of the obstacle contour is proposed to reduce redundant nodes and edges in the v-graph. Thirdly, a bidirectional breadth-first search is combined into the path searching process in the proposed fast path planner algorithm in order to reduce the waste of exploring space. Finally, the simulation results indicate that the proposed sparse v-graph planner can significantly improve the efficiency of building the v-graph and reduce the time of path search. In highly convoluted unknown or partially known environments, our method is 40% faster than the FAR Planner and produces paths 25% shorter than it. Moreover, the physical experiment shows that the proposed path planner is faster than the FAR Planner in both the v-graph update process and laser process. The method proposed in this paper performs faster when seeking paths than the conventional method based on the occupancy grid.

Keywords: visibility graph; computational geometry; path planning; mapping

Citation: Li, Q.; Xie, F.; Zhao, J.; Xu, B.; Yang, J.; Liu, X.; Suo, H. FPS: Fast Path Planner Algorithm Based on Sparse Visibility Graph and Bidirectional Breadth-First Search. *Remote Sens.* **2022**, *14*, 3720. https://doi.org/10.3390/rs14153720

Academic Editor: Andrzej Stateczny

Received: 16 June 2022
Accepted: 31 July 2022
Published: 3 August 2022

1. Introduction

With the popularity of the robotics industry, simultaneous localization and mapping (SLAM) technology has developed rapidly. SLAM technology can be divided into three categories, i.e., LiDAR-SLAM [1–4], visual-SLAM [5–8] and LiDAR fusion visual SLAM [9–11], and it is widely used for robot navigation tasks. In the application of robot navigation, a by-product of SLAM is the map, including metric and topological maps. The metric map emphasizes accurately representing the positional relationships of objects, while the topological map emphasizes the relationships between map elements.

In smaller spaces, such as corridors and houses, occupancy grid maps [12] are preferred over topological maps. The topology maps, on the other hand, are more appropriate for path planning in large areas where the occupancy grid maps are computationally expensive. In this paper, the visibility graph [13], a topology-based type of map, will be constructed for route planning and navigation.

Path planning has been an emerging trend in research nowadays to cater to the needs of autonomous systems. The visibility graph (v-graph) is an efficient map representation for path planning, which allows the robot to move from one node to another, but it has some drawbacks. Firstly, it is not only hard to map the visibility graph completely in the 3D world [14], but it is also difficult to extract the outlines of obstacles. Secondly, when the number of nodes in the graph increases, the associated edges will be doubled, which causes increased computational cost in terrain where complex obstacles exist [15]. In this case, it is necessary to simplify complex obstacles to reduce vertices and edges for path planning.

Most existing v-graph generation methods [16–21] store the pointcloud explored by the LiDAR into the local grid and then perform plane mapping to extract the vertices of the polygon. However, these algorithms, such as [17–20], face the following problems. Firstly, if the local grid is too sparse, the sampling accuracy will be decreased. On the other hand, if the local grid is dense, the shape of a polygon will be more accurately determined, but the amount of calculation will increase dramatically. Secondly, the quantity of vertices and edges affects how complicated the v-graph-based path search algorithm is. There will be a lot of redundant vertices and edges in maps with a lot of intricate barriers, which makes path search a time-consuming task. Finally, the large number of vertices and edges in the v-graph slows down its traversal speed, which leads to a decrease in the maintenance speed of the v-graph. Although WonheeLee et al. [21] proposed a v-graph-based obstacle avoidance strategy, they did not address the issue of dense v-graph in complicated scenes. A sparse v-graph-based path planner is proposed as a solution to these issues, which lowers the cost of v-graph maintenance, increases the effectiveness of v-graph building, and decreases space waste during the path search. Overall, the main contributions of the paper are summarized as follows:

- Compared to existing methods for storing a laser pointcloud, this paper proposes a complementary holed structure for iteratively updating the local grid. Basically, only half of the pointcloud data needs to be processed in each data frame to update the map. The pointcloud data are subjected to image blurring after planar mapping, and then key vertices are extracted from the blurred image.
- For obstacles with complex contours, this paper proposes a filtering method based on the edge length and the number of vertices of the obstacle contour. The method effectively reduces the number of vertices in the v-graph and the maintenance cost of the v-graph by performing vertex filtering on large complex obstacles. Since the v-graph and the path search algorithm are tightly coupled, the efficiency of the path search algorithm will also be improved.
- A bidirectional breadth-first search algorithm was introduced since exploring uncharted territory requires a lot of search space. In this paper, the edge between the goal point and the existing vertices in the v-graph is established by geometric checking. Therefore, the bidirectional breadth-first search algorithm could reduce the waste of exploration space in navigation.

2. Related Work

The current mainstream of path planning research is divided into the following categories: search-based, sampling (probability)-based, genetic algorithm (GA)-based, and learning-based. According to the planning results, it is further divided into complete planning algorithms and probabilistic complete planning algorithms.

Search-based planning methods: These methods mainly include Dijkstra [22] and its variants, such as A* [23], D* [24], etc. The Dijkstra and A* algorithms are often used to search on discrete grids. Such algorithms are re-initialized for each search cycle, thus taking a long time to plan routes. An incremental version of the Dijkstra-derived algorithm was proposed to reduce re-planning time by adjusting the local information to the planning result in the previous cycle. However, similar to D* Lite [25], when encountering complex environments, the computational load of the incremental algorithm to re-evaluate the current environment is even greater than that of A* without increment. Many improved A* algorithms have been proposed to decrease the memory space and achieve a better trajectory [26,27], but they are all based on the occupancy grids and thus cannot avoid the shortcomings of the occupancy grids.

Random sampling planning methods: These methods mainly include rapidly expanding random tree (RRT)[28] and a series of its variants, such as RRT* [29], informed RRT* [30], and RRT-Connect [31]. These methods are designed for a known environment. Some algorithms derived from RRT are used for planning in an unknown or partially known environment. These methods must perform maintenance or regeneration of random trees frequently to account for newly observed environments.

Genetic algorithm methods (GA): This algorithm performs path planning through crossover and mutation of chromosomes. Tu et al. [32] proposed a method for generating collision-free paths based on GA. Chen et al. [33] proposed an improved GA to minimize the total distance of the UAV. The method in [34] combined deep learning and GA to design a followable path for multi-robots. Moreover, there are many different improvements to GA [35–37] to achieve better results. In the genetic algorithm, the choice of parameters such as crossover rate and mutation rate has a significant impact on the quality of the solution, but these values are typically chosen based on experience.

Planning by deep learning: These methods [38–40] require a large number of ground-truth labels for training. For example, the essence of training based on deep learning is to encode and decode environmental information. During testing, these methods can handle scenarios similar to the training environment; the essence of training based on reinforcement learning is to encode and map the information of the state of motion every time. During testing, the results are output according to the encoded motion state information. GuichaoLin et al. [41] proposed a collision-free model based on deep reinforcement learning to allow robots to avoid obstacles. Tutsoy et al. [42] provided a minimum time path for a balancing task through reinforcement learning, and [43] considered the energy consumption to design a general dynamic control model based on deep reinforcement learning. However, learning-based methods are inherently data-driven, and the magnitude of the data limits their ability to scale to different environments.

The majority of the aforementioned path planning techniques, including common search-based algorithms, RRT algorithms, and evolutionary algorithms, are based on occupancy grids. When the scope of the searched scene is large, the drawbacks of occupancy grids will constrain the speed of these algorithms, i.e., as shown in Figure 1, in a 10×10 grid map, the algorithm based on the occupancy grid map traversed 20 grids to reach the goal, while the algorithm based on the v-graph only traversed five vertices of the obstacle. This paper mainly studies the visibility graph-based and path-planning methods. Although the literature [13,44–46] studied the use of the visibility graph for robot navigation, and FAR Planner [19] applied the theoretical part to the exploration and navigation of the actual 3D world, the maintenance of a graph in complex situations is still expensive, and the way-point generated in unknown environments is easy to detour. In order to address these issues, a sparse v-graph path planner is proposed. This path planner enhances the efficiency of v-graph building while decreasing space waste and v-graph maintenance costs.

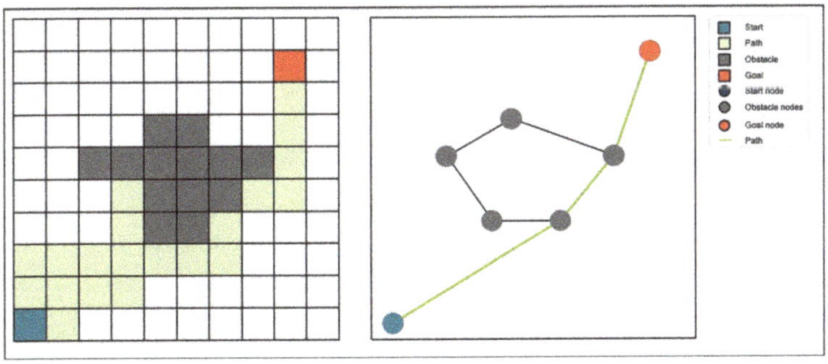

Figure 1. Comparison of the occupancy grids and the visibility graph.

Similar to FAR Planner [19], the pointcloud is extracted from obstacles and mapped into polygons, from which vertices and edges are extracted to construct the v-graph for navigation. The improvement of our approach is that each data frame of pointcloud in the local area does not fully participate in the construction of the global layer. The complementary hole structure for iteratively updating the local grid is used to store the pointcloud information in the current local area, which means that only half of the points in each data frame need to be processed each time. The method continuously updates the pointcloud information on each data frame until a global map is formed. Compared with the original algorithm of FAR Planner, the proposed algorithm can reach the target point within a shorter distance and take less time.

In simulation experiments, the feasibility of the method is evaluated through the simulated physical environment. The environment of the simulation experiment includes medium-scale, complex-scale, and large-scale environments in the Autonomous Exploration Development Environment provided by CMU [47], and medium-scale indoor environments and complex large-scale indoor environments provided by Matterport3D [48]. In the physical experiment, the LiDAR, and an Inertial Measurement Unit (IMU) are coupled to generate state estimation of the mobile robot [4], and the proposed replacednavigationnavgiation algorithm will be tested in a real garage.

3. Sparse Visibility Graph-Based Path Planner

Define $\mathcal{Q} \subset \mathbb{R}^3$ as the robot navigation space, and $\mathcal{S} \subset \mathcal{Q}$ as the sensor data from obstacles. A down-sample strategy is used to update and maintain the v-graph, denoted as $\mathcal{G}$, and the grid to store pointcloud is denoted as $\mathcal{L}$. Define the position of robot as $P_{robot} \in \mathcal{Q}$, the goal $P_{goal} \in \mathcal{Q}$.

The flow chart of the path planner proposed in the paper is shown in Figure 2. And the process consists of three parts: (1) generating the geometric contours of obstacles by LiDAR-to-plane mapping; (2) aggregating and simplifying complex obstacle information to maintain the v-graph at a low cost; and (3) searching for nodes and edges to generate the path from the start point to the goal through the v-graph.

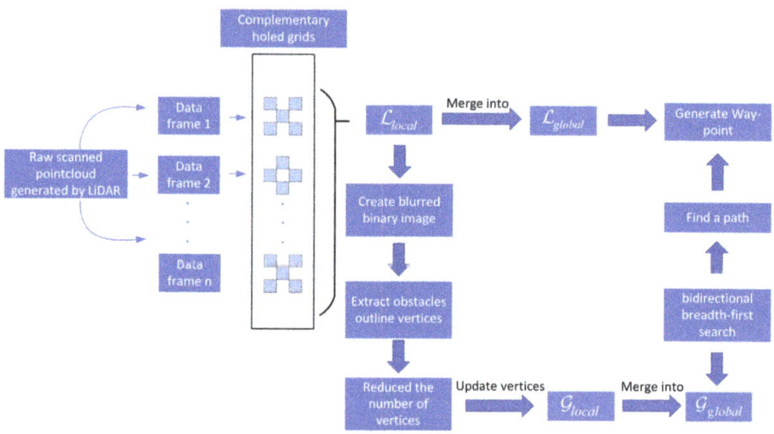

Figure 2. The main flow chart of the path planner based on the v-graph.

3.1. *Pointcloud Extraction Structure*

We denote the process of extracting and mapping the pointcloud to geometric contours as extract $\left\{P^k_{cloud} \subset \mathcal{Q} \,|k \in Z^+\right\}$, and the grid as *Grid*, respectively. In most laser-based SLAM, grids are used for accessibility analysis, which means that pointcloud information needs to be recorded in the global layer $\mathcal{L}_{global}$ and local $\mathcal{L}_{local}$. Although an incremental method of updating the pointcloud is proposed, in the case of complex terrain and high-resolution grids, the computational resources used to update the pointcloud are still very high. Therefore, a general sparsification module denoted as $\mathcal{F}$ is used to create the holed-structure local grid and incrementally update the pointcloud.

The dilated convolutional module [49] is usually used in neural networks to enlarge the receptive field in the picture, and its whole structure gives another way to deal with the pointcloud in the local grid: as shown in Figure 3, the pointcloud will be stored in the complementary hole grid.

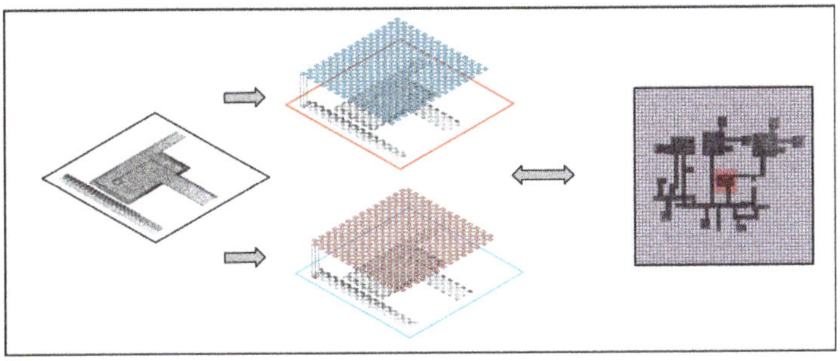

Figure 3. A schematic diagram of the holed grid structure used for the local update. The pointcloud is updated through the complementary holed grids between every two data frames and merged into the global layer.

The holed-structure local grid is defined as $Grid_d$, and the $\mathcal{F}$ contains $Sub \subset Grid_d$; when obstacles are detected by the LiDAR, the sensor data $\mathcal{S}$ is transferred to Sub, all Sub forms the $Grid_d$. We denote the voxel size as V_S, and this value will affect the density of pointcloud. In this paper, the value of V_S is set to 0.15 m. When the $Grid_d$ is formed, a PCL filter with a kernel size of (V_S, V_S, V_S) will be applied to reduce the size of the pointcloud. $\mathcal{S}'$ is the remaining pointcloud in the $Grid_d$. After the local pointcloud is formed, the $\mathcal{S}'$ are classified as obstacles or free, and we denote the classified $\mathcal{S}'$ as $\left\{P^k_{cloud}\right\}$. At this step, the fully classified $Grid_d$ is denoted as $\mathcal{L}_{local}$, then integrates $\mathcal{L}_{local}$ to $\mathcal{L}_{global}$. The complementary hole grids will be generated respectively in different data frames, so that the final $\mathcal{L}_{global}$ still contains all the information of the pointcloud. The module $\mathcal{F}$ is shown in Algorithm 1.

Algorithm 1 Module $\mathcal{F}$.

Input: Sensor data $\mathcal{S}$

Output: $\mathcal{S}'$

$input\ \mathcal{S}$
for $every\ data\ frame$ **do**
    Generate $Grid_d$
    for $each\ Sub \subset Grid_d$ **do**
        $store\ pointcloud \in \mathcal{S}$
    end for
end for
$Apply\ PCL\ filter\ to\ point \in Grid_d$
$classify\ point \in Grid_d$
$\mathcal{S}' = remain\ pointcloud\ in\ Grid_d$
$Update\ Grid_d\ to\ \mathcal{L}_{local}$

As shown in Figure 4, the Sub is a cell in the grid, and it stores a part of the pointcloud information in the 3D space. The standard practice is to form a grid from all cells, but in this paper, a grid with holed structure, as shown in Figure 5a,b, is used to let only part of the cells participate in the calculation. In fact, for a grid of a certain size, the number of cells depends on its resolution. As shown in Figure 6, the higher the resolution, the more cells there are, and the denser the grid, the better its mapping effect. However, the amount of computation increases dramatically. Therefore, this kind of holed-structure grid can save the calculation cost very well because it mainly requires half cells to participate in the calculation in each data frame.

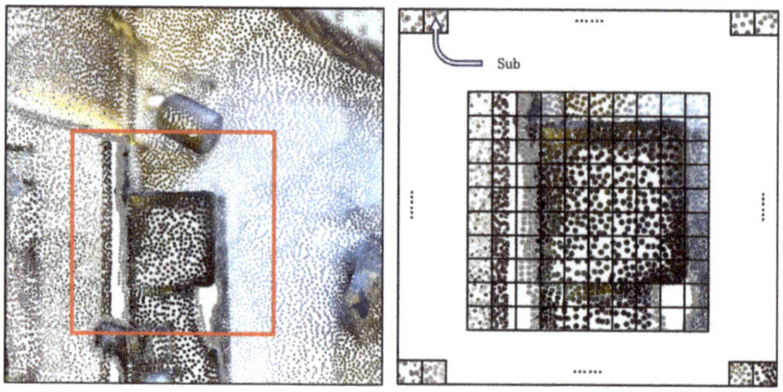

(**a**) Point cloud (**b**) Local grid

Figure 4. (**a**) is the spatial 3D pointcloud; (**b**) is the mapping of the pointcloud information in the local grid.

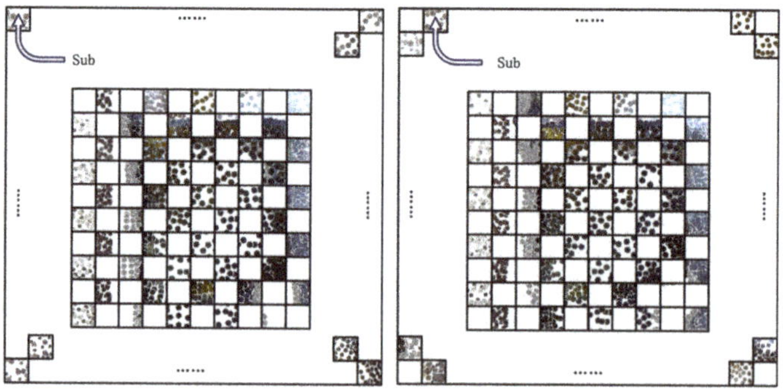

(**a**) Holed structure-based grid1 (**b**) Holed structure-based grid2

Figure 5. The complementary grids with holed structure.

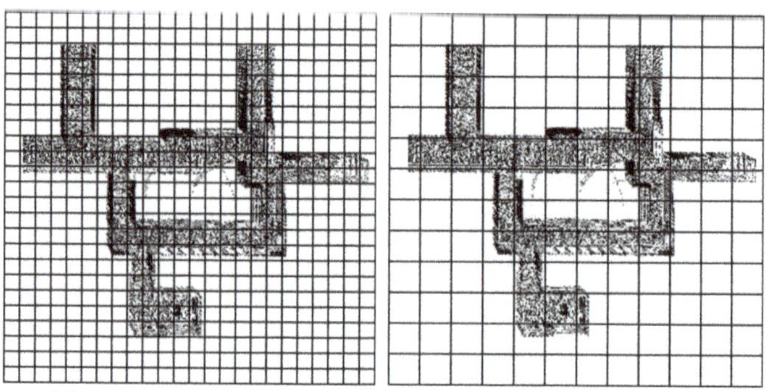

(**a**) Grid with higher resolution (**b**) Grid with lower resolution

Figure 6. Grid with (**a**) higher resolution, and (**b**) lower resolution.

For the obstacle extraction process and obstacle vertices reconstruction process in the v-graph, the sensor information $\mathcal{S}$ is gridded and stored by the module $\mathcal{F}$ to obtain $\mathcal{S}'$. After that, the $\mathcal{S}'$ will be converted to a binary image $\mathcal{I}$. To enhance robustness, the image $\mathcal{I}$ will be blurred, then obstacle vertices will be extracted through image processing [50] to generate polygons $\left\{P^k_{contour} \subset \mathcal{Q} \mid k \in Z^+\right\}$. The polygon extraction algorithm is shown in Algorithm 2.

To define the kernel size of the box filter in Algorithm 2, the equation is as follows, where R_W and R_L are the width and length of the robot, respectively, and V_S is the voxel size. In this paper, V_S is set to 0.15 m:

$$(kernel\ width,\ kernal\ height) = max(\lfloor \frac{\frac{max(R_W, R_L)}{2} + V_S}{V_S} \rceil, 5) \tag{1}$$

Figure 7 demonstrates the blurred picture of LiDAR-mapped obstacle geometry and the time consumption of the laser process. As can be seen from Figure 7a, the hollow structure (using module $\mathcal{F}$) does not affect the image after blurring. Compared with the original (without using a hollow structure), it can be found that our generated contour approximates the original image. The projected outline of the obstacle is thicker because of the blurred image, and the details inside the outline are lost.

The time required for the laser process, according to Figure 7b, includes gathering the raw pointcloud and downsampling. For the same area, the time consumed by using the holed structure is more gentle, while the processing process without the holed structure is steeper and its curve fluctuates greatly. The time it takes to process an image is depicted in Figure 7c. The total time of the image process includes mapping the pointcloud in $\mathcal{L}_{local}$ into the image $\mathcal{I}$, blurring the image, and initially extracting the obstacle contour points. About 20% of the total image processing time is spent on blurring the outline of obstacles.

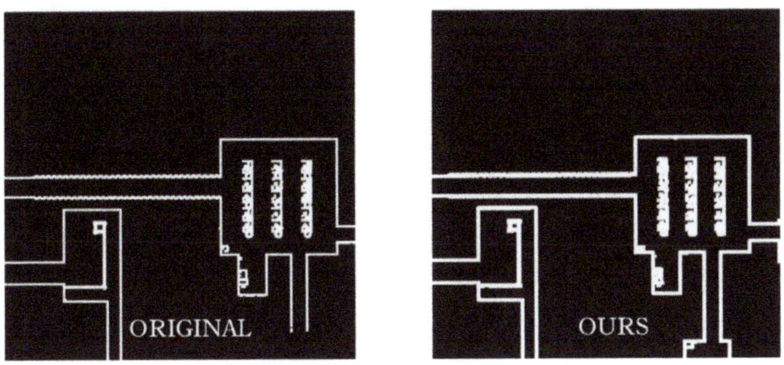

(**a**) Blurred image comparison

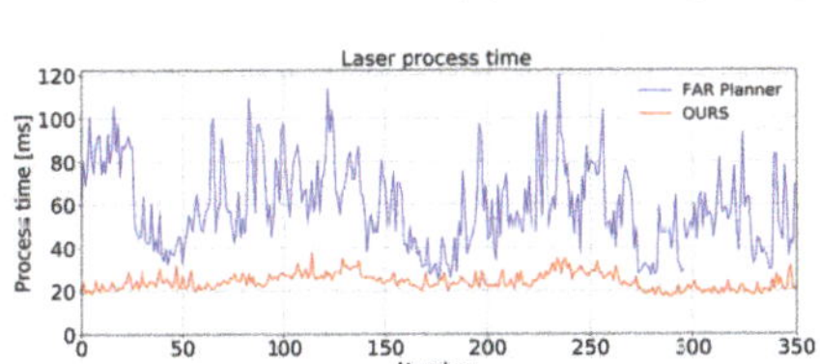

(**b**) Laser process consumption

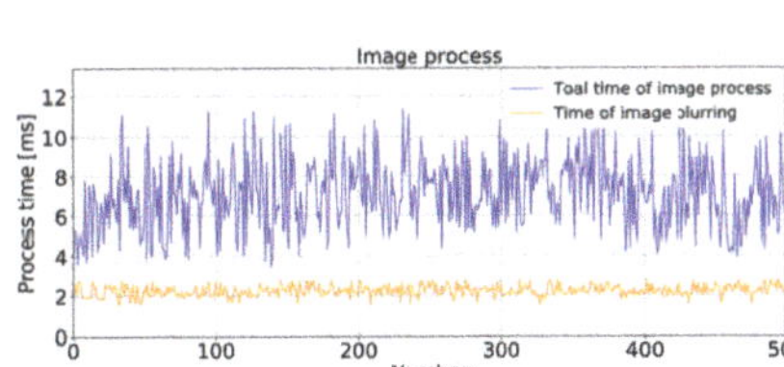

(**c**) Image process

Figure 7. (**a**) shows blurred images, the vertices of the obstacle will be extracted through the blurred image. The FAR Planner generates images without holed structure, and ours generates pictures through the holed structure. (**b**) shows the consumption of the laser process and (**c**) shows the consumption of the image process.

Algorithm 2 Polygon Extraction.

Input: $\mathcal{S}' \in \mathcal{L}_{local}$
Output: $Polygons : \left\{ P^k_{contour} \right\}$

1: $input\ \mathcal{S}'$
2: $Create\ binary\ image\ \mathcal{I}\ from\ points\ in\ \mathcal{S}'$
3: $Apply\ box\ filter\ with\ kernal\ size\ of\ (kernel\ width,\ kernal\ height)\ to\ blur\ image\ \mathcal{I}$
4: $Extract\ polygons\ \left\{ P^k_{contour} \right\}\ based\ on$ [50]
5: **for** $each\ P^k_{contour}$ **do**
6: $Downsample\ vertices\ in\ P^k_{contour}\ based\ on$ [51]
7: **end for**

3.2. Simplified Complex Contours Algorithm

In the previous Section 3.1, the $\mathcal{L}$ was constructed to obtain the pointcloud from LiDAR. In this section, the polygons will be extracted and simplified based on $\mathcal{L}$.

For the graph update method of the two-layer architecture, shown in Figure 8, we define $\mathcal{G}_{local}$ and $\mathcal{G}_{global}$. Between them, $\mathcal{G}_{local}$ is the local layer around the robot, and $\mathcal{G}_{global}$ is the layer set of the entire observation environment. $\mathcal{G}_{local}$ will be generated by the sensor information $\mathcal{S}$ for each data frame and then merged into $\mathcal{G}_{global}$ For each data frame, the sensor information $\mathcal{S}$ will generate $\mathcal{G}_{local}$ and then be merged with $\mathcal{G}_{global}$, noting that since the module $\mathcal{F}$ is used to store the sensor information $\mathcal{S}$, now $\mathcal{S}' \in \mathcal{L}_{local}$ can be used to merge $\mathcal{G}_{global}$ at a lower cost.

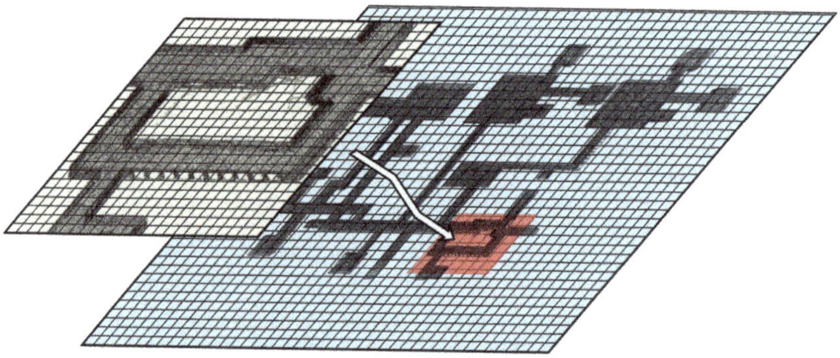

Figure 8. A two-layer update structure, where the blue grid is the global map, and the turquoise grid is the local map. The local map is located in the red boxed area in the global map.

It is known that the computational complexity involved in constructing a v-graph is $O(n^2 \log n)$ [52], where n is the number of vertices in the graph. In normal case, the cost of building a local graph in the environment is small enough so that computational resources can be allocated to each data frame in an incremental update manner. However, redundant nodes will also be generated during each v-graph update if the environment has numerous complex obstacles, leading to a significant increase in the number of edges connecting the nodes. To ensure the effectiveness of the v-graph update, a method for further sparse operation on complex contours is required.

Constructing local layers: The $\mathcal{S}' \in \mathcal{L}_{local}$ will be converted into local polygons $\left\{ P^k_{contour} \right\}$, and use $\left\{ P^k_{contour} \right\}$ to construct a local visualization graph $\mathcal{G}_{local}$. Note that for complex polygons, as shown in Figure 9, the polygon contains many vertices composed of short edges. Adding redundant vertices will construct more useless edges; thus, a lot of computing resources are wasted on unnecessary vertices and edges in the process of path search in complex terrain.

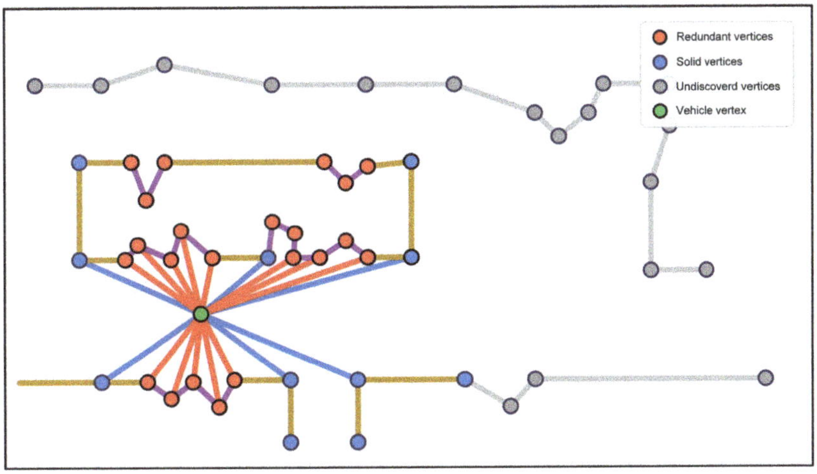

Figure 9. The red solid lines are redundant edges which connect with the robot, and the purple lines are the redundant edges from obstacles themselves.

A threshold η is set to control the number of vertices for complex large local polygons. When the number of vertices of the polygon $\left\{P^k_{contour}\right\}$ is greater than η, the vertices will be reduced, which not only optimizes the geometric outline of large and complex obstacles, but also retains the geometric characteristics of small obstacles. As shown in Figure 10, the continuous vertices inside those red circles in Figure 10a should be eliminated, but the current method does not eliminate them well, resulting in more vertices and edges in the v-graph. Compared to Figure 10a, the optimized version in Figure 10b has fewer vertices.

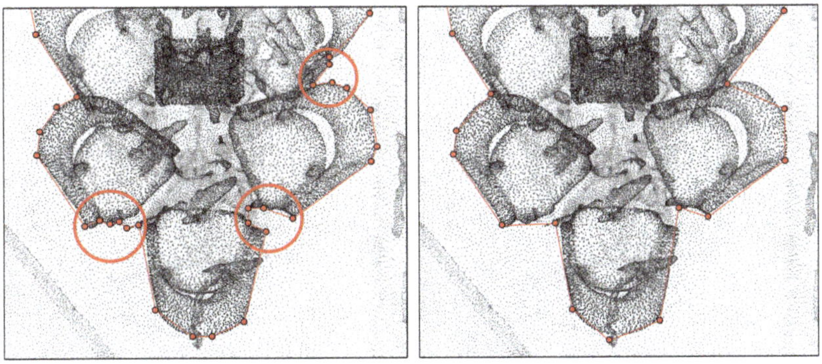

(a) Extract vertices **(b)** Optimization after extracting vertices

Figure 10. (**a**) shows the obstacle vertices extracted after pointcloud mapping. (**b**) shows the remaining obstacle vertices after optimizing the (**a**).

When the number of vertices of an obstacle is greater than η in the local layer, the algorithm preferentially records the distance between the two longest vertices in the obstacle. For example, the distance between the longest two vertices is $dist_{max}$. The algorithm traverses the three consecutive vertices $vertex_{i-1}$, $vertex_i$, $vertex_{i+1}$ in the obstacle and calculates the length between the two vertices, respectively. The distance between them is denoted as $dist_{(i,i-1)}$ and $dist_{(i,i+1)}$. If both $dist_{(i,i-1)}$ and $dist_{(i,i+1)}$ are less than $0.1\times$ $dist_{max}$, it means that $vertex_i$ is an invalid vertex (excess vertex), in which case we delete $vertex_i$ and destroy its connection edges $edge_{(i-1,i)}$ and $edge_{(i,i+1)}$.

Since the simplified complex contours algorithm only works on the $\mathcal{G}_{local}$, the v-graph update process will not be slowed down by the accumulation of the number of nodes in the $\mathcal{G}_{global}$.

Update the global layer: After $\mathcal{G}_{local}$ is constructed, the $\mathcal{G}_{local}$ and the $\mathcal{G}_{global}$ are fused. The strategy is: take out the overlapping parts of $\mathcal{G}_{local}$ in $\mathcal{G}_{global}$, and associate the vertex position in the $\mathcal{G}_{local}$ to the $\mathcal{G}_{global}$. The Euclidean distance is used to associate vertices in two layers, and the associated vertices are recorded. The entire graph updating algorithm is as follows in Algorithm 3, and the final obstacle contours and edges are shown in Figure 11.

For the given two points a(a_x, a_y, a_z) and b(b_x, b_y, b_z), the $distance(a, b)$ in Algorithm 3 is defined as followed:

$$distance(a,b) = \sqrt{(a_x - b_x)^2 + (a_y - b_y)^2 + (a_z - b_z)^2} \tag{2}$$

Algorithm 3 Visibility Graph Update.

Input: $\mathcal{S}' \in \mathcal{L}_{local}$, *graph* $\mathcal{G}$
Output: *Update graph* $\mathcal{G}$

$\left\{P^k_{contour}\right\} \leftarrow Polygon\ Extraction(\mathcal{S}');\ //\ from\ Algorithm2$
for *each* $P^k_{contour}$ **do**
  if *the number of vertices* $> \eta$ **then**
    for *each vertex in contour* **do**
      $dist_{max} = max(distance(vertex_i, vertex_{i+1}))$
    end for
    while *true* **do**
      for *each vertex in vertices* **do**
        $dist_{(i,i-1)} = distance(vertex_{i-1}, vertex_i)$
        $dist_{(i,i+1)} = distance(vertex_i, vertex_{i+1})$
        if $dist_{(i,i\pm1)} < 0.1 \times dist_{max}$ **then**
          Eliminate $vertex_i$
          Eliminate unnecessary $edge_{(i-1,i)}$ *and* $edge_{(i,i+1)}$
        end if
      end for
      if *all* $dist_{(i,i\pm1)} \geq 0.1 \times dist_{max}$ **then**
        break
      end if
    end while
  else
    continue
  end if
end for
Associate vertices $P^k_{contour}$ *in the* $\mathcal{G}_{global}$
Upate to visibility graph $\mathcal{G}$

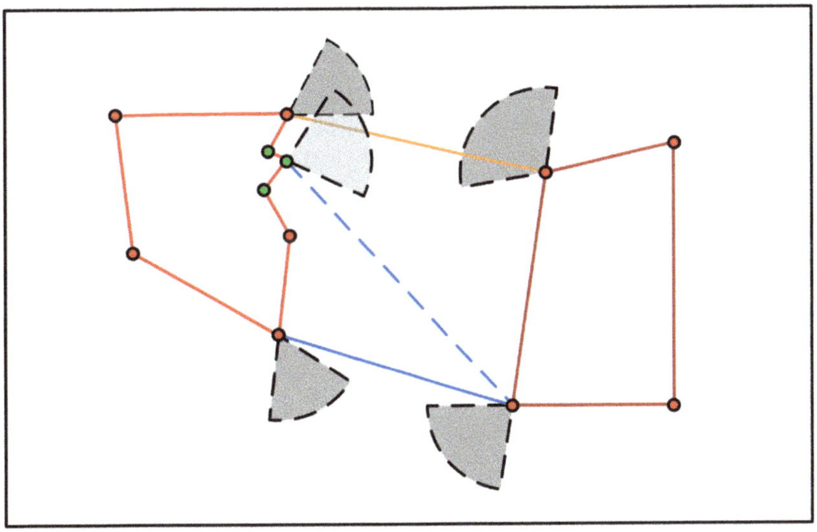

Figure 11. An illustration of sparse v-graph. The edge (orange) that head into at least one polygon from the shaded angle are eliminated, and the blue one will be kept. After eliminating those green vertices, the dot blue edge will be removed from the $\mathcal{G}_{local}$.

3.3. Path-Planning Based on Bidirectional Breadth-First Search

In FAR Planner, the goal point P_{goal} is used as a vertex, and the Euclidean distance is used as the score to update the parent node of P_{goal}. Although the path can be found in an unknown environment, its spatial search range is obviously wasted. The strategy adopted by the robot in many cases is to explore many unnecessary spaces until it finally reaches the goal. As shown in Figure 12, the robot travels from position 0 (start) to position 1, resulting in unnecessary exploration space.

A bidirectional breadth-first search (bidirectional BFS) structure is combined with the v-graph to search for a path, selecting a vertex of a connecting edge of the robot in the forward search while simultaneously beginning a backward search from the P_{goal} to find the path to the robot's current position. This minimizes the amount of unnecessary exploration space.

In the planning, assume that there are no obstacles in the unknown area where the P_{goal} is located. The P_{goal} uses geometric collision checking to establish edges with existing vertices $\left\{P_{node} \mid P_{node} \subset \mathcal{G}_{global}\right\}$ in the v-graph, and then the P_{goal} will be connected to the vertices of the discovered obstacles in the v-graph as shown in Figure 13a. The one-way BFS usually wastes some search space in unknown or partially known environments. This is because the one-way BFS starts from the nodes connected to the robot, calculates the target point according to the cost, and then iterates to the robot position according to the parent node of the target point.

As shown in Figure 13b,c, the one-way BFS enters a fork in the planning of the global path from the starting point to the ending point, resulting in an increase in the search space. The result is shown in Figure 13d, from the red point to the green point, the one-way search wastes a huge amount of space. Therefore, this paper embeds the goal in the v-graph and associates it with the existing vertices in the graph, and adopts a bidirectional breadth-first algorithm for path search.

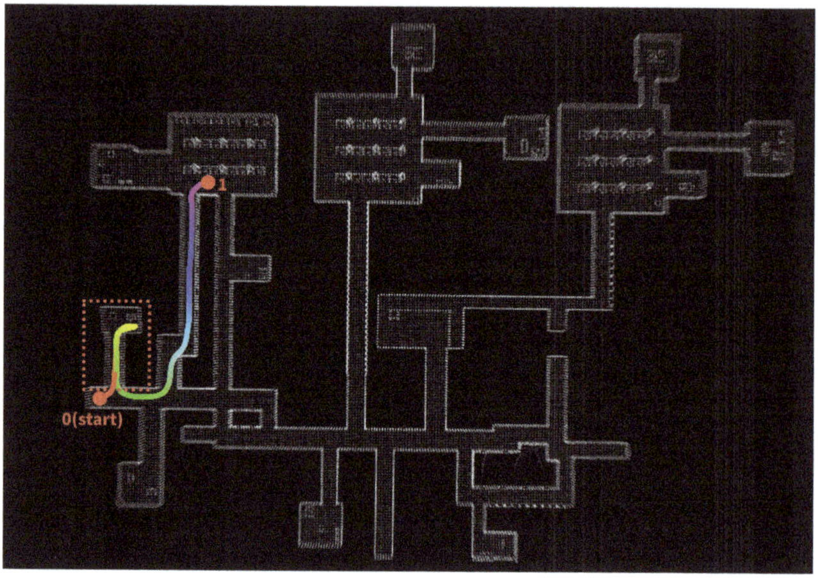

Figure 12. The robot travels from position 0 to position 1, and the red dotted box represents the wasted exploration space during navigation.

The bidirectional BFS structure shows in Algorithm 4, in which the two BFS are divided into forward and backward according to the direction of the search (forward searches from the robot position to the P_{goal}, and backward searches from the P_{goal} to the robot position).

In the Algorithm 4, $parent_F(\cdot)$ and $parent_B(\cdot)$ are the functions returning the forward and backward parent of a node. Q_F and Q_B are the min-priority queues in forward and backward ones, respectively, and Q_F is ordered by g_F, Q_B is ordered by g_B. μ is the cost of the best path found so far (initially, μ is set to ∞). Whenever the robot reaches a node and expands in the other search, μ will be updated if a better path goes through the node. g_F is the current distance from start and g_B is the current distance from P_{goal}. top_F and top_B are the distances to the top nodes in the forward and backward queues, respectively. The STEP function in Algorithm 5 is responsible for advancing the search through the v-graph and updating μ.

One of the benefits of bidirectional search is that it can, to some extent, avoid the wasted search space caused by entering invalid forks. As shown in Figure 13e, the globally planned path no longer passes through the fork, thus avoiding excessive searching. As a result, as shown in Figure 13f, compared to Figure 13d, the distance traveled by the robot is greatly reduced.

It is not possible to use a vertex that has just been extracted from the v-graph $\mathcal{G}$ as a point of navigation directly; instead, a transform is used to turn the vertex into a way-point. As Figure 14 shows, in the obstacle where the point is located, the vertices connected to the point at the polygon will be extracted to calculate the direction vector of the point (in Algorithm 6, they are $\overrightarrow{dir}_{front}$ and $\overrightarrow{dir}_{back}$, respectively, and the direction vector of the point is the $\overrightarrow{surf}_{dir}$).

A detailed description is shown in Algorithm 6. In Algorithm 6, the parameter of $search_{dist}$ is set to constrain the searching area, and the $near_{dist}$ is a step parameter which extends the way-point from $\overrightarrow{surf}_{dir}$ direction, and R_W and R_L are the width and length of the robot, respectively. When the way-point extends, the max_{extend} is set to constrain the length of the extension. The $NearOBS(\cdot)$ function is to obtain a range of obscloud from $\mathcal{L}_{global}$, with $P_{way-point}$ as center and $search_{dist}$ as the radius. *Check_Collision* is used to detect if the expanded $P_{way-point}$ collides with surrounding obstacles, and the detail of the *Check_Collision* function is shown in Algorithm 7.

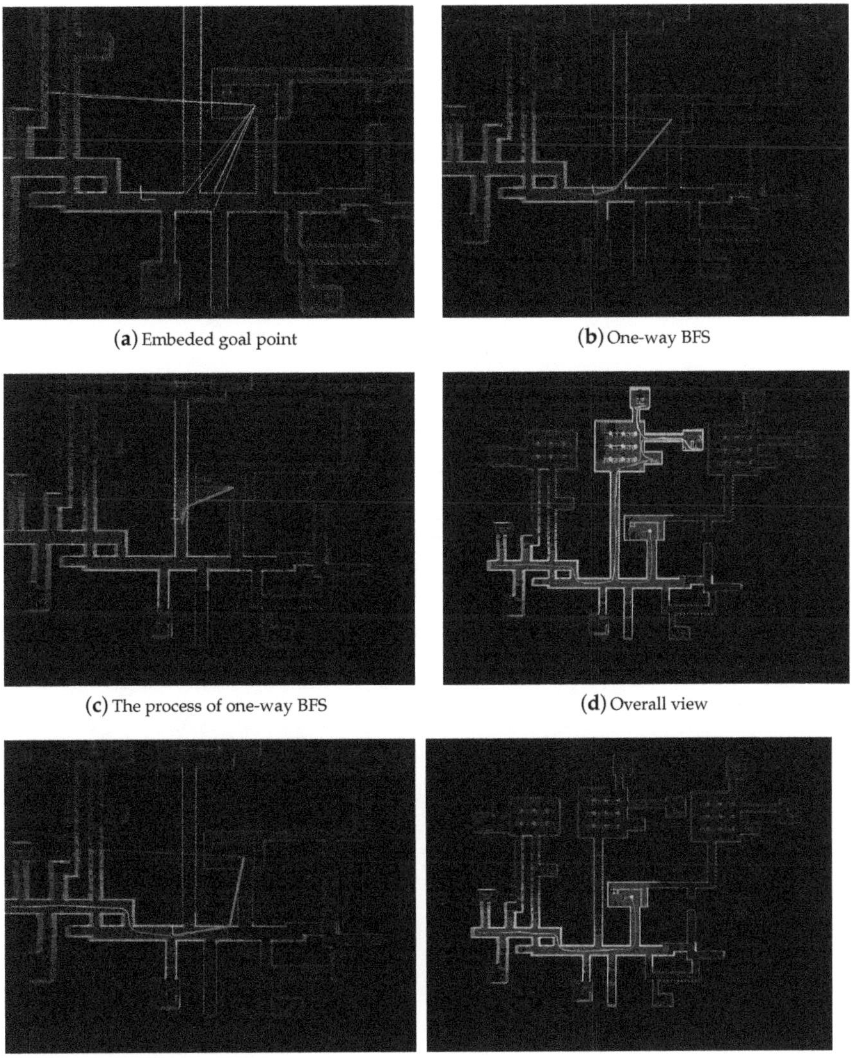

(**a**) Embeded goal point
(**b**) One-way BFS
(**c**) The process of one-way BFS
(**d**) Overall view
(**e**) bidirectional BFS
(**f**) Overall view

Figure 13. The blue area of the map represents the part that has been explored by LiDAR and is considered known. (**a**) shows the connection between the endpoint and the existing node, represented by a solid yellow line. (**b**,**c**) show the path planning using one-way BFS. (**d**) is the result of one-way BFS. The red dot is the starting point, and the green dot is the endpoint. (**e**,**f**), respectively, show the path planning and the final result using bidirectional BFS. The globally planned path appears on the map as a thick blue line.

Algorithm 4 Bidirectional BFS.

Input: P_{start}, P_{goal}, $Visibility\ Graph$: $\mathcal{G}$
Output: $path : \left\{P_{path}\right\}$

$Q_F, Q_B \leftarrow make\ min - priority\ queues\ for\ the\ nodes\ initially\ containing\ only\ P_{start}\ and\ P_{goal}$
$expanded_F, expanded_B \leftarrow make\ forward\ and\ backward\ lookup\ tables$
$search_F \leftarrow (Q_F, expanded_F, parent_F)$
$search_B \leftarrow (Q_B, expanded_B, parent_B)$
$\mu \leftarrow \infty$
$initialize\ \mathcal{G}, associate\ P_{goal}\ in\ \mathcal{G}$
$path \leftarrow none$
while $top_F + top_B < \mu$ **do**
  if $at\ least\ one\ queue\ is\ non - empty$ **then**
    $choose\ the\ search\ to\ advance$
  else
    return $path$
  end if
  if $forward\ search\ was\ chosen$ **then**
    $(path, \mu) \leftarrow$ **STEP**$(search_F, search_R, path, \mu)$
  else
    $(path, \mu) \leftarrow$ **STEP**$(search_R, search_F, path, \mu)$
  end if
end while
return $path$

Algorithm 5 STEP $(search_1, search_2, solution, \mu)$.

// 1 denotes the chosen direction and 2 is other direction
// $c(\cdot)$ is the cost function, in this paper, manhattan distance is used
// $c(u,v) = |u_x - v_x| + |u_y - v_y| + |u_z - v_z|$
$\mu \leftarrow pop\ the\ min\ g_1\ node\ from\ Q_1$
for $v \in parent_1(u)$ **do**
  if $v \notin expanded_1 \cup Q_1\ or\ g_1(u) + c(u,v) < g_1(v)$ **then**
    $g_1(v) \leftarrow g_1(u) + c(u,v)$
    Add v to Q_1
    if $v \in expanded_2\ and\ g_1(v) + g_2(v) < \mu$ **then**
      $path \leftarrow reconstruct\ the\ path\ through\ u\ and\ v$
      $\mu \leftarrow g_1(v) + g_2(v)$
    end if
  end if
  return $(path, \mu)$
end for

In Algorithm 6, the $normalize(P_a, P_b)$ and $normalize(\cdot)$ are defined as followed:

$$normalize(P_a, P_b) = \frac{P_b - P_a}{||P_b - P_a||} \tag{3}$$

$$normalize(\vec{V}) = \frac{\vec{V}}{||\vec{V}||} \tag{4}$$

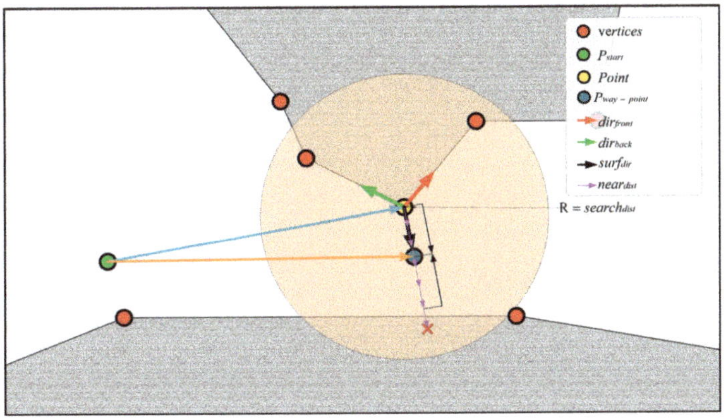

Figure 14. The schematic diagram of generating a way-point.

Algorithm 6 Way-point generation.

Input: $P_{start}, Path, \mathcal{L}_{global}$
Output: $P_{way-point}$

$vertex = Path[0]$
$Path = Path[1:]$
$search_{dist} = max(2.5 \times max(R_L, R_W), 5)$
$near_{dist} = min(min(R_L, R_W), 0.5)$
$P_{way-point} = vertex$
$max_{extend} = min(search_{dist}, distance(P_{start}, P_{way-point})$
$\overrightarrow{dir}_{front} = normalize(polygon_k(i-1), polygon_k(i))$
$\overrightarrow{dir}_{back} = normalize(polygon_k(i+1), polygon_k(i))$
if $P_{way-point}$ *is a convex point* **then**
  $\overrightarrow{surf}_{dir} = -normalize(\overrightarrow{dir}_{front} + \overrightarrow{dir}_{back})$
else
  $\overrightarrow{surf}_{dir} = normalize(\overrightarrow{dir}_{front} + \overrightarrow{dir}_{back})$
end if
$obscloud = NearOBS(\mathcal{L}_{global}, search_{dist}, P_{way-point})$
$obscloud = setInputCloud(obscloud)$ //setInputCloud is a pcl library function to build the KDTree of a set of pointcloud
$temp = P_{way-point} + \overrightarrow{surf}_{dir} \times near_{dist}$
$is\_collide = Check\_Collision(temp, obscloud)$ //from Algorithm 6
$extend_{dist} = near_{dist}$
while $is\_collide$ *is false and* $extend_{dist} < search_{dist}$ **do**
  $temp += \overrightarrow{surf}_{dir} \times near_{dist}$
  $extend_{dist} += near_{dist}$
  $is\_collide = Check\_Collision(temp, obscloud)$
  if $extend_{dist} < max_{extend}$ **then**
    $P_{way-point} = temp$
  end if
end while
if $is\_collide$ *is true and* $extend_{dist} > max(R_W, R_L)$ **then**
  $P_{way-point} = \frac{P_{way-point} + vertex - \overrightarrow{surf}_{dir} \times near_{dist}}{2}$
  return $P_{way-point}$
else
  return *drop this vertex and re* − *search path*
end if

Algorithm 7 Check_Collision.

Input: *point P, obscloud*
Output: *false or true*
1: $radius = near_{dist}/2 + V_S$
2: $thre = near_{dist}/V_S$
3: $KDTree \rightarrow radiusSearch(P, radius, indices, dis)$ // search points with P as center and radius as the search radius, return the number of points in indices and the distance between each point and P.
4: **if** $indices > thre$ **then**
5: **return** *true*
6: **else**
7: **return** *false*
8: **end if**

4. Experiments and Results

The paper uses the same experimental parameters as FAR Planner [19], (uniform sensor parameters, the robot speed is set to 2 m/s). A highly complex channel network tunnel, a parking garage with multiple floors, and a forest of trees with many irregularly shaped trees are all included in the simulated experimental environment. The indoor is moderately complex but easy to detour. Additionally, Matterport3D [48] provides a simple environment 17DRP5sb8fy (denoted as 17DR), a slightly complex environment PX4nDJXEHrG (denoted as PX4n), and a large complex environment 2azQ1b91cZZ (denoted as 2azQ).

In the simulation environment, all methods run on a 2.6Ghz i7 laptop, and the v-graph-based methods use images at 0.2 m/pixel resolution to extract points to form polygons. The local layer on the v-graph is in a 40 m × 40 m area with the robot in the center. The threshold of the length of each visibility edge is set to 5 m. Finally, the simulated mobile robot is 0.6 m long, 0.5 m wide, and 0.6 m high.

In the physical environment, ours runs on an embedded device, and the robot speed is set to 0.7 m/s. To adapt to the real environment, the local layer of the v-graph is set to 20 m × 20 m. The length, width, and height of the mobile robot are set as 0.32 m, 0.25 m, and 0.31 m, respectively.

4.1. Simulation Experiment

4.1.1. Laser Process Simulation Experiment

In the laser process simulation experiment, seven different types of environments are used to compare the holed structure with the original one. The experimental results are shown in Figure 15. The robot moves according to a fixed route, and the experiment analyzes the time of the laser processing process. The laser processing process refers to the whole process of extracting and storing the pointcloud information of the local layer into the grid and classification (obstacle pointcloud or free pointcloud).

As shown in Figure 15, for example, in the indoor environment, the robot will start from 0 (start) and pass through the target points 1, 2, 3, 4, 5, and 6 in sequence. The initial state of the robot is set to be in an unknown environment, and the known information in the environment is continuously accumulated through exploration. The program records the time of receiving and processing the pointcloud data from LiDAR during the movement of the robot.

Figure 16a is a summary of the average time of laser processing in different environments. It can be clearly seen from Figure 16a that our method used less time in the processing of the pointcloud information in every data frame. Figure 16b–h show that using the grid with the holed-structure leads to the smooth processing of the laser pointcloud. Compared with the original grid, the use of the holed structure can improve the processing speed of the pointcloud by 30.5∼44.5%.

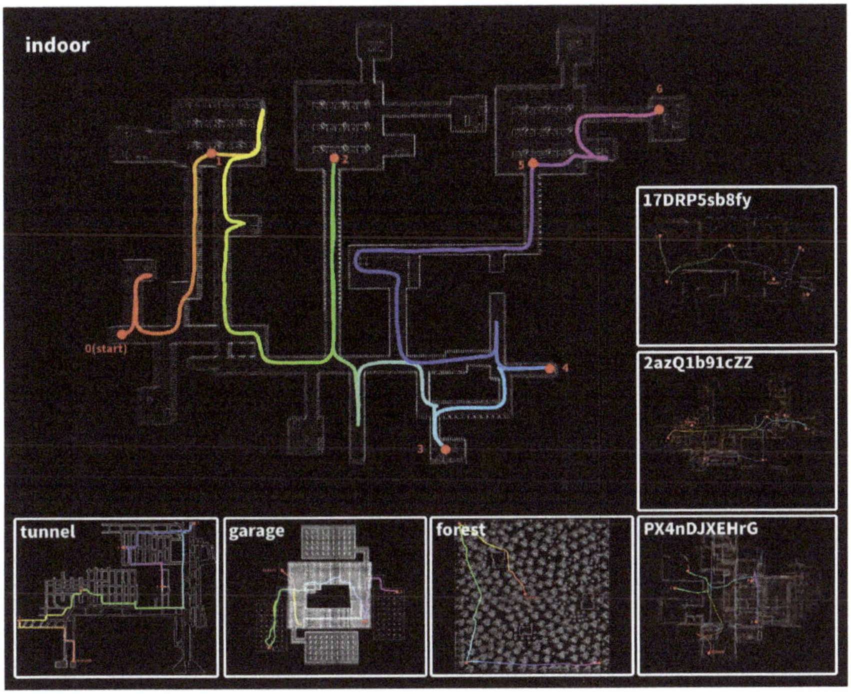

Figure 15. The overall view of seven different environments.

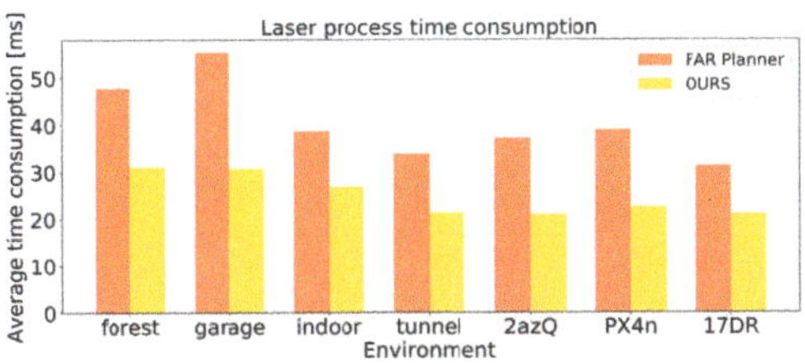

(**a**) Overall average process consumption

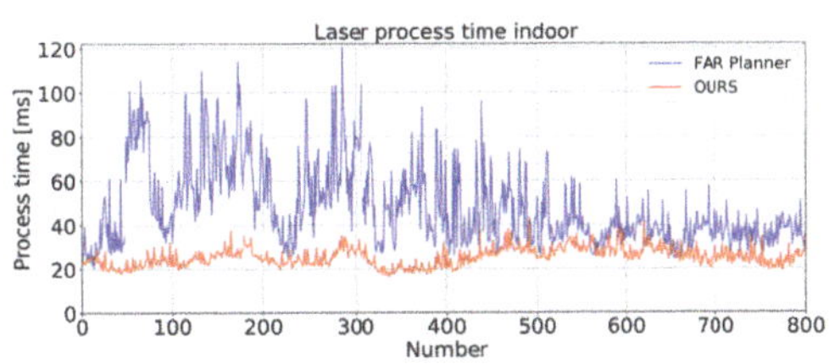

(**b**) Laser process(indoor)

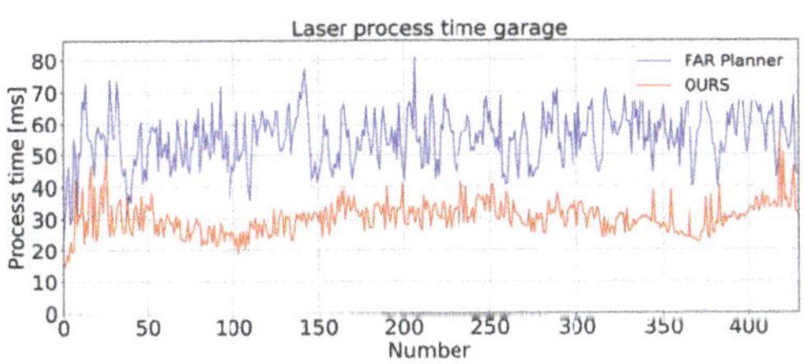

(**c**) Laser process(garage)

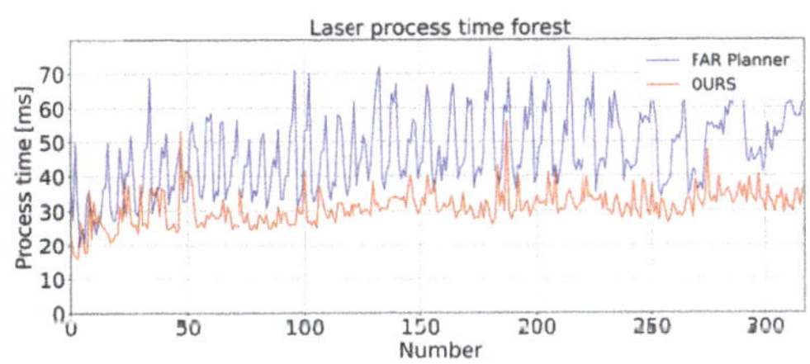

(**d**) Laser process(forest)

Figure 16. *Cont.*

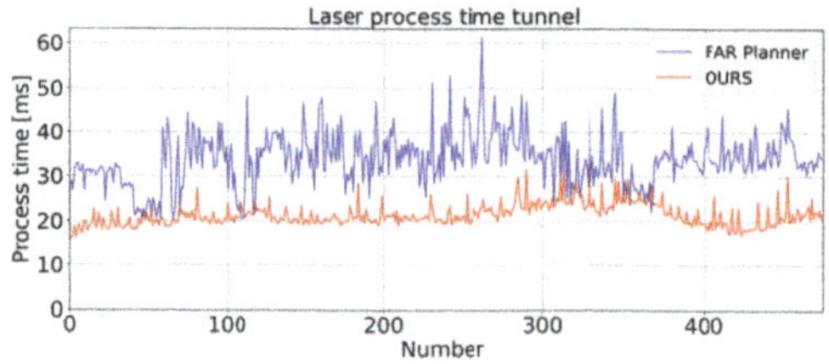

(**e**) Laser process(tunnel)

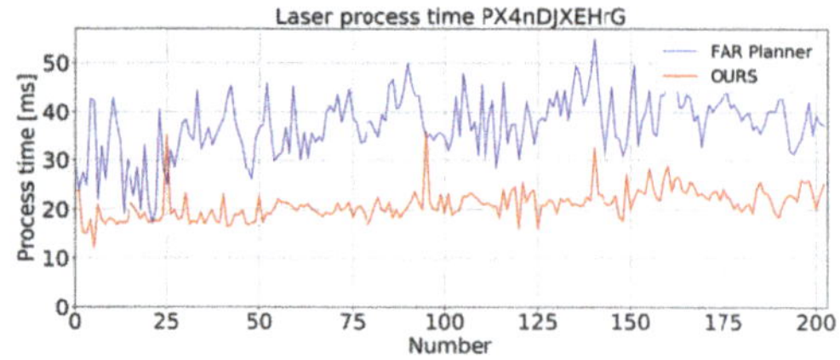

(**f**) Laser process(PX4nDJXEHrG)

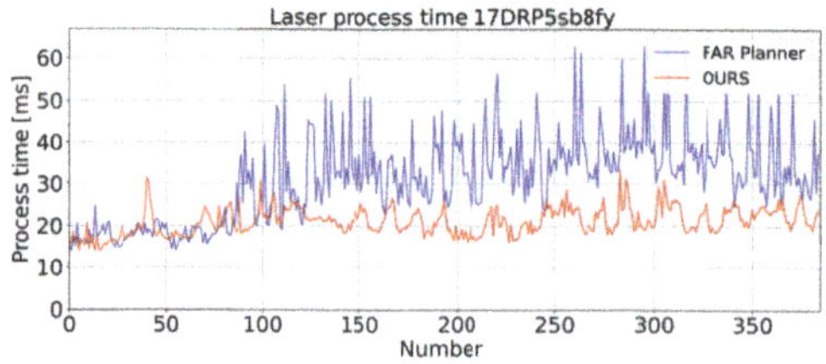

(**g**) Laser process(17DRP5sb8fy)

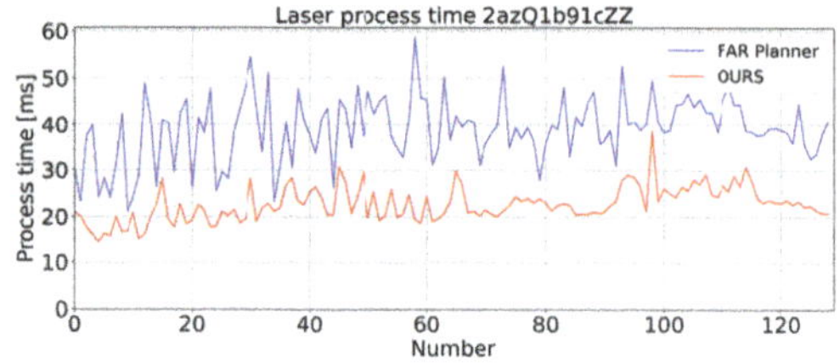

(**h**) Laser process(2azQ1b91cZZ)

Figure 16. (**a**) is the summary of the average time of laser processing in different environments. (**b–h**) show the process time of pointcloud in each data frame.

4.1.2. Visibility Graph Update Simulation Experiment

Different values of η, ranging from 5 to 25, are set in representative indoor and outdoor environments in order to select an appropriate value. The robot in the simulation experiment travels throughout the entire environment to count all of the vertices in the v-graph and logs the average update speed. As the Table 1 shows, it can be seen that the number of vertices in the v-graph and the update speed of the v-graph are positively correlated with the value of η, but η is not as small as possible.

Table 1. Relevances among η, total vertices besides the speed of v-graph update.

Test	Without η	$\eta = 25$	$\eta = 20$	$\eta = 15$	$\eta = 10$	$\eta = 5$
total vertices (Indoor)	904	873	725	714	695	630
v-graph update (Indoor)	21.48 ms	20.64 ms	14.13 ms	13.88 ms	13.82 ms	12.81 ms
total vertices (Forest)	3976	3589	2893	2838	2743	2334
v-graph update (Forest)	71.25 ms	54.23 ms	39.12 ms	38.97 ms	36.82 ms	35.64 ms

Theoretically, a smaller η value should lead to fewer vertices in the v-graph. However, a value of η that is too small will have some drawbacks. The function of collision detection may be affected when some minor obstacles are ignored, as shown in Figure 17a, and this phenomenon also occurs in forest environments. When η is equal to 5, there are two small trees missing from the v-graph in Figure 17b. When η is greater than or equal to 15 and less than or equal to 20, with the increase of η, the outline of the obstacle is well guaranteed, and the update speed of the v-graph is also relatively fast. When η is greater than 25, the number of vertices in the v-graph gradually increases toward the direction of none η. After several rounds of testing, a preliminary conclusion can be drawn that when the value of η is between 15 and 20, the algorithm is most suitable for generating obstacle vertices.

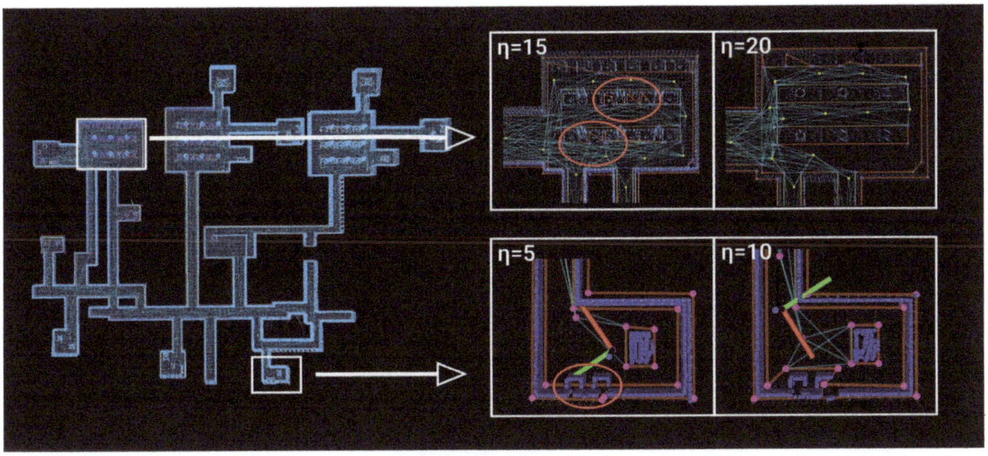

(a) Indoor environment

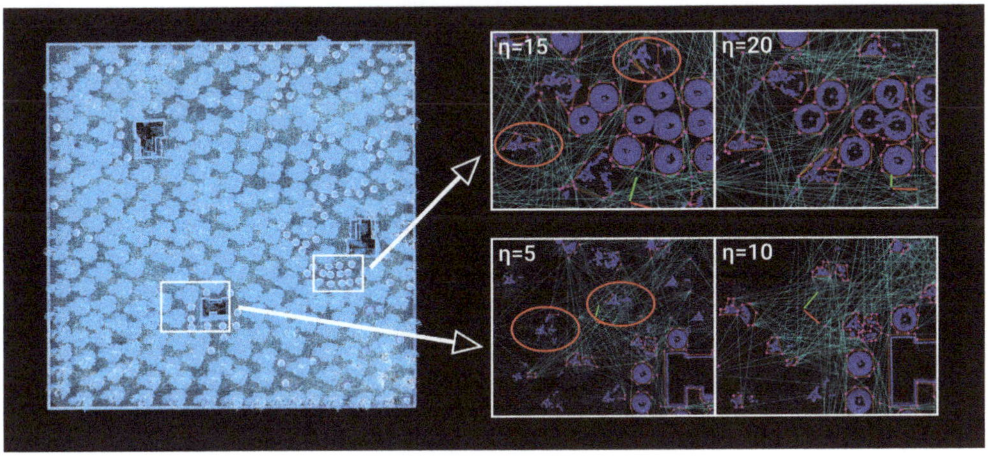

(b) Forest environment

Figure 17. Different η in the indoor and forest environment.

In the v-graph update simulation experiment, seven different environments are used to compare our v-graph update method with FAR Planner's, and the parameter of η is set to 20. In different environments, a series of target points are established for the mobile robot to travel. These points are fixed, and both methods let the mobile robot pass through them in sequence. For example, in the indoor environment, the robot will pass through the six target points 1, 2, 3, 4, 5, and 6 in sequence. During the driving process of the robot, the update speed of the v-graph will be recorded, and similarly, all the vertices in the v-graph will be recorded. The experimental results are shown in Figure 18.

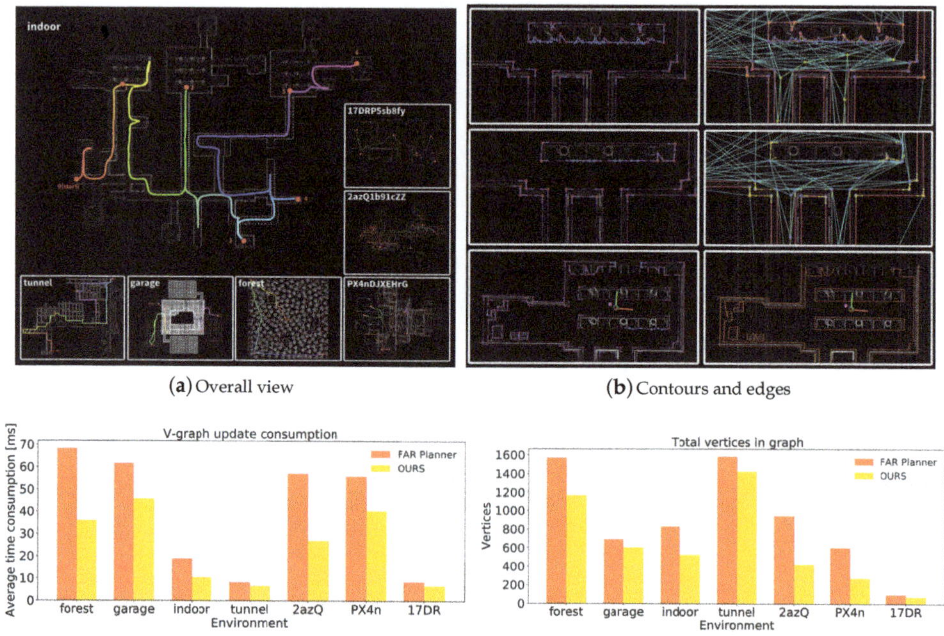

(**a**) Overall view

(**b**) Contours and edges

(**c**) Visibility graph update

(**d**) Total vertices in graph

Figure 18. (**a**) shows that the robot runs in different environments and passes through a series of target points. (**b**) shows the geometric outline of obstacles and the connection of edges, red points are valid vertices, and cyan lines represent effective edges. The last one in (**b**) is the optimized global map. (**c**) shows the update speed of our method and the original method in different environments. (**d**) shows the number of vertices in the v-graph after running the same trajectory.

In each environment in Figure 18a, the robot passes through a series of target points, and the known environment information is reset after reaching each target point. Figure 18b shows the optimized nodes and edges for complex irregular objects. For complex obstacles, ours simplifies the vertex information of the obstacle. When the redundant vertices are reduced, the redundant edges will also be correspondingly reduced. As shown in Figure 18b, the cyan and blue edges and red nodes in the optimized v-graph are significantly reduced, respectively.

Figure 18c shows the average speed of the v-graph update, and Figure 18d shows the total number of vertices in the v-graph. For simple terrain and most of the obstacles with simple shapes and corners, such as the tunnel and 17DRP5sb8fy, our method obtains similar results to FAR Planner, and the improvement is only 10∼20%; for large-scale maps containing complex obstacles, such as the forest, 2azQ1b91cZZ, PX4nDJXEHrG, indoor, etc., our method significantly improves the efficiency by 40∼60% compared to FAR Planner.

It is evident from Figure 19 that for environments with complex terrain, our method greatly reduces the number of vertices for obstacles in the v-graph. In Figure 19e,f, the original method has a total of 73 vertices while our method has only 49 vertices.

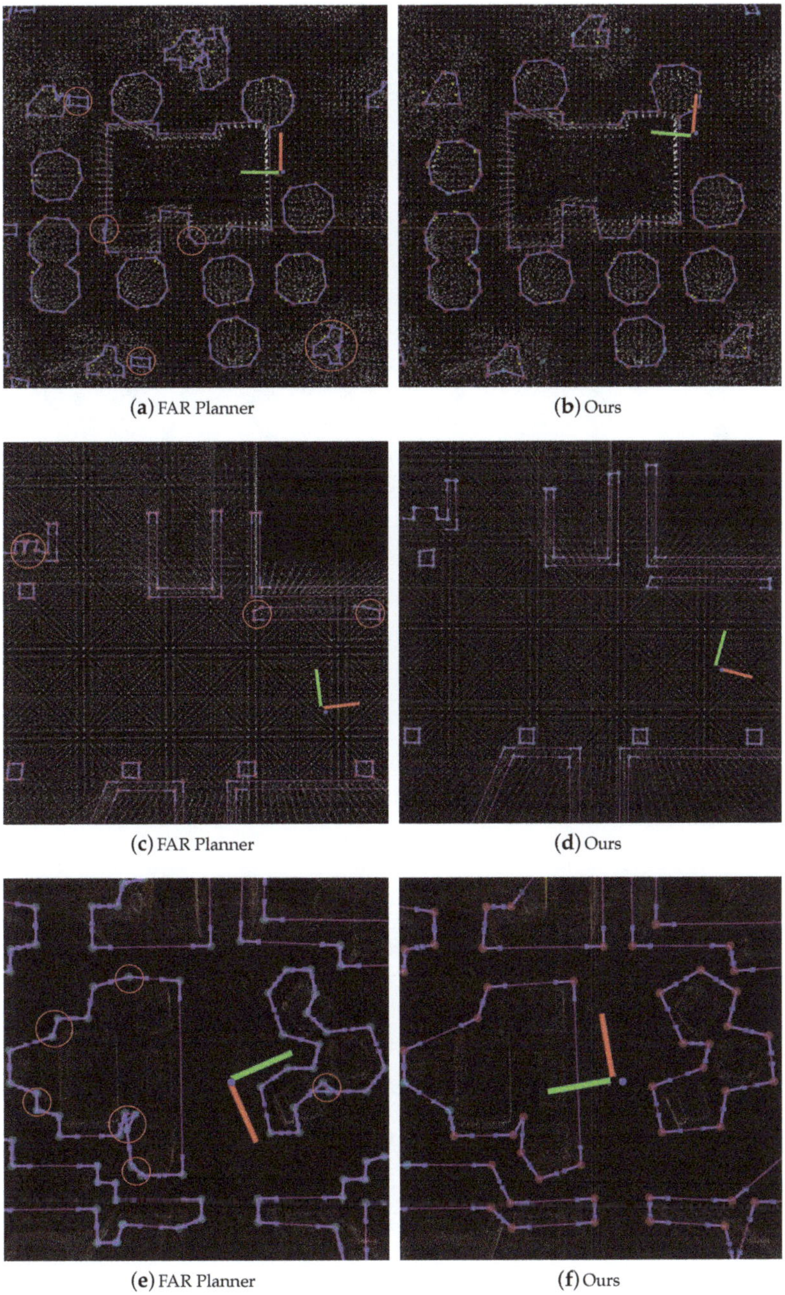

Figure 19. From (**a**–**f**) is the comparison result of the extracted obstacle vertices.

4.1.3. Path Planning Simulation Experiment

For the path exploration in the unknown environment, we compared the slightly complex indoor environment and the tunnel environment with complex network structures, respectively. Similar to the graph update simulation experiment, a series of waypoints are set up in each environment, allowing the robot to pass one by one. Define the robot to accumulate environmental information while exploring the unknown environment.

FAR Planner (v-graph-based), A*, and D* Lite (occupancy grid-based) are all added for comparison with our algorithm. Figure 20 shows the trajectory paths generated by our algorithm and other algorithms in navigation. In the case of reaching the same navigation point, our algorithm can avoid unnecessary exploration space to a greater extent than the FAR Planner and A*, D* Lite, so as to achieve a shorter distance and less time.

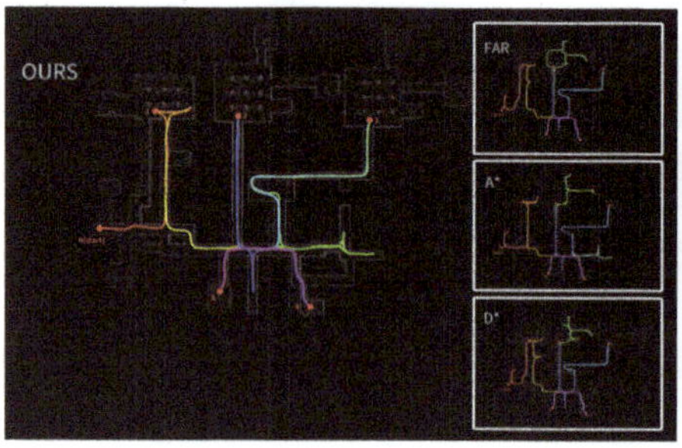

(a) Indoor

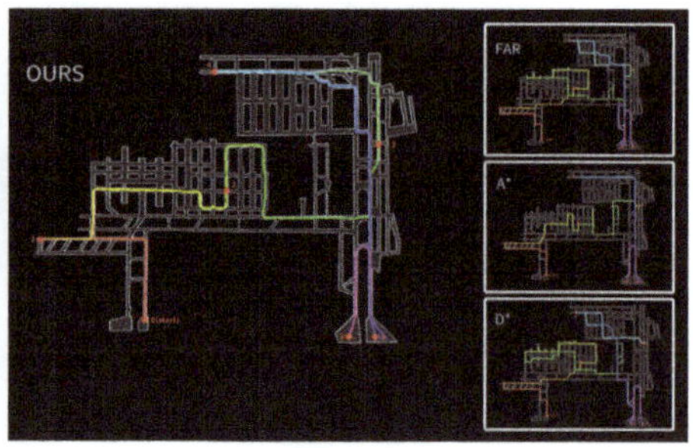

(b) Tunnel

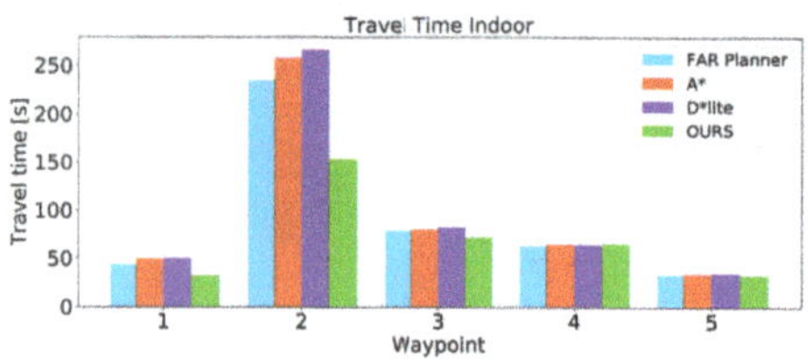

(c) Travel Time (Indoor)

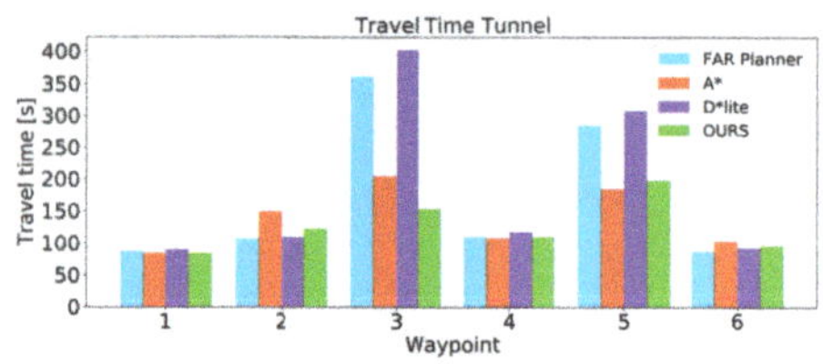

(d) Travel Time (Tunnel)

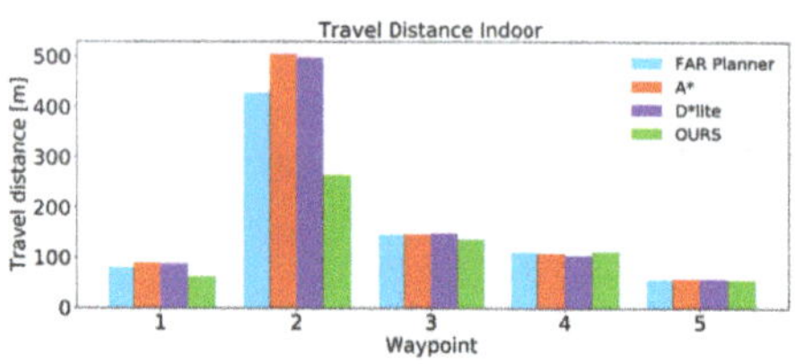

(e) Travel Distance (Indoor)

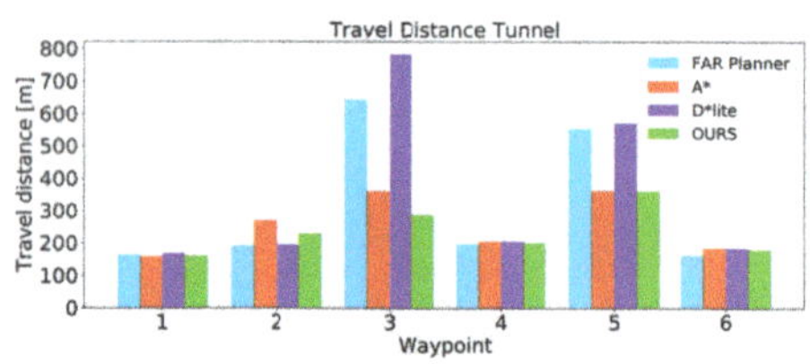

(f) Travel Distance (Tunnel)

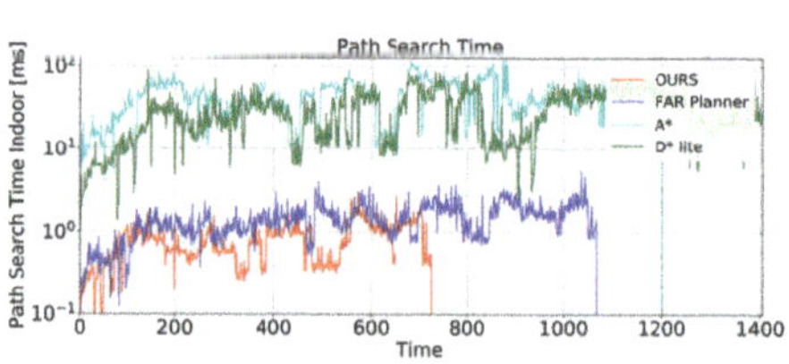

(g) Path Search Time (Indoor)

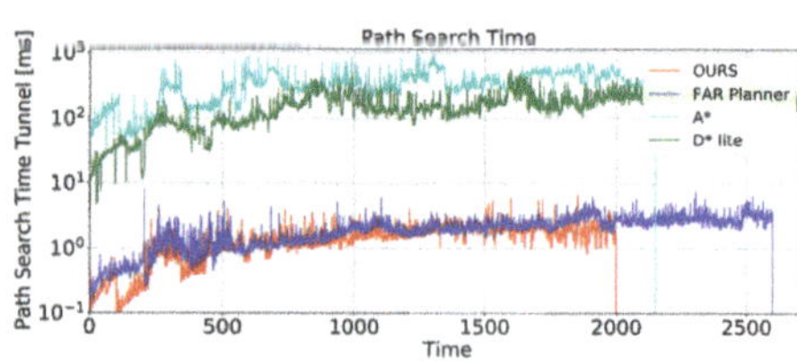

(h) Path Search Time (Tunnel)

Figure 20. (**a**,**b**) show that the trajectory of the robot using different algorithms to navigate in indoor and tunnel simulation environments respectively and passes through a series of target points. (**c**,**d**) show the time consumption for each target point. (**e**,**f**) show the distance robot takes to travel to each target point. (**g**,**h**) show different path-searching algorithms' time consumption.

Figure 20c–f show the time and distance it takes for the robot to travel from one navigation point to another, and the robot accumulates map information during driving.

Tables 2 and 3 (FAR Planner denoted as FAR) gives a summary of the overall travel time, distance, and search time used for each map robot navigation.

As can be seen from Figure 20a for the indoor environment, the search space is wasted to varying degrees when using FAR Planner, A*, and D* Lite to navigate from point 1 to point 2. They are more inclined to explore the place where point 3 is located and then turn back after finding that there is no passage leading to point 2. The possible reason for this is that most of their cost functions only refer to the cost of the current node itself and the cost of the Euclidean distance between the endpoint and the current node. This causes the robot to tend to drive towards the node with the lowest total cost in a single direction, even though that node may not be able to reach the goal.

For adjacent navigation points with relatively short distances, such as from navigation point 4 to navigation point 5 in an indoor environment, the time and distance consumed by all algorithms are not much different.

The A* and D* Lite are known for their search integrity in finding the optimal path. However, those methods are difficult to scale as the computational cost increases significantly when environments are large and complex [19]. For the tunnel environment, although the environment contains a series of complex #- and *T*-shaped structures, there are almost no dead ends, that is, the goal can be reached in any direction, so the robot hardly needs to be turned back during the running process. For traditional A* and D* Lite, due to the increase in scene scale, the number of grids that need to be calculated increases sharply, and the cost of algorithm operation increases significantly in large-scale scenes. However, for us, the use of bidirectional BFS allows us to avoid some bifurcations and travel a shorter distance to the destination point.

Table 2. The overall time spent by the robot using different algorithms in [s] in Indoor environment.

Test	Overall Time [s]	Overall Distance [m]	Average Search Time [ms]
FAR(baseline)	455.31	824.1	1.43
A*	486.4	906.3	44.21
Compare to FAR	+6.7%	+9%	+2991%
D* Lite	498.1	896.6	29.6
Compare to FAR	+9.4%	+8.8%	+1967%
OURS	357.2	633.1	0.81
Compare to FAR	−21.5%	−19.5%	−43.3%

Table 3. The overall time spent by the robot using different algorithms in [s] in Tunnel environment.

Test	Overall Time [s]	Overall Distance [m]	Average Search Time [ms]
FAR(baseline)	1038.2	1914.4	2.05
A*	833.6	1543.8	332.95
Compare to FAR	−19.7%	−19.3%	+16,141%
D* Lite	1119.8	2109.8	162.3
Compare to FAR	+7.8%	+10.2%	+7817%
OURS	763.8	1420.4	1.44
Compare to FAR	−26.4%	−25.8%	−29.7%

As shown in Figure 20a,b, our planner is able to search for shorter paths and generate effective trajectories. Table 2 shows that in the indoor case, our method reduces travel time by 23% compared to A*, 28% compared to D* Lite, and 21% compared to FAR Planner's original algorithm. In terms of increased time, FAR Planner has the most time wasted due to ineffective exploration, while A* and D* Lite are time-consuming when the robot is constantly swinging back and forth in a certain position, but what all three have in common is wasted search space between some navigation points. Table 3 shows that in the tunnel case, our method produces the shortest distance, which is 8% shorter than A*, 32% shorter than D* Lite, and 25% shorter than FAR Planner.

Tables 2 and 3 show that our planning algorithm can run faster because of the use of a hole-structured mesh to update the graph and vertex optimization for complex obstacles. Compared to FAR Planner, our search algorithm update rate is 43% faster.

4.2. Physical Experiment

The physical experiment uses the mobile robot platform in Figure 21 with the speed set to 0.7 m/s. The mobile robot is equipped with a Velodyne-16 LiDAR and an Inertial Measurement Unit (IMU) 9250. The entire autonomous system is built on Robot Operating System [53], and the Raspberry Pi 4B is the master and the laptop is the slave. In the autonomous system, the master is used to transmit LiDAR data, run the CMU autonomous exploration interface [47], and drive the stm32. The navigation algorithm runs on the slave. The camera at 640 $\times$ 480 resolution in the mobile robot is only used to obtain pictures of the environment. The LiDAR and IMU are coupled to generate the state estimation of the robot through Lego-LOAM [4]. The main system structure of the mobile robot is shown in Figure 22, the autonomy system incorporates navigation modules from CMU autonomous development interface, e.g., terrain analysis and way-point following as fundamental navigation modules.

Figure 21. The mobile robot.

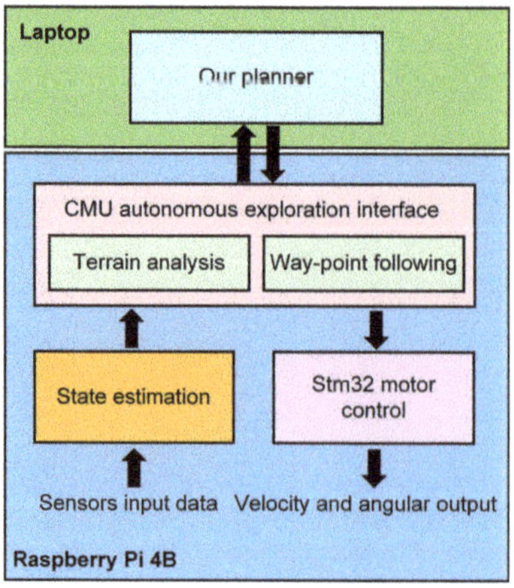

Figure 22. The main structure of autonomous system.

As shown in Figure 23a,b, obstacles are mapped as polygons in exploration, and solid edges are formed from each relevant vertex. In Figure 23c, colored pointclouds of obstacles are displayed to better show the details; orange lines denote the obstacles' outlines, while cyan lines denote the relationships between the edges of various obstacles.

In the navigation, as shown on the left side of Figure 23d, the mobile robot started from point 1 and arrived at points 2, 3, and 4 in sequence and on the right side of the Figure 23d shows pictures when the mobile robot navigates to the corresponding position.

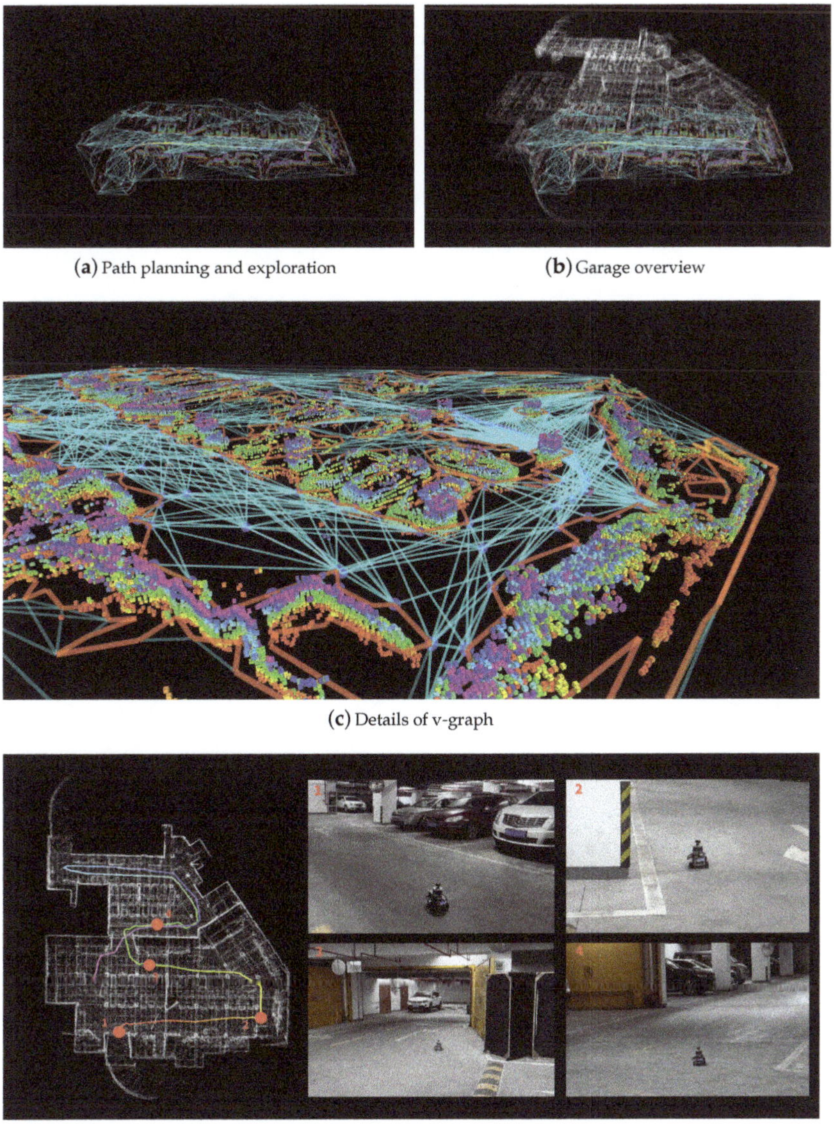

(**a**) Path planning and exploration (**b**) Garage overview

(**c**) Details of v-graph

(**d**) Navigate in the garage

Figure 23. (**a**,**b**) show the path planning and exploration of the garage. (**c**) shows the details of exploration, and (**d**) shows the entire navigation in the garage.

As shown in Figure 24, our algorithm is compared with FAR Planner in the physical experiment. The processing speed of the laser data, the update speed of the v-graph, and the operation speed of the search algorithm are recorded every 0.5 seconds.

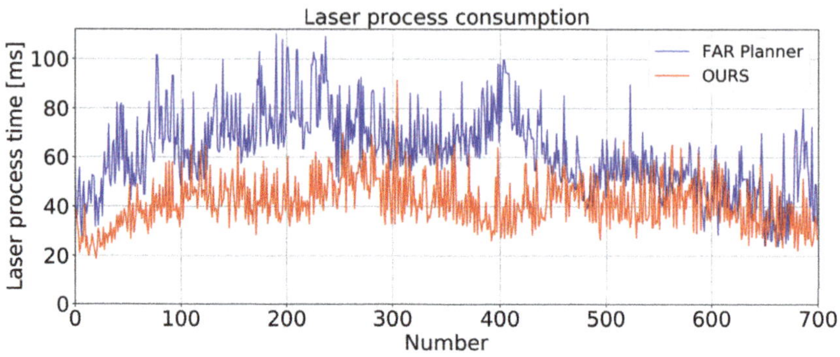

(**a**) Laser process time in the physical experiment

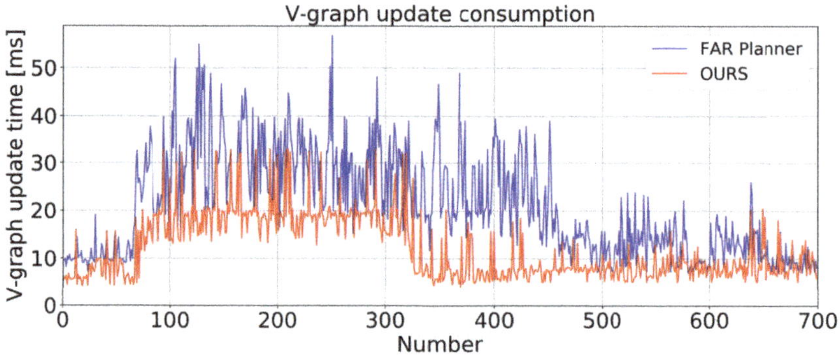

(**b**) V-graph update time in the physical experiment

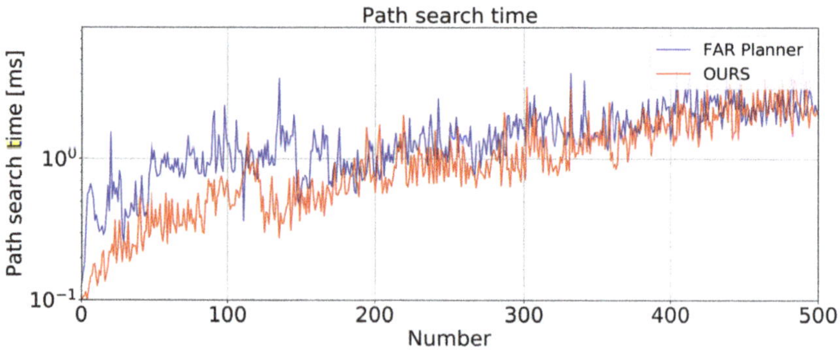

(**c**) Path search time in the physical experiment

Figure 24. Data recorded when the path planning algorithm runs.

In Figure 24a, The mean and standard deviation for the laser processing of FAR Planner are 60.95 ms and 16.59 ms, respectively, while our method's mean and standard deviation of the laser processing are 41.59 ms and 9.93 ms, respectively. Not only is ours 31.76% faster on average than FAR Planner, but the time taken by the algorithm is also more stable. As shown in Figure 24b, both approaches will take longer to update the v-graph because of the unexpected rise in obstacles, but ours is on average 37.12% quicker than FAR Planner. In Figure 24c, ours is on average 18.06% faster than the FAR Planner on the search algorithm.

5. Discussion

5.1. Grid in Mapping

In the process of robot simultaneous localization and mapping, grids can be used to store pointcloud data [12,54–58], and can also be used as occupancy grid maps for robot navigation [56,59–63]. As mentioned in Lau, B et al. [58], the processing speed of a computer is limited by the resolution of the grid. When the grid resolution is higher, the information represented by each cell is more accurate, but the calculation time is longer. In Homm et al. [61], they use a Graphics Processing Unit (GPU) to speed up the formation of fine grids, and in A. Birk et al. [56], they use multiple robots to jointly maintain grid maps, an approach that indirectly speeds up the construction of the grid map.

The aim of the work in this paper is to use a discrete hollow grid to convert pointcloud information into a binary image and extract obstacle vertices. Therefore, the focus of this paper is on how to quickly extract the grid to generate the vertices of obstacles. For a 30 m × 30 m local grid, the processing speed of a high-resolution grid, such as 0.5 m × 0.5 m per cell, is much slower than that of a low-resolution grid, such as 1 m × 1 m per cell. Therefore, a grid with spaced hollow structures is designed to speed up the processing of high-resolution grids. Since the use of hollow structures will affect subsequent images, such as the discontinuous edges of obstacles, it is necessary to blur the generated images. Blurring the image can cover empty spots caused by hollow structures. Additionally, a complementary hollow-structured grid is also considered, storing a set of complementary hollow-structured grids under adjacent data frames and integrated into the local map.

It is foreseeable that the grid application with this sparse structure can significantly reduce the amount of computation and shorten the computation time in three-dimensional space. In future work, the authors hope to use this sparse structure for 3D grids.

5.2. The Reduced Visibility Graph

The v-graph is a topology map that is widely used for path planning since it is constructed using the vertices of obstacles. The difficulty of calculating and maintaining the v-graph mainly depends on the number of vertices in the graph; thus, many researchers focus on how to simplify the v-graph [19,64–70].

In Nguyet et al. [69], he proposed a method for clustering small obstacles according to their volume, which can well reduce the number of vertices in the v-graph, thereby improving the efficiency of the path search algorithm. However, this method needs to calculate the total area of the global map for each iteration, which wastes a lot of time for multiple small-volume obstacles. In Yang et al. [19], an angle ζ is set to limit the visibility of each obstacle vertex, which can well reduce the number of edges of the obstacle vertex The method used in this paper combined Yang's method [19] and proposed a method of simplifying obstacle vertices that only act on the local map. Compared with [69], the speed of the algorithm proposed in this paper is not affected by the global map, and it effectively reduces invalid vertices and redundant edges.

5.3. Uncertainties in the Path Planning

The path planning is based on the produced map (they can be occupancy maps, topological maps, or semantic maps), and the localization of the robot is very important for constructing the map. In the simulation experiment, we can easily obtain the relevant data of the robot, such as the simulated IMU sensor information and the simulated LiDAR information, and this information is accurate and unbiased in the simulated environment to estimate the state of the robot. How well the robot is positioned determines whether the map used for navigation is available. However, in real life, we cannot obtain such unbiased data; therefore, we used Lego-LOAM for the state estimation of the robot. If the robot's state estimation data has a large error, the v-graph it builds will deviate from the real world.

As shown in Figure 25, in the real world, the mobile robot made an error in the state estimation, and the white pointcloud newly scanned by the LiDAR was obviously offset from the colored obstacle pointcloud. This will lead to the establishment of an unreliable v-graph, thus affecting the effect of path planning.

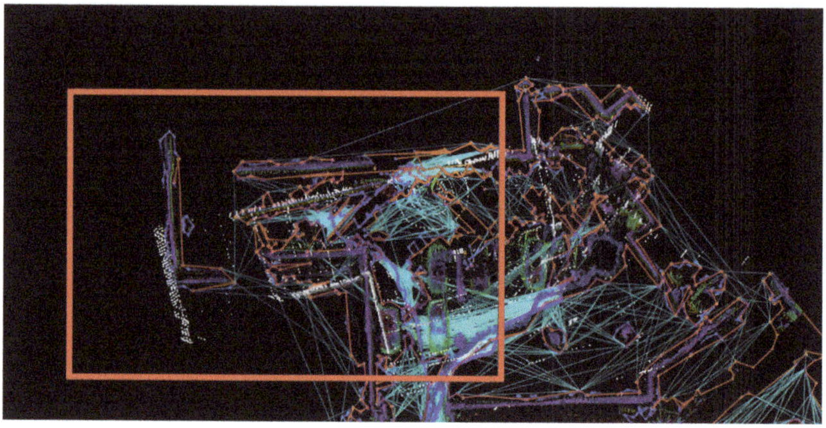

Figure 25. The unreliable v-graph.

6. Conclusions

This paper proposed a sparse visibility graph-based path planner based on the FAR Planner framework. Our method is far superior to the FAR Planner in terms of the efficiency of v-graph maintenance and generation. Our method can be used for navigation in known environments and exploration in unknown environments. Moreover, a complementary hollow grid is designed for local layer updates and merges the local layer into the global layer. For complex obstacles in the environment, a method was proposed to reduce the cost of maintaining the v-graph by simplifying the vertices and edges of polygons. For small obstacles, their information is still preserved in the graph. Moreover, the paper proposed a path planning method based on a bidirectional breadth-first search combined with the v-graph. By comparing the original algorithm of FAR Planner and the traditional search algorithms A* and D* Lite, ours achieves the optimal path planning in unknown environments, and the speed of the path search algorithm is faster than that of FAR Planner.

Author Contributions: Conceptualization, Q.L. and F.X.; methodology, Q.L., F.X., J.Z. and B.X.; software, Q.L., F.X. and J.Z.; validation, Q.L., F.X. and B.X.; formal analysis, F.X., J.Z.; investigation, J.Y.; resources, X.L.; data curation, H.S. and J.Z.; writing—original draft preparation, Q.L. and F.X.; writing—review and editing, Q.L., F.X. and B.X.; visualization, H.S. and J.Z.; supervision, J.Y.; project administration, F.X. and X.L.; funding acquisition, J.Y. All authors have read and agreed to the published version of the manuscript.

Funding: This work was partially supported by the National Natural Science Foundation of China (Grant Nos. 41974033 and 61803208), the Scientific and technological achievements transformation project of Jiangsu Province (BA2020004), 2020 Industrial Transformation and Upgrading Project of Industry and Information Technology Department of Jiangsu Province, Postgraduate Research & Practice Innovation Program of Jiangsu Province.

Data Availability Statement: The CMU simulation environment can be acquired through https://www.cmu-exploration.com. The Matterport3D simulation environment can be acquired through https://niessner.github.io/Matterport. The original framework used in the paper can be acquired through https://github.com/MichaelFYang/far_planner (all accessed on 14 June 2022).

Acknowledgments: The authors would like to sincerely thank all the funders.

Conflicts of Interest: The authors declare no conflict of interest. The funders had no role in the design of the study; in the collection, analyses, or interpretation of data; in the writing of the manuscript, or in the decision to publish the results.

Abbreviations

The following abbreviations are used in this manuscript:

FAR Planner	Fast, Attemptable Route Planner using Dynamic Visibility Update [19]
SLAM	Simultaneous Localization And Mapping
IMU	Inertial Measurement Unit
GPU	Graphics Processing Unit

References

1. Hess, W.; Kohler, D.; Rapp, H.; Andor, D. Real-time loop closure in 2D LIDAR SLAM. In Proceedings of the 2016 IEEE International Conference on Robotics and Automation (ICRA), Stockholm, Sweden, 16–21 May 2016; pp. 1271–1278. [CrossRef]
2. Grisetti, G.; Kümmerle, R.; Stachniss, C.; Burgard, W. A Tutorial on Graph-Based SLAM. *IEEE Intell. Transp. Syst. Mag.* **2010**, *2*, 31–43. [CrossRef]
3. Li, T.; Pei, L.; Xiang, Y.; Wu, Q.; Xia, S.; Tao, L.; Guan, X.; Yu, W. P3-LOAM: PPP/LiDAR Loosely Coupled SLAM With Accurate Covariance Estimation and Robust RAIM in Urban Canyon Environment. *IEEE Sens. J.* **2021**, *21*, 6660–6671. [CrossRef]
4. Shan, T.; Englot, B. Lego-loam: Lightweight and ground-optimized lidar odometry and mapping on variable terrain. In Proceedings of the 2018 IEEE/RSJ International Conference on Intelligent Robots and Systems (IROS), Madrid, Spain, 1–5 October 2018; pp. 4758–4765.
5. Li, J.; Pei, L.; Zou, D.; Xia, S.; Wu, Q.; Li, T.; Sun, Z.; Yu, W. Attention-SLAM: A Visual Monocular SLAM Learning From Human Gaze. *IEEE Sens. J.* **2021**, *21*, 6408–6420. [CrossRef]
6. Mur-Artal, R.; Montiel, J.M.M.; Tardós, J.D. ORB-SLAM: A Versatile and Accurate Monocular SLAM System. *IEEE Trans. Robot.* **2015**, *31*, 1147–1163. [CrossRef]
7. Mur-Artal, R.; Tardós, J.D. ORB-SLAM2: An Open-Source SLAM System for Monocular, Stereo, and RGB-D Cameras. *IEEE Trans. Robot.* **2017**, *33*, 1255–1262. [CrossRef]
8. Qin, T.; Li, P.; Shen, S. VINS-Mono: A Robust and Versatile Monocular Visual-Inertial State Estimator. *IEEE Trans. Robot.* **2018**, *34*, 1004–1020. [CrossRef]
9. Chan, S.H.; Wu, P.T.; Fu, L.C. Robust 2D Indoor Localization Through Laser SLAM and Visual SLAM Fusion. In Proceedings of the 2018 IEEE International Conference on Systems, Man, and Cybernetics (SMC), Miyazaki, Japan, 7–10 October 2018; pp. 1263–1268. [CrossRef]
10. Debeunne, C.; Vivet, D. A Review of Visual-LiDAR Fusion based Simultaneous Localization and Mapping. *Sensors* **2020**, *20*, 2068. [CrossRef]
11. Nguyen, T.M.; Cao, M.; Yuan, S.; Lyu, Y.; Nguyen, T.H.; Xie, L. VIRAL-Fusion: A Visual-Inertial-Ranging-Lidar Sensor Fusion Approach. *IEEE Trans. Robot.* **2022**, *38*, 958–977. [CrossRef]
12. Thrun, S. Learning Occupancy Grid Maps with Forward Sensor Models. *Auton. Robot.* **2003**, *15*, 111–127. [CrossRef]
13. Lozano-Pérez, T.; Wesley, M.A. An algorithm for planning collision-free paths among polyhedral obstacles. *Commun. ACM* **1979**, *22*, 560–570. [CrossRef]
14. Kitzinger, J.; Moret, B. The Visibility Graph among Polygonal Obstacles: A Comparison of Algorithms. Ph.D. Thesis, University of New Mexico, Albuquerque, NM, USA, 2003.
15. Alt, H.; Welzl, E. Visibility graphs and obstacle-avoiding shortest paths. *Z. Für Oper. Res.* **1988**, *32*, 145–164. [CrossRef]
16. Shen, X.; Edelsbrunner, H. A tight lower bound on the size of visibility graphs. *Inf. Process. Lett.* **1987**, *26*, 61–64. [CrossRef]
17. Sridharan, K.; Priya, T.K. An Efficient Algorithm to Construct Reduced Visibility Graph and Its FPGA Implementation. In Proceedings of the VLSI Design, International Conference, Mumbai, India, 9 January 2004; IEEE Computer Society: Los Alamitos, CA, USA, 2004; p. 1057. [CrossRef]
18. Wu, G.; Atilla, I.; Tahsin, T.; Terziev, M.; Wang, L. Long-voyage route planning method based on multi-scale visibility graph for autonomous ships. *Ocean Eng.* **2021**, *219*, 108242. [CrossRef]
19. Yang, F.; Cao, C.; Zhu, H.; Oh, J.; Zhang, J. FAR Planner: Fast, Attemptable Route Planner using Dynamic Visibility Update. *arXiv* **2021**, arXiv:2110.09460.

20. Pradhan, S.; Mandava, R.K.; Vundavilli, P.R. Development of path planning algorithm for biped robot using combined multi-point RRT and visibility graph. *Int. J. Inf. Technol.* **2021**, *13*, 1513–1519. [CrossRef]
21. D'Amato, E.; Nardi, V.A.; Notaro, I.; Scordamaglia, V. A Visibility Graph approach for path planning and real-time collision avoidance on maritime unmanned systems. In Proceedings of the 2021 International Workshop on Metrology for the Sea; Learning to Measure Sea Health Parameters (MetroSea), Reggio Calabria, Italy, 4–6 October 2021; pp. 400–405. [CrossRef]
22. Dijkstra, E.W. A note on two problems in connexion with graphs. *Numer. Math.* **1959**, *1*, 269–271. [CrossRef]
23. Hart, P.E.; Nilsson, N.J.; Raphael, B. A formal basis for the heuristic determination of minimum cost paths. *IEEE Trans. Syst. Sci. Cybern.* **1968**, *4*, 100–107. [CrossRef]
24. Stentz, A. Optimal and efficient path planning for partially-known environments. In Proceedings of the 1994 IEEE International Conference on Robotics and Automation, San Diego, CA, USA, 8–13 May 1994; Volume 4, pp. 3310–3317. [CrossRef]
25. Koenig, S.; Likhachev, M. Fast replanning for navigation in unknown terrain. *IEEE Trans. Robot.* **2005**, *21*, 354–363. [CrossRef]
26. XiangRong, T.; Yukun, Z.; XinXin, J. Improved A-star algorithm for robot path planning in static environment. *J. Phys. Conf. Ser.* **2021**, *1792*, 012067. [CrossRef]
27. Zhang, L.; Li, Y. Mobile Robot Path Planning Algorithm Based on Improved A Star. *J. Phys. Conf. Ser.* **2021**, *1848*, 012013. [CrossRef]
28. LaValle, S.M.; Kuffner, J.J.; Donald, B. Rapidly-exploring random trees: Progress and prospects. *Algorithmic Comput. Robot. New Dir.* **2001**, *5*, 293–308.
29. Karaman, S.; Frazzoli, E. Sampling-based algorithms for optimal motion planning. *Int. J. Robot. Res.* **2011**, *30*, 846–894. [CrossRef]
30. Gammell, J.D.; Srinivasa, S.S.; Barfoot, T.D. Informed RRT*: Optimal sampling-based path planning focused via direct sampling of an admissible ellipsoidal heuristic. In Proceedings of the 2014 IEEE/RSJ International Conference on Intelligent Robots and Systems, Chicago, IL, USA, 14–18 September 2014; pp. 2997–3004.
31. Kuffner, J.J.; LaValle, S.M. RRT-connect: An efficient approach to single-query path planning. In Proceedings of the 2000 ICRA, Millennium Conference, IEEE International Conference on Robotics and Automation, Symposia Proceedings (Cat. No. 00CH37065), San Francisco, CA, USA, 24–28 April 2000; Volume 2, pp. 995–1001.
32. Tu, J.; Yang, S. Genetic algorithm based path planning for a mobile robot. In Proceedings of the 2003 IEEE International Conference on Robotics and Automation (Cat. No.03CH37422), Taipei, Taiwan, 14–19 September 2003; Volume 1, pp. 1221–1226. [CrossRef]
33. Chen, J.; Ye, F.; Li, Y. Travelling salesman problem for UAV path planning with two parallel optimization algorithms. In Proceedings of the 2017 Progress in Electromagnetics Research Symposium—Fall (PIERS - FALL), Singapore, 19–22 November 2017; pp. 832–837. [CrossRef]
34. Pan, Y.; Yang, Y.; Li, W. A Deep Learning Trained by Genetic Algorithm to Improve the Efficiency of Path Planning for Data Collection With Multi-UAV. *IEEE Access* **2021**, *9*, 7994–8005. [CrossRef]
35. Hao, K.; Zhao, J.; Wang, B.; Liu, Y.; Wang, C. The application of an adaptive genetic algorithm based on collision detection in path planning of mobile robots. *Comput. Intell. Neurosci.* **2021**, *2021*, 5536574. [CrossRef] [PubMed]
36. Rahmaniar, W.; Rakhmania, A.E. Mobile Robot Path Planning in a Trajectory with Multiple Obstacles Using Genetic Algorithms. *J. Robot. Control (JRC)* **2022**, *3*, 1–7. [CrossRef]
37. Zhang, T.W.; Xu, G.H.; Zhan, X.S.; Han, T. A new hybrid algorithm for path planning of mobile robot. *J. Supercomput.* **2022**, *78*, 4158–4181. [CrossRef]
38. Richter, C.; Vega-Brown, W.; Roy, N., Bayesian Learning for Safe High-Speed Navigation in Unknown Environments. In *Robotics Research: Volume 2*; Bicchi, A., Burgard, W., Eds.; Springer: Cham, Switzerland, 2018; pp. 325–341. [CrossRef]
39. Zeng, J.; Ju, R.; Qin, L.; Hu, Y.; Yin, Q.; Hu, C. Navigation in unknown dynamic environments based on deep reinforcement learning. *Sensors* **2019**, *19*, 3837. [CrossRef]
40. Guo, X.; Fang, Y. Learning to Navigate in Unknown Environments Based on GMRP-N. In Proceedings of the 2019 IEEE 9th Annual International Conference on CYBER Technology in Automation, Control, and Intelligent Systems (CYBER), Suzhou, China, 29 July–2 August 2019; pp. 1453–1458.
41. Lin, G.; Zhu, L.; Li, J.; Zou, X.; Tang, Y. Collision-free path planning for a guava-harvesting robot based on recurrent deep reinforcement learning. *Comput. Electron. Agric.* **2021**, *188*, 106350. [CrossRef]
42. Tutsoy, O.; Brown, M. Reinforcement learning analysis for a minimum time balance problem. *Trans. Inst. Meas. Control* **2016**, *38*, 1186–1200. [CrossRef]
43. Tutsoy, O.; Barkana, D.E.; Balikci, K. A novel exploration-exploitation-based adaptive law for intelligent model-free control approaches. *IEEE Trans. Cybern.* **2021**, 1–9. [CrossRef]
44. Oommen, B.; Iyengar, S.; Rao, N.; Kashyap, R. Robot navigation in unknown terrains using learned visibility graphs. Part I: The disjoint convex obstacle case. *IEEE J. Robot. Autom.* **1987**, *3*, 672–681. [CrossRef]
45. Wooden, D.; Egerstedt, M. Oriented visibility graphs: Low-complexity planning in real-time environments. In Proceedings of the 2006 IEEE International Conference on Robotics and Automation, Orlando, FL, USA, 15–19 May 2006; pp. 2354–2359.
46. Rao, N.S. Robot navigation in unknown generalized polygonal terrains using vision sensors. *IEEE Trans. Syst. Man Cybern.* **1995**, *25*, 947–962. [CrossRef]

47. Cao, C.; Zhu, H.; Yang, F.; Xia, Y.; Choset, H.; Oh, J.; Zhang, J. Autonomous Exploration Development Environment and the Planning Algorithms. In Proceedings of the 2022 International Conference on Robotics and Automation (ICRA), Philadelphia, PA, USA, 23–27 May 2022.
48. Chang, A.; Dai, A.; Funkhouser, T.; Halber, M.; Niessner, M.; Savva, M.; Song, S.; Zeng, A.; Zhang, Y. Matterport3D: Learning from RGB-D Data in Indoor Environments. *arXiv* **2017**, arXiv:1709.06158.
49. Wei, Y.; Xiao, H.; Shi, H.; Jie, Z.; Feng, J.; Huang, T.S. Revisiting dilated convolution: A simple approach for weakly-and semi-supervised semantic segmentation. In Proceedings of the IEEE Conference on Computer Vision and Pattern Recognition, Salt Lake City, UT, USA, 18–23 June 2018; pp. 7268–7277.
50. Suzuki, S. Topological structural analysis of digitized binary images by border following. *Comput. Vision Graph. Image Process.* **1985**, *30*, 32–46. [CrossRef]
51. Teh, C.H.; Chin, R.T. On the detection of dominant points on digital curves. *IEEE Trans. Pattern Anal. Mach. Intell.* **1989**, *11*, 859–872. [CrossRef]
52. Lee, D.T. Proximity and Reachability in the Plane. Ph.D. Thesis, University of Illinois at Urbana-Champaign, Champaign, IL, USA, 1978.
53. Quigley, M.; Conley, K.; Gerkey, B.; Faust, J.; Foote, T.; Leibs, J.; Wheeler, R.; Ng, A.Y. ROS: An open-source Robot Operating System. In Proceedings of the ICRA Workshop on Open Source Software, Kobe, Japan, 12–17 May 2009; Volume 3, p. 5.
54. Will, P.; Pennington, K. Grid coding: A preprocessing technique for robot and machine vision. *Artif. Intell.* **1971**, *2*, 319–329. [CrossRef]
55. Agarwal, P.K.; Arge, L.; Danner, A., From Point Cloud to Grid DEM: A Scalable Approach. In *Progress in Spatial Data Handling: 12th International Symposium on Spatial Data Handling*; Riedl, A., Kainz, W., Elmes, G.A., Eds.; Springer: Berlin/Heidelberg, Germany, 2006; pp. 771–788. [CrossRef]
56. Birk, A.; Carpin, S. Merging Occupancy Grid Maps From Multiple Robots. *Proc. IEEE* **2006**, *94*, 1384–1397. [CrossRef]
57. Meyer-Delius, D.; Beinhofer, M.; Burgard, W. Occupancy grid models for robot mapping in changing environments. In Proceedings of the Twenty-Sixth AAAI Conference on Artificial Intelligence, Toronto, ON, Canada, 22–26 July 2012.
58. Lau, B.; Sprunk, C.; Burgard, W. Efficient grid-based spatial representations for robot navigation in dynamic environments. *Robot. Auton. Syst.* **2013**, *61*, 1116–1130. [CrossRef]
59. Kim, B.; Kang, C.M.; Kim, J.; Lee, S.H.; Chung, C.C.; Choi, J.W. Probabilistic vehicle trajectory prediction over occupancy grid map via recurrent neural network. In Proceedings of the 2017 IEEE 20th International Conference on Intelligent Transportation Systems (ITSC), Yokohama, Japan, 16–19 October 2017; pp. 399–404. [CrossRef]
60. Elfes, A. Using occupancy grids for mobile robot perception and navigation. *Computer* **1989**, *22*, 46–57. [CrossRef]
61. Homm, F.; Kaempchen, N.; Ota, J.; Burschka, D. Efficient occupancy grid computation on the GPU with lidar and radar for road boundary detection. In Proceedings of the 2010 IEEE Intelligent Vehicles Symposium, La Jolla, CA, USA, 21–24 June 2010; pp. 1006–1013. [CrossRef]
62. Li, H.; Tsukada, M.; Nashashibi, F.; Parent, M. Multivehicle Cooperative Local Mapping: A Methodology Based on Occupancy Grid Map Merging. *IEEE Trans. Intell. Transp. Syst.* **2014**, *15*, 2089–2100. [CrossRef]
63. Li, Y.; Ruichek, Y. Occupancy Grid Mapping in Urban Environments from a Moving On-Board Stereo-Vision System. *Sensors* **2014**, *14*, 10454–10478. [CrossRef]
64. Kneidl, A.; Borrmann, A.; Hartmann, D. Generation and use of sparse navigation graphs for microscopic pedestrian simulation models. *Adv. Eng. Inform.* **2012**, *26*, 669–680. [CrossRef]
65. Welzl, E. Constructing the visibility graph for n-line segments in O(n2) time. *Inf. Process. Lett.* **1985**, *20*, 167–171. [CrossRef]
66. Majeed, A.; Lee, S. A Fast Global Flight Path Planning Algorithm Based on Space Circumscription and Sparse Visibility Graph for Unmanned Aerial Vehicle. *Electronics* **2018**, *7*, 375. [CrossRef]
67. Oleynikova, H.; Taylor, Z.; Siegwart, R.; Nieto, J. Sparse 3D Topological Graphs for Micro-Aerial Vehicle Planning. In Proceedings of the 2018 IEEE/RSJ International Conference on Intelligent Robots and Systems (IROS), Madrid, Spain, 1–5 October 2018; pp. 1–9. [CrossRef]
68. Himmel, A.S.; Hoffmann, C.; Kunz, P.; Froese, V.; Sorge, M. Computational complexity aspects of point visibility graphs. *Discret. Appl. Math.* **2019**, *254*, 283–290. [CrossRef]
69. Nguyet, T.T.N.; Hoai, T.V.; Thi, N.A. Some Advanced Techniques in Reducing Time for Path Planning Based on Visibility Graph. In Proceedings of the 2011 Third International Conference on Knowledge and Systems Engineering, Hanoi, Vietnam, 14–17 October 2011; pp. 190–194. [CrossRef]
70. Ben-Moshe, B.; Hall-Holt, O.; Katz, M.J.; Mitchell, J.S.B. Computing the Visibility Graph of Points within a Polygon. In Proceedings of the Twentieth Annual Symposium on Computational Geometry, Brooklyn, NY, USA, 8–11 June 2004; Association for Computing Machinery: New York, NY, USA, 2004; p. 27–35. [CrossRef]

remote sensing

Article

3D LiDAR Aided GNSS/INS Integration Fault Detection, Localization and Integrity Assessment in Urban Canyons

Zhipeng Wang [1], Bo Li [1], Zhiqiang Dan [1], Hongxia Wang [2,*] and Kun Fang [3]

1 National Key Laboratory of CNS/ATM, School of Electronics and Information Engineering, Beihang University, Beijing 100191, China
2 Beihang School, Beihang University, Beijing 100191, China
3 Research Institute for Frontier Science, Beihang University, Beijing 100191, China
* Correspondence: 10455@buaa.edu.cn; Tel.: +86-010-61716783

Abstract: The performance of Global Navigation Satellite System (GNSS) and Inertial Navigation System (INS) integrated navigation can be severely degraded in urban canyons due to the non-line-of-sight (NLOS) signals and multipath effects. Therefore, to achieve a high-precision and robust integrated system, real-time fault detection and localization algorithms are needed to ensure integrity. Currently, the residual chi-square test is used for fault detection in the positioning domain, but it has poor sensitivity when faults disappear. Three-dimensional (3D) light detection and ranging (LiDAR) has good positioning performance in complex environments. First, a LiDAR aided real-time fault detection algorithm is proposed. A test statistic is constructed by the mean deviation of the matched targets, and a dynamic threshold is constructed by a sliding window. Second, to solve the problem that measurement noise is estimated by prior modeling with a certain error, a LiDAR aided real-time measurement noise estimation based on adaptive filter localization algorithm is proposed according to the position deviations of matched targets. Finally, the integrity of the integrated system is assessed. The error bound of integrated positioning is innovatively verified with real test data. We conduct two experiments with a vehicle going through a viaduct and a floor hole, which, represent mid and deep urban canyons, respectively. The experimental results show that in terms of fault detection, the fault could be detected in mid urban canyons and the response time of fault disappearance is reduced by 70.24% in deep urban canyons. Thus, the poor sensitivity of the residual chi-square test for fault disappearance is improved. In terms of localization, the proposed algorithm is compared with the optimal fading factor adaptive filter (OFFAF) and the extended Kalman filter (EKF). The proposed algorithm is the most effective, and the Root Mean Square Error (RMSE) in the east and north is reduced by 12.98% and 35.1% in deep urban canyons. Regarding integrity assessment, the error bound can overbound the positioning errors in deep urban canyons relative to the EKF and the mean value of the error bounds is reduced.

Keywords: GNSS/INS; 3D LiDAR; fault detection; localization; integrity assessment

Citation: Wang, Z.; Li, B.; Dan, Z.; Wang, H.; Fang, K. 3D LiDAR Aided GNSS/INS Integration Fault Detection, Localization and Integrity Assessment in Urban Canyons. *Remote Sens.* **2022**, *14*, 4641. https://doi.org/10.3390/rs14184641

Academic Editors: Yuwei Chen, Changhui Jiang, Qian Meng, Bing Xu, Wang Gao, Panlong Wu, Lianwu Guan and Zeyu Li

Received: 28 July 2022
Accepted: 14 September 2022
Published: 16 September 2022

1. Introduction

A reliable positioning system is the basis of autonomous driving [1]. Integrity is an important indicator for ensuring the driving safety of vehicles. The integration of the Global Satellite Navigation System (GNSS) and the Inertial Navigation System (INS) could provide real-time and high-precision positioning and this approach is widely used in military and civil fields [2,3]. However, positioning and navigation in urban canyons is still a challenge [4] because multipath GNSS signals are received due to reflection or non-line-of-sight (NLOS) signals are received due to diffraction; this eventually leads to unacceptable GNSS measurement errors [5–7]. The comprehensive positioning performance of the GNSS/INS integrated navigation system is seriously degraded. Therefore, to meet the positioning performance requirements in urban canyons, such as viaducts, floor holes and tunnels, the real-time detection of GNSS/INS integrated system positioning performance is essential. However, in urban canyons, there are a lot of satellites affected by NLOS and multipath. With all four core GNSS constellations of the world in operation, the number of affected satellites is ten or more. It is difficult to detect all the affected satellites and exclude them from the localization solution by traditional algorithm such as Multiple Solution Separation (MSS). Therefore, we detect faults in the positioning domain. After faults are detected, a GNSS/INS integration algorithm can be designed based on the detection results to improve the positioning accuracy and ensure the integrity of the system.

GNSS fault detection methods are mainly divided based on whether to detect specific faulty satellites [8]. Blanch et al. only used GNSS measurement information to propose a method combining a greedy search and L1 norm minimization to detect satellites with pseudo range errors greater than 20 m that were affected by multipath and NLOS signals [9]. However, in urban canyons, rapid changes in observation satellites reduced the performance of the method. Sun et al. proposed a dynamic detection and multiple fault elimination algorithm based on pseudo range comparison [10]. Using inertial measurement unit (IMU) and GNSS pseudo range data, a parallel fault detection and exclusion scheme consisting of a sliding window and a detector was designed, which can detect multiple faulty satellites in real-time. This scheme is suitable for the tightly and loosely coupled architecture. Groves et al. proposed a likelihood-based 3D-mapping-aided (3DMA) GNSS ranging algorithm and signals, which are predicted to be non-line-of-sight (NLOS) to contribute to the position solution without explicitly computing the additional path delay due to NLOS reception, which is computationally expensive [11]. Sun et al. proposed a new measurement noise covariance update scheme, with the adaptive indicator generated from pseudo range error prediction results, for a tightly coupled GNSS/IMU navigation system in urban areas [12]. Shytermeja and Attia et al. proposed the method of using a fish-eye camera to suppress NLOS signals and eliminate influenced satellites [13,14]. Fisheye images are divided into sky and non-sky regions. The GNSS fault detection can be realized by identifying GNSS satellites in the non-sky region received by the receiver. However, satellites affected by multipath signals cannot be excluded by this method. Moreover, the processing technique used for camera data is complex. Wen et al. proposed a similar approach using a real-time 3D point cloud or a sliding window map for identifying edges on the top of the given building [15,16]. Based on the relationship between the edges on the top of the building or double-decker bus and all observed satellites, the satellites influenced by the NLOS signals are detected and pseudo ranges are corrected. However, only the satellites affected by NLOS signals are detected, and the environmental conditions are highly demanding. All four of the above methods need to detect specific affected satellites and exclude them from the positioning solution. However, in urban canyons, lots of satellites are affected by NLOS and multipath signals, so it is difficult to exclude them individually. The residual chi-square test is widely used for fault detection; it constructs a fault detector via the innovation of the Kalman filter (KF) [17]. The test is performed at the positioning domain to detect whether a positioning fault is present. The algorithm is efficient and works in real-time, but the sensitivity of the residual chi-square test is poor when faults disappear. As mentioned above, although the effectiveness of the existing fault detection algorithms for detecting satellites affected by multipath and NLOS signals in complex environments has been proven, challenges remain concerning the real-time performance of dynamic applications [18,19]. Therefore, it is necessary to propose a new real-time fault detection algorithm to improve the sensitivity of the residual chi-square test to fault disappearances, and the integrity of this approach should be assessed in terms of its false alarm rate and missed detection rate. LiDAR is an important sensor in autonomous driving localization, which has good performance in ranging and perception [20]. As the cost of solid-state LiDAR decreases, LiDAR has a wider range of applications [21]. LiDAR has robust performance in building high-precision map and object detection for autonomous driving [22,23]. In this paper, we carry out fault detection algorithm aided by LiDAR.

The KF is a famous recursive algorithm for discrete linear systems and has been widely used in many fields [24]. When the given system is linear and the observed noise follows a Gaussian distribution, the KF can be proven to be optimal. However, Chang et al. pointed out that the noise probability does not obey a Gaussian distribution in practice, and it is difficult to determine the dynamic model and statistical information of the noise distribution [25]. Many scholars have studied a series of filters for non-Gaussian-distributed noise. Gordon et al. developed a particle filter (PF) to address the problem of non-Gaussian Bayesian estimation [26]. However, high-dimensional state estimation may result in a heavy computational burden. The H∞ filter was proposed for the uncertainty of measurement noise, but Rigatos pointed out that it could not solve the problem that GNSS-measured values were outliers [27]. The fading filter and adaptive filter are two widely used filters that have been proposed to solve the uncertainty of the noise distribution characteristics of dynamic systems. Chen et al. analyzed the difference between the adaptive filter and fading filter in detail. The fading filter mainly sets the weight of the state covariance matrix, while the adaptive filter mainly sets the weight of the measured noise [28]. Fagin and Sorenson proposed the fading filter algorithm in the 1960s [29]. Lee and Sun et al. proposed utilizing the fading filter method in the field of integrated navigation [30,31]. Although these methods improve the performance of the KF and the resultant positioning accuracy, many matrix multiplication and inversion operations are involved, resulting in poor practicability. Li et al. proposed a dynamic fading filter algorithm to solve problems with difficult matrix multiplication operations, greatly improving the performance of the algorithm [2]. Li et al. proposed a fading filter to defend against outliers. When a GNSS fault occurred, the innovation of the last n epochs was used to estimate the innovation of the current epoch, but the weights of the last n epochs were not considered [32]. Wang et al. developed the Sage–Husa adaptive filter, designed a sliding window and determined the covariance matrix of the current epoch based on an iterative method, but it was quite difficult to choose the window length [33]. Zhou et al. proposed a new adaptive unscented KF (N-AUKF) algorithm [34]. A covariance matching criterion was designed to judge the filtering divergence, and an adaptive weighted coefficient was applied to restrain the divergence. However, this approach requires adequate measurements. Li et al. extended this definition to Kalman filtering to detect gross errors, explained its nature and its relation with the currently adopted Chi-square variables of Kalman filtering in model and data spaces and compared them with the predictive residual statistics, starting by defining an incremental chi-square method of recursive least squares [35]. All of the above algorithms are effective in complex environments. However, the measurement noise is estimated by prior modeling with a certain error. Therefore, it is necessary to estimate the measurement noise with multi-sensor redundant measurement information.

Integrity is an important indicator of GNSS positioning in the aviation field [36]. Integrity evaluation from the GNSS to the GNSS/INS integrated positioning. Protection level is an important index of integrity monitoring, which represents the upper limit of positioning error [37]. However, in the existing research, the protection level in GNSS/INS integrated positioning is mainly based on simulation experiments [38], so it is necessary to carry out the verification protection level using the real test data. The protection level can meet the application requirement of autonomous driving.

The main contributions of this paper are as follows. (1) Regarding the problem that the residual chi-square test has low sensitivity when faults disappear, a LiDAR aided real-time GNSS/INS integrated fault detection algorithm based on the characteristics of high-precision 3D LiDAR is proposed. The test statistic is constructed based on the mean position deviation of the matched targets, and the dynamic threshold is constructed by a sliding window based on the power series. (2) To solve the problem that the measurement noise is estimated by prior modeling with a certain error, a LiDAR aided measurement noise estimation and filtering algorithm is proposed. The mean position deviations of the matched targets for the last n epochs are normalized, and an adaptive weight sequence is built. The adaptive measurement noise factor is joined with the Extended Kalman

Filter (EKF) innovation covariance. (3) For the problem that the integrated navigation error bounds are calculated by simulations in the existing research, we innovatively verify the error bounds of the integrated navigation system on real test data and optimize the algorithm of error bounds by using the filter in (2).

The subsequent sections of this paper are organized as follows. An overview of the proposed algorithm is given in Section 2. The proposed LiDAR aided real-time fault detection algorithm is presented in Section 3. In Section 4, the LiDAR aided measurement noise estimation adaptive filter algorithm is provided. Two experiments are performed to verify the effectiveness of the proposed algorithm, and its integrity is assessed in Section 5. Finally, Section 6 presents the conclusions and directions for future studies.

2. Overview of the Proposed Algorithm

An overview of the algorithm proposed in this paper is shown in Figure 1. First, based on LiDAR's high-precision ranging and its perception of the surrounding environment, a LiDAR aided real-time GNSS/INS integrated positioning fault detection algorithm is proposed. A 3D LiDAR global point cloud map based on the LiDAR odometry and mapping (LOAM) [39] algorithm is established in this paper, and the map is processed in Cloud Compare software to mark the detected targets and achieve target detection. The targets in this study are mainly rods, including tree trunks and lampposts. Onboard the LiDAR, real-time scanning of the surrounding environment is carried out to detect the targets of interest based on the single frame point cloud. The targets detected based on this single frame point cloud are matched with the targets marked on the global point cloud map. A test statistic is constructed based on the mean position deviation of the matched targets, and a dynamic threshold is constructed by a sliding window based on the power series. A detector is used for the detection of GNSS/INS integrated positioning faults affected by NLOS and multipath signals. The specific implementation and detailed description of the fault detection algorithm is in Section 3.

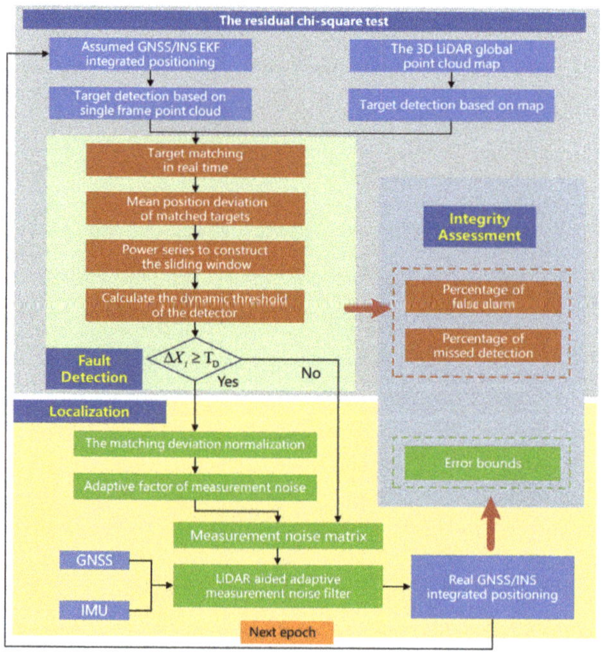

Figure 1. System framework of the proposed algorithm.

Second, a LiDAR aided real-time measurement noise estimation adaptive filter positioning algorithm is proposed. Based on the proposed fault detection algorithm, the proposed positioning algorithm is used to improve the performance of GNSS/INS integrated positioning. The adaptive measurement noise factor is joined with the EKF innovation covariance. The mean position deviations of the matched targets for the last n epochs are normalized, and an adaptive weight sequence is built. Then, the innovation of the current epoch is calculated. Thus, the adaptive measurement noise factor for the current epoch is determined to achieve the proposed filtering and localization algorithm. The specific implementation and detailed description of the localization algorithm are in Section 4.

Finally, the integrity of the proposed algorithm is assessed based on real test data; the evaluation metrics include the missed detection and false alarm rates for fault detection and the error bounds of the localization algorithm. The integrity assessment is realized for the proposed LiDAR aided fault detection and localization algorithm under the influence of NLOS and multipath signals.

3. LiDAR Aided Real-Time Fault Detection Algorithm

3.1. KF Architecture and the Residual Chi-Square Test

3.1.1. KF Architecture

The KF is composed of state prediction and measurement update equations, as shown in Equations (1) and (2), respectively [35]. In these equations, $\mathbf{X}_k$ represents the state vector of the system at the kth epoch, $\mathbf{\Phi}_{k/k-1}$ denotes the state transition matrix of the system from the $(k-1)$th epoch to the kth epoch, $\mathbf{W}_k$ represents the system noise, $\mathbf{Z}_k$ is the measurement vector of the system at the kth epoch, $\mathbf{H}_k$ represents the system measurement matrix at the kth epoch, $\mathbf{V}_k$ is the measurement noise vector and $\mathbf{\Gamma}_k$ is the system noise distribution matrix.

$$\mathbf{X}_k = \mathbf{\Phi}_{k/k-1}\mathbf{X}_{k-1} + \mathbf{\Gamma}_k\mathbf{W}_k \tag{1}$$

$$\mathbf{Z}_k = \mathbf{H}_k\mathbf{X}_k + \mathbf{V}_k \tag{2}$$

In our research, the KF is used for integrating the GNSS and the INS and the loosely coupled architecture is used. According to [40], we estimate a 15-dimensional state vector $\mathbf{X}$, which includes the position error δp, velocity error δv^n, attitude error ϕ^n, static accelerometer bias δb_a and static gyroscope bias δb_g, each of which possesses a 3D vector. The state vector $\mathbf{X}$ is given by

$$\mathbf{X} = \left[\delta p\ \delta v^n \phi^n \delta b_a \delta b_g\right]^T \tag{3}$$

The measurement vector $\mathbf{Z}$ is 3-dimensional, which includes the deviation between the position of GNSS and INS. p_{GNSS} and p_{INS} are the GNSS position and INS position, respectively. The p_{GNSS} is the GNSS position of the output of Newton-M2 receiver with onboard RTK.

$$\mathbf{Z} = [p_{GNSS} - p_{INS}] = \begin{bmatrix} p^X_{GNSS} - p^X_{INS} \\ p^Y_{GNSS} - p^Y_{INS} \\ p^Z_{GNSS} - p^Z_{INS} \end{bmatrix} \tag{4}$$

The system noise vector $\mathbf{W}$ is 6-dimensional, which includes the gyroscope and accelerometer noise. The matrix $\mathbf{Q}$ is the covariance of $\mathbf{W}$.

$$\mathbf{Q} = diag\left[\sigma^2_{gR}\quad \sigma^2_{gF}\quad \sigma^2_{gU}\quad \sigma^2_{aR}\quad \sigma^2_{aF}\quad \sigma^2_{aU}\right] \tag{5}$$

The measurement noise vector $\mathbf{V}$ is 3-dimensional, which includes the GNSS uncertainty. The matrix $\mathbf{R}$ is the covariance of $\mathbf{V}$.

$$\mathbf{R} = diag\left[\sigma^2_X\quad \sigma^2_Y\quad \sigma^2_Z\right] \tag{6}$$

In the GNSS/INS integrated positioning, the $\mathbf{Q}$ is set by IMU sensor information and the $\mathbf{R}$ is set by the GNSS uncertainty of the output of Newton-M2 receiver, which changes with time. When the GNSS positioning quality is degraded, the uncertainty becomes significantly larger. The uncertainty of the output is in the Earth-North-Up (ENU) frame and it is transformed to Earth-Centered and Earth-Fixed (ECEF) frame.

When the initial state information, the state vector $\mathbf{X}_0$ and the initial value of the one-step prediction covariance matrix $\mathbf{P}_0$ is given, the optimal estimation of the state parameters can be achieved according to the time prediction and measurement update processes by using the state prediction and measurement update equations of the system in [31].

3.1.2. The Residual Chi-Square Test

According to [17], the test statistic of the residual chi-square test q_k at the kth epoch is constructed as

$$q_k = \sum_{i=1}^{k} \gamma_i^T \mathbf{S}_i^{-1} \gamma_i \tag{7}$$

γ_k is the innovation of KF and $\mathbf{S}_i$ is the covariance matrix of γ_k at the ith epoch.

In the residual chi-square test, the mean value of innovation is 0 under the no-fault hypothesis $\mathrm{H_0}$. The test statistic follows the central chi-square distribution. In contrast, under the fault hypothesis $\mathrm{H_1}$, the mean KF innovation value is not 0 and the test statistic obeys the non-central chi-square distribution.

$$\begin{cases} \mathrm{H_0}: \mathrm{E}(\gamma_k) = 0, q_k \sim \chi^2(n) \\ \mathrm{H_1}: \mathrm{E}(\gamma_k) \neq 0, q_k \sim \chi^2(n, \lambda) \end{cases} \tag{8}$$

For a determined false alarm probability P_{FA}, the residual chi-square test threshold $\mathrm{T_D}$ is determined by the inverse of the chi-square cumulative distribution function (CDF) [3], which is given by

$$P(q_k \leq \mathrm{T_D}/\mathrm{H_0}) = \int_0^T f_{\chi^2(n+1)}(x)\mathrm{d}x = 1 - P_{FA} \tag{9}$$

The residual chi-square test algorithm is used to detect faults in this study. The performance of this algorithm is compared with that of the proposed algorithm.

3.2. *LiDAR Aided Real-Time Fault Detection*

3.2.1. The Theory of the Fault Detection Algorithm

The fault detection algorithm is realized by calculating the mean position deviation of all matched targets. The positions of the detected targets can be determined by establishing a global point cloud map. When no fault occurs, the vehicle can be located by GNSS/INS integrated positioning. The observed position of each target can be obtained by jointly exploiting the single frame point cloud target detection process of LiDAR. The position deviation of the matched targets between the real-time observed positions and the detected targets based on the 3D LiDAR map is very small. When the GNSS signals are affected by NLOS and multipath signals, the GNSS/INS integrated positioning results are biased; thus, large position deviations occur among the matched targets. A schematic diagram of the fault detection procedure aided by LiDAR is shown in Figure 2.

3.2.2. Three-Dimensional LiDAR Mapping and Target Detection Based on a Map

An a priori 3D LiDAR global point cloud map is established based on IMU and LiDAR data produced by LOAM [39]. First, the feature points are extracted, including the corner points and flat points. Thus, the odometry is obtained by interframe matching. Then, the map is obtained by matching the current frame with the built 10 Hz odometry to obtain more accurate mileage information and update the 3D LiDAR map. Rods, including lampposts and tree trunks, are extracted from the 3D LiDAR map using Cloud Compare software [41]. They are used as the targets to be detected based on the 3D LiDAR map.

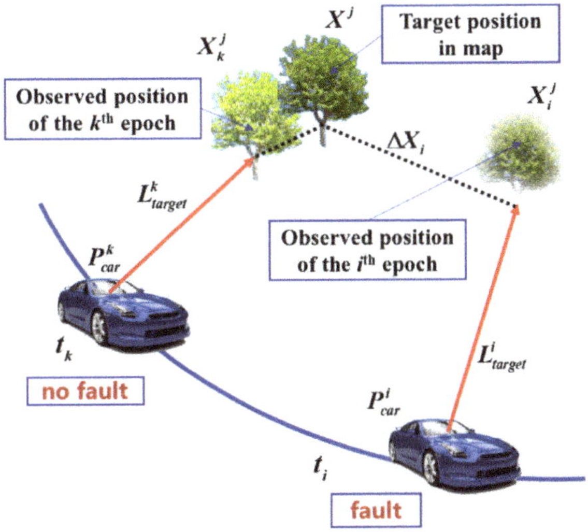

Figure 2. Schematic diagram of the fault detection process aided by LiDAR.

3.2.3. Single Frame Point Cloud Target Detection

Real-time single frame point cloud target detection is mainly achieved based on a point cloud library (PCL) [42]. First, the detection area of the single frame point cloud is set, and then, a voxel grid filter is used to filter the single frame point cloud. Second, the point cloud is segmented based on the road plane and obstacles by random sample consensus (RANSAC). Third, the obstacles on the road are clustered by Euclidean clustering, and a K-dimensional (KD) tree is used to conduct a nearest-neighbor search. Finally, the bounding boxes are associated with the detected targets based on the single frame point cloud [43].

3.2.4. The Construction of the Test Statistic

The targets detected by the single frame point cloud are matched with the targets detected by the global map, and the test statistic is constructed based on the mean position deviation of the matched targets, where the number of matched targets is f.

The real position of the jth matched target X^j can be obtained from the 3D LiDAR map at the kth epoch without faults. The observed position X_k^j can be obtained through the vehicle position P_{car}^k, which is estimated by GNSS/INS integrated positioning, and the relative positioning L_{target}^k, which is estimated by single frame point cloud target detection. The real position X^j and the observed position X_k^j are approximately equal, as shown below:

$$X^j \approx X_k^j = P_{car}^k + L_{target}^k \tag{10}$$

The observed position X_i^j of the jth matched target at the ith epoch with GNSS faults can also be obtained, and a large error is present between the 3D LiDAR map position X^j and the matched position X_i^j.

$$X^j \neq X_i^j = P_{car}^i + L_{target}^i \tag{11}$$

The test statistic is then constructed by calculating the mean horizontal difference ΔX_i between the real target position X^j and the observed target position X_i^j at each epoch for all f matched targets, i.e.,

$$\Delta X_i = \frac{1}{f}\sum_{j=1}^{f}\left(X^j - X_i^j\right)_{horizontal} \tag{12}$$

With the equations described in this section, the test statistic is constructed. In the next section, a threshold calculation approach is provided.

3.2.5. The Threshold Constructed by an Adaptive Sliding Window

An adaptive sliding window approach is proposed to construct a dynamic threshold for fault detection. An adaptive weight sequence is set up to determine the weight of each value in the sliding window.

The threshold of the ith epoch is constructed by selecting the mean position deviation ΔX over the previous m epochs to construct a sliding window, which is defined as $\{\Delta X_{i-m} \cdots \Delta X_{i-2}, \Delta X_{i-1}\}$. The adaptive weight sequence is selected as $\{\alpha_n\} = \{\alpha_1, \alpha_2, \cdots \alpha_m\}$, which satisfies the condition $\sum_{i=1}^{m} \alpha_n = 1$. Each value in the adaptive weight sequence $\{\alpha_n\}$ is determined by a geometric sequence:

$$b^1 + b^2 + b^3 \cdots + b^m = \left(1 - b^{m+1}\right) / (1 - b) \tag{13}$$

As $0 < b < 1$, Equation (13) can be transformed into

$$\left(b^1 + b^2 + b^3 \cdots + b^m\right)(1 - b) / \left(1 - b^{m+1}\right) = 1 \tag{14}$$

Assuming that $(1 - b)/(1 - b^{m+1}) = \delta$, Equation (14) is redefined as

$$\left(b^1 + b^2 + b^3 \cdots + b^m\right)\delta = \delta b + \delta b^2 \cdots + \delta b^m = 1 \tag{15}$$

Thus, the weight sequence can be determined by

$$\{\alpha_n\} = \{\alpha_1, \alpha_2 \cdots, \alpha_m\} = \left\{\delta b, \delta b^2 \cdots, \delta b^m\right\} \tag{16}$$

The threshold can be constructed as

$$\mathrm{T_D} = \sum_{n=1}^{m} \alpha_n \Delta X_{i-n} \tag{17}$$

To avoid false alarms when the threshold is too small while considering the measurement errors of LiDAR and the lateral and longitudinal positioning accuracy standards stated in [44], we set a threshold of 0.5 m when there are no NLOS and multipath effects. Therefore, the final threshold is given by

$$\mathrm{T_D} = \max\left\{0.5, \sum_{n=1}^{m} \alpha_n \Delta X_{i-n}\right\} \tag{18}$$

Finally, with the test statistic and threshold, the hypothesis test is constructed as

$$\begin{cases} \mathrm{H_0} : no\ fault, & \Delta X_i \le \mathrm{T_D} \\ \mathrm{H_1} : fault, & \Delta X_i > \mathrm{T_D} \end{cases} \tag{19}$$

With the hypothesis test given by Equation (19), GNSS positioning faults are detected in this study.

4. LiDAR Aided Measurement Noise Estimation Adaptive Filter Algorithm

4.1. Existing Adaptive Filter Algorithms

4.1.1. Single Fading Factor Adaptive Filter

Li et al. proposed an adaptive filter against outliers in [32] and designed a sliding window with length n to save the innovation of the filter γ_k. If the GNSS fault is detected at the kth epoch, the innovation γ_k is replaced by the average value of the last n innovation sequences and calculated as follows:

$$\gamma_k = (\gamma_{k-1} + \gamma_{k-2} \cdots + \gamma_{k-n})/n \tag{20}$$

γ_k is brought into the KF calculation, and the modified innovation of the filter is saved in the sliding data window.

4.1.2. Optimal Fading Factor Adaptive Filter

When the GNSS of an integrated navigation system has abnormal errors, the observation noise increases, which leads to an increase in the state covariance matrix $\mathbf{P}$. Therefore, Geng et al. proposed an optimal fading adaptive filter algorithm, adding a fading factor S_k to the state prediction step of the state covariance matrix $\mathbf{P}$ [45]:

$$\mathbf{P}_{k/k-1} = S_k \mathbf{\Phi}_{k/k-1} \mathbf{P}_{k-1} \mathbf{\Phi}^T_{k/k-1} + \mathbf{Q}_{k-1} \tag{21}$$

The convergence criterion of the KF is:

$$\gamma_k^T \gamma_k \leq \kappa \mathrm{tr}\left(\mathrm{E}\left(\gamma_k \gamma_k^T\right)\right) \tag{22}$$

where γ_k is the KF innovation, κ is the regulating coefficient and $\kappa \geq 1$. The strictest convergence condition is $\kappa = 1$, and $\gamma_k^T \gamma_k$ takes the minimum value satisfying Equation (22).

To realize the optimal fading factor, it must meet the strictest convergence condition:

$$\kappa = 1,\ \gamma_k^T \gamma_k = \mathrm{tr}\left(\mathrm{E}\left(\gamma_k \gamma_k^T\right)\right) \tag{23}$$

This factor can be obtained from the properties of the innovation sequence:

$$\mathrm{E}\left(\gamma_k \gamma_k^T\right) = \mathbf{H}_k \mathbf{P}_{k/k-1} \mathbf{H}_k^T + \mathbf{R}_k \tag{24}$$

We substitute Equation (21) into Equations (23) and (24) to obtain:

$$\gamma_k^T \gamma_k = S_k \mathbf{H}_k \mathbf{\Phi}_{k/k-1} \mathbf{P}_{k-1} \mathbf{\Phi}^T_{k/k-1} \mathbf{H}_k^T + \mathbf{H}_k \mathbf{Q}_{k-1} \mathbf{H}_k^T + \mathbf{R}_k \tag{25}$$

Equation (25) is traced and simplified, and the formula of S_k is obtained:

$$S_k = \frac{\gamma_k^T \gamma_k - \mathrm{tr}\left(\mathbf{H}_k \mathbf{Q}_{k-1} \mathbf{H}_k^T + \mathbf{R}_k\right)}{\mathrm{tr}\left(\mathbf{H}_k \mathbf{\Phi}_{k/k-1} \mathbf{P}_{k-1} \mathbf{\Phi}^T_{k/k-1} \mathbf{H}_k^T\right)} \tag{26}$$

It can be seen from Equation (26) that when the observed data are abnormal, the sum squares of the innovation $\gamma_k^T \gamma_k$ increase. Since the other parameters of the system remain unchanged in Equation (26), the current S_k is increased correspondingly, which is equivalent to increasing the state error covariance matrix $\mathbf{P}$ in Equation (21).

4.2. LiDAR Aided Real-Time Measurement Noise Estimation of Adaptive Filter

Single fading factor adaptive filter and optimal fading factor adaptive filter algorithms are used to solve the uncertainty of measurement noise when GNSS is affected by multipath and NLOS. The above algorithms can realize the optimal estimate in theory, but are based on prior modeling. The real measurement noise is not considered.

Taking advantage of the above two algorithms and considering the specific problem, we propose a LiDAR aided real-time measurement noise estimation adaptive filter algorithm. In the proposed algorithm, a sliding window with a length of n is constructed. The data in the sliding window are determined by the normalization of the mean position deviation of the matched targets. The innovation of the filter at the fault epoch is estimated in real-time. The value of the adaptive measurement noise factor is determined based on the strict theoretical derivation of the filter with the optimal fading factor. Real-time measurement noise estimation is realized. Finally, the accuracy and robustness of GNSS/INS integrated positioning are improved.

When faults are detected by the LiDAR aided real-time fault detection algorithm, the LiDAR aided real-time measurement noise estimation adaptive filter is applied for GNSS/INS integration. When faults are not detected, the EKF is applied. The algorithm flow chart is shown in Figure 3.

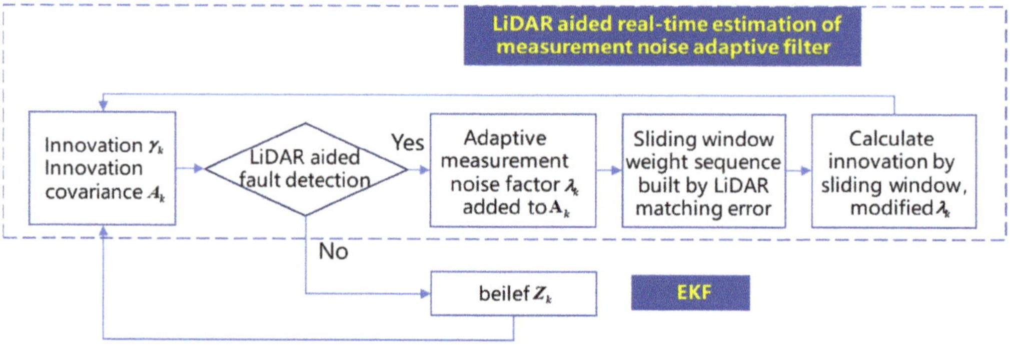

Figure 3. Schematic diagram of the LiDAR aided measurement noise estimation adaptive filter.

The specific implementation of the LiDAR aided real-time measurement noise estimation adaptive filter algorithm is shown in the following steps.

(1) An adaptive measurement noise factor λ_k is added to the innovation covariance $\mathbf{A}_k$ to produce Equation (27):

$$\mathbf{A}_k = \mathrm{E}\left(\gamma_k \gamma_k^T\right) = \mathbf{H}_k \mathbf{P}_{k/k-1} \mathbf{H}_k^T + \lambda_k \mathbf{R}_k \tag{27}$$

(2) According to the filter convergence conditions for Equations (23) and (24), the adaptive measurement noise factor λ_k can be calculated [43].

$$\lambda_k = \frac{\gamma_k^T \gamma_k - \mathrm{tr}\left(\mathbf{H}_k \mathbf{P}_{k/k-1} \mathbf{H}_k^T\right)}{\mathrm{tr}(\mathbf{R}_k)} \tag{28}$$

(3) The filter innovation in a fault epoch is calculated based on the sliding window. λ_k is only related to γ_k from Equation (28), and γ_k in the kth epoch is constructed by selecting the γ obtained with the previous n epochs to construct a sliding window, which is defined as $\{\gamma_{k-1}, \gamma_{k-2} \cdots, \gamma_{k-n}\}$. The adaptive weight sequence is selected as $\{\beta_i\} = \{\beta_1, \beta_2, \cdots \beta_n\}$. The mean value $\Delta\overline{\mathbf{X}}$ of the corresponding epochs $\left\{\Delta X_{target}^{k-1}, \Delta X_{target}^{k-2} \cdots, \Delta X_{target}^{k-n}\right\}$ is normalized, and the adaptive weight sequence is determined.

$$\beta_i = \frac{\Delta X_{target}^i - \Delta\overline{X}}{\max\left\{\Delta X_{target}^i\right\} - \Delta\overline{X}} \tag{29}$$

$$\gamma_k = \beta_1 \gamma_{k-1} + \beta_2 \gamma_{k-2} \cdots + \beta_n \gamma_{k-n} \tag{30}$$

(4) Intelligent switching between the LiDAR aided real-time measurement noise estimation adaptive filter algorithm and the EKF algorithm is realized to adapt to real-time environmental changes.

(5) Finally, the proposed algorithm is compared with the optimal fading factor adaptive filter (OFFAF) and the EKF.

5. Experimental Results and Discussion

5.1. Introduction to the Experiment

5.1.1. Sensor Setups

In both experiments, Newton-M2, a low-accuracy GNSS/INS integrated navigation system, was used to collect real-time kinematic (RTK) GNSS data at a frequency of 1 Hz and INS data at a frequency of 100 Hz. A 3D LiDAR sensor (Velodyne 16) was to collect raw 3D point clouds at a frequency of 10 Hz. In addition, the NovAtel SPAN-CPT7, a high-accuracy GNSS (GPS, Global Navigation Satellite System (GLONASS), and Beidou) RTK/INS integrated navigation system, was used to provide the ground truth positioning data. The RTK is onboard and runs in real-time. The NovAtel SPAN-CPT7 is also affected by multipath and NLOS, but the accuracy is higher and our research focuses on low-precision and cheap integrated navigation devices. Therefore, the SPAN-CPT7 was used for providing the ground truth. The coordinate systems between all the sensors were calibrated before conducting the experiments. The experimental equipment is shown in Figure 4.

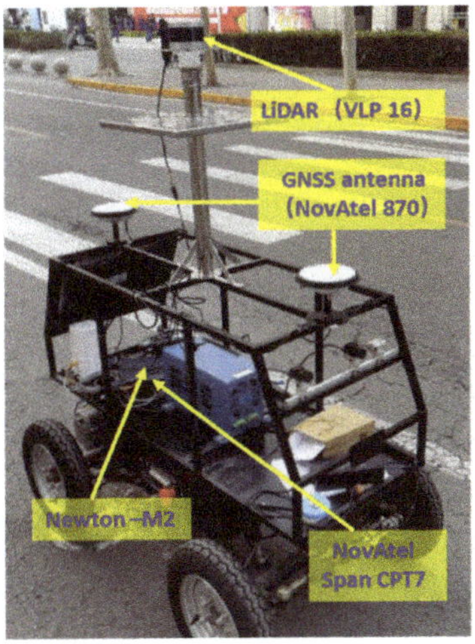

(a)

(b)

Figure 4. The experimental equipment for data collection. (**a**) The vehicle used for data collection and the installation positions of the sensors. (**b**) Local magnification of the Newton-M2 and NovAtel Span CPT7.

The Apollo 5.5 was used to collect GNSS and INS data from Newton-M2 and LiDAR data from Velodyne 16. The data format is Protobuf. The Novatel SPAN-CPT7 saves the data through the serial port. The data format is OEM7. The Newton-M2 and the Novatel SPAN-CPT7 are synchronized to the GNSS and INS through GPS timing, and the LiDAR is synchronized by the pulse per second (PPS) of the GPS timing.

5.1.2. Experimental Scenes

Since the residual chi-square test is widely used to detect faults in the positioning domain and the traditional EKF is the most commonly used GNSS/INS integrated navigation algorithm, the proposed algorithm was compared with the EKF to verify its performance. Two experiments were carried out in typical urban canyons in Beijing. The scenarios are shown in Figure 5a,b, representing mid and deep urban canyons, respectively. We first tested the situation in which the vehicle went through the narrow viaduct and the GNSS signals were slightly affected for a short time, as shown in Figure 5a. Then, we carried out another experiment in which the vehicle went through a wide floor hole, and the GNSS signals were seriously affected for a long time, as shown in Figure 5b.

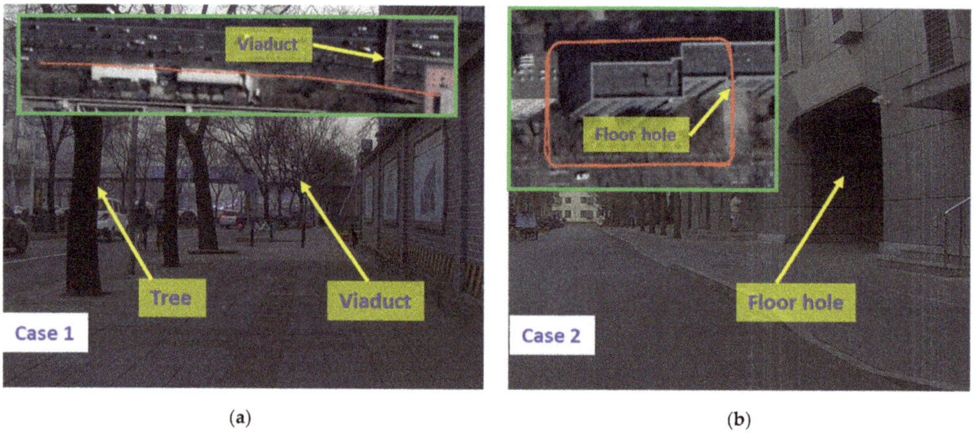

Figure 5. The experimental scenes of cases 1 and 2. (**a**) Case 1: The GNSS was slightly affected when the vehicle passed the narrow viaduct. (**b**) Case 2: The GNSS was seriously affected when the vehicle passed the wide floor hole.

5.2. Case One: The Narrow Viaduct

In case one, the GNSS signal was slightly affected by multipath and NLOS signals when the vehicle went through the narrow viaduct. The trajectory of the vehicle is shown in Figure 6a. In our analysis, the positioning error is higher in the period from 149th s to 164th s. It happens to be similar that according to the onboard RTK data, the solution type of GNSS positioning in the period from 149th s to 164th s was not "NARROW INT" (the "NARROW INT" indicates that the multi-frequency RTK is a fixed solution in the NovAtel receiver). Therefore, we needed to perform fault detection in this period. The positioning errors in the east and north are shown in Figure 6b. The maximum errors in the east and north were 1.1156 m and −0.84 m, respectively.

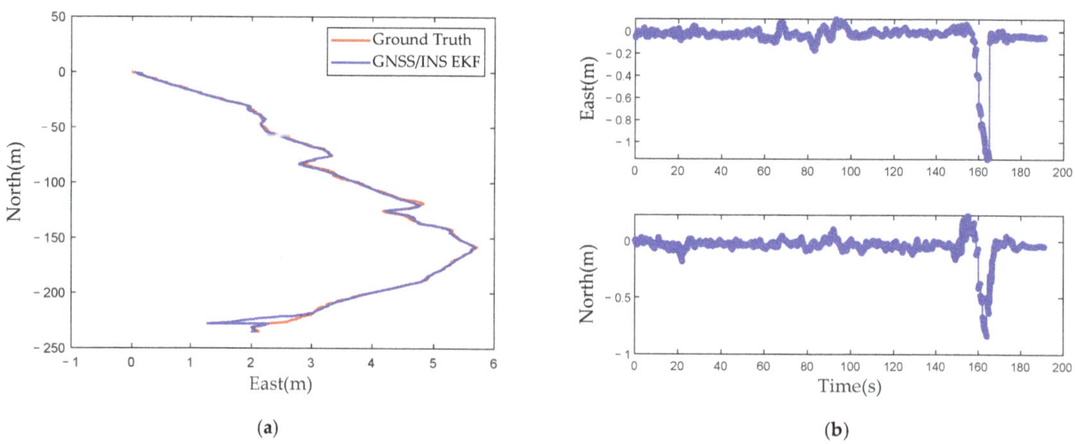

Figure 6. Trajectory and positioning errors in the east and north. (**a**) The trajectory of the vehicle in case 1. (**b**) The positioning errors in the east and north.

The top view of the 3D LiDAR global point cloud map built by the LOAM algorithm is shown in Figure 7. Figure 8 shows 31 tree trunks and lampposts marked in the 3D LiDAR global point cloud map as detection targets. These targets are marked with colored points in Figure 8 and numbered 0–30 from left to right in the east view.

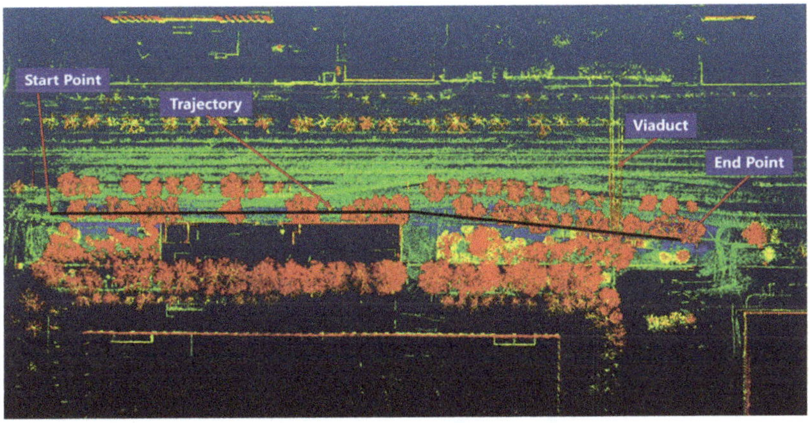

Figure 7. Three-Dimensional LiDAR global point cloud map in case one.

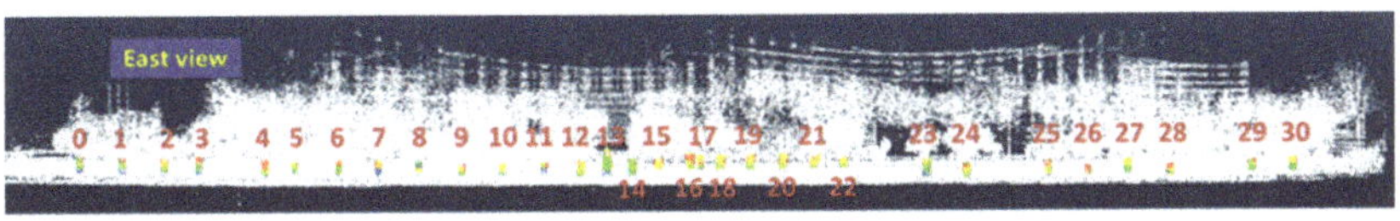

Figure 8. Target detection in map numbered 0–30, including tree trunks and lampposts, derived from the east view.

Two frame point clouds are randomly selected to present the point cloud target detection results, as shown in Figure 9. Figure 9a shows the 645th frame, and a total of 10 targets were detected, including targets No. 8–10 marked in Figure 8. Figure 9b shows the 1023rd frame, and a total of four targets were detected, including targets No. 17–18 marked in Figure 8.

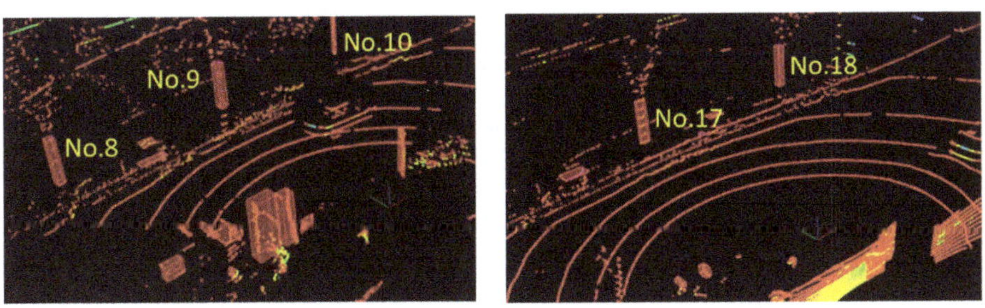

(**a**) (**b**)

Figure 9. Single frame point cloud target detection. (**a**) The 645th frame and No. 8–10 targets in Figure 8. (**b**) The 1023rd frame and No. 17–18 targets in Figure 8.

Figure 10a shows the matched targets between the single frame target detection outputs and the global map-based target detection results. Figure 10b shows the number of detected targets and matched targets in each frame. It can be seen that at least one matched target is present in each frame. This meets the time continuity requirement of matched targets and can support subsequent fault detection processes. Table 1 summarizes the number of detected targets and matched targets obtained with single frame detection.

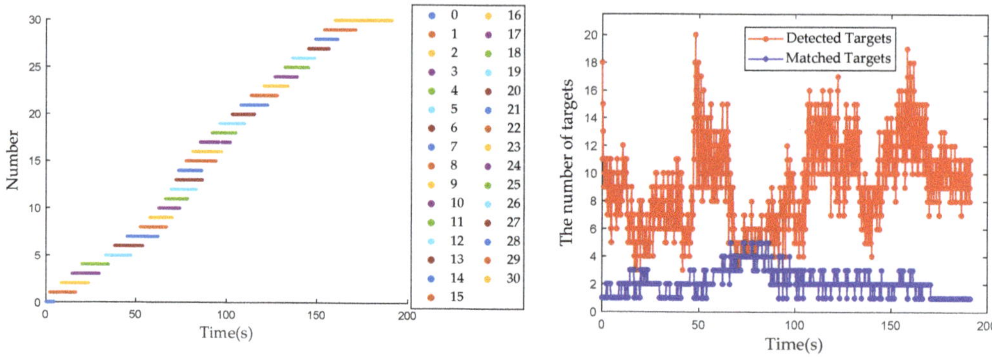

(**a**) (**b**)

Figure 10. Matched targets between the single frame and map-based target detection results. (**a**) The No. 0–30 of matched targets for target detection in global map. (**b**) The number of detected and matched targets in each frame.

Table 1. The number of detected and matched targets for single frame target detection.

	Mean	Max	Min
The number of detected targets	9.297	20	2
The number of matched targets	2.092	5	1

The position deviations of all matched targets in each frame in the east and north directions are shown in Figure 11a,b, respectively. Figure 12 shows the mean position deviation of the matched targets in each frame, which is used as the test statistic for the proposed fault detection algorithm.

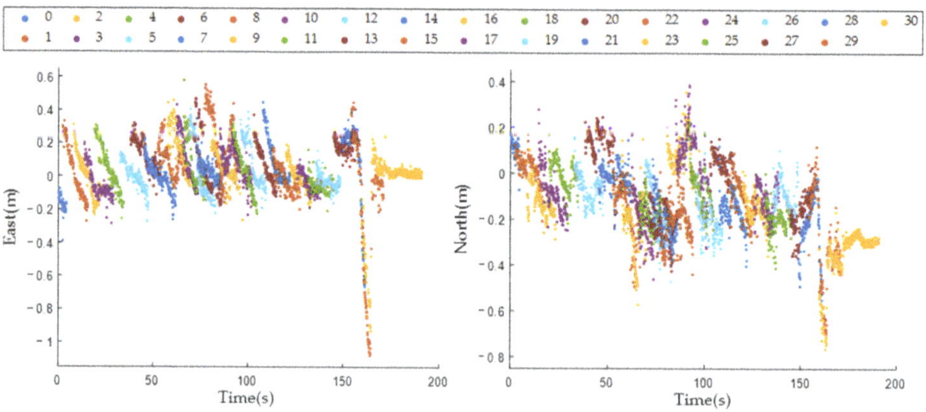

(**a**) (**b**)

Figure 11. The position deviations of the matched targets in the east and north. (**a**) The position deviation of the matched targets in the east. (**b**) The position deviation of the matched targets in the north.

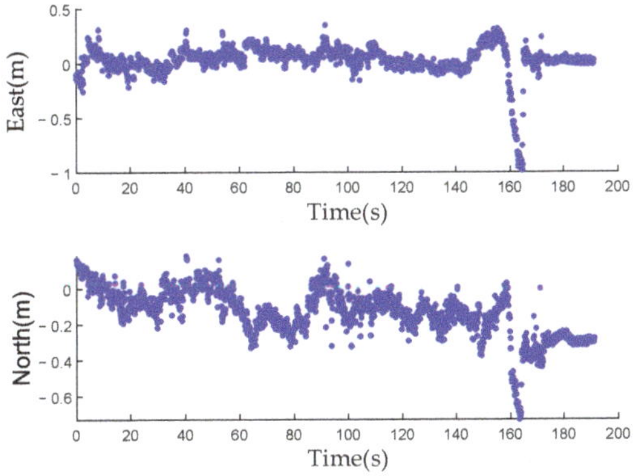

Figure 12. The mean position deviations of the matched targets in the east and north in case one.

Sensitivity is an important fault detection performance evaluation index, and the missed detection rate and false alarm rate are important integrity assessment indices. Figure 13a shows the fault detection result produced by the residual chi-square test, and Figure 13b is the result of the proposed LiDAR aided real-time fault detection algorithm. The fault detection performance of the two algorithms is compared in Table 2. The fault could not be detected by the residual chi-square test, while the proposed algorithm could detect the fault from 160.02th to 164.75th s. Therefore, it can be proven that the proposed algorithm has higher sensitivity. The percentages of false alarms and missed detections were reduced by 42.67% and 31.2%, respectively, relative to those of the residual chi-square test.

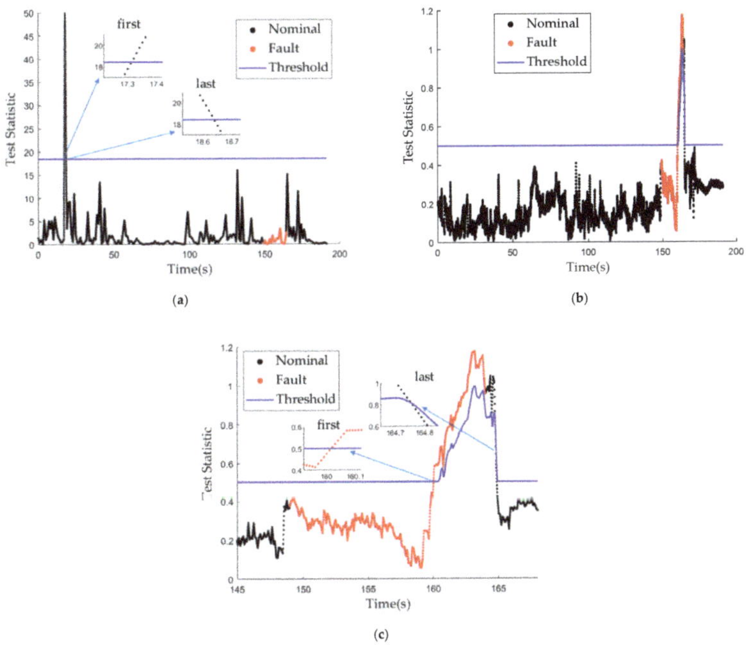

Figure 13. Fault detection results of the residual chi-square test and the proposed algorithm. (**a**) The fault detection result of the residual chi-square test. (**b**) The LiDAR aided real-time fault detection result. (**c**) Local magnification for fault detection of proposed algorithm.

Table 2. The fault detection performance of the chi-square test and the proposed algorithm.

	The Residual Chi-Square Test	Proposed Algorithm
The period of the fault	From 149th to 164th s	
Time of the first detected fault	-	160.02nd s
Time of the last detected fault	-	164.75nd s
Missed detection epochs	1600	1102
False alarm epochs	132	75
Response time of fault occurrence	-	11.02 s
Response time of fault disappearance	-	0.75 s
Percentage of false alarm	0.75%	0.43%
Percentage of missed detection	100%	68.88%

The proposed algorithm can mainly be used to solve the existing problems faced by the EKF algorithm during the fault period. The positioning errors of the EKF, OFFAF and the proposed algorithm in the east and north are shown in Figure 14. The results are analyzed in Table 3. Compared with the EKF, the mean, maximum and root mean square error (RMSE) positioning errors of the proposed algorithm were reduced by 67.66%, 51.9% and 71.58% in the east and 12.93%, 27.02% and 33.6% in the north, respectively. Compared with the OFFAF, the mean, maximum and root mean square error (RMSE) positioning errors of the proposed algorithm were reduced by 51%, 39.1% and 60.3% in the east and −5.2%, 21.1% and 19.2% in the north, respectively. The proposed algorithm is more effective than the OFFAF.

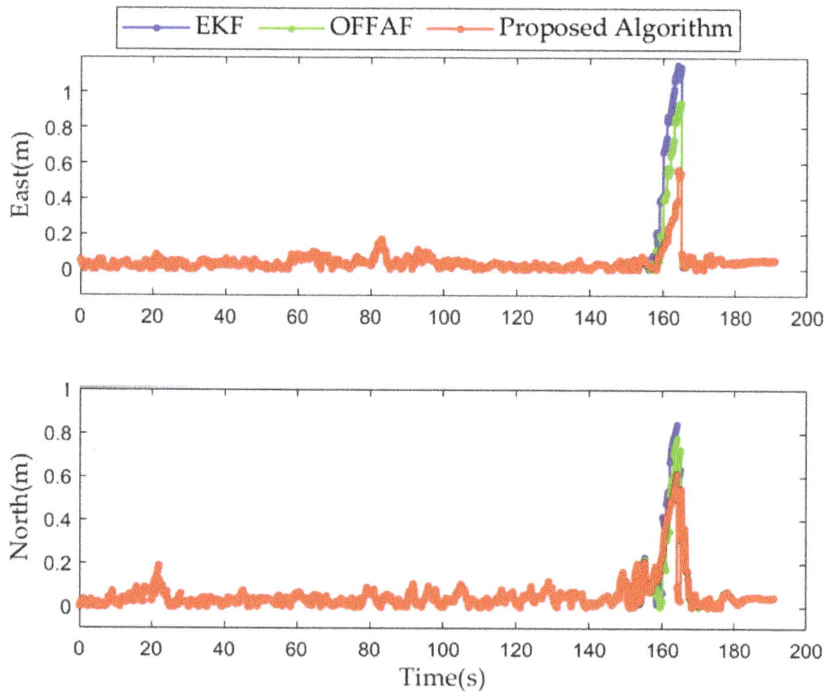

Figure 14. Positioning errors in the east and north for the EKF, OFFAF and the proposed algorithm.

Table 3. The GNSS/INS integrated positioning performance of the EKF and the proposed algorithm.

	East			North		
	Mean (m)	Max (m)	RMSE (m)	Mean (m)	Max (m)	RMSE (m)
GNSS/INS EKF	0.303	1.156	0.387	0.232	0.84	0.247
OFFAF	0.2	0.9132	0.2768	0.192	0.777	0.203
Proposed Algorithm	0.098	0.556	0.11	0.202	0.613	0.164

The error bound, which is the upper limit of positioning error, is an important integrity assessment index. When the positioning error exceeds the error bounds, an alarm is triggered. The existing research on integrated navigation error bounds were mainly performed by simulation experiments in [38]. In this study, the error bound of integrated navigation is innovatively verified with real test data. The error bounds and horizontal positioning error for the EKF and the proposed algorithm are shown in Figure 15a,b, respectively. In the two algorithms, the error bounds could overbound the horizontal error. From Table 4, the mean value and maximum value of the error bounds of the proposed algorithm were reduced by 53.03% and 57.88%, respectively, relative to the EKF during the fault period.

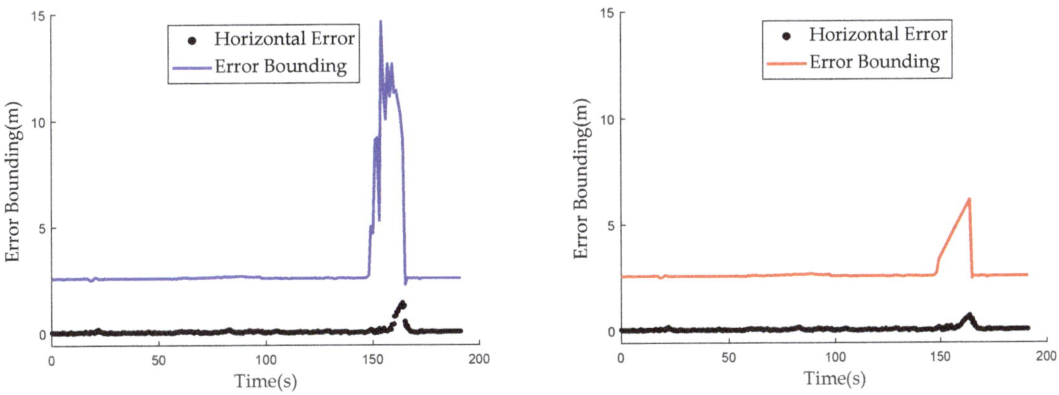

(a) (b)

Figure 15. The error bounds and horizontal error of the EKF and the proposed algorithm in case one. (a) The error bounds and horizontal error of the EKF. (b) The error bounds and horizontal error of the proposed algorithm.

Table 4. The error bounds of the EKF and proposed algorithm during the fault period.

Error Bounds	GNSS/INS EKF	Proposed Algorithm
Fail to overbound epochs	0	0
Mean (m)	10.1346	4.76
Max (m)	14.6365	6.165

5.3. Case Two: The Wide Floor Hole

In case two, the GNSS signal was seriously affected by multipath and NLOS signals when the vehicle went through the wide floor hole. The trajectory of the vehicle during data collection is shown in Figure 16a. The vehicle circled twice around the red trajectory in Figure 5b. In our analysis, the positioning error is higher in the period from the 179th to 198th s and from the 377th to 401st s. It happens to be similar according to the onboard RTK data, the solution types of the GNSS positioning results in this period were not "NARROW INT". It is believed that the GNSS was affected by multipath and NLOS signals. The positioning errors induced in the east and north during data collection are shown in Figure 16b. The maximum errors in the east and north were −4.2601 m and −19.5848 m, respectively.

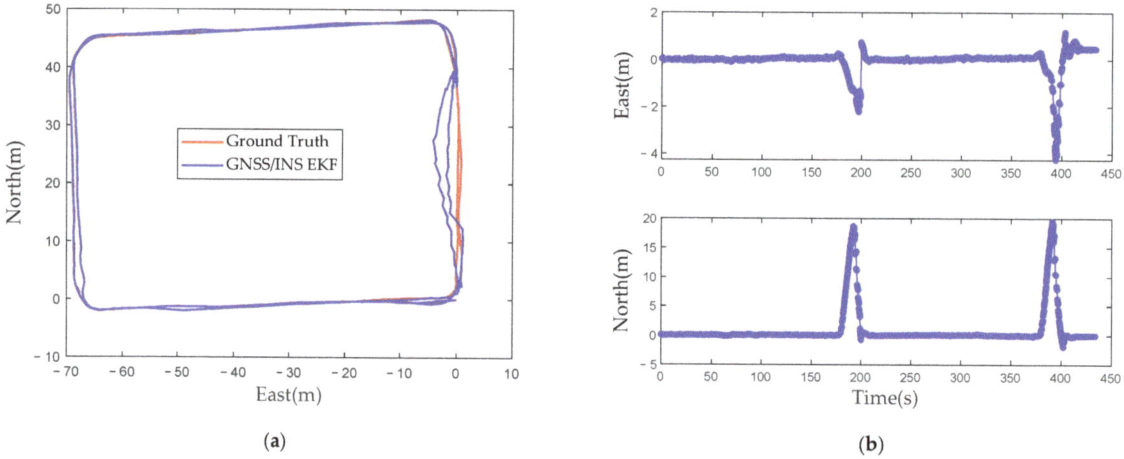

Figure 16. Trajectory and positioning errors in the east and north. (**a**) The trajectory of the vehicle in case 2. (**b**) The positioning errors in the east and north.

Figure 17 shows the top view of the 3D LiDAR global point cloud map built by the LOAM algorithm. Figure 18 shows 26 tree trunks and lampposts marked as detection targets in the 3D LiDAR global point cloud map. They are marked with colored points in Figure 18 and numbered from 0–25.

Figure 17. Three-Dimensional LiDAR global point cloud map in case two.

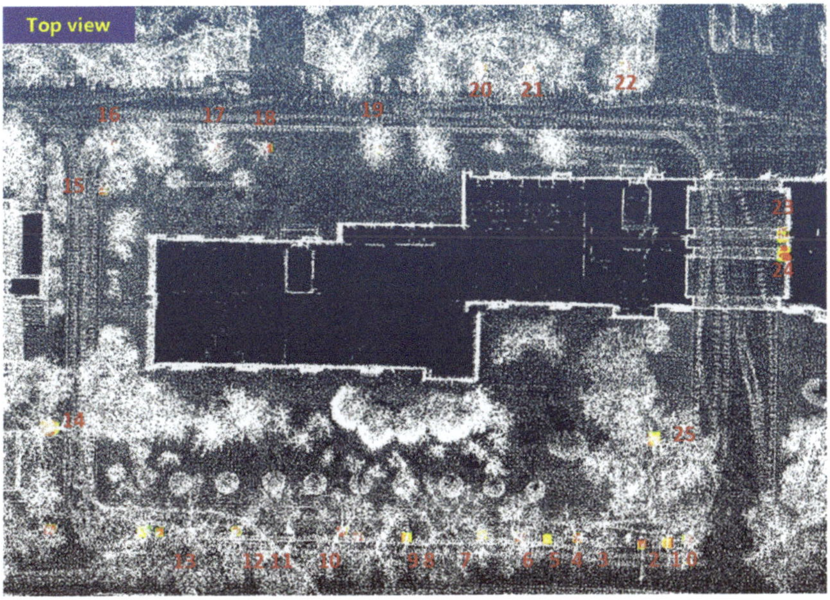

Figure 18. Detection targets in the map numbered 0–25, including tree trunks and lampposts.

Figure 19 shows the results of single frame target detection. We took two frames as the vehicle went through the floor hole. Figure 19a shows the 1742nd frame, and a total of nine targets were detected, including targets numbered twenty-three to twenty-four marked in Figure 18. Figure 19b shows the 1856th frame, and a total of twelve targets were detected, including targets numbered twenty-three to twenty-five marked in Figure 18.

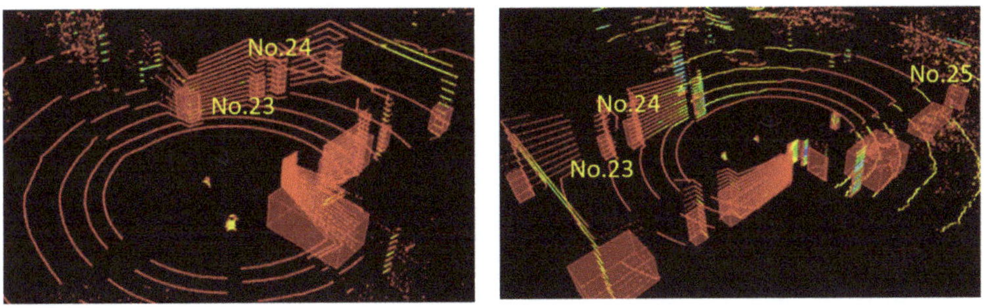

(a) (b)

Figure 19. Single frame point cloud target detection results. (**a**) The 1742nd frame and No. 23–24 targets in Figure 18. (**b**) The 1856th frame and No. 23–25 targets in Figure 18.

Figure 20a shows the matched targets between the single frame target detection outputs and the global map-based target detection results. Figure 20b shows the number of detected targets and matched targets in each frame. It can be seen that at least one matched target was observed in any epoch. This outcome meets the time continuity requirement of matched targets and can support subsequent fault detection processes. Table 5 summarizes the number of detected targets and matched targets obtained with single frame target detection.

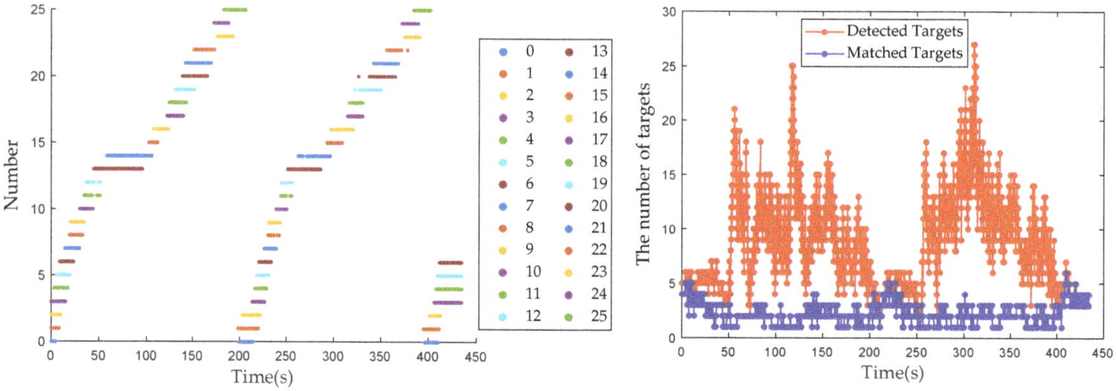

(**a**) (**b**)

Figure 20. Matched targets between the single frame and map-based target detection results. (**a**) The No. 0–25 of matched targets for target detection in the global map. (**b**) The number of detected and matched targets in each frame.

Table 5. The number of detected targets and matched targets.

	Mean	**Max**	**Min**
The number of detected targets	8.40	27	2
The number of matched targets	2.19	6	1

The position deviations of all matched targets in each frame in the east and north directions are shown in Figure 21a,b, respectively. Figure 22 shows the mean position deviation of the matched targets in each frame, which is used as the test statistic for the proposed fault detection algorithm.

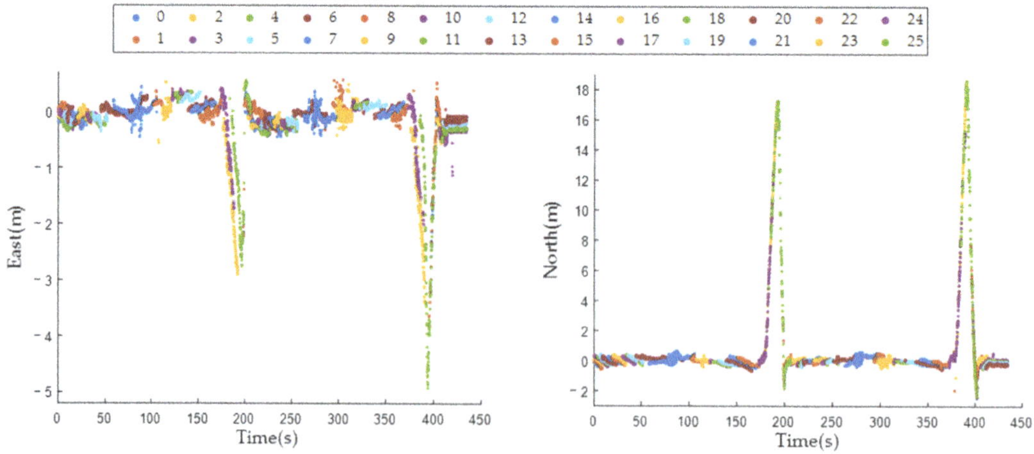

(**a**) (**b**)

Figure 21. The position deviations of the single frame matched targets in the east and north. (**a**) The position error of the matched targets in the east. (**b**) The position error of the matched targets in the north.

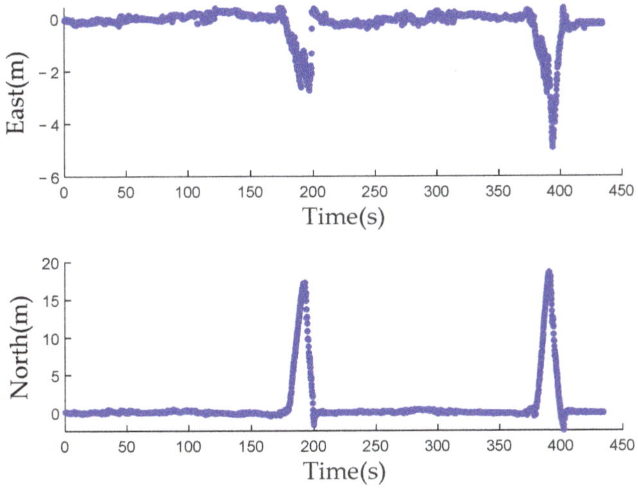

Figure 22. The mean position deviations of the matched targets in the east and north in case two.

Similar with Figure 13, Figure 23a shows the fault detection results produced by the residual chi-square test. The results of the two fault detection algorithms are locally magnified in Figure 24a,b. Figure 23b is the result of the proposed LiDAR aided real-time fault detection algorithm. The results of the two fault detection approaches are locally magnified in Figure 24c,d. The fault detection performances of the two algorithms are compared in Table 6. The response time of fault disappearance was 8.94 s and 6.62 s for the residual chi-square test, but they were reduced by 5.465 s on average with the proposed algorithm. Therefore, the low-sensitivity problem of the residual chi-square test for the fault disappearance case is effectively ameliorated. Furthermore, the percentages of false alarm and missed detection were 6.26% and 0.69% for the proposed LiDAR aided real-time fault detection algorithm, which are reductions of 76.49% and 79.03% relative to the residual chi-square test results, respectively.

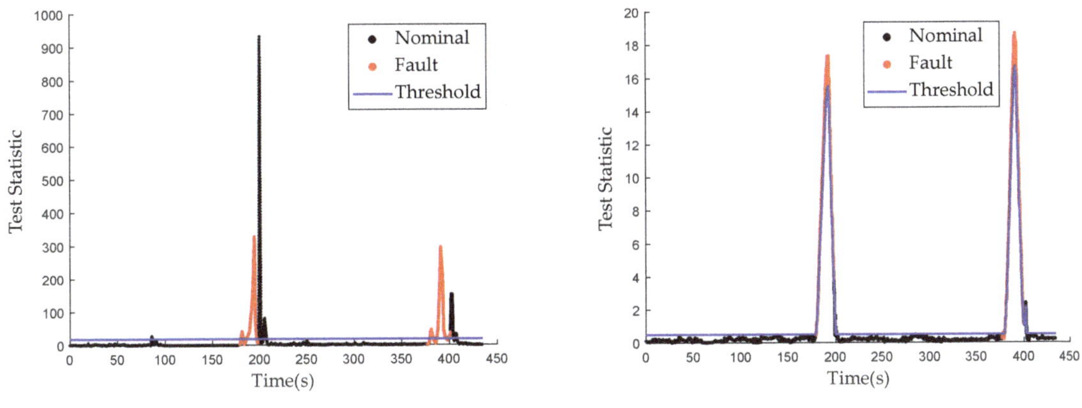

Figure 23. Fault detection results of the residual chi-square test and the proposed algorithm. (**a**) The fault detection results of the residual chi-square test. (**b**) The LiDAR aided real-time fault detection results.

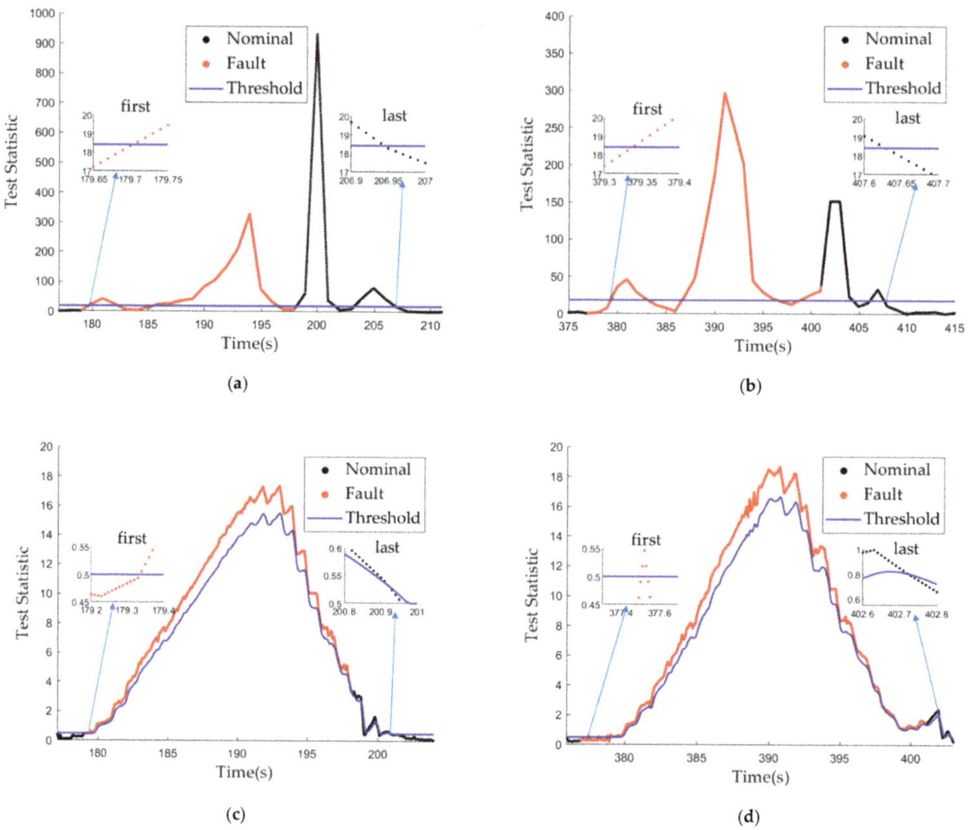

Figure 24. Local magnifications of Figure 23 for the fault detection algorithm. (**a**) The first fault detected by the residual chi-square test. (**b**) The second fault detected by the residual chi-square test. (**c**) The first fault detected by the LiDAR aided fault detection algorithm. (**d**) The second fault detected by the LiDAR aided fault detection algorithm.

Table 6. The fault detection performance of the residual chi-square test and the proposed algorithm.

	The Residual Chi-Square Test		**Proposed Algorithm**	
The fault period	From 179th to 198th s	From 377th to 401th s	From 179th to 198th s	From 377th to 401th s
Time of the first detected fault	179.71th s	379.34th s	179.34th s	377.51th s
Time of the last detected fault	206.94th s	407.62th s	200.91th s	402.72th s
Missed detection epochs	1145		269	
False alarm epochs	1287		271	
Response time of fault occurrence	0.71 s	2.34 s	0.34 s	0.51 s
Response time of fault disappearance	8.94 s	6.62 s	2.91 s	1.72 s
Percentage of false alarm	26.63%		6.26%	
Percentage of missed detection	3.29%		0.69%	

Similar with Figure 14, the positioning errors of the EKF, OFFAF and the proposed algorithm in the east and north for case two are shown in Figure 25. The results are analyzed in Table 7. Compared with the EKF, the mean, maximum and RMSE positioning errors of the proposed algorithm were reduced by 7.4%, 20.49% and 12.81% in the east and 79.12%, 68.22% and 73.31% in the north, respectively, during the fault period. Compared with the OFFAF, the mean, maximum and RMSE positioning errors of the proposed algorithm were reduced by 26.2%, 3.7% and 12.98% in the east and 51.7%, 18.8% and 35.1% in the north, respectively, during the fault period. The proposed algorithm is more effective than the OFFAF.

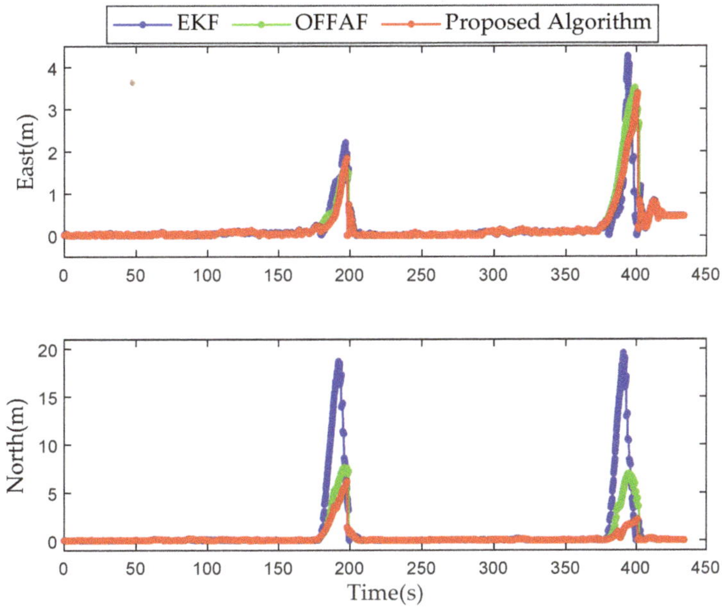

Figure 25. Positioning errors of the proposed algorithm in the east and north in case two.

Table 7. The GNSS/INS integrated positioning performance of the EKF, OFFAF and the proposed algorithm.

	East			North		
	Mean (m)	**Max (m)**	**RMSE (m)**	**Mean (m)**	**Max (m)**	**RMSE (m)**
GNSS/INS EKF	1.094	4.26	1.007	8.876	19.5848	6.137
OFFAF	1.372	3.517	1.009	3.84	7.669	2.524
Proposed Algorithm	1.013	3.387	0.878	1.853	6.224	1.638

Similar with Figure 15, the error bounds and positioning errors induced by the EKF and the proposed algorithm in the horizontal direction in case two are shown in Figure 26a,b, respectively. The error bounds of the EKF could not overbound the horizontal errors in some epochs. However, the error bound of the proposed algorithm could overbound the horizontal error in all epochs. From Table 8, the error bounds failed to overbound the horizontal error in 1490 epochs. The mean value and maximum value of the error bounds yielded by the proposed algorithm were reduced by 56.35% and 73.2%, respectively, relative to the EKF during the fault period. Therefore, the integrity of the system could be guaranteed by the proposed algorithm.

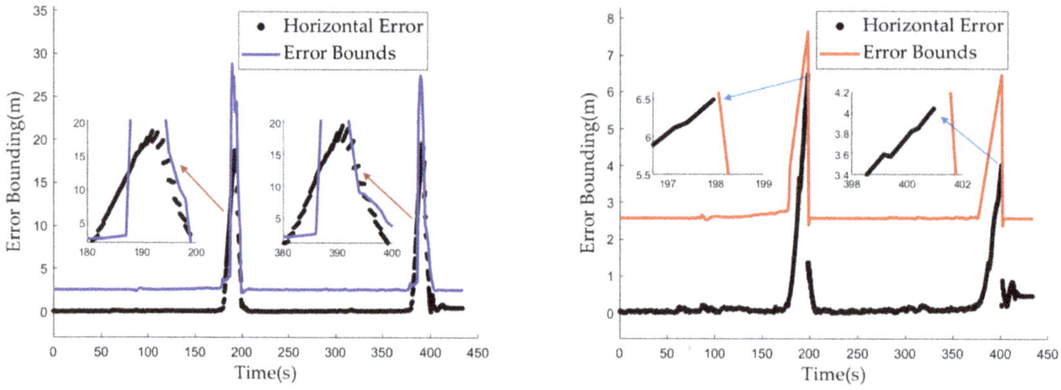

(**a**) (**b**)

Figure 26. The error bounds and horizontal error of the EKF and the proposed algorithm in case 2. (**a**) The error bounds and horizontal error of the EKF. (**b**) The error bounds and horizontal error of the proposed algorithm.

Table 8. The error bounds of the EKF and the proposed algorithm during the fault period.

Error Bounds	GNSS/INS EKF	Proposed Algorithm
Fail to overbound epochs	1490	0
Mean (m)	11.771	5.137
Max (m)	28.546	7.649

5.4. Discussion

In our research, the LiDAR aided GNSS/INS integration fault detection and localization algorithm is proposed. The proposed fault detection algorithm could effectively improve the sensitivity of residual chi-square test when the fault disappears. The response time for fault disappearance is an important indicator. The localization algorithm could reduce the positioning error compared to EKF. Finally, the integrity of the proposed algorithm is evaluated including false alarm rate, missed detection rate and error bounds. The algorithms proposed in this paper are oriented to the autonomous driving for level four (L4) or level five (L5). However, due to the limitation of experimental conditions, some problems should be considered.

Firstly, in terms of prior point cloud map, the map constructed by LOAM has errors. However, in our experiments, the map is built on a small scene and the error of the map established by LOAM is less than 0.3 m, which is very small compared with the error caused by multipath and NLOS. Therefore, the error of LOAM is acceptable and it has little effect on the performance of the algorithm. The cost of construction of a prior global point cloud map is high only for the proposed algorithm. However, a high-precision point cloud map is an indispensable part of autonomous localization in the existing research. In the future, the HD map can be produced in the industry which is cheaper and more accurate. The proposed algorithm can be applied with the help of HD map and produce more effect. The map is established offline and will not affect the real-time performance of our algorithm. Storing maps takes up a lot of memory, but is not a major direction in our research. We believe this problem of memory will be solved with the autonomous driving implementation.

Secondly, in terms of point cloud processing being involved in target detection, in this paper, the target detection of a single frame point cloud is carried out by clustering. Object detection of single frame is required in the perception module of autonomous driving. Our algorithm can reuse relevant results in the application process. Therefore,

14. Attia, D.; Meurie, C.; Ruichek, Y.; Marais, J.; Flancquart, A. Image analysis based real time detection of satellites reception state. In Proceedings of the 13th International IEEE Annual Conference on Intelligent Transportation Systems, Funchal, Portugal, 19–22 September 2010; pp. 1651–1656.
15. Wen, W.; Zhang, G.; Hsu, L.-T. Correcting NLOS by 3D LiDAR and building height to improve GNSS single point positioning. *Navigation* **2019**, *66*, 705–718. [CrossRef]
16. Wen, W.; Hsu, L.-T. 3D LiDAR Aided GNSS NLOS Mitigation in Urban Canyons. In Proceedings of the IEEE Transactions on Intelligent Transportation Systems, Macau, China, 8–12 October 2022; pp. 1–13.
17. Liu, B.; Gao, Y.; Gao, Y.; Wang, S. HPL calculation improvement for Chi-squared residual-based ARAIM. *GPS Solut.* **2022**, *26*, 45. [CrossRef]
18. Qian, C.; Liu, H.; Tang, J.; Chen, Y.; Kaartinen, H.; Kukko, A.; Zhu, L.; Liang, X.; Chen, L.; Hyyppä, J. An Integrated GNSS/INS/LiDARSLAM Positioning Method for Highly Accurate Forest Stem Mapping. *Remote Sens.* **2017**, *9*, 3. [CrossRef]
19. Feng, S.; Ochieng, W.Y. A difference test method for early detection of slowly growing errors in GNSS positioning. *J. Navig.* **2007**, *60*, 427. [CrossRef]
20. Aldibaja, M.; Suganuma, N.; Yoneda, K.; Yanase, R. Challenging Environments for Precise Mapping Using GNSS/INS-RTK Systems: Reasons and Analysis. *Remote Sens.* **2022**, *14*, 4058. [CrossRef]
21. Wang, Y.; Lou, Y.; Zhang, Y.; Song, W.; Huang, F.; Tu, Z. A Robust Framework for Simultaneous Localization and Mapping with Multiple Non-Repetitive Scanning Lidars. *Remote Sens.* **2021**, *13*, 2015. [CrossRef]
22. Wang, W.; Liu, J.; Wang, C.; Luo, B.; Zhang, C. DV-LOAM: Direct Visual LiDAR Odometry and Mapping. *Remote Sens.* **2021**, *13*, 3340. [CrossRef]
23. Fiorucci, M.; Verschoof-van der Vaart, W.B.; Soleni, P.; Le Saux, B.; Traviglia, A. Deep Learning for Archaeological Object Detection on LiDAR: New Evaluation Measures and Insights. *Remote Sens.* **2022**, *14*, 1694. [CrossRef]
24. Jiang, H.; Li, T.; Song, D.; Shi, C. An Effective Integrity Monitoring Scheme for GNSS/INS/Vision Integration Based on Error State EKF Model. *IEEE Sens. J.* **2022**, *22*, 7063–7073. [CrossRef]
25. Chang, G. Robust Kalman filtering based on Mahalanobis distance as outlier judging criterion. *J. Geodesy.* **2014**, *88*, 391–401. [CrossRef]
26. Gordon, N.J.; Salmond, D.J.; Smith, A.F.M. Novel approach to non-linear/non-Gaussian Bayesian state estimation. *IEE Proc. F-Radar Signal Process* **1993**, *140*, 107–113. [CrossRef]
27. Rigatos, G.; Siano, P.; Wira, P.; Busawon, K.; Binns, R. A nonlinear H-infinity control approach for autonomous truck and trailer systems. *Unmanned Syst.* **2020**, *8*, 49–69. [CrossRef]
28. Jiang, C.; Zhang, S.; Li, H. Performance evaluation of the filters with adaptive factor and fading factor for GNSS/INS integrated systems. *GPS Solut.* **2021**, *25*, 130. [CrossRef]
29. Fagin, S.L. Recursive linear regression theory, optimal filter theory and error analysis of optimal systems. *IEEE Int.* **1964**, *12*, 216–240.
30. Lee, T. Theory and application of adaptive fading memory Kalman filters. *IEEE Trans. Circuits Syst.* **1988**, *35*, 474–477. [CrossRef]
31. Sun, J.; Ye, Q.; Lei, Y. In-Motion Alignment Method of SINS Based on Improved Kalman Filter under Geographic Latitude Uncertainty. *Remote Sens.* **2022**, *14*, 2581. [CrossRef]
32. Li, W. Research on Adaptive Kalman Filter and Fault-Tolerant Algorithm Used in in-Vehicle Integrated Navigation System. Master's Thesis, National University of Defense Technology, Changsha, China, 2008.
33. Wang, Y.; Liu, J.; Wang, J.; Zeng, Q.; Shen, X.; Zhang, Y. Micro Aerial Vehicle Navigation with Visual-Inertial Integration Aided by Structured Light. *J. Navig.* **2020**, *73*, 16–36. [CrossRef]
34. Zhou, H.; Huang, H.; Zhao, H.; Zhao, X.; Yin, X. Adaptive Unscented Kalman Filter for Target Tracking in the Presence of Nonlinear Systems Involving Model Mismatches. *Remote Sens.* **2017**, *9*, 657. [CrossRef]
35. Li, B.; Chen, W.; Peng, Y.; Dong, D.; Wang, Z.; Xiao, T.; Yu, C.; Liu, M. Robust Kalman Filtering Based on Chi-square Increment and Its Application. *Remote Sens.* **2020**, *12*, 732. [CrossRef]
36. Gao, Z.; Fang, K.; Wang, Z.; Guo, K.; Liu, Y. An Error Overbounding Method Based on a Gaussian Mixture Model with Uncertainty Estimation for a Dual-Frequency Ground-Based Augmentation System. *Remote Sens.* **2022**, *14*, 1111. [CrossRef]
37. Fang, X.; Song, D.; Shi, C.; Fan, L.; Hu, Z. Multipath Error Modeling Methodology for GNSS Integrity Monitoring Using a Global Optimization Strategy. *Remote Sens.* **2022**, *14*, 2130. [CrossRef]
38. Lee, J.; Kim, M.; Lee, J.; Pullen, S. Integrity assurance of Kalman-filter based GNSS/IMU integrated systems against IMU faults for UAV applications. In Proceedings of the 31st International Technical Meeting of the Satellite Division of The Institute of Navigation (ION GNSS+ 2018), Miami, FL, USA, 24–28 September 2018; pp. 2484–2500.
39. Zhang, J.; Singh, S. Low-drift and real-time lidar odometry and mapping. *Auton. Robots* **2017**, *41*, 401–416. [CrossRef]
40. Wan, G.; Yang, X.; Cai, R.; Li, H.; Zhou, Y.; Wang, H.; Song, S. Robust and Precise Vehicle Localization Based on Multi-Sensor Fusion in Diverse City Scenes. In Proceedings of the 2018 IEEE International Conference on Robotics and Automation (ICRA), Brisbane, QLD, Australia, 21–25 May 2018; pp. 4670–4677.
41. CloudCompare. Available online: https://www.cloudcompare.org/doc/qCC (accessed on 22 August 2022).
42. Rusu, R.; Cousins, S. 3D is here: Point Cloud Library (PCL). In Proceedings of the 2011 IEEE International Conference on Robotics and Automation (ICRA), Shanghai, China, 9–13 May 2011; pp. 1–4.

43. Teng, D. Resaeach on Dynamic Path Planning for Driverless Vehicles Based on LiDAR and Camera. Master's Thesis, Nanjing University of Aeronautics and Astronautics, Nanjing, China, 2020.
44. Groves, P.D. *Principles of GNSS, Inertial, and Multi Sensor Integrated Navigation Systems*; Artech House: London, UK, 2013.
45. Geng, Y.; Wang, J. Adaptive estimation of multiple fading factors in Kalman filter for navigation applications. *GPS Solut.* **2008**, *12*, 273–279. [CrossRef]
46. Zhou, Z.; Mertikopoulos, P.; Bambos, N.; Glynn, P.; Ye, Y. Distributed Stochastic Optimization with Large Delays. *Math. Oper. Res.* **2021**. *ahead of print*. [CrossRef]
47. Zhou, Z.; Mertikopoulos, P.; Bambos, N.; Glynn, P.; Ye, Y.; Li, L.; Li, F. Distributed Asynchronous Optimization with Unbounded Delays: How Slow Can You Go? In Proceedings of the 35th International Conference on Machine Learning Conference (ICML 2018), Stockholm, Sweden, 10–15 July 2018; pp. 597–5979.
48. Wright, J.; Ma, Y. *High-Dimensional Data Analysis with Low-Dimensional Models: Principles, Computation, and Applications*, 1st ed.; Cambridge University Press: Cambridge, UK, 2020; pp. 54–80.

remote sensing

Article

A Robust Nonlinear Filter Strategy Based on Maximum Correntropy Criterion for Multi-GNSS and Dual-Frequency RTK

Jian Liu [1], Tong Liu [1], Yuanfa Ji [2], Mengfei Sun [1], Mingyang Lyu [3], Bing Xu [3], Zhiping Lu [1] and Guochang Xu [1,*]

[1] Institute of Space Science and Applied Technology, Harbin Institute of Technology (Shenzhen), Shenzhen 518055, China
[2] Guangxi Key Laboratory of Wireless Broadband Communication and Signal Processing, Guilin 541004, China
[3] Department of Aeronautical and Aviation Engineering, The Hong Kong Polytechnic University, Hong Kong SAR, China
* Correspondence: xuguochang@hit.edu.cn

Abstract: The multi-constellation, multi-frequency Global Navigation Satellite System (GNSS) has the potential to empower precise real-time kinematics (RTK) with higher accuracy, availability, continuity, and integrity. However, to enhance the robustness of the nonlinear filter, both the measurement quality and efficiency of parameter estimation need consideration, especially for GNSS challenging or denied environments where outliers and non-Gaussian noise exist. This study proposes a nonlinear Kalman filter with adaptive kernel bandwidth (KBW) based on the maximum correntropy criterion (AMC-KF). The proposed method excavates data features of higher order moments to enhance the robustness against noise. With the wide-lane and ionosphere-free combination, a dual frequency (DF) data-aided ambiguity resolution (AR) method is also derived to improve the measurement quality. The filtering strategy based on the DF data-aided AR method and AMC-KF is applied for multi-GNSS and DF RTK. To evaluate the proposed method, the short baseline test, long baseline test, and triangle network closure test are conducted with DF data from GPS and Galileo. For the short baseline test, the proposed filter strategy could improve the positioning accuracy by more than 30% on E and N components, and 60% on U. The superiority of the proposed adaptive KBW is validated both in efficiency and accuracy. The triangle network closure test shows that the proposed DF data-aided AR method could achieve a success rate of more than 93%. For the long baseline test, the integration of the above methods gains more than 40% positioning accuracy improvement on ENU components. This study shows that the proposed nonlinear strategy could enhance both robustness and accuracy without the assistance of external sensors and is applicable for multi-GNSS and dual-frequency RTK.

Keywords: multi-GNSS; real-time kinematic; maximum correntropy criterion; Kalman filter; wide-lane; ionosphere-free

Citation: Liu, J.; Liu, T.; Ji, Y.; Sun, M.; Lyu, M.; Xu, B.; Lu, Z.; Xu, G. A Robust Nonlinear Filter Strategy Based on Maximum Correntropy Criterion for Multi-GNSS and Dual-Frequency RTK. *Remote Sens.* **2022**, *14*, 4578. https://doi.org/10.3390/rs14184578

Academic Editor: Giuseppe Casula

Received: 30 July 2022
Accepted: 9 September 2022
Published: 13 September 2022

1. Introduction

The current operating Global Navigation Satellite System (GNSS) includes the Global Positioning System (GPS), Globalnaya Navigazionnaya Sputnikovaya Sistema (GLONASS), BeiDou (BDS), and Galileo. The precise position, velocity, and time (PVT) generated from real-time kinematics (RTK) are crucial for engineering surveys, fleet monitoring, intelligent transportation systems, geographical information systems, guidance and control, etc. [1,2]. By 2024, more than 110 satellites with different frequencies are expected to be accessible for multi-GNSS. Compared with a single constellation, multiple constellations and frequencies could improve the accuracy, continuity, availability, and integrity significantly [3], while enhancing the robustness against noise and outliers [4].

Multi-GNSS applications usually apply loosely and tightly coupled models for data curation [5,6]. Both of them could achieve similar performance [7] in precise point positioning [8] and multi-sensor fusion such as integrating GNSS with the Inertial Navigation

System, Simultaneous Location and Mapping, Lidar, and 5G [9–15]. Although the tightly coupled model shares one common pivot satellite for each constellation, it is challenging to improve the model strength due to the presence of double difference inter-system bias (DISB) originating from receivers [16]. In this study, the loosely coupled model is used due to its usability [17].

Even if dual-frequency (DF) data is available [18–20], the ambiguity resolution is still impacted in urban environments due to signal blockage and interference [18–21]. The traditional extended Kalman filter is sensitive to outliers and noise [22], as it is based on the second-order statistics (e.g., variance, correlation, etc.) that exist in the Gaussian noise assumption and minimum mean square error (MSE) criterion [23]. Thus, the following limitations still need to be addressed: (1) The unmodeled non-Gaussian noise and heavy-tail noise originated not only from the outliers and gross errors but also the missed data due to signal blockage and deformation [24]. (2) The lack of complete prior knowledge of system dynamics and observations may eventually cause divergence [25]. Extensive attempts have been made to address these problems. One solution is the particle filter (PF), which is capable of estimating the posterior probability density function (PDF) by massive particles [26]. Multi-model filters such as the Gaussian sum filter (GSF) are another solution, which parallelly implement and interactively combine several filters to estimate the system states. A computationally economical filter was studied by integrating the robust M estimation into the GSF framework [27]. Heavy-tailed distribution-based filters and the H_∞ filter have also been studied in [28]. However, attention still needs to be paid, as most of them are deficient in universality and efficiency [29].

Recently, correntropy has received growing attention in signal processing, posture estimation, and machine learning [30–33]. It is a measurable metric of local similarity and is established on Gaussian kernel functions. Specifically, higher dimensional data matching and error detection can be realized, as Gaussian kernel functions enrich data features by transforming the observation into the Hibert space with higher dimensions. Thus, the maximum correntropy between the input and output is robust against various types of noise and even arbitrarily large outliers, and can be achieved according to the maximum correntropy criterion (MCC). Although the kernel bandwidth (KBW) is the key parameter for implementing the MCC, existing studies mostly treat it as an exogenous parameter from empirical experiments, rather than an endogenous variable of the system [33–36]. In this paper, a nonlinear adaptive Kalman filter (KF) based on MCC with adaptive KBW is proposed. The Gaussian hypothesis and MSE criterion are further relaxed, which aims to improve the adaptability and robustness [37]. The main contributions of the proposed method are: (1) The AMC-KF is proposed as a new robust nonlinear filter to improve the precision and robustness of multi-GNSS DF RTK [38]. (2) The DF data-aided AR method is proposed to fix ambiguities with the wide-lane and ionosphere-free combinations. The wide-lane pseudo-range is introduced for medium and long baseline RTK. (3) A nonlinear filter strategy is designed by integrating the DF data-aided AR method into the proposed AMC-KF. The test results show the significant superiority of the proposed strategy with various baselines.

The remainder of this paper is organized as follows: In Section 2, the double difference (DD) RTK model is introduced, followed by the derivation of the proposed DF data-aided AR method. The loosely coupled model for multi-GNSS is also presented. In Section 3, correntropy is introduced and the AMC-KF is elaborated in detail. Moreover, the derivation of the adaptive KBW is outlined. In Section 4, the performance of the proposed filter strategy is demonstrated. Datasets collected from the continuously operating reference stations (CORS) network of Australia are used for the short baseline test, adaptive strategy test, and the long baseline test. Finally, some conclusions are given in Section 5.

2. RTK Mathematic Model

2.1. Constrained Loosely Coupled Model

The DD carrier phase measurement φ (in cycles) with wavelength λ and code measurement P (in meters) is defined as follows [1]:

$$\begin{cases} \lambda\nabla\Delta\varphi = \nabla\Delta\rho - \lambda\nabla\Delta N + \nabla\Delta T - \nabla\Delta I + \nabla\Delta\varepsilon \\ \nabla\Delta \boldsymbol{P} = \nabla\Delta\rho + \nabla\Delta T + \nabla\Delta I + \nabla\Delta\xi \end{cases} \tag{1}$$

where $\nabla\Delta$ is the DD operator; ρ, T, I, and N are the receiver-satellite geometric range, tropospheric delay, ionospheric delay, and carrier phase ambiguity, respectively; and ξ and ε are the unmodeled errors including multipath noise, system noise, etc. The covariance matrix of the system state, system noise, measurement noise, and design matrices can be defined as $\boldsymbol{P}$, $\boldsymbol{Q}$, $\boldsymbol{R}$, and $\boldsymbol{H}$. The filter-state vector consists of positioning information and the DD ambiguities can be expressed as [7]:

$$\boldsymbol{x} = \begin{bmatrix} x_n^G & x_u^G & x_e^G & x_e^G & x_n^G & x_u^G & \nabla\Delta N^G & \nabla\Delta N^E \end{bmatrix}^T$$

where the superscripts 'G' and 'E' represent GPS and Galileo, and the subscripts e, n, and u represent different directions. The $diag(\cdot)$ is a diagonal matrix. The KF recursive process is defined as [1]:

$$\begin{cases} \left.\begin{array}{l} \hat{\boldsymbol{x}}_{k|k-1} = \boldsymbol{F}_{k,k-1}\hat{\boldsymbol{x}}_{k-1} \\ \boldsymbol{P}_{k|k-1} = \boldsymbol{F}_{k,k-1}\boldsymbol{P}_{k-1}\boldsymbol{F}_{k,k-1}^{\mathrm{T}} + \boldsymbol{\Gamma}_{k-1}\boldsymbol{Q}_{k-1}\boldsymbol{\Gamma}_{k-1}^{\mathrm{T}} \end{array}\right\} prediction \\ \left.\begin{array}{l} \boldsymbol{P}_k = (\boldsymbol{I} - \boldsymbol{K}_k\boldsymbol{H}_k)\boldsymbol{P}_{k|k-1} \\ \boldsymbol{K}_k = \boldsymbol{P}_{k|k-1}\boldsymbol{H}_k^{\mathrm{T}}(\boldsymbol{H}_k\boldsymbol{P}_{k|k-1}\boldsymbol{H}_k^{\mathrm{T}} + \boldsymbol{R}_k)^{-1} \\ \hat{\boldsymbol{x}}_k = \hat{\boldsymbol{x}}_{k|k-1} + \boldsymbol{K}_k(\boldsymbol{z}_k - \boldsymbol{H}_k\hat{\boldsymbol{x}}_{k|k-1}) \end{array}\right\} update \end{cases} \tag{2}$$

where $\boldsymbol{K}$, $\boldsymbol{F}$ are the gain and state transform matrix; and $\boldsymbol{z}_k$ and $\boldsymbol{x}_k$ are the measurements and state that needs to be estimated at the kth epoch. As $\boldsymbol{x}_G = \boldsymbol{x}_E$ once successfully located, the following constraint can be attached [39]:

$$\underbrace{\begin{bmatrix} \boldsymbol{I}_{3\times3} & -\boldsymbol{I}_{3\times3} \end{bmatrix}}_{D} \underbrace{\begin{bmatrix} \boldsymbol{x}_G \\ \boldsymbol{x}_E \end{bmatrix}}_{X} = \underbrace{\begin{bmatrix} \boldsymbol{0}_{3\times3} \end{bmatrix}}_{M} \tag{3}$$

where $\boldsymbol{I}$ represents the unit matrix, and $\boldsymbol{D}$ and $\boldsymbol{M}$ are the constraint matrices. The prediction step in (2) can be developed to constrain KF as follows:

$$\begin{cases} \boldsymbol{x}_{k|k-1} = \boldsymbol{x}_{k|k-1} - \boldsymbol{D}_k^T\left(\boldsymbol{D}_k\boldsymbol{D}_k^T\right)^{-1}\left(\boldsymbol{D}_k\boldsymbol{D}_k^T\right)^{-1}\left(\boldsymbol{D}_k\boldsymbol{x}_{k|k-1} - \boldsymbol{M}_k\right) \\ \boldsymbol{P}_{k|k-1} = \left(\boldsymbol{I} - \boldsymbol{D}_k^T\left(\boldsymbol{D}_k\boldsymbol{D}_k^T\right)^{-1}\boldsymbol{D}_k\right)^T \boldsymbol{P}_{k|k-1}\left(\boldsymbol{I} - \boldsymbol{D}_k^T\left(\boldsymbol{D}_k\boldsymbol{D}_k^T\right)^{-1}\boldsymbol{D}_k\right) \end{cases} \tag{4}$$

2.2. DF Data-Aided AR

Defining N_1 and N_2 as the carrier ambiguities on frequencies f_1 and f_2, respectively, the ρ on different frequencies can be formed by $\nabla\Delta\varphi$ as follows:

$$\begin{cases} \nabla\Delta\rho_1 = (\nabla\Delta\varphi_1 + \nabla\Delta N_1)\lambda_1 - \frac{A}{f_1^2} \\ \nabla\Delta\rho_2 = (\nabla\Delta\varphi_2 + \nabla\Delta N_2)\lambda_2 - \frac{A}{f_2^2} \end{cases} \tag{5}$$

where Ne is the number of electrons in unit area $A = 40.3\int_S Neds$. The wide-lane ambiguity $\nabla\Delta N_w$ with wavelength λ_w can be expressed as:

$$\nabla\Delta N_w = \nabla\Delta\varphi_1 - \nabla\Delta\varphi_2 - \frac{1}{\lambda_w}\left(\nabla\Delta\rho - \nabla\Delta T - \frac{f_1}{f_2}\nabla\Delta I - \nabla\Delta\varepsilon\right) \tag{6}$$

For the atmosphere errors in the above equation, $\nabla\Delta I$ cannot be ignored for long baselines due to the spatial difference. $\nabla\Delta T$ varies from 2 to 20 m depending on the satellite elevation, but a higher cutoff angle reduces data utilization [40]. Thus, the timeliness of the fixed solution could not be guaranteed, which is crucial for dynamic RTK once the satellites are available [41–44]. The proposed DF data-aided AR method is summarized as follows:

$$\begin{cases} \nabla\Delta N_w = (\nabla\Delta\varphi_1 - \nabla\Delta\varphi_2) - \frac{(f_1\nabla\Delta P_1 + f_2\nabla\Delta P_2)}{\lambda_w(f_1+f_2)} + \nabla\Delta\varepsilon \\ \nabla\Delta N_1 = \frac{1}{m\lambda_1 - n\lambda_2}[\nabla\Delta\varphi_{IF} - m\nabla\Delta\varphi_1 + n\nabla\Delta\varphi_2 - n\lambda_2\nabla\Delta N_w] \\ \nabla\Delta N_2 = \nabla\Delta N_1 - \nabla\Delta N_w \end{cases} \tag{7}$$

where $\nabla\Delta\varphi_{\mathrm{IF}}$ is the ionosphere-free combination for carrier observations, $m = \frac{f_1^2}{f_1^2 - f_2^2}$ and $n = \frac{f_2^2}{f_1^2 - f_2^2}$, and $\nabla\Delta\varepsilon$ represents the noise. The above method introduces the code wide-lane combination to invert $\nabla\Delta N_1$ and the derivation can be found in Appendices A and B. The proposed method shows the following merits: (1) The positioning accuracy and AR success rate are improved as the influence of $\nabla\Delta I$ error is eliminated. (2) The method is applicable for medium and long baselines as the limitation of geometry distance ρ is eliminated. (3) The $\nabla\Delta N$ on each frequency can be inverted directly from the high precise ionosphere-free and wide-lane measurements. Moreover, the following moving average with n epochs is also adopted to reduce the effect result from noise in $\nabla\Delta P$ [45]:

$$\nabla\Delta\hat{N}_w(n) = \frac{\sum_{i=k}^{k+n} Z^i \nabla\Delta N_w^i}{\sum_{i=k}^{k+n} Z^i} \tag{8}$$

where k is the start epoch of the observation arc without cycle slip. Z^i is the weight of ith epoch. The $\nabla\Delta\hat{N}_w(n)$ could be fixed to $round(\nabla\Delta\hat{N}_w(n))$ by the integer rounding method [46] if the differential residual of $\nabla\Delta\hat{N}_w$ between adjacent epochs meets the following constraint:

$$\left|\nabla\Delta\hat{N}_w(n) - \nabla\Delta\hat{N}_w(n-1)\right| < 0.1 \text{ cycles}$$

Then, the corresponding $\nabla\Delta N_1$, $\nabla\Delta N_2$ and precise carrier measurements could be formed by Equation (7). In the following test, the moving average window width n is set to be five epochs.

3. AMC-KF

The obtained $\nabla\Delta N_1$ and $\nabla\Delta N_2$ could be applied to form precise $\nabla\Delta\varphi$ in Equation (1), and used for state estimating in the nonlinear filter.

3.1. KF Based on MCC Derivation

Assuming the joint PDF of random variables $\boldsymbol{X}$ and $\boldsymbol{Y}$ as $F_{XY}(x,y)$, the correntropy is defined as follows [47]:

$$\begin{aligned} V(X,Y) &= E[G_\sigma(X-Y)] \\ &= \int G_\sigma(x-y)dF_{XY}(x,y) \\ &= \int \exp\left(-\frac{(X-Y)^2}{2\sigma^2}\right)dF_{XY}(x,y) \end{aligned} \tag{9}$$

where σ is the KBW, E is the expectation operator, and $\boldsymbol{G_\sigma}$ is the non-negative Gaussian kernel function. Furthermore, the Taylor expansion of the above equation is:

$$V(X,Y) = \frac{1}{\sqrt{2\pi}\sigma}\sum_{n=0}^{\infty}\frac{(-1)^n}{2^n n!}E\left[\frac{(X-Y)^{2n}}{\sigma^{2n}}\right] \tag{10}$$

Here, V is essentially a correlation function in the local kernel space controlled by σ, as it is the weighted sum of all even order moments of (X-Y). This localization proves meaningful in measuring the similarity between X and Y [48,49]. Then, the KF based on MCC (MCC-KF) can be established by optimizing the following loss function [50–52]:

$$J_C = G_\sigma(\|z_k - H\hat{x}_k\|) + G_\sigma(\|\hat{x}_k - \Phi\hat{x}_{k-1}\|) \tag{11}$$

where $\|\ \|$ denotes the Euclidean norm. J_c is only a function of σ [25]. Let $\frac{\partial J_C}{\partial \hat{x}_k} = 0$; the estimated state can be obtained as follows [53]:

$$\hat{x}_k = \Phi\hat{x}_{k-1} + \frac{G_\sigma(\|z_k - H\hat{x}_k\|)}{G_\sigma(\|\hat{x}_k - F\hat{x}_{k-1}\|)} H^T(z_k - H\hat{x}_k) \tag{12}$$

It tells that the MCC will be achieved if $X = Y$, as G_σ reaches the upper bound and the PDF of the predicted value and the measured value matched to the maximum extent [54]. The further results can be obtained while $x_k \approx F\hat{x}_{k-1}$:

$$\hat{x}_k = \hat{x}_k^- + G_\sigma(\|z_k - H\hat{x}_k\|)H^T(z_k - H\hat{x}_k) \tag{13}$$

3.2. AMC-KF Derivation

The KBW of the originally proposed MCC shown in Equation (13) is usually predefined empirically, which results in the compromise between fast learning initially and fast learning near the optimum point. To derive an adaptive KBW, the loss function is further enhanced as follows:

$$\begin{aligned} J_C = &\ G_\sigma\left(\|z_k - H\hat{x}_k\|_{R_k^{-1}}\right) + G_\sigma\left(\|\hat{x}_k - \Phi\hat{x}_{k-1}\|_{P_{k|k-1}^{-1}}\right) \\ &\le G_\sigma\left(\|z_k - H\hat{x}_k\|_{R_k^{-1}}\right) + \Lambda \\ &= \frac{1}{N\sqrt{2\pi}\sigma}\sum_{i=n-N+1}^{n} \exp\left(\frac{-\|z_k - H\hat{x}_k\|_R^2}{2\sigma^2}\right) + \Lambda \end{aligned} \tag{14}$$

where Λ is a constant overbounded by $\lim_{\|\hat{x}_k - \Phi\hat{x}_{k-1}\|_{P_{k|k-1}^{-1}} \to 0} G_\sigma\left(\|\hat{x}_k - \Phi\hat{x}_{k-1}\|_{P_{k|k-1}^{-1}}\right)$ [47,55]. To search for the proper σ, the gradient ascent approach is applied by taking a small step μ along the positive gradient, then the nth iteration can be expressed as $\sigma_{n+1} = \sigma_n + \mu\nabla J_c$ [47]. The J_c can be minimized as follows:

$$\begin{aligned} \nabla J_C &= \frac{\partial J_C}{\partial\sigma} = -\frac{1}{N\sqrt{2\pi}\sigma^2}\sum_{i=n-N+1}^{n}\exp\left(-\frac{\|z_k - H\hat{x}_k\|_{R^{-1}}}{2\sigma^2}\right) + \frac{1}{N\sqrt{2\pi}\sigma}\sum_{i=n-N+1}^{n}\left(\frac{\|z_k - H\hat{x}_k\|_{R^{-1}}}{\sigma^3}\right)\exp\left(-\frac{\|z_k - H\hat{x}_k\|_{R^{-1}}}{2\sigma^2}\right) \\ &= -\frac{1}{\sqrt{2\pi\sigma^2}}\exp\left(-\frac{\|z_k - H\hat{x}_k\|_{R^{-1}}}{2\sigma^2}\right) + \frac{\|z_k - H\hat{x}_k\|_{R^{-1}}}{\sqrt{2\pi\sigma^4}}\exp\left(-\frac{\|z_k - H\hat{x}_k\|_{R^{-1}}}{2\sigma^2}\right) \\ &= 0 \end{aligned}$$

Then, the closed-formed KBW, which can adaptively adjust according to R, is expressed as follows:

$$\sigma = \sqrt{\frac{\|z_k - H\hat{x}_k\|_{R^{-1}}}{2}} + \sigma' \tag{15}$$

Through Equations (4), (12), (13) and (15), the proposed AMC-KF is finally obtained. It should be noted that the exponential part of AMC-KF reduces to constant and is no longer correntropy-based if Equation (15) is applied without the small penalty term σ'. The penalty term is artificially added and can be determined according to [56].

3.3. Filter Implementation

The procedure of the proposed nonlinear strategy is summarized as follows: (1) Removing the ambiguities $\nabla\Delta N$ on each frequency by the DF data-aided AR method. A threshold and moving average operation are applied to ensure stability and reduce noise. (2) The precise $\nabla\Delta\varphi$ without ambiguities is fed into the AMC-KF for multi-GNSS float solution. The adaptive KBW is used for the prediction and update step during the filtering. Both of them help improve the robustness and accuracy of the float solution. (3) To keep consistency with other RTK structures, the least square ambiguity decorrelation adjustment (LAMBDA) is adopted to transfer the float solution to the fixed solution.

To initialize the proposed filter, the variance-covariance matrix is deduced by the least square method (LS) at the initial epoch. The $\boldsymbol{F}$ and $\boldsymbol{Q}$ can be defined as an identity matrix and a zero matrix without cycle slips. The noise level for non-difference code and carrier measurements are set to 3 m and 3 cm, respectively [1]. The framework of the proposed nonlinear strategy is depicted in Figure 1. As shown in Equation (8), the window width of the moving average and the threshold in the 'DF Data-aid AR Stage' is usually set to 5 epochs and 0.1 cycles, which implies the influence on the first time to fix ambiguity is tiny and controllable.

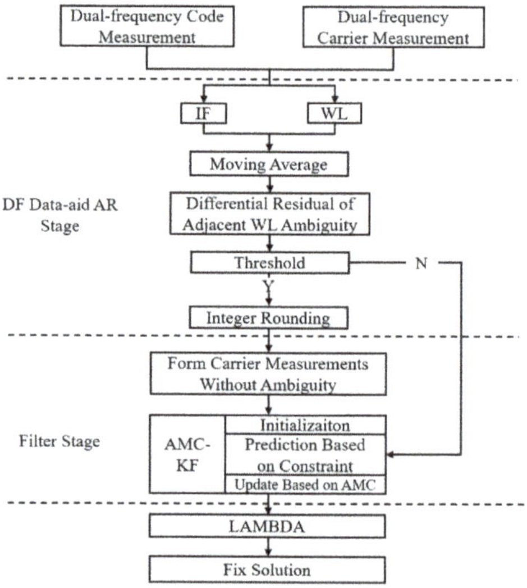

Figure 1. The procedure of the filter implementation.

4. Test and Result

To validate the proposed filter strategy, short and long baseline tests are conducted and the traditional DD KF (DD-KF) model mentioned in [1] is also used for comparison. The DD-KF is established in Equations (1)–(4) without the 'DF Data-aid AR Stage'. As the noise of original observations is significantly less than those of wide-lane and ionosphere-free combined measurements, thus, the DF Data-aid AR method is not enabled for the short baseline test. The improvements illustrated in the short baseline test are only beneficial from the AMC-KF method. For the long baseline, the DF Data-aid AR stage is enabled to eliminate atmosphere errors; thus, the improvements in the long baseline test are beneficial both from the AMC-KF and the DF Data-aid AR method. All results obtained are based on post-processing performed on an Intel Core i7 2.30 GHz notebook with 16 GB RAM running on Windows 10.

As Figure 2 shows, the dataset is collected from seven Australia CORS stations (BONE, QCLF, ANGS, STNY, NEWH, GSBN, WBEE) on January 1, 2021, and all formed baselines are elaborated in Table 1. The first six baselines range from 19–60 km and are used for the short baseline test. The last two baselines are formed by (BONE, GSBN, WBEE) and used for long baselines. The sample interval and cut-off elevation for all tests are 30 s and 10°, respectively.

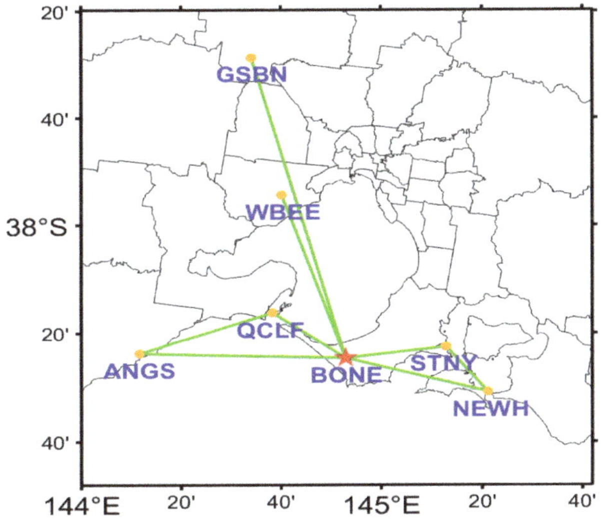

Figure 2. Distribution of CORS station BONE, QCLF, ANGS, STNY, NEWH, GSBN, WBEE. (https://gnss.ga.gov.au/network (accessed on 1 January 2021)).

Table 1. Information for different baselines.

No.	Baseline	Distance (km)	Sample Interval	Cut-Off Elevation	Processing Model
1	STNY-NEWH	19.562	30 s	10°	AMC-KF, DD-KF
2	BONE-QCLF	26.363	30 s	10°	AMC-KF, DD-KF
3	BONE-STNY	29.231	30 s	10°	AMC-KF, DD-KF
4	QCLF-ANGS	41.335	30 s	10°	AMC-KF, DD-KF
5	NEWH-BONE	42.606	30 s	10°	AMC-KF, DD-KF
6	ANGS-BONE	60.213	30 s	10°	AMC-KF, DD-KF
7	BONE-WBEE	58.942	30 s	10°	AMC-KF, DD-KF
8	BONE-GSBN	106.877	30 s	10°	AMC-KF, DD-KF

4.1. Position Accuracy Test

Figure 3 depicts the available satellites and the relative dilution of the precision (RDOP) on BONE. The other stations will obtain similar indicators for the short baseline. The average available satellites for filtering in the whole day are 6 for GPS, 6 for Galileo, and 10 for GPS+Galileo. The RDOP for GPS+Galileo is 0.7228 and indicates an ideal environment for relative positioning [1]. If the available satellite number of Gallileo is less than four, the position result is obtained only by GPS.

The root mean square (RMS) and standard deviation (STD) of positioning errors on the East (E), North (N), and Up (U) components are shown in Table 2, where the positive and negative values represent improvement and degradation.

For AMC-KF, the RMS on BONE-QCLF, QCLF-ANGS and ANGS-BONE is (0.13417 m, 0.20254 m, 0.30294 m), (0.09421 m, 0.17402 m, 0.27439 m), (0.16395 m, 0.18714 m, 0.26284 m), respectively. While the RMS for DD-KF is (0.28258 m, 0.43671 m, 0.80099 m), (0.23641 m, 0.23035 m, 0.96756 m), (0.53574 m, 0.15190 m, 0.65454 m). The accuracy improvement by (+52.52%, +53.62%, +62.18%), (+60.15%, +24.45%, +71.64%), (+69.40%, −23.20%, +59.84%) is achieved for AMC-KF. Similarly, the STD on each baseline for AMC-KF is (0.10886 m, 0.18606 m, 0.30278 m), (0.08226 m, 0.16659 m, 0.26107 m) and (0.16027 m, 0.16694 m, 0.24867 m) while for DD-KF is (0.10066 m, 0.33062 m, 0.40077 m), (0.11478 m, 0.17340 m, 0.52163 m), (0.23124 m, 0.10958 m, 0.53847 m). AMC-KF ameliorates the performance by (−8.15%, +43.72%, +24.45%), (+28.33%, +3.93%, +49.95%), (+30.69%, −52.35%, +53.82%).

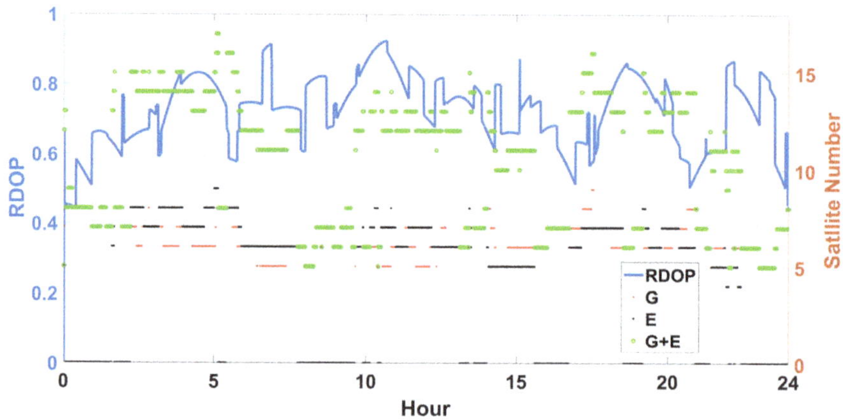

Figure 3. RODP and available satellites of BONE.

For BONE-STNY, STNY-NEWH, NEWH-BONE, the AMC-KF possesses an RMS improvement by (+55.90%, +64.82%, +85.24%), (−22.94%, +60.42%, +83.79%), (+53.43%, +82.67%, +67.43%), and an STD improvement by (+31.55%, +46.827%, +78.57%), (−11.22%, +35.33%, +53.71%) and (+47.26%, +73.50%, +66.16%), respectively.

Compared to DD-KF in each direction, the proposed AMC-KF obtains an improvement by (32.24%, 34.48%, 63.07%) on average, despite negative values existing. Thus, AMC-KF is a beneficial scheme for short baseline RTK as it retains a low-level positioning error. Particularly, the positioning errors on U are reduced by more than 60%. One possible reason for the negative values is that AMC-KF also redistributes the positioning residual since errors in different directions are coupled [57].

Table 2. RMS and STD improvement on ENU for BONE-QCLF-ANGS-BONE.

Baseline		E		N		U		Improvement on ENU
		AMC-KF	DD-KF	AMC-KF	DD-KF	AMC-KF	DD-KF	
RMSE (m)	BONE-QCLF	0.13417	0.28258	0.20254	0.43671	0.30294	0.80099	+52.52%, +53.62%, +62.18%
	QCLF-ANGS	0.09421	0.23641	0.17402	0.23035	0.27439	0.96756	+60.15%, +24.45%, +71.64%
	ANGS-BONE	0.16395	0.53574	0.18714	0.15190	0.26284	0.65454	+69.40%, −23.20%, +59.84%
STD (m)	BONE-QCLF	0.10886	0.10066	0.18606	0.33062	0.30278	0.40077	−8.15%, +43.72%, +24.45%
	QCLF-ANGS	0.08226	0.11478	0.16659	0.17340	0.26107	0.52163	+28.33%, +3.93%, +49.95%
	ANGS-BONE	0.16027	0.23124	0.16694	0.10958	0.24867	0.53847	+30.69%, −52.35%, +53.82%
RMSE (m)	BONE-STNY	0.15830	0.35895	0.15017	0.42681	0.19493	1.32102	+55.90%, +64.82%, +85.24%
	STNY-NEWH	0.13166	0.10709	0.13620	0.34410	0.19730	1.21688	−22.94%, +60.42%, +83.79%
	NEWH-BONE	0.17273	0.37092	0.13468	0.77698	0.19800	0.60796	+53.43%, +82.67%, +67.43%
STD (m)	BONE-STNY	0.13460	0.19663	0.14762	0.27757	0.18249	0.85175	+31.55%, +46.827%, +78.57%
	STNY-NEWH	0.11695	0.10515	0.13590	0.21013	0.19655	0.42460	−11.22%, +35.33%, +53.71%
	NEWH-BONE	0.10496	0.19900	0.13333	0.50321	0.19746	0.58359	+47.26%, +73.50%, +66.16%

4.2. Adaptive Strategy Test

The variation and statistics of KBW are shown in Figure 4. The AMC-KF is proved to be effective as the KBW increases rapidly after initialization to respond to the input GNSS measurements and varies epoch by epoch. In Figure 4b, the mean and standard deviation found for (STNY-NEWH, QCLF-ANGS, NEWH-BONE, BONE-STNY, BONE-

QCLF) are (16.1284, 14.9827, 20.6101, 19.1174, 13.1624, 17.0062) and (4.7120, 5.1061, 6.1030, 5.4980, 3.9002, 5.38731), respectively. Although the KBW is different from each other as all baselines are spatially separated, the similar variation trend verified that the adaptive KBW is sensitive to the environment.

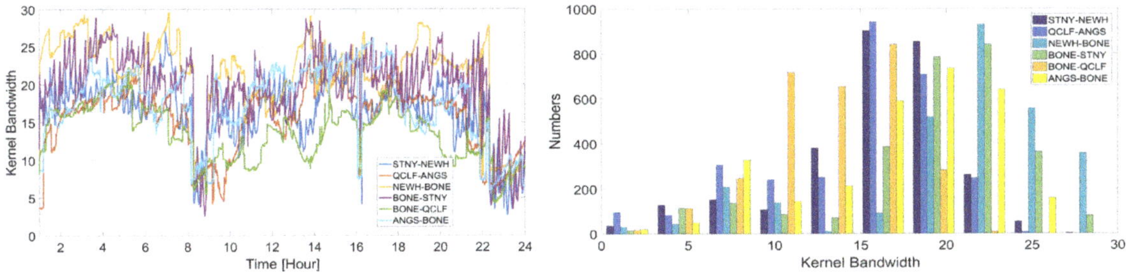

(**a**) (**b**)

Figure 4. KBW time series and statistics for each baseline. (**a**) Time series for each baseline; (**b**) the statistics of KBW time series.

The filter time consumption with the proposed adaptive KBW and the original fixed KBW in MCC is illustrated in Figures 5 and 6. The fixed KBW used here is set to be 1, 5, 25, and 30, as all KBW has shown in Figure 4b fall in [0, 30].

It could be found that the adaptive KBW owns smoother and more stable processing results. It means that the embedded devices and on-chip modules may benefit from power conservation [1]. For adaptive KBW, the average time consumption at each epoch is (0.0683 s, 0.0535 s, 0.0520 s, 0.0674 s, 0.0641 s, 0.0495 s) on ANGS-BONE, BONE-QCLF, BONE- STNY, NEWH-BONE, QCLF-ANGS, STNY-NEWH. While for fixed KBW (1, 5, 25, 30) are (0.0619 s, 0.0599 s, 0.0717 s, 0.0659 s), (0.0528 s, 0.0549 s, 0.0517 s, 0.0544 s), (0.0515 s, 0.0543 s, 0.0581 s, 0.0582 s), (0.0542 s, 0.0681 s, 0.0680 s, 0.0543 s), (0.0658 s, 0.0594 s, 0.0679 s, 0.0662 s) and (0.0581 s, 0.0560 s, 0.0574 s, 0.0499 s), respectively.

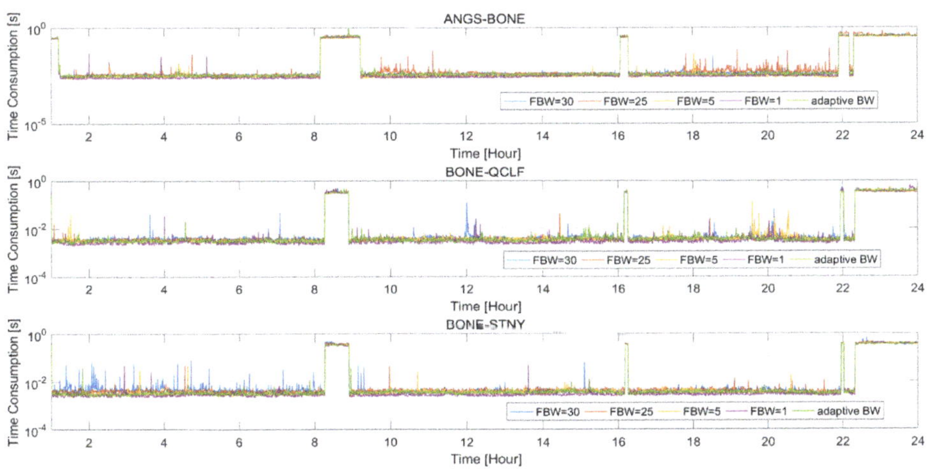

Figure 5. Time consumption with different KBW strategies on ANGS-BONE, BONE-QCLF, BONE-STNY.

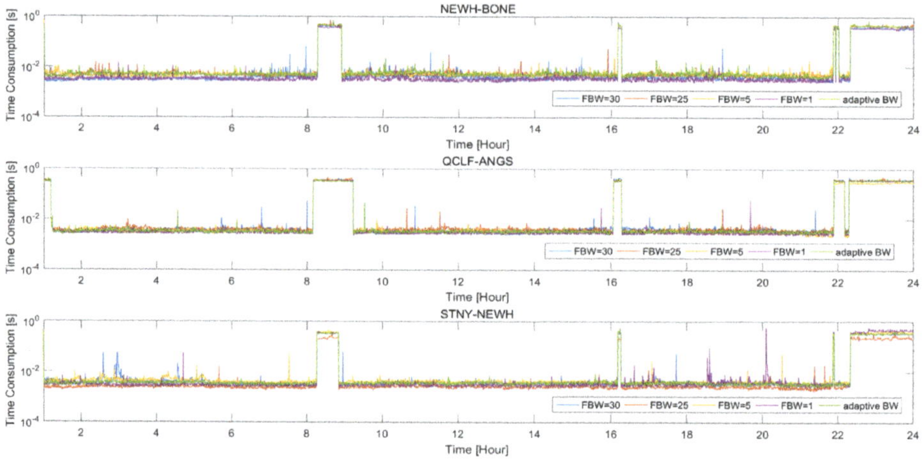

Figure 6. Time consumption with different KBW strategies on NEWH-BONE, QCLF-ANGS, STNY-NEWH.

Treating adaptive KBW as the benchmark, the efficiency improvement is demonstrated in Figure 7. For all 24 cases, negative values (shown in nine cases) indicate a longer time consumption than the benchmark, and the positive values (shown in 15 cases) indicate the opposite results. In general, degradation exists in most cases; the calculation load increased by 6.54% in the other 15 cases, and 5 of them take 10% more time. Only three cases achieved more than a 10% improvement, and the remaining six cases averagely improved by 5.41%. The superiority of the adaptive KBW strategy is the most obvious in STNY-NEWH and BONE-STNY. Thus, the proposed AMC-KF and adaptive KBW strategy can generally improve filtering efficiency.

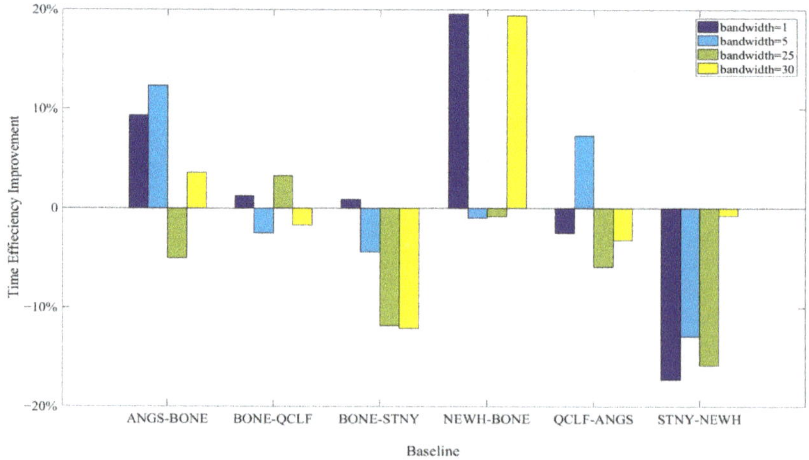

Figure 7. Time efficiency improvement for different KBW strategies.

The RMS improvement of the proposed AMC-KF compared to the original fixed KBW is shown in Figure 8. Here, the negative values mean a positioning-accuracy degradation compared to the adaptive KBW strategy.

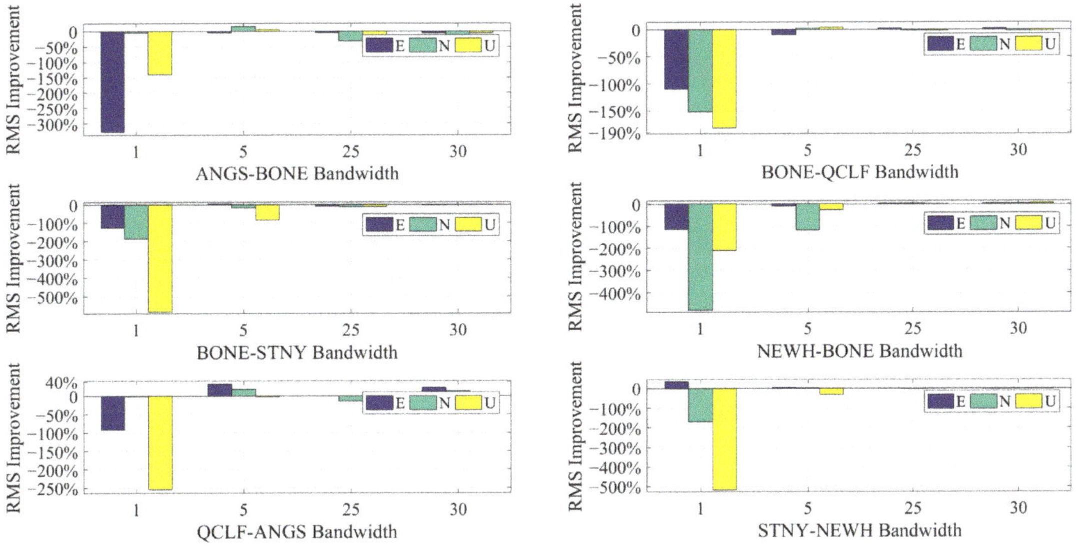

Figure 8. RMS improvement on each baseline.

The RMS increases significantly at least on one direction component while KBW = 1 and 5. Especially, the RMS on U deteriorated by almost six times (−584.261%) compared to the adaptive KBW. However, no significant performance fluctuation appears when the fixed KBW = 25 and 30, except for the −43.13% degradation on U (KBW = 25) and the 23.3% improvement on E (KBW = 30), which both occur on QCLF-ANGS.

Although the large KBW seems better, the improvement is hardly permitted as the increase in KBW amplifies the time consumption and positioning errors. In conclusion, the proposed AMC-KF method takes both efficiency and accuracy into account and is more progressive than the traditional methods with KBW fixed.

4.3. Ambiguity Resolution Analysis

In the triangle network formed by BONE-STNY-NEWH, all baselines shared the common pivot satellite, and the closure residual of DD ambiguities should meet the following constraint: $\nabla\Delta N_{closure} = \nabla\Delta N_{BONE\text{-}STNY} + \nabla\Delta N_{STNY\text{-}NEWH} + \nabla\Delta N_{NEWH\text{-}BONE} = 0$. Thus, $\nabla\Delta N_{closure}$ could be used to verify the proposed DF data-aided AR method. It should be noted that only DF data from GPS is used for a long baseline test since only GPS signal is stably received.

To illustrate the reasonability for setting the window width of the moving average to be five epochs, Figure 9 gives the differential residuals of adjacent $\nabla\Delta N_w$ on each baseline. The statics of results are shown in Table 3 and the results within ±1 cycle are shown for easy observation. The threshold of ±0.1 cycles, ±0.15 cycles, and ±0.5 cycles are also illustrated as limitations bounds.

Table 3. $\nabla\Delta N_w$ statistics of the triangle closure network.

Baseline	(−0.1, 0.1)	(−0.15, 0.15)	(0.5, 0.5)	Others
BONE_STNY	98.6679%	98.7700%	99.0339%	0.9661%
NEWH_BONE	98.3097%	98.5279%	98.9901%	1.0099%
STNY_NEWH	98.8000%	98.8726%	99.1373%	0.8627%

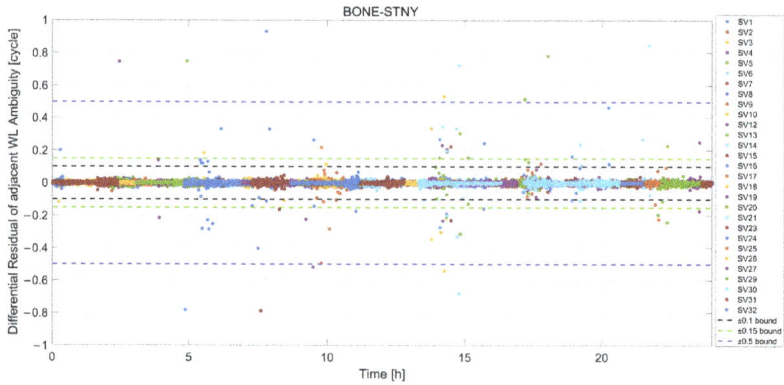

(**a**)

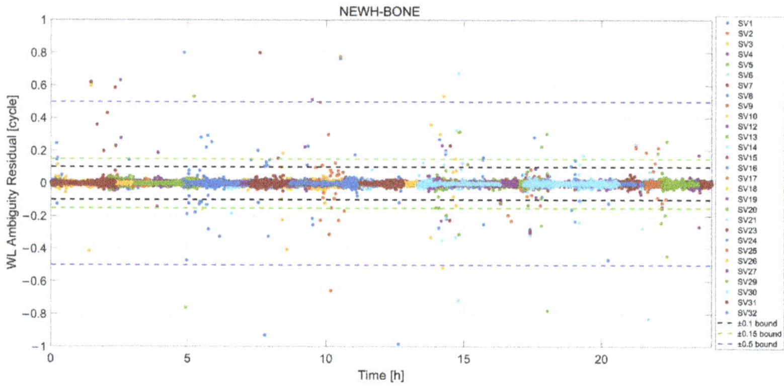

(**b**)

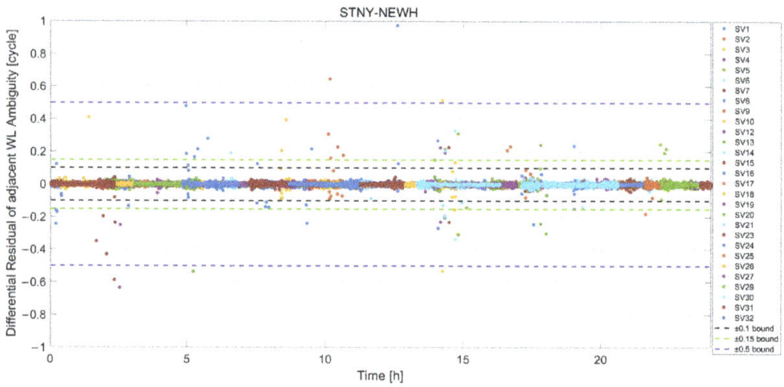

(**c**)

Figure 9. The $\nabla\Delta N_{\mathrm{w}}$ for each baseline of the triangle closure network. (**a**) BONE-STNY; (**b**) NEWH-BONE; (**c**) STNY-NEWH.

In Table 3, the proportion of residuals suppressed within ±0.1 cycles is 98.6679% for BONE-STNY, 98.3097% for NEWH-BONE, and 98.8000% for STNY-NEWH, respectively. Meanwhile, the results increase slightly when the threshold bounds increase. Thus, setting the threshold to 0.1 cycles is reasonable and conservative, as most validated epochs are included.

The $\nabla\Delta N_{closure}$ of all available satellites is shown in Figure 10. It could be found that, for most satellites, the $\nabla\Delta N_{closure}$ converges to 0 once they are used and $\nabla\Delta N_{closure} = 0$ accounts for the majority. This means that $\nabla\Delta N_{BONE\text{-}STNY}$, $\nabla\Delta N_{STNY\text{-}NEWH}$, $\nabla\Delta N_{NEWH\text{-}BONE}$ are fixed correctly without initialization. The outliers usually appeared at discrete epochs contaminated by cycle slip and could be further eliminated by refined data synchronization and cycle-slip detection.

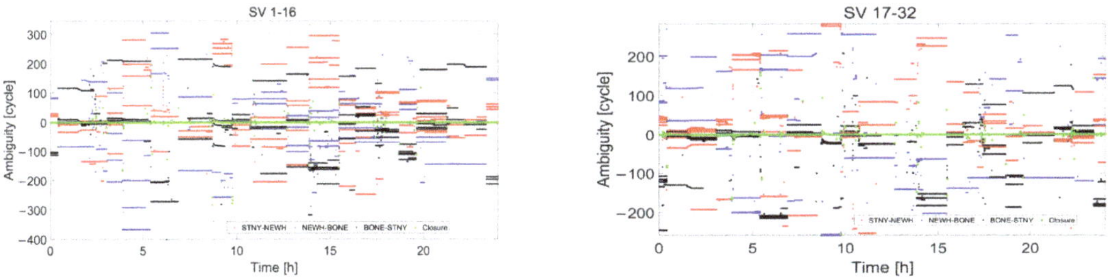

(**a**) (**b**)

Figure 10. $\nabla\Delta N_{closure}$ in the network. (**a**) $\nabla\Delta N_{closure}$ for SV1-SV16; (**b**) $\nabla\Delta N_{closure}$ for SV17-SV32.

Figure 11 shows the detailed results of $\nabla\Delta N_{closure}$ with a total of 23377 effective epochs used. The minimum and maximum outliers accompanied by the cycle slip are (−283.56 cycles, 365.86 cycles). According to Table 4, the $\nabla\Delta N_{closure} < 0.5$ cycles in most epochs, meaning that the $\nabla\Delta N_w$ can be correctly fixed by integer rounding with a success rate of not less than 93%. The 1.7154% epochs fall into 0.5–1 cycles and 2.7848% fall into 1–5 cycles are treated as small residuals and could be improved by synchronization and cycle-slip repair. However, at least one of the three baselines fails to fix its ambiguity for the remaining 1.6041% of epochs that include residuals larger than 10 cycles. Once the $\nabla\Delta N_w$ is fixed, the corrected $\nabla\Delta\varphi$ is used for the float solution, which is expected to be with a small variance.

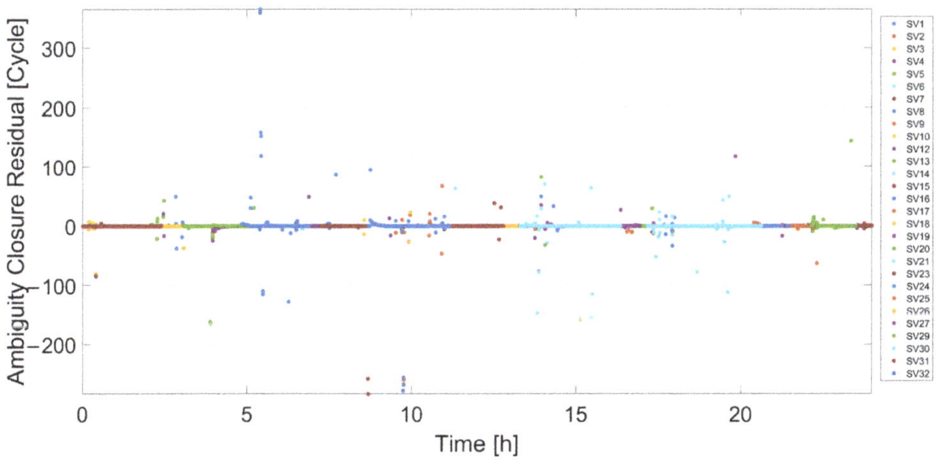

Figure 11. Ambiguity closure residual in the network.

Table 4. The statistics of $\nabla\Delta N_{\text{closure}}$.

$\nabla\Delta N_{\text{closure}}$	Residual Range (Cycle)				
	[0, 0.5)	[0.5, 1)	[1, 5)	[5, 10)	[10, +∞)
Proportion	93.1043%	1.7154%	2.7848%	0.7914%	1.6041%

The BONE-GSBN and BONE-GSBN with the distance of 106.877 km and 58.942 km, respectively, are used for the long baseline test. The improvements are both benefiting from the 'DF Data-aid AR stage' and the 'filter stage'. Figure 12 and Table 5 show the positioning error on ENU components. The AMC-KF maintains the positioning error around 0 and no obvious difference occurs in all directions.

It can be inferred the proposed filter strategy suppresses the noise in DD measurements on the whole, as the correntropy can measure the similarity between the random variables through PDF.

For AMC-KF, the RMS is improved by (+78.60%, +88.85%, +77.74%) at BONE-WBEE and (+57.49%, +69.52%, +42.31%) at BONE-GSBN than DD-KF. The STD is improved by (+64.97%, +66.26%, +60.81%) at BONE-WBEE and (+51.10%, +46.89%, +40.34%) at BONE-GSBN, respectively. The proposed filter strategy can reduce the positioning error significantly for the long baseline.

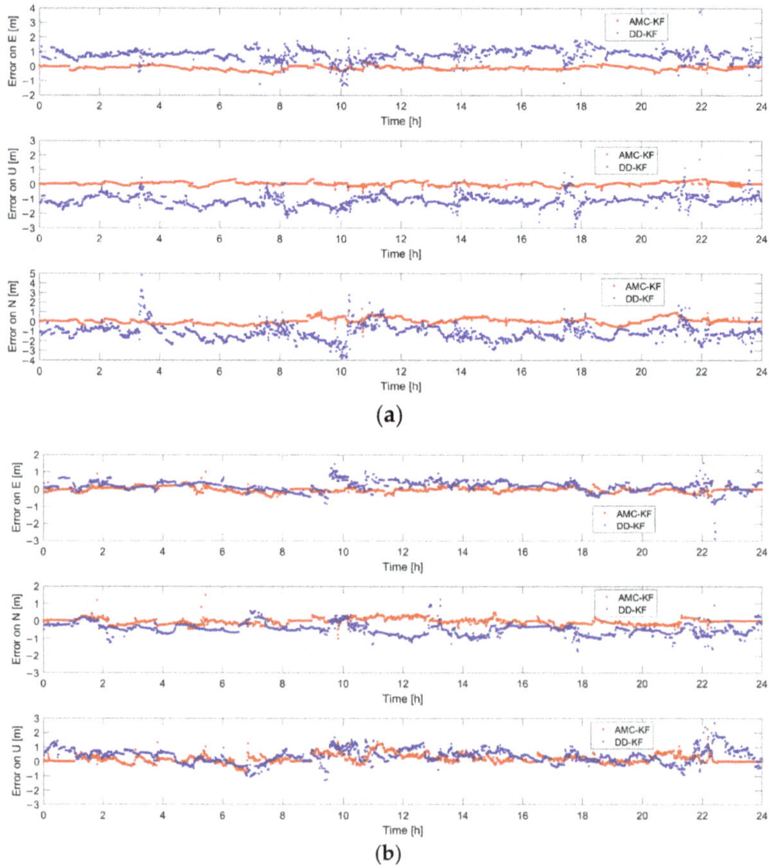

Figure 12. Positioning error on ENU. (**a**) Positioning error of BONE-WBEE; (**b**) positioning error of BONE-GSBN.

Table 5. The RMS and STD of position error on the ENU component.

Baseline	Model	RMS (m)			STD (m)		
		E	N	U	E	N	U
BONE-WBEE	DD-KF	0.87648	1.17600	1.38090	0.37734	0.37925	0.76978
	AMC-KF	0.18754	0.13117	0.30743	0.13217	0.12794	0.30166
BONE-GSBN	DD-KF	0.34605	0.58798	0.62407	0.29501	0.33371	0.50501
	AMC-KF	0.14710	0.17921	0.36000	0.14426	0.17722	0.30130

5. Conclusions

In terms of the timeliness and accuracy of RTK in harsh environments, both the measurement quality and the filter robustness need to improve, especially with the presence of non-Gaussian noise. This paper focus on multi-GNSS DF applications and a new nonlinear filter strategy is proposed. It consists of the DF data-aided AR method and the AMC-KF based on MCC and adaptive KBW. The superiorities are verified through tests with various baselines. First of all, ionosphere-free and wide-lane measurements are used for the DF data-aided AR method. The ambiguities on each frequency are directly converted without searching. Then, the corrected carrier measurements are used for the float solution by the proposed AMC-KF. The AMC-KF is robust to non-Gaussian noise and sparking noise as it employs MCC and adaptive KBW to measure the similarity between the input and output. Compared to the conventional DD-KF, the proposed strategy achieves higher accuracy and efficiency. The following conclusions are obtained:

(1) For the positioning accuracy of short baselines, the RMS and STD of positioning error improved by more than 30%, 30%, and 60% on the E, N, and U components, respectively. The applicability and flexibility of the proposed AMC-KF are validated significantly.
(2) For the proposed adaptive KBW, the efficiency and accuracy are compared and validated with fixed KBW (1, 5, 25, 30). The results show that large KBWs increase the computational load, while the small KBWs gain a worse positioning accuracy. The proposed optimization strategy can change the KBW adaptively according to the measurements and is validated to be effective as it considers both efficiency and accuracy.
(3) A total of 29 satellites were involved in the ambiguity resolution test with long baselines. It validates that the proposed DF data-aided AR method achieves a success rate of more than 93%. The results are expected to be further improved with stringent data synchronization and cycle-slip detection.
(4) For the long baseline test, the proposed filter strategy obtains an improvement of more than 40% in all directions as the noise is effectively suppressed. For the longer baseline BONE-WBEE, the RMS of positioning error is reduced by more than 75% on E, N, and U, which shows that the proposed method plays a better role in long baseline RTK when the multi-GNSS multi-frequency data is stably available.

Our future work focuses on deriving the sequential form of the proposed nonlinear filter strategy and applying it to smartphone RTK applications. To improve the precision and reliability of dynamic navigation in urban environments, the integration of the proposed method with vector-tracking GNSS receivers will also be explored.

Author Contributions: Conceptualization, J.L.; methodology, J.L.; software, J.L. and T.L.; investigation, Y.J., B.X., and Z.L.; data curation, J.L., M.S. and T.L.; writing—original draft preparation, J.L.; writing—review and editing, B.X., M.S., M.L., Z.L. and G.X.; supervision, Y.J., B.X., Z.L., and G.X. All authors have read and agreed to the published version of the manuscript.

Funding: This study is supported by the Guangdong Basic and Applied Basic Research Foundation (No. 2021A1515012600); the Opening Project of Guangxi Wireless Broadband Communication and Signal Processing Key Laboratory (No. GXKL06200217); the Open Fund of Key Laboratory of Urban Land Resources Monitoring and Simulation, Ministry of Natural Resources (No. KF-2021-06-104).

Data Availability Statement: The authors are grateful to the Global Navigation Satellite System Data Centre of Australia for publicly sharing their GNSS data.

Acknowledgments: We are grateful to the anonymous reviewers and editors for their helpful and constructive suggestions, which significantly improved the quality of the paper.

Conflicts of Interest: The authors declare no conflict of interest.

Appendix A

Defining the common geometry distance $\rho + c(\delta t_u - \delta t^s) + T$ without considering ionosphere delay as Θ. The carrier phase measurement φ (in cycles) with wavelength λ and code measurements P (in meters) are defined as follows:

$$\begin{cases} \varphi_1 = \frac{f_1}{c}\Theta - \frac{A}{cf_1} + N_1 + \xi_{\varphi_1} \\ \varphi_2 = \frac{f_2}{c}\Theta - \frac{A}{cf_2} + N_2 + \xi_{\varphi_2} \\ P_1 = \Theta + \frac{A}{f_1^2} + \varepsilon_{P_1} \\ P_2 = \Theta + \frac{A}{f_2^2} + \varepsilon_{P_2} \end{cases} \tag{A1}$$

where f_1 and f_2 represent different frequencies, t_u and t^s are the clock errors from the user receiver and satellite and ξ and ε are the unmodelled noise on φ and P. The c denotes the speed of light. The Θ and A can be expressed by P_1 and P_2 as follows:

$$A = \frac{f_1^2 f_2^2[(P_1 - P_2) - (\varepsilon_{P_1} - \varepsilon_{P_2})]}{f_2^2 - f_1^2} = \frac{f_1^2 f_2^2 (P_1 - P_2)}{f_2^2 - f_1^2} + \varepsilon_A \tag{A2}$$

$$\Theta = \frac{(f_1^2 P_1 - f_2^2 P_2) - (f_1^2 \varepsilon_{P_1} - f_2^2 \varepsilon_{P_2})}{f_1^2 - f_2^2} = \frac{f_1^2 P_1 - f_2^2 P_2}{f_1^2 - f_2^2} + \varepsilon_\Theta \tag{A3}$$

where, $\varepsilon_A = \frac{f_1^2 f_2^2(\varepsilon_{P_1} - \varepsilon_{P_2})}{f_2^2 - f_1^2}$ and $\varepsilon_\Theta = \frac{(f_2^2 \varepsilon_{P_2} - f_1^2 \varepsilon_{P_1})}{f_1^2 - f_2^2}$ are the noise on A and Θ, respectively. The wide-lane combination of φ is expressed as:

$$\varphi_{WL} = \varphi_1 - \varphi_2 = \left(\frac{f_1}{c} - \frac{f_2}{c}\right)\Theta - \left(\frac{f_2 - f_1}{cf_1 f_2}\right)A + N_w + \xi_w \tag{A4}$$

where N_w is the wide-lane ambiguity, $\xi_w = (\xi_1 - \xi_2)$. Then, the following expression can be obtained:

$$\begin{aligned} \varphi_{WL} &= \frac{f_1 - f_2}{c} \cdot \frac{f_1^2 P_1 - f_2^2 P_2}{f_1^2 - f_2^2} - \left(\frac{f_2 - f_1}{cf_1 f_2}\right) \cdot \frac{f_1^2 f_2^2 (P_1 - P_2)}{f_2^2 - f_1^2} + N_w + \varepsilon \\ &= \frac{f_1^2 P_1 - f_2^2 P_2}{\lambda_w (f_1^2 - f_2^2)} + \left(\frac{f_1 - f_2}{cf_1 f_2}\right) \cdot \frac{f_1^2 f_2^2 (P_1 - P_2)}{f_2^2 - f_1^2} + N_w + \varepsilon \\ &= \frac{f_1^2 P_1 - f_2^2 P_2}{\lambda_w (f_1^2 - f_2^2)} + \frac{f_1 f_2 (P_1 - P_2)}{\lambda_w (f_2^2 - f_1^2)} + N_w + \varepsilon \\ &= \frac{(f_1^2 P_1 - f_2^2 P_2) - f_1 f_2 (P_1 - P_2)}{\lambda_w (f_1^2 - f_2^2)} + N_w + \varepsilon \\ &= \frac{(f_1 P_1 + f_2 P_2)}{\lambda_w (f_1 + f_2)} + N_w + \varepsilon \end{aligned}$$

where ε is the combination of the noise terms which can be expressed as $\varepsilon = \xi_w + \left(\frac{f_1}{c} - \frac{f_2}{c}\right)\varepsilon_\theta - \frac{f_2 - f_1}{cf_1 f_2}\varepsilon_A$. Finally, the N_w can be obtained as follows:

$$N_{WL} = (\varphi_1 - \varphi_2) - \frac{(f_1 P_1 + f_2 P_2)}{\lambda_w (f_1 + f_2)} + \varepsilon \tag{A5}$$

In addition, the corresponding DD wide-lane ambiguity can be obtained by:

$$\nabla\Delta N_w = (\nabla\Delta\varphi_1 - \nabla\Delta\varphi_2) - \frac{(f_1\nabla\Delta P_1 + f_2\nabla\Delta P_2)}{\lambda_w(f_1+f_2)} + \nabla\Delta\varepsilon \tag{A6}$$

Appendix B

According to Equation (5), the ionosphere-free measurement is defined as $\nabla\Delta\varphi_{IF} = m\nabla\Delta\rho_1 - n\nabla\Delta\rho_2$. Where $m = \frac{f_1^2}{f_1^2-f_2^2}$ and $n = \frac{f_2^2}{f_1^2-f_2^2}$, the definition of symbols and variables stay the same as those above. Thus, we have the following expansion:

$$\begin{aligned}\nabla\Delta\varphi_{IF} &= m\left[\nabla\Delta\varphi_1 + \nabla\Delta N_1\lambda_1 - \frac{A}{f_1^2}\right] - n\left[\nabla\Delta\varphi_2 + \nabla\Delta N_2\lambda_2 - \frac{A}{f_2^2}\right]\\ &= m\nabla\Delta\varphi_1 + m\nabla\Delta N_1\lambda_1 - n\nabla\Delta\varphi_2 - n\nabla\Delta N_2\lambda_2\\ &= m\nabla\Delta\varphi_1 - n\nabla\Delta\varphi_2 + m\lambda_1\nabla\Delta N_1 - n\lambda_2(\nabla\Delta N_1 - \nabla\Delta N_w)\\ &= m\nabla\Delta\varphi_1 - n\nabla\Delta\varphi_2 + m\lambda_1\nabla\Delta N_1 - n\lambda_2\nabla\Delta N_1 + n\lambda_2\nabla\Delta N_w\\ &= m\nabla\Delta\varphi_1 - n\nabla\Delta\varphi_2 + (m\lambda_1 - n\lambda_2)\nabla\Delta N_1 + n\lambda_2\nabla\Delta N_w\end{aligned}$$

In addition, then, we have

$$\begin{aligned}(m\lambda_1 - n\lambda_2)\nabla\Delta N_1 &= \nabla\Delta\varphi_{IF} - m\nabla\Delta\varphi_1 + n\nabla\Delta\varphi_2 - n\lambda_2\nabla\Delta N_w\\ \nabla\Delta N_1 &= \frac{1}{m\lambda_1 - n\lambda_2}[\nabla\Delta\varphi_{IF} - m\nabla\Delta\varphi_1 + n\nabla\Delta\varphi_2 - n\lambda_2\nabla\Delta N_w]\end{aligned}$$

References

1. Liu, J.; Zhang, B.; Liu, T.; Xu, G.; Ji, Y.; Sun, M.; Nie, W.; He, Y. An efficient UD factorization implementation of Kalman filter for RTK based on equivalent principle. *Remote Sens.* **2022**, *14*, 967. [CrossRef]
2. Zhang, K.; Jiao, W.; Wang, L.; Li, Z.; Zhou, K. Smart-RTK: Multi-GNSS kinematic positioning approach on Android smart devices with doppler-smoothed-code filter and constant acceleration model. *Adv. Space Res.* **2019**, *64*, 1662–1674. [CrossRef]
3. Deng, C.; Tang, W.; Liu, J.; Shi, C. Reliable single-epoch ambiguity resolution for short baselines using combined GPS/BeiDou system. *GPS Solut.* **2014**, *18*, 375–386. [CrossRef]
4. Sun, Q.; Xia, J.; Foster, J.; Falkmer, T.; Lee, H. Pursuing precise vehicle movement trajectory in urban residential area using multi-GNSS RTK tracking. *Transp. Res. Procedia* **2017**, *25*, 2361–2376. [CrossRef]
5. Atz, M.F.; Konukseven, C.; Sermet, T.C.; Alay, S. Comparative analysis of the performance of Multi-GNSS RTK: A case study in Turkey. *Int. J. Eng. Geosci.* **2022**, *7*, 67–80. [CrossRef]
6. Al-Shaery, A.; Zhang, S.; Lim, S.; Rizos, C. *Multi-GNSS Opportunities and Challenges;* Pres at the EGU General Assembly; European Geosciences Union: Vienna, Austria, 2012.
7. Deng, C.; Liu, Q.; Zou, X.; Tang, W.; Cui, J.; Wang, Y.; Guo, C. Investigation of tightly combined single-frequency and single-epoch precise positioning using multi-GNSS data. *Remote Sens.* **2020**, *12*, 285. [CrossRef]
8. Chen, C.; Chang, G. PPPLib: An open-source software for precise point positioning using GPS, BeiDou, Galileo, GLONASS, and QZSS with multi-frequency observations. *GPS Solut.* **2021**, *25*, 18. [CrossRef]
9. Ibrahim, A.S.; Yacoub, A.M.; Aloi, D.N. A 3-dimensional multiband antenna for vehicular 5G sub-6 GHz/GNSS/V2X applications. *Int. J. Antennas Propag.* **2022**, *2022*, 5609110. [CrossRef]
10. Li, X.; Wang, H.; Li, S.; Feng, S.; Wang, X.; Liao, J. GIL: A tightly coupled GNSS PPP/INS/LiDAR method for precise vehicle navigation. *Satell. Navig.* **2021**, *2*, 26. [CrossRef]
11. Bai, L.; Sun, C.; Dempster, A.G.; Zhao, H.; Cheong, J.W.; Feng, W. GNSS-5G hybrid positioning based on multi-rate measurements fusion and proactive measurement uncertainty prediction. *IEEE Trans. Instrum. Meas.* **2022**, *71*, 8501415. [CrossRef]
12. Xin, S.; Geng, J.; Zeng, R.; Zhang, Q.; Ortega-Culaciati, F.; Wang, T. In situ real time seismogeodesy by integrating multi-GNSS and accelerometers. *Measurement* **2021**, *179*, 109453. [CrossRef]
13. Vagle, N.; Broumandan, A.; Lachapelle, G. Multiantenna GNSS and inertial sensors/odometer coupling for robust vehicular navigation. *IEEE Internet Things J.* **2018**, *5*, 4816–4828. [CrossRef]
14. Bonnor, N. Principles of GNSS, inertial, and multisensor integrated navigation systems—Second EditionPaul D. Groves Artech House, 2013, 776 pp ISBN-13: 978-1-60807-005-3. *J. Navig.* **2014**, *67*, 191–192. [CrossRef]
15. Broumandan, A.; Lachapelle, G. Spoofing detection using GNSS/INS/Odometer coupling for vehicular navigation. *Sensors* **2018**, *18*, 1305. [CrossRef]
16. Odijk, D.; Teunissen, P. Characterization of between-receiver GPS-Galileo inter-system biases and their effect on mixed ambiguity resolution. *GPS Solut.* **2013**, *17*, 521–533. [CrossRef]
17. Gao, W.; Pan, S.; Gao, C.; Wang, Q.; Shang, R. Tightly combined GPS and GLONASS for RTK positioning with consideration of differential inter-system phase bias. *Meas. Sci. Technol.* **2019**, *30*, 054001. [CrossRef]

18. Li, X.; Wang, B.; Li, X.; Huang, J.; Lyu, H.; Han, X. Principle and performance of multi-frequency and multi-GNSS PPP-RTK. *Satell. Navig.* **2022**, *3*, 7. [CrossRef]
19. Liu, T.; Zhang, B. Estimation of code observation-specific biases (OSBs) for the modernized multi-frequency and multi-GNSS signals: An undifferenced and uncombined approach. *J. Geod.* **2021**, *95*, 97. [CrossRef]
20. Psychas, D. Fast and Reliable Multi-GNSS Precise Point Positioning with Integer Ambiguity Resolution. Ph.D. Thesis, Delft University of Technology, Delft, The Netherlands, January 2022.
21. Zhang, G.; Xu, P.; Xu, H.; Hsu, L.-T. Prediction on the urban GNSS measurement uncertainty based on deep learning networks with long short-term memory. *IEEE Sens. J.* **2021**, *21*, 20563–20577. [CrossRef]
22. Smola, A.; Gretton, A.; Le, S.; Schlkopf, B. A hilbert space embedding for distributions. In Proceedings of the International Conference on Algorithmic Learning Theory, Sendai, Japan, 1–4 October 2007.
23. Li, T.; Zhang, H.; Gao, Z.; Chen, Q.; Niu, X. High-accuracy positioning in urban environments using single-frequency multi-GNSS RTK/MEMS-IMU integration. *Remote Sens.* **2018**, *10*, 205. [CrossRef]
24. Gong, X.; Zheng, F.; Gu, S.; Zhang, Z.; Lou, Y. The long-term characteristics of GNSS signal distortion biases and their empirical corrections. *GPS Solut.* **2022**, *26*, 52. [CrossRef]
25. Chen, B.; Wang, X.; Li, Y.; Principe, J.C. Maximum correntropy criterion with variable center. *IEEE Signal Process. Lett.* **2019**, *26*, 1212–1216. [CrossRef]
26. Zhang, J.; Liu, Y.; Liu, H.; Wang, J. Learning local–global multiple correlation filters for robust visual tracking with Kalman filter redetection. *Sensors* **2021**, *21*, 1129. [CrossRef]
27. Valipour, M.; Ricardez-Sandoval, L.A. Abridged Gaussian sum extended Kalman filter for nonlinear state estimation under non-Gaussian process uncertainties. *Comput. Chem. Eng.* **2021**, *155*, 107534. [CrossRef]
28. Wang, H.; Zhang, W.; Zuo, J.; Wang, H. Outlier-robust Kalman filters with mixture correntropy. *J. Franklin Inst.* **2020**, *357*, 5058–5072. [CrossRef]
29. Sahmoudi, M.; Landry, R. A nonlinear filtering approach for robust multi-GNSS RTK positioning in presence of multipath and ionospheric delays. *IEEE J. Sel. Top. Signal Process.* **2009**, *3*, 764–776. [CrossRef]
30. Flores, T.K.S.; Villanueva, J.M.M.; Gomes, H.P.; Catunda, S.Y.C. Adaptive pressure control system based on the maximum correntropy criterion. *Sensors* **2021**, *21*, 5156. [CrossRef] [PubMed]
31. Hu, C.; Wang, G.; Ho, K.; Liang, J. Robust ellipse fitting with Laplacian kernel based maximum correntropy criterion. *IEEE Trans. Image Process.* **2021**, *30*, 3127–3141. [CrossRef]
32. Xiong, W.; Schindelhauer, C.; So, H.C.; Wang, Z. Maximum correntropy criterion for robust TOA-based localization in NLOS environments. *Circuits Syst. Signal Process.* **2021**, *40*, 6325–6339. [CrossRef]
33. Zhao, H.; Liu, D.; Lv, S. Robust maximum correntropy criterion subband adaptive filter algorithm for impulsive noise and noisy input. *IEEE Trans. Circuits Syst. II Express Briefs* **2021**, *69*, 604–608. [CrossRef]
34. Fakoorian, S.; Santamaria-Navarro, A.; Lopez, B.T.; Simon, D.; Agha-mohammadi, A.-A. Towards robust state estimation by boosting the maximum correntropy criterion Kalman filter with adaptive behaviors. *IEEE Robot. Autom. Lett.* **2021**, *6*, 5469–5476. [CrossRef]
35. Fan, X.; Wang, G.; Han, J.; Wang, Y. Interacting multiple model based on maximum correntropy Kalman filter. *IEEE Trans. Circuits Syst. II Express Briefs* **2021**, *68*, 3017–3021. [CrossRef]
36. Zhao, W.; Zhao, H.; Liu, L. A single beacon-aided cooperative localization algorithm based on maximum correntropy criterion. In Proceedings of the 2021 International Wireless Communications and Mobile Computing (IWCMC), Harbin, China, 28 June–2 July 2021; pp. 1835–1839.
37. Chen, B.; Xing, L.; Zhao, H.; Xu, B.; Principe, J.C. Robustness of maximum correntropy estimation against large outliers. *arXiv* **2017**, arXiv:1703.08065.
38. Odolinski, R.; Teunissen, P.; Odijk, D. Combined BDS, Galileo, QZSS and GPS single-frequency RTK. *GPS Solut.* **2015**, *19*, 151–163. [CrossRef]
39. Wang, J.; Xu, T.; Nie, W.; Xu, G. A simplified processing algorithm for multi-baseline RTK positioning in urban environments. *Measurement* **2021**, *179*, 109446. [CrossRef]
40. Chen, P.; Ma, Y.; Liu, H.; Zheng, N. A new global tropospheric delay model considering the spatiotemporal variation characteristics of ZTD with altitude coefficient. *Earth Space Sci.* **2020**, *7*, e2019EA000888. [CrossRef]
41. Lai, C.; Guo, R. Extend the RTK survey to GNSS-denied areas using a low-cost inertial-aided positioning pole. In Proceedings of the 34th International Technical Meeting of the Satellite Division of The Institute of Navigation (ION GNSS+ 2021), St. Louis, MO, USA, 20–24 September 2021; pp. 3212–3226.
42. Liu, Y.; Gao, Z.; Xu, Q.; Li, Y.; Chen, L. Assessing partial ambiguity resolution and WZTD-constraint multi-frequency RTK in an urban environment using new BDS signals. *GPS Solut.* **2022**, *26*, 88. [CrossRef]
43. Miwa, M.; Ushiroda, T. Precision flight drones with RTK-GNSS. *J. Robot. Mechatron.* **2021**, *33*, 371–378. [CrossRef]
44. Xue, X.; Qin, H.; Lu, H. High-precision time synchronization of kinematic navigation system using GNSS RTK differential carrier phase time transfer. *Measurement* **2021**, *176*, 109132. [CrossRef]
45. Roberts, G.W.; Tang, X.; Brown, C.J. Measurement and correlation of displacements on the Severn Suspension Bridge using GPS. *Appl. Geomat.* **2018**, *11*, 161–176. [CrossRef]
46. Teunissen, P. Towards a unified theory of GNSS ambiguity resolution. *J. Glob. Position. Syst.* **2003**, *2*, 1–12. [CrossRef]

47. Singh, A.; Principe, J.C. Using Correntropy as a cost function in linear adaptive filters. In Proceedings of the 2009 International Joint Conference on Neural Networks, Atlanta, GA, USA, 14–19 June 2009.
48. Chen, B.; Xing, L.; Zhao, H.; Du, S.; Príncipe, J.C. Effects of outliers on the maximum correntropy estimation: A robustness analysis. *IEEE Trans. Syst. Man Cybern. Syst.* **2019**, *51*, 4007–4012. [CrossRef]
49. Lu, L.; Zhao, H. Active impulsive noise control using maximum correntropy with adaptive kernel size. *Mech. Syst. Signal Process.* **2017**, *87*, 180–191. [CrossRef]
50. Cinar, G.T.; Principe, J.C. Hidden state estimation using the Correntropy Filter with fixed point update and adaptive kernel size. In Proceedings of the 2012 International Joint Conference on Neural Networks (IJCNN), Brisbane, Australia, 10–15 June 2012.
51. Chen, B.; Liu, X.; Zhao, H.; Principe, J.C. Maximum correntropy Kalman filter. *Automatica* **2017**, *76*, 70–77. [CrossRef]
52. Liu, W.; Pokharel, P.P.; Principe, J.C. Correntropy: Properties and applications in non-Gaussian signal processing. *IEEE Trans. Signal Process.* **2007**, *55*, 5286–5298. [CrossRef]
53. Izanloo, R.; Fakoorian, S.A.; Yazdi, H.S.; Dan, S. Kalman filtering based on the maximum correntropy criterion in the presence of non-Gaussian noise. In Proceedings of the 2016 Annual Conference on Information Science and Systems (CISS), Princeton, NJ, USA, 16–18 March 2016.
54. Heravi, A.R.; Hodtani, G.A. A new information theoretic relation between minimum error entropy and maximum correntropy. *IEEE Signal Process. Lett.* **2018**, *25*, 921–925. [CrossRef]
55. Fakoorian, S.; Izanloo, R.; Shamshirgaran, A.; Dan, S. Maximum correntropy criterion Kalman filter with adaptive kernel size. In Proceedings of the 2019 IEEE National Aerospace and Electronics Conference (NAECON), Dayton, OH, USA, 15–19 July 2019.
56. Botev, Z.I.; Grotowski, J.F.; Kroese, D.P. Kernel density estimation via diffusion. *Ann. Stat.* **2010**, *38*, 2916–2957. [CrossRef]
57. Hou, P.; Zhang, B.; Yasyukevich, Y.V.; Liu, T.; Zha, J. Multi-frequency phase-only PPP-RTK model applied to BeiDou data. *GPS Solut.* **2022**, *26*, 76. [CrossRef]

remote sensing

MDPI

Article

A Novel Method for AI-Assisted INS/GNSS Navigation System Based on CNN-GRU and CKF during GNSS Outage

Shuai Zhao, Yilan Zhou * and Tengchao Huang

State Key Laboratory of Modern Optical Instrumentation, Zhejiang University, Hangzhou 310027, China
* Correspondence: naliyuohz@zju.edu.cn

Abstract: In the fields of positioning and navigation, the integrated inertial navigation system (INS)/global navigation satellite systems (GNSS) are frequently employed. Currently, high-precision INS typically utilizes fiber optic gyroscopes (FOGs) and quartz flexural accelerometers (QFAs) rather than MEMS sensors. But when GNSS signals are not available, the errors of high-precision INS also disperse rapidly, similar to MEMS-INS when GNSS signals would be unavailable for a long time, leading to a serious degradation of the navigation accuracy. This paper presents a new AI-assisted method for the integrated high-precision INS/GNSS navigation system. The position increments during GNSS outage are predicted by the convolutional neural network-gated recurrent unit (CNN-GRU). In the process, the CNN is utilized to quickly extract the multi-dimensional sequence features, and GRU is used to model the time series. In addition, a new real-time training strategy is proposed for practical application scenarios, where the duration of the GNSS outage time and the motion state information of the vehicle are taken into account in the training strategy. The real road test results verify that the proposed algorithm has the advantages of high prediction accuracy and high training efficiency.

Keywords: INS/GNSS integrated navigation; CNN-GRU; CKF; GNSS outage

Citation: Zhao, S.; Zhou, Y.; Huang, T. A Novel Method for AI-Assisted INS/GNSS Navigation System Based on CNN-GRU and CKF during GNSS Outage. *Remote Sens.* **2022**, *14*, 4494. https://doi.org/10.3390/rs14184494

Academic Editors: Yuwei Chen, Changhui Jiang, Qian Meng, Bing Xu, Wang Gao, Panlong Wu, Lianwu Guan and Zeyu Li

Received: 31 July 2022
Accepted: 6 September 2022
Published: 9 September 2022

1. Introduction

Global Navigation Satellite systems (GNSS) can provide accurate position and velocity information in outdoor environments, and its errors do not accumulate over time [1]. The disadvantages are that it can only provide less accuracy attitude information, the output frequency is low (1–20 Hz), and it is vulnerable to environmental interference. In contrast, the Inertial Navigation System (INS) is less dependent on the environment, and relies entirely on the angular velocity and acceleration information that is measured by the Inertial Measurement Unit (IMU), which can provide high-frequency navigation information [2,3]. But the position error will disperse over time due to the integral acquisition of positional information, resulting large errors in navigation results. Therefore, combining the respective advantages of GNSS and INS to obtain the navigation results with high accuracy, high interference immunity, and high frequency is a hot topic of research in the field of navigation at present [4,5].

The Kalman filter (KF) and its upgrade variants are the most widely utilized algorithms for INS and GNSS information fusion [6,7]. The traditional Kalman filter algorithm can only be applied to linear systems, but most of the information in real navigation systems are nonlinear. Bucy et al. [8] proposed the extended Kalman filter (EKF), which linearizes the nonlinear function around the current estimate, and truncates the first-order linearization of the Taylor expansion of the nonlinear function. The remaining higher-order terms are ignored, and their performance depends on the degree of local nonlinearity. The unscented Kalman filter (UKF) was proposed to further improve the performance under nonlinear systems by making the nonlinear system equations applicable to linear assumptions through lossless transformations [9,10]. By approximating the posterior probability density of the state with a series of deterministic samples, the problem of the EKF accuracy dispersion under a highly nonlinear system is avoided. But the UKF has low accuracy in the high-dimensional case, the cubature Kalman filter (CKF) that is based on the spherical radial volume criterion is applied to data fusion, which can effectively approximate the Gaussian density function with higher accuracy, convenient parameter selection, and good convergence effect [11–13]. In order to improve the fusion accuracy in complex measurement environments, robust Kalman algorithms have also started to attract the attention of researchers [14,15]. To solve the problem of error model that is caused by measurement anomalies, Chen et al. [16] proposed a new cardinal maximum correlation entropy Kalman filter, which uses the robust maximum correlation entropy criterion (MCC) as the optimality criterion to solve the state estimation problem under outlier interference by maximizing the correlation entropy between states and measurements. Yun et al. [17] proposed a variational Bayesian-based state estimation algorithm to improve the CKF accuracy under dynamic model mismatch and outlier disturbance.

When the GNSS signals are unavailable, KF operates in predictive mode and corrects INS measurements according to the system error model. At this time, the accuracy of data fusion that relies only on the KF is not effective and navigation performance deteriorates rapidly. To improve the integrated navigation accuracy during GNSS outage, machine learning has started to be applied to integrated navigation systems. Ning et al. [18] proposed an optimal radial basis function (RBF)-based neural network that can improve the overall positioning accuracy during short-term GNSS signal outages. Hang et al. [19] proposed a new hybrid intelligence algorithm combining a discrete gray predictor (DGP) and a multilayer perceptron (MLP) neural network that provides pseudo-GPS positions during GNSS failures and uses GNSS position information from the last few moments to predict positions for future moments. Compared with traditional artificial neural networks, recurrent neural networks are more advantageous in combined navigation systems and can make full use of historical information [20–22]. Liu et al. [23] proposed a multi-channel long-short term memory (LSTM) network to predict the increments of vehicle position, which reduces the navigation error in case of GNSS outages by an order of magnitude. In practical applications, a large amount of historical data before the GNSS outage needs to be trained when the GNSS outage occurs, so the training efficiency of neural networks also has high requirements. Tang et al. [24] proposed a hybrid algorithm that was based on the gated recurrent unit (GRU) and adaptive Kalman filter (AKF), and the experimental results showed that GRU outperformed LSTM in terms of prediction accuracy and training efficiency. Zhi et al. [25] proposed a convolutional neural network-long short-term memory (CNN-LSTM) model, which uses convolutional neural network (CNN) to quickly extract the features of the input and LSTM network to output the pseudo-GPS signal, further improving the training efficiency. However, most of the current articles use the offline simulation, assuming that the GNSS failure time is known and do not consider the time that is required to train the model online. Al Bitar et al. [26] proposed a novel real-time training strategy for regular training on the past one minute data, with the disadvantage that only short historical data are used and the accuracy is poor when the time of GNSS outage is long.

To overcome the shortcomings of the traditional methods, our paper proposes a new AI-assisted method for the integrated high-precision INS/GNSS navigation system. The method consists of two parts: first, CKF is used to provide more accurate neural network training samples. Then, by building a CNN-GRU network to predict the position increments during GNSS outage, the CNN is utilized to quickly extract the multi-dimensional sequence features, and GRU is used to model the time series. In addition, a new real-time training strategy is proposed for practical application scenarios, where the duration of the GNSS outage time and the motion state information of the vehicle are taken into account in the training strategy. The experiments verify that the proposed algorithm has the advantages of high prediction accuracy and high training efficiency.

The rest of the paper is organized as follows: Section 2 introduces the INS error propagation model and the integrated navigation model that is based on CKF, Section 3 introduces the proposed CNN-GRU network, Section 4 performs the road test and result analysis, and the conclusion is presented in Section 5.

2. Mathematical Integrated System Model

The INS and GNSS are loosely coupled as shown in Figure 1. The INS and GNSS can complete position and velocity independently. Then, the position and velocity errors of INS are estimated using CKF, and these errors are used to correct the navigation output of INS and achieve the correction of gyro and accelerometer drift, thus reducing the impact of INS errors. In this section, we derive the error propagation model of INS, and then introduce the integrated navigation model that is based on CKF.

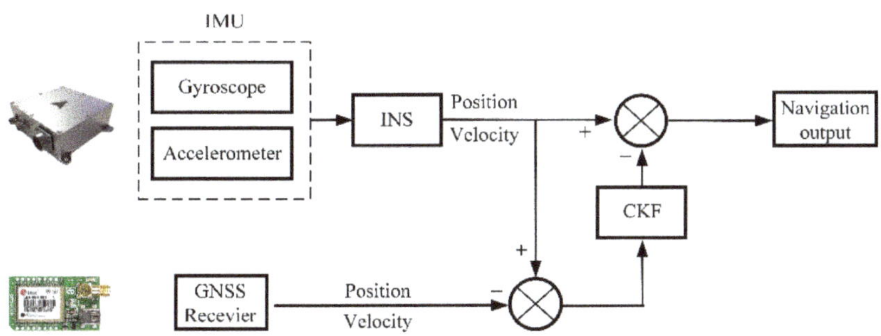

Figure 1. Flowchart of the loosely coupled integrated system.

2.1. The Error Propagation Model

By integrating the angular velocity that is measured by gyro, the attitude direction of the inertial component can be obtained. By using the directional cosine matrix of the attitude direction, the specific force components of the accelerometer observations along each axis of the carrier system can be converted to the navigation coordinate system, the velocity and position can be calculated.

In the process of attitude calculation, the navigation coordinate system that is obtained is regarded as the real navigation coordinate system. In practice, due to the interference of various factors, the calculated navigation coordinate system will have deviations compared to the real navigation coordinate system, and the attitude error equation can be expressed as:

$$\dot{\phi} = \phi \times \omega_{in}^{n} + \delta\omega_{in}^{n} - \delta\omega_{ib}^{n}, \tag{1}$$

where $\delta\omega_{in}^{n} = \delta\omega_{ie}^{n} + \delta\omega_{en}^{n}$ is the angular velocity error in the navigation coordinate system, $\delta\omega_{ie}^{n}$ is the angular velocity error of the Earth's rotation, $\delta\omega_{en}^{n}$ is the rotation error of the navigation system, $\delta\omega_{ib}^{n} = C_b^n \delta\omega_{ib}^{b} = C_b^n \left[(\delta K_G + \delta G)\omega_{ib}^{b} + \varepsilon^b \right]$ is the gyro measurement error, ε^b is the gyro drift error, and δK_G and δG are the gyro scale factor error and non-orthogonal installation error, which can be expressed as:

$$\begin{aligned} \delta K_G &= \mathrm{diag}(\begin{bmatrix} \delta K_{Gx} & \delta K_{Gy} & \delta K_{Gz} \end{bmatrix}) \\ \delta G &= \begin{bmatrix} 0 & \delta G_z & -\delta G_y \\ -\delta G_z & 0 & \delta G_x \\ \delta G_y & -\delta G_x & 0 \end{bmatrix} \end{aligned} \tag{2}$$

Both calibration residuals and installation errors can be considered as constants, while random drift can be expressed as a cumulative model of random constants and a first-order Markov model. In the presence of attitude error and specific force measurement error, the velocity differential error can be expressed as:

$$\delta\dot{v}^n = f_{sf}^n \times \phi + (2\delta\omega_{ie}^n + \delta\omega_{en}^n) \times v^n - (2\omega_{ie}^n + \omega_{en}^n) \times \delta v^n + \delta f_{sf}^n + \delta g^n \tag{3}$$

where δf_{sf}^{n} is the accelerometer measurement error, which can be expressed as the cumulative model of accelerometer zero deviation and white noise.

$$\delta g^{n} = \begin{bmatrix} 0 \\ -\beta_3(\sin 2L\cdot\delta h + 2h\cos 2L\cdot\delta L) \\ -[g_e \sin 2L(\beta - 4\beta_1 \cos 2L)\delta L - \beta_2 \delta h] \end{bmatrix} \tag{4}$$

where β is the gravity flattening, $\beta_1 = 2.32718 \times 10^{-5} s^{-2}$, $\beta_2 = 3.08 \times 10^{-6} s^{-2}$, $\beta_3 = 8.08 \times 10^{-9} s^{-2}$.

The position error is [27]:

$$\begin{cases} \delta\dot{L} = \frac{1}{R_M+h}\delta v_N - \frac{v_N}{(R_M+h)^2}\delta h \\ \delta\dot{\lambda} = \frac{\sec L}{R_N+h}\delta v_E + \frac{v_E \sec L \tan L}{R_N+h}\delta L - \frac{v_E \sec L}{(R_N+h)^2}\delta h \\ \delta\dot{h} = \delta v_U \end{cases}, \tag{5}$$

2.2. The Integrated Navigation Model Based on CKF

In the integrated navigation system, the CKF estimation is adopted to estimate the system state vector which is a 15 array vector:

$$X = \begin{bmatrix} \varphi_e\ \varphi_n\ \varphi_u\ \delta V_e\ \delta V_n\ \delta V_u\ \delta L\ \delta\lambda\ \delta h\ \nabla_x\ \ \nabla_y\ \nabla_z\ \varepsilon_x\ \varepsilon_y\ \varepsilon_z \end{bmatrix}^T \tag{6}$$

where φ_e, φ_n, and φ_u denote the attitude error angles of the INS in the east, north, and zenith directions. δV_e, δV_n, and δV_u denote the velocity error of INS in the three directions. δL, $\delta\lambda$, and δh denote the latitude error, longitude error, and altitude error. ∇_x, ∇_y, and ∇_z are the accelerometer bias vectors, and ε_x, ε_y, and ε_z are the gyro bias vectors.

The equation of state and measurement equation of the system can be expressed as:

$$\begin{cases} \dot{X} = FX + GW \\ Z = HX + V \end{cases} \tag{7}$$

where F is the system state transfer matrix, which consists of the INS system error. G is the system noise matrix. Z is the measurement vector, which is set to the position error vector and velocity error vector of INS and GNSS. H is the observation matrix. W is the process noise vector. V denotes the measured noise vector.

The CKF is based on the 3rd order volume rule and is suitable for high-dimensional filtering problems. The computation consists of two steps: the time update and measurement update phase, as shown in Algorithm 1.

Algorithm 1: Cubature Kalman Filter.

Require: $\hat{\mathbf{x}}_{k-1|k-1}, \mathbf{P}_{k-1|k-1}, \mathbf{Q}_{k-1}, \mathbf{R}_k$
for i = 1, . . . , N **do**
  $\mathbf{x}^i_{k-1} = S_{k-1|k-1}\xi_i + \hat{\mathbf{x}}_{k-1|k-1}$
  $\mathbf{w}^i_{k-1} = 1/N$
end for
Prediction phase:
$\hat{\mathbf{x}}_{k|k-1} = \sum_{i=1}^{N} \mathbf{w}^i_{k-1} \mathrm{f}\left(\mathbf{x}^i_{k-1}\right)$
$\mathbf{P}_{k|k-1} = \sum_{i=1}^{N} \mathbf{w}^i_{k-1} \left(\mathrm{f}\left(\mathbf{x}^i_{k-1}\right) - \hat{\mathbf{x}}_{k|k-1}\right)\left(\mathrm{f}\left(\mathbf{x}^i_{k-1}\right) - \hat{\mathbf{x}}_{k|k-1}\right)^T + \mathbf{Q}_{k-1}$
for i = 1, . . . , N **do**
  $\mathbf{x}^i_{k|k-1} = S_{k|k-1}\xi_i + \hat{\mathbf{x}}_{k|k-1}$
  $\mathbf{w}^i_{k|k-1} = 1/N$
end for
Update phase:
  $\hat{\mathbf{z}}_{k|k-1} = \sum_{i=1}^{N} \mathbf{w}^i_{k|k-1} \mathrm{h}\left(\mathbf{x}^i_{k|k-1}\right)$
  $\mathbf{P}^{zz}_{k|k-1} = \sum_{i=1}^{N} \mathbf{w}^i_{k|k-1} \left(\mathrm{h}\left(\mathbf{x}^i_{k|k-1}\right) - \hat{\mathbf{z}}_{k|k-1}\right)\left(\mathrm{h}\left(\mathbf{x}^i_{k|k-1}\right) - \hat{\mathbf{z}}_{k|k-1}\right)^T + \mathbf{R}_k$
  $\mathbf{P}^{xz}_{k|k-1} = \sum_{i=1}^{N} \mathbf{w}^i_{k|k-1} \left(\mathbf{x}^i_{k|k-1} - \hat{\mathbf{x}}_{k|k-1}\right)\left(\mathrm{h}\left(\mathbf{x}^i_{k|k-1}\right) - \hat{\mathbf{z}}_{k|k-1}\right)^T$

3. CNN-GRU

The overall architecture of our method is shown in Figure 2. In this paper, we use a loosely coupled integrated navigation scheme that is based on the combination of CKF and CNN-GRU. The CKF module provides highly accurate position, velocity, attitude, and IMU error information. The inputs and outputs of the network are shown in Figure 2, where W_I and F_I are the angular velocity and specific force that are provided by the IMU, A_{INS} and V_{INS} are the attitude and velocity information that is calculated by the INS. The outputs of the network are the position increments δP output by the CKF module, which are integrated as the pseudo-GNSS position information. When the GNSS signals are available, the CNN-GRU module operates in learning mode. When the GNSS signals become unavailable, the CNN-GRU module operates in prediction mode, and the pseudo-GNSS position increments are predicted to ensure navigation accuracy. Specifically, three operating modes are included: learning mode when the GNSS signals are available, prediction mode and learning mode during GNSS short-term outage, and prediction mode during GNSS long-term outage. When the GNSS signals are available, the length of each learning sample is 2 min and the learning interval is controlled in 1 min. The reason for choosing – min is that most GNSS interruption scenarios last less than 2 min, and the purpose of the learning interval is to reset the CKF filter. When the GNSS signals are unavailable, the model that was trained in the previous phase is used for prediction, while the previous historical data are used for training the new fine model, and the model is switched to the new fine model to improve the prediction accuracy when the GNSS signals that are interrupted exceed 2 min. In order to ensure that the training of historical data can be completed within 2 min, we consider the differences of the model under different motion states of the vehicle and reduce the length of the training data. The decision is done using the vehicle motion state according to the output data of INS, which are zero speed, zero angular speed, zero lateral speed, and zero vertical speed, and stop saving data when a period of continuous motion state exceeds five minutes, thus improving the training efficiency.

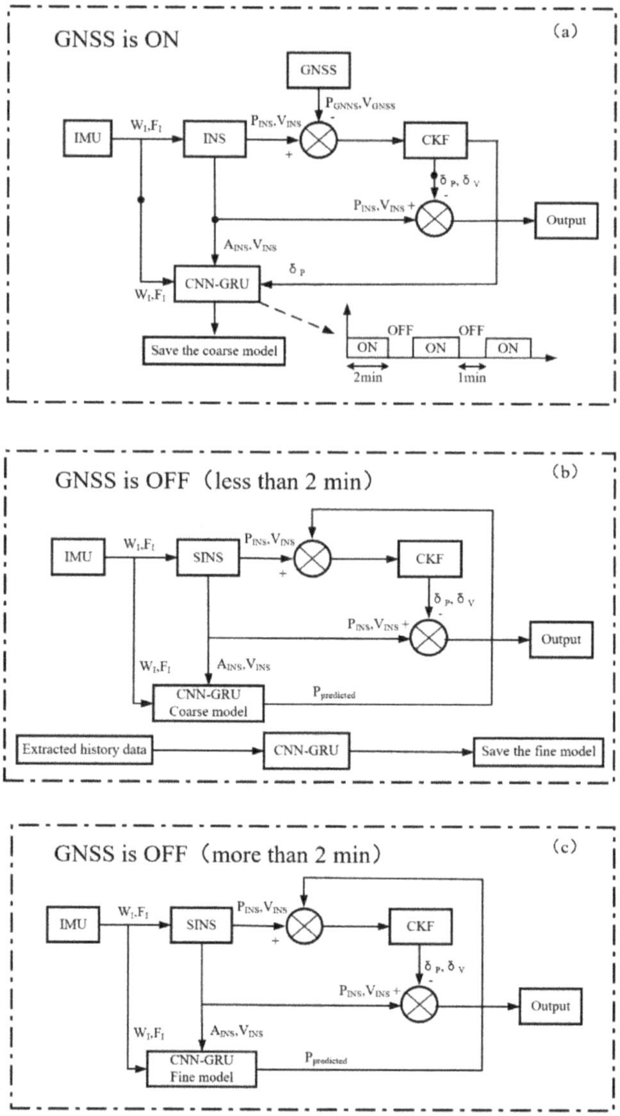

Figure 2. The overall architecture of the proposed method (**a**) GNSS is available; (**b**) GNSS ioutage (less than 2 min); (**c**) GNSS outage (more than 2 min).

The CNN-GRU networks consist of a one-dimensional version of CNN, GRU, and a fully connected layer. Since the inputs involve multiple sensors and the coupling of multi-dimensional motion information, the intake features need to be extracted more accurately, so the CNN is used to quickly extract features from the sensor sequences. Since the vehicle motion and IMU sensor errors are time-dependent, the GRU is adopted to extract deeper hidden information from the sensor history data. Finally, a fully connected layer is used to obtain the final navigation information.

The structure of CNN is shown in Figure 3. CNN is one of the common network models in the field of deep learning, which is a multi-layer feedforward neural network with high generalization ability and robustness by local connectivity, weight sharing, and pooling operation [28]. The pooling layer is used to compress the high-dimensional features of the input after processing in the convolutional layer to reduce the parameter matrix dimension, which reduces the computational workload by reducing the parameters of the network. The fully connected layer can combine all of the local features into global features.

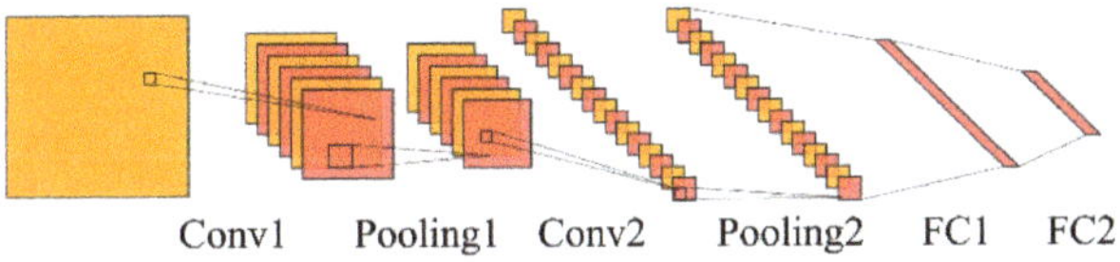

Figure 3. The structure of CNN.

The GRU neural network is an improvement on the LSTM neural network. The LSTM neural network provides a new solution to the short-term memory problem and solves the problem of gradient disappearance and gradient explosion when the RNN exists to handle longer sequences. The GRU is simplified for the same estimation accuracy and higher training efficiency compared to the LSTM. The GRU model consists of two gates: the update gate and the reset gate. The update gate determines how much information from the previous state is brought into the current state. The larger the update gate is, the more the previous state is brought into the current state. The reset gate determines how the new input information is combined with the previous memory. The smaller the reset gate is, the more information about the previous state is ignored. The schematic diagram of GRU is shown in Figure 4.

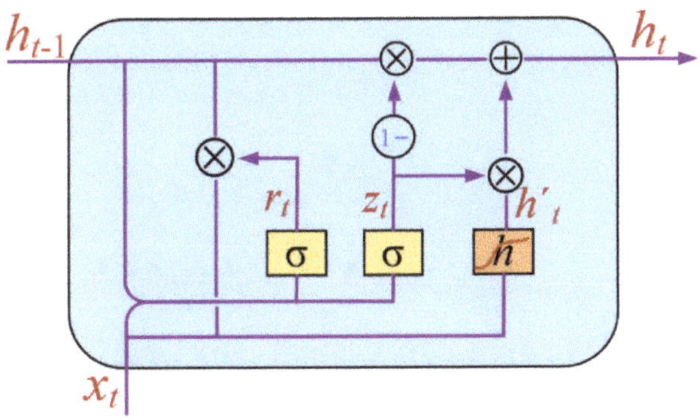

Figure 4. The structure of GRU.

GRU undertakes the most important task of sequence analysis, and the time-dependent nature of GRU makes training more difficult. Therefore, the core parameters are the size and structure of the GRU network. The hyperparameters that have the greatest impact on the performance include the number of GRU layers, the number of neurons, and the step size. Too many neurons lead to an overfitting phenomenon and degradation of the generalization performance, while insufficient number of neurons cannot fully extract the relationship between the input and output sequences, and too many GRU layers also lead to instability of the model. Considering the prediction accuracy and computation time, the number of GRU layers, the number of neurons, and the step size are set to 2, 48, and 2, respectively. The training time of the network increases with the above three hyperparameters. Two layers of GRU are sufficient to extract the hidden information, and too many layers can lead to overfitting.

4. Experiment Results

The road experiments were carried out on a vehicle platform using an INS/GNSS system, as shown in Figure 5. The INS system uses high-precision fiber optic gyroscope and quartz flexible accelerometer that was developed by our group. The GNSS receiver is Ublox NEO-M8T. The sampling frequency of the INS is set to 400 Hz, and the sampling frequency of the GNSS is 10 Hz. RTK GNSS provides the ground truth values. The specific parameters are shown in Table 1. Two typical road experiments were carried out. After the initial alignment, the INS/GNSS was started in a loosely coupled integrated navigation mode. The experimental locations were in Zhejiang Province, China:

(1) Experiment 1: Urban roads as shown in Figure 6: the duration of the experiment is 1 h, and the road conditions include straight lines, turns, and lane changes. There were three segments of simulated GNSS signals interruptions that were introduced, and the signal interruption durations were 60 s, 180 s, and 300 s, respectively.

(2) Experiment 2: Expressway including tunnels as shown in Figure 7: the duration of the experiment is 1 h, the road conditions are mainly long straight lines, and the driving trajectory contains multiple tunnels of different lengths to verify the performance of the proposed method in the case of real GNSS signals outage scenarios.

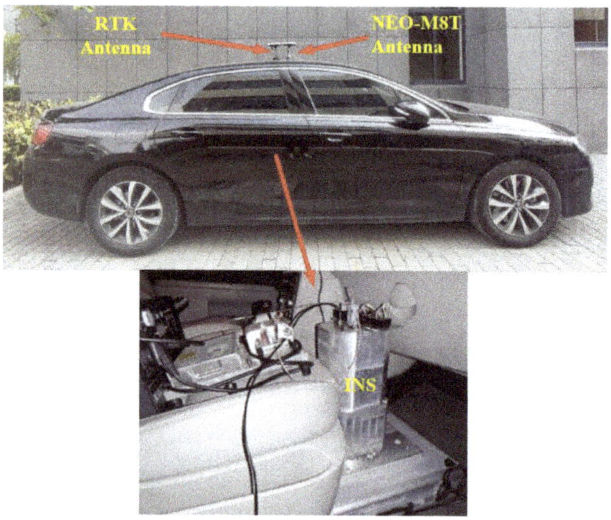

Figure 5. Data acquisition vehicle platform.

Table 1. The parameters of the sensors.

Fiber optic gyroscope	Bias Scale factor Angular Random walk	0.02°/h 10 ppm $0.002°/\sqrt{h}$
Quartz Flexible Accelerometer	Bias Scale factor	10 ug 20 ppm
GNSS(Ublox NEO-M8T)	Position accuracy Velocity accuracy	2.5 m 0.05 m/s
RTK GNSS (NovAtel ProPak6)	Position accuracy	3 cm

Figure 6. The vehicle trajectory in Experiment 1.

Figure 7. The vehicle trajectory in Experiment 2.

In order to verify the performance of the CNN-GRU-CKF that is proposed in our paper, we selected two typical road experiments, Experiment 1 focuses on urban roads, simulating GNSS interruptions by artificially turning off GNSS, while RTK is still working normally and can provide the ground truth of the position for verifying the algorithm accuracy. The GNSS signals interruption duration is 1 min, 3 min, and 5 min, respectively. In order to better reflect the effectivity of the algorithm, the trajectory containing the turn is deliberately chosen. When the GNSS signals are unavailable, the position information that is obtained by CNN-GRU prediction is used instead of the true GNSS information, and the measurement update process of CKF is carried out. Meanwhile, the performance of pure INS, MLP, and GRU is compared.

Due to the short GNSS outage time in the first period, the INS that is based on high precision fiber optic gyro shows high accuracy, as shown in Figures 8 and 9, the position errors in the east and north direction during the 60 s GNSS outage are within 2 m. The horizontal direction error of 60 s outage is shown in Figure 10. It can be seen that when the horizontal error is at its maximum, its east and north errors are not necessarily the maximum. Due to the high accuracy of INS, the overall horizontal error is within 2 m, and the difference between the accuracy of different algorithms is not significant, and the turning point of error dispersion mainly occurs at the vehicle corners. The trajectories during this period are shown in the Figure 11, and it can be seen that the position errors are mainly generated at the corners. The maximum position errors are shown in Table 2. The proposed method in our paper reduces the maximum position error in the east direction by 58.61%, 67.05%, and 63.35% compared to pure INS, MLP, and GRU, respectively. The maximum position error in the north direction is similar, increasing by 4.44% and 12.74% compared to pure INS and GRU, reducing by 46.34% compared to MLP, and the maximum position in the horizontal direction error is reduced by 16.86%, 45.29%, and 35.34% compared to pure INS, MLP, and GRU, respectively. It can be seen that the accuracy of the RNN is significantly better than that of the MLP.

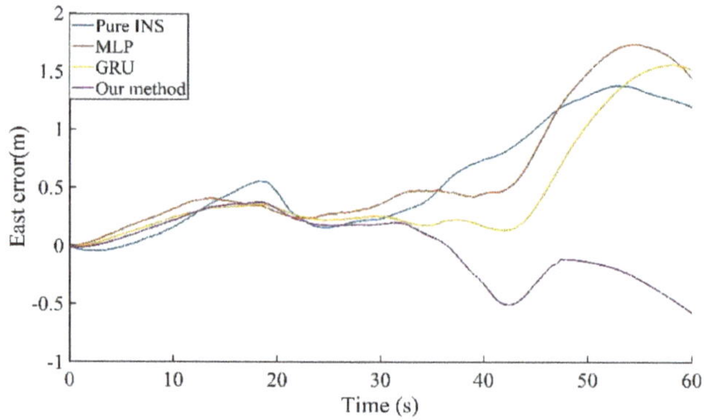

Figure 8. The east position error result of 60 s outage of Experiment 1.

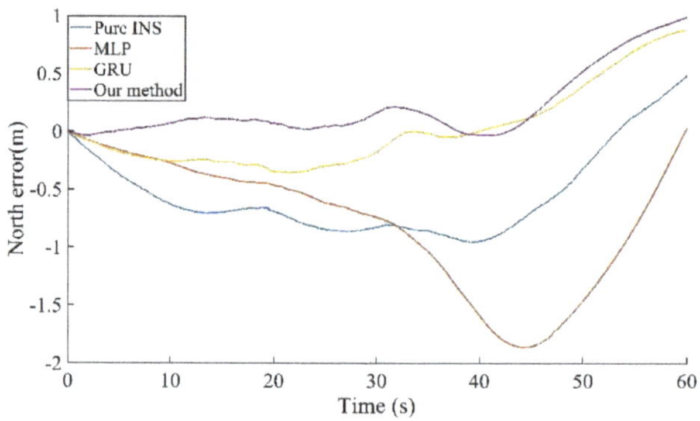

Figure 9. The north position error result of 60 s outage of Experiment 1.

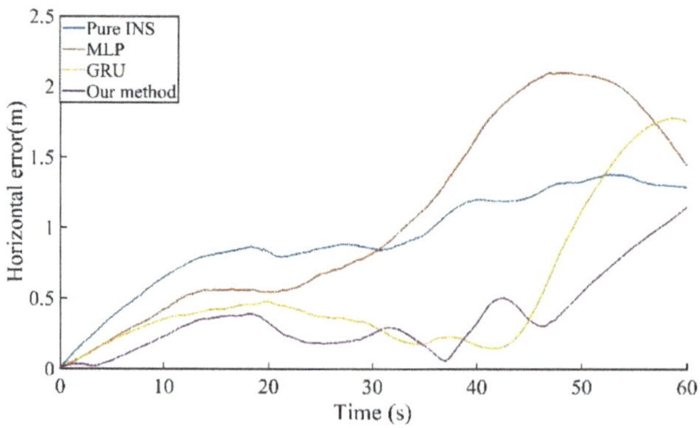

Figure 10. The horizontal position error result of 60 s outage of Experiment 1.

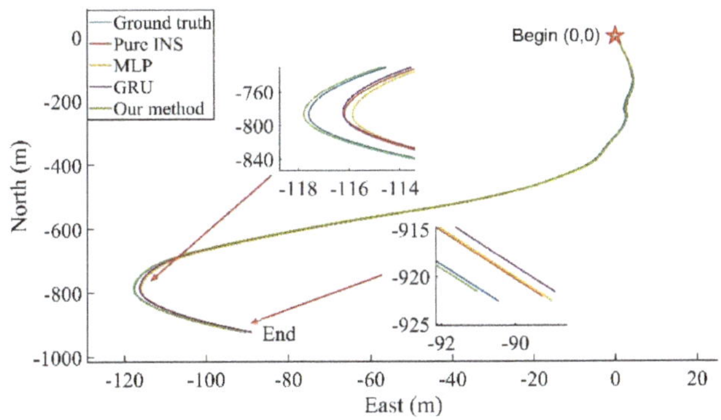

Figure 11. The trajectories of 60 s outage of Experiment 1.

Table 2. Maximum position error of 60 s outage of Experiment 1.

	East Error (m)	North Error (m)	Horizontal Error (m)
Pure INS	1.38	0.96	1.38
MLP	1.74	1.86	2.10
GRU	1.56	0.89	1.78
Our method	0.57	1.00	1.15

Figures 12 and 13 show the position errors of the different algorithms in the east and north directions during the 120 s GNSS outage. It can be seen that the north position error of the pure INS has started to decrease. The horizontal direction error of 180 s outage is shown in Figure 14, and it can be seen that as the GNSS outage time increases to 3 min, the accuracy of the pure INS starts to dissipate, and the MLP method does not perform well, starting to dissipate from around 100 s. The trajectories during this period are shown in Figure 15. The maximum position errors are shown in Table 3. Compared to the pure INS, MLP, and GRU, the proposed method in this paper reduces the maximum position errors in the east direction by 92.00%, 89.95%, and 81.10%, in the north direction by 37.39%, 80.45%, and 56.96%, and in the horizontal direction by 86.66%, 86.08%, and 72.18%.

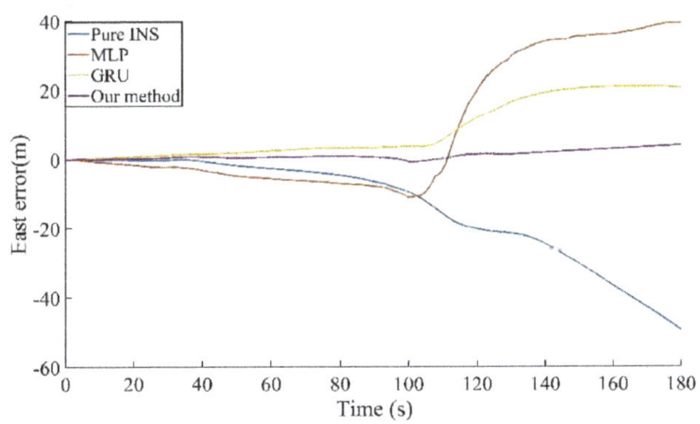

Figure 12. The east position error result of 180 s outage of Experiment 1.

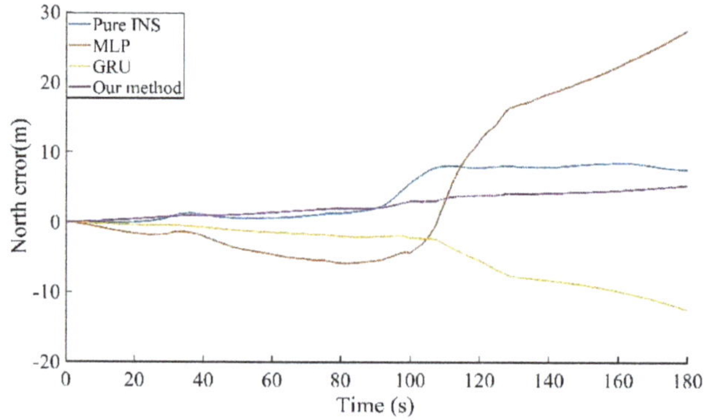

Figure 13. The north position error result of 180 s outage of Experiment 1.

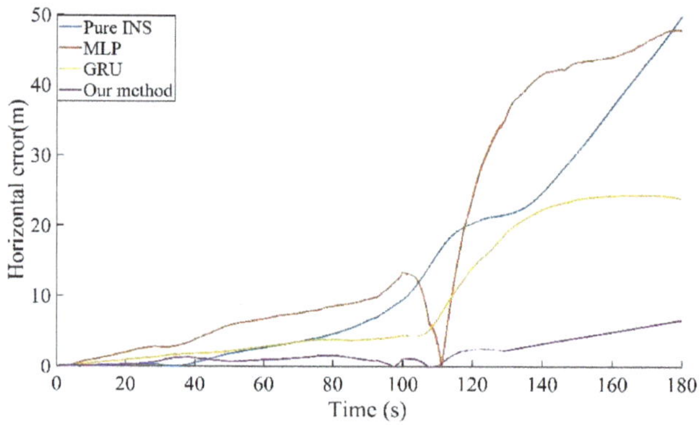

Figure 14. The horizontal position error result of 180 s outage of Experiment 1.

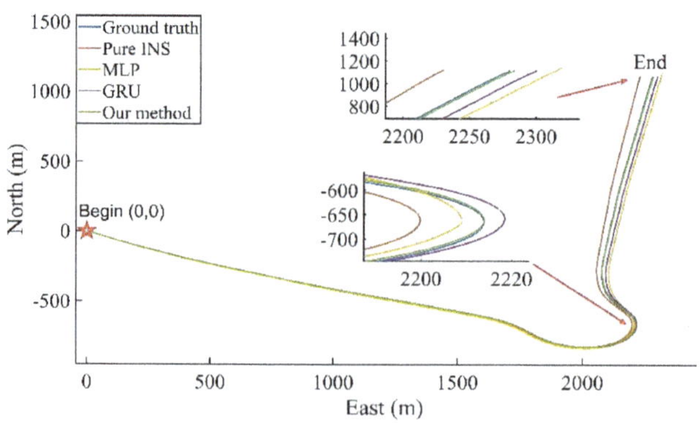

Figure 15. The trajectories of 180 s outage of Experiment 1.

Table 3. Maximum position error of 180 s outage of Experiment 1.

	East Error (m)	North Error (m)	Horizontal Error (m)
Pure INS	49.42	8.58	50.03
MLP	39.35	27.49	47.93
GRU	20.92	12.49	23.98
Our method	3.95	5.37	6.67

Since the GNSS outage time reached 300 s, the model was switched to the fine model when the GNSS outage time increase to more than two minutes. Figures 16 and 17 show the position errors of the different algorithms in the east and north directions during the 300 s GNSS outage. It can be seen that the prediction performance of MLP for the north position error was unsatisfactory, while the effect of the direction that was proposed in our paper is obvious. Horizontal direction error of 300 s outage is shown in Figure 18. It can be seen that as the GNSS outage time increases to 5 min, the accuracy of both the pure INS and the MLP method begin to diverge, while the GRU and CNN-GRU accuracy is better maintained. The trajectories during this period are shown in the Figure 19. The trajectory that is predicted by the method that is proposed in our paper is close to the real trajectory. The maximum position errors are shown in Table 4. Compared to the pure INS, MLP, and GRU, the proposed method in our paper reduces the maximum position errors in the east direction by 93.96%, 77.60%, and 61.27%, in the north direction by 86.34%, 82.87%, and 57.67%, and in the horizontal direction by 93.36%, 84.58%, and 66.81%.

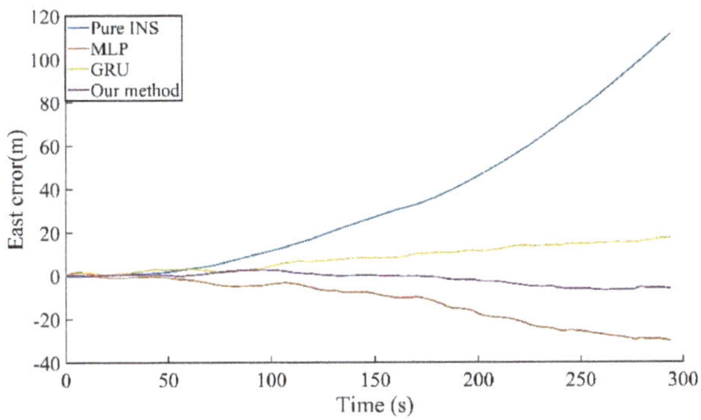

Figure 16. The east position error result of 300 s outage of Experiment 1.

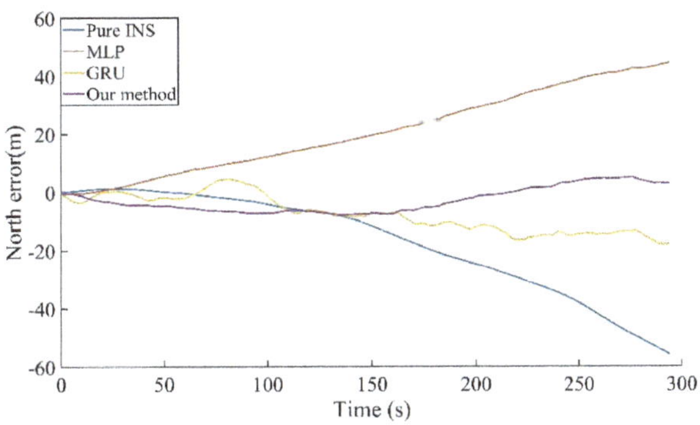

Figure 17. The north position error result of 300 s outage of Experiment 1.

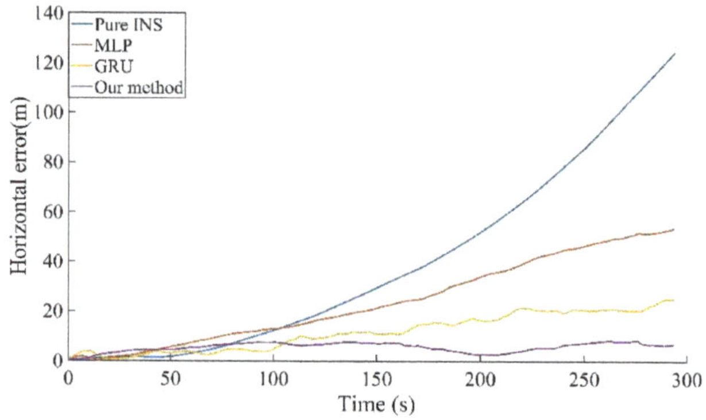

Figure 18. The horizontal position error result of 300 s outage of Experiment 1.

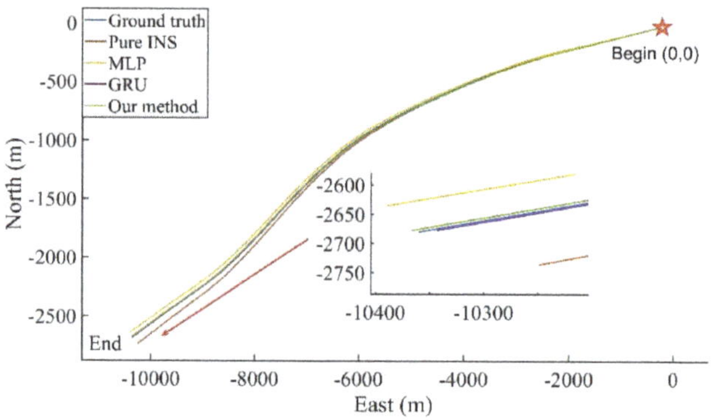

Figure 19. The trajectories of 300 s outage of Experiment 1.

Table 4. Maximum position error of 300 s outage of Experiment 1.

	East Error (m)	North Error (m)	Horizontal Error (m)
Pure INS	111.48	55.81	124.67
MLP	30.08	44.51	53.72
GRU	17.40	18.01	24.96
Our method	6.74	7.62	8.28

As shown in Figure 20, Experiment 2 contains five tunnels, with lengths of 1.3 km, 1.7 km, 3.2 km, 5.2 km, and 1.7 km, respectively. The first three sections of the tunnel are closely spaced which is specifically designed to more accurately show the algorithms' effectiveness. As shown in Figure 21, the compensation effect of different algorithms for the first three tunnel sections can be seen. Since RTK cannot obtain position information in the tunnel, we choose the horizontal position error at the end of the tunnel as the evaluation index, and the results are shown in Table 5. Compared with pure INS, MLP, and GRU, the method that was proposed in this paper reduces the average horizontal position error at the end of the tunnel by 66.07%, 59.85%, and 36.50%.

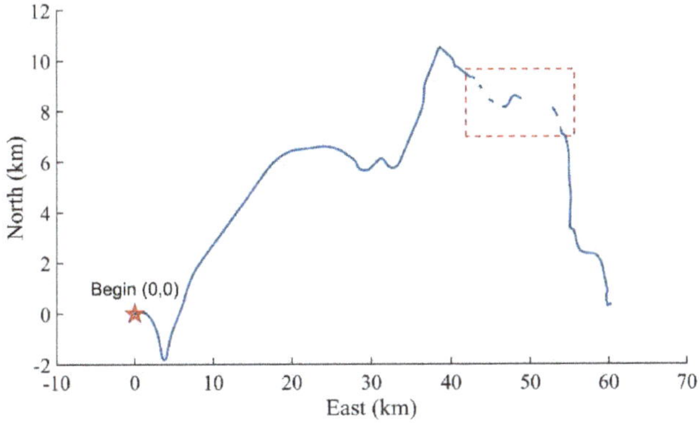

Figure 20. The trajectories of Experiment 2.

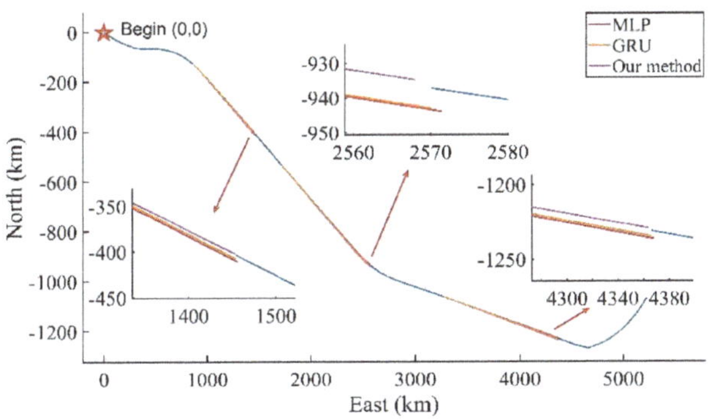

Figure 21. The trajectories of periods #1-3 of Experiment 2.

Table 5. Horizontal position error at the end of the different tunnels.

	#1 (m)	#2 (m)	#3 (m)	#4 (m)	#5 (m)
Pure INS	2.56	5.27	6.82	18.91	6.57
MLP	7.46	6.94	5.77	9.62	4.12
GRU	3.96	5.82	3.49	5.52	2.65
Our method	1.90	3.02	2.52	4.36	1.81

5. Conclusions

In order to improve the positioning accuracy of integrated INS/GNSS navigation during GNSS outage, our paper proposes a new AI assisted method. The method consists of two parts: first, CKF is used to provide more accurate positioning results. Then, by building a CNN-GRU network we can predict the position increments during GNSS outage. In the process, the CNN is utilized to quickly extract the multi-dimensional sequence features, and GRU is used to model the time series. In addition, a new real-time training strategy is proposed for practical application scenarios, where the duration of the GNSS outage time and the motion state information of the vehicle are taken into account in the training strategy. The experimental results show that compared with pure INS, MLP, and GRU, the proposed method reduces the maximum position error in the horizontal direction by 93.36%, 84.58%, and 66.81% in the 5 min simulated GNSS disruption experiments compared to the pure INS, MLP, and GRU, respectively. In the real GNSS failure scenario, the average horizontal position error at the end of the tunnel using our method is reduced by 66.07%, 59.85%, and 36.50%. The algorithm can provide real-time high-precision navigation results with high efficiency and has a good reduction effect on the error dispersion that is caused by prolonged GNSS failure.

Author Contributions: Methodology, writing—original draft preparation, software, S.Z.; Investigation, writing—reviewing and editing, Y.Z.; Supervision, conceptualization, T.H. All authors have read and agreed to the published version of the manuscript.

Funding: This research received no external funding.

Data Availability Statement: Not applicable.

Conflicts of Interest: The authors declare no conflict of interest.

References

1. Hightower, J.; Borriello, G. Location Systems for Ubiquitous Computing. *Computer* **2001**, *34*, 57–66. [CrossRef]
2. Barbour, N.; Schmidt, G. Inertial Sensor Technology Trends. *IEEE Sens. J.* **2001**, *1*, 332–339. [CrossRef]
3. Lee, B. Review of the Present Status of Optical Fiber Sensors. *Opt. Fiber Technol.* **2003**, *9*, 57–79. [CrossRef]
4. Godha, S.; Cannon, M.E. GPS/MEMS INS Integrated System for Navigation in Urban Areas. *GPS Solut.* **2007**, *11*, 193–203. [CrossRef]
5. Zhang, Q.; Niu, X.; Shi, C. Assessment of the Effect of GNSS Sampling Rate on GNSS/INS Relative Accuracy on Different Time Scales for Precision Measurements. *Measurement* **2019**, *145*, 583–593. [CrossRef]
6. Falco, G.; Pini, M.; Marucco, G. Loose and Tight GNSS/INS Integrations: Comparison of Performance Assessed in Real Urban Scenarios. *Sensors* **2017**, *17*, 255. [CrossRef]
7. Li, Q.; Li, R.; Ji, K.; Dai, W. Kalman Filter and Its Application. In Proceedings of the 2015 8th International Conference on Intelligent Networks and Intelligent Systems (ICINIS), Tianjin, China, 1–3 November 2015; pp. 74–77.
8. Kalman, R.E.; Bucy, R.S. New Results in Linear Filtering and Prediction Theory. *J. Basic Eng.* **1961**, *83*, 95–108. [CrossRef]
9. Costanzi, R.; Fanelli, F.; Meli, E.; Ridolfi, A.; Caiti, A.; Allotta, B. UKF-Based Navigation System for AUVs: Online Experimental Validation. *IEEE J. Ocean. Eng.* **2019**, *44*, 633–641. [CrossRef]
10. Al Bitar, N.; Gavrilov, A. A Novel Approach for Aiding Unscented Kalman Filter for Bridging GNSS Outages in Integrated Navigation Systems. *Navigation* **2021**, *68*, 521–539. [CrossRef]
11. Shen, C.; Zhang, Y.; Guo, X.; Chen, X.; Cao, H.; Tang, J.; Li, J.; Liu, J. Seamless GPS/Inertial Navigation System Based on Self-Learning Square-Root Cubature Kalman Filter. *IEEE Trans. Ind. Electron.* **2021**, *68*, 499–508. [CrossRef]
12. Song, R.; Fang, Y. Vehicle State Estimation for INS/GPS Aided by Sensors Fusion and SCKF-Based Algorithm. *Mech. Syst. Signal Process.* **2021**, *150*, 107315. [CrossRef]
13. Cui, B.; Wei, X.; Chen, X.; Li, J.; Li, L. On Sigma-Point Update of Cubature Kalman Filter for GNSS/INS Under GNSS-Challenged Environment. *IEEE Trans. Veh. Technol.* **2019**, *68*, 8671–8682. [CrossRef]
14. Cui, B.; Chen, X.; Tang, X.; Huang, H.; Liu, X. Robust Cubature Kalman Filter for GNSS/INS with Missing Observations and Colored Measurement Noise. *ISA Trans.* **2018**, *72*, 138–146. [CrossRef] [PubMed]
15. Wei, X.; Li, J.; Zhang, D.; Feng, K. An Improved Integrated Navigation Method with Enhanced Robustness Based on Factor Graph. *Mech. Syst. Signal Process.* **2021**, *155*, 107565. [CrossRef]
16. Chen, B.; Liu, X.; Zhao, H.; Principe, J.C. Maximum Correntropy Kalman Filter. *Automatica* **2017**, *76*, 70–77. [CrossRef]
17. Yun, P.; Wu, P.; He, S.; Li, X. A Variational Bayesian Based Robust Cubature Kalman Filter under Dynamic Model Mismatch and Outliers Interference. *Measurement* **2022**, *191*, 110063. [CrossRef]
18. Ning, Y.; Wang, J.; Han, H.; Tan, X.; Liu, T. An Optimal Radial Basis Function Neural Network Enhanced Adaptive Robust Kalman Filter for GNSS/INS Integrated Systems in Complex Urban Areas. *Sensors* **2018**, *18*, 3091. [CrossRef]
19. Zhang, Y.; Wang, L. A Hybrid Intelligent Algorithm DGP-MLP for GNSS/INS Integration during GNSS Outages. *J. Navig.* **2019**, *72*, 375–388. [CrossRef]
20. LeCun, Y.; Bengio, Y.; Hinton, G. Deep Learning. *Nature* **2015**, *521*, 436–444. [CrossRef]
21. Hochreiter, S.; Schmidhuber, J. Long Short-Term Memory. *Neural Comput.* **1997**, *9*, 1735–1780. [CrossRef]
22. Zhao, S.; Zhou, Y.; Shu, X. Study on Nonlinear Error Calibration of Fiber Optical Gyroscope Scale Factor Based on LSTM. *Measurement* **2022**, *190*, 110783. [CrossRef]
23. Liu, J.; Guo, G. Vehicle Localization During GPS Outages with Extended Kalman Filter and Deep Learning. *IEEE Trans. Instrum. Meas.* **2021**, *70*, 7503410. [CrossRef]
24. Tang, Y.; Jiang, J.; Liu, J.; Yan, P.; Tao, Y.; Liu, J. A GRU and AKF-Based Hybrid Algorithm for Improving INS/GNSS Navigation Accuracy during GNSS Outage. *Remote Sens.* **2022**, *14*, 752. [CrossRef]
25. Zhi, Z.; Liu, D.; Liu, L. A Performance Compensation Method for GPS/INS Integrated Navigation System Based on CNN–LSTM during GPS Outages. *Measurement* **2022**, *188*, 110516. [CrossRef]

26. Al Bitar, N.; Gavrilov, A. A New Method for Compensating the Errors of Integrated Navigation Systems Using Artificial Neural Networks. *Measurement* **2021**, *168*, 108391. [CrossRef]
27. Savage, P.G. Strapdown Inertial Navigation Integration Algorithm Design Part 2: Velocity and Position Algorithms. *J. Guid. Control Dyn.* **1998**, *21*, 208–221. [CrossRef]
28. Alom, M.Z.; Taha, T.M.; Yakopcic, C.; Westberg, S.; Sidike, P.; Nasrin, M.S.; Hasan, M.; Van Essen, B.C.; Awwal, A.A.S.; Asari, V.K. A State-of-the-Art Survey on Deep Learning Theory and Architectures. *Electronics* **2019**, *8*, 292. [CrossRef]

remote sensing

MDPI

Article

An Aerial and Ground Multi-Agent Cooperative Location Framework in GNSS-Challenged Environments

Haoyuan Xu [1], Chaochen Wang [1,*], Yuming Bo [1], Changhui Jiang [2], Yanxi Liu [1], Shijie Yang [1] and Weisong Lai [3]

[1] School of Automation, Nanjing University of Science and Technology, No. 200, Xiaolingwei Street, Xuanwu District, Nanjing 210094, China
[2] Department of Remote Sensing and Photogrammetry, Finnish Geospatial Research Institute, 02430 Masala, Finland
[3] School of Materials Science and Engineering, Nanjing University of Science and Technology, No. 200, Xiaolingwei Street, Xuanwu District, Nanjing 210094, China
* Correspondence: wangchaochen@njust.edu.cn

Abstract: In order to realize the cooperative localization of multi-unmanned platforms in the GNSS-denied environment, this paper proposes a collaborative SLAM (simultaneous localization and mapping, SLAM) framework based on image feature point matching. Without GNSS, a single unmanned platform UGV and UAV (unmanned ground vehicle, UGV; unmanned aerial vehicle, UAV) equipped with vision and IMU (inertial measurement unit, IMU) sensors can exchange information through data communication to jointly build a three-dimensional visual point map, and determine the relative position of each other through visual-based position re- identification and PnP (Perspective-n-Points, PnP) methods. When any agent can receive reliable GNSS signals, GNSS positioning information will greatly improve the positioning accuracy without changing the positioning algorithm framework. In order to achieve this function, we designed a set of two-stage position estimation algorithms. In the first stage, we used the modified ORB-SLAM3 algorithm for position estimation by fusing visual and IMU information. In the second stage, we integrated GNSS positioning and cooperative positioning information using the factor graph optimization (FGO) algorithm. Our framework consists of an UGV as the central server node and three UAVs carried by the UGV, that will collaborate on space exploration missions. Finally, we simulated the influence of different visibility and lighting conditions on the framework function on the virtual simulation experiment platform built based on ROS (robot operating system, ROS) and Unity3D. The accuracy of the cooperative localization algorithm and the single platform localization algorithm was evaluated. In the two cases of GNSS-denied and GNSS-challenged, the error of co-location reduced by 15.5% and 19.7%, respectively, compared with single-platform independent positioning.

Keywords: differential GNSS; SLAM; cooperation SLAM; multi robot system; UAV; UGV

Citation: Xu, H.; Wang, C.; Bo, Y.; Jiang, C.; Liu, Y.; Yang, S.; Lai, W. An Aerial and Ground Multi-Agent Cooperative Location Framework in GNSS-Challenged Environments. *Remote Sens.* **2022**, *14*, 5055. https://doi.org/10.3390/rs14195055

Academic Editor: Xiaogong Hu

Received: 2 August 2022
Accepted: 1 October 2022
Published: 10 October 2022

1. Introduction

The aerial and ground collaborative unmanned systems are a heterogeneous cross-domain collaborative unmanned system composed of unmanned aerial vehicles (UAVs) and unmanned ground vehicles (UGVs), with complex functions such as perception, positioning, control and navigation. It can not only perform tasks independently, but also interacting with multiple aircraft across domains. A heterogeneous team of UAVs and UGVs can compensate for the lack of mobility, payload, and observation conditions between different platforms. UAVs can quickly cover large areas and have a good point of view for situational assessment. Ground vehicles have longer battery life, can carry large payloads, and actively interact with the environment.

In recent years, single-platform SLAM technology has been developed significantly. Early SLAM framework of sensor fusion mostly adopted the extended Kalman filter (EKF). For example, MSCKF [1] proposed a multi-sensor location information fusion method loosely coupled with visual-inertial odometry (VIO). ORB-SLAM3 [2] proposed by Campos e.g., revealed its potential in the aspects of high precision and high robustness. In the aspect of map fusion, the ORBSLAM-Atlas [3] module used camera pose error covariance to estimate the observability of the camera pose to determine whether to retain the camera pose or create a new map. In the field of multi-sensor fusion SLAM, VINS-FUSION [4] proposed a method that fused loosely-coupled global positioning information and VO/VIO positioning result. Some excellent collaborative SLAM frameworks also emerged from this foundation. When lacking external measurements, the relative position measurement between unmanned platforms mainly relies on visual position re-identification. A common way to obtain loop closures is to use visual place recognition methods, based on image or keypoint descriptors and a bag of words model, such as [4,5]. Some recent works have also studied closed-loop detection between distributed robots [6,7]. This method finds closed-loops through local communication between robots [8], collects observation data in a central server, and obtains a motion trajectory estimation of each robot through pose graph optimization (PGO). Different from the above, Yun Chang et al. proposed a collaborative SLAM method based on deep learning semantic description features [9].

There are two communication modes of multi-robot collaborative SLAM: distributed and centralized. [9–11] are representative works of a distributed framework. In related research of a centralized framework, Zou et al. introduced CoSLAM [12] in the early years, which demonstrated considerable potential of the centralized collaborative SLAM, CCM-SLAM [13] that deploys resource-consuming computations on servers, while still ensuring each agent's autonomy at low computational resource requirements by running a visual odometry system onboard. CVISLAM [14] was the first collaborative SLAM framework for bidirectional communication and extended visual-inertial odometry to the collaborative SLAM domain. It achieved higher accuracy and metric scale estimation. However, this study did not integrate GPS positioning information and thus lacks flexibility. Jialing Liu et al. proposed a collaborative monocular inertial SLAM system for smart phones. This was the first multi-agent collaborative SLAM system to run on a mobile phone [15], supporting cross-device collaboration. Similar work has reported CoVins [16] which can perform collaborative SLAM tasks on a larger scale. This study showed advantages in removing redundant information and reducing the coordination overhead.

All the above research only provides some thought to solving the ground-air collaborative navigation problems. They did not evaluate the specific application of various methods in aerial and ground collaborative navigation problems under GNSS-challenged environments. Even so, many challenges of this application still exist. For example, how to overcome the place recognition of crossing platforms under the aerial-ground difference of visual angles, or how to correct drift errors of GNSS positioning information for different platforms. The previous research mostly considered that a single platform did not need to run a complete SLAM optimization process during collaborative SLAM and only needed to execute visual odometry or visual-inertia odometry. However, with the rapid development of the terminal equipment computing power, we considered that a deploy communication interface, loosely-coupled with two-stage optimization and a complete single-platform SLAM process on a terminal device at the same time will not only improve the positioning accuracy of the single-platform but also improve the robustness of the single-platform positioning algorithm of the whole system in the case of communication disorder. The key to collaborative positioning in GNSS-challenging environments is to ensure system initialization and positioning without GNSS signals, and to improve overall positioning accuracy with GNSS positioning when GNSS positioning information is available.

In order to solve ground-air collaborative positioning problems, Moafipoor et al. proposed a method that used UGV and UAV collaboration to navigate [17]. When GPS is not available, the constraints of the external measurements provided by the extended Kalman filter and tracking filter are used to ensure the normal operation of the navigation function under adverse GPS conditions. In this paper, graph optimization was adopted to solve similar problems, assuming that GPS signals of each agent may be interfered with at any time. Peter Fankhauser et al. proposed a completely integrated method of relative observation between robots, independent of external positioning, and without initial guesstimates about the robot's posture [18]. This method was applied to the mutual positioning between a hexacopter and a quadrupedal robot. J Park et al. studied the map point registration between UAV and UGV through feature points and realized the work of spatial data collection by multiple agents in a decentralized environment [19]. Hailong Qin proposed a two layer shared novel optimized exploration path planning and navigation framework, which provided optimal exploration paths and integrated the collaborative exploration and mapping efforts through an OctoMap-based volumetric motion planning interface [20]. They only considered GNSS-denied environments, not GNSS-challenged environments. In practical applications, on the one hand, we hope to use GNSS positioning information when receiving GNSS signals, and on the other hand, we hope to maintain a certain navigation function in the absence of GPS.

In this paper, we proposed an algorithm framework based on feature point matching and graph optimization for ground-air collaborative positioning in GNSS-challenged environments and verify its function in the virtual simulation system. Compared with previous related work, the main contributions of this paper are summarized as follows:

1. GNSS information is used to eliminate drift errors of the SLAM process in a loosely-coupled way. This method can work normally even if the GNSS signal is disturbed or missing;
2. The cooperative location process is divided into several stages, and we try our best to balance the server computing power, communication bandwidth and algorithm performance. This allows the system to perform a complete global map initialization in the environment without GNSS signals;
3. The function of the proposed method is verified by using a virtual simulation system, and the positioning accuracy of the algorithm is analyzed.

2. Materials and Methods

2.1. System Overview

The architecture of the framework is depicted in Figure 1. At the early stages of system startup, all the UAV take off from a platform on the top of a UGV. After the unmanned platform completes IMU initialization and performs the first global bundle adjustment (BA), map fusion and relative pose solutions are completed between unmanned platforms through the local map. A flowchart of the co-location program executed on the server is shown below:

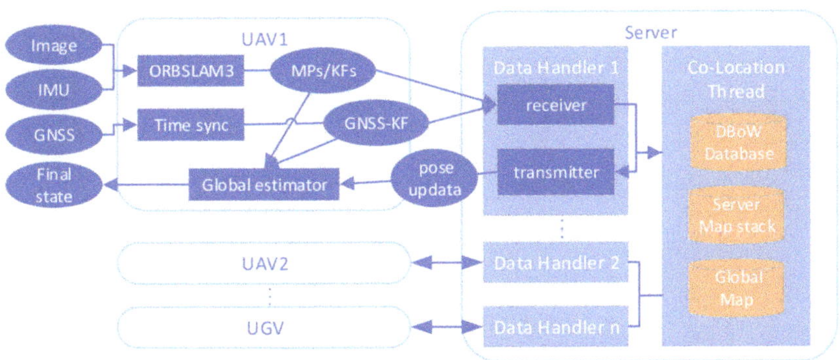

Figure 1. Overall algorithm framework. "MPs/KFs" is short for map points and keyframes. In the figure, the oval block represents all kinds of data, and the square block represents the processing

of data. Input for each platform includes images, IMU, and GNSS location information. Each data handler thread is responsible for handling a corresponding agent. Pose estimation and global optimization on a single platform and server are realized by factor graph optimization. Finally, the final state after global optimization is output to the subsequent program for path planning, task allocation or map construction.

For convenience of explanation, the first stage of the position estimation coordinate system for a single platform is called the VIO frame. The coordinate system where each agent's local map is located after a single platform has completed the second stage of position estimation is called the local frame. The coordinate system where the global map is located after map fusion is completed on the server is called the global frame. The co-location program on the server is shown in Figure 2. When there is no GNSS positioning or collaborative positioning information, the VIO frame will overlap with the local frame. The global frame overlaps with the local frame of the UGV before initialization, and the process of global map initialization is to obtain the relative positioning relation between each UAV and UGV through PnP solutions and convert the local map of each agent to a global map based on UGV's local map. After initialization, the local frame of each agent will overlap with the global frame. As the unmanned platform continues to move, each agent generates a new keyframe during the local SLAM process and sends these new keyframes and map points to the server through wireless data communication. The server stack will cache map information from each agent. These keyframes and map points will be added to the global map through the initialized relative position changes between platforms. After the program discovers place recognition among platforms through detecting feature points, loop-closure and map fusion of the global map will be executed, as well as optimization. The optimized position and pose will be used to estimate that in the second stage, together with the GNSS positioning from each platform. Finally, the new pose of the new keyframes after collaborative positioning in the closed-loop position will be sent back to the corresponding agent. And the agent that accepts the collaborative positioning information will adjust their pose during the second stage of local optimization.

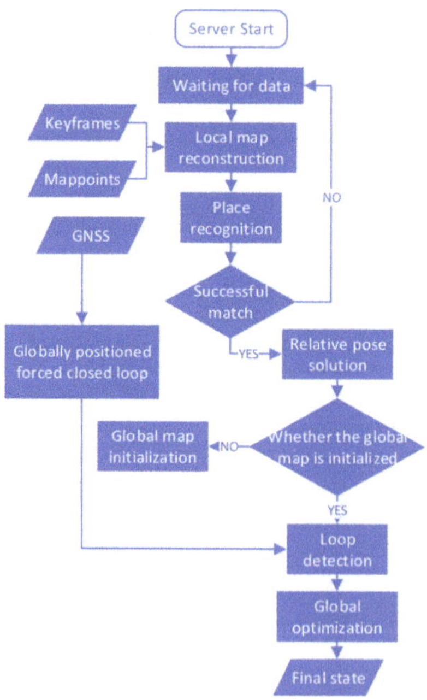

Figure 2. The running process of the co-location module in Figure 1 is explained in detail. The function

of this module is to restore the received MPs/KFs data from each agent to a local map. The local map is further stitched into a global map through visual position re-identification, or a new closed-loop is added to the global map that has been initialized. Finally, the GNSS positioning information is integrated to perform global optimization and output the result of co-location.

2.2. *Keyframe-Based SLAM Algorithm Fusing GNSS Positioning Information and Co-Location Information*

We deployed a SLAM framework modified and based on ORB-SLAM3 on each agent to run independently, to ensure each agent can complete independent navigation tasks in the situation where communication links are lost. The algorithm's flow is shown in Figure 3. ORB-SLAM3 was modified as follows:

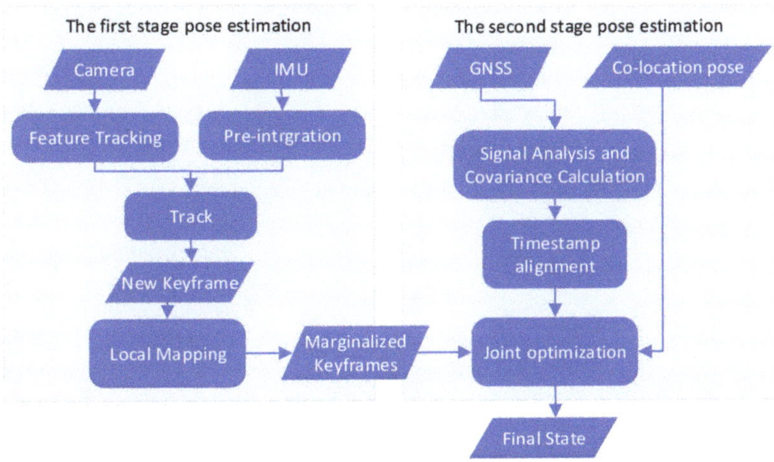

Figure 3. The tracking thread processes sensor information in real time and initially estimates the pose of the current frame. In the process of tracking the local map, the track thread determines whether the current frame is used as a keyframe. Different from ORBSLAM3's keyframe addition strategy, when valid GNSS positioning information completes timestamp alignment, frames aligned with GNSS information are also inserted as keyframes.

- A communication module to the server was added to exchange keyframe and map point information with the server instantly. The details about this part will be introduced in Section 2.3.1;
- The position estimation of the second stage was added, and both GNSS positioning information and collaborative positioning information were fused.

Each node represents a keyframe position and pose in the world frame. In Figure 4, the line between two nodes is called an "edge", which represents the constraint of the amount to be optimized in the optimization of the factor graph. The edge between two consecutive nodes is a local constraint, which comes from ORB-SLAM3's pose estimation. The other edge is the constraint from the co-location results of the server and GNSS satellite positioning information. We used the VIO factor as the local constraints, and GNSS location information and co-location information as global constraints. An illustration of the second-stage global pose graph structure is shown in Figure 4.

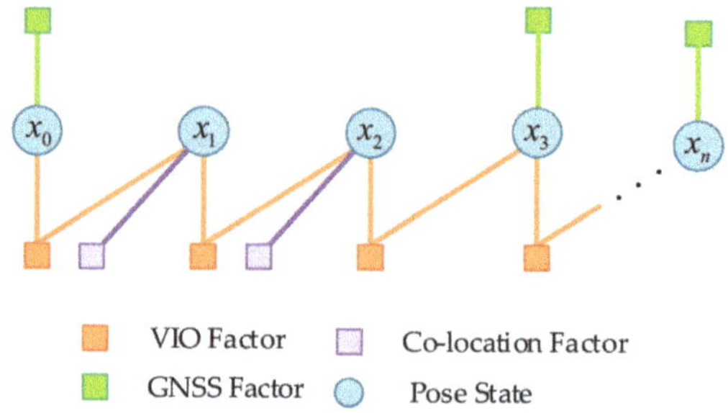

Figure 4. Schematic diagram of the pose graph structure to be optimized in the second stage. The circle is the state quantity, and the yellow square is the constraint of local observation, that is, the relative pose transformation of VIO from ORBSLAM3. The other colored squares are constrained by global observations.

Construction of residuals with local constraints refer to ORBSLAM3 [1], supposing that there are two continuous keyframes KF_{t_i} and KF_{t_j}, we define q_t^v as the attitude quaternion KF_t under the VIO coordinate system, and p_t^v as the three-dimensional coordinate under the VIO coordinate system of KF_t. Similarly, q_t^l and p_t^l are the quaternion and coordinates under the global coordinate system. The VIO local factor is derived as:

$$z^L - s^L(\chi) = z^L - s^L(x_{t_j}, x_{t_i}) = \begin{bmatrix} q_{t_j}^{v\,-1}\left(P_{t_i}^v - P_{t_j}^v\right) \\ q_{t_j}^{v\,-1} q_{t_i}^v \end{bmatrix} \odot \begin{bmatrix} q_{t_j}^{l\,-1}\left(P_{t_i}^l - P_{t_j}^l\right) \\ q_{t_j}^{l\,-1} q_{t_i}^l \end{bmatrix} \quad (1)$$

where z is the measured value, provided by the results of the one-stage position estimation; χ is the state prediction; and s is the observation equation. χ is calculated from the pose transformation between the two moments. The specific method used in this paper uses the relative pose of the current moment and the previous moment obtained from the position estimation in the first stage to add to the position coordinate x_{t_j} at the previous moment to obtain the position coordinate x_{t_i} at the current moment. $\odot$ is the minus operation on the error state of quaternion. We take the pose covariance matrix generated during SLAM as the covariance of local measurements. The essence of the local factor is the relative change in the pose in two keyframes.

Consider the following two situations where GNSS does not work: the GNSS signal is poor or has interference, resulting in a large error in the GNSS positioning information, resulting in a large drift in an agent's navigation trajectory; the GNSS signal is completely disabled, and there is no GNSS positioning information. Upon receipt of a valid GNSS location, the longitudinal dimension heights of the original measurements will be transformed into x, y, z coordinates in the local Cartesian coordinates system (ENU) during GNSS data preprocessing. $P_t^G = [x_t^w, y_t^w, z_t^w]^T$ is the coordinate of the GNSS information in the transformed ENU coordinate system. P_t^G is the measured value of GNSS at that moment, t. The uncertainty of measurement is assumed to be a Gaussian distribution with mean and covariance. $P_t{}^w$ is the assumed GNSS estimate. The GNSS factor is derived as:

$$z_t^G - s_t^G(\chi) = z_t^G - s_t^G(x_t) = P_t^G - P_t^w \quad (2)$$

When each agent receives a new pose about a keyframe's collaborative positioning from the server, we introduce the result of collaborative positioning as a new measured value into the process of secondary optimization. We assume that the result obtained by KF_t in the first-stage pose estimation at that moment in the VIO coordinate system is $\{q_v, p_v\}$.

The pose obtained by KF_t after the co-location is transformed into $\{q_c, p_c\}$ on the local framework of the agent. The co-location factor is derived as:

$$z_t^C - s_t^C(\chi) = z_t^C - s_t^C(x_t) = \begin{bmatrix} q_v^{l^{-1}}\left(P_c^l - P_v^l\right) \\ q_v^{l^{-1}} q_c^l \end{bmatrix} \odot \begin{bmatrix} q_v^{w-1}(P_c^w - P_v^w) \\ q_v^{w-1} q_c^w \end{bmatrix} \tag{3}$$

Lastly, in order to prevent VIO from drifting too much, the transformation of the VIO coordinate system to the global coordinate system should be updated after each optimization. The operation must make the position after fusion overlap with the results of GNSS positioning and cooperative positioning as much as possible, and the difference between the two frames should be as equal as possible with VIO. The data from VIO does not impose constraints on the absolute position after fusion, and only requires that the incremental error of the position after fusion and the incremental error of VIO should be as small as possible.

2.3. *Communication*

The communication modules on the agents and server was based on The ROS communication infrastructure [21]. It was used for message passing over a wireless network. The communication interface supports two-way communication between the agents and the server, we applied it to update the keyframe pose at a given moment for single-platform SLAM. The server uses it to receive map information from each agent. With this module, the objects waiting to be sent were first serialized/deserialized. Since ROS is not a real-time communication system, time stamps needed to be attached to packets.

2.3.1. Agent-to-Server Communication

In order to realize map data sharing from agent to server and save data bandwidth as much as possible, we added several state variables into the ORB-SLAM3 program to monitor the running status of SLAM on a single platform, including: IMU initialization process, whether visual-inertial bundle adjustment (VIBA) is completed, and tracking the running status of threads. After the IMU completes initialization and VIBA, all the previous keyframes and map point information is serialized and sent. The keyframe information includes the unique ID of this keyframe on the agent, its pose and time stamp, the information of feature points it contains, and the numbers of all map points observed by this keyframe. The map point information contains the ID of this map point on the agent, its three-dimensional coordinates, its descriptors, and the ID of all keyframes that have observed this map point so far. Unique IDs ensure that these messages are not sent repeatedly to reduce the required network bandwidth. The communication module distinguishes between dynamic information (such as the pose of a keyframe that may be adjusted according to the running state of SLAM) and static information (descriptors of feature points). In the process of communicating with the server, each agent will send the smallest packet at the highest frequency (including at least one newly added keyframe information).

2.3.2. Server-to-Agent Communication

Each agent handler thread is responsible for processing keyframes and map points from the corresponding agent on the server. Each keyframe is not processed until all of its observed map points have been deserialized by an agent handler and inserted into the map stack. The server map stack contains separate maps that contain data from each agent. When the server maps complete location identification and merge, the two maps are removed from the map stack and replaced with their merged map until there is only one global map in the entire server map stack. When the server detects the new closed-loop and completes the global pose optimization, it sends the ID and pose of the keyframe back to the corresponding agent to be updated.

2.4. Global Map Initialization and Intra-Agent Measurements

In order to enable the system to complete the positioning of each agent without GNSS at the initial stage of operation, we let the UAVs and UGVs start from the same position with the same angle of view. In this paper, UAVs are set to take off from the top platform of the UGV. After the IMU initialization is completed (which commonly needs 15–35 keyframes and takes 5–10 s), the agents perform a visual-inertial bundle adjustment. The aim of this action is to improve the precision of the initial map. Then, each agent sends its initial optimized local map to the server. For a certain terminal $Agent_c$, the server will start the alignment from the keyframe KF^c at the most recent moment sent in the Map_c. First, a subset of maps Map_m from the server map stack and brute-force descriptors are chosen that match all keyframes contained within that map with KF_t^c through DBoW2, and obtained the most likely matched candidate $\left\{KF_{t_1}^m, KF_{t_2}^m \ldots KF_{t_i}^m\right\}$ (We set i to four times that of the number of agents). For each keyframe, among all the feature points contained in it and its five best common-view keyframes, the 2D-2D feature point matching provided by DBoW2 is used to obtain the 3D-3D matching between their corresponding map points. The map points of candidacy $KF_{t_n}^m$ will be converted to KF_t^c through T_{cm}^n. If the reprojection error is less than the threshold, a vote is cast for the corresponding $KF_{t_n}^m$, and the frame with the most votes is selected as the matching frame. For further improvement of the closed-loop's robustness, the matching frame and its common-view frame together need to achieve three successful matches to be considered as the complete place recognition. If, at this moment, $Agent_c$ is a UAV, $Agent_m$ is a UGV, then the UAV's local map will be fused into the UGV's through $T_{UAV_c\_UGV_m} = T_{cm}$. If both $Agent_c$ and $Agent_m$ are UAVs, the relative pose transformation will be saved until one of them is fused with the UGV's map, and another will complete fusion through $T_{UAV_c\_UAV_m} \times T_{UAV_m\_UGV}$. After the above pairwise pairing of the unmanned platforms, the local map of each UAV is finally converted into the coordinate system of the local map of the unmanned vehicle. New incremental keyframes and map points received by the server during subsequent operations will be transferred to the global map with the corresponding poses until a new loop closure or GNSS location provides a new location constraint.

2.5. Loop Closure and Global Optimization

The place recognition of loop closure is similar to the global map initialization, which depends on the feature point matching provided by DBoW2, and the initial matching through RANSAC and the PnP algorithm. Different from the global map initialization, considering the long-term drift of single-platform SLAM, a finer relative position transition is required. As above, we suppose that KF^c of $Agent_c$ and KF^m of KF^m have complete initial matching and $T_{cm} = \begin{pmatrix} R & t \\ 0 & 1 \end{pmatrix} \in SE(3)$. All converted map points contained within the common-view frame of KF^m, and those found matching to the key points in KF^c. After the above intra-agent measurements have been completed, we need to adjust the map information according to the new closed-loop and fuse all the positioning information from each agent to perform global optimization. The details are as follows.

2.5.1. Refinement of Transformation Matrix

T_{cm} is used to convert all map points contained within the common-view keyframe of KF^m, and matching map points are located in the feature points of KF^c. To obtain as many matches as possible, the map points are found that matches KF^c in KF^m and all of its keyframes at the same time. The nonlinear optimization of $T_{cm}^* = \begin{pmatrix} sR & t \\ 0 & 1 \end{pmatrix} \in sim(3)$ is carried out by using all the map point matching relationships found (the initial value of s is 1), and the goal function is the bidirectional reprojection error

2.5.2. Pose Graph Optimization

After obtaining the optimized transformation matrix T^*_{cm}, the redundant map points need to be eliminated in the above matching process for each pair of matching map points describing the same feature point. Then propagation of the corrections must be disseminated to the rest of the map through pose graph optimization. To eliminate duplicate points, for each pair of matches the points from $Agent_c$ are removed, and the points from $Agent_m$ inherit all the observations of the removed points. By adding edges to the co-visibility and essential graphs, the observability is propagating between the common-view frames of KF^m and KF^m.

As it has sufficient computing power on the server and is not sensitive to the time cost, global BA optimization can be directly carried out on the adjusted visibility and essential graphs.

2.5.3. Global the Second-Stage Pose Estimation

The factor graph structure of the global two-stage optimization is similar to that of the single platform, the difference is that in the global optimization process only the GNSS positioning information (if any) is needed to be taken into consideration. This is due to the GNSS positioning information from different agents will also generate new position constraints on the global map making the track change significantly and intersect, even if there is no visual re-location at that position. In this case, we would lower the threshold in line with the RANSAC algorithm. If it can pass the detection after reducing the number of inner points, a new closed-loop will be established here. The subsequent processing is consistent with the closed-loop process described above.

3. Simulation and Experimental Result

3.1. Virtual Simulation Experiment Platform

The overall architecture of the multi-unmanned platform simulation system based on Unity3D and ROS architecture is shown in Figure 5. The UAV flight control and visual simulation is based on Flightmare [22]. The ROS Gazebo [23] simulation environment was run on Computer A to constrain the movements of UAVs and UGVs through dynamic models, and to generate the true values of the position and motion velocity. Position and velocity errors were superimposed to generate virtual IMU and GNSS data. Among them, GPS positioning information in GPGGA format was chosen as GNSS satellite positioning. The location and timestamp of an agent sent to the visual simulation module was passed through ROS-Unity3D interface. The visual simulation module moves the agent's model to the corresponding coordinates, renders the photo, and finally sends the most up-to-date image and timestamp to the algorithm verification program on Computer B. The configuration of the two computers in the figure is shown in the Table 1.

3.1.1. Engine Dynamics Modeling and Simulation for a Type of UAV and UGV

The rigid body appearance, physical properties, joint types and other aspects of the unmanned platform are described through URDF files. The motion of the unmanned platform model in the ROS environment can be viewed through Gazebo. Its position in Gazebo is taken as the true value of the pose of the unmanned platform. hybrid A* algorithm to UGV is applied to realize the path planning, and TEB algorithm to realize trajectory tracking. The UAV adopts a four-rotor model, using the quadrotor dynamics to design the control algorithm, and the Euler method and fourth-order Runge-Kutta method to integrate the UAV dynamics equation.

3.1.2. Visual Simulation Based on Unity3D

The visual simulation module includes scene resources and unmanned platform models. The scene resource module is responsible for the construction and optimization of 3D scenes in the visual simulation software, and makes the scenes restore the real environment as much as possible to run smoothly through the 3D model import and level of detail (LOD) optimization. Multiple unmanned platforms can be loaded in the simulation scene according to the demand, and the objects can be driven to move according to the immediate feedback of the data in ROS. Particle effect components are used to simulate weather, such as rain and fog. The light was set in the Unity scene to simulate a sunny day, dusk and night. Various meteorological and light environments were combined to simulate different visibilities in the real world. Lastly, the virtual RGB images were obtained through the Unity camera module.

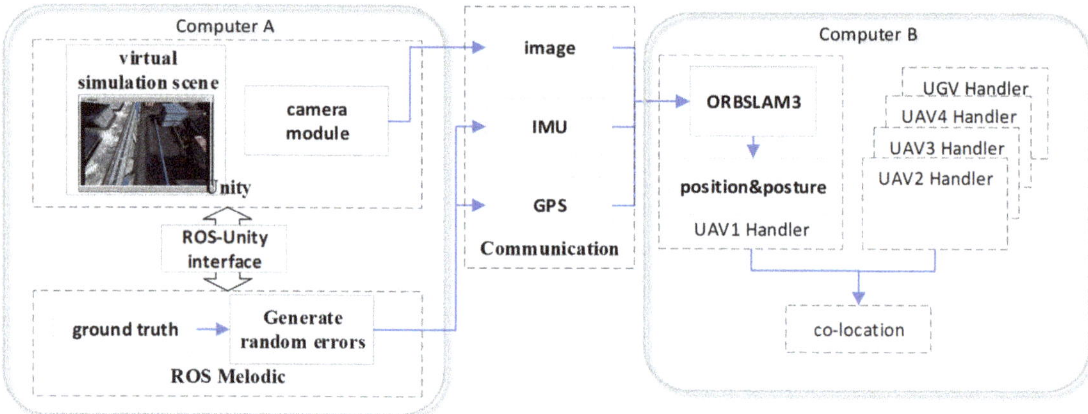

Figure 5. The virtual data generation program is deployed on Computer A, and the agent handler and cooperative location program is deployed on Computer B. They are connected using a gigabit network.

Table 1. The hardware configuration of the computers running the virtual simulation platform and the positioning algorithm.

Platform	Type	Characteristic
Computer A	Desktop computer	Intel9-11700k + 32 Gb + 2080Ti
Computer B	Desktop computer	AMD3990x + 64 Gb + 2080Ti

3.1.3. Virtual Sensor Data Generation

Based on the modeling of the unmanned platform in Section 3.1.2, two sensor components were added: an inertial measurement unit module and a GPS module. IMU and GPS data were obtained by adding noise to the true value of the sensor model in the Gazebo coordinate system. The IMU-related parameters are shown in Table 2:

Table 2. IMU data production module parameter configuration table.

Parameter	Value
Gyroscope noise density	0.0003394 [Hz]
Gyro deviation random walk	0.000038785 [Hz]
Gyro deviation related time constant	1000.0 [s]
Gyroscope opening deviation standard deviation	0.0087 [rad/s]
Accelerometer noise density	0.004 [Hz]
Accelerometer deviation random walk	0.006 [Hz]
Accelerometer deviation related time constant	300.0 [s]
Update frequency	100 [Hz]

The parameters related to GPS are shown in Table 3.

Table 3. GPS data production module parameter configuration table.

Parameter	Value
Standard deviation of Gaussian noise at horizontal position	0.05 [m]
Standard deviation of Gaussian noise in vertical position	0.15 [m]
Standard deviation of horizontal velocity Gaussian noise	0.05 [m/s]
Standard deviation of vertical velocity Gaussian noise	0.05 [m/s]
Update frequency	5 [Hz]

3.2. Experimental Results

This section introduces the simulation and experimental results of the autonomous positioning in an unknown environment by using the heterogeneous UAV and UGV system proposed. In the following experiment, we generated specified waypoints for the three UAVs and a UGV through the virtual simulation and saved the resulting sensor data and time stamps in the ROS bag for subsequent repeating tests. We examined the robustness of the method we proposed in an actual engineering scenario by modifying the ambient light, weather in the scene, and the down view angle of the UAV camera. Then we analyzed and compared the influence of the collaborative positioning framework on the positioning accuracy of each platform with single-platform positioning. The detailed settings for each dataset are shown in the Table 4.

Table 4. Parameter Settings for each dataset.

Dataset	Lighting Conditions	Meteorology	Agent	Camera Angle	Path Length
Factory 01	bright	High visibility	UAV1	45°	162.47 m
			UAV2	45°	161.29 m
			UAV3	45°	168.96 m
			UGV	0°	106.43 m
Factory 02	bright	thick-foggy	UAV1	45°	162.47 m
			UAV2	45°	161.29 m
			UAV3	45°	168.96 m
			UGV	0°	106.43 m
Factory 03	somber	High visibility	UAV1	45°	162.47 m
			UAV2	45°	161.29 m
			UAV3	45°	168.96 m
			UGV	0°	106.43 m
Factory 04	somber	thick-foggy	UAV1	45°	162.47 m
			UAV2	45°	161.29 m
			UAV3	45°	168.96 m
			UGV	0°	106.43 m
Factory 05	bright	mist	UAV1	45°	162.47 m
			UAV2	45°	161.29 m
			UAV3	45°	168.96 m
			UGV	0°	106.43 m
Factory 06	somber	mist	UAV1	45°	162.47 m
			UAV2	45°	161.29 m
			UAV3	45°	168.96 m
			UGV	0°	106.43 m
Factory 07	bright	High visibility	UAV1	25°	162.47 m
			UAV2	25°	161.29 m
			UAV3	25°	168.96 m
			UGV	0°	106.43 m
Factory 08	bright	High visibility	UAV1	25°	162.47 m
			UAV2	45°	161.29 m
			UAV3	75°	168.96 m
			UGV	0°	106.43 m

3.2.1. The Influence of External Factors on the Function of Algorithm Framework

Our algorithm depends on visual graph feature points to complete place recognition, which is the foundation of subsequent map fusion and loop closure. Considering that visual angle difference and observation conditions are the main factors affecting feature point matching, we set three meteorological environments in the virtual simulation system: no occlusion, mist and fog, and two lighting environments: normal light and high light ratio under sunset. There was no cover, mist and fog representing the visibility of infinity, 150 m and 50 m, respectively. A schematic of the scenario is shown in Figure 6.

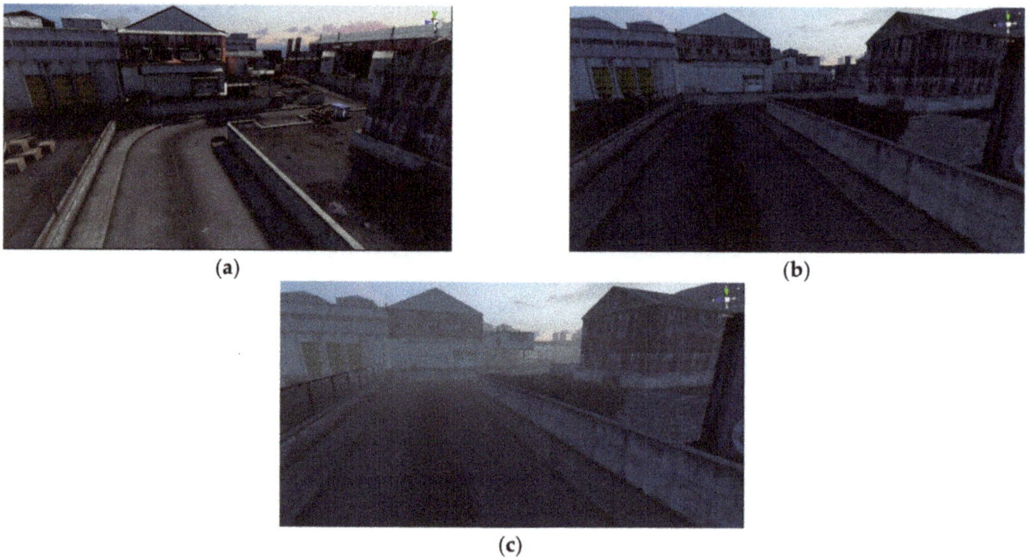

Figure 6. Schematic diagram of the virtual simulation scene (**a**) bright scene with high visibility; (**b**) a dimly lit but fog-free scene; (**c**) foggy and dim scene. Different observation conditions will affect the extraction and matching of feature points.

The first thing worth paying attention to is the network bandwidth occupied by each agent and server for exchanging data in the virtual communication network environment. In Table 5, we list the average bandwidth and instantaneous traffic peaks of upload and downlink data between each agent and server. The peak of uploaded data occurs during map initialization, and the bandwidth usage is relatively smooth thereafter. There is a positive correlation between bandwidth usage and keyframe generation speed.

Table 5. This table shows the network bandwidth usage after all simulation datasets were run. The results are obtained from seven experiments.

Agent	Upload		Downlink	
	Average	Maximum	Average	Maximum
UAV1	0.35 MB/s	1.1 MB/s	0.4 KB/s	0.8 KB/s
UAV2	0.34 MB/s	1.1 MB/s	0.6 KB/s	0.9 KB/s
UAV3	0.40 MB/s	1.3 MB/s	0.4 KB/s	0.7 KB/s
UGV	0.31 MB/s	0.9 MB/s	0.2 KB/s	0.4 KB/s

Here, we compared the number of closed-loops generated with that of the feature points matched on the closed-loop frame during the operation process of the collaborative positioning algorithm on different datasets. From the results in Table 6, we found that the feature point matching had similar performance under two similar viewing angles when the viewing angle of UAV was the same. As many feature point pairs as possible are beneficial not only to achieve robust place recognition, but also to improve the positioning accuracy of SLAM. However, the number of feature point matching between UAV and UGV will decrease with an increase in the angle of view difference, which is more obvious when a UAV flies at low altitude. We believe that setting the camera angle of the UAV to 45° in the downward view can balance the need for visual repositioning among UAVs and between UAVs and UGVs better.

Table 6. The number of feature points matching and successful location re-identification under different angles of view and illuminations.

Camera Angle	Lighting Conditions	Number of Successful Relocations	Average Number of Matching Point Pairs
25–25°	bright	28	123.83
	somber	27.3	110
45–45°	bright	27	139
	somber	27	113
75–75°	bright	28	128.33
	somber	27	129.5
25–75°	bright	5	36
	somber	4	35
25–45°	bright	11	47
	somber	10	44.83
45–75°	bright	14	55.5
	somber	11	49
0–25°	bright	12	47
	somber	10	42
0–45°	bright	17	51.5
	somber	17	38.83
0–75°	bright	4	23.33
	somber	FAIL	18

As can be seen from the data in Table 7, the change of light affected the detection of feature points, but the number of feature points in the picture was still enough to produce the correct closed-loop. In the mist mode, 59% of feature point pairs were lost, and the number of correct closed-loops produced also dropped dramatically. Moreover, dense fog interfered the feature point detection and place recognition thoroughly. In the dense fog mode, except when the map initialization was completed when all the UAVs and UGVs started from the same position at the first stage of the operation, all the position re-identifications failed at the position where the closed-loop should be generated due to insufficient matching feature points in subsequent operations.

Table 7. The number of feature points matching and successful location reidentification under different angles of view and illuminations.

Lighting Conditions	Meteorology	Average Number of Successful Relocations	Average Number of Matching Point Pairs
bright	high visibility	27	125.83
	mist	11	58
	thick-foggy	FAIL	17.5
somber	high visibility	27	112
	mist	7	39.83
	thick-foggy	FAIL	15.33

3.2.2. Collaborative SLAM Estimation Accuracy

To estimate the positioning accuracy of the system, in this section the influence of external observation conditions on the experiment results were ignored and the camera view that generated the largest average number of matching point pairs (45°) was adopted. The experiment for this section was run on Factory 01, 03, 05, 06 on the datasets described in Table 8.

Table 8. The accuracy of each agent running in different modes.

Agent	Mode	ATE_RMSE (m)/Mean Error (m)
UAV1	GPS-challenged Co-location 1	0.262/0.170
	GPS-challenged Co-location 2	0.311/0.197
	GPS-Denied Co-location	0.930/0.842
	ORB-SLAM3	1.110/0.911
	ORB-SLAM3 fusion GPS	0.371/0.223
UAV2	GPS-challenged Co-location 1	0.416/0.293
	GPS-challenged Co-location 2	0.472/0.401
	GPS-Denied Co-location	1.298/1.025
	ORB-SLAM3	1.523/1.465
	ORB-SLAM3 fusion GPS	0.488/0.317
UAV3	GPS-challenged Co-location 1	0.368/0.273
	GPS-challenged Co-location 2	0.397/0.285
	GPS-Denied Co-location	0.782/0.502
	ORB-SLAM3	0.979/0.735
	ORBSLAM3 fusion GPS	0.396/0.299
UGV	GPS-challenged Co-location 1	0.172/0.134
	GPS-challenged Co-location 2	0.173/0.137
	GPS-Denied Co-location	0.219/0.171
	ORB-SLAM3	0.405/0.297
	ORB-SLAM3 fusion GPS	0.186/0.175

We estimated the absolute trajectory error (ATE) of each agent. We tested the positioning performance in GPS-denied and GPS-challenged modes separately. In the GPS-challenged 1 mode, all agents obtained valid GPS location information every 20 s. In the GPS-challenged 2 mode, the UAV could only randomly obtain four valid GPS positions during the operation, and the UGV masked all GPS location information. In the GPS-denied mode, all agents would not use the GPS location data in the dataset. The effect trajectory of the three UAVs and the unmanned vehicle in the collaborative positioning is shown in Figure 7a.

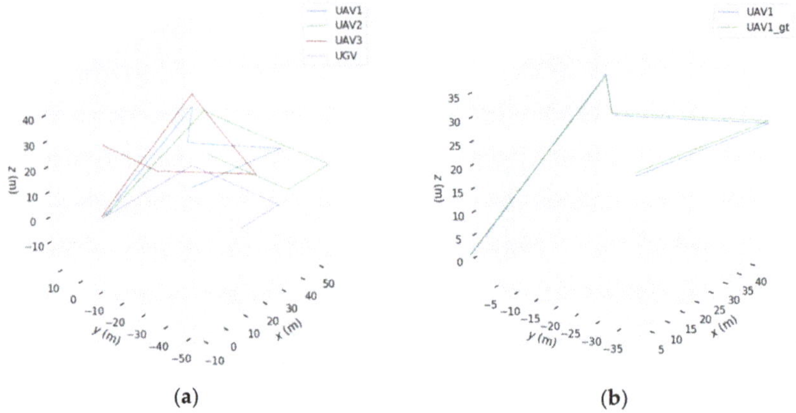

Figure 7. *Cont.*

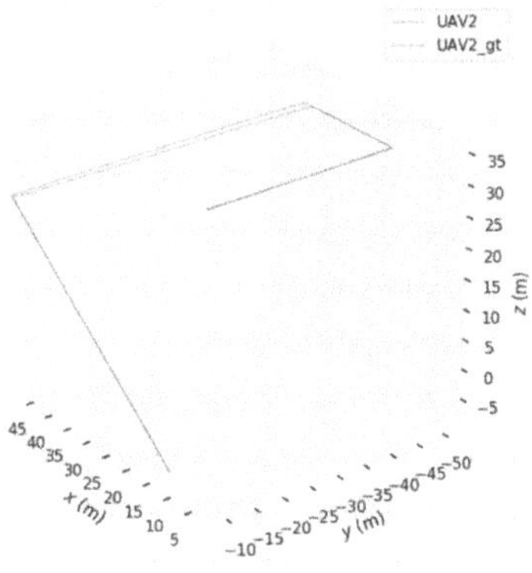

(c)

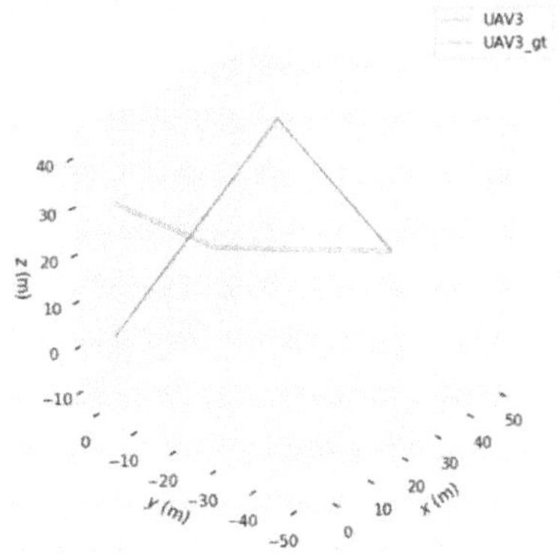

(d)

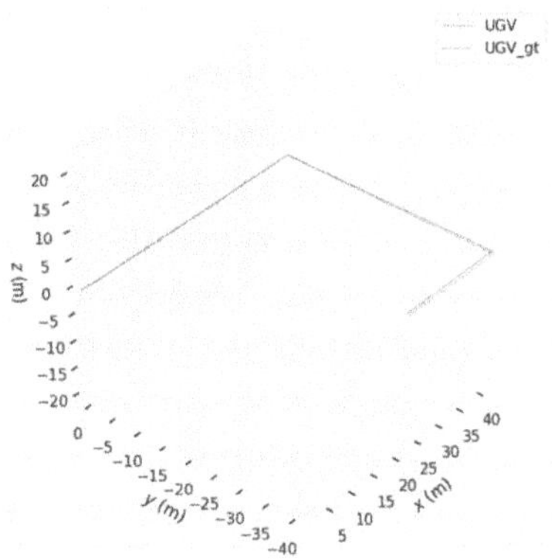

(e)

Figure 7. (**a**) The rendering of the co-location of three UAVs and a UGV; (**b–e**) trajectories and ground truth for a single-platform operation of each agent.

4. Discussion

From the experiments described in Section 3 we draw the following three conclusions.

1. Experiments in the virtual simulation platform demonstrated that the co-location framework we designed can maintain usable positioning accuracy over hundreds of meters of trajectory when the GNSS signal is rejected. At the same time, the co-location framework outperforms the single platform in terms of accuracy. This is essentially attributed to the new position constraints arising from place recognition between the different platforms.
2. The SLAM front-ends we currently employ were not sufficiently stable under poor observation conditions. For example, poor visibility environments or poor near orthogonal viewing angles affected the proper function and positioning accuracy of the air-ground co-location system. The main reason is that the ORB feature-based SLAM front-end we adopted often cannot provide a sufficient number of feature points for matching to achieve robust location re-identification. Other common front-end methods, such as optical flow method, have more stringent requirements on observation conditions and viewing angles. In the SLAM front-end process, if the threshold of hamming distance for ORB detection is lowered in order to increase the number of matched pairs, it will lead to excessive false matches affecting the accuracy of VIO. Therefore, place recognition that relies on ORB features can only be adapted to

the images generated under more optimum observation conditions. The visual front-end used in this manuscript fulfills the basic requirements for cooperative air-ground localization but is not yet sufficiently robust for weak observation conditions.

3. The method described in this manuscript allows the location information of each agent in the entire unmanned cluster to be propagated to the other agents in the cluster. This process improves the positioning accuracy while also allowing the server and the individual unmanned platforms to acquire relative positions to each other. We designed a comparison of four cases. They were GPS-challenged co-location, GPS-denied co-location, single-platform independent operation of ORBSLAM3, and single-platform independent operation of ORBSLAM3 fusion GPS. In GPS-denied mode, the simulation was performed on a solar body with no GNSS fix except of Earth, or on Earth but the GNSS signal had interference. In this case, the positioning of each agent completely depended on the camera and IMU. It can be seen from Table 8 that the accuracy of co-location was significantly improved compared with the positioning accuracy of ORBSLAM3 running on a single platform. Assuming that some agents in the system obtained satellite positioning information, the positioning accuracy of all agents was improved. It is worth noting that even if we assume in the experiment that the UGV cannot obtain satellite positioning signals at all, its positioning accuracy still benefits from the co-location algorithm. This is due to the fact that the new constraints brought by the satellite positioning information are propagated to the local maps of all agents through the global map. Through the comparison of the two GPS-challenged modes, we found that only a few GPS points were enough to greatly improve the positioning accuracy. The GPS-challenged 1 mode used several times more GPS positioning points than the GPS-challenged 2 mode, but the positioning accuracy was not significantly improved compared with the latter.

In the experiments designed in this manuscript, we set the UAV to use high-precision GNSS positioning information, while the UGV could not obtain GNSS signals. There were two main reasons for this assumption. The first point was that, in practice, UAVs located in the sky can often obtain GNSS positions with good accuracy through RTK or relative positioning measurements. Whereas vehicles located on the ground may temporarily lose their GNSS position due to obstruction by reinforced concrete buildings, or by entering tunnels and interiors. The second point was that ORBSLAM3 itself is an excellent single-platform SLAM algorithm, which can have a trajectory drift error of less than 1% with a closed-loop bottleneck. This makes it difficult for the meter-level errors inherent in GNSS to contribute to the improvement in positioning accuracy if differential-free GNSS positioning is used in a virtual scene of limited size. If the scale of the motion trajectory reaches several kilometers, however, even differential-free GNSS positioning information can greatly optimize the positioning accuracy.

In our paper, we focused on collaborative localization. However, in engineering applications, the movement of each agent in the system was not infinite due to the limitation of communication bandwidth and communication distance. Due to the lack of prior information of the global map, the pose information of each agent and the local map information received by the server had time delays, and the global positioning pose updated through the visual closed-loop had no gradient information. In this case, if a motion strategy for convex online learning to train unmanned clusters is required, delayed mirror descent (DMD) [24] would be a good choice. We will use this approach in the subsequent work. In the aspect of multi-agent motion control, our method only used the most basic D* Lite algorithm. Only the trajectory of a single agent was considered to fit the route generated in advance. However, in practical applications, there may be obstacles in the waypoints planned in advance, and the route itself is unreachable. Cooperation among multiple agents can be difficult to achieve. Collaborative algorithms such as [25] realize the decentralized and real-time cooperative pursuit of a single evader in the planar domain. By improving the method proposed in this paper, the control of the prior map obtained by the UAV would be improved.

Finally, it is worth mentioning that in the statistics of network bandwidth usage, we found that the peak bandwidth usage occurs in the map initialization phase, which is about three times the average value. In the case of long distance or interference, this will put higher requirements on the performance of wireless communication devices. When there is a prior map, it will greatly relieve the bandwidth pressure during system initialization. During system initialization, at least four high-quality GNSS positioning points are needed to convert the VIO coordinate system to the northeast sky coordinate system. Meanwhile, ORBSLAM3′s IMU initialization requires initial zero-bias estimation and gravity direction estimation to obtain scale information, which makes the initialization of the global map take too long. How to complete global map initialization more efficiently and quickly or to make better use of prior map information is the next urgent problem to be solved.

5. Conclusions

This manuscript uses ORB-based feature point re-identification to improve the positioning accuracy of all unmanned platforms in a cluster by fusing the local maps of each unmanned platform with new position constraints and global GNSS positioning information. Through a centralized collaborative positioning service, this can provide low latency positioning information for subsequent collaborative path planning and task allocation algorithms. The co-location algorithm proposed in this manuscript has better accuracy in both GNSS-challenged and GNSS-denied modes than the ORB-SLAM3 algorithm running on a single platform. The two-stage position estimation method used can also be combined with other positioning sensors such as UWBs and barometers in addition to the GNSS global positioning information applied in the manuscript. Current techniques for the localization of air-ground unmanned clusters in complex environments present new demands in the direction of visual front-ends, optimization methods and multi-sensor fusion. The approach proposed in this manuscript can be extended to other unmanned clusters in areas such as UAVs, logistics, agriculture and military. In the future, based on the current results, our subsequent work will further investigate the impact of different vision front-end techniques on location re-identification, a key aspect that critically affects map fusion and closed-loop detection. We will consider the use of feature detection and matching techniques based on deep learning or point and line features to improve the robustness of the algorithmic framework and to improve the localization accuracy based on this. Finally, our proposed approach in this manuscript has high computational resource requirements for map fusion and global optimization, and we will aim to mitigate the computational power and communication bandwidth required for global map maintenance on a server.

Author Contributions: Conceptualization, H.X.; methodology, H.X.; software, H.X.; validation, Y.L. and S.Y.; formal analysis, C.W.; resources, Y.B.; data curation, C.W.; writing—original draft preparation, H.X.; writing—review and editing, C.J.; visualization, W.L.; supervision, W.L.; project administration, Y.B. All authors have read and agreed to the published version of the manuscript.

Funding: This research received no external funding.

Data Availability Statement: Not applicable.

Conflicts of Interest: The authors declare no conflict of interest.

References

1. Mourikis, A.I.; Roumeliotis, S.I. A multi-state constraint kalman filter for vision-aided inertial navigation. In Proceedings of the IEEE International Conference on Robotics and Automation, Orlando, FL, USA, 15–19 May 2006; pp. 456–478.
2. Campos, C.; Elvira, R.; Rodriguez, J.J.G.; Montiel, J.M.M.; Tardos, J.D. ORB-SLAM3: An Accurate Open-Source Library for Visual, Visual–Inertial, and Multimap SLAM. *IEEE Trans. Robot.* **2021**, *37*, 1874–1890. [CrossRef]
3. Elvira, R.; Tardos, J.D.; Montiel, J. ORBSLAM-Atlas: A robust and accurate multi-map system. *arXiv* **2019**, arXiv:1908.11585.
4. Cadena, C.; Galvez-López, D.; Tardos, J.D.; Neira, J. Robust Place Recognition with Stereo Sequences. *IEEE Trans. Robot.* **2012**, *28*, 871–885. [CrossRef]

5. Pumarola, A.; Vakhitov, A.; Agudo, A.; Sanfeliu, A.; Moreno-Noguer, F. PL-SLAM: Real-time monocular visual SLAM with points and lines. In Proceedings of the 2017 IEEE International Conference on Robotics and Automation (ICRA), Singapore, 29 May–3 June 2017; pp. 4503–4508.
6. Tian, Y.; Khosoussi, K.; How, J.P. A resource-aware approach to collaborative loop-closure detection with provable performance guarantees. *Int. J. Robot. Res.* **2021**, *40*, 1212–1233. [CrossRef]
7. Giamou, M.; Khosoussi, K.; How, J.P. Talk Resource-Efficiently to Me: Optimal Communication Planning for Distributed Loop Closure Detection. In Proceedings of the 2018 IEEE International Conference on Robotics and Automation (ICRA), Brisbane, Australia, 21–25 May 2018; pp. 3841–3848.
8. Cieslewski, T.; Scaramuzza, D. Efficient Decentralized Visual Place Recognition Using a Distributed Inverted Index. *IEEE Robot. Autom. Lett.* **2017**, *2*, 640–647. [CrossRef]
9. Chang, Y.; Tian, Y.; How, J.P.; Carlone, L. Kimera-Multi: A System for Distributed Multi-Robot Metric-Semantic Simultaneous Localization and Mapping. In Proceedings of the 2021 IEEE International Conference on Robotics and Automation (ICRA), Xi'an, China, 30 May–5 June 2021; pp. 11210–11218.
10. Lajoie, P.-Y.; Ramtoula, B.; Chang, Y.; Carlone, L.; Beltrame, G. DOOR-SLAM: Distributed, Online, and Outlier Resilient SLAM for Robotic Teams. *IEEE Robot. Autom. Lett.* **2020**, *5*, 1656–1663. [CrossRef]
11. Choudhary, S.; Carlone, L.; Nieto-Granda, C.; Rogers, J.; Christensen, H.I.; Dellaert, F. Distributed mapping with privacy and communication constraints: Lightweight algorithms and object-based models. *Int. J. Robot. Res.* **2017**, *36*, 1286–1311. [CrossRef]
12. Zou, D.; Tan, P. Coslam: Collaborative visual slam in dynamic environments. *IEEE Trans. Pattern Anal. Mach. Intell.* **2012**, *35*, 354–366. [CrossRef] [PubMed]
13. Schmuck, P.; Chli, M. CCM-SLAM: Robust and efficient centralized collaborative monocular simultaneous localization and mapping for robotic teams. *J. Field Robot.* **2019**, *36*, 763–781. [CrossRef]
14. Karrer, M.; Schmuck, P.; Chli, M. CVI-SLAM—Collaborative visual-inertial SLAM. *IEEE Robot. Autom. Lett.* **2018**, *3*, 2762–2769. [CrossRef]
15. Liu, J.; Liu, R.; Chen, K.; Zhang, J.; Guo, D. Collaborative Visual Inertial SLAM for Multiple Smart Phones. In Proceedings of the 2021 IEEE International Conference on Robotics and Automation (ICRA), Xi'an, China, 30 May–5 June 2021; pp. 11553–11559.
16. Schmuck, P.; Ziegler, T.; Karrer, M.; Perraudin, J.; Chli, M. COVINS: Visual-Inertial SLAM for Centralized Collaboration. In Proceedings of the 2021 IEEE International Symposium on Mixed and Augmented Reality Adjunct (ISMAR-Adjunct), Bari, Italy, 4–8 October 2021; pp. 171–176.
17. Moafipoor, S.; Bock, L.; Fayman, J.A.; Conroy, E. Vision-Based Collaborative Navigation for UAV-UGV-Dismounted Units in GPS Challenged Environments. In Proceedings of the 33rd International Technical Meeting of the Satellite Division of the Institute of Navigation (ION GNSS+ 2020), Online, 22–25 September 2020; pp. 573–584.
18. Fankhauser, P.; Bloesch, M.; Krüsi, P.; Diethelm, R.; Wermelinger, M.; Schneider, T.; Dymczyk, M.; Hutter, M.; Siegwart, R. Collaborative navigation for flying and walking robots. In Proceedings of the 2016 IEEE/RSJ International Conference on Intelligent Robots and Systems (IROS), Daejeon, Korea, 9–14 October 2016; pp. 2859–2866.
19. Park, J.; Kim, P.; Cho, Y.K.; Kang, J. Framework for automated registration of UAV and UGV point clouds using local features in images. *Autom. Constr.* **2019**, *98*, 175–182. [CrossRef]
20. Qin, H.; Meng, Z.; Meng, W.; Chen, X.; Sun, H.; Lin, F.; Ang, M.H. Autonomous Exploration and Mapping System Using Heterogeneous UAVs and UGVs in GPS-Denied Environments. *IEEE Trans. Veh. Technol.* **2019**, *68*, 1339–1350. [CrossRef]
21. Quigley, M.; Conley, K.; Gerkey, B.; Faust, J.; Foote, T.; Leibs, J.; Ng, A.Y. ROS: An open-source Robot Operating System. In Proceedings of the ICRA Workshop on Open Source Software, Kobe, Japan, 12–17 May 2009; Volume 3, p. 5.
22. Song, Y.; Naji, S.; Kaufmann, E.; Loquercio, A.; Scaramuzza, D. Flightmare: A flexible quadrotor simulator. *arXiv* **2020**, arXiv:2009.00563.
23. Koenig, N.; Howard, A. Design and use paradigms for gazebo, an open-source multi-robot simulator. In Proceedings of the 2004 IEEE/RSJ International Conference on Intelligent Robots and Systems (IROS), (IEEE Cat. No. 04CH37566). Sendai, Japan, 28 September–2 October 2004; Volume 3, pp. 2149–2154.
24. Zhou, Z.; Mertikopoulos, P.; Bambos, N.; Glynn, P.W.; Tomlin, C. Countering feedback delays in multi-agent learning. *Adv. Neural Inf. Process. Syst.* **2017**, *30*.
25. Zhou, Z.; Zhang, W.; Ding, J.; Huang, H.; Stipanović, D.M.; Tomlin, C.J. Cooperative pursuit with Voronoi partitions. *Automatica* **2016**, *72*, 64–72. [CrossRef]

remote sensing

MDPI

Article

An Improved Pedestrian Navigation Method Based on the Combination of Indoor Map Assistance and Adaptive Particle Filter

Zhengchun Wang [1,2], Li Xing [3,*], Zhi Xiong [1,2], Yiming Ding [1,2], Yinshou Sun [1,2] and Chenfa Shi [1,2]

1 College of Automation Engineering, Nanjing University of Aeronautics and Astronautics, Nanjing 210016, China
2 Navigation Research Center, School of Automation Engineering, Nanjing University of Aeronautics and Astronautics, Nanjing 211106, China
3 School of Railway Transportation, Shanghai Institute of Technology, Shanghai 201418, China
* Correspondence: xingli393855907@gmail.com

Abstract: At present, the traditional indoor pedestrian navigation methods mainly include pedestrian dead reckoning (PDR) and zero velocity update (ZUPT), but these methods have the problem of error divergence during long time navigation. To solve this problem, under the condition of not relying on the active sensing information, combined with the characteristics of particles "not going through the wall" in the indoor map building structure, an improved adaptive particle filter (PF) based on the particle "not going through the wall" method is proposed for pedestrian navigation in this paper. This method can restrain the error divergence of the navigation system for a long time. Compared to the traditional pedestrian navigation method, based on the combination of indoor map assistance (MA) and particle filter, a global search method based on indoor MA is used to solve the indoor positioning problem under the condition of the unknown initial position and heading. In order to solve the problem of low operation efficiency caused by the large number of particles in PF, a calculation method of adaptively adjusting the number of particles in the process of particle resampling is proposed. The results of the simulation data and actual test data show that the proposed indoor integrated positioning method can effectively suppress the error divergence problem of the navigation system. Under the condition that the total distance is more than 415.44 m in the indoor environment of about 2600 m^2, the average error and the maximum error of the position are less than two meters relative to the reference point.

Keywords: map assistance; particle filter; global search algorithm; pedestrian navigation

Citation: Wang, Z.; Xing, L.; Xiong, Z.; Ding, Y.; Sun, Y.; Shi, C. An Improved Pedestrian Navigation Method Based on The Combination of Indoor Map Assistance and Adaptive Particle Filter. *Remote Sens.* **2022**, *14*, 6282. https://doi.org/10.3390/rs14246282

Academic Editors: Yuwei Chen, Changhui Jiang, Qian Meng, Bing Xu, Wang Gao, Panlong Wu, Lianwu Guan and Zeyu Li

Received: 26 September 2022
Accepted: 8 December 2022
Published: 11 December 2022

1. Introduction

With the progress of urbanization, the indoor environment has become an important place for human production and life. Indoor pedestrian navigation technology has been widely considered and studied by scholars in disaster relief and rescue, medical search and rescue, public security, anti-terrorism and other fields. Providing accurate navigation and positioning capabilities for pedestrians in indoor working environments is the basis for achieving indoor rescue work. In indoor working environments, the signals of global positioning system (GPS) [1], Beidou and other global navigation satellite systems (GNSS) [2] are seriously blocked, which makes it difficult to play the role of normal navigation and positioning and the incapacity to provide accurate navigation and positioning function for pedestrians. Therefore, it is necessary to carry out research on the pedestrian navigation method in the indoor satellite failure environment.

The current indoor pedestrian navigation technology mainly includes active navigation and passive navigation. Active navigation means that navigation and positioning must be carried out with the help of sensors other than itself, including ultra wide band

(UWB) [3], wireless fidelity (Wi-Fi) [4], bluetooth (BT) [5], ZigBee [6], radio frequency identification (RFID) [7], near field communication (NFC) [8] and other methods. While the positioning error of the active navigation algorithm does not accumulate over time, it is greatly affected by the indoor environment, obstacles, multipath propagation [9,10] and other environmental factors. It is necessary to arrange the source base station in advance and build a fingerprint database. The cost of construction and maintenance is high. In addition, the indoor rescue site is often accompanied by problems such as the unavailability of beacons caused by power system interruption. The relevant navigation and positioning technology is difficult to meet the availability requirements of indoor rescue positioning. Passive navigation refers to a navigation and positioning method that only relies on its own sensors, without relying on external sensor information sources. It mainly includes the methods of simultaneous localization and mapping (SLAM) based on the laser radar sensor [11] and visual sensor [12] (monocular camera, binocular camera, depth camera), and the methods based on inertial measurement unit (IMU). For the indoor rescue navigation and positioning system with high real-time requirements, there are shortcomings, such as the laser radar remaining unchanged, and the visual sensor may be affected by the indoor environment. As it is inconvenient for pedestrians to carry lidars, and visual sensors may be affected by indoor environment, these sensors cannot meet the needs of pedestrian navigation and positioning system for indoor rescue in actual use. With the development of technology, the micro electro–mechanical system (MEMS) [13] has been continuously improved and developed. As a kind of autonomous navigation and positioning equipment, a wearable inertial sensor based on MEMS technology has been widely studied in indoor pedestrian navigation. It only needs to fix the IMU on the human body and calculate pedestrian navigation parameters by collecting IMU data to realize autonomous navigation and positioning.

The pedestrian navigation algorithms based on IMU are mainly divided into two categories: The PDR algorithm and ZUPT algorithm. The pedestrian navigation method based on PDR estimates the step length, step number, heading and other parameters of the pedestrian in the walking process by collecting the acceleration, angular velocity and other data of the pedestrian, and calculates the pedestrian motion trace. In 2017, Dina [14] proposed a method to estimate the step length through the information of leg and foot inertial sensors of two navigation systems. In 2018, Xu [15] studied the PDR navigation algorithm based on handheld mobile phones. In order to improve the step length estimation accuracy of the algorithm for different users, she proposed a step length detection method based on state transition and a step length estimation method based on neural networks. In 2020, Ding [16] proposed a PDR navigation algorithm based on the relationship between waist inertial data and step length. Based on the ZUPT algorithm, the inertial sensor is installed on the foot. According to the algorithm, the velocity of the foot is zero in theory during the period of time when the foot contacts the ground during periodic movement, and the velocity during this period is used as the observation quantity to periodically correct the position and velocity of the human. In 2016, Ruppelt [17] proposed a navigation and positioning technology about ZUPT detection based on the finite state machine. It was used to analyze the gait cycle of human foot mounted IMU, which could detect the zero velocity interval more accurately. In 2017, Hsu [18] proposed a sensor fusion technology based on a two-stage quaternion extended Kalman filter for the inertial sensor cumulative error, and the error between the starting point and the end point was 2.01%. In 2018, Suresh [19] proposed the method of combining the ZUPT and the high pass filter. He applied the high pass filter to the complementary filtering, reducing the error drift of the angular velocity. In 2021, Abdallah [20] proposed a foot-mounted and synthetic aperture indoor navigation method based on inertial/ZUPT/depth neural network, which reduced the accumulated error of inertial navigation system through the integrated navigation algorithm based on ZUPT. From the working principle, the pedestrian navigation system (PNS), based only on inertial sensors, will diverge after a long time of operation.

The error of PNS relying only on inertial sensors will diverge over time. To solve this problem, scholars have studied a variety of methods to correct the navigation error. In 2016, Ilyas [21] studied the indoor geomagnetic assisted pedestrian navigation. A large amount of information interferes with the magnetometer, resulting in magnetic information distortion. In this regard, a magnetic anomaly detection method is proposed to compensate the abnormal data. In 2018, Song [22] proposed a two-stage Kalman filter, in which a magnetic sensor is installed at the waist and an inertial sensor is installed at the foot. Compared to the traditional pedestrian navigation method based on ZUPT, the error is reduced by 30%. In 2016, Diez [23] proposed an improved heuristic drift elimination algorithm (iHDE) to install the inertial sensor on the wrist. Compared to the algorithm, iHDE reduces the error by 95%. In 2018, Muhammad [24] used the indoor corridor for heading correction and proposed an HDE algorithm based on waist heading. The author divided the 360-degree heading into 16 equal sectors. When pedestrians moved along the orthogonal corridor direction or the main heading, the algorithm corrected the heading. If the motion trace is a curve or not moving along the main heading, no course correction will be made. In 2021, Kim [25] proposed a topological map construction method based on tge architectural plan and sensors to solve the problem that it took a lot of time to create indoor maps in real time. This method can provide a safe path, and the indoor plan can be updated more easily in the future, even if the internal structure of the building changes. Since 2009, the German Aerospace Center has studied the navigation method based on Foot SLAM [26,27], which is only based on inertial sensors and can maintain the navigation accuracy for a long time. References [28,29] combined indoor map and PF to modify the pedestrian navigation results obtained by the inertial sensor solution. This method greatly improves the navigation accuracy, but the computational efficiency is low. As the indoor geomagnetic interference is large, and the auxiliary navigation effect is not good, the HDE method needs to obtain indoor environmental constraint features in advance. The Foot SLAM method needs to form a closed loop of motion trajectory, which has great limitations in practical applications. For the MA method, the indoor architectural plan is relatively easy to obtain, and navigation and positioning are realized by combining the indoor map with PF.

For the PNS relying only on inertial sensors, in order to effectively solve the needs of pedestrian autonomous navigation under special tasks such as indoor rescue, this paper proposes an improved pedestrian navigation positioning method based on the combination of indoor MA and adaptive PF (IMAPF). In order to solve the problem of high precision localization when pedestrians enter an unfamiliar environment with unknown initial position and heading, a global search method based on MA is proposed. Aiming at the problem that a large number of particle operations are required under the unknown initial position and heading, which leads to low computational efficiency, an adaptive particle number calculation method is proposed. It solves the problem of high-precision navigation and positioning of indoor pedestrians for a long time and the positioning problem under the unknown initial position and heading, and improves the computational efficiency, accuracy and reliability of the indoor pedestrian navigation system.

The structure of this paper is as follows: Section 2 describes the algorithm in detail. In Section 3, the proposed algorithm is verified by the simulation and experiment. Section 4 discusses the results of the experimental activities proposed. Finally, Section 5 concludes the paper.

2. Materials and Methods

2.1. Proposed System Scheme

In order to solve the error divergence problem of pedestrian navigation and positioning system for a long time, on the premise of no other external sensors, a pedestrian navigation and positioning method based on the combination of indoor IMAPF is proposed to constrain and correct the position and heading change information calculated by the navigation system through the constraint relationship between the indoor map information and the position and heading calculated by the inertial navigation. The frame of the designed integrated navigation and positioning method is shown in Figure 1.

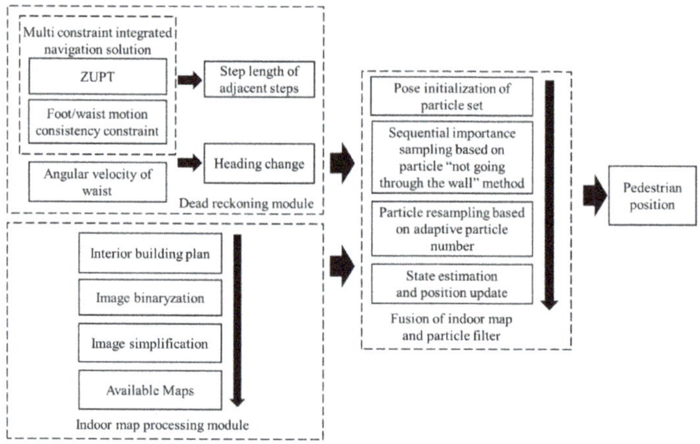

Figure 1. General scheme of the indoor pedestrian navigation and location method based on the IMAPF method.

The algorithm proposed in this paper is as follows: Firstly, through the dead reckoning module [30], a PNS based on foot-mounted ZUPT is established, and the navigation system is corrected by taking the difference between the heading change of foot and heading change of waist in one step as the observation, taking the physical distance between two adjacent zero velocity intervals as the step length, and then the heading change is obtained by integrating the waist angular velocity with time. Then, through the indoor map processing module, the indoor architectural plan is binarized, and then the image is simplified to obtain the available map. Establish a PF model, input the step length, heading change and indoor map obtained from the above two modules into the PF. First, initialize the position and heading of the particle set respectively according to the known or unknown initial position and heading of pedestrian navigation, detect whether particles "going through the wall", delete "illegal particles" or retain "legal particles", and calculate the particle normalization weight according to the sequential importance sampling. Secondly, the effective value of the particles is calculated to determine whether resampling is required. When resampling is required, the adaptive particle number is calculated to update the current state estimation value. Then the pedestrian position at the current time is solved, and the motion trace obtained from the solution is projected into the indoor map.

2.2. *Pedestrian Navigation Method Based on Indoor MA and PF*

2.2.1. Theoretical Model of Algorithm

(1) Pedestrian navigation model based on dead reckoning

Figure 2 is the schematic diagram of the pedestrian motion trace update.
The state transition equation of particle filter of PDR can be obtained as

$$\begin{cases} \psi_k^i = \psi_{k-1}^i + \Delta\psi_k^i \\ u_k^i = u_{k-1}^i + l_k^i \sin\psi_k^i \\ v_k^i = v_{k-1}^i + l_k^i \cos\psi_k^i \end{cases} \tag{1}$$

In Equation (1), the system state quantity is the pedestrian position coordinate $\left(u_k^i, v_k^i\right)$ in the two-dimensional plane. The control quantity of the system is $\Delta\psi_k^i$ and l_k^i, where $\Delta\psi_k^i$ is the heading change, and l_k^i is the step length. Subscript k is step k-th step, and superscript i is the i-th particle. The particle contains the possible two-dimensional position and heading at the current time. The position of the particle is corrected by constantly adjusting the particle weight and the two-dimensional position.

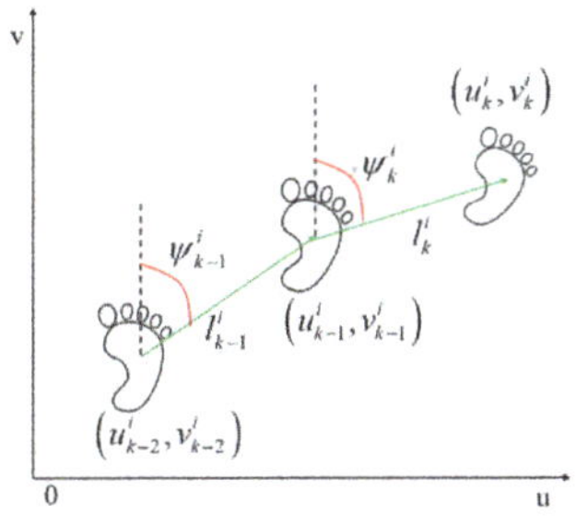

Figure 2. Schematic diagram of the pedestrian motion trace update.

The control quantity equation of the particle filter based on PDR is:

$$\begin{cases} \hat{l}_k = l_k + \gamma_{lk} \\ \Delta\hat{\psi}_k = \Delta\psi_k + \gamma_{\psi k} \end{cases} \tag{2}$$

In Equation (2), γ_{lk} and $\gamma_{\psi k}$ represent step noise and heading change noise, respectively, which all obey the Gaussian distribution.

In this paper, the step length and heading change are calculated with the method of multiple constraints for indoor navigation (MCIN) [30] based on multiple constraints, as the input of the IMAPF method proposed in this paper. This method consists of the following two parts: (a) A pedestrian navigation model based on ZUPT is constructed; (b) based on the feature that the difference between heading change of foot and heading change of waist is small in one step motion, a navigation correction model based on the consistency constraint of heading change of waist during human motion is constructed as a virtual observation. This method corrects the pedestrian navigation error in a period of time, but the error will increase when moving for a long time. As the motion state of adjacent steps does not change much in the motion process, the distance between the adjacent zero velocity intervals can be used as the step length control value through the MCIN method. In addition, since the waist IMU motion is relatively stable, the heading change obtained from the waist angular velocity integral is taken as the control value.

The step length is calculated as follows:

$$l = \sqrt{(g_k - g_{k-1})^2 + (h_k - h_{k-1})^2} \tag{3}$$

In Equation (3), (g_k, h_k) and (g_{k-1}, h_{k-1}) are the position coordinates of the current step and the previous step calculated by the MCIN method, respectively.

(2) Observation model based on particle "not going through the wall" method

In the navigation and positioning method based on the IMAPF, the "not going through the wall" method is to judge whether the current particle position is valid according to the position of the particle at the previous time after one step of movement.

To determine whether a particle is a "valid particle", it is based on inaccessible areas in the map (such as patios, elevators, walls, etc.) or impossible paths in reality (such as going from one room to another without going through a door). Figure 3 is a schematic diagram of particle motion. The particles representing the human at the last moment is marked as a white circle, and he is currently walking in the corridor.

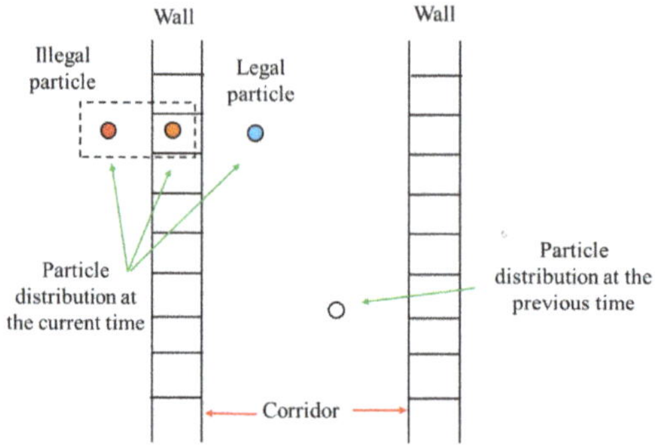

Figure 3. Schematic diagram of the update of the particle position.

According to the state equation, the particle coordinates at the current time can be obtained, including the following possible positions: Corridor, inside the wall, going through the wall into another room, etc. It is compared with the indoor accessible area (the accessible area is obtained through the architectural plan). If a particle position coordinate P_k^i belongs to the accessible region P_e of the blue circle, it is regarded as "legal particle", and the particle weight w_k^i will not be changed. When a particle turns into an orange circle through state transition and enters the inaccessible area P_{m1} such as the wall, it can be judged that this type of particle is an "illegal particle", and then the weight value of this type of particle is set to 0. If a particle turns into a red circle and directly enters another room after the state transfer, there is no connectivity between the new particle and the particle at the previous moment, which belongs to an inaccessible area P_{m2}. The particle weight value is set to 0, and the "illegal particle" is deleted. The equation is as follows:

$$w_k^i = \begin{cases} 1 & P_k^i \subset P_e \\ 0 & P_k^i \subset P_{m1} \\ 0 & P_k^i \subset P_{m2} \end{cases} \tag{4}$$

The thickness of the wall is determined by the pixel length of the wall after the architectural plan is converted to the binary map. Generally, the physical length represented by one pixel is less than the thickness of the wall, that is, the wall is represented by at least several consecutive pixels. Even when the wall has only one pixel, it is possible to calculate that the particle is inside or through the wall, and then proceed to the next steps. Therefore, the accuracy of the method is independent of the thickness of the wall.

2.2.2. Algorithm Process Design

(1) Optimization of initial position and heading of particle set

The core idea of PF is that it is composed of a finite number of random samples (particles) with weight. Each particle represents the estimation of the current state. The integral operation of the posterior probability density distribution $p(x_k|y_{1:k-1})$ is approximately expressed as the sum operation of the finite samples. The posterior probability distribution of the system is expressed by the density of the particle distribution. The value of the particle weight w represents the possibility of the state, which is used to represent the probability distribution of the state variable to approximate the true probability distribution of the system. Select n particles $\{x_k^i, w_k^i\}$ $(i = 1, 2, \ldots, n)$, with

$$p(x_{0:k}|y_{1:k}) \approx \sum_{i=0}^{N} \tilde{w}_k^i \delta\left(x_{0:k} - x_{0:k}^i\right) \tag{5}$$

$$\sum_{i=0}^{N} \widetilde{w}_k^i = 1 \tag{6}$$

In Equations (5) and (6), N represents the number of particles, i represents the number of each particle, δ represents the Dirac function, $x_{0:k}$ represents the historical system state in time interval $0 \sim k$, the observation quantity of the system is marked as y, representing the probability distribution, $y_{1:k}$ represents the historical observation in time interval $1 \sim k$, and $\widetilde{w}_k^i$ represents the normalized importance weight of the ith particle at moment k.

At the initial moment, set $k = 0$, and randomly generate N sample particle groups $\{x_0^i, i = 1, 2, \ldots, N\}$ according to a priori probability $p(x_0)$. The weights of all particles are $w_0^i = 1/N$, and each particle sampled is recorded as $\{x_k^i, 1/N\}$.

The current research mainly focuses on the of navigation and positioning methods for indoor pedestrians when the initial positions and heading are known in a special working environment. The initial particles are added with Gaussian white noise, as shown in Equation (2), and the initial particles obey

$$p(x_0) \sim N(\mu, \sigma) \tag{7}$$

In Equation (7), $p(x_0)$ is the prior distribution of the initial position, $N(\mu, \sigma)$ is the Gaussian distribution, μ is the expectation, and σ is the mean square error.

However, in the actual environment, due to the special emergency of the work site and the inability to provide the absolute position information in a short time, when pedestrians must enter the work environment, this paper proposes a method based on global search to solve the pedestrian's position and heading information. The initial particles are distributed in the entire indoor map in a uniform way, and the following equation holds:

$$p(x_0) \sim U(a, b) \tag{8}$$

In Equation (8), $p(x_0)$ is the prior distribution of the initial position, $U(a, b)$ is the uniform distribution, and a and b are the minimum and maximum values of the pixel coordinates in the picture, respectively.

(2) One-step prediction of particle states

According to the state equation, state x_k at the current moment is estimated by the prior probability density of state x_{k-1} at the previous moment. The prior distribution $p(x_0)$ of the initial state is known.

$$\begin{aligned} p(x_k|y_{1:k-1}) &= \int p(x_k, x_{k-1}|y_{1:k-1})dx_{k-1} \\ &= \int p(x_k|x_{k-1}, y_{1:k-1})p(x_{k-1}|y_{1:k-1})dx_{k-1} \\ &= \int p(x_k|x_{k-1})p(x_{k-1}|y_{1:k-1})dx_{k-1} \end{aligned} \tag{9}$$

(3) Particle weight updating based on prior map information

Using the observation y_k at the current moment, modify $p(x_k|y_{1:k-1})$ to obtain a posterior probability density $p(x_k|y_{1:k})$.

$$\begin{aligned} p(x_k|y_{1:k}) &= \frac{p(y_k|x_k, y_{1:k-1})p(x_k|y_{1:k-1})}{p(y_k|y_{1:k-1})} \\ &= \frac{p(y_k|x_k)p(x_k|y_{1:k-1})}{p(y_k|y_{1:k-1})} \\ &= \frac{p(y_k|x_k)p(x_k|y_{1:k-1})}{\int p(y_k|x_k)p(x_k|y_{1:k-1})dx_k} \end{aligned} \tag{10}$$

At moment $k + 1$, y_{k+1} is updated, and the importance weight value of the whole state sequence needs to be recalculated, so the amount of calculation increases greatly over time. This problem is solved through Sequential Importance Sampling (SIS). A group of known random samples with weights is used to represent the posterior probability density, and the state estimation value is calculated based on the known random samples and weights.

The prior probability is selected as the importance density function, as follows

$$q(x_k^i \middle| x_{k-1}^i, y_k) = p(x_k^i \middle| x_{k-1}^i) \tag{11}$$

The particles sampled according to this function are as follows:

$$x_k^i = q(x_k | x_{k-1}, y_{1:k}) \tag{12}$$

The normalized importance weight is

$$w_k^i = w_{k-1}^i \frac{p(y_k | x_k^i) p\left(x_k^i \middle| x_{k-1}^i\right)}{q\left(x_k^i \middle| x_{k-1}^i, y_k\right)} \tag{13}$$

Substituting Equation (11) into Equation (13) has

$$w_k^i = w_{k-1}^i p\left(y_k \middle| x_k^i\right) \tag{14}$$

In Equation (14), $p(y_k | x_k^i)$ represents the detection result of the particle "not going through the wall" method.

The normalized weight calculation equation is

$$\widetilde{w}_k^i = \frac{w_{k-1}^i}{\sum_{j=1}^{N} w_k^i} \tag{15}$$

(4) Particle resampling based on adaptive particle number

In order to reduce the amount of calculation and avoid unnecessary resampling, judge whether resampling is required according to the effective particle number N_{eff}. The equation for calculating the effective value of particles is

$$N_{eff} = 1/\sum_{i=1}^{N} \left(w_k^i\right)^2 \tag{16}$$

The smaller the number of effective particles is, the more serious the particle weight degradation is, and the there is more need to resample. Set the resampling threshold N_{th}, when $N_{eff} \leq N_{th}$, particle set $\{x_k^i, w_k^i\}$ needs resampling.

Theoretically, the larger the number of particles, the more accurate the results of PF. However, as the number of particles increases, the calculation time increases exponentially, and the improvement of navigation and positioning accuracy is not obvious, especially for the navigation and positioning results under the condition of the known initial position and heading. For position under the condition of unknown initial position and heading, a large number of particles are required at the beginning. Using a large number of particles after finding the location of a pedestrian in the map will lead to computational inefficiency. Therefore, this paper proposes a method to adaptively adjust the particle number during resampling, which can greatly improve the computational efficiency of PF and ensure the high-precision navigation and positioning results of pedestrian navigation system. The calculation equation of the adaptive particle number is

$$pNum = floor(e \cdot c_1) + c_2 \tag{17}$$

In Equation (17), $floor$ is the integral function, and e is the sum of the main diagonal elements of the covariance matrix of the position coordinates of all particles at the current moment. c_1 and c_2 are empirical constants. c_1 is the proportional coefficient, and the value of c_2 ensures that the particle filter function can be achieved when $pNum$ is at the minimum value. In this paper, c_1 and c_2 are 1000 and 2000, respectively.

The updated particle set is recorded as $\{x_k^i, 1/pNum\}$.

2.2.3. Algorithm Flow

To sum up, the flow chart of pedestrian navigation and positioning method of IMAPF is shown in Figure 4.

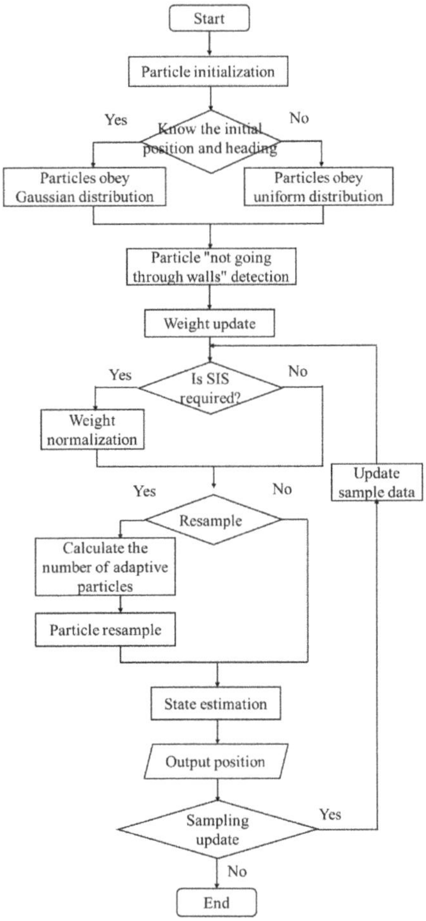

Figure 4. Flow chart of the improved pedestrian navigation and positioning method based on IMAPF.

The algorithm flow is as follows:

Step 1: Initialization of position and heading of particle set: For the known initial position and heading, the initialization adopts Equation (7) to satisfy the Gaussian distribution. In the case of the unknown initial position and heading, the initialization adopts Equation (8) to meet the uniform distribution.

Step 2: Sequential importance sampling based on particle "not going through the wall" method: Update the particle weight according to the particle "not going through the wall" method, and then calculate the normalized weight value through the Equations (11)–(15).

Step 3: Particle resampling based on adaptive particle number: The particle number is adaptively calculated by Equations (16) and (17) and resample.

Step 4: State estimation and location update: The state $\hat{x}_k$ at the current moment is estimated by the updated adaptive number of particle sets and weights as

$$\hat{x}_k = \sum_{i=1}^{N} \widetilde{w}_k^i x_k^i \tag{18}$$

The position P_k at the current moment is

$$P_k = \sum_{i=1}^{N} \widetilde{w}_k^i \left(x_k^i - \hat{x}_k\right)\left(x_k^i - \hat{x}_k\right)^T \tag{19}$$

At moment k, when a particle is in an inaccessible area, the particle is resampled. Copy the navigation parameters (step length and heading at all times of 0~k) of the valid particle to the particle. The weighted average method is used to calculate the position of the pedestrian at the moment $k, k-1, \ldots \ldots, 0$ in turn.

Project the updated position onto the indoor map.

Step 5: $k = k + 1$, go to Step 2.

3. Results

3.1. Verification of Simulation

3.1.1. Conditions of Simulation

In order to verify the effectiveness of the improved pedestrian navigation and location method based on the indoor map assistance and particle filter, a series of simulation experiments were carried out. The simulation data is processed on the desktop computer, and the computer platform parameters are shown in Table 1.

Table 1. Computer platform parameters.

Characteristic	Parameter
Computer operating system	Windows10
CPU	Intel(R) Core(TM) i7-8700, Dominant frequency 3.20 GHz
Memory	32 GB
Software	Matlab2020

The simulation environment is based on the architectural plan of the fifth floor of no. 1 Building and no. 2 Building of the College of Automation Engineering, as shown in Figure 5. In Figure 5a, the red line indicates the corridor path. The four black dots (①②③④) represent the reference point of the relative position.

The simulation movement trace is shown in Figure 6a. The blue point at the lower left corner is the starting point/end point. A complete closed loop path ①→②→③→④→① has a distance of 207.72 m. The simulation moves two circles in a counterclockwise direction. The parameters are set as follows: The mean square error of the step noise is 0.1 m, and the mean square error of heading change noise is 1°. The simulation data of pedestrian position and course change obtained are also saved. The four reference coordinates in the trace are shown in Figure 6b. Due to the process of entering the room, the total distance cannot be measured accurately, and the total distance exceeds 415.44 m.

Define the positioning error as

$$Err = \sqrt{\Delta x^2 + \Delta y^2} \tag{20}$$

In Equation (20), Err is the Euclidean Distance between the reference point and the measuring point. Δx is the difference between the abscissa calculated by the proposed method and the abscissa calculated by the standard path; Δy is the difference between the ordinate calculated by the proposed method and the ordinate calculated by the standard path.

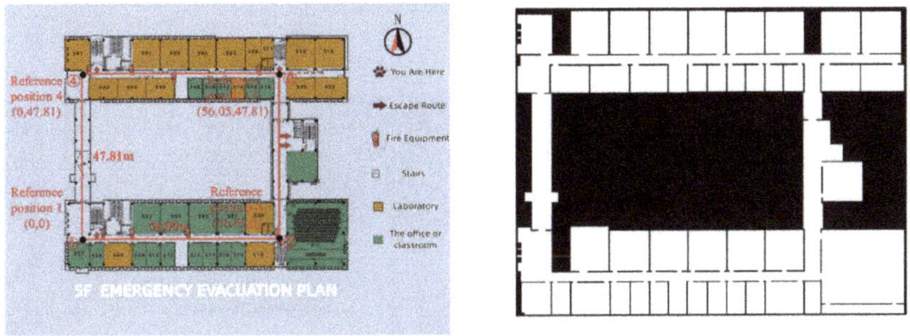

(a) (b)

Figure 5. Digital map on the fifth floor. (**a**) Architectural plan. (**b**) Available binary maps.

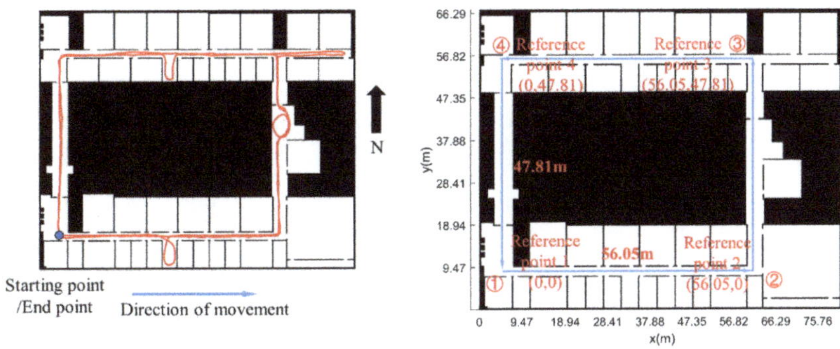

(a) (b)

Figure 6. Simulation trace and coordinates of reference point. (**a**) Simulation trace. (**b**) Schematic diagram of the coordinates of the reference point.

3.1.2. Analysis of Simulation Results

This section first analyzes the error comparison results between the IMAPF method and the PDR algorithm when the initial position and heading are known, then analyzes the positioning effect of the proposed method and PDR algorithm when the initial position and heading are unknown, and finally analyzes the navigation error and calculation efficiency when the initial position and heading are unknown and the adaptive particle number and fixed particle number.

(1) The initial position and heading of pedestrian are known

The distribution of the sampled particles is shown in Figure 7. The noise of step length and heading change of particles conform to Gaussian distribution. At the beginning, the particles are distributed within a certain range, with red representing "illegal particles" and blue representing passable "legal particles".

Figure 8 shows the navigation trace comparison and its positioning error CDF curve with an known initial position and heading. In Figure 8a, the red line represents the ideal trace without noise, the green line represents the trace with noise calculated by PDR, and the blue line represents the trace diagram of the IMAPF method proposed in this paper. It can be seen that the method proposed in this paper can well correct the position of the navigation system. Figure 8b shows the CDF curve, which shows that the algorithm proposed in this paper is better than the PDR method in error correction. Figure 8c shows the error of each step.

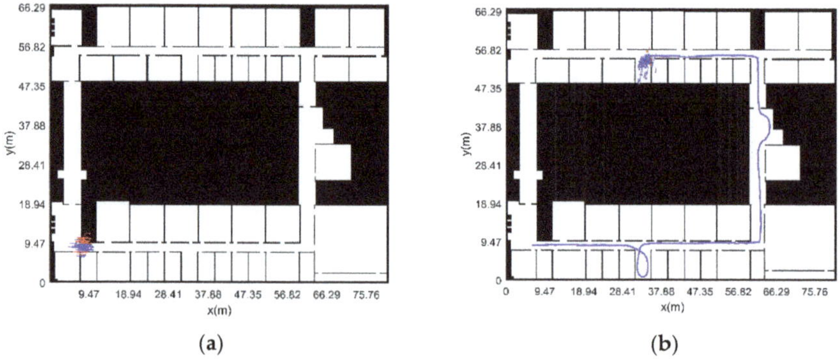

Figure 7. Sampling particle distribution and motion trace with known position and heading. (**a**) Particle distribution at initial moment. (**b**) Particle distribution in motion.

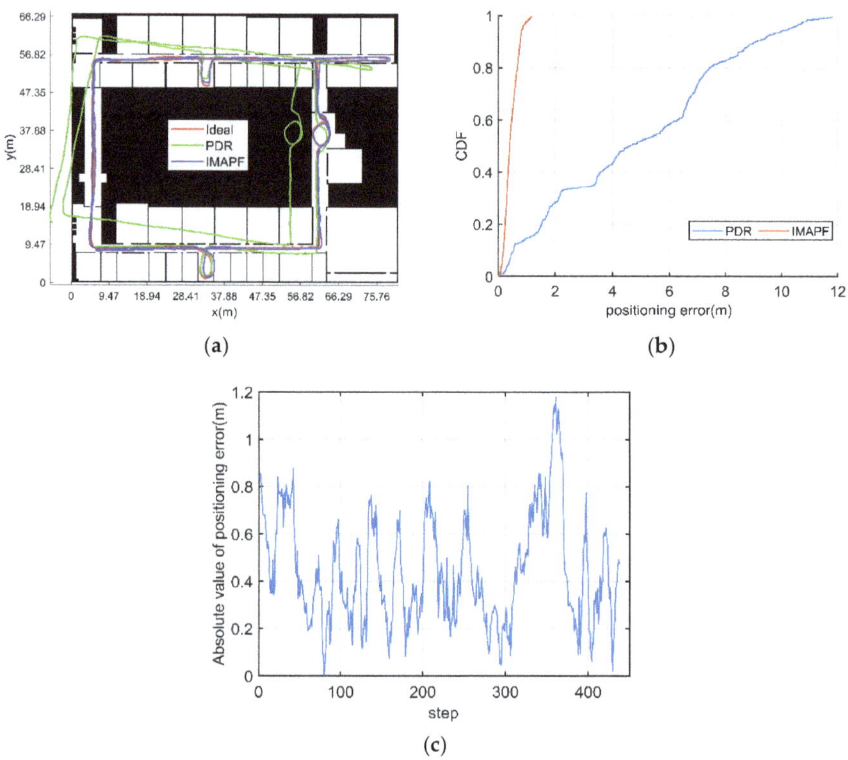

Figure 8. Navigation trace comparison and positioning error CDF curve with known initial position and heading. (**a**) Comparison diagram of positioning trace. (**b**) CDF curve of positioning error. (**c**) Absolute position error of each step.

Table 2 shows the error comparison under the condition that the initial position and heading are known. It can be seen from the data in the table that the error of PDR algorithm increases with the increase in motion time. The method proposed in this paper can effectively restrain error divergence.

Table 2. Error comparison under the condition that the initial position and heading are known.

Navigation Method	Mean Error (m)	Maximum Error (m)
PDR	4.78	11.81
IMAPF	0.44	1.18

(2) The initial position and heading of pedestrian are unknown (adaptive particle number)

The distribution of sampling particles is shown in Figure 9. Using the global search method, the particles are evenly distributed in the whole map at the beginning, with red indicating "illegal particles" and blue indicating passable "legal particles". With continuous movement, the approximate position of pedestrians can be found at the 74th step, and then the positioning coordinates will be modified for pedestrians, finally providing navigation and positioning functions for pedestrians. Due to the structural features, such as rooms and corridors, the path complexity can be increased by increasing the number of room entry and exit to achieve global search as soon as possible.

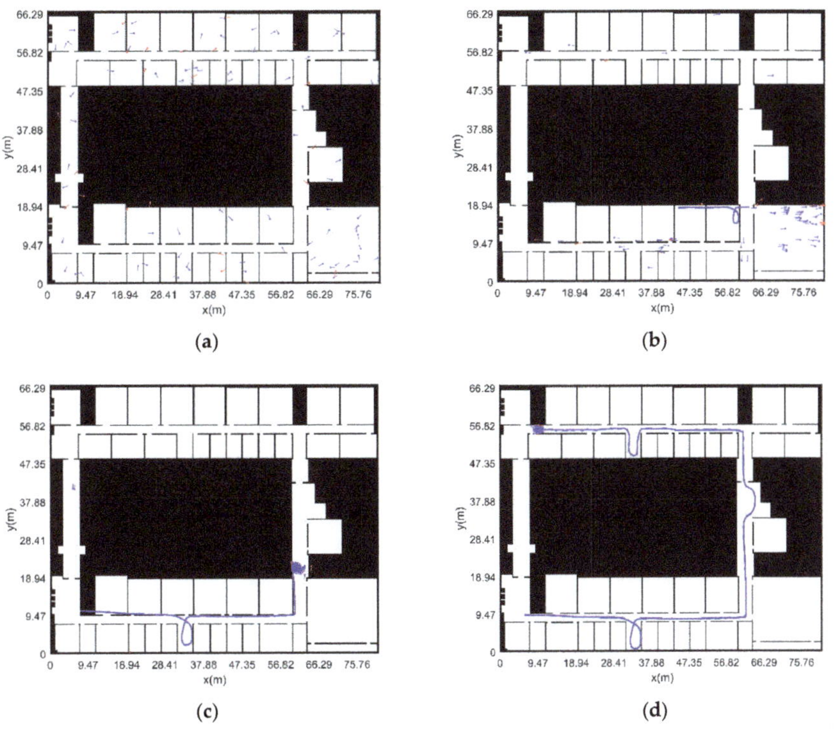

Figure 9. Sampling particle distribution and motion trace with unknown initial position and heading. (**a**) Particle distribution at initial moment; (**b**) particle distribution and motion trace of the 44th step in the motion process; (**c**) particle distribution and motion trace of the 74th step in the motion process; and (**d**) particle distribution and motion trace of the 162th step in the motion process.

Figure 10 shows the navigation trace comparison and its positioning error CDF curve under the condition of unknown initial position and heading, while the PDR algorithm is not applicable to this condition. In Figure 10a, the red line represents the real motion trace without noise, and the blue line represents the trace diagram of the IMAPF method proposed in this paper. It can be seen that the method proposed in this paper can provide accurate positioning function for pedestrians when the initial position and heading are unknown. Figure 10b shows the CDF curve, which clearly shows that the algorithm proposed in this paper can provide accurate positioning function for pedestrians under the condition of unknown initial positon and heading. Figure 10c shows the error of each step. The average error of IMAPF algorithm is 0.36 m, and the maximum error is 0.84 m.

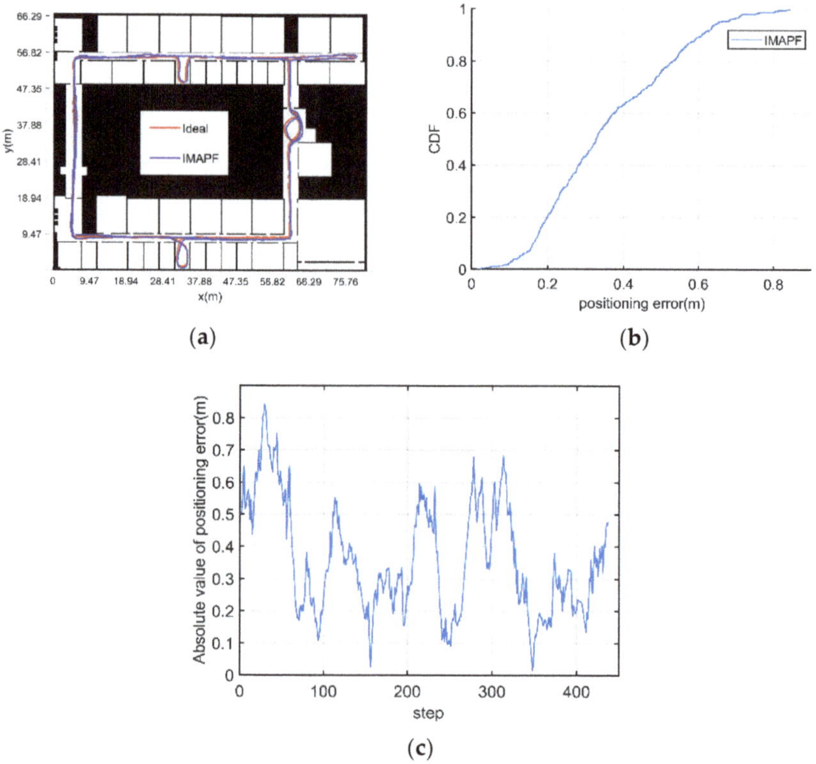

Figure 10. Navigation trace comparison and positioning error CDF curve with unknown initial position and heading. (**a**) Comparison diagram of positioning trace; (**b**) CDF curve of positioning error; and (**c**) absolute position error of each step.

It can be seen that the IMAPF algorithm studied in this paper performs better than the PDR algorithm for pedestrian navigation, whether the initial position and heading are known or not.

(3) The initial position and heading of pedestrian are unknown (fixed particle number)

In order to compare the calculation efficiency and error value between the fixed particle number and the adaptive particle number, under the condition of unknown initial position and heading of pedestrians, the fixed particle number of 2000, 10,000, 50,000 and 100,000 are compared, respectively. Table 3 shows statistics of positioning errors with different particle numbers.

Figure 11 shows the pedestrian motion trajectory obtained by solving with different particle numbers and compares the navigation results of four different particle numbers. The navigation and positioning error value of the adaptive particle number method proposed in this paper is smaller than that of the fixed particle number of 100,000. However, the calculation time of the adaptive particle number method is reduced by about 20 times lower compared to the calculation time of 100,000 fixed particles.

Table 3. Statistics of positioning errors with different particle numbers under simulation conditions.

Particle Number	Mean Error (m)	Maximum Error (m)	Calculation Time (s)
2 thousand fixed particles	10.34	15.99	16.64
10 thousand fixed particles	7.74	12.07	103.05
50 thousand fixed particles	0.75	3.01	909.04
100 thousand fixed particles	0.50	1.91	2624.36
adaptive particle numbers	0.36	0.84	116.14

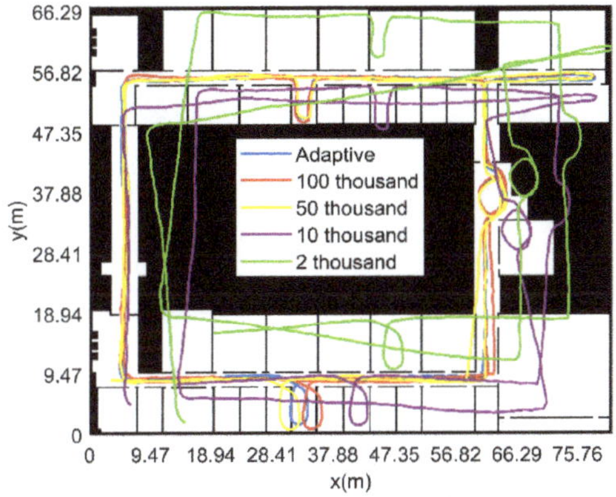

Figure 11. Pedestrian motion trace calculated with different particle numbers under simulation conditions.

3.2. Experiment and Verification

3.2.1. Experimental Conditions

In order to further verify the practicability and effectiveness of the proposed method, this paper verifies its effect through experiments. In the experiment, five IMUs are used to collect pedestrian movement data, of which four IMUs use FSS-IMU6132 independently designed by Forsense Technology Company to collect accelerometer and gyroscope data with a sampling frequency of 100 Hz; One MTI-G-710 inertial device of XSENS Company is used to collect air pressure data with a sampling frequency of 50 hz. The above five sensors are all connected to Raspberry Pie 4B through data lines, and the mobile phone app controls Raspberry Pie to collect data through Bluetooth communication. The sensor installation method is shown in Figure 12.

The performance parameters of FSS-IMU6132 are shown in Table 4.

Table 4. Performance parameters of FSS-IMU6132.

/	Sensor Range	Bias Stability
Accelerometer	±6 g	10 μg
Gyroscope	±500 deg/s	1.0 deg/h

The performance parameters of MTI-G-710 are shown in Table 5 below.

Table 5. Performance parameter of XSENS MTI-G-710.

/	Sensor Range	Total Root Mean Square Noise
Barometer	300–1100 mBar	3.6 Pa

The computer performance parameters are shown in Table 1.

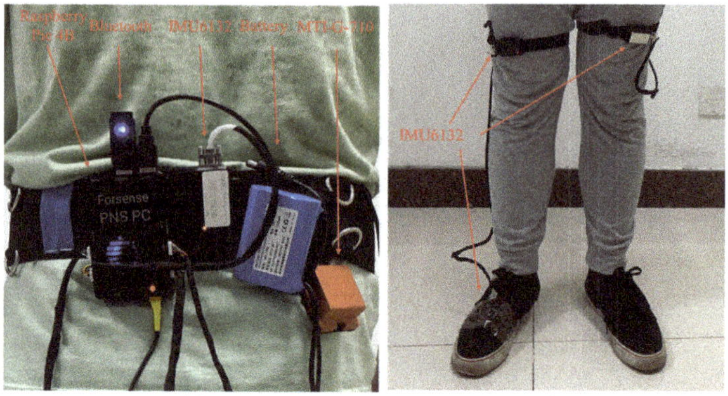

(a) **(b)**

Figure 12. Schematic diagram of sensor installation. (**a**) Installation diagram of sensors on waist. (**b**) Installation diagram of sensors on leg and foot.

3.2.2. Experimental Verification Analysis

(1) MCIN method

The experimental site is the fifth floor of No.1 Building and No.2 Building of the College of Automation Engineering. The experimental trace and indoor experimental scene are shown in Figure 13. The experimenter started from position ① (0, 0) and walked counterclockwise to collect two rounds of data. The experimenter experiences the reference point in the order of ①②③④⑤⑥⑦⑧①③④⑤⑨⑦⑧①. Additionally, the trace included multiple "entering the house" behaviors. The actual movement time in the experiment is 451.74 s.

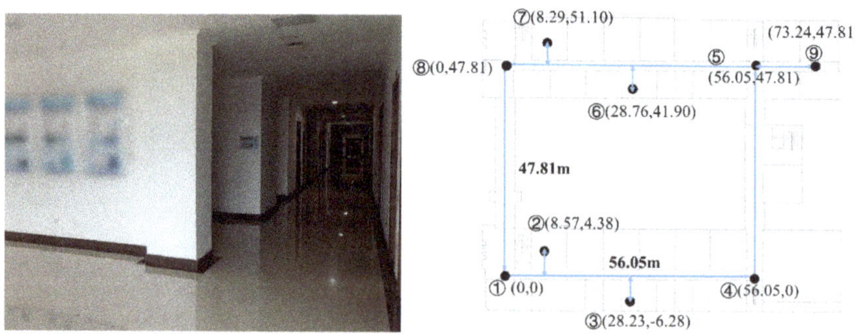

(a) **(b)**

Figure 13. Indoor experiment scene and experiment trace. (**a**) Indoor experiment scene. (**b**) Reference points.

The trace solved by the MCIN method and the positioning error of the reference point experienced 15 times are shown in Figure 14. The mean error of the reference point position is 1.98 m, and the maximum error is 4.16 m. For Figure 14b, if the MCIN method based on inertial navigation only is adopted, the navigation error will gradually diverge. Due to the uncertainty of course error divergence, the navigation solution results of some paths in a closed path will show smaller errors.

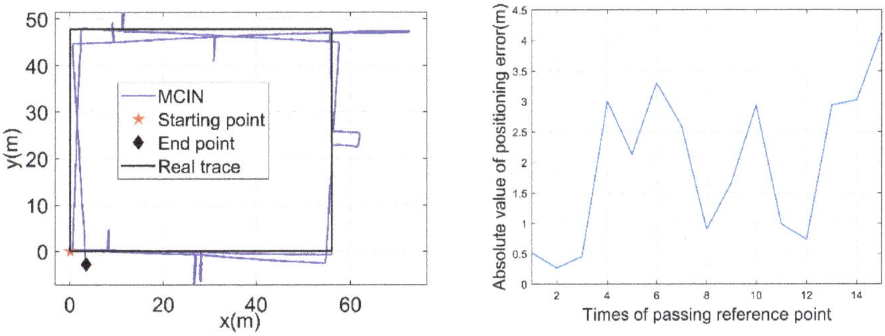

(a) (b)

Figure 14. Two dimensional trace and absolute value of positioning error based on MCIN method. (**a**) Two dimensional trace. (**b**) Absolute value of positioning error.

(2) The initial position and heading of pedestrian are known

The distribution of sampled particle is shown in Figure 15. The step length and heading change of the particles obey the Gaussian distribution. Initially, the particles are distributed in the range near the starting point, where red represents "illegal particles" and blue represents passable "legal particles".

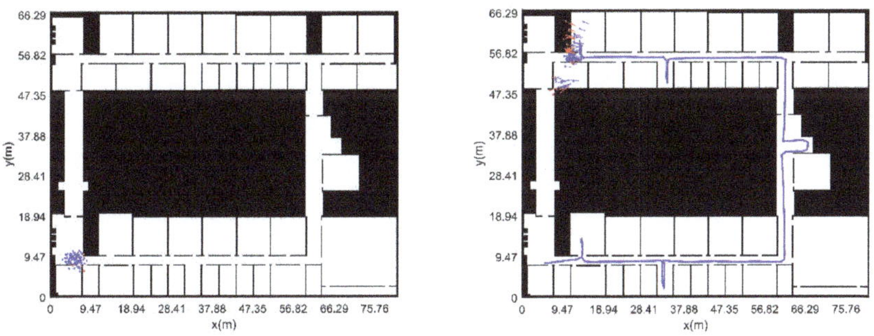

(a) (b)

Figure 15. Sampling particle distribution and motion trace under the condition of known initial position and heading. (**a**) Particle distribution at the initial moment. (**b**) Particle distribution and trace during motion.

Figure 16 shows the motion trace and positioning error under the condition that the initial position and heading are known. In Figure 16a, the red line represents the motion trace solved by the MCIN method, and the blue line represents the trace with IMAPF method. It can be seen that the IMAPF can well correct the position of the navigation system over time. Figure 16b is the absolute value curve of the positioning error under the condition of known initial position and heading, which clearly shows the effectiveness of the algorithm in correcting the error. The MCIN method can effectively restrain error divergence.

Table 6 shows the error comparison under the condition that the initial position and heading are known. It can be seen from the data in the table that the method proposed in this paper can effectively restrain error divergence compared to the MCIN method.

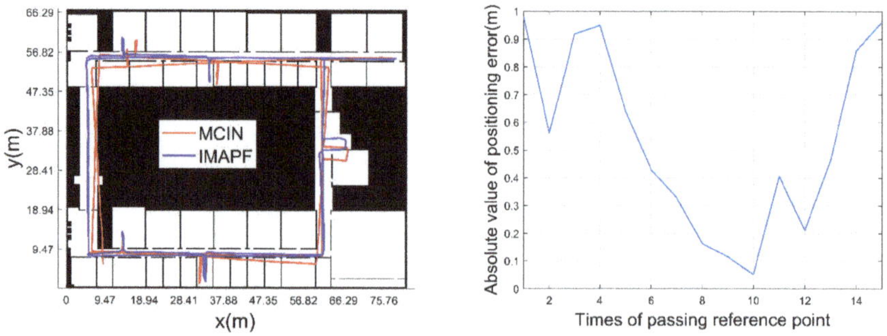

(**a**) (**b**)

Figure 16. Navigation trace comparison and curve of absolute value of positioning error with known initial position and heading. (**a**) Positioning trace comparison diagram. (**b**) Absolute value of positioning error.

Table 6. Error comparison under the condition that the initial position and heading are unknown.

Navigation Method	Mean Error (m)	Maximum Error (m)
MCIN	1.98	4.16
IMAPF	0.54	0.98

(3) The initial position and heading of pedestrian are unknown (adaptive particle number)

The distribution of sampled particles is shown in Figure 17. Initially, particles are uniformly distributed throughout the map, with red representing "illegal particles" and blue representing passable "legal particles". Starting from the 114th step, the general location of the pedestrian is searched. With the continuous movement of the pedestrian, the navigation and positioning function is finally provided for the pedestrian. There is a possibility of symmetry in the indoor structure, and pedestrians may not know in advance. According to the structural characteristics of indoor rooms, corridors, etc., the global search can be realized as soon as possible by increasing the path complexity.

Figure 18 is the comparison diagram of positioning trace under the condition of unknown initial position and heading. In Figure 18a, the red line represents the trace of the MCIN method, and the blue line represents the trace of IMAPF method. It can be seen that with the increase of time, IMAPF method can well correct the position of the navigation system. Figure 18b shows the absolute value curve of the positioning error under the condition that the initial position and heading are unknown. As the pedestrian keeps moving, navigation and positioning functions are gradually provided with map constraints. After 114 steps, functions of navigation and positioning can be provided for pedestrians in the map. Figure 18b shows the absolute value of the positioning error calculated according to the final navigation and positioning results, so the errors of the first two reference points are not considered. It can be seen that the algorithm is effective in correcting errors, which can be limited in a certain range and do not diverge over time. The mean error of IMAPF method is 1.06 m, and the maximum error is 1.33 m.

Figures 16b and 18b show the absolute value of navigation error using the IMAPF method in this paper. In general, the navigation error of IMAPF method is constrained in a small range, which is better than MCIN.

It can be seen that the IMAPF method studied in this paper performs better than the MCIN method for pedestrian navigation, whether the initial position and heading are known or not.

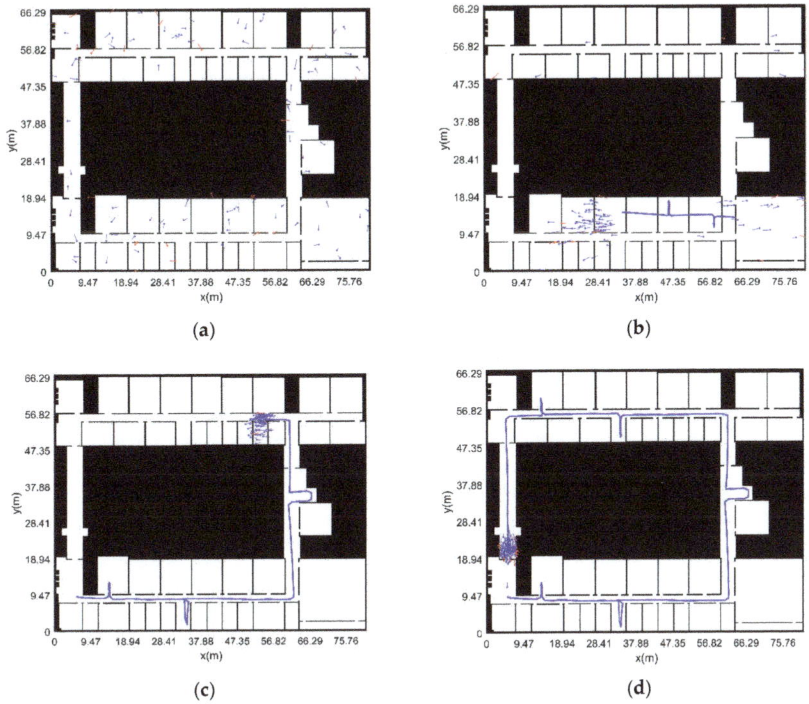

Figure 17. Distribution of sampled particles and motion trace with unknown initial position and heading. (**a**) Particle distribution at initial moment. (**b**) Particle distribution and motion track of the 58th step in the motion process. (**c**) Particle distribution and motion track of the 114th step in the motion process. (**d**) Particle distribution and motion track of the 190th step in the motion process.

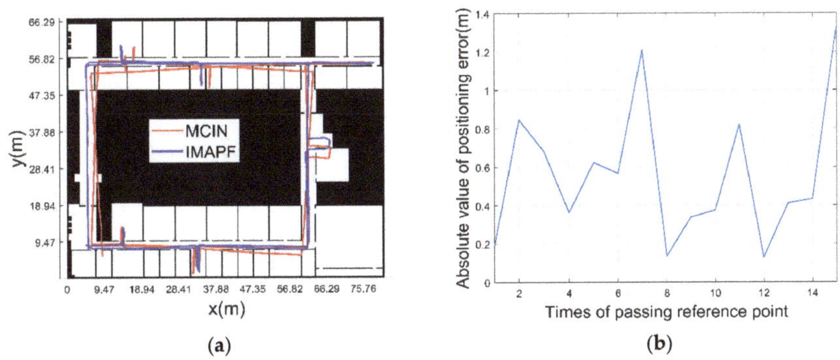

Figure 18. Navigation trace comparison and curve of absolute value of positioning error with unknown initial position and heading. (**a**) Positioning trace comparison diagram. (**b**) Absolute value of the positioning error.

(4) The initial position and heading of pedestrian are unknown (fixed particle number)

In order to compare the computational efficiency and error value between the fixed particle number and the adaptive particle number, under the condition of unknown pedestrian initial position and heading, the fixed number of particles is 2000, 10,000, 50,000 and 100,000, respectively. Table 7 is the error statistics table of different particle numbers.

Table 7. Statistics of positioning errors with different particle numbers under experimental conditions.

Particle number	Mean Error (m)	Maximum Error (m)	Calculation Time (s)
2 thousand fixed particles	3.89	7.03	16.37
10 thousand fixed particles	2.74	9.21	93.02
50 thousand fixed particles	1.13	1.40	755.33
100 thousand fixed particles	1.04	1.11	2229.13
adaptive particle number	1.06	1.33	131.59

Figure 19 shows the pedestrian motion trace obtained by different particle numbers. Comparing the navigation results of five different particle numbers, the adaptive particle number method proposed in this paper is close to the navigation positioning error value when the fixed number of particles is 50,000. The error is small when the number of fixed particles is smaller, but the calculation time of the adaptive particle number method is reduced about 4.7 times lower compared to the calculation time of 50,000 fixed particles.

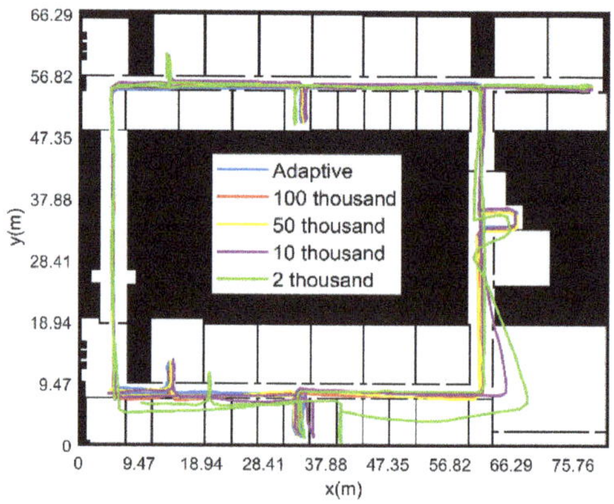

Figure 19. Pedestrian motion trace calculated with different particle numbers under experimental conditions.

4. Discussion

The purpose of this paper is to study an improved pedestrian navigation and location method based on the combination of indoor map assistance and adaptive particle filter. For the multi constraint integrated navigation method that only relies on the wearable inertial sensor node network for indoor pedestrians, there is an unavoidable problem of the accumulation and divergence of navigation errors. It is urgent to further improve the accuracy of the navigation system based on the available indoor auxiliary information. Considering that the indoor architectural plan is the most basic and accessible information source in the indoor rescue process, a navigation and positioning method based on the combination of indoor map assistance and particle filter is proposed. This method makes full use of the existing indoor map constraint information to assist in improving the performance of the pedestrian navigation and positioning system based on inertial sensor network. In this paper, the algorithm framework of indoor pedestrian navigation based on map assistance is designed. Combined with the characteristics that pedestrians cannot actually go through the walls and other obstacles when moving in buildings. By establishing the filtering algorithm under the property of particle "not going through the wall", the effective constraints on navigation error are realized; In view of the pedestrian's initial entry into an unfamiliar indoor environment, a map aided localization algorithm based on global search is proposed under the condition of unknown initial position and heading; In order to solve the problem that a large number of particles are required to complete the global search, which leads to low computational efficiency, a particle resampling method based on adaptive particle number is proposed. While maintaining the accuracy of navigation and positioning, it also improves the computing efficiency and achieves accurate indoor positioning in unfamiliar environments. On this basis, based on indoor map constraints, the problem of inertial accumulation error divergence is well suppressed, which provides a strong support for pedestrian indoor navigation and positioning with high precision and reliability.

Through the verification and analysis of simulation data and measured data, the pedestrian navigation and positioning method based on the combination of improved indoor map assistance and adaptive particle filter proposed in this paper is suitable for the conditions of known and unknown initial position and heading. In the indoor environment of about 2600 m^2, when the total distance exceeds 415.44 m, the mean error and the maximum error of the position relative to the reference point are both less than 2 m. It effectively suppresses the pedestrian navigation error based on inertial devices, and greatly improves the calculation efficiency, which can meet the needs of indoor pedestrians for a long time.

In fact, in the process of motion, both lateral and longitudinal errors are derived from step length error and heading errors. In a one-step correction process, if it is calculated that the coordinates of a particle in the lateral or longitudinal direction are in the inaccessible area, the particle is in the inaccessible area and needs to be resampled and given new navigation parameters. The pedestrian position is then calculated by the weighted average method.

The method proposed in this paper also has some limitations. It is based on the indoor building plan, and combines the characteristics of particles "not going through the wall" to constrain and modify the pedestrian trace. When the indoor structure is simple and the environment is open, then the distance between the walls on both sides of the walkway is very far, it is difficult to correct the pedestrian trace through this method.

5. Conclusions

In this paper, an improved pedestrian navigation method based on indoor map assistance and particle filter is proposed. Based on the fact that particles cannot "going through the wall", this method limits the pedestrian navigation positioning error to a low range for a long time. In addition, a global search algorithm is proposed to solve the problem of high-precision localization of pedestrians in unfamiliar environments with unknown initial position and heading; An adaptive particle number calculation method in the process of particle resampling is also proposed, which can improve the calculation efficiency and achieve long-term high-precision navigation and positioning for indoor pedestrians.

The method proposed in this paper can be used in indoor environments such as disaster relief and rescue, medical search and rescue. With the building plan and inertial sensors, navigation and positioning accuracy can be maintained for a long time. This method is of great significance to the practical application of pedestrian inertial navigation.

In the future, we will further study the method of multi person cooperative navigation in a complex environment according to the method in this paper to obtain higher navigation and positioning accuracy in a longer period.

Author Contributions: Conceptualization, Z.W., L.X. and Z.X.; methodology, Z.W.; validation, Z.W., Y.D., Y.S. and C.S.; formal analysis, Z.W.; software, Z.W.; data curation, Y.D. and Y.S.; supervision, L.X. and Z.X., writing—original draft, Z.W., L.X. and Z.X.; writing—review and editing, Z.W., L.X., Z.X., Y.D., Y.S. and C.S. All authors have read and agreed to the published version of the manuscript.

Funding: This research was funded by the National Natural Science Foundation of China, grant number 61873125; The National Natural Science Foundation of China, grant number 62073163; The National Natural Science Foundation of China, grant number 62103285; Support for projects in special zones for national defense science and technology innovation; The advanced research project of the equipment development grant number 30102080101; National Basic Research Program, grant number JCKY2020605C009; The Natural Science Fund of Jiangsu Province, grant number BK20181291; The Aeronautic Science Foundation of China, grant number ASFC-2020Z071052001; The

Aeronautic Science Foundation of China, grant number 202055052003; The Fundamental Research Funds for the Central Universities, grant number NZ2020004; Shanghai Aerospace Science and Technology Innovation Fund, grant number SAST2019-085; Introduction plan of high end experts, grant number G20200010142. Foundation of Key Laboratory of Navigation, Guidance and Health-Management Technologies of Advanced Aerocraft (Nanjing Univ. of Aeronautics and Astronautics), Ministry of Industry and Information Technology, Jiangsu Key Laboratory "Internet of Things and Control Technologies" & the Priority Academic Program Development of Jiangsu Higher Education Institutions, Science and Technology on Avionics Integration Laboratory. Supported by the 111 Project, grant number B20007.

Data Availability Statement: The study did not report any data. I choose to exclude this statement.

Conflicts of Interest: The authors declare no conflict of interest.

References

1. Jwo, D.J.; Lee, J.T. Kernel Entropy Based Extended Kalman Filter for GPS Navigation Processing. *Comput. Mater. Contin.* **2021**, *68*, 857–876.
2. Cao, X.; Shen, F.; Zhang, S.; Li, J. Time delay bias between the second and third generation of BeiDou Navigation Satellite System and its effect on precise point positioning. *Measurement* **2021**, *168*, 108346. [CrossRef]
3. Kröll, H.; Steiner, C. Indoor ultra-wideband location fingerprinting. In Proceedings of the 2010 International Conference on Indoor Positioning and Indoor Navigation, Zurich, Switzerland, 15–17 September 2010.
4. Li, Y.; Zhuang, Y.; Zhang, P.; Lan, H.; Niu, X.; Naser, E.-S. An improved inertial/wifi/magnetic fusion structure for indoor navigation. *Inf. Fusion* **2017**, *34*, 101–119.
5. Kriz, P.; Maly, F.; Kozel, T. Improving Indoor Localization Using Bluetooth Low Energy Beacons. *Mob. Inf. Syst.* **2016**, *2016 Pt 2*, 1–11. [CrossRef]
6. Bianchi, V.; Ciampolini, P.; De Munari, I. RSSI-Based Indoor Localization and Identification for ZigBee Wireless Sensor Networks in Smart Homes. *IEEE Trans. Instrum. Meas.* **2019**, *68*, 566–575. [CrossRef]
7. Chawla, K.; Thomason, W.; Mcfarland, C.; Robins, G. An accurate real-time RFID-based location system. *Int. J. Radio Freq. Identif. Technol. Appl.* **2018**, *5*, 48–76. [CrossRef]
8. Nadarajah, V.R.; Singh, M.M. Privacy-by-Design(PbD) IoT Framework: A Case of Location Privacy Mitigation Strategies for Near Field Communication (NFC) Tag Sensor. *Adv. Sci. Technol. Eng. Syst. J.* **2017**, *2*, 134–138. [CrossRef]
9. Tian, D.; Ming, D.; Mao, G.; Lin, Z.; David, L.P. Uplink Performance Analysis of Dense Cellular Networks with LoS and NLoS Transmissions. *IEEE Trans. Wirel. Commun.* **2017**, *16*, 2601–2613.
10. Mao, K.J.; Wu, J.B.; Jin, H.B.; Miao, C.Y.; Chen, Q.Z. Indoor localization algorithm for NLOS environment. *Acta Electron. Sin.* **2016**, *44*, 1174–1179.
11. Marek, P.; Philippe, G.; Rasmus, A. Mapping forests using an unmanned ground vehicle with 3D LiDAR and graph-SLAM. *Comput. Electron. Agric.* **2018**, *145*, 217–225.
12. Liu, J.; Gao, W.; Hu, Z. Bidirectional Trajectory Computation for Odometer-Aided Visual-Inertial SLAM. *IEEE Robot. Autom. Lett.* **2021**, *99*, 1670–1677. [CrossRef]
13. Qureshi, U.; Golnaraghi, F. An algorithm for the in-field calibration of a MEMS IMU. *IEEE Sens. J.* **2017**, *17*, 7479–7486.
14. Ahmed, D.B.; Diaz, E.M. Loose Coupling of Wearable-Based INSs with Automatic Heading Evaluation. *Sensors* **2017**, *17*, 2534. [CrossRef] [PubMed]
15. Xu, L.; Xiong, Z.; Liu, J.; Wang, Z.; Ding, Y. A Novel Pedestrian Dead Reckoning Algorithm for Multi-Mode Recognition Based on Smartphones. *Remote Sens.* **2019**, *11*, 294. [CrossRef]
16. Ding, Y.; Xiong, Z.; Li, W.; Cao, Z.; Wang, Z. Pedestrian Navigation System with Trinal-IMUs for Drastic Motions. *Sensors* **2020**, *20*, 5570.
17. Ruppelt, J.; Kronenwett, N.; Scholz, G.; Trommer, G.F. High-precision and robust indoor localization based on foot-mounted inertial sensors. In Proceedings of the 2016 IEEE/ION Position, Location and Navigation Symposium (PLANS), Savannah, GA, USA, 11–14 April 2016; pp. 67–75.
18. Hsu, Y.L.; Wang, J.S.; Chang, C.W. A Wearable Inertial Pedestrian Navigation System with Quaternion-Based Extended Kalman Filter for Pedestrian Localization. *IEEE Sens. J.* **2017**, *17*, 3193–3206. [CrossRef]
19. Suresh, R.P.; Sridhar, V.; Pramod, J.; Talasila, V. Zero velocity potential update (ZUPT) as a correction technique. In Proceedings of the 2018 3rd International Conference on Internet of Things: Smart Innovation and Usages (IoT-SIU), Bhimtal, India, 23–24 February 2018; pp. 1–8.
20. Abdallah, A.A.; Jao, C.S.; Kassas, Z.M.; Shkel, A.M. A pedestrian indoor navigation system using deep-learning-aided cellular signals and ZUPT-aided foot-mounted IMUs. *IEEE Sens. J.* **2021**, *22*, 5188–5198. [CrossRef]
21. Ilyas, M.; Cho, K.; Baeg, S.-H.; Park, S. Drift reduction in pedestrian navigation system by exploiting motion constraints and magnetic field. *Sensors* **2016**, *16*, 1455. [CrossRef]
22. Song, J.W.; Park, C.G. Enhanced pedestrian navigation based on course angle error estimation using cascaded Kalman filters. *Sensors* **2018**, *18*, 1281. [CrossRef]

23. Diez, L.E.; Bahillo, A.; Bataineh, S.; Masegosa, A.D.; Perallos, A. Enhancing improved heuristic drift elimination for wrist-worn PDR systems in buildings. In Proceedings of the 2016 IEEE 84th Vehicular Technology Conference (VTC-Fall), Montreal, QC, Canada, 18–21 September 2016; pp. 1–5.
24. Muhammad, M.N.; Salcic, Z.; Wang, K.I.-K. Detecting turns and correcting headings using low-cost INS. *J. Navig.* **2018**, *71*, 189–207. [CrossRef]
25. Kim, S.-Y.; Park, Y.-Y.; Lee, K.-O.; Jung, J.-B.; Kye, J.-E. A Study on the Making Indoor Topology Map for the Autonomous Driving of Wearable Robots. *Ilkogr. Online* **2021**, *20*, 1153–1159.
26. Angermann, M.; Robertson, P. FootSLAM: Pedestrian Simultaneous Localization and Mapping Without Exteroceptive Sensors—Hitchhiking on human perception and cognition. *Proc. IEEE* **2012**, *100*, 1840–1848. [CrossRef]
27. Kaiser, S.; Puyol, M.G.; Robertson, P. Measuring the Uncertainty of Probabilistic Maps Representing Human Motion for Indoor Navigation. *Mob. Inf. Syst.* **2016**, *2016*, 9595306. [CrossRef]
28. Wang, Q.; Luo, H.; Men, A.; Zhao, F.; Huang, Y. An Infrastructure-Free Indoor Localization Algorithm for Smartphones. *Sensors* **2018**, *18*, 3317. [CrossRef]
29. Yu, C.; El-Sheimy, N.; Lan, H.; Liu, Z. Map-based indoor pedestrian navigation using an auxiliary particle filter. *Micromachines* **2017**, *8*, 225. [CrossRef]
30. Wang, Z.; Xiong, Z.; Xing, L.; Ding, Y.; Sun, Y. A method for autonomous multi-motion modes recognition and navigation optimization for indoor pedestrian. *Sensors* **2022**, *22*, 5022. [CrossRef]

remote sensing

Review

Simultaneous Localization and Mapping (SLAM) for Autonomous Driving: Concept and Analysis

Shuran Zheng [1,*], Jinling Wang [1], Chris Rizos [1], Weidong Ding [1] and Ahmed El-Mowafy [2]

1 School of Civil and Environmental Engineering, UNSW Sydney, Sydney 2052, Australia
2 School of Earth and Planetary Sciences, Curtin University, Perth 6845, Australia
* Correspondence: shuran.zheng@unsw.edu.au

Abstract: The Simultaneous Localization and Mapping (SLAM) technique has achieved astonishing progress over the last few decades and has generated considerable interest in the autonomous driving community. With its conceptual roots in navigation and mapping, SLAM outperforms some traditional positioning and localization techniques since it can support more reliable and robust localization, planning, and controlling to meet some key criteria for autonomous driving. In this study the authors first give an overview of the different SLAM implementation approaches and then discuss the applications of SLAM for autonomous driving with respect to different driving scenarios, vehicle system components and the characteristics of the SLAM approaches. The authors then discuss some challenging issues and current solutions when applying SLAM for autonomous driving. Some quantitative quality analysis means to evaluate the characteristics and performance of SLAM systems and to monitor the risk in SLAM estimation are reviewed. In addition, this study describes a real-world road test to demonstrate a multi-sensor-based modernized SLAM procedure for autonomous driving. The numerical results show that a high-precision 3D point cloud map can be generated by the SLAM procedure with the integration of Lidar and GNSS/INS. Online four–five cm accuracy localization solution can be achieved based on this pre-generated map and online Lidar scan matching with a tightly fused inertial system.

Keywords: Simultaneous Localization and Mapping; autonomous driving; localization; high definition map

Citation: Zheng, S.; Wang, J.; Rizos, C.; Ding, W.; El-Mowafy, A. Simultaneous Localization and Mapping (SLAM) for Autonomous Driving: Concept and Analysis. *Remote Sens.* **2023**, *15*, 1156. https://doi.org/10.3390/rs15041156

Academic Editors: Yuwei Chen, Changhui Jiang, Qian Meng, Bing Xu, Wang Gao, Panlong Wu, Lianwu Guan and Zeyu Li

Received: 27 October 2022
Revised: 6 January 2023
Accepted: 10 January 2023
Published: 20 February 2023

1. Introduction

Autonomous (also called self-driving, driverless, or robotic) vehicle operation is a significant academic as well as an industrial research topic. It is predicted that fully autonomous vehicles will become an important part of total vehicle sales in the next decades. The promotion of autonomous vehicles draws attention to the many advantages, such as service for disabled or elderly persons, reduction in driver stress and costs, reduction in road accidents, elimination of the need for conventional public transit services, etc. [1,2].

A typical autonomous vehicle system contains four key parts: localization, perception, planning, and controlling (Figure 1). Positioning is the process of obtaining a (moving or static) object's coordinates with respect to a given coordinate system. The coordinate system may be a local coordinate system or a geodetic datum such as WGS84. Localization is a process of estimating the carrier's pose (position and attitude) in relation to a reference frame or a map. The perception system monitors the road environment around the host vehicle and identifies interested objects such as pedestrians, other vehicles, traffic lights, signage, etc.

By determining the coordinates of objects in the surrounding environment a map can be generated. This process is known as Mapping.

Path planning is the step that utilizes localization, mapping, and perception information to determine the optimal path in subsequent driving epochs, guiding the automated vehicle from one location to another location. This plan is then converted into action using the controlling system components, e.g., brake control before the detected traffic lights, etc.

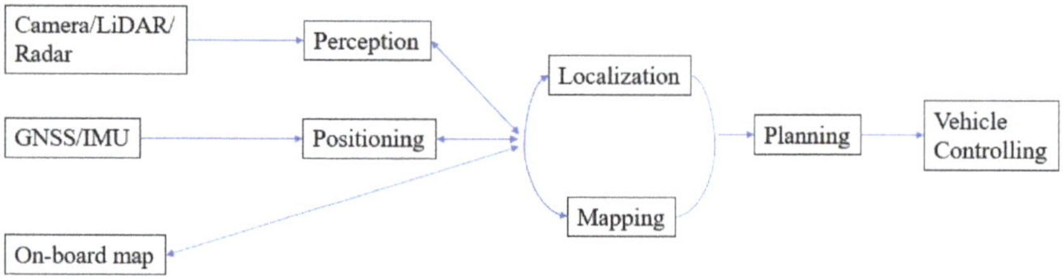

Figure 1. Functional components of an autonomous driving system.

All these parts are closely related. The location information for both vehicle and road entities can be obtained by combining the position, perception, and map information. In contrast, localization and mapping can be used to support better perception. Accurate localization and perception information is essential for correct planning and controlling.

To achieve fully automated driving, there are some key requirements that need to be considered for the localization and perception steps. The first is accuracy. For autonomous driving, the information about where the road is and where the vehicle is within the lane supports the planning and controlling steps. To realize these, and to ensure vehicle safety, there is a stringent requirement for position estimation at the lane level, or even the "where-in-lane" level (i.e., the sub-lane level). Recognition range is important because the planning and controlling steps need enough processing time for the vehicle to react [3]. Robustness means the localization and perception should be robust to any changes while driving, such as driving scenarios (urban, highway, tunnel, rural, etc.), lighting conditions, weather, etc.

Traditional vehicle localization and perception techniques cannot meet all of the aforementioned requirements. For instance, GNSS error occurs as the signals may be distorted, or even blocked, by trees, urban canyons, tunnels, etc. Often an inertial navigation system (INS) is used to support navigation during GNSS signal outages, to continue providing position, velocity, and altitude information. However, inertial measurement bias needs frequently estimated corrections or calibration, which is best achieved using GNSS measurements. Nevertheless, an integrated GNSS/INS system is still not sufficient since highly automated driving requires not only positioning information of the host vehicle, but also the spatial characteristics of the objects in the surrounding environment. Hence perceptive sensors, such as Lidar and Cameras, are often used for both localization and perception. Lidar can acquire a 3D point cloud directly and map the environment, with the aid of GNSS and INS, to an accuracy that can reach the centimeter level in urban road driving conditions [4]. However, the high cost has limited the commercial adoption of Lidar systems in vehicles. Furthermore, its accuracy is influenced by weather (such as rain) and lighting conditions. Compared to Lidar, Camera systems have lower accuracy but are also affected by numerous error sources [5,6]. Nevertheless, they are much cheaper, smaller in size, require less maintenance, and use less energy. Vision-based systems can provide abundant environment information, similar to what human eyes can perceive, and the data can be fused with other sensors to determine the location of detected features.

A map with rich road environment information is essential for the aforementioned sensors to achieve accurate and robust localization and perception. Pre-stored road information makes autonomous driving robust to the changing environment and road dynamics. The recognition range requirement can be satisfied since an onboard map can provide timely information on the road network. Map-based localization and navigation have been studied using different types of map information. Google Map is one example as it provides worldwide map information including images, topographic details, and satellite images [7], and it is available via mobile phone and vehicle apps. However, the use of maps will be limited by the accuracy of the maps, and in some selected areas the map's resolution may be inadequate. In [8], the authors considered low-accuracy maps for navigation by combining data from other sensors. They detected moving objects using Lidar data and utilized a GNSS/INS system with a coarse open-source GIS map. Their results show their fusion technique can successfully detect and track moving objects. A precise curb-map-based localization method that uses a 3D-Lidar sensor and a high-precision map is proposed in [9]. However, this method will fail when curb information is lacking, or obstructed.

Recently, so-called "high-definition" (HD) maps have received considerable interest in the context of autonomous driving since they contain very accurate, and large volumes of, road network information [10]. According to some major players in the commercial HD map market, 10–20 cm accuracy has been achieved [11,12], and it is predicted that in the next generation of HD maps, a few centimeters of accuracy will be reached. Such maps contain considerable information on road features, not only the static road entities and road geometry (curvature, grades, etc.), but also traffic management information such as traffic signs, traffic lights, speed limits, road markings, and so on. The autonomous car can use the HD map to precisely locate the host-car within the road lane and to estimate the relative location of the car with respect to road objects by matching the landmarks which are recognized by onboard sensors with pre-stored information within the HD map.

Therefore, maps, especially HD maps, play several roles in support of autonomous driving and may be able to meet the stringent requirements of accuracy, precision, recognition ranging, robustness, and information richness. However, the application of the "map" for autonomous driving is also facilitated by techniques such as Simultaneous Localization and Mapping (SLAM). SLAM is a process by which a moving platform builds a map of the environment and uses that map to deduce its location at the same time. SLAM, which is widely used in the robotic field, has been demonstrated [13,14] as being applicable for autonomous vehicle operations as it can support not only accurate map generation but also online localization within a previously generated map.

With appropriate sensor information (perception data, absolute and dead reckoning position information), a high-density and accurate map can be generated offline by SLAM. When driving, the self-driving car can locate itself within the pre-stored map by matching the sensor data to the map. SLAM can also be used to address the problem of DATMO (detection and tracking of moving objects) [15] which is important for detecting pedestrians or other moving objects. As the static parts of the environment are localized and mapped by SLAM, the dynamic components can concurrently be detected and tracked relative to the static objects or features. However, SLAM also has some challenging issues when applied to autonomous driving applications. For instance, "loop closure" can be used to reduce the accumulated bias within SLAM estimation in indoor or urban scenarios, but it is not normally applicable to highway scenarios.

This paper will review some key techniques for SLAM, the application of SLAM for autonomous driving, and suitable SLAM techniques related to the applications. Section 2 gives a brief introduction to the principles and characteristics of some key SLAM techniques. Section 3 describes some potential applications of SLAM for autonomous driving. Some challenging issues in applying the SLAM technique for autonomous driving are discussed in Section 4. A real-world road test to show the performance of a multi-sensor-based SLAM procedure for autonomous driving is described in Section 5. The conclusions are given in Section 6.

2. Key SLAM Techniques

Since its initial introduction in 1986 [16], a variety of SLAM techniques have been developed. SLAM has its conceptual roots in geodesy and geospatial mapping [17].

In general, there are two types of approaches to SLAM estimation: filter-based and optimization-based. Both approaches estimate the vehicle pose states and map states at the same time. The vehicle pose includes 3D or 2D vehicle position, but sometimes also velocity, orientation or attitude, depending on the sensor(s) used and on the application(s).

2.1. Online and Offline SLAM

Figures 2 and 3 illustrate two general SLAM implementations: online SLAM and offline SLAM (sometimes referred to as full SLAM). According to [18], full SLAM seeks to calculate variables over the entire path along with the map, instead of just the current pose, while the online SLAM problem is solved by removing past poses from the full SLAM problem.

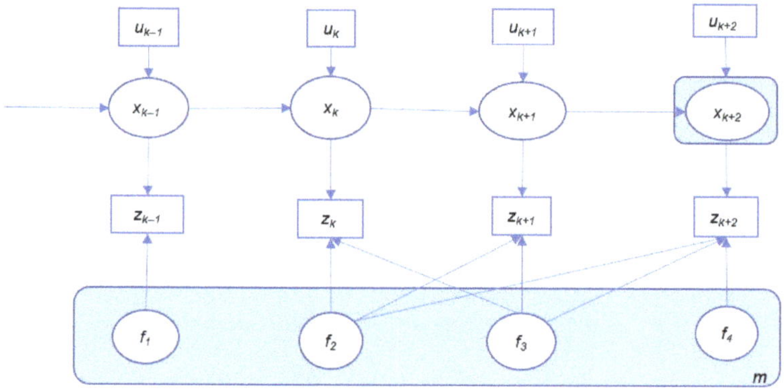

Figure 2. Description of online SLAM.

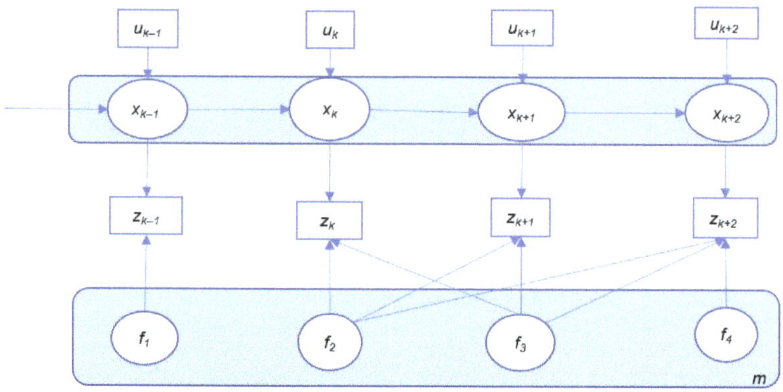

Figure 3. Description of offline SLAM.

Here, x_k represents the vehicle pose (position, attitude, velocity, etc.) at time k. m is the map that consists of stored landmarks (f_1–f_4) with their position states. u_k is the control inputs that represent the vehicle motion information between time epochs $k-1$ and k, such as acceleration, turn angle, etc., which can be acquired from vehicle motion sensors such as wheel encoders or an inertial sensor. At some epoch k, the onboard sensors (such as Camera, Lidar, and Radar) will perceive the environment and detect one or more landmarks. The relative observations between the vehicle and all the observed landmarks are denoted as z_k. With this information, the variables (including the vehicle pose and the map states) can be estimated.

The rectangle with blue background in Figures 2 and 3 represents the state variables that are estimated in these two implementations. In most cases, for online SLAM, only the current vehicle pose x_{k+2} is estimated while the map is generated and updated with the most recent measurements (u_{k+2} and z_{k+2}), whereas in the case of the offline SLAM implementation, the whole trajectory of the vehicle is updated together with the whole map. All the available control and observation measurements will be utilized together for the offline SLAM implementation.

However, with the development of SLAM algorithms and increased computational capabilities, the full SLAM solution may be obtained in real-time with an efficient SLAM algorithm, which can also be treated as an online problem. Therefore, implementing a SLAM method online or offline may be dependent on whether the measurement inputs (control and observation) it requires are from current/history or from future epochs, and on its processing time (real-time or not).

2.2. Filter-Based SLAM

Filter-based SLAM recursively solves the SLAM problem in two steps. Firstly, the vehicle and map states are predicted with processing models and control inputs. In the next step, a correction of the predicted state is carried out using the current sensor observations. Therefore, the filter-based SLAM is suitable for online SLAM.

Extended Kalman Filter-based SLAM (EKF-SLAM) represents a standard solution for the SLAM problem. It is derived from Bayesian filtering in which all variables are treated as Gaussian random variables. It consists of two steps: time update (prediction) and measurement update (filtering). At each time epoch the measurement and motion models are linearized (using the current state with the first-order Taylor expansion). However, since the linearization is not made around the true value of the state vector, but around the estimated value [19], the linearization error will accumulate and could cause a divergence of the estimation. Therefore, inconsistencies can occur.

Another issue related to EKF-SLAM is the continuous expansion of map size which makes the quadratic calculation process of large-scale SLAM impractical. For autonomous driving, the complex road environment and long driving period will introduce a large number of features, which makes real-time computation not feasible. A large number of algorithms have been developed in order to improve computational efficiency. For example, the Compressed Extended Kalman Filter (CEKF) [20] algorithm can significantly reduce computations by focusing on local areas and then extending the filtered information to the global map. Algorithms with sub-maps have also been used to address the computation issues [21–24]. A new blank map is used to replace the old map when the old one reaches a predefined map size. A higher-level map is maintained to track the link between each sub-map.

There are some other filter-based SLAM approaches, such as some variants of the Kalman Filter. One of them, the Information Filter (IF), is propagated with the inverse form of the state error covariance matrix, which makes this method more stable [25]. This method is more popular in multi-vehicle SLAM than in single-vehicle systems.

Another class of filter-based SLAM techniques is the Particle Filter (PF) which has become popular in recent years. PF executes Sequential Monte-Carlo (SMC) estimation by a set of random point clusters (or particles) representing the Bayesian aposteriori. The Rao–Blackwellized Particle Filter was proposed in [26]. Fast-SLAM is a popular implementation that treats the robot position distribution as a set of Rao–Blackwellized particles, and uses an EKF to maintain local maps. In this way, the computational complexity of SLAM is greatly reduced. Real-time application is possible with Fast-SLAM [27], making online SLAM possible for autonomous driving. Another advantage over EKF is that the particle filters can cope with non-linear motion models [28]. However, according to [29,30], Fast-SLAM suffers from degeneration since it cannot forget the past. If marginalizing the map and when resampling is performed, statistical accuracy is lost.

2.3. Optimization-Based SLAM

Full SLAM estimates all the vehicle pose and map states using the entire sensor data, and it is mostly optimization based. Similar to filter-based SLAM, the optimization-based SLAM system consists of two main parts: the frontend and the backend. In the frontend step, the SLAM system extracts the constraints of the problem with the sensor data, for example, by performing feature detection and matching, motion estimation, loop closure detection, etc. Nonlinear optimization is then applied to acquire the maximum likelihood estimation at the backend.

Graph SLAM is one of the main classes of full SLAM which uses a graphical structure to represent the Bayesian SLAM. All the platform poses along the whole trajectory and all the detected features are treated as nodes. Spatial constraints between poses are encoded in the edges between the nodes. These constraints result from observations, odometry measurements, and from loop closure constraints. After the graph construction, graph optimization is applied in order to optimize the graph model of the whole trajectory and map. To solve the full optimization and to calculate the Gaussian approximation of the aposteriori, a number of methods can be used, such as Gauss–Newton or Levenberg–Marquardt [31].

For graph-based SLAM, the size of its covariance matrix and update time are constant after generating the graph, therefore graph SLAM has become popular for building large-scale maps. Reducing the computational complexity of the optimization step has become one of the main research topics for practical implementations of the high-dimensional SLAM problem. The key to solving the optimization step efficiently is the sparsity of the normal matrix. The fact that each measurement is only associated with a very limited number of variables makes the matrix very sparse. With Cholesky factorization and QR factorization methods, the information matrix and measurement Jacobian matrix can be factorized efficiently, and hence the computational cost can be significantly reduced. Several algorithms have been proposed, such as TORO and g2o. The sub-map method is also a popular strategy for solving large-scale problems [32–36]. The sub-maps can be optimized independently and are related to a local coordinate frame. The sub-map coordinates can be treated as pose nodes, linked with motion constraints or loop closure constraints. Thus, a global pose graph is generated. In this way the computational complexity and update time will be improved.

Smoothing and Mapping (SAM), another optimization-based SLAM algorithm, is a type of nonlinear least squares problem. Such a least squares problem can be solved incrementally by Incremental Smoothing and Mapping (iSAM) [37] and iSAM2 [38]. Online SLAM can be obtained with incremental SAMs as they avoid unnecessary calculations with the entire covariance matrix. iSAM2 is more efficient as it uses a Bayes tree to obtain incremental variable re-ordering and fluid re-linearization.

SLAM++ is another incremental solution for nonlinear least squares optimization-based SLAM which is very efficient. Moreover, for online SLAM implementations, fast state covariance recovery is very important for data association, obtaining reduced state representations, active decision-making, and next best-view [39,40]. SLAM++ has an advantage as it allows for incremental covariance calculation which is faster than other implementations [40].

Table 1 is a summary of the characteristics of some typical SLAM techniques. Note that Graph SLAM utilizes all available observations and control information and can achieve very accurate and robust estimation results. It is suitable for offline applications and its performance relies on a good initial state guess. Filter-based SLAM is more suitable for small-scale environments when used for online estimation, but for the complex environment, a real-time computation may be difficult with the traditional EKF-SLAM. Other variants or fastSLAM should be considered. The incremental optimization method can do incremental updating, so as to provide an optimal estimation of a large-scale map with very high efficiency and in real-time.

Table 1. Characteristics of some typical SLAM techniques.

SLAM	Type	Advantages	Disadvantages	Typical Studies
EKF SLAM	Bayesian filter	• Mature method, widely studied; • Uncertainty is estimated.	• Suffers from linearization errors; • No re-linearization step; • Needs huge memory and computational resources for large maps.	[29,41,42]

Table 1. *Cont.*

SLAM	Type	Advantages	Disadvantages	Typical Studies
IF SLAM	Bayesian filter	• Already inversed covariance matrix; • Faster and more stable than EKF; • Suitable for multi-vehicle systems.	• Suffers from linearization errors; • No re-linearization step.	[25,43]
CEKF SLAM	Bayesian filter	• Cost-effective; • Outliers/errors only affect local maps; • Auxiliary coefficient matrix is used for inactive parts.	• Needs correct link between local and global maps.	[20,44]
Fast SLAM	Particle filter	• Capable of updating with unknown data association; • Less computation and memory cost than EKF; • Suitable for nonlinear cases; • Robust in cases where motion noise is high relative to measurement noise.	• Loses accuracy when marginalizing the map and resampling is performed.	[26,28,30,45]
Graph SLAM	Batch Least Squares optimization	• Suitable for nonlinear cases; • More accurate; • Can handle a large number of features.	• Not suitable for online applications; • Relies on good initial value.	[46–51]
iSAM2	Incremental optimization	• Very fast; • Suitable for nonlinear cases; • Allows re-linearization and data association correction.	• Complexity grows when graph become dense.	[37,38]
SLAM++	Incremental optimization	• Suitable for nonlinear cases; • Very fast estimation (faster than iSAM2); • Efficient uncertainty estimation; • Suitable for large-scale mapping.	• Complexity grows with increasing number of observations.	[39,40]

2.4. Sensors and Fusion Method for SLAM

New SLAM methods have appeared thanks to advances in sensor and computing technology. These methods are also optimization-based or filtered-based at the backend estimation step while the frontend step is highly dependent on the application of different sensor modalities. Two of the major sensors used for SLAM are Lidar and Camera. The Lidar method has become popular due to its simplicity and accuracy compared to other sensors [52]. The core of Lidar-based localization and mapping is scan-matching, which recovers the relative position and orientation of two scans or point clouds. Popular approaches for scan matching include the Iterative Closet Point (ICP) algorithm and its variants [53–55], and the normal distribution transform (NDT) [56]. These methods are highly dependent on good initial guess, and are impacted by local minimums [57,58]. Some other matching methods include probabilistic methods such as correlative scan matching (CSM) [59], feature-based methods [57,60], and others. Many of the scan-matching methods focus on initial free of or robust to, initialization error, but they still face the computation efficiency challenge.

Some range sensors that can be used for SLAM estimation are Radar and Sonar/ultrasonic sensors. Radar works in a similar manner to Lidar, but the system emits radio waves instead of light to measure the distance to objects. Furthermore, since Radar can observe the relative velocity between the sensor and the object using the measured Doppler shift [61], it is suitable for distinguishing between stationary and moving objects, and can be used to discard moving objects during the map-building process [62]. Some research on using Radar for SLAM can be found in [42,62–66]. When compared to Lidar, lower price, lower power consumption, and less sensitivity to atmospheric conditions make it well suited for outdoor applications. However, Radar has lower measurement resolution, and its detections are more sparse than Lidar. Thus, it is harder to match Radar data and deal with the data association problem, which results in its 3D mapping being less accurate.

Sonar/ultrasonic sensors also measure the time-of-flight (TOF) to determine the distance to objects, by sending and receiving sound waves. Sonar-based SLAM was initially used for underwater [67,68], and indoor [69] applications. It has become popular due to its low cost and low power consumption. It is not affected by visibility restrictions and can be used with multiple surface types [70]. However, similar to Radar, it obtains sparse information and suffers from inaccurate feature extraction and long processing time. Thus, it is of limited use for high-speed vehicle applications. Moreover, Sonar/ ultrasonic sensors have limited sensing range and may be affected by environmental noise and other platforms using ultrasound with the same frequency [71].

Camera is another popular sensor for SLAM. Different techniques have been developed, such as monocular [72,73], stereo [74–77], and multi-camera [78–81]. These techniques can be used in a wide range of environments, both indoor and outdoor. The single-camera system is easy to deploy, however, it suffers from scale uncertainty [82]. Stereo-camera systems can overcome the scale factor problem and can retrieve 3D structural information by comparing the same scene from two different perspectives [61]. Multi-camera systems have gained increasing interest, particularly as they achieve a large field of view [78] or are even capable of panoramic vision [81]. This system is more robust in complicated environments, while single sensor system may be very vulnerable to environmental interference [81]. However, the integration of Cameras requires additional software and hardware, and requires more calibration and synchronization effort [71,83]. Another special Camera, the RGB-D Camera, has been studied by the SLAM and computer vision communities [84–91] since it can directly obtain depth information. However, this system is mainly applicable in indoor environments because it uses infrared spectrum light and is therefore sensitive to external illumination [70].

The Visual SLAM can also be classified as feature-based or direct SLAM depending on how the measurements are used. The feature-based SLAM repeatedly detects features in images and utilizes descriptive features for tracking and depth estimation [92]. Some fundamental frameworks for this feature-based system include MonoSLAM [72,93], PTAM [94], ORB-SLAM [95], and ORB-SLAM2 [96]. Instead of using any feature detectors and descriptors, the direct SLAM method uses the whole image. Examples of direct SLAM include DTAM [97], LSD-SLAM [73], and SVO [98]. A dense or semi-dense environment model can be acquired by these methods, which makes them more computationally demanding than feature-based methods. Engel et al. [74] extended the LSD-SLAM from a monocular to a stereo model while Caruso et al. [99] extended the LSD-SLAM to an omnidirectional model. A detailed review of Visual SLAM can be found in [5] and [70,92,100,101].

Each of these perceptive sensors has its advantages and limitations. Lidar approaches can provide precise and long-range observations, but with limitations such as being sensitive to atmospheric conditions, being expensive, and currently rather bulky. Radar systems are relatively low cost, but are more suitable for object detection than for 3D map building. Sonar/ultrasonic sensors are not suitable for high-speed platform applications. Cameras are low-cost, even when multiple Cameras are used. Cameras can also provide rich visual information. However, they are sensitive to environment texture and light, and in general, have high computational demands. Therefore, a popular strategy is to combine a variety of sensors, making the SLAM system more robust.

There are several strategies to integrate data from different sensors for SLAM. One is fusing independently processed sensor results to then obtain the final solution. In [102], a mapping method that merged two grid maps, which were generated individually from laser and stereo camera measurements, into a single grid map was proposed. In this method, the measurements of the different sensors need to be mapped to a joint reference system. In [103], a multi-sensor SLAM system that combined the 3-DoF pose estimation from laser readings, the 6-DoF pose estimation from a monocular visual system, and the inertial-based navigation estimation results to generate the final 6-DoF position using an EKF processing scheme was proposed. For this type of strategy, the sensors can provide redundancy and the system will be robust to possible single-sensor failure. A decision-making step may be needed to identify whether the data from each sensor is reliable, and to decide whether to adopt the estimation from that sensor modality or to ignore it. Another fusion strategy is using an assistant sensor to improve the performance of other sensor-based SLAM algorithms. The main sensor could be Lidar or Camera, while the assistant sensor could be any other type of sensor. In this strategy, the assistant sensor is used to overcome the limitations of the main sensor. The work in [104] incorporated visual information to provide a good initial guess on the rigid body transformation, and then used this initial transformation to seed the ICP framework. Huang et al. [105] extracted the depth of point-based and line-based landmarks from the Lidar data. The proposed system used this depth information to guide camera tracking and also to support the subsequent point-line bundle adjustment to further improve the estimation accuracy.

The above two strategies can be combined. In the work of [106], the fusing consists of two models, one deals with feature fusion that utilizes line feature information from an image to remove any "pseudo-segments", which result from dynamic objects, in the laser segments. Another is a modified EKF SLAM framework that incorporates the state estimates obtained from the individual monocular and laser SLAM in order to reduce the pose estimation covariance and improve localization accuracy. This modified SLAM framework can run even when one sensor fails since the sensor SLAM processes are parallel to each other.

Some examples of more tight fusion can also be found in the literature. The work of [107] combined both the laser point cloud data and image feature point data as constraints and conducted a graph optimization with both of these constraints using a specific cost function. Furthermore, an image feature-based loop closure was added to this system to remove accumulation errors.

Inertial SLAM incorporates an inertial measurement unit (IMU) as an assistant sensor. The IMU can be fused with the Camera or Lidar to support pose (position, velocity, attitude) estimation. With an IMU, the attitudes, especially the heading, are observable [108]. The integration of IMU measurements can also improve the motion tracking performance during the gaps of observations. For instance, for a Visual SLAM, illumination change, texture-less area, or motion blur will cause losses of visual tracks [108]. For a Lidar system, the raw Lidar scan data may suffer from skewing caused by high-acceleration motion, such as moving fast or shaking suddenly, resulting in sensing error that is difficult to account for [109]. The work of [110] used an IMU sensor to deal with fast velocity changes and to initialize motion estimates for scan-matching Lidar odometry to support their LOAM system. The high-frequency IMU data between two Lidar scans can be used to de-skew Lidar point clouds and improve their accuracy [109].

The fusion of inertial sensors can be as a simple assistant [111,112] or more tightly coupled [108,113–115]. For the simple assistant case, the IMU is mainly used to provide orientation information, such as to support the system initialization. The IMU is used as prior for the whole system, and the IMU measurements are not used for further optimization. For the tightly coupled case, IMU data is fused with Camera/Lidar states to build up measurement models, and then perform state estimation and feedback to the inertial navigation system to improve navigation performance [116]. Therefore the former method is more efficient than the latter, however, it is less accurate [117]. For the tightly coupled case, a Kalman filter could be used to correct the IMU states, even during GNSS outages [118].

2.5. Deep Learning-Based SLAM

Most of the aforementioned SLAM methods are geometric model-based, which build up models of platform motion and the environment based on geometry. These methods have achieved great success in the past decade. However, they still face many challenging issues. For instance, Visual SLAM (VSLAM) is limited under extreme lighting conditions. For large-scale applications, the model-based methods need to deal with large amounts of information, such as features and dynamic obstacles. Recently, deep learning techniques, such as data-driven approaches developed in the computer vision field, have attracted more attention. Many researchers have attempted to apply deep learning methods to SLAM problems.

Most of the current research activities focus on utilizing learning-based methods for VSLAM problems since deep learning techniques have made breakthroughs in the areas of image classification, recognition, object detection, and image segmentation [119]. For instance, deep learning has been successfully applied to the visual odometry (VO) problem, which is an important element of VSLAM. Optical flow estimation is utilized in some learned VO models as inputs [120–124]. The application of learning approaches can be applied in an end-to-end manner without adopting any module in the conventional VO pipeline [125,126]. Wang et al. [125] introduced an end-to-end VO algorithm with deep Recurrent Convolutional Neural Networks (RCNNs) by combining CNNs with the RNNs. With this algorithm, the pose of the camera is directly estimated from raw RGB images, and neither prior knowledge nor parameters are needed to recover the absolute scale [125]. Li et al. [127] proposed an Unsupervised Deep Learning based VO system (UnDeepVO) which is trained with stereo image pairs and then performs both pose estimation and dense depth map estimation with monocular images. Unlike the one proposed by Wang et al. [125], ground truth is not needed for UnDeepVO since it operates in an unsupervised manner.

The learning-based methods can be combined with the VSLAM system to replace or add on an individual or some modules of traditional SLAM, such as image depth estimation [128–130], pose estimation [131–133], and loop closure [134–137], etc., to improve the traditional method. Li et al. [138] proposed a fully unsupervised deep learning-based VSLAM that contains several components, including Mapping-net, Tracking-net, Loop-net, and a graph optimization unit. This DeepSLAM method can achieve accurate pose estimation and is robust in some challenging scenarios, combining the important geometric models and constraints into the network architecture and the loss function.

Sematic perception of the environment and semantic segmentation are current research topics in the computer vision field. They can provide a high-level understanding of the environment and are extremely important for autonomous applications. The rapid development of deep learning can assist in the introduction of semantic information into VSLAM [139] for semantic segmentation [140–142], localization and mapping [143–147], and dynamic object removal [148–151]. Some detailed reviews of deep learning-based VSLAM can be found in [92,139,152–154].

Fusion with an inertial sensor can also benefit from deep learning techniques, especially the RNN, which has an advantage in integrating temporal information and helping to establish consistency between nearby frames [139]. The integration of visual and inertial data with RNN or Long Short-Term Memory (LSTM), a variant of RNN that allows RNN to learn long-term trends [155], has been proven to be more effective and convenient than traditional fusion [156–158]. According to Clark et al. [157], the data-driven approach eliminates the need for manual synchronization of the camera and IMU, and the need for manual calibration between the camera and IMU. It outperforms the traditional fusion method since it is robust to calibration errors and can mitigate sensor drifts. However, to deal with the drift problem, a further extension of the learning-based visual-inertial odometry system to a larger SLAM-like system with loop-closure detection and global relocalization still needs to be investigated.

Compared to the visual-based SLAM, the applications of deep learning techniques for laser scanners or Lidar-based SLAM are still in the early stages and can be considered a new challenge [159]. Velas et al. [160] used CNN for Lidar odometry estimation by using the IMU sensor to support rotation parameter estimation. The results are competitive with state-of-the-art methods such as LOAM. Li et al. [161] introduced an end-to-end Lidar odometry, LO-Net, which has high efficiency, and high accuracy, and can handle dynamic objects. However, this method is trained with ground truth data, which limits its application to large-scale outdoor scenarios. Li et al. [162] designed a visual-Lidar odometry framework, which is self-supervised, without using any ground truth labels. The results indicate that this VLO method outperforms other current self-supervised visual or Lidar odometry methods, and performs better than fully supervised VOs. Data-driven approaches also make semantic segmentation of Lidar data more accurate and faster, making it suitable for supporting autonomous vehicles [163–165]. Moving objects can be distinguished from static objects by LMNet [166] based on CNNs of 3D Lidar scans. One limitation of some cost-effective 3D Lidar applications for autonomous driving in challenging dynamic environments is its relatively sparse point clouds. In order to overcome this drawback, high-resolution camera images were utilized by Yue et al. [167] to enrich the raw 3D point cloud. ERFNet is employed to segment the image with the aid of sparse Lidar data. Meanwhile, the sparsity invariant CNN (SCNN) is employed to predict the dense point cloud. Then the enriched point clouds can be refined by combining these two outputs using a multi-layer convolutional neural network (MCNN). Finally, Lidar SLAM can be performed with this enriched point cloud. Better target segmentation can be achieved with this Lidar data enrichment neural network method. However, due to the small training dataset, this method did not show improvement in SLAM accuracy with the enriched point cloud when compared to the original sparse point cloud. More training and further investigation of dynamic objects may be needed to satisfy autonomous driving application requirements [167].

The generation of complex deep learning architectures has contributed to achieving more accurate, robust, adaptive, and efficient computer vision solutions, confirming the great potential for their application to SLAM problems. The availability of large-scale datasets is still the key to boosting these applications. Moreover, with no need for ground truth, unsupervised learning is more promising for SLAM applications in autonomous driving. Compared to the traditional SLAM algorithms, data-driven SLAM is still in the development stage, especially for Lidar SLAM. In addition, combining multiple sensing modalities may overcome the shortcomings of individual sensors, for which the learning methods-based integration system still needs further investigation.

3. Application of SLAM in Autonomous Driving

Depending on the different characteristics of SLAM techniques, there could be different applications for autonomous driving. One classification of the applications is whether they are offline or online. A map satisfying a high-performance requirement is typically generated offline, such as the High Definition (HD) map [10]. For this kind of 3D point cloud map, an offline map generation process ensures the accuracy and reliability of the map. Such maps can be pre-generated to support the real-time operations of autonomous vehicles.

3.1. High Definition Map Generation and Updating

As stated earlier, SLAM can be used to generate digital maps used for autonomous driving, such as the HD map [10]. Due to the stringent requirements, high quality sensors are used. Lidar is one of the core sensors for automated cars as it can generate high-density 3D point clouds. High-end GNSS and INS technology are also used to provide accurate position information. Cameras can provide information that is similar to the information detected by human eyes. The fusion of sensor data and analysis of road information to generate HD maps needs considerable computational power, which is not feasible in current onboard vehicle systems. Therefore the HD map is built-up offline, using techniques such as optimization-based SLAM. The offline map creation can be performed by driving the road network several times to collect information, and then all the collected perceptive sensor information and position information is processed together to improve the accuracy of the final map. An example of a HD map is shown in Figure 4 [11].

Figure 4. An image from a high definition map (https://here.com/) [11].

The road environment and road rules may change, for instance, the speed limit may be reduced due to road work, road infrastructure may be changed due to building development, and so on. Therefore the HD map needs frequent updates. Such updates can utilize the online data collected from any autonomous car. For example, the data is transmitted to central (cloud) computers where the update computations are performed. Other cars can receive such cloud-based updates and make a timely adjustment to driving plans. Jo et al. [168] proposed a SLAM change update (SLAMCU) algorithm, utilizing a Rao–Blackwellized PF approach for online vehicle position and (new) map state estimation. In the work of [169], a new feature layer of HD maps can be generated using Graph SLAM when a vehicle is temporarily stopped or in a parking lot. The new feature layer from one vehicle can then be uploaded to the map cloud and integrated with that from other vehicles into a new feature layer in the map cloud, thus enabling more precise and robust vehicle localization. In the work of Zhang et al. [170], real-time semantic segmentation and Visual SLAM were combined to generate semantic point cloud data of the road environment, which was then matched with a pre-constructed HD map to confirm map elements that have not changed, and generate new elements when appearing, thus facilitating crowdsource updates of HD maps.

3.2. Small Local Map Generation

SLAM can also be used for small local areas. One example is within parking areas. The driving speed in a parking lot is low, therefore the vision technique will be more robust than in other high-speed driving scenarios. The parking area could be unknown (public parking lot or garage), or known (home zone)–both cases can benefit from SLAM. Since SLAM can be used without GNSS signals, it is suitable for vehicles in indoor or underground parking areas, using just the perceptive sensor and odometry measurements (velocity, turn angle) or IMU measurements. For unknown public parking areas, the position of the car and the obstacles, such as pillars, sidewalls, etc., can be estimated at the same time, guiding the parking system. For home zone parking, the pre-generated map and a frequent parking trajectory can be stored within the automated vehicle system. Each time the car returns home, re-localization using the stored map can be carried out by matching detected features with the map. The frequent trajectory could be used for the planning and controlling steps.

An approach that utilizes multi-level surface (MLS) maps to locate the vehicle, and to calculate and plan the vehicle path within indoor parking areas was proposed in [171]. In this study, graph-based SLAM was used for mapping, and the MLS map is then used to plan a global path from the start to the destination, and to robustly localize the vehicle with laser range measurements. In the work of [172], a grid map and an EKF SLAM algorithm were used with W-band radar for autonomous back-in parking. In this work, an efficient EKF SLAM algorithm was proposed to enable real-time processing. In [173], the authors proposed an around-view monitor (AVM)/ Lidar sensor fusion method to recognize the parking lane and to provide rapid loop closing performance. The above studies have demonstrated that both filter-based SLAM and optimization-based SLAM can be used to support efficient and accurate vehicle parking assistance (local area mapping and localization), even without GNSS. In the work of Qin et al. [174], pose graph optimization is performed so as to achieve an optimized trajectory and a global map of a parking lot, with semantic features such as guide signs, parking lines, and speed bumps. These kinds of features are more (long-term) stable and robust than traditional geometrical features, especially in underground parking environments. An EKF was then used to complete the localization system for autonomous driving.

3.3. Localization within the Existing Map

In map-based localization, a matching method is used to match "live" data with map information, using methods such as Iterative Closest Point (ICP), Normal Distribution Transform (NDT), and others [10,175]. These algorithms can be linked to the SLAM problem since SLAM executes loop closing and re-localization using similar methods. For a SLAM problem, the ability to recognize a previously mapped object or feature and to relocate the vehicle within the environment is essential for correcting the maps [13]. Therefore, the reuse of a pre-generated map to localize the vehicle can be considered an extension of a SLAM algorithm. In other words, the pre-generated and stored map can be treated as a type of "sensor" to support localization.

However, matching live data with a large-scale pre-prepared map requires substantial computational resources. Hence, some methods have been proposed to increase computational efficiency. One method is to first narrow down the possible matching area from the map with position estimated from GNSS or GNSS/INS, and then carry out detailed matching of the detected features with the map [176].

Due to the current limited installation of Lidar systems in commercial vehicles (high price of sensor and high power consumption), localization of a vehicle with a low-cost sensor (e.g., vision sensor) in a pre-generated HD map is of considerable practical interest. For instance, the work in [177] located a vehicle within a dense Lidar-generated map using vision data and demonstrated that a similar order of magnitude error rate can be achieved to traditional Lidar localization but with several orders of magnitude cheaper sensor technology. Schreiber et al. [178] proposed to first generate a highly accurate map with road markings and curb information using a high-precision GNSS unit, Velodyne laser scanner, and cameras. Then during the localization process, a stereo camera system was used to detect road information and match it with the pre-generated map to achieve lane-level real-time localization. Jeong et al. [179] utilized road markings obtained from camera images for global localization. A sub-map that contained road information, such as 3D road marking points, was generated and utilized to recognize a revisited place and to support accurate loop detection. The pose graph-based approach was then used to eliminate the drift. Qin et al. [146] proposed a semantic localization system to provide a light-weight localization solution for low-cost cars. In this work, a local semantic map was generated by combining the CNN-based semantic segmentation results and the optimized trajectory after pose graph optimization. A compacted global map was then generated (or updated) in the cloud server for further end-user localization based on the ICP method and within an EKF framework. The average size of the semantic map was 36 kb/km. This proposed camera-based localization framework is reliable and practical for autonomous driving.

In addition to the aforementioned applications, moving objects within the road environment will cause a drift of perception, localization, and mapping for autonomous driving. SLAM can be used to address the problem of DATMO (detection and tracking of moving objects) [15] because one of the assumptions of SLAM is that the detected features are stationary. As the static parts of the environment are localized and mapped by SLAM, the dynamic parts can be concurrently detected and tracked. Some approaches have dealt with dynamic obstacles [180–182].

4. Challenges of Applying SLAM for Autonomous Driving and Suggested Solutions

4.1. Ensuring High Accuracy and High Efficiency

Localization and mapping for automated vehicles need to be accurate and robust to any changes in the environment and executed with high efficiency. With rapidly developing sensor technology, the combination of different sensors can compensate for the limitations of a particular sensor. Examples include GNSS/INS + Lidar/Camera SLAM, Radar SLAM, and some others. There is considerable research and development associated with low-cost and/or miniaturized Lidar sensors. New Lidar sensor concepts promise a significant reduction in the cost of Lidar systems, with the potential for real-time implementation in future autonomous vehicles. For instance, RoboSense has unveiled a new $200 Lidar sensor combining MEMS sensors and an AI-based deep-learning algorithm to support high-performance autonomous driving applications [183].

Choosing a SLAM approach should take into consideration different application scenarios with different level of requirements. Optimization-based SLAM can provide more accurate and robust estimation, however, it is more suitable for offline estimation. EKF SLAM suffers from the quadratic increase in the number of state variables, which restricts its online application in large-scale environments. Although high-resolution map generation can be offline, real-time, or near-real-time, solutions are essential for map updating and map-based localization applications.

Any change in the road environment should be quickly updated on the map and transmitted to other road users. Emerging 5G wireless technology can make the communication between vehicle-to-vehicle (V2V), vehicle-to-infrastructure (V2I), and vehicle-to-cloud more reliable and with higher throughput [14].

4.2. Representing the Environment

There are different types of maps that can be used to represent the road environment. Three major types of maps in the robotic field for SLAM applications are occupancy grid maps, feature-based maps, and topological maps [184]. They are also applicable to road environments. Each of them has its own advantages and limitations for autonomous driving applications. The grid map divides the environments into many fixed-size cells, and each cell contains its own unique property, such as whether the grid is occupied, free or unknown [185,186]. The obstacle occupancy information can be directly fed to the planning algorithms. This kind of map can be merged easily and has flexibility in incorporating data from numerous types of sensors [184]. Mentasti and Matteucci [185] proposed an occupancy grid creation method that utilized data from all the available sensors on an autonomous vehicle, including Lidar, Camera, Laser, and Radar. The grid map also shows the potential for detecting moving objects [187]. Mutz et al. [188] compare the performance of mapping and localization with different grid maps, including occupancy, reflectivity, color, and semantic grid maps, for self-driving car applications in diverse driving environments, including under challenging conditions. GraphSLAM was used for mapping, while localization was based on particle filter solutions. According to their results, the occupancy showed more accurate localization results, followed by the reflectivity grid map. Semantic grid maps kept the position tracking without losses in most scenarios, however with bigger errors than the first two map approaches. Colorized grid maps were most inconsistent and inaccurate for use in localization, which may be due to the influence of illumination conditions. One shortcoming that limits the occupancy grip map for large-scale autonomous driving is its dense representation, which needs big storage space and high computation power [189]. Thus Li [186] suggested applying this technique for real-time local mapping with a controlled size instead of for global mapping.

The feature-based map is a popular map type for autonomous driving. It represents the map with a group of features extracted from sensor data. For outdoor road environments, the typical features are traffic lanes, kerbs, road markings and signs, buildings, trees, etc. For indoor areas, especially in parking areas, the features are mainly the parking lane, sidewalls, etc. These features can be represented by points, lines, and planes, tagged with coordinate information. The point feature represents the environment as dense point clouds. The high-density point cloud maps generated using Lidar and/or vision sensors can provide abundant features and 3D structure information of the area surrounding the vehicle. However, the transmission, updating, and processing of this volume of data is burdensome for complex road environments. The sparser line and plane features are suitable for structured environments, such as indoor environments, urban areas, or highways, with clear markings. These features are more sophisticated than the point features, with lower memory requirements [186], and are less susceptible to noise [189]. Im et al. [173] proposed a parking line-based SLAM approach which extracted and analyzed parking line features to achieve rapid loop closure and accurate localization in a parking area. Javanmardi et al. [190] generated a city road map with 2D lines and 3D planes to represent the buildings and the ground along the road. However, for autonomous driving, the application environment is variable. A specific landmark-based algorithm may not be suitable for other driving scenarios. Furthermore, in some rural areas, the road may be unpaved and there is no road lane marking. Thus the related feature-based map approaches may not be feasible due to the lack of road markings and irregular road curve geometry [191].

The topological map represents the environment with a series of nodes and edges. The nodes indicate the important objects, such as corners, intersections, and feature points; while the edges denote the topological relationships between them [192,193]. One typical topological map is OpenStreetMAP (OSM) [194] which contains the coordinates of features as well as road properties such as road direction, lane numbers, etc. This kind of map significantly reduces the storage and computational requirements. However, it loses some useful information about the nature and structure of the actual environment [184]. Thus, some approaches combine topological maps with other types of maps. Bernuy and Ruiz-del-Solar [195] proposed the use of a topological map based on semantic information to provide robust and efficient mapping and localization solutions for large-scale outdoor scenes for autonomous vehicles and ADAS systems. According to Bernuy and Ruiz-del-Solar [195], the graph-based topological semantic mapping method was suitable for large-scale driving tasks on highways, rural roads, and city areas, with less computational expense than metrics maps. Bender et al. [196] introduced a highly detailed map, Lanelets, which combines both geometrical and topological representations, and includes information on traffic regulations and speed limits.

The semantic map is becoming increasingly important in autonomous fields as it contains semantic information that allows the robot or vehicle to better understand the environment, and to complete higher-lever tasks, such as human-robot interaction. For outdoor applications, the labeled objects could be statistic background (e.g., 'building', 'tree', 'traffic sign') or dynamic entities (e.g., 'vehicle', 'pedestrian'). Therefore, this kind of map can facilitate complex tasks for autonomous vehicles, such as planning and navigation [195,197]. Associating semantic concepts with geometric entities has become a popular research topic and semantic SLAM approaches have been investigated that combine geometric and semantic information [139,143,149]. The semantic SLAM approaches can contribute to making localization and mapping more robust [174], to supporting re-localization at revisited areas [143], and very importantly, to tracking moving objects detected in dynamic environments [149,151,198]. One critical problem faced by semantic map generation and utilization is that some modules within them, such as semantic segmentation, are very computationally demanding, which makes them unsuitable for real-time applications [199], especially for large-scale outdoor scenarios. Thus, some investigations seek to solve this problem. Ros et al. [199] proposed an offline-online strategy that generates a dense 3D semantic map offline without sacrificing accuracy. Afterward, real-time self-localization can be performed by matching the current view to the 3D map, and the related geometry and semantics can be retrieved accordingly. Meanwhile, the new dynamic objects can be detected online to support instantaneous motion planning. With the advent of deep learning, the efficiency and reliability of semantic segmentation and semantic SLAM have been vastly improved [147,200–203]. However, as previously mentioned, when applying deep learning-based semantic SLAM to autonomous driving, there are still some challenges, such as the need for large amounts of training data, or the lack of ground truth that makes unsupervised learning methods necessary.

The different map representations are essential to support a highly automated vehicle operating in a challenging and complex road environment. Therefore, a detailed digital map, such as the HD map, which contains different layers of data, has been increasingly adopted. In addition to the most basic 3D point cloud map layer, the HD map may also contain layers with information on road topology, geometry, occupancy, lane features, road furniture, road regulation, real-time knowledge, and more. The storing, updating, and utilizing of such dense data without losing accuracy is a challenge. Some researchers have proposed the concept of "Road DNA" to represent the road environment and to deal with the Big Data problem [12,204]. Road DNA converts a 3D point cloud road pattern into a compressed, 2D view of the roadway without losing details [12], with the objective to reduce processing requirements.

4.3. Issue of Estimation Drifts

SLAM estimation drifts may be caused by accumulated linearization error, the presence of dynamic obstacles, noisy sensor data, wrong data association, etc.

In most SLAM algorithms, nonlinear models are used to represent the vehicle motion pattern and the environment. EKF SLAM suffers from a divergence problem due to the accumulation of linearization errors. Biases may occur when linearization is performed using values of state variables that are far from their true values. For optimization-based SLAM, a poor initial guess of variables will lead to poor convergence performance. Rotation may be the cause of nonlinearity and has a strong impact on the divergence of estimation [205,206], thus the accumulated vehicle orientation error will cause the inconsistency of the SLAM problem. One solution to the linearization challenge is the Linear SLAM algorithm proposed in [205], which modifies the relative state vector and carries out "map joining". Sub-map joining, which involves solving a linear least squares problem and performing nonlinear coordinate transformations, does not require an initial guess or iteration. In the work of [207], a robocentric local map sequencing approach was presented which can bound location uncertainty within each local map and improve the linearization accuracy with sensor uncertainty level constraints. Many variants of the classical EKF-SLAM have been proposed to overcome the divergence of the filter. The study of [208] demonstrated that the Unscented SLAM can improve the online consistency for large-scale outdoor applications. Huang et al. [209] proposed two alternatives for EKF-SLAM, Observability Constrained EKF, and First-Estimates Jacobian EKF, both of which significantly outperform the EKF in terms of accuracy and consistency. A linear time varying (LTV) Kalman filtering was introduced in [210] which avoids linearization error by creating virtual measurements. Some nonparametric approaches which are mainly based on the PF, such as fastSLAM [28], Unscented fastSLAM [211–214], show better performance than the EKF-SLAM.

For the nonlinear optimization-based SLAM approach, computing a good initial guess (solving the initialization problem), will lead to faster convergence and reduce the risk of convergence to local minima. Olson et al. [215] presented a fast iterative algorithm for optimizing pose graphs using a variant of Stochastic Gradient Descent (SGD), which is robust against local minima and converges quickly even with a bad initial guess. Then in the work of [50], an extension of Olson's algorithm was proposed which uses a tree-based parameterization for the nodes in the graph. This algorithm was demonstrated to be more efficient than Olson's and robust to the initial configuration. An approximation solution for 2D pose-graphs, called Linear Approximation for a pose Graph Optimization (LAGO), can be used as an exact solution or for bootstrapping nonlinear techniques [216,217]. This method first solves a linear estimation problem to obtain the suboptimal orientation estimate, and then uses it to estimate the relative position measurements in the global reference frame. Finally, the position and orientation solution is obtained by solving another linear estimation problem. This solution can then be treated as an initial guess for a Gauss-Newton iteration. This method can provide a good initial guess, however, it is limited to 2D pose-graph, and is sensitive to noisy measurements. An algorithm with more complex initialization was proposed in [218] that uses the M-estimator, in particular the Cauchy function, as a bootstrapping technique. Similar to approaches that use the M-estimator to make estimation robust to outliers, the M-estimator proved to also be robust to a bad initial guess. In contrast to LAGO and TORO, this method can be applied to different variants of SLAM (pose-graphs and feature-based) in both 2D and 3D [218]. Carlone et al. [219] surveyed different 3D rotation estimation techniques and demonstrated the importance of good rotation estimate to bootstrap iteration pose graph solvers. More recent research presented a heuristic method called Multi-Ancestor Spatial Approximation Tree (MASAT), which has low complexity and is computationally efficient without needing a preliminary optimization step [220]. This method is still for the pose graph. Other studies seek to obtain a good initial guess by introducing inertial measurements to support initialization [221,222] or conducting parameter calibration [223–225].

Dynamic objects such as pedestrians, bicycles, other vehicles, etc., may cause estimation drifts since the system may wrongly identify them as static road entities. There are some methods to avoid this. Probabilistic Maps that use probabilistic infrared intensity values have been proposed in [226]. In this study, GNSS/INS and a 64-beam Lidar sensor were combined to achieve robust position RMS errors of 9 cm in dynamic environments. However, this system suffers from high costs and a high computational burden. The 3D Object Tracker [227] can be used to track moving objects in Visual SLAM methods. Another algorithm proposed in [228] uses Canny's edge detector to find dominant edges in the vertical direction of a tree trunk and to select these tree trunks as typical salient features. Deep learning methods are increasingly investigated to deal with the dynamic environment as aforementioned [148–151,166,198].

Another source of drifts is the outlier within the sensor observations. Each sensor has its own error sources. For example, in the case of a camera, the fuzzy image due to high speed and poor light conditions may cause incorrect identification of landmarks. Lidar sensors are sensitive to weather conditions (such as rainfall), and large changes in the road environment. GNSS may suffer from signal blockage. FDI (Fault Detection and Isolation system) techniques can be used to detect measurement outliers and reject the influence of these outliers on positioning and localization [229].

The aforementioned SLAM error sources may also result in incorrect data association, which is an important process to associate measurement(s) to a specific landmark. Wrong data association may happen due to not only the noisy sensor data, inconsistency, wrong detection of dynamic objects, etc., but also to some specific road environments. For instance, the highway environment is sometimes visually repetitive and contains many similar features, which makes it difficult to recognize a previously explored area.

Some researchers avoid the challenge of wrong data association directly at the frontend step of SLAM by using RANSAC [230], which is commonly used in Visual SLAM to reject outliers. In [231], the authors proposed a middle layer, referred to as Graph-Tinker (GTK), that can detect and remove false-positive loop closures. Artificial loop closures are then injected into the pose graph when using an Extended Rauch–Tung–Striebel smoother framework.

The data association challenge can also be addressed at the backend step since there is still a chance that outliers are not totally eliminated. The concept of Switchable Constraints (SC) was introduced in [232], such that a switchable variable is introduced into each loop closure constraint. Once a constraint is considered as an outlier, it can be turned off during optimization. In [233], the authors introduced an algorithm known as Realizing, Reversing, and Recovering (RRR), which is a consistency-based loop closure verification method. More recently, Carlone et al. [234] used $\ell 1$_relaxation to select "reliable" measurements, and Carlone and Calafiore [235] use convex relaxations to solve the nonconvex problem without the need for an initial guess of unknown poses. The potential causes of SLAM drifts and the corresponding suggested solutions are summarized in Table 2.

Table 2. Potential causes of SLAM drifts and solutions.

SLAM Drift	Possible Solutions
Linearization error	• Variants of EKF-SLAM [208–210]; • Nonparametric approaches [28,211–214]; • Local map joining [205,207]; • Gradient Descent based optimization scheme [50,215]; • Improved initialization method [216–218,220]; • Aiding with inertial measurements [221–225].
Sensor outliers	• Fault Detection and Isolation [229]; • Sensor fusion to compensate for different sensor errors [236,237].
Dynamic objects	• Probabilistic maps [226]; • 3D Object Tracker [227]; • Salient feature detection [228]; • Deep learning-based methods [148–151,166,198].
Wrong data association	• RANSAC [230]; • Graph-Tinker [231]; • Switchable Constraints [232]; • RRR [233]; • ℓ1_relaxation [234], convex relaxations [235].

4.4. *Lack of Quality Control*

The quantitative evaluation of the SLAM algorithms is another important challenge. There are some criteria to evaluate SLAM algorithms, such as their accuracy, scalability, availability, recovery (which is the ability to localize the vehicle inside a large-scale map), and updatability. Quantitative analysis of the performance of SLAM algorithms is essential since they can provide numerical evaluation and a basis for comparison of different SLAM algorithms.

Estimation accuracy is a widely used quality analysis metric, but it can be difficult in practice for autonomous driving. Most approaches evaluate the performance of SLAM algorithms by comparing the results to ground truth using, for example, an accurate map. However, a suitable ground truth map is seldom available. Sometimes the estimated map is evaluated by overlaying it onto the floor plan and searching for differences [238], which is harder for outdoor applications and needs human operator intervention [239]. Two popular accuracy metrics, relative pose error (RPE) and absolute trajectory error (ATE), were proposed by Sturm et al. [240] which evaluate a Visual SLAM system by comparing the estimated camera motion against the true trajectory, instead of doing complex map comparison. The RPE measures the local accuracy of trajectory over a fixed time interval, while ATE compares the absolute distances between the estimated and the ground truth trajectory and thus estimates the global consistency. These two trajectories should first be aligned using the Horn method [240]. According to [240], the RPE considers both translational and rotational errors, while the ATE only considers the translational error. These metrics have been widely used by the SLAM community for evaluating and comparing different SLAM approaches. However, similar to maps, the precise location of the vehicle trajectory on the actual road surface may not always be available. In [239], the authors proposed a framework for analyzing the accuracy of SLAM by measuring the error of the corrected trajectory. Root Mean Square Error (RMSE) of vehicle poses is normally used to indicate the accuracy of the SLAM trajectory estimation result. Another widely used quality analysis method is the Chi-squared (χ^2) test. According to [241], the χ^2 test is a statistic test to quantify the quality of the provided covariance matrices for landmark measurements and odometry error. When the minimum χ^2 error is nearly equal to the difference of the dimension of the measurement vector and the size of the state vector, the measure would be considered as being of good quality [241].

Some researchers [242–244] have considered the consistency of their SLAM algorithms. According to [242], the major reason for SLAM inconsistency is the accumulated error caused by the incorrect odometry model and inaccurate linearization of the SLAM nonlinear functions. When the estimation error is beyond the uncertainty, it can be assumed that the estimation results are inconsistent. EKF-SLAM suffers from such an inconsistency problem unless the Jacobians of observation/odometry functions are evaluated around the true system state. In [30] and [245], the consistency of fastSLAM and EKF-SLAM algorithms was quantitatively determined using the measure indicator normalized estimation error squared (NEES). In [246], observability properties of the filter's error state model were analyzed to investigate the fundamental causes of the inconsistency of EKF-SLAM. In the work of [247], the consistency of an incremental graph SLAM was checked by applying a χ^2 test to the weighted sum of measurement residuals. Whether inconsistency can be tolerated ultimately depends on the application of the SLAM results [19].

The reliability of the output of the localization, mapping, and navigation system should also be checked. However, few studies have been made on the quantitative analysis of the reliability of SLAM. Some reliability studies for other localization systems (such as GNSS, GNSS/INS) can be used as a reference to guide the SLAM community. System reliability can be considered as having two components: internal reliability and external reliability. The former identifies the ability of the system to detect faults, which is quantified by the Minimal Detectable Bias (MDB), and is indicated by the lower bound for detectable faults. The latter estimates the influence of undetected faults on the final solution [175,248–251]. When the MDB value is low, the system is more reliable. Similarly, the reliability of the SLAM system feature observation model and vehicle motion model can also be evaluated with these approaches.

Integrity is very important, as it is an indicator of the "trustworthiness" of the information supplied by the localization system, and can provide timely warning of the risks caused by inaccuracy [252]. Integrity measures are used to quantify the requirements for localization safety. The concept was first established in aviation and is also applicable to land vehicle localization [253]. Due to the strict safety requirement of autonomous driving, there is increasing attention to integrity by autonomous driving researchers. The localization and navigation of a self-driving car are based on the use of multiple sensors, therefore the traditional integrity analysis methods for GNSS should be extended. Fault detection and isolation (FDI) is one of the most popular alert generation approaches for GNSS-based localization [229,254–256].

5. Lidar/GNSS/INS Based Mapping and Localization: A Case Study

The Lidar-based Simultaneous Localization and Mapping (SLAM) technology approach is widely studied and used in the robotics field because Lidar can generate a very dense 3D point cloud with a fast sensing rate and high accuracy. Normally the SLAM system experiences estimation error which increases with the travel distance, thus it needs "loop closure" to correct the errors. However, the closed loop is hard to achieve in some large-scale outdoor applications of autonomous driving, such as driving on a highway, or a complex trajectory in urban areas. Furthermore, the Lidar-only SLAM will only provide the relative localization information. Therefore, the combination of GNSS/INS with Lidar SLAM will effectively reduce the dependence on loop closures and provide absolute positioning information.

Furthermore, a Lidar system can also support localization using existed HD maps when GNSS signals are not available. A modernized SLAM procedure that combines Lidar, GNSS, and INS is tested here. This procedure contains two parts: Lidar/GNSS/INS-based offline mapping part, and Lidar/HD map-based online localization and mapping part.

5.1. Experiment Setup

Land vehicle tests were conducted in some urban areas of Sydney, Australia, to test the proposed Lidar/GNSS/INS multi-sensor system. The vehicle was equipped with a VLP-16

LiDAR sensor, a tactical-grade IMU sensor, and two GNSS antennas from PolyExplore, Inc., San Jose, CA, USA (Figure 5). The second antenna can be used to provide a dual-antenna-aided heading update for the online localization system. The sampling rate of the Lidar was 10-Hz, the sample rate of GNSS was 1-Hz, and for the IMU it was 100-Hz.

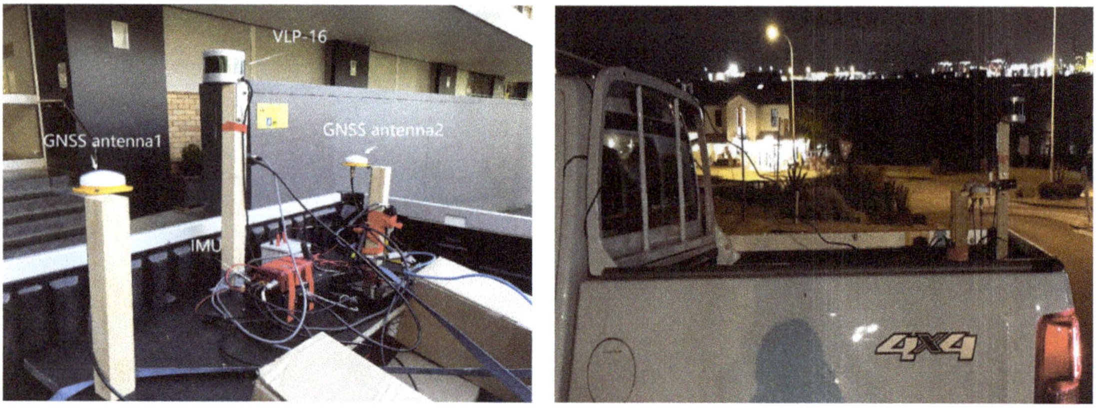

(**a**) (**b**)

Figure 5. Experimental platform: (**a**) The multi-sensor system, (**b**) side view of the system installed within a vehicle.

The trajectory of the road test is shown in Figure 6a. The vehicle was driven from the campus of the University of New South Wales (UNSW) in Kensington to La Perouse (Section A), and then back to UNSW (Section B). In this study, the forward journey (from UNSW to La Perouse) was used to a produce high precision 3D point cloud map of the road, and the backward journey (from La Perouse to UNSW) was used to test the performance of the Lidar/3D point cloud map-based localization method.

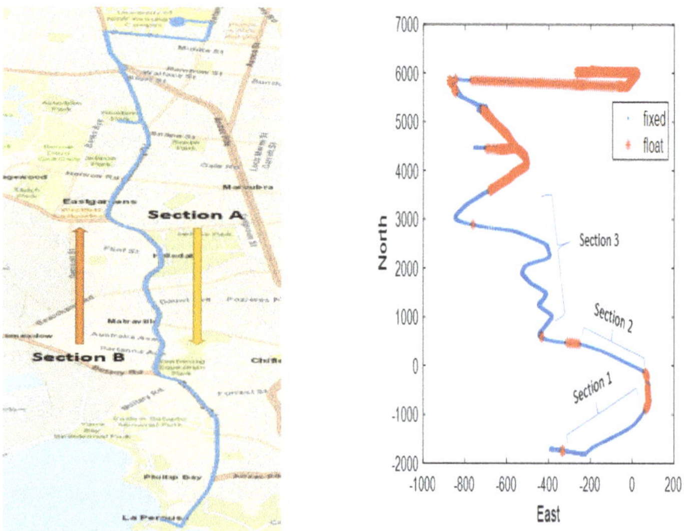

(**a**) (**b**)

Figure 6. (**a**) The road test trajectory (in blue) on Google Maps; (**b**) GNSS/INS localization of the whole trajectory with RTK positioning status in a local coordinate system.

In order to conduct a quantitative analysis of the localization performance, three sections of the trajectory were selected (Figure 6b). For each of the selected sections on the driving trajectory, the GNSS-RTK status for the forward journey and backward journey was "integer-ambiguity fixed". Hence, the offline mapping results are expected to be accurate at about the 5 cm level. For the backward journey (from La Perouse to UNSW), the selected sections will have accurate GNSS/INS positioning results as a reference to evaluate the performance of the Lidar/3D point cloud map-based localization method.

5.2. Lidar/GNSS/INS Mapping

The acquired dataset of the forward journey (from UNSW to La Perouse) was used to generate a georeferenced point cloud map of the road environment. The georeferenced map was generated using Lidar odometry frame-to-frame matching and GNSS/INS positioning/attitude. Figure 7 shows an overview of the offline mapping system architecture.

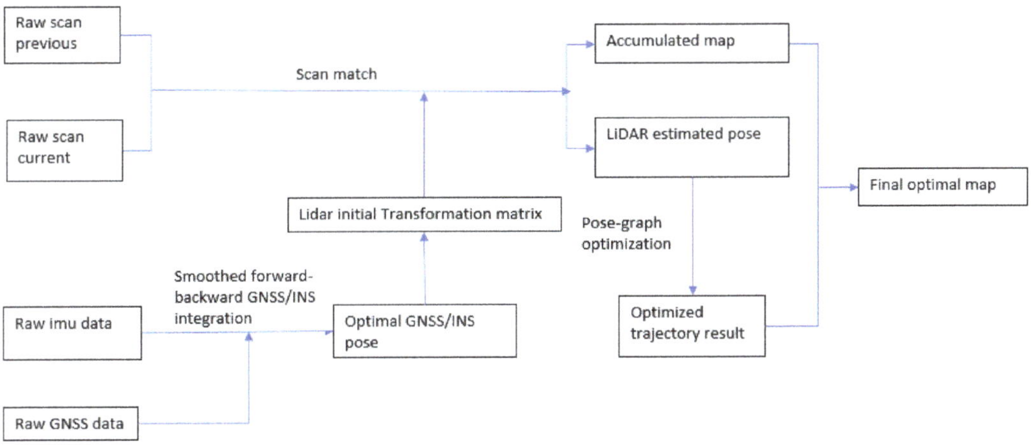

Figure 7. Overview of the Lidar/GNSS/INS mapping system architecture.

The GNSS/INS system can provide the geodetic positioning and attitude information. Since this map generation was performed offline, an optimal GNSS/INS trajectory can be obtained. The GNSS/INS-derived position and attitude results were used as initial values for the frame-by-frame matching to transfer the newly merged point cloud to the referenced frame. In this way, the point cloud can be georeferenced. When GNSS results are unavailable the inertial navigation 6-DOF pose results can be used to generate the initial transformation before the GNSS signals are reacquired. When conducting Lidar odometry, each current frame was matched to the previous frame with Normal Distributions Transform (NDT) scan matching algorithm, with the initial transformation information provided by GNSS/INS. The point clouds were firstly pre-processed to remove the ground plane point (Figure 8), before matching by NDT to improve the accuracy of registration.

Figure 9 shows two scan views before being matched. It appears that these two scan views have slight differences in features. The matched point cloud from the two scan views can be generated (Figure 10).

By conducting Lidar odometry sequentially with all the available Lidar scans, the newly matched point cloud can be merged with the previously generated point cloud maps, and the accumulated map of the whole trajectory can be obtained and georeferenced, as shown in Figure 11.

By enlarging Figure 11, details of the road map can be seen, and its corresponding real-world road view can be compared with Google Earth images (since this map is georeferenced). Figure 12 shows a comparison of one zoomed-in section of this generated map and the corresponding view in Google Earth.

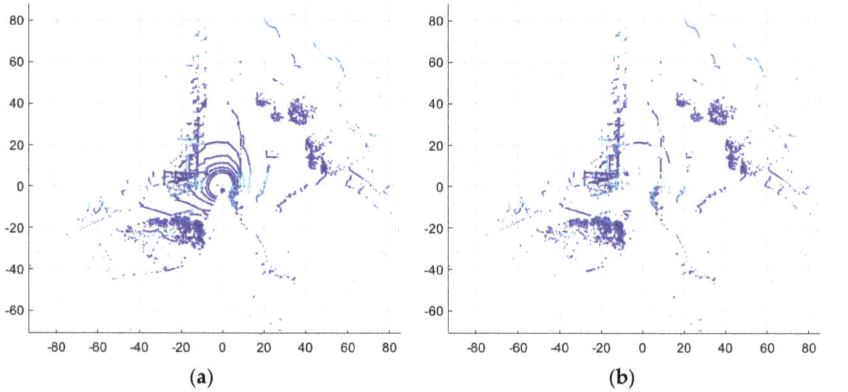

Figure 8. Scan view of a Lidar scan frame (**a**) the original scan view; (**b**) the view after pre-processing.

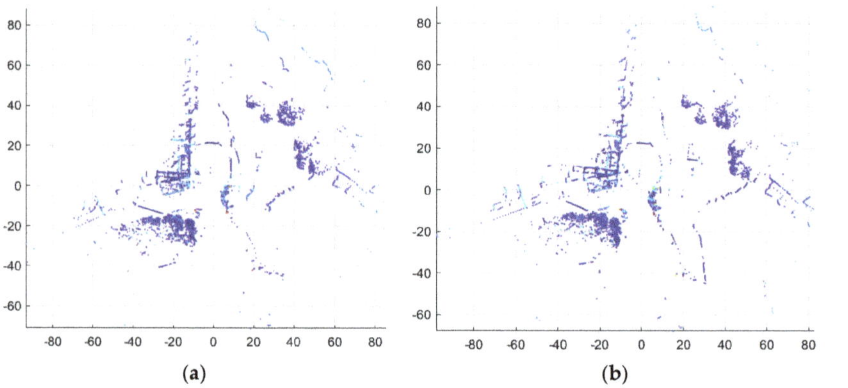

Figure 9. Scan views of two sequenced Lidar scan frames ((**a**) previous scan frame; (**b**) current scan frame) for scan matching.

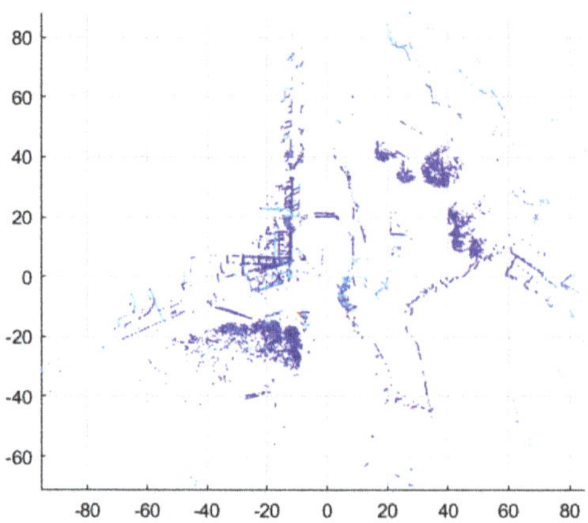

Figure 10. Generated map point cloud after matching two sequenced Lidar scan frames.

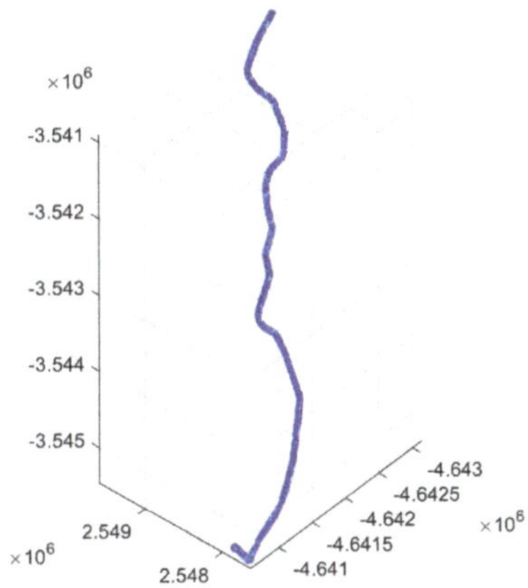

Figure 11. Global georeferenced road map from UNSW to La Perouse (frame: ECEF, unit: meter) from the 3D point cloud-based map.

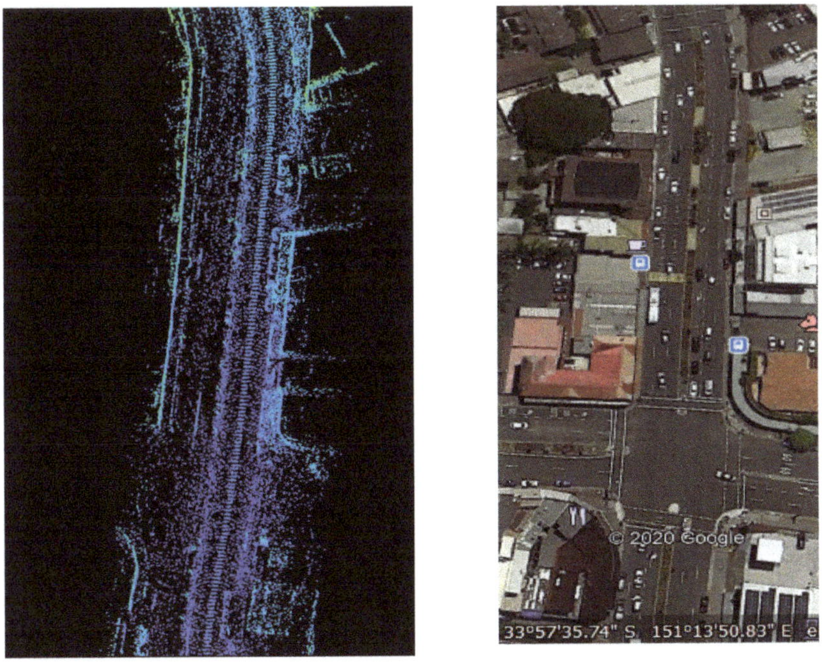

(**a**) (**b**)

Figure 12. A section of (**a**) generated map, and (**b**) the Google Earth view for the same location.

This generated map shows a good structure of the road environment, including the road edge, buildings, trees, and parked vehicles along the road.

Three control points with known coordinates are placed around the UNSW Scientia Lawn. These control points can be used to evaluate the accuracy of the generated point cloud. By comparing the coordinates of the identified control points within the map to the real known position, it is found the difference at each axis X, Y, and Z is around 2–8 cm. Therefore, the offline-generated map accuracy is considered to be 5 cm.

5.3. Localization with Lidar Scans and the GeoReferenced 3D Point Cloud Map Matching

The georeferenced 3D point cloud map produced from the data of Section A (the forward journey from UNSW to La Perouse) can then be used to support the Lidar-based localization for Section B (the backward journey from La Perouse to UNSW) by matching the Lidar scans to the map. The procedure of the online Lidar/3D map matching-based localization method is shown in Figure 13. An INS is used to support the Lidar/3D map-based online localization. In order to show the performance of different fusion levels, two fusion methods were investigated. The first method simply utilizes the IMU as an assistant sensor that directly uses the INS solution as initial information for scan/map matching. The second fusion method is a tightly coupled one that not only uses the INS solution to support matching, but also contains an EKF-based error state update step that enhances the inertial navigation performance.

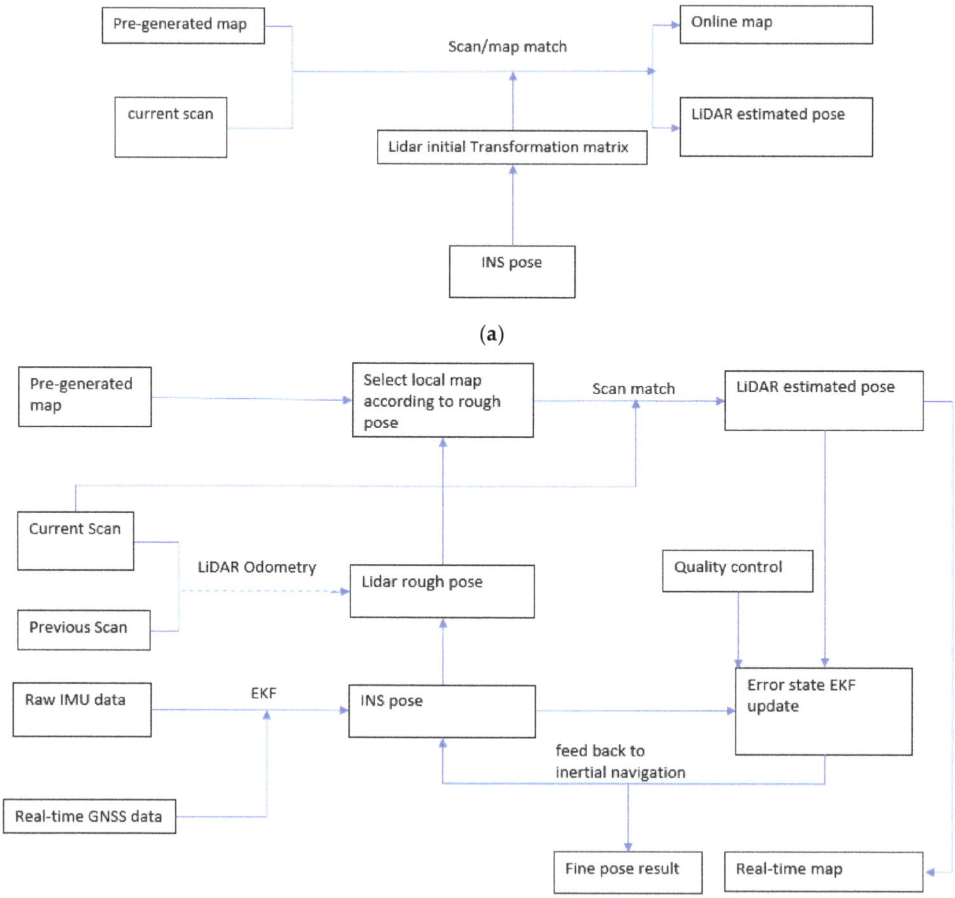

(b)

Figure 13. Overview of the proposed Lidar/3D map matching based localization system architecture. (**a**) Method 1: fusing IMU as an assistant sensor; (**b**) Method 2: fusing IMU using the EKF-based tightly coupled method.

Method 2 consists of two parts: scan matching and EKF fusing. Firstly, if the inertial navigation information is not available, the frame-to-frame Lidar odometry can be used to support localization. After initializing the error-state EKF, the estimated pose from the inertial navigation will provide a rough pose for the current Lidar scan frame, and the Lidar odometry can be shut down to lower the computation load. With the rough position provided by the INS, a local map is searched and selected from the pre-generated global map to improve the matching efficiency. NDT-based scan matching between the current Lidar frame and the local map is undertaken with the inertial-based initial Transformation Matrix. The Lidar-estimated vehicle pose can be obtained. A new real-time road map can also be generated if needed.

After obtaining the Lidar pose, the difference between the Lidar pose and the inertial propagation pose can be obtained, and the error within the inertial navigation information is estimated by the error-state EKF and then fed back to the inertial system to improve the pose results and bias estimation. When GNSS information is available, such as the RTK position results, these can also be used to correct the inertial navigation information to improve the accuracy and reliability of the localization system.

For data fusion of Lidar, INS, and GNSS, some current work has proposed using the graph optimization-based method to generate optimal localization and mapping solutions [179,257–259]. However, some of them are post-processed or highly dependent on GNSS data to mitigate the navigation drift, or even ignore the IMU bias. Since for our online Lidar/map matching based localization method, a reliable inertial navigation solution is essential to provide a good initialization for the scan/map matching process, and to increase the efficiency and accuracy of local map searching and selection, in-time IMU bias correction is critical and is more easily achieved using an EKF. As our test is undertaken within an urban area where GNSS signals are frequently lost, the feedback to the IMU states should also depend on the Lidar data, especially during GNSS outages. Moreover, the estimation uncertainty, which is an important parameter for the analyzing system solution, is seldom estimated in graph-based methods but can be directly estimated through the EKF method. Therefore in our current Method 2, the EKF method is used to fuse the Lidar/map localization, GNSS, and inertial navigation results. A comparison of Method 1 and Method 2 also highlights the difference between using an IMU sensor as a separate aiding sensor and as a tightly-coupled aiding sensor.

5.3.1. Estimation Results of Lidar/3D Map-Based Localization System

Since there is no ground truth information for this urban road test, the Lidar/map matching-based solutions are compared with the GNSS/INS solution within the three selected trajectory sections (Figure 14), during which the RTK status is "ambiguity-fixed" (Figure 6b).

Table 3 shows the comparison between the Lidar/map matching-based localization and the reference GNSS/INS localization results. For Method 1, the result differences fluctuate around zero, and their mean values are at the centimeter to decimeter level. The standard deviation for all three sections is around 0.1–0.2 m, therefore we treat the difference of the coordinates larger than 0.6 m as indicating possible outliers. The epochs that have outliers are about 1.7% of the total test data, which means the presence of outliers is rare. The possible reason for the outlier will be discussed in the next Section. For Method 2, the result has better accuracy. The standard deviation is around 0.05 m, much lower than that of Method 1. It can be seen from Figure 14, that Method 2 has a lower difference to the reference during the periods that Method 1 shows possible outliers, indicating that the tightly coupled method is more robust to outliers than simply using the INS solution for initialization.

5.3.2. Quality Analysis of the Numerical Results

The details of measurements during the epochs with big jumps (such as during the "red and green boxes" in Figure 14) are checked to investigate possible causes of the detected outliers. For Trajectory Section 1 in Figure 14 (red box), it is found that when driving around a roundabout, there were some big outliers by Method 1. The trajectory of the Lidar/map system and the GNSS/INS solution around this roundabout and their views in Google Maps are shown in Figure 15. It can be seen that the GNSS/INS solutions (Figure 15 yellow line) are smoother in this area since the GNSS integer ambiguities are "fixed", while the Lidar/map solution has some differences to the reference trajectory if only using the IMU as a simple assistant (Method 1, Figure 15 blue line).

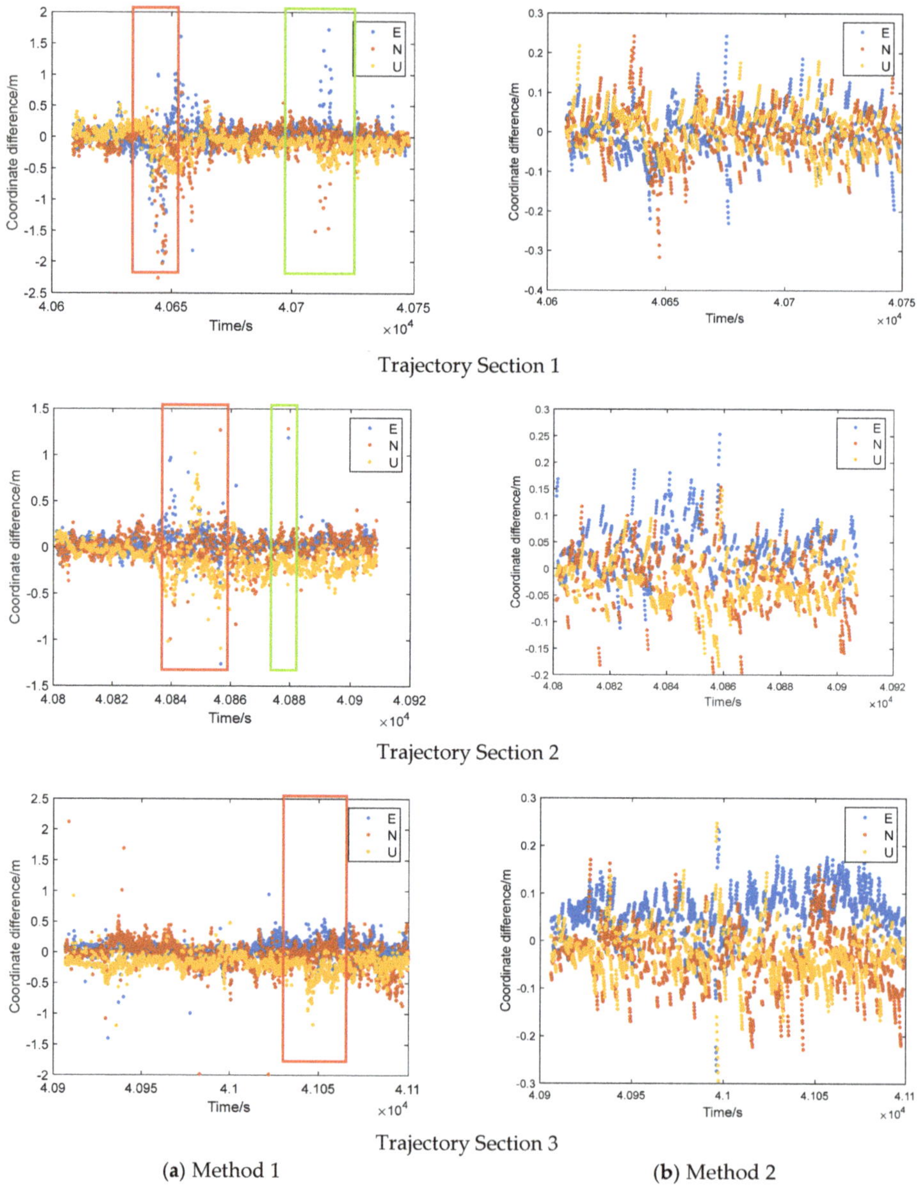

Figure 14. Coordinate difference between the proposed Lidar/map matching-based localization method and the reference GNSS/INS localization method at three trajectory sections. (**a**) Method 1: fusing IMU as an assistant sensor; (**b**) Method 2: fusing IMU using the EKF-based tightly coupled method. The red and green boxes indicate epochs with big coordinate difference with different timestamps.

Table 3. Mean and standard deviation for the difference between Lidar/map matching-based localization and the reference GNSS/INS localization results, for Trajectory Sections 1, 2 and 3 in Figure 14. (a) Method 1: fusing IMU as an assistant sensor; (b) Method 2: fusing IMUs using the EKF-based tightly coupled method.

		Trajectory Section 1	Trajectory Section 2	Trajectory Section 3
Mean (m)		Method 1		
	East	0.020	−0.036	0.051
	North	−0.035	0.0031	−0.048
	Up	−0.084	0.140	−0.189
		Method 2		
	East	−0.0026	0.0358	0.0571
	North	−0.0052	−0.0221	−0.0371
	Up	0.0041	−0.0250	−0.0228
Stdev (m)		Method 1		
	East	0.142	0.099	0.128
	North	0.162	0137	0.188
	Up	0.182	0.151	0.123
		Method 2		
	East	0.0556	0.0466	0.0503
	North	0.0605	0.0530	0.0574
	Up	0.0481	0.0410	0.0486

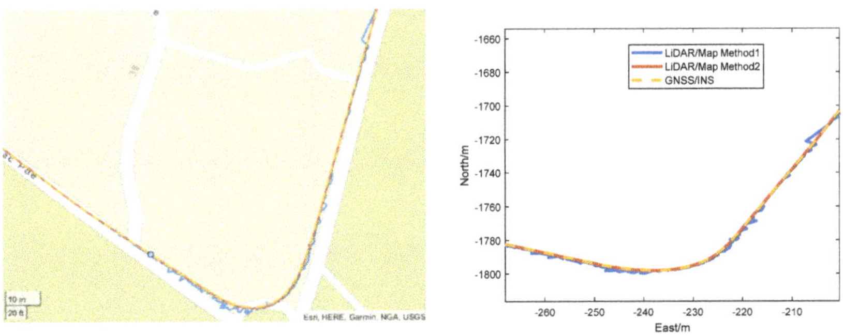

(**a**) (**b**)

Figure 15. Trajectory of Lidar/map matching-based localization Method 1 (blue), Method 2 (Red) and GNSS/INS localization (Yellow): (**a**) view in Google Map; (**b**) view in local coordinate system around the roundabout.

Figure 16 shows the map view at this roundabout. It can be seen that the structure of the pre-generated map on the driving side of the road is not very clear since it lacks features around the trajectory. The roundabout is located at a parking area of a tourist attraction. There is no building and very few trees around this area. Since the testing was undertaken in the evening, there were not many parked vehicles that could be used as features. Therefore the quality of the matching step may be poorer, which results in degraded localization accuracy.

Figure 17 shows the Lidar scan view at this point with a range threshold of 20 m, and it can be seen that this Lidar scan does not have many usable features, especially after pre-processing. Similar road environments with fewer features can be found when another cluster of outliers appeared in Trajectory Section 1 (green box), shown in Figure 14. In this situation, extending the range threshold may enhance the accuracy by including more features, however, it will increase the computational burden and be impacted by more outlier sources. Incorporating the inertial motion model by fusing the IMU more tightly may make the localization system more robust to this featureless condition (Figure 15 red line).

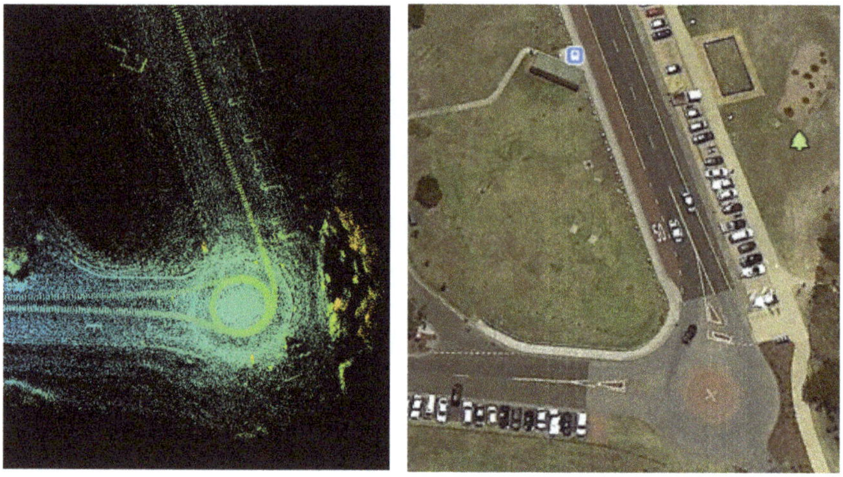

(**a**) (**b**)

Figure 16. A section of (**a**) the generated map, and (**b**) the Google Earth view for the same location around the roundabout.

(**a**) (**b**)

Figure 17. Scan frame at epoch 40,647 s with big outlier in Trajectory Section 1 around the roundabout ((**a**) the original scan view; (**b**) the view after pre-processing).

Another major source of outliers is the other moving entities around the host vehicle. Figures 18–20 show the Lidar views when there are big outliers in the localization stage.

No matter whether the other moving vehicle is on the same side of the road or the other side of the road, such a moving vehicle will influence the quality of the Lidar/map matching-based localization. When the moving vehicle is on the same side of the road as the host vehicle, it will result in bad estimation when it is initially detected or has a different speed from the host vehicle, or when it turns and drives onto another road. This will make this vehicle no longer detectable (red box in Figure 14 Trajectory Section 2 and Figure 18).

(**a**) (**b**)

Figure 18. Scan frame at epoch 40,854 s with big outlier in Trajectory Section 2: a following vehicle is driving to another road, (**a**) Lidar scan; (**b**) Google Earth view.

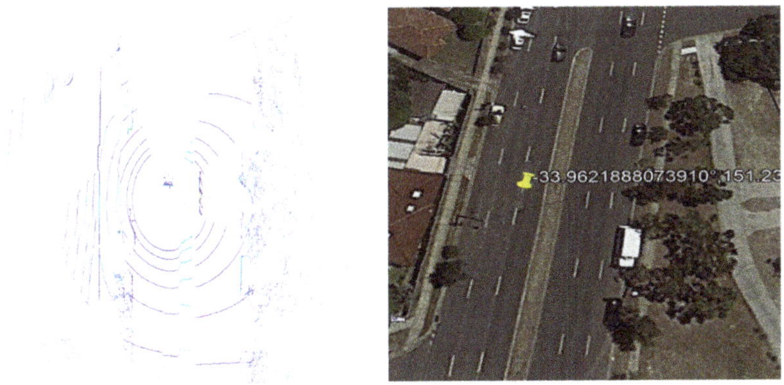

(**a**) (**b**)

Figure 19. Scan frame at epoch 40,888 s with big outlier in Trajectory Section 2, an opposite driving vehicle detected, (**a**) Lidar scan; (**b**) Google Earth view.

(**a**) (**b**)

Figure 20. Scan frame at epoch 41,103 s with big outlier in Trajectory Section 3, with one tall bus driving past (**a**) Lidar scan; (**b**) Google Earth view.

Once the host vehicle detected an opposite driving vehicle, the localization estimation errors could reach 1–1.5 m (green box in Figure 14 Trajectory Section 2 and Figure 19).

The type of moving elements will also impact the presence of outliers. Most of the time, the vertical position estimation is less influenced by the moving elements. However, when checking the red box in Trajectory Section 3 (Figure 14), it can be found that the differences in the vertical direction are much higher than in other sections. By looking at the details of the Lidar view it is found that, at that section, a tall bus was driving past the host vehicle (Figure 20), which means vertical differences between the current Lidar scan and the pre-generated map might have caused some systematic vertical biases.

The moving objects within the road environment will be a major source of measurement outliers because the system treated the pre-generated map as a fixed reference map. Therefore, if there are any moving objects that cause the structures of the pre-generated map and the current scan frame to be different, outliers will occur. The moving objects, such as other vehicles, may exist in both the previous road mapping stage and in the current road scans for use in localization. For the step of offline HD map generation, such moving objects should be carefully identified and removed from the static 3D point cloud maps. For the online step, the moving objects could be identified and removed based on the cleaned pre-generated map, or directly achieved semantic segmentation with sensor data. Some researchers have developed some moving object segmentation methods, such as LMNet [166] which can distinguish moving and static objects based on CNN. Therefore, the possible detected moving objects could be removed, or the possible road environment change could be updated to the global map to enhance the accuracy of future driving around the same road path. These methods may be undertaken during the perception step. The aid of some numerical quality control methods may also contribute to this task at the localization and mapping steps, such as the FDI method, or outlier detection and identification methods, which can directly estimate and mitigate the influence of outliers from all kinds of resources, not only the moving outlier, and also other sensor or model faults.

Some FDI methods or integrity monitoring methods [255,256] have already been successfully applied to the GNSS/INS integration system under an EKF framework. Since in this case study the EKF method is used to fuse Lidar/map localization results and INS pose to generate high-frequency precise pose solutions, these quality control methods also indicate the potential for this proposed localization system. This will be a future research topic.

6. Conclusions

This paper gives a brief review of different SLAM approaches and their characteristics. SLAM has become a key approach for localization, mapping, planning, and controlling in automated driving. It shows promising progress in generating high-resolution maps for autonomous driving and for vehicle localization in road environments. The advantages and disadvantages of different SLAM techniques have been identified and their applications for autonomous driving have been discussed.

The trustworthiness of localization and navigation algorithms is an important issue for autonomous driving. There are many challenges that limit the performance of the SLAM techniques, which affect the safety of the localization and navigation results. These challenging issues, and possible solutions, are mentioned in this review. Furthermore, in order to ensure safety, the performance of the algorithms should be quantitatively evaluated with respect to such measures as accuracy, consistency, precision, reliability, and integrity. The methods to evaluate these qualities are briefly discussed.

A real-world road test was conducted to demonstrate the application of SLAM for autonomous driving with multi-sensor integration. The numerical results show that a GNSS/INS-aided-Lidar system can generate a georeferenced high-density point cloud map. This pre-generated map can then be used to support online localization, which has achieved about centimeter-level accuracy. This Lidar/map matching-based localization method may also be useful to support an autonomous driving system during periods when GNSS signals are unavailable, which makes it suitable for urban area driving. A more tightly coupled fusion of IMU measurements will make the Lidar/map-based localization more accurate and robust to outliers than simply utilizing the inertial solution as assistant information.

Future studies should be focused on how to detect moving entities and mitigate their impact in the 3D point cloud mapping and localization process. In addition, integrity monitoring procedures for such Lidar/GNSS/INS-based vehicle localization and mapping system should be investigated.

Author Contributions: J.W. and S.Z. had the idea for the article. S.Z. performed the literature search, conducted the experiments and corresponding analysis, and wrote and revised the manuscript. J.W. reviewed and commented on the draft manuscript. W.D. supported the experiment setup. C.R. and A.E.-M. provided critical feedback. All authors have read and agreed to the published version of the manuscript.

Funding: This work is supported by the Australian Research Council (ARC) Project No. DP170103341.

Data Availability Statement: The data presented in this study are available on request from the corresponding author. The data are not publicly available due to the size of the data.

Conflicts of Interest: The authors declare no conflict of interest.

References

1. Litman, T. Autonomous Vehicle Implementation Predictions: Implications for Transport Planning. In Proceedings of the 2015 Transportation Research Board Annual Meeting, Washington, DC, USA, 11–15 January 2015.
2. Katrakazas, C.; Quddus, M.; Chen, W.H.; Deka, L. Real-time motion planning methods for autonomous on-road driving: State-of-the-art and future research directions. *Transp. Res. C-EMER* **2015**, *60*, 416–442. [CrossRef]
3. Seif, H.G.; Hu, X. Autonomous driving in the iCity- HD maps as a key challenge of the automotive industry. *Engineering* **2016**, *2*, 159–162. [CrossRef]
4. Suganuma, N.; Yamamto, D.; Yoneda, K. Localization for autonomous vehicle on urban roads. *J. Adv. Control Autom. Robot.* **2015**, *1*, 47–53.
5. Fuentes-Pacheco, J.; Ruiz-Ascencio, J.; Rendon-Mancha, J.M. Visual simultaneous localization and mapping: A survey. *Aritf. Intell. Rev.* **2015**, *43*, 55–81. [CrossRef]
6. Pupilli, M.; Calway, A. Real-time visual SLAM with resilience to erratic motion. In Proceedings of the 2006 IEEE Computer Society Conference on Computer Vision and Pattern Recognition (CVPR'06), 2006, New York, NY, USA, 17–22 June 2006; pp. 1244–1249. [CrossRef]
7. He, L.; Lai, Z. The study and implementation of mobile GPS navigation system based on Google maps. In Proceedings of the International Conference on Computer and Information Application (ICCIA), Tianjin, China, 3–5 December 2010; pp. 87–90. [CrossRef]
8. Hosseinyalamdary, S.; Balazadegan, Y.; Toth, C. Tracking 3D moving objects based on GPS/IMU navigation solution, Laser Scanner Point Cloud and GIS Data. *ISPRS Int. J. Geo.-Inf.* **2015**, *4*, 1301–1316. [CrossRef]
9. Wang, L.; Zheng, Y.; Wang, J. Map-based Localization method for autonomous vehicle using 3D-LiDAR. *IFCA Pap.* **2017**, *50*, 278–281.
10. Liu, R.; Wang, J.; Zhang, B. High definition map for automated driving, overview and analysis. *J. Navig.* **2020**, *73*, 324–341. [CrossRef]
11. HERE. HERE HD Live Map—The Most Intelligent Vehicle Sensor. 2017. [Online]. Available online: https://here.com/en/products-services/products/here-hd-live-map (accessed on 19 March 2017).
12. TomTom. 2017, RoadDNA, Robust and Scalable Localization Technology. Available online: http://download.tomtom.com/open/banners/RoadDNA-Product-Info-Sheet-1.pdf (accessed on 25 May 2019).
13. Bresson, G.; Alsayed, Z.; Yu, L.; Glaser, S. Simultaneous localization and mapping: A survey of current trends in autonomous driving. *IEEE Trans. Intell. Veh.* **2017**, *2*, 194–220. [CrossRef]
14. Kuutti, S.; Fallah, S.; Katsaros, K.; Dianati, M.; Mccullough, F.; Mouzakitis, A. A survey of the state-of-the-art localization techniques and their potentials for autonomous vehicle applications. *IEEE Internet Things* **2018**, *5*, 829–846. [CrossRef]
15. Wang, C.C.; Thorpe, C.; Thrun, S.; Hebert, M.; Durrant-Whyte, H. Simultaneous Localization, Mapping and Moving Object Tracking. *Int. J. Robot. Res.* **2007**, *26*, 889–916. [CrossRef]
16. Smith, R.; Cheeseman, P. On the representation and estimation of spatial uncertainty. *Int. J. Robot. Res.* **1986**, *5*, 56–68. [CrossRef]
17. Agarwal, P.; Burgard, W.; Stachniss, C. Geodetic approaches to mapping and relationship to graph-based SLAM. *IEEE Robot. Autom. Mag.* **2014**, *21*, 63–80. [CrossRef]
18. Thrun, S.; Burgard, W.; Fox, D. *Probabilistic Robotics*; MIT Press: Cambridge, MA, USA, 2005; p. 246.

19. Dissanayake, G.; Huang, S.; Wang, Z.; Ranasinghe, R. A review of recent developments in simultaneous localization and mapping. In Proceedings of the 2011 6th International Conference on Industrial and Information Systems, Kandy, Sri Lanka, 16–19 August 2011; pp. 477–482.
20. Guivant, J.E.; Nebot, E.M. Optimization of the Simultaneous Localization and Map-building Algorithm for Real-time Implementation. *IEEE Trans. Robot. Autom.* **2001**, *17*, 242–257. [CrossRef]
21. Williams, S.B. Efficient Solutions to Autonomous Mapping and Navigation Problems. Ph.D. Thesis, University of Sydney, Australian Centre for Field Robotics, Sydney, Australia, 2001.
22. Bailey, T. Mobile Robot Localisation and Mapping in Extensive Outdoor Environments. Ph.D. Thesis, University of Sydney, Australian Centre for Field Robotics, Sydney, Australia, 2002.
23. Paz, L.M.; Tardos, J.D.; Neira, J. Divide and Conquer: EKF SLAM in O(n). *IEEE Trans. Robot.* **2008**, *24*, 1107–1120. [CrossRef]
24. Chli, M.; Davison, A.J. Automatically and efficiently inferring the hierarchical structure of visual maps. In Proceedings of the 2009 IEEE International Conference on Robotics and Automation, Kobe, Japan, 12–17 May 2009; pp. 387–394. [CrossRef]
25. Thrun, S.; Liu, Y. Multi-robot SLAM with Sparse Extended Information Filers. In *Robotics Research. The Eleventh International Symposium, Siena, Italy, 19–22 October 2005*; Dario, P., Chatila, R., Eds.; Springer Tracts in Advanced Robotics; Springer: Berlin/Heidelberg, Germany, 2005; Volume 15, pp. 254–266.
26. Grisetti, G.; Stachniss, C.; Burgard, W. Improved techniques for grid mapping with Rao-Blackwellized Particle Filters. *IEEE Trans. Robot.* **2007**, *23*, 34–46. [CrossRef]
27. Eade, E.; Drummond, T. Scalable monocular SLAM. In Proceedings of the IEEE Computer Society Conference on Computer Vision and Pattern Recognition, New York, NY, USA, 17–22 June 2006; IEEE: New York, NY, USA, 2006; Volume 1, pp. 469–476. [CrossRef]
28. Thrun, S.; Montemerlo, M.; Koller, D.; Wegbreit, B.; Nieto, J.; Nebot, E.M. Fastslam: An efficient solution to the simultaneous localization and mapping problem with unknown data association. *J. Mach. Learn. Res.* **2004**, *4*, 380–407.
29. Durrant-Whyte, H.; Bailey, T. Simultaneous localization and mapping: Part I. *IEEE Robot. Autom. Mag.* **2006**, *13*, 99–110. [CrossRef]
30. Bailey, T.; Nieto, J.; Nebot, E. Consistency of the FastSLAM algorithm. In Proceedings of the 2006 IEEE International Conference on Robotics and Automation, 2006, ICRA 2006, Orlando, FL, USA, 15–19 May 2006; pp. 424–429. [CrossRef]
31. Press, W.; Teukolsky, S.; Vetterling, W.; Flannery, B. *Numerical Recipes*, 2nd ed.; Cambridge University Press: Cambridge, UK, 1992.
32. Wagner, R.; Frese, U.; Bauml, B. Graph SLAM with signed distance function maps on a humanoid robot. In Proceedings of the 2014 IEEE/RSJ International Conference on Intelligent Robots and Systems, Chicago, IL, USA, 14–18 September 2014; pp. 2691–2698.
33. Ni, K.; Steedly, D.; Dellaert, F. Tectonic SAM: Exact, out-of-core, submap-based SLAM. In Proceedings of the 2007 IEEE International Conference on Robotics and Automation, Rome, Italy, 12 April 2007; pp. 1678–1685.
34. Huang, S.; Wang, Z.; Dissanayake, G.; Frese, U. Iterated D-SLAM map joining: Evaluating its performance in terms of consistency, accuracy and efficiency. *Auton. Robot.* **2009**, *27*, 409–429. [CrossRef]
35. Pinies, P.; Paz, L.M.; Tardos, J.D. CI-Graph: An efficient approach for large scale SLAM. In Proceedings of the 2009 IEEE International Conference on Robotics and Automation, Kobe, Japan, 12–17 May 2009; pp. 3913–3920.
36. Ho, B.; Sodhi, P.; Teixeira, P.; Hsiao, M.; Kusnur, T.; Kaess, M. Virtual Occupancy Grid Map for Submap-based Pose Graph SLAM and Planning in 3D Environments. In Proceedings of the 2018 IEEE/RSJ International Conference on Intelligent Robots and Systems (IROS), Madrid, Spain, 1–5 October 2018; pp. 2175–2182.
37. Kaess, M.; Ranganathan, A.; Dellaert, F. iSAM: Incremental Smoothing and Mapping. *IEEE Trans. Robot.* **2008**, *24*, 1365–1378. [CrossRef]
38. Kaess, M.; Johannsson, H.; Roberts, R.; Ila, V.; Leonard, J.J.; Dellaert, F. iSAM2: Incremental smoothing and mapping using the Bayes tree. *Int. J. Robot. Res.* **2012**, *31*, 216–235. [CrossRef]
39. Ila, V.; Polok, L.; Šolony, M.; Smrz, P.; Zemcik, P. Fast covariance recovery in incremental nonlinear least square solvers. In Proceedings of the 2015 IEEE International Conference on Robotics and Automation (ICRA), Seattle, WA, USA, 26–30 May 2015; pp. 4636–4643.
40. Ila, V.; Polok, L.; Solony, M.; Svoboda, P. SLAM++-A highly efficient and temporally scalable incremental SLAM framework. *Int. J. Robot. Res.* **2017**, *36*, 210–230. [CrossRef]
41. Leonard, J.J.; Durrant-Whyte, H.F. Mobile robot localization by tracking geometric beacons. *IEEE Trans. Robot. Autom.* **1991**, *7*, 376–382. [CrossRef]
42. Dissanayake, M.W.M.G.; Newman, P.; Clark, S.; Durrant-Whyte, H.F.; Csorba, M. A solution to the simultaneous localization and map building (SLAM) problem. *IEEE Trans. Robot. Autom.* **2001**, *17*, 229–241. [CrossRef]
43. Thrun, S.; Koller, D.; Ghahramani, Z.; Durrant-Whyte, H.; Ng, A.Y. Simultaneous mapping and localization with sparse extended information filters. In Proceedings of the Fifth International Workshop on Algorithmic Foundations of Robotics, Nice, France, 15–17 December 2002.
44. Guivant, J.E.; Nebot, E.M. Solving computational and memory requirements of feature-based simultaneous localization and mapping algorithms. *IEEE Trans. Robot. Autom.* **2003**, *19*, 749–755. [CrossRef]
45. Montemerlo, M.; Thrun, S.; Koller, D.; Wegbreit, B. FastSLAM: A factored solution to the simultaneous localization and mapping problem. In Proceedings of the AAAI National Conference on Artiflcial Intelligence, Edmonton, AB, Canada, 28 July–1 August 2002; pp. 593–598.

46. Gutmann, J.-S.; Konolige, K. Incremental mapping of large cyclic environments. In Proceedings of the IEEE International Symposium on Computational Intelligence in Robotics and Automation (CIRA), Monterey, CA, USA, 8–9 November 1999; pp. 318–325.
47. Frese, U.; Larsson, P.; Duckett, T. A multilevel relaxation algorithm for simultaneous localisation and mapping. *IEEE Trans. Robot.* **2005**, *21*, 196–207. [CrossRef]
48. Folkesson, J.; Christensen, H. Graphical SLAM—A self-correcting map. In Proceedings of the IEEE International Conference on Robotics and Automation, ICRA '04, New Orleans, LA, USA, 26 April—1 May 2004; Volume 1, pp. 383–390. [CrossRef]
49. Kummerle, R.; Grisetti, G.; Strasdat, H.; Konolige, K.; Burgard, W. g2o: A General Framework for Graph Optimization. In Proceedings of the IEEE International Conference on Robotics and Automation (ICRA), Shanghai, China, 9–13 May 2011.
50. Grisetti, G.; Stachniss, C.; Grzonka, S.; Burgard, W. A tree parameterization for efficiently computing maximum likelihood maps using gradient descent. In *Robotics: Science and Systems*; Georgia Institute of Technology: Atlanta, GA, USA, 2007.
51. Grisetti, G.; Stachniss, C.; Burgard, W. Non-linear Constraint Network Optimization for Efficient Map Learning. *IEEE Trans. Intell. Transp. Syst.* **2009**, *10*, 428–439. [CrossRef]
52. Debeunne, C.; Vivet, D. A Review of Visual-LiDAR Fusion based Simultaneous Localization and Mapping. *Sensors* **2020**, *20*, 2068. [CrossRef]
53. Besl, P.J.; McKay, N.D. A method for registration of 3-D shapes. *IEEE Trans. Pattern Anal. Mach. Intell.* **1992**, *14*, 239–256. [CrossRef]
54. Segal, A.; Haehnel, D.; Thrun, S. Generalized-ICP. In Proceedings of the Robotic: Science and Systems V Conference, Seattle, WA, USA, 28 June–1 July 2009.
55. Censi, A. An ICP variant using a point-to-line metric. In Proceedings of the 2008 IEEE International Conference on Robotics and Automation, Pasadena, CA, USA, 19–23 May 2008; pp. 19–25. [CrossRef]
56. Biber, P.; Strasser, W. The normal distributions transform: A new approach to laser scan matching. In Proceedings of the 2003 IEEE/RSJ International Conference on Intelligent Robots and Systems (IROS 2003) (Cat. No.03CH37453), Las Vegas, NV, USA, 27–31 October 2003; Volume 3, pp. 2743–2748. [CrossRef]
57. Li, J.; Zhong, R.; Hu, Q.; Ai, M. Feature-Based Laser Scan Matching and Its Application for Indoor Mapping. *Sensors* **2016**, *16*, 1265. [CrossRef]
58. Wolcott, R.W.; Eustice, R.M. Fast LIDAR localization using multiresolution Gaussian mixture maps. In Proceedings of the 2015 IEEE International Conference on Robotics and Automation (ICRA), Seattle, WA, USA, 26–30 May 2015; pp. 2814–2821. [CrossRef]
59. Olson, E.B. Real-Time Correlative Scan Matching. In Proceedings of the 2009 IEEE International Conference on Robotics and Automation, Kobe, Japan, 12–17 May 2009.
60. Ramos, F.T.; Fox, D.; Durrant-Whyte, H.F. Crf-Matching: Conditional Random Fields for Feature-Based Scan Matching. In Proceedings of the Robotics: Science and Systems, Atlanta, GA, USA, 27–30 June 2007.
61. Woo, A.; Fidan, B.; Melek, W.W.; Zekavat, S.; Buehrer, R. Localization for Autonomous Driving. In *Handbook of Position Location*; Zekavat, S.A., Buehrer, R.M., Eds.; Wiley: Hoboken, NJ, USA, 2018; pp. 1051–1087. [CrossRef]
62. Holder, M.; Hellwig, S.; Winner, H. Real-Time Pose Graph SLAM based on Radar. In Proceedings of the 2019 IEEE Intelligent Vehicles Symposium (IV), Paris, France, 9–12 June 2019; pp. 1145–1151. [CrossRef]
63. Jose, E.; Adams, M. Relative radar cross section based feature identification with millimeter wave radar for outdoor SLAM. In Proceedings of the IEEE/RSJ International Conference on Intelligent Robots and Systems (IROS), Sendai, Japan, 28 September–2 October 2004; Volume 1, pp. 425–430.
64. Jose, E.; Adams, M. An augmented state SLAM formulation for multiple lineof-sight features with millimetre wave radar. In Proceedings of the IEEE International Conference on Intelligent Robots and Systems (IROS), Edmonton, AB, Canada, 2–6 August 2005; pp. 3087–3092.
65. Rouveure, R.; Faure, P.; Monod, M. Radar-based SLAM without odometric sensor. In Proceedings of the ROBOTICS 2010: International Workshop of Mobile Robotics for Environment/Agriculture, Clermont Ferrand, France, 6–8 September 2010.
66. Vivet, D.; Checchin, P.; Chapuis, R. Localization and Mapping Using Only a Rotating FMCW Radar Sensor. *Sensors* **2013**, *13*, 4527–4552. [CrossRef]
67. Leonard, J.J.; Feder, H.J.S. A Computationally Efficient Method for Large-Scale Concurrent Mapping and Localization. *Int. Symp. Robot. Res.* **2000**, *9*, 169–178.
68. Newman, P.; Leonard, J.J. Pure Range-Only Sub-Sea SLAM. In Proceedings of the IEEE International Conference on Robotics and Automation, Taipei, Taiwan, 14–19 September 2003; Volume 2, pp. 1921–1926.
69. Tardos, J.D.; Neira, J.; Newman, P.M.; Leonard, J.J. Robust Mapping and Localization in Indoor Environments Using Sonar Data. *Int. J. Robot. Res.* **2002**, *21*, 311–330. [CrossRef]
70. Chong, T.J.; Tang, X.J.; Leng, C.H.; Yogeswaran, M.; Ng, O.E.; Chong, Y.Z. Sensor Technologies and Simultaneous Localization and Mapping (SLAM). *Procedia Comput. Sci.* **2015**, *76*, 174–179. [CrossRef]
71. Aqel, M.O.A.; Marhaban, M.H.; Saripan, M.I.; Ismail, N.B. Review of visual odometry: Types, approaches, challenges, and applications. *SpringerPlus* **2016**, *5*, 1897. [CrossRef]
72. Davison, A.J.; Reid, I.D.; Molton, N.D.; Stasse, O. Monoslam: Real-time single camera SLAM. *Pattern Anal. Mach. Intell. IEEE Trans.* **2007**, *29*, 1052–1067. [CrossRef] [PubMed]

73. Engel, J.; Schöps, T.; Cremers, D. LSD-SLAM: Large-scale direct monocular SLAM. In Proceedings of the European Conference on Computer Vision, Zurich, Switzerland, 6–12 September 2014; pp. 834–849.
74. Engel, J.; Stückler, J.; Cremers, D. Large-scale direct SLAM with stereo cameras. In Proceedings of the 2015 IEEE/RSJ International Conference on Intelligent Robots and Systems (IROS), Hamburg, Germany, 28 September–2 October 2015; pp. 1935–1942. [CrossRef]
75. Mei, C.; Sibley, G.; Cummins, M.; Newman, P.; Reid, I. RSLAM: A System for Large-Scale Mapping in Constant-Time Using Stereo. *Int. J. Comput. Vis.* **2011**, *94*, 198–214. [CrossRef]
76. Paz, L.; Piniés, P.; Tardós, J.; Neira, J. Large-scale 6-DoF SLAM with stereo-inhand. *IEEE Trans. Robot.* **2008**, *24*, 946–957.
77. Bellavia, F.; Fanfani, M.; Pazzaglia, F.; Colombo, C. Robust Selective Stereo SLAM without Loop Closure and Bundle Adjustment. In *Image Analysis and Processing—ICIAP 2013. ICIAP 2013*; Petrosino, A., Ed.; Lecture Notes in Computer Science; Springer: Berlin/Heidelberg, Germany, 2013; Volume 8156. [CrossRef]
78. Harmat, A.; Sharf, I.; Trentini, M. Parallel tracking and mapping with multiple cameras on an unmanned aerial vehicle. In *International Conference on Intelligent Robotics and Applications*; Springer: Berlin/Heidelberg, Germany, 2012; pp. 421–432.
79. Urban, S.; Hinz, S. MultiCol-SLAM-A modular real-time multi-camera SLAM system. *arXiv* **2016**, arXiv:1610.07336.
80. Yang, S.; Scherer, S.A.; Yi, X.; Zell, A. Multi-camera visual SLAM for autonomous navigation of micro aerial vehicles. *Robot. Auton. Syst.* **2017**, *93*, 116–134. [CrossRef]
81. Yang, Y.; Tang, D.; Wang, D.; Song, W.; Wang, J.; Fu, M. Multi-camera visual SLAM for off-road navigation. *Robot. Auton. Systs.* **2020**, *128*, 103505. [CrossRef]
82. Kitt, B.M.; Rehder, J.; Chambers, A.D.; Schonbein, M.; Lategahn, H.; Singh, S. Monocular visual odometry using a planar road model to solve scale ambiguity. In Proceedings of the European Conference on Mobile Robots; Örebro University: Örebro, Sweden, 2011; pp. 43–48.
83. Heng, L.; Lee, G.H.; Pollefeys, M. Self-calibration and visual SLAM with a multi-camera system on a micro aerial vehicle. *Auton. Robot.* **2015**, *39*, 259–277. [CrossRef]
84. Khoshelham, K.; Elberink, S.O. Accuracy and Resolution of Kinect Depth Data for Indoor Mapping Applications. *Sensors* **2012**, *12*, 1437–1454. [CrossRef] [PubMed]
85. Endres, F.; Hess, J.; Sturm, J.; Cremers, D.; Burgard, W. 3-D Mapping with an RGB-D Camera. *IEEE Trans. Robot.* **2014**, *30*, 177–187. [CrossRef]
86. Kahler, O.; Prisacariu, V.; Ren, C.; Sun, X.; Torr, P.; Murray, D. Very high frame rate volumetric integration of depth images on mobile devices. *IEEE Trans. Vis. Comput. Graph.* **2015**, *21*, 1241–1250. [CrossRef] [PubMed]
87. Henry, P.; Krainin, M.; Herbst, E.; Ren, X.; Fox, D. RGB-D mapping: Using Kinect-style depth cameras for dense 3D modeling of indoor environments. *Int. J. Robot. Res.* **2012**, *31*, 647–663. [CrossRef]
88. Kerl, C.; Sturm, J.; Cremers, D. Dense visual SLAM for RGB-D cameras. In Proceedings of the 2013 IEEE/RSJ International Conference on Intelligent Robots and Systems, Tokyo, Japan, 3–8 November 2013; pp. 2100–2106.
89. De Medeiros Esper, I.; Smolkin, O.; Manko, M.; Popov, A.; From, P.J.; Mason, A. Evaluation of RGB-D Multi-Camera Pose Estimation for 3D Reconstruction. *Appl. Sci.* **2022**, *12*, 4134. [CrossRef]
90. Salas-Moreno, R.F.; Newcombe, R.A.; Strasdat, H.; Kelly, P.H.J.; Davison, A.J. SLAM++: Simultaneous localisation and mapping at the level of objects. In Proceedings of the IEEE Conference on Computer Vision and Pattern Recognition, Portland, OR, USA, 23–28 June 2013; pp. 1352–1359.
91. Tateno, K.; Tombari, F.; Navab, N. When 2.5D is not enough: Simultaneous reconstruction, segmentation and recognition on dense SLAM. In Proceedings of the IEEE International Conference on Robotics and Automation (ICRA), Stockholm, Sweden, 16–21 May 2016; pp. 2295–2302.
92. Milz, S.; Arbeiter, G.; Witt, C.; Abdallah, B.; Yogamani, S. Visual SLAM for Automated Driving: Exploring the Applications of Deep Learning. In Proceedings of the 2018 IEEE/CVF Conference on Computer Vision and Pattern Recognition Workshops (CVPRW), Salt Lake City, UT, USA, 18–22 June 2018; pp. 360–36010. [CrossRef]
93. Davison, A.J. Real-time simultaneous localisation and mapping with a single camera. In Proceedings of the International Conference on Computer Vision, Nice, France, 13–16 October 2003; pp. 1403–1410.
94. Klein, G.; Murray, D.W. Parallel tracking and mapping for small AR workspaces. In Proceedings of the International Symposium on Mixed and Augmented Reality, Washington, DC, USA, 13–16 November 2007; pp. 225–234.
95. Mur-Artal, R.; Montiel, J.M.M.; Tardos, J.D. ORB-SLAM: A versatile and accurate monocular SLAM system. *IEEE Trans. Robot.* **2015**, *31*, 1147–1163. [CrossRef]
96. Mur-Artal, R.; Tardós, J.D. Orb-slam2: An open-source slam system for monocular, stereo, and rgb-d cameras. *IEEE Trans. Robot.* **2017**, *33*, 1255–1262. [CrossRef]
97. Newcombe, R.A.; Lovegrove, S.J.; Davison, A.J. DTAM: Dense tracking and mapping in real-time. In Proceedings of the 2011 International Conference on Computer Vision, Barcelona, Spain, 6–13 November 2011; pp. 2320–2327.
98. Forster, C.; Pizzoli, M.; Scaramuzza, D. SVO: Fast semi-direct monocular visual odometry. In Proceedings of the 2014 IEEE International Conference on Robotics and Automation (ICRA), Hong Kong, China, 31 May–7 June 2014; pp. 15–22.
99. Caruso, D.; Engel, J.; Cremers, D. Large-scale direct SLAM for omnidirectional cameras. In Proceedings of the 2015 IEEE/RSJ International Conference on Intelligent Robots and Systems (IROS), Hamburg, Germany, 28 September–2 October 2015; pp. 141–148. [CrossRef]

100. Taketomi, T.; Uchiyama, H.; Ikeda, S. Visual SLAM algorithms: A survey from 2010 to 2016. *IPSJ Trans. Comput. Vis. Appl.* **2017**, *9*, 16. [CrossRef]
101. Azzam, R.; Taha, T.; Huang, S.; Zweiri, Y. Feature-based visual simultaneous localization and mapping: A survey. *SN Appl. Sci.* **2020**, *2*, 224. [CrossRef]
102. Valente, M.; Joly, C.; Fortelle, A. Fusing Laser Scanner and Stereo Camera in Evidential Grid Maps. In Proceedings of the 2018 15th International Conference on Control, Automation, Robotics and Vision (ICARCV), Singapore, 18–21 November 2018; pp. 990–997. [CrossRef]
103. López, E.; García, S.; Barea, R.; Bergasa, L.M.; Molinos, E.J.; Arroyo, R.; Romera, E.; Pardo, S. A Multi-Sensorial Simultaneous Localization and Mapping (SLAM) System for Low-Cost Micro Aerial Vehicles in GPS-Denied Environments. *Sensors* **2017**, *17*, 802. [CrossRef] [PubMed]
104. Pandey, G.; Savarese, S.; McBride, J.R.; Eustice, R.M. Visually bootstrapped generalized ICP. In Proceedings of the 2011 IEEE International Conference on Robotics and Automation, Shanghai, China, 9–13 May 2011; pp. 2660–2667. [CrossRef]
105. Huang, S.S.; Ma, Z.Y.; Mu, H.; Fu, T.J.; Hu, S.-M. Lidar-Monocular Visual Odometry using Point and Line Features. In Proceedings of the 2020 IEEE International Conference on Robotics and Automation (ICRA), Paris, France, 31 May–31 August 2020; pp. 1091–1097. [CrossRef]
106. Zhang, X.; Rad, A.B.; Wong, Y.-K. Sensor Fusion of Monocular Cameras and Laser Rangefinders for Line-Based Simultaneous Localization and Mapping (SLAM) Tasks in Autonomous Mobile Robots. *Sensors* **2012**, *12*, 429–452. [CrossRef] [PubMed]
107. Jiang, G.; Lei, Y.; Jin, S.; Tian, C.; Ma, X.; Ou, Y. A Simultaneous Localization and Mapping (SLAM) Framework for 2.5D Map Building Based on Low-Cost LiDAR and Vision Fusion. *Appl. Sci.* **2019**, *9*, 2105. [CrossRef]
108. Qin, T.; Li, P.; Shen, S. VINS-Mono: A Robust and Versatile Monocular Visual-Inertial State Estimator. *arXiv* **2017**, arXiv:1708.03852. [CrossRef]
109. He, L.; Jin, Z.; Gao, Z. De-Skewing LiDAR Scan for Refinement of Local Mapping. *Sensors* **2020**, *20*, 1846. [CrossRef] [PubMed]
110. Zhang, J.; Singh, S. LOAM: Lidar Odometry and Mapping in Real-time. In Proceedings of the 2014 Robotics: Science and Systems (RSS2014), Berkeley, CA, USA, 12–16 July 2014; Volume 2.
111. Falquez, J.M.; Kasper, M.; Sibley, G. Inertial aided dense & semi-dense methods for robust direct visual odometry. In Proceedings of the IEEE/RSJ International Conference on Intelligent Robots and Systems, Daejeon, Republic of Korea, 9–14 October 2016; pp. 3601–3607.
112. Fang, W.; Zheng, L.; Deng, H.; Zhang, H. Real-Time Motion Tracking for Mobile Augmented/Virtual Reality Using Adaptive Visual-Inertial Fusion. *Sensors* **2017**, *17*, 1037. [CrossRef] [PubMed]
113. Lynen, S.; Sattler, T.; Bosse, M.; Hesch, J.; Pollefeys, M.; Siegwart, R. Get Out of My Lab: Large-scale, Real-Time Visual-Inertial Localization. In Proceedings of the Robotics: Science and Systems, Rome, Italy, 13–17 July 2015.
114. Zhang, Z.; Liu, S.; Tsai, G.; Hu, H.; Chu, C.C.; Zheng, F. PIRVS: An Advanced Visual-Inertial SLAM System with Flexible Sensor Fusion and Hardware Co-Design. *arXiv* **2017**, arXiv:1710.00893.
115. Chen, C.; Zhu, H. Visual-inertial SLAM method based on optical flow in a GPS-denied environment. *Ind. Robot Int. J.* **2018**, *45*, 401–406. [CrossRef]
116. Chen, C.; Zhu, H.; Li, M.; You, S. A review of visual-inertial simultaneous localization and mapping from filtering-based and optimization-based perspectives. *Robotics* **2018**, *7*, 45. [CrossRef]
117. Ye, H.; Chen, Y.; Liu, M. Tightly coupled 3D lidar inertial odometry and mapping. *arXiv* **2019**, arXiv:1904.06993.
118. Hemann, G.; Singh, S.; Kaess, M. Long-range gps-denied aerial inertial navigation with lidar localization. In Proceedings of the Intelligent Robots and Systems (IROS), 2016 IEEE/RSJ International Conference on IEEE, Daejeon, Republic of Korea, 9–14 October 2016; pp. 1659–1666. [CrossRef]
119. Zhang, H. Deep Learning Applications in Simultaneous Localization and Mapping. *J. Phys. Conf. Ser.* **2022**, *2181*, 012012. [CrossRef]
120. Costante, G.; Mancini, M.; Valigi, P.; Ciarfuglia, T.A. Exploring Representation Learning with CNNs for Frame-to-Frame Ego-Motion Estimation. *IEEE Robot. Autom. Lett.* **2016**, *1*, 18–25. [CrossRef]
121. Costante, G.; Ciarfuglia, T.A. LS-VO: Learning Dense Optical Subspace for Robust Visual Odometry Estimation. *IEEE Robot. Autom. Lett.* **2018**, *3*, 1735–1742. [CrossRef]
122. Muller, P.; Savakis, A. Flowdometry: An Optical Flow and Deep Learning Based Approach to Visual Odometry. In Proceedings of the 2017 IEEE Winter Conference on Applications of Computer Vision (WACV), Santa Rosa, CA, USA, 24–31 March 2017; pp. 624–631. [CrossRef]
123. Saputra, M.R.; Gusmão, P.P.; Wang, S.; Markham, A.; Trigoni, A. Learning Monocular Visual Odometry through Geometry-Aware Curriculum Learning. In Proceedings of the 2019 International Conference on Robotics and Automation (ICRA), Montreal, QC, Canada, 20–24 May 2019; pp. 3549–3555.
124. Zhu, R.; Yang, M.; Liu, W.; Song, R.; Yan, B.; Xiao, Z. DeepAVO: Efficient pose refining with feature distilling for deep Visual Odometry. *Neurocomputing* **2022**, *467*, 22–35. [CrossRef]
125. Wang, S.; Clark, R.; Wen, H.; Trigoni, N. DeepVO: Towards end-to-end visual odometry with deep Recurrent Convolutional Neural Networks. In Proceedings of the 2017 IEEE International Conference on Robotics and Automation (ICRA), Singapore, 29 May–3 June 2017; pp. 2043–2050. [CrossRef]

126. Xue, F.; Wang, Q.; Wang, X.; Dong, W.; Wang, J.; Zha, H. Guided Feature Selection for Deep Visual Odometry. In *Computer Vision—ACCV 2018, ACCV 2018, Lecture Notes in Computer Science*; Jawahar, C., Li, H., Mori, G., Schindler, K., Eds.; Springer: Cham, Switzerland, 2018; Volume 11366. [CrossRef]
127. Li, R.; Wang, S.; Long, Z.; Gu, D. UnDeepVO: Monocular Visual Odometry Through Unsupervised Deep Learning. In Proceedings of the 2018 IEEE International Conference on Robotics and Automation (ICRA), Brisbane, QLD, Australia, 21–25 May 2018; pp. 7286–7291. [CrossRef]
128. Tateno, K.; Tombari, F.; Laina, I.; Navab, N. CNN-SLAM: Real-Time Dense Monocular SLAM with Learned Depth Prediction. In Proceedings of the 2017 IEEE Conference on Computer Vision and Pattern Recognition (CVPR), Honolulu, HI, USA, 21–26 July 2017; pp. 6565–6574. [CrossRef]
129. Ma, F.; Karaman, S. Sparse-to-Dense: Depth Prediction from Sparse Depth Samples and a Single Image. In Proceedings of the 2018 IEEE International Conference on Robotics and Automation (ICRA), Brisbane, QLD, Australia, 21–25 May 2018; pp. 4796–4803. [CrossRef]
130. Bloesch, M.; Czarnowski, J.; Clark, R.; Leutenegger, S.; Davison, A.J. CodeSLAM—Learning a Compact, Optimisable Representation for Dense Visual SLAM. In Proceedings of the 2018 IEEE/CVF Conference on Computer Vision and Pattern Recognition, Salt Lake City, UT, USA, 18–23 June 2018; pp. 2560–2568.
131. DeTone, D.; Malisiewicz, T.; Rabinovich, A. Toward geometric deep slam. *arXiv* **2017**, arXiv:1707.07410.
132. Yang, N.; Stumberg, L.V.; Wang, R.; Cremers, D. D3VO: Deep Depth, Deep Pose and Deep Uncertainty for Monocular Visual Odometry. In Proceedings of the 2020 IEEE/CVF Conference on Computer Vision and Pattern Recognition (CVPR), Seattle, WA, USA, 13–19 June 2020; pp. 1278–1289.
133. Li, Y.; Ushiku, Y.; Harada, T. Pose Graph optimization for Unsupervised Monocular Visual Odometry. In Proceedings of the 2019 International Conference on Robotics and Automation (ICRA), Montreal, QC, Canada, 20–24 May 2019; pp. 5439–5445. [CrossRef]
134. Gao, X.; Zhang, T. Unsupervised learning to detect loops using deep neural networks for visual SLAM system. *Auton. Robot* **2017**, *41*, 1–18. [CrossRef]
135. Memon, A.R.; Wang, H.; Hussain, A. Loop closure detection using supervised and unsupervised deep neural networks for monocular SLAM systems. *Robot. Auton. Syst.* **2020**, *126*, 103470. [CrossRef]
136. Wang, Z.; Peng, Z.; Guan, Y.; Wu, L. Manifold Regularization Graph Structure Auto-Encoder to Detect Loop Closure for Visual SLAM. *IEEE Access* **2019**, *7*, 59524–59538. [CrossRef]
137. Qin, C.; Zhang, Y.; Liu, Y.; Lv, G. Semantic loop closure detection based on graph matching in multi-objects scenes. *J. Vis. Commun. Image Represent.* **2021**, *76*, 103072. [CrossRef]
138. Li, R.; Wang, S.; Gu, D. DeepSLAM: A Robust Monocular SLAM System with Unsupervised Deep Learning. *IEEE Trans. Ind. Electron.* **2021**, *68*, 3577–3587. [CrossRef]
139. Chen, W.; Shang, G.; Ji, A.; Zhou, C.; Wang, X.; Xu, C.; Li, Z.; Hu, K. An Overview on Visual SLAM: From Tradition to Semantic. *Remote Sens.* **2022**, *14*, 3010. [CrossRef]
140. Zhang, H.; Geiger, A.; Urtasun, R. Understanding High-Level Semantics by Modeling Traffic Patterns. In Proceedings of the 2013 IEEE International Conference on Computer Vision, Sydney, NSW, Australia, 1–8 December 2013; pp. 3056–3063. [CrossRef]
141. Long, J.; Shelhamer, E.; Darrell, T. Fully convolutional networks for semantic segmentation. In Proceedings of the 2015 IEEE Conference on Computer Vision and Pattern Recognition (CVPR), Boston, MA, USA, 7–12 June 2015; pp. 3431–3440. [CrossRef]
142. Chen, L.C.; Papandreou, G.; Schroff, F.; Adam, H. Rethinking Atrous Convolution for Semantic Image Segmentation. *arXiv* **2017**. [CrossRef]
143. Zhao, Z.; Mao, Y.; Ding, Y.; Ren, P.; Zheng, N. Visual-Based Semantic SLAM with Landmarks for Large-Scale Outdoor Environment. In Proceedings of the 2019 2nd China Symposium on Cognitive Computing and Hybrid Intelligence (CCHI), Xi'an, China, 21–22 September 2019; pp. 149–154.
144. Li, R.; Gu, D.; Liu, Q.; Long, Z.; Hu, H. Semantic Scene Mapping with Spatio-temporal Deep Neural Network for Robotic Applications. *Cogn. Comput.* **2018**, *10*, 260–271. [CrossRef]
145. Rosinol, A.; Abate, M.; Chang, Y.; Carlone, L. Kimera: An Open-Source Library for Real-Time Metric-Semantic Localization and Mapping. In Proceedings of the 2020 IEEE International Conference on Robotics and Automation (ICRA), Virtually, 31 May–31 August 2020; pp. 1689–1696.
146. Qin, T.; Zheng, Y.; Chen, T.; Chen, Y.; Su, Q. A Light-Weight Semantic Map for Visual Localization towards Autonomous Driving. In Proceedings of the 2021 IEEE International Conference on Robotics and Automation (ICRA), Xi'an, China, 30 May–5 June 2021; pp. 11248–11254.
147. McCormac, J.; Handa, A.; Davison, A.; Leutenegger, S. SemanticFusion: Dense 3D semantic mapping with convolutional neural networks. In *IEEE International Conference on Robotics and Automation(ICRA)*; IEEE: New York, NY, USA, 2017; pp. 4628–4635.
148. Schörghuber, M.; Steininger, D.; Cabon, Y.; Humenberger, M.; Gelautz, M. SLAMANTIC—Leveraging Semantics to Improve VSLAM in Dynamic Environments. In Proceedings of the 2019 IEEE/CVF International Conference on Computer Vision Workshop (ICCVW), Seoul, Republic of Korea, 27–28 October 2019; pp. 3759–3768.
149. Yu, C.; Liu, Z.; Liu, X.; Xie, F.; Yang, Y.; Wei, Q.; Qiao, F. DS-SLAM: A Semantic Visual SLAM towards Dynamic Environments. In Proceedings of the 2018 IEEE/RSJ International Conference on Intelligent Robots and Systems (IROS), Madrid, Spain, 1–5 October 2018; pp. 1168–1174.
150. Sualeh, M.; Kim, G.-W. Semantics Aware Dynamic SLAM Based on 3D MODT. *Sensors* **2021**, *21*, 6355. [CrossRef]

151. Han, S.; Xi, Z. Dynamic Scene Semantics SLAM Based on Semantic Segmentation. *IEEE Access* **2020**, *8*, 43563–43570. [CrossRef]
152. Duan, C.; Junginger, S.; Huang, J.; Jin, K.; Thurow, K. Deep Learning for Visual SLAM in Transportation Robotics: A review. *Transp. Saf. Environ.* **2019**, *1*, 177–184. [CrossRef]
153. Li, R.; Wang, S.; Gu, D. Ongoing Evolution of Visual SLAM from Geometry to Deep Learning: Challenges and Opportunities. *Cogn. Comput.* **2018**, *10*, 875–889. [CrossRef]
154. Chen, C.; Wang, B.; Lu, C.X.; Trigoni, A.; Markham, A. A Survey on Deep Learning for Localization and Mapping: Towards the Age of Spatial Machine Intelligence. *arXiv* **2020**, arXiv:abs/2006.12567.
155. Hochreiter, S.; Schmidhuber, J. Long short-term memory. *Neural Comput.* **1997**, *9*, 1735–1780. [CrossRef]
156. Han, L.; Lin, Y.; Du, G.; Lian, S. DeepVIO: Self-supervised Deep Learning of Monocular Visual Inertial Odometry using 3D Geometric Constraints. In Proceedings of the 2019 IEEE/RSJ International Conference on Intelligent Robots and Systems (IROS), Venetian Macao, Macau, China, 3–8 November 2019; pp. 6906–6913.
157. Clark, R.; Wang, S.; Wen, H.; Markham, A.; Trigoni, N. VINet: Visual-Inertial Odometry as a Sequence-to-Sequence Learning Problem. In Proceedings of the AAAI Conference on Artificial Intelligence, San Francisco, CA, USA, 4–9 February 2017; Volume 31. [CrossRef]
158. Gurturk, M.; Yusefi, A.; Aslan, M.F.; Soycan, M.; Durdu, A.; Masiero, A. The YTU dataset and recurrent neural network based visual-inertial odometry. *Measurement* **2021**, *184*, 109878. [CrossRef]
159. Da Silva, M.A.V. SLAM and Data Fusion for Autonomous Vehicles: From Classical Approaches to Deep Learning Methods. Machine Learning [cs.LG]. Ph.D. Thesis, Université Paris Sciences et Lettres, Paris, France, 2019. (In English)
160. Velas, M.; Spanel, M.; Hradis, M.; Herout, A. Cnn for imu assisted odometry estimation using velodyne lidar. In Proceedings of the 2018 IEEE International Conference on Autonomous Robot Systems and Competitions (ICARSC), Torres Vedras, Portugal, 25–27 April 2018; IEEE: New York, NY, USA, 2018; pp. 71–77.
161. Li, Q.; Chen, S.; Wang, C.; Li, X.; Wen, C.; Cheng, M.; Li, J. LO-Net: Deep Real-Time Lidar Odometry. In Proceedings of the 2019 IEEE/CVF Conference on Computer Vision and Pattern Recognition (CVPR), Long Beach, CA, USA, 15–20 June 2019; pp. 8465–8474.
162. Li, B.; Hu, M.; Wang, S.; Wang, L.; Gong, X. Self-supervised Visual-LiDAR Odometry with Flip Consistency. In Proceedings of the 2021 IEEE Winter Conference on Applications of Computer Vision (WACV), Waikoloa, HI, USA, 3–8 January 2021; pp. 3843–3851. [CrossRef]
163. Milioto, A.; Vizzo, I.; Behley, J.; Stachniss, C. RangeNet++: Fast and Accurate LiDAR Semantic Segmentation. In Proceedings of the 2019 IEEE/RSJ International Conference on Intelligent Robots and Systems (IROS), Macau, China, 3–8 November 2019; pp. 4213–4220. [CrossRef]
164. Wu, B.; Wan, A.; Yue, X.; Keutzer, K. SqueezeSeg: Convolutional Neural Nets with Recurrent CRF for Real-Time Road-Object Segmentation from 3D LiDAR Point Cloud. In Proceedings of the IEEE International Conference on Robotics & Automation (ICRA), Brisbane, QLD, Australia, 21–25 May 2018.
165. Wu, B.; Zhou, X.; Zhao, S.; Yue, X.; Keutzer, K. SqueezeSegV2: Improved Model Structure and Unsupervised Domain Adaptation for Road-Object Segmentation from a LiDAR Point Cloud. In Proceedings of the IEEE International Conference on Robotics & Automation (ICRA), Montreal, QC, Canada, 20–24 May 2019.
166. Chen, X.; Li, S.; Mersch, B.; Wiesmann, L.; Gall, J.; Behley, J.; Stachniss, C. Moving Object Segmentation in 3D LiDAR Data: A Learning-Based Approach Exploiting Sequential Data. *IEEE Robot. Autom. Lett.* **2021**, *6*, 6529–6536. [CrossRef]
167. Yue, J.; Wen, W.; Han, J.; Hsu, L. LiDAR data enrichment using Deep Learning Based on High-Resolution Image: An Approach to Achieve High-Performance LiDAR SLAM Using Low-cost LiDAR. *arXiv* **2020**, arXiv:2008.03694.
168. Jo, K.; Kim, C.; Sunwoo, M. Simultaneous localization and map change update for the high definition map-based autonomous driving car. *Sensors* **2018**, *18*, 3145. [CrossRef] [PubMed]
169. Kim, C.; Cho, S.; Sunwoo, M.; Jo, K. Crowd-sourced mapping of new feature layer for high-definition map. *Sensors* **2018**, *18*, 1472. [CrossRef] [PubMed]
170. Zhang, P.; Zhang, M.; Liu, J. Real-Time HD Map Change Detection for Crowdsourcing Update Based on Mid-to-High-End Sensors. *Sensors* **2021**, *21*, 2477. [CrossRef] [PubMed]
171. Kummerle, R.; Hahnel, D.; Dolgov, D.; Thrun, S.; Burgard, W. Autonomous driving in a multi-level parking structure. In Proceedings of the 2009 IEEE International Conference on Robotics and Automation, Kobe, Japan, 12–17 May 2009; pp. 3395–3400.
172. Lee, H.; Chun, J.; Jeon, K. Autonomous back-in parking based on occupancy grid map and EKF SLAM with W band radar. In Proceedings of the 2018 International Conference on Radar (RADAR), Brisbane, QLD, Australia, 27–31 August 2018; pp. 1–4.
173. Im, G.; Kim, M.; Park, J. Parking line based SLAM approach using AVM/LiDAR sensor fusion for rapid and accurate loop closing and parking space detection. *Sensors* **2019**, *19*, 4811. [CrossRef] [PubMed]
174. Qin, T.; Chen, T.; Chen, Y.; Su, Q. AVP-SLAM: Semantic Visual Mapping and Localization for Autonomous Vehicles in the Parking Lot. In Proceedings of the 2020 IEEE/RSJ International Conference on Intelligent Robots and Systems (IROS), Las Vegas, NV, USA, 24 October 2020–24 January 2021; pp. 5939–5945. [CrossRef]
175. Zheng, S.; Wang, J. High definition map based vehicle localization for highly automated driving. In Proceedings of the 2017 International Conference on Localization and GNSS (ICL-GNSS), Nottingham, UK, 27–29 June 2017; pp. 1–8. [CrossRef]

176. Li, H.; Nashashibi, F. Multi-vehicle cooperative localization using indirect vehicle-to-vehicle relative pose estimation. In Proceedings of the IEEE International Conference on Vehicular Electronics and Safety, Istanbul, Turkey, 24–27 July 2012; pp. 267–272.
177. Wolcott, R.W.; Eustice, R.M. Visual Localization within LiDAR maps for automated urban driving. In Proceedings of the Intelligent Robots and Systems (IROS2014), 2014 IEEE/RSJ International Conference on Intelligent Robots and Systems, Chicago, IL, USA, 14–18 September 2014; pp. 176–183.
178. Schreiber, M.; Knöppel, C.; Franke, U. LaneLoc: Lane marking based localization using highly accurate maps. In Proceedings of the 2013 IEEE Intelligent Vehicles Symposium (IV), Gold Coast City, QLD, Australia, 23 June 2013; pp. 449–454.
179. Jeong, J.; Cho, Y.; Kim, A. Road-SLAM: Road marking based SLAM with lane-level accuracy. In Proceedings of the 2017 IEEE Intelligent Vehicles Symposium (IV), Los Angeles, CA, USA, 11–14 June 2017; p. 1736-1473. [CrossRef]
180. Vu, T.D. Vehicle Perception: Localization, Mapping with Detection, Classification and Tracking of Moving Objects. Ph.D. Thesis, Institut National Polytechnique de Grenoble-INPG, Grenoble, France, 2009.
181. Wang, C.C.; Thorpe, C.; Thrun, S. Online simultaneous localization and mapping with detection and tracking of moving objects: Theory and results from a ground vehicle in crowded urban areas. In Proceedings of the IEEE International Conference on Robotics and Automation, Taipei, Taiwan, 14–19 September 2003; pp. 842–849.
182. Fei, Y. *Real-Time Detecting and Tracking of Moving Objects Using 3D LIDAR*; Zhejian University: Hangzhou, China, 2012.
183. Miller, W. RoboSense Develops $200 LiDAR System for Autonomous Vehicles. Electronic Products. Available online: https://www.electronicproducts.com/Automotive/RoboSense_develops_200_LiDAR_system_for_autonomous_vehicles.aspxMorales (accessed on 1 October 2022).
184. Rone, W.; Ben-Tzvi, P. Mapping, localization and motion planning in mobile multi-robotic systems. *Robotica* **2013**, *31*, 1–23. [CrossRef]
185. Mentasti, S.; Matteucci, M. Multi-layer occupancy grid mapping for autonomous vehicles navigation. In Proceedings of the 2019 AEIT International Conference of Electrical and Electronic Technologies for Automotive (AEIT AUTOMOTIVE), Turin, Italy, 2–4 July 2019; pp. 1–6. [CrossRef]
186. Li, H.; Tsukada, M.; Nashashibi, F.; Parent, M. Multi-Vehicle Cooperative Local Mapping: A Methodology Based on Occupancy Grid Map Merging. *IEEE Trans. Intell. Transp. Syst.* **2014**, *15*, 12. [CrossRef]
187. Wang, X.; Wang, W.; Yin, X.; Xiang, C.; Zhang, Y. A New Grid Map Construction Method for Autonomous Vehicles. *IFAC-PapersOnLine* **2018**, *51*, 377–382. [CrossRef]
188. Mutz, F.; Oliveira-Santos, T.; Forechi, A.; Komati, K.S.; Badue, C.; França, F.M.G.; De Souza, A.F. What is the best grid-map for self-driving cars localization? An evaluation under diverse types of illumination, traffic, and environment. *Expert Syst. Appl.* **2021**, *179*, 115077. [CrossRef]
189. Yu, S.; Fu, C.; Gostar, A.K.; Hu, M. A Review on Map-Merging Methods for Typical Map Types in Multiple-Ground-Robot SLAM Solutions. *Sensors* **2020**, *20*, 6988. [CrossRef]
190. Javanmardi, E.; Gu, Y.; Javanmardi, M.; Kamijo, S. Autonomous vehicle self-localization based on abstract map and multi-channel LiDAR in urban area. *IATSS Res.* **2019**, *43*, 1–13. [CrossRef]
191. Ort, T.; Paull, L.; Rus, D. Autonomous Vehicle Navigation in Rural Environments without Detailed Prior Maps. In Proceedings of the 2018 IEEE International Conference on Robotics and Automation (ICRA), Brisbane, QLD, Australia, 21–25 May 2018; pp. 2040–2047. [CrossRef]
192. Badue, C.; Guidolini, R.; Carneiro, R.V.; Azevedo, P.; Cardoso, V.B.; Forechi, A.; Jesus, L.; Berriel, R.; Paixão, T.M.; Mutz, F.; et al. Self-driving cars: A survey. *Expert Syst. Appl.* **2021**, *165*, 113816. [CrossRef]
193. Li, L.; Yang, M.; Wang, B.; Wang, C. An overview on sensor map based localization for automated driving. In Proceedings of the 2017 Joint Urban Remote Sensing Event (JURSE), Dubai, United Arab Emirates, 6–8 March 2017; pp. 1–4. [CrossRef]
194. Haklay, M.; Weber, P. OpenStreetMap: User-Generated Street Maps. *IEEE Pervasive Comput.* **2008**, *7*, 12–18. [CrossRef]
195. Bernuy, F.; Ruiz-del-Solar, J. Topological Semantic Mapping and Localization in Urban Road Scenarios. *J. Intell. Robot. Syst.* **2018**, *92*, 19–32. [CrossRef]
196. Bender, P.; Ziegler, J.; Stiller, C. Lanelets: Efficient map representation for autonomous driving. In Proceedings of the 2014 IEEE Intelligent Vehicles Symposium Proceedings, Dearborn, MI, USA, 8–11 June 2014; pp. 420–425. [CrossRef]
197. Dube, R.; Cramariuc, A.; Dugas, D.; Nieto, J.; Siegwart, R.; Cadena, C. SegMap: 3d segment mapping using data-driven descriptors. Robotics: Science and Systems (RSS). *arXiv* **2018**, arXiv:1804.09557/.
198. Zhong, F.; Wang, S.; Zhang, Z.; Chen, C.; Wang, Y. Detect-slam: Making object detection and slam mutually beneficial. In Proceedings of the 2018 IEEE Winter Conference on Applications of Computer Vision (WACV), Lake Tahoe, NV, USA, 12–15 March 2018; pp. 1001–1010. [CrossRef]
199. Ros, G.; Ramos, S.; Granados, M.; Bakhtiary, A.; Vazquez, D.; Lopez, A.M. Vision-Based Offline-Online Perception Paradigm for Autonomous Driving. In Proceedings of the 2015 IEEE Winter Conference on Applications of Computer Vision, Waikoloa, HI, USA, 5–9 January 2015; pp. 231–238. [CrossRef]
200. Hempel, T.; Al-Hamadi, A. An online semantic mapping system for extending and enhancing visual SLAM. *Eng. Appl. Artif. Intell.* **2022**, *111*, 104830. [CrossRef]
201. Nakajima, Y.; Saito, H. Efficient object-oriented semantic mapping with object detector. *IEEE Access* **2019**, *7*, 3206–3213. [CrossRef]

202. Paz, D.; Zhang, H.; Li, Q.; Xiang, H.; Christensen, H.I. Probabilistic Semantic Mapping for Urban Autonomous Driving Applications. In Proceedings of the 2020 IEEE/RSJ International Conference on Intelligent Robots and Systems (IROS), Las Vegas, NV, USA, 24 October 2020–24 January 2021; pp. 2059–2064. [CrossRef]
203. Chen, X.; Milioto, A.; Palazzolo, E.; Giguere, P.; Behley, J.; Stachniss, C. SuMa++: Efficient LiDAR-based semantic slam. In Proceedings of the 2019 IEEE/RSJ International Conference on Intelligent Robots and Systems (IROS), Macau, China, 3–8 November 2019; pp. 4530–4537. [CrossRef]
204. Li, L.; Yang, M. Road dna based localization for autonomous vehicles. In Proceedings of the 2016 IEEE Intelligent Vehicles Symposium(IV), Gothenburg, Sweden, 19–22 June 2016; pp. 883–888. [CrossRef]
205. Huang, S.; Lai, Y.; Frese, U.; Dissanayake, G. How far is SLAM from a linear least squares Problems? In Proceedings of the 2010 IEEE/RSJ International Conference on Intelligent Robots and Systems, Taipei, Taiwan, 18–22 October 2010; pp. 3011–3016.
206. Bresson, G.; Aufrere, R.; Chapuis, R. A general consistent decentralized SLAM solution. *Robot Auton. Syst.* **2015**, *74*, 128–147. [CrossRef]
207. Martinez-Cantin, R.; Castellanos, J.A. Bounding uncertainty in EKF-SLAM: The robocentric local approach. In Proceedings of the 2006 IEEE International Conference on Robotics and Automation, 2006. ICRA 2006, Orlando, FL, USA, 15–19 May 2006; pp. 430–435. [CrossRef]
208. Martinez-Cantin, R.; Castellanos, J.A. Unscented SLAM for large-scale outdoor environments. In Proceedings of the 2005 IEEE/RSJ International Conference on International Conference on Intelligent Robots and Systems, Edmonton, AB, Canada, 2–6 August 2005; pp. 3427–3432.
209. Huang, G.P.; Mourikis, A.I.; Roumeliotis, S.I. Observability-based rules for designing consistent EKF SLAM estimators. *Int. J. Robot. Res.* **2010**, *29*, 502–528. [CrossRef]
210. Tan, F.; Lohmiller, W.; Slotine, J. Simultaneous Localization and Mapping without Linearization. *arXiv* **2015**, arXiv:1512.08829.
211. Cugliari, M.; Martinelli, F. A FastSLAM Algorithm Based on the Unscented Filtering with Adaptive Selective Resampling. In *Field and Service Robotics*; Laugier, C., Siegwart, R., Eds.; Springer Tracts in Advanced Robotics; Springer: Berlin/Heidelberg, Germany, 2008; Volume 42. [CrossRef]
212. Wang, H.; Wei, S.; Che, Y. An improved rao-blackwellized particle filter for slam. In Proceedings of the International Symposium on Intelligent Information Technology Application Workshops 2008, II'AW'08, Shanghai, China, 21–22 December 2008; pp. 515–518.
213. He, B.; Ying, L.; Zhang, S.; Feng, X.; Yan, T. Autonomous navigation based on unscented-FastSLAM using particle swarm optimization for autonomous underwater vehicles. *Measurement* **2015**, *71*, 89–101. [CrossRef]
214. Zhang, F.; Li, S.; Yuan, S.; Sun, E.; Zhao, L. Algorithms analysis of mobile robot SLAM based on Kalman and particle filter. In Proceedings of the 2017 9th International Conference on Modelling, Identification and Control (ICMIC), Kunming, China, 10–12 July 2017; pp. 1050–1055. [CrossRef]
215. Olson, E.; Leonard, J.; Teller, S. Fast iterative alignment of pose graphs with poor initial estimates. In Proceedings of the 2006 IEEE International Conference on Robotics and Automation, 2006, ICRA 2006, Orlando, FL, USA, 15–19 May 2006; pp. 2262–2269. [CrossRef]
216. Carlone, L.; Aragues, R.; Castellanos, J.; Bona, B. A linear approximation for graph-based simultaneous localization and mapping. In Proceedings of the Robotics: Science and Systems, Los Angeles, CA, USA, 27–30 June 2011.
217. Carlone, L.; Aragues, R.; Castellanos, J.; Bona, B. A fast and accurate approximation for planar pose graph optimization. *Int. J. Robot. Res.* **2014**, *33*, 965–987. [CrossRef]
218. Hu, G.; Khosoussi, K.; Huang, S. Towards a reliable SLAM back-end. In Proceedings of the 2013 IEEE/RSJ International Conference on Intelligent Robots and Systems, Tokyo, Japan, 3–7 November 2013; pp. 37–43. [CrossRef]
219. Carlone, L.; Tron, R.; Daniilidis, K.; Dellaert, F. Initialization techniques for 3D SLAM: A survey on rotation estimation and its use in pose graph optimization. In Proceedings of the 2015 IEEE International Conference on Robotics and Automation (ICRA), Seattle, WA, USA, 26–30 May 2015; pp. 4597–4604. [CrossRef]
220. Harsányi, K.; Kiss, A.; Szirányi, T.; Majdik, A. MASAT: A fast and robust algorithm for pose-graph initialization. *Pattern Recognit. Lett.* **2020**, *129*, 131–136. [CrossRef]
221. Campos, C.; Montiel, J.M.M.; Tardos, J.D. Inertial-Only Optimization for Visual-Inertial Initialization, 2020 International Conference on Robotics and Automation. *arXiv* **2020**, arXiv:2003.0576.
222. Skoglund, M.A.; Sjanic, Z.; Gustafsson, F. *Initialisation and Estimation Methods for Batch Optimization of Inertial/Visual SLAM*; Linköping University: Linköping, Sweden, 2013.
223. Dong-Si, T.-C.; Mourikis, A.I. Estimator initialization in vision-aided inertial navigation with unknown camera-IMU calibration. In Proceedings of the 2012 IEEE/RSJ International Conference on Intelligent Robots and Systems, Vilamoura-Algarve, Portugal, 7–12 October 2012; pp. 1064–1071. [CrossRef]
224. Mu, X.; Chen, J.; Zhou, Z.; Leng, Z.; Fan, L. Accurate Initial State Estimation in a Monocular Visual-Inertial SLAM System. *Sensors* **2018**, *18*, 506. [CrossRef] [PubMed]
225. Cheng, J.; Zhang, L.; Chen, Q. An Improved Initialization Method for Monocular Visual-Inertial SLAM. *Electronics* **2021**, *10*, 3063. [CrossRef]

226. Levinson, L.; Thrun, S. Robust Vehicle Localization in Urban Environments Using Probabilistic Maps. In Proceedings of the 2010 IEEE International Conference on Robotics and Automation (ICRA), Anchorage, AK, USA, 3–7 May 2010; pp. 4372–4378. [CrossRef]
227. Wangsiripitak, S.; Murray, D.W. Avoiding moving outliers in visual SLAM by tracking moving objects. In Proceedings of the 2009 IEEE International Conference on Robotics and Automation, Kobe, Japan, 12–17 May 2009.
228. Asmar, D. Vision-Inertial SLAM Using Natural Features in Outdoor Environments. Ph.D. Thesis, University of Waterloo, Waterloo, ON, Canada, 2006.
229. Morales, Y.; Takeuchi, E.; Tsubouchi, T. Vehicle localization in outdoor woodland environments with sensor fault detection. In Proceedings of the IEEE International Conference on Robotics and Automation, Pasadena, CA, USA, 19–23 May 2008; pp. 449–454. [CrossRef]
230. Kitt, B.; Geiger, A.; Lategahn, H. Visual odometry based on stereo image sequences with RANSAC-based outlier rejection scheme. In Proceedings of the 2010 IEEE Intelligent Vehicles Symposium, University of Califomia, San Diego, CA, USA, 21–24 June 2010; pp. 486–492.
231. Xie, L.; Wang, S.; Markham, A.; Trigoni, N. GraphTinker: Outlier rejection and inlier injection for pose graph SLAM. In Proceedings of the 2017 IEEE/RSJ International Conference on Intelligent Robots and Systems (IROS), Vancouver, BC, Canada, 24–28 September 2017; pp. 6777–6784.
232. Sünderhauf, N.; Protzel, P. Switchable constraints for robust pose graph SLAM. In Proceedings of the 2012 IEEE/RSJ International Conference on Intelligent Robots and Systems, Vilamoura, Portugal, 7–12 October 2012; pp. 1879–1884.
233. Latif, Y.; Cadena, C.; Neira, J. Robust loop closing over time for pose graph SLAM. *Int. J. Robot. Res.* **2013**, *32*, 1611–1626. [CrossRef]
234. Carlone, L.; Censi, A.; Dellaert, F. Selecting good measurements via $\ell 1$ relaxation: A convex approach for robust estimation over graphs. In Proceedings of the 2014 IEEE/RSJ International Conference on Intelligent Robots and Systems, Chicago, IL, USA, 14–18 September 2014; pp. 2667–2674. [CrossRef]
235. Carlone, L.; Calafiore, G.C. Convex Relaxations for Pose Graph Optimization with Outliers. *IEEE Robot Autom. Let.* **2018**, *3*, 1160–1167. [CrossRef]
236. Wei, L.; Cappelle, C.; Ruichek, Y. Camera/laser/GPS fusion method for vehicle positioning under extended NIS-based sensor validation. *IEEE Trans. Instrum. Meas.* **2013**, *62*, 3110–3322. [CrossRef]
237. Zhang, J.; Singh, S. Visual-LiDAR odometry and mapping, low-drift, robust and fast. In Proceedings of the IEEE International Conference on Robotics and Automation, Seattle, WA, USA, 26–30 May 2015.
238. Sturm, J.; Burgard, W.; Cremers, D. Evaluating Egomotion and Structure-from-Motion Approaches Using the TUM RGB-D Benchmark. In Proceedings of the IEEE/RSJ International Conference Intelligent Robots & Systems, Vilamoura, Portugal, 7–12 October 2012.
239. Kummerle, R.; Steder, B.; Dornhege, C.; Ruhnke, M.; Grisetti, G.; Stachniss, C.; Kleiner, A. On measuring the accuracy of SLAM algorithms. *Auton. Robot.* **2009**, *27*, 387–407. [CrossRef]
240. Sturm, J.; Engelhard, N.; Endres, F.; Burgard, W.; Cremers, D. A benchmark for the evaluation of RGB-D SLAM systems. In Proceedings of the 2012 IEEE/RSJ International Conference on Intelligent Robots and Systems, Vilamoura, Portugal, 7–12 October 2012; pp. 573–580. [CrossRef]
241. Kurlbaum, J.; Frese, U. A Benchmark Data Set for Data Association. Technical Report. University of Bremen, 2009. Available online: https://www.dfki.de/fileadmin/user_upload/import/4432_kurlbaum_tr_09.pdf (accessed on 1 October 2022).
242. Li, Q.; Song, Y.; Hou, Z. Neural network based FastSLAM for autonomous robots in unknown environments. *Neurocomputing* **2015**, *165*, 99–110. [CrossRef]
243. Cadena, C.; Neira, J. SLAM in O(log n) with the Combined Kalman—Information filter. In Proceedings of the 2009 IEEE/RSJ International Conference on Intelligent Robots and Systems, St. Louis, MO, USA, 10–15 October 2009; pp. 2069–2076.
244. Huang, S.; Dissanayake, G. Convergence and consistency analysis for extended Kalman Filter based SLAM. *IEEE Trans. Robot.* **2007**, *23*, 1036–1049. [CrossRef]
245. Bailey, T.; Nieto, J.; Guivant, J.; Stevens, M.; Nebot, E. Consistency of the EKF-SLAM Algorithm. In Proceedings of the 2006 IEEE/RSJ International Conference on Intelligent Robots and Systems, Beijing, China, 9–15 October 2006; pp. 3562–3568. [CrossRef]
246. Huang, G.P.; Mourikis, A.I.; Roumeliotis, S.I. A First-Estimates Jacobian EKF for Improving SLAM Consistency. In *Experimental Robotics*; Khatib, O., Kumar, V., Pappas, G.J., Eds.; Springer Tracts in Advanced Robotics; Springer: Berlin/Heidelberg, Germany, 2009; Volume 54. [CrossRef]
247. Graham, M.C.; How, J.P.; Gustafson, D.E. Robust incremental SLAM with consistency-checking. In Proceedings of the 2015 IEEE/RSJ International Conference on Intelligent Robots and Systems (IROS), Hamburg, Germany, 28 September–2 October 2015; pp. 117–124. [CrossRef]
248. Wang, J.; Knight, N.L. New outlier separability test and its application in GNSS positioning. *J. Glob. Position. Syst.* **2012**, *11*, 6–57. [CrossRef]
249. Yang, L.; Wang, J.; Knight, N.L.; Shen, Y. Outlier separability analysis with a multiple alternative hypotheses test. *J. Geodesy* **2013**, *87*, 591–604. [CrossRef]

250. Baarda, W. *A Testing Procedure for Use in Geodetic Networks;* New Series 2(4); Netherlands Geodetic Commission: Apeldoorn, The Netherlands, 1968.
251. Li, Z.; Wang, J.; Alquarashi, M.; Chen, K.; Zheng, S. Geometric analysis of reality-based indoor 3D mapping. *J. Glob. Position Sys.* **2016**, *14*, 1. [CrossRef]
252. El-Mowafy, A.; Imparato, D.; Rizos, C.; Wang, J.; Wang, K. On hypothesis testing in RAIM algorithms: Generalized likelihood ratio test, solution separation test and a possible alternative. *Meas. Sci. Technol.* **2019**, *30*, 2019. [CrossRef]
253. Wang, J.; Ober, P.B. On the availability of Fault Detection and Exclusion in GNSS receiver autonomous integrity monitoring. *J. Navig.* **2009**, *62*, 251–261. [CrossRef]
254. Sundvall, P.; Jensfelt, P. Fault detection for mobile robots using redundant positioning systems. In Proceedings of the IEEE International Conference on Robotics and Automation, Orlando, FL, USA, 15–19 May 2006; pp. 3781–3786. [CrossRef]
255. Hewitson, S.; Wang, J. Extended receiver autonomous integrity monitoring (eRAIM) for GNSS/INS integration. *J. Surv. Eng.* **2010**, *136*, 13–22. [CrossRef]
256. El-Mowafy, A. On Detection of Observation Faults in the Observation and Position Domains for Positioning of Intelligent Transport Systems. *J. Geod.* **2019**, *93*, 2109–2122. [CrossRef]
257. Kukko, A.; Kaijaluoto, R.; Kaartinen, H.; Lehtola, V.V.; Jaakkola, A.; Hyyppä, J. Graph SLAM correction for single scanner MLS forest data under boreal forest canopy. *ISPRS J. Photogramm. Remote Sens.* **2017**, *132*, 199–209. [CrossRef]
258. Hess, W.; Kohler, D.; Rapp, H.; Andor, D. Real-time loop closure in 2D LIDAR SLAM. In Proceedings of the IEEE International Conference on Robotics and Automation, Stockholm, Sweden, 16–21 May 2016; pp. 1271–1278.
259. Chang, L.; Niu, X.; Liu, T.; Tang, J.; Qian, C. GNSS/INS/LiDAR-SLAM Integrated Navigation System Based on Graph Optimization. *Remote Sens.* **2019**, *11*, 1009. [CrossRef]

MDPI
St. Alban-Anlage 66
4052 Basel
Switzerland
Tel. +41 61 683 77 34
Fax +41 61 302 89 18
www.mdpi.com

Remote Sensing Editorial Office
E-mail: remotesensing@mdpi.com
www.mdpi.com/journal/remotesensing